T0075245

FOUNDATIONS AND FRONTIERS IN COMPUTER, COMMUNICATION
AND ELECTRICAL ENGINEERING

PROCEEDINGS OF THE 3RD INTERNATIONAL CONFERENCE ON FOUNDATIONS AND FRONTIERS IN COMPUTER, COMMUNICATION AND ELECTRICAL ENGINEERING (C2E2–2016), MANKUNDU, WEST BENGAL, INDIA, 15–16 JANUARY 2016

Foundations and Frontiers in Computer, Communication and Electrical Engineering

Editor

Aritra Acharyya
Department of Electronics and Communication,
Supreme Knowledge Foundation Group of Institutions (SKFGI),
Mankundu, West Bengal, India

CRC Press
Taylor & Francis Group
Boca Raton   London   New York   Leiden

CRC Press is an imprint of the
Taylor & Francis Group, an **informa** business

A BALKEMA BOOK

CRC Press/Balkema is an imprint of the Taylor & Francis Group, an informa business

© 2016 Taylor & Francis Group, London, UK

Typeset by V Publishing Solutions Pvt Ltd., Chennai, India

All rights reserved. No part of this publication or the information contained herein may be reproduced, stored in a retrieval system, or transmitted in any form or by any means, electronic, mechanical, by photocopying, recording or otherwise, without written prior permission from the publisher.

Although all care is taken to ensure integrity and the quality of this publication and the information herein, no responsibility is assumed by the publishers nor the author for any damage to the property or persons as a result of operation or use of this publication and/or the information contained herein.

Published by: CRC Press/Balkema
        P.O. Box 11320, 2301 EH Leiden, The Netherlands
        e-mail: Pub.NL@taylorandfrancis.com
        www.crcpress.com – www.taylorandfrancis.com

ISBN: 978-1-138-02877-7 (Hbk)
ISBN: 978-1-315-65791-2 (eBook PDF)

Table of contents

Preface

The 3rd International Conference on "Foundations and Frontiers in Computer, Communication and Electrical Engineering–2016" (C2E2–2016), hosted by Supreme Knowledge Foundation Group of Institutions (SKFGI), India, in association with CSIR-CEERI, Pilani, Rajasthan, India, followed the success of previous two International Conferences hosted by SKFGI in 2012 and 2015.

C2E2–2016 was a notable event which brings academia, researchers, engineers and students of Electronics and Communication, Computer and Electrical Engineering together. The conference was a perfect platform to share experience, foster collaborations across industry and academia, and evaluate emerging technologies across the globe. The conference was technically sponsored by IEEE Kolkata Section, IET Kolkata Network along with several IEEE chapters under Kolkata Section such as Electron Devices Society, Power and Energy Society, Dielectrics and Electrical Insulation Society and Computer Society. The conference was sponsored by CSIR-CEERI, Pilani, Rajasthan, Defence Research and Development Organization (DRDO) and Indian National Science Academy (INSA).

C2E2–2016 celebrated the historical moment of the four-way call between Alexander Graham Bell in New York, his assistant Thomas Watson in San Francisco, President Woodrow Wilson in Washington and Theodore Vail, President of American Telephone and Telegraph company (AT&T) in Jekyll Island, i.e. the first transcontinental phone call made on January 25, 1915. The theme of C2E2–2016 rejoiced a walk down telephone memory lane; a remarkable footstep of human being in the field of electrical communication. Many scopes from the earlier International Conferences hosted by SKFGI are still relevant today and therefore remain unchanged for the 3rd C2E2–2016; these include electron devices, ion integrated circuits, interconnects, semiconductors, quantum-effect structures, microwave and millimeter-wave vacuum devices, emerging materials with applications in bioelectronics, biomedical electronics, computation, communications, displays, microelectromechanics, imaging, micro-actuators, nano-electronics, optoelectronics, photovoltaics, power ICs, micro-sensors, digital and analog VLSI, photonics, plasma devices, microwave/millimeter-wave components, devices, circuits and systems, antennas, millimeter-wave and sub-millimeter-wave techniques, antenna signal processing and control, tubes, missile tracking and guided systems, radio astronomy, propagation and radiation aspects of terrestrial and space based communication, satellite and mobile communication systems, radar, high power microwave systems (HPMS), optical communications, information processing science and technology, machine learning and artificial intelligence, networking, image processing, soft computing, cloud computing, data mining & data warehousing, generation, transmission, distribution, storage and usage of electric energy, dielectric phenomena and measurements, high voltage engineering, electrical machines, power systems, control systems, non-conventional energies, power electronics and drives, etc.

The 3rd C2E2–2016 retained the same format as the previous International Conferences at SKFGI, specifically a two-day programme with plenary session format for the forenoon sessions followed by parallel technical session format for the afternoon sessions of each day. Only the oral format had been kept for the authors to present their research works during the parallel technical sessions. Professor (Dr.) Akhtar Kalam of Victoria University, Melbourne, Australia being the Chief Guest, inaugurated the Conference. Professor Kalam also delivered the opening keynote address of the conference at the forenoon session on the first day. He explored the state-of-the-art of the smart power grids in the 21st century during his talk. Dr. Guillermo Carpintero of Universidad Carlos III de Madrid, Spain, delivered the next keynote address on photonic integrated circuit technology for ultra-high speed wireless communications. Dr. Sekhar Bhattacharya of SSN Research Centre, Chennai, India, Er. Tulika Mehta, Director, Fortune consultancy services, India, and Mr. Rajdeep Chowdhury, Department of Computer Applications, JIS College of Engineering, India, delivered notable lectures during the forenoon plenary session of the second day of the conference. Dr. Bhattacharyya discussed regarding the atomic layer deposition technique in nanoelectronics and plasma enhanced chemical vapour deposition for photovoltaics during his lecture. Er. Mehta talked about the advanced applied cognition neuroscience education technology to eradicate drug addiction. The last invited speaker Mr. Chowdhury delivered his lecture on data warehouse performance enhancement employing minimized query processing proposal and implementation of security modus operandi. All broad scopes of the conference were covered by the keynote addresses as well as invited lectures. We are sincerely grateful to the keynote and invited speakers who generously have contributed their time, expertise and experience to these comprehensive lectures.

More than 140 research articles were submitted by the authors from all over the world. Each paper has been peer-reviewed by at least two reviewers, drawn from academic institutions and industry from around the world. And finally, 106 papers were accepted for oral presentation during the conference. All those accepted papers were presented by the respective authors during the parallel technical sessions during the afternoon sessions of each day of the conference. At each technical session, five parallel sessions were simultaneously held; out of which two sessions were designated for electronics and communication engineering related papers, two for electrical engineering related papers and one session was designated for the papers fall under the scope of computer science and engineering. The presented papers collected in these proceedings which provide a comprehensive reference of the current state-of-the-art in computer, communication and electrical engineering. We are indebted to the efforts of all the reviewers, who undoubtedly have raised the quality of the proceedings. We are also earnestly thankful to all the authors who have contributed their valuable research works to the conference.

We convey our heartiest thank to the managerial bodies of SKFGI for their immense help and supports for organizing the event. We also thank Professor (Dr.) Abhijit Lahiri, Campus Director, SKFGI for his enormous contribution in overall coordination with the contributing authors, invited speakers, publisher and intense supervision during the programme. We are thankful to Mr. Biswadeepam Pal for his assistance with graphic design and webpage development. We are also grateful for the cooperative advice from Professor (Dr.) B. N. Basu with respect to planning of the conference. Finally, SKFGI is grateful to the industrial and academic sponsors for providing financial support, the members of our Local Organizing Committee, National and International Advisory Committees, and all the technical sponsors under whose crucial auspices the C2E2–2016 experienced the utmost success.

Aritra Acharyya
March 2016

Advisory committees

INTERNATIONAL ADVISORY COMMITTEE

Dr. Tadao Nagatsuma, *Professor, Department of Systems Information, Graduate School of Engineering Science, Osaka University, Japan.*

Dr. Tzyh-Ghuang Ma, *Distinguished Professor, Group, Communication and Electromagnetic Engineering, National Taiwan University of Science and Technology. Taiwan.*

Dr. Dipak Ranjan Poddar, *Emeritus Professor, Department of Electronics and Telecommunication Engineering, Jadavpur University, Kolkata, India.*

Dr. V. Rodolfo García Colón H., *Registrar, IIE Centro de Posgrado, IEEE DEIS Adcom Member and DEIS Chapters Chair*

Dr. Sivaji Chakravorti, *Professor, Electrical Engineering Department, Jadavpur University, Kolkata, India.*

Dr. Nikhil Ranjan Pal, *INAE Chair Professor, ECSU, Indian Statistical Institute, Kolkata, India.*

Dr. Dharma P. Agrawal, *Ohio Board of Regents Distinguished Professor, Department of EECS University of Cincinnati, Ohio.*

NATIONAL ADVISORY COMMITTEE

Dr. Chandan Sarkar, *Professor, Department of Electronics and Telecommunication Engineering, Jadavpur University, Kolkata, India.*

Dr. Monojit Mitra, *Professor, Department of Electronics and Telecommunication Engineering Indian Institute of Engineering Science and Technology, Shibpur, Howrah, India.*

Dr. Nandini Gupta, *Professor, Department of Electrical Engineering, Indian Institute of Technology, Kanpur, India.*

Dr. Saibal Chatterjee, *Professor, Department of Electrical Engineering, North Eastern Regional Institute of Science & Technology, Arunachal Pradesh, India.*

Dr. Mita Nasipuri, *Professor, Department of Computer Science & Engineering, Jadavpur University, Kolkata, India.*

Dr. Soumya Pandit, *Assistant Professor, Institute of Radio Physics and Electronics, University of Calcutta, Kolkata, India.*

Dr. Ujjwal Maulik, *Professor, Department of Computer Science & Engineering, Jadavpur University, Kolkata, India.*

Dr. Suranjan Ghose, *Professor, Department of Computer Science & Engineering, Jadavpur University, Kolkata, India.*

Organizing committees

CONFERENCE PATRONS

Prof. D. K. Basu, *Chairman, B.O.G., SKFGI*
Mr. B. G. Mallick, *Chairman Trustee*
Mr. D. K. Mondal, *Secretary, Trustee*
Mr. C. K. Bhattacharya, *CEO, Trustee*
Mr. K. C. Mondal, *Treasurer, Trustee*

CONFERENCE CHAIR

Professor B. N. Biswas, *Chairman, Education Division, SKFGI*

CONFERENCE CONVENER

Professor T. K. Sengupta, *Chief Technical Director, SKFGI*

CONFERENCE CO-CONVENER

Professor Abhijit Lahiri, *Campus Director, SKFGI*

CONFERENCE SECRETARY

Professor T. K. Dey, *Additional Chief Technical Director, SKFGI*

CONFERENCE COORDINATOR

Mr. Aritra Acharyya, *Department of ECE, SKFGI*

CONFERENCE LIAISON

Dr. S. N. Joshi, *CSIR-CEERI, Pilani, Rajasthan*
Professor B. N. Basu, *Research Coordinator, SKFGI*
Mr. Subhradeep Pal (On lien), *SKFGI*
Mr. Subhadip Chowdhury (On lien), *SKFGI*

CONFERENCE HOST

Ms. Srima Nandi, *SKFGI*

PUBLICATION CHAIR

Dr. Rajib Bag, *Dean of Student Affairs, SKFGI*

TREASURER

Mr. Sourav Koley, *Department of CSE, SKFGI*

EDITOR IN CHIEF

Mr. Aritra Acharyya, *Department of ECE, SKFGI*

Frontiers in Computer, Communication and Electrical Engineering – Acharyya (Ed.)
© 2016 Taylor & Francis Group, London, ISBN: 978-1-138-02877-7

A microscopic view on the effect of anisotropy in the breakdown phenomenon of the 4H-SiC power diodes

Subhashri Chatterjee, Adrija Das, Alka Singh, Tripti Guin Biswas & Aritra Acharyya
Supreme Knowledge Foundation Group of Institutions, Mankundu, Hooghly, West Bengal, India

ABSTRACT: In this paper, the authors have studied the effect of anisotropy on the static characteristics, such as breakdown voltage, avalanche region voltage drop, avalanche width, etc., of 4H-SiC power diodes. A sample of n^+-n-p-p^+ structured 4H-SiC power diodes having different structural and doping parameters have been taken into consideration and a simulation study based on the drift-diffusion model has been carried out by taking into account the material parameters of both <0001> and <11$\bar{2}$0> oriented 4H-SiC to investigate the influence of anisotropy exhibited by 4H-SiC due to those material parameters depending on the orientation of the wafer growth on the aforementioned static characteristics. Results show that the <0001> oriented 4H-SiC power diodes have a larger breakdown voltage as compared to their <11$\bar{2}$0> counterparts. The reasons behind the better breakdown characteristics of <0001> oriented 4H-SiC power diodes have also been discussed from a microscopic insight of the breakdown phenomena.

1 INTRODUCTION

Since the last two and half decades, one of the most promising wide bandgap semiconductor material, silicon carbide (SiC), has attracted the attention of researchers for high-power and high-frequency device applications (Weitzel *et al.* 1995, Acharyya *et al.* 2013). Due to the excellent electrical, thermal, and other physical properties of SiC, such as high breakdown field, high saturation drift velocity and mobility of charge carriers, high thermal conductivity, etc., it turns out to be superior to conventional semiconductor materials, like Si, GaAs, InP, etc., for high-power, high-temperature, and high-frequency operational conditions. The 4H-SiC excels other polytypes of SiC (e.g., 6H-SiC, 3C-SiC, etc.) as the potential candidate to be the base material of high-power and high-frequency devices due to its large bandgap (Eg ≈ 3.26 eV at 300 K), most favorable carrier transportation properties, and the availability of the most advanced epitaxial growth.

Some of the properties, which are indispensible for high-power solid-state devices are (a) high breakdown voltage, (b) negligible leakage current, (c) very small on resistance, (d) high switching speed, (e) high reliability throughout the operational lifetime, capacity of withstanding very high temperature, and (f) cost effectiveness. In this regard, the most favorable electrical, thermal, and other physical properties of 4H-SiC enables it to excel other polytypes of SiC as well as other conventional semiconductor materials mentioned earlier. Several solid-state power devices, such as

Schottky diodes, *p-n* junction diodes, *p-i-n* diodes, thyristors, UMOSFETs, SITs, RF MOSFETs, and RF JFETs, based on 4H-SiC have already been fabricated and their performances are found to be superior to those based on Si, GaAs, InP, etc. (Weitzel *et al.* 1995).

Experimental studies show that some important material parameters, such as impact ionization rates, mobilities, and drift velocities of charge carriers, in <0001> and <11$\bar{2}$0> oriented 4H-SiC epitaxial wafers exhibit large anisotropy (Hatakeyama *et al.* 2004). Since the abovementioned material parameters have a major role in the breakdown mechanism of *p-n* junctions based on both <0001> and <11$\bar{2}$0> oriented 4H-SiC, the orientation becomes crucial in the design of power devices. Both experimental and simulation results of breakdown phenomena in 4H-SiC diodes have shown a large anisotropy. In the present paper, the authors have studied the effect of anisotropy on the static characteristics of 4H-SiC power diodes. Simulation is carried out to obtain the spatial variations of electric field, particle current densities, carrier densities as well as breakdown voltage, avalanche region voltage drop, and avalanche width in n^+-n-p-p^+ structured 4H-SiC power diodes having different structural and doping parameters as a function of bias current density. Simulation study has been carried out by taking into account the material parameters of both <0001> and <11$\bar{2}$0> oriented 4H-SiC to investigate the influence of anisotropy exhibited by 4H-SiC due to those material parameters depending on the orientation of the wafer growth on the aforementioned static characteristics of the power diodes under consideration. Results show that the <0001> oriented 4H-SiC power diodes have a larger breakdown voltage as compared to their <11$\bar{2}$0> counterparts. The reason behind the better breakdown characteristics of <0001> oriented 4H-SiC power diodes has also been discussed from microscopic insights of the breakdown phenomena.

2 MATERIAL PARAMETERS

The material parameters of the base semiconductor play a major role in the breakdown mechanism of reverse biased power diodes. Important material parameters of 4H-SiC, such as energy bandgap (E_g in eV), effective density of the states in the conduction band (N_c in m^{-3}), effective density of states in the valance band (N_v in m^{-3}), density of the state effective mass of charge carriers (m_d^* in Kg), saturated drift velocities of electrons and holes ($v_{n(sat)}$ and $v_{p(sat)}$ in m s^{-1}), diffusivity of electrons and holes (D_n and D_p m^2 s^{-1}), dielectric constant (ε_r), thermal conductivity (k in W m^{-1} K^{-1}), etc., at room temperature (i.e., at 300 K) have been taken for simulation from the published literature (Electronic Archive 2015, Lee 2008). In the year 2004, Hatakeyama et al. experimentally measured the electron and hole ionization rates (α_n and α_p in m^{-1}) in the 4H-SiC diodes in two different directions, i.e., <0001> and <11$\bar{2}$0> directions (Hatakeyama et al. 2004). They fit their experimental data with the well-known empirical relation of the field dependence of the ionization rates of the charge carriers in a semiconductor, i.e., $\alpha_{n,p}(\xi) = A_{n,p} \exp(-B_{n,p}/\xi)$; where ξ is the electric field in V m^{-1} and $A_{n,p}$ (m^{-1}) and $B_{n,p}$ (V m^{-1}) are the ionization coefficients. According to their report, the fitting parameters', A_n, A_p, B_n, B_p, values are found to be different for <0001> and <11$\bar{2}$0> oriented 4H-SiC. These values of $A_{n,p}$ and $B_{n,p}$ and the expression of $\alpha_{n,p}(\xi)$ have been incorporated in the simulation program to introduce anisotropy in 4H-SiC in terms of the ionization rates of the charge carriers.

The low-field mobility model used in the present simulation is taken to be

$$\mu_{n,p} = \mu_{n,p(min)} \left(T/300 \right)^{a_{n,p}}$$
$$+ \left\{ \mu_{n,p(max)} \Big/ \left[1 + \left((N_A + N_D)/N_{n,p(ref)} \right) \right]^{b_{n,p}} \right\}$$
$$\left(T/300 \right)^{a_{n,p}},$$

where N_A and N_D are the acceptor and donor concentrations, respectively. Other parameters, such as $\mu_{n,p(min)}$, $\mu_{n,p(max)}$, $a_{n,p}$, $b_{n,p}$, have been taken from the published literature for 4H-SiC oriented in <0001> and <11$\bar{2}$0> directions (Schaffer et al. 1994). It can be observed from the experimental results reported

in (Schaffer et al. 1994) that the electron mobility of <0001> oriented 4H-SiC is around 1.20 times of that of the <11$\bar{2}$0> oriented 4H-SiC above 200 K. The widely used parallel field mobility model for 4H-SiC has been taken into account in the simulation as a high-field mobility, which can be expressed as

$$\mu_{n,p}(\xi) = \mu_{n,p}^0 \left/ \left[1 + \left(\mu_{n,p}^0 \xi / v_{n,p(sat)} \right)^{c_{n,p}} \right]^{1/c_{n,p}} \right. ,$$

where $c_{n,p} = 2$ and $v_{n,p(sat)} = 2 \times 10^5$ m s^{-1} have been chosen for a high field mobility.

3 SIMULATION TECHNIQUE

A one-dimensional (1-D) model of reverse biased n^+-n-p-p^+ structured Power Diode (PD) shown in Figure 1 has been considered for the simulation. In the present paper, simulation has been carried out based on a drift-diffusion model to study the breakdown characteristics of PDs under consideration. The spatial variations of the electric field ($\xi(x)$ vs. x), electron and hole current densities ($J_n(x)$, $J_p(x)$ vs. x), and electron and hole concentrations ($n(x)$, $p(x)$ vs. x) within the depletion layer of the reverse biased PD can be obtained from the simultaneous numerical solution of fundamental device equations, such as (a) Poisson's equation, (b) combined carrier continuity equations at steady-state, (c) current density equations, and (d) mobile space charge equation subject to appropriate boundary conditions at the depletion layer edges.

Now the appropriate boundary conditions associated with $\xi(x)$, $J_n(x)$, and $J_p(x)$ to be imposed at depletion layer edges, i.e., at $x = 0$ and $x = W$ to solve the aforementioned equations can be formulated as follows. Due to the high conductivity of n^+- and p^+-layers, the electric field within those regions must be zero. Thus, the boundary conditions for the electric field at $x = 0$ and $x = W$ are given by:

$$\xi(x = 0) = \xi(x = W) = 0. \qquad (1)$$

The normalized current density parameter ($P(x)$) may be defined as $P(x) = (J_p(x) - J_n(x))/J_0$, where $J_0 = J_n(x) + J_p(x)$ is the total bias current density.

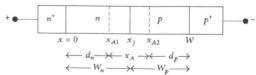

Figure 1. 1-D model of the n^+-n-p-p^+ structured power diode.

The boundary conditions for $J_n(x)$ and $J_p(x)$ at $x = 0$ and $x = W$ can be defined in terms of $P(x)$ as:

$$\left.\begin{array}{l} P(x=0) = \left(\dfrac{2J_p(x=0)}{J_0} - 1\right) \\[3mm] P(x=W) = \left(1 - \dfrac{2J_n(x=W)}{J_0}\right) \end{array}\right\}, \qquad (2)$$

where $J_p(x = 0) = J_{ps}$ and $J_n(x = W) = J_{ns}$; J_{ps} and J_{ns} are the electron and hole component of the thermally generated reverse saturation current, respectively, where $J_{ns,ps} = (qD_{n,p}n_i^2/L_{n,p}N_{D,A})$, L_n and L_p are the diffusion lengths of the electrons and holes, and n_i is the intrinsic carrier concentration of the base semiconductor at room temperature. For both the cases, $J_0 \gg J_{ns,ps}$ after the breakdown of the p-n junction; therefore, the ratios $(2J_p(x = 0)/J_0)$ and $(2J_n(x = 0)/J_0)$ tend to become zero after the breakdown. Thus, the boundary conditions in equation (6) reduce to $P(x = 0) = -1$ and $P(x = W) = +1$.

4 RESULTS AND DISCUSSION

Two power diodes having different structural and doping parameters have been chosen for the simulation study. The n- and p-side epitaxial layer lengths (W_n and W_p) have been taken to be 2.50 μm in the first diode (i.e., PD1) and those are taken to be 2.00 μm each in the second one (i.e., PD2). The doping concentrations of the epitaxial layers have been chosen ($N_D = N_A = 3.00 \times 10^{22}$ m^{-3} in PD1 and $N_D = N_A = 5.00 \times 10^{22}$ m^{-3} in PD2) keeping in mind that both of the diodes do not behave as high punch-through diodes even at high bias current densities. The thickness and doping concentrations of the highly doped n^+- and p^+-contact layers (W_{n+}, W_{p+} and N_{n+}, N_{p+}) are taken to be 0.50 μm and 5.00×10^{25} m^{-3} each in both PD1 and PD2, respectively. The structural and doping parameters of both PD1 and PD2 are given in Table 1.

The numerical integration of $\xi(x)$-profile by using Simpson's –1/3 rule within $x = 0$ and $x = W$ and $x = x_{A1}$ and $x = x_{A2}$ provide the breakdown voltage (V_B) and avalanche zone voltage drop (V_A) for a particular J_0. The V_B and V_A values in both PD1 and PD2 based on both <0001> and <11$\bar{2}$0> oriented 4H-SiC have been obtained for different bias current densities ranging from 0.7×10^8 to 1.4×10^8 A m^{-2} and plotted in Figures 2 and 3, respectively. It is observed that both the breakdown voltage and avalanche region voltage drop are greater in PD1s based <0001> oriented 4H-SiC PDs as compared to their <11$\bar{2}$0> counterparts. The reason behind it can be easily explained in the following way. Figure 4 shows the variations

Table 1. Structural and doping parameters.

Parameters	PD1	PD2
W_n (μm)	2.50	2.00
W_p (μm)	2.50	2.00
W_{n+} (μm)	0.50	0.50
W_{p+} (μm)	0.50	0.50
N_D ($\times 10^{22}$ m^{-3})	3.00	5.00
N_A ($\times 10^{22}$ m^{-3})	3.00	5.00
N_{n+} ($\times 10^{25}$ m^{-3})	5.00	5.00
N_{p+} ($\times 10^{25}$ m^{-3})	5.00	5.00

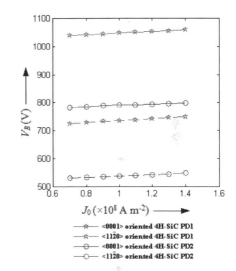

Figure 2. Variation of breakdown voltage in <0001> and <11$\bar{2}$0> oriented 4H-SiC power diodes with bias current densities.

of electron and hole ionization rates in <0001> and <11$\bar{2}$0> oriented 4H-SiC in an electric field. It is very much clear from Figure 4 that within the operating field range of both PDs as indicated by the grey box, both the α_n and α_p values of <0001> oriented 4H-SiC are smaller than its <11$\bar{2}$0> counterparts. Lower ionization rates signified that a higher electric field is required to cause impact ionization within the multiplication region of PDs based in <0001> oriented 4H-SiC as compared to those based in <11$\bar{2}$0> oriented 4H-SiC. That is why the electric field at each space point is found to be greater in <0001> oriented 4H-SiC PDs than those in <11$\bar{2}$0> oriented 4H-SiC PDs. A higher electric field at each space point obviously leads to higher V_B as well as V_A in <0001> oriented 4H-SiC PDs than in <11$\bar{2}$0> oriented 4H-SiC PDs. Moreover, the inset of the Figure 4 shows the variations of (α_n/α_p) in <0001> and <11$\bar{2}$0> oriented 4H-SiC

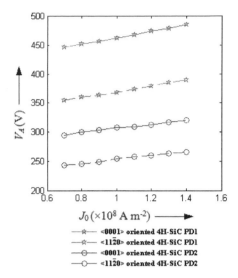

Figure 3. The variation of avalanche zone voltage drop in <0001> and <11$\bar{2}$0> oriented 4H-SiC power diodes with bias current densities.

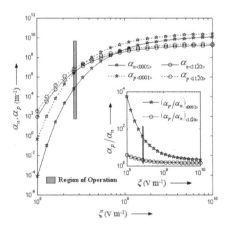

Figure 4. Variations of electron and hole ionization rates in <0001> and <11$\bar{2}$0> oriented 4H-SiC with electric field. Inset of the figure shows the variations of the ratio of the hole to the electron ionization rates in <0001> and <11$\bar{2}$0> oriented 4H-SiC with electric field.

with the electric field. It is observed that the values of the ratios (α_n/α_p) are around 70 and 1.5 within the operating field range (indicated by the grey box) of PD1 and PD2. Higher ionization rates of holes as compared to electrons cause a narrower p-side avalanche width $x_{Ap} = (x_{A2} - x_j)$ than that in the n-side, i.e., $x_{An} = (x_j - x_{A1})$. Moreover, narrower p-side avalanche width leads to smaller avalanche region voltage drop in the p-side, i.e., $V_{Ap} < V_{An}$ (where $V_A = V_{Ap} + V_{An}$). Since (α_n/α_p) ratio is much larger in <0001> oriented 4H-SiC within the operating field range of the PDs, the values of (x_{Ap}/x_{An})

and (V_{Ap}/V_{An}) are found to be much smaller in <00$\underline{01}$> oriented 4H-SiC PDs as compared to their <11$\bar{2}$0> counterparts for any bias current density. It is observed that the values of (x_{Ap}/x_{An}) and (V_{Ap}/V_{An}) vary between 1.30 and 1.52 and $\underline{1}$.46 and 1.55, respectively, in PDs based on <11$\bar{2}$0> oriented 4H-SiC, while they vary between 3.40 and 3.60 and 4.00 and 4.20, respectively, in PDs based on <0001> oriented 4H-SiC for the variation of J_0 from 0.7×10^8 to 1.4×10^8 A m^{-2}.

5 CONCLUSION

The authors have studied the influence of anisotropy on the static characteristics of 4H-SiC power diodes. Numerical simulation study based on drift-diffusion model on the n^+-n-p-p^+ structured 4H-SiC power diodes having different structural and doping parameters has been carried out by taking into account the material parameters of both <0001> and <11$\bar{2}$0> oriented 4H-SiC to investigate the influence of anisotropy exhibited by 4H-SiC due to those material parameters depending on the orientation of the wafer growth on the aforementioned static characteristics. Results show that the <0001> oriented 4H-SiC power diodes have a larger breakdown voltage as compared to their <11$\bar{2}$0> counterparts. The results presented in this paper will be extremely helpful for future experimentalists for choosing the appropriate orientation of the crystal growth of 4H-SiC for the fabrication of power diodes having high voltage ratings.

REFERENCES

Acharyya, A., & Banerjee, J.P. (2013). Potentiality of IMPATT devices as terahertz source: An avalanche response time based approach to determine the upper cut-off frequency limits. *IETE Journal of Research*, 59(2), 118–127.

Electronic Archive: New Semiconductor Materials, Characteristics and Properties (2015). Retrieved from http://www.ioffe.ru/SVA/NSM/Semicond/SiC/index.html.

Hatakeyama, T., Watanabe, T., Shinohe, T. (2004). Impact ionization coefficients of 4H silicon carbide. *Applied Physics Letters*, 85(8), 1380–1382.

Lee, H.S. (2008). Fabrication and Characterization of Silicon Carbide Power Bipolar Junction Transistors. PhD Dissertation, KTH.

Schaffer, W.J., Kong, H.S., Negley, G.H., and Palmour, J.W. (1994). Hall Effect and C-V Measurement on Epitaxial 6H—and 4H-SiC," In Proc. of Inst. Phys. Conf. Ser., 137 (pp. 155–159).

Weitzel, C.E., Palmour, J.W., Carter, C.H., Moore, K., Nordquist, K.J., Allen, S., Thero, C., and Bhatnagar, M. (1995) Silicon Carbide High-Power Devices. *IEEE Transactions on Electron Devices*, 43(10), 1732–1741.

Frontiers in Computer, Communication and Electrical Engineering – Acharyya (Ed.)
© 2016 Taylor & Francis Group, London, ISBN: 978-1-138-02877-7

Wireless power transmission—part I: A brief history

Nirvik Patra, Debodyuti Banerjee, Subhashri Chatterjee, T.K. Sengupta,
Subhendu Chakraborty & Aritra Acharyya
Supreme Knowledge Foundation Group of Institutions, Mankundu, Hooghly, West Bengal, India

ABSTRACT: In this part of the paper, authors have discussed the historical background of the wireless power transmission since the inception of it to the current state-of-the-art. This initial part of the paper may be regarded as the introductory part of authors' experimental studies on wireless power transfer via near-field coupling of electrically-small transmitting and receiving loop antennas which have been recently carried out at the Antenna and Propagation Laboratory, SKFGI, Mankundu. The theoretical modeling of the transmitting and receiving loop antennas as the wireless power transfer module and the corresponding experimental studies have been discussed in the successive parts later on as the continuation of the same paper.

1 INTRODUCTION

After the invention of electrical energy, the efficient energy transportation became an imperative concern for mankind. The biggest and sophisticated energy transportation network is the electrical grid. Distribution grids transport the electrical energy from the point of its generation to the far ends in a very efficient way. However, during the transmission, a loss of around 30% of energy remains unavoidable due to several reasons. Some potential modern-day applications significantly promote the concept of electrical energy transportation and distribution without the copper cables, i.e. wireless energy transfer. Some of those prospective applications of wireless energy/power transfer are (a) charging mobile devices, electrical cars, unmanned aircrafts, etc., (b) driving home appliances like irons, vacuum cleaners, television, etc., (c) supplying power to the biomedical implants such as pacemakers, cochlear implants, subcutaneous drug supplier, etc. mankind has desired to transfer electrical energy without use of wires from as far back as the mid 1800s. In the year 1864, James C. Maxwell predicted the existence of electromagnetic waves by means of his famous set of equations, known as Maxwell's equations (Maxwell 1964). Later in 1884, John H. Poynting realized that the Poynting vector would play an important role in quantifying the energy/power of electromagnetic wave (Poynting 1884). Reinforced by the Maxwell's equations, Heinrich Hertz first showed the experimental evidence of radio waves by his spark-gap radio transmitter in the year 1988 (Hertz 1888); and this evidence of radio wave (i.e. the experimental validation of Maxwell's equations) in the end of 19th century was the inception of wireless power transmission.

2 INITIATION BY NIKOLA TESLA

Nikola Tesla was the pioneer of wireless electrical power transmission (Tesla 1904, Tesla 1914). He initiated his efforts on wireless transmission of electrical power on 1989, just after the Hertz's successful experiment. During 1889–1900, Tesla carried out experiments in Colorado Springs, CO, USA, related to wireless power transmission via the electric field and capacitive coupling, as well as transmission line or waveguide-like effects (Tesla 1891). In the year 1901, Tesla began the construction of Wardenclyffe Tower near Long Island Sound, New York, UA (Figure 1) for broadcasting, wireless communication and transmission of wireless power. However, earlier than Tesla, Sir J.C. Bose already demonstrated experimentally the first millimeter-wave radio communication link in the year 1894 at Presidency College, Kolkata, India (Bondyopadhyay 1998). Later, Tesla believed that the capability of his own invention for wireless power transmission was even more significant than its abilities as a way of wireless communication, it has been said that Tesla's experiments achieved to light lumps several kilometers away from the source without any wire connection. Nonetheless, due to very low power transfer efficiency, depletion of financial resources and most importantly the dangerous nature of his experiments, Tesla finally discarded his experimentations. He left this bequest in the form of a patent that was never commercially exploited (Tesla 1914). Thus, undoubtedly,

Figure 1. Tesla's Wardenclyffe Tower Facility. It was dismantled during World War I for scrap metal.

if Sir J. C. Bose was the pioneer of wireless transmission of information (i.e. wireless communication), the Nikola Tesla was the same of wireless transmission of electrical power.

3 DIFFERENT METHODS OF WIRELESS POWER TRANSMISION

Electromagnetic radiation which has been comprehensively used for wireless transmission of information (voice and data) can also be used for transportation of electrical energy/power to distant end. Highly efficient transfer of electrical power is possible in a directional way by means of microwaves (Glaser 1973). Though the said method is extremely efficient, it has some significant disadvantages such as it requires line of sight path, it is a dangerous mechanism for living beings, etc. more viable option for efficient and harmless short and medium range wireless energy transfer is the use of electromagnetic resonance (Karalis et al. 2008, Kurs 2007). However a number of methods are available for wireless energy transportation at present days. Some of those are: (a) by means of high power laser beam targeted to distant solar cell of relevant detector (NASA 2003), (b) by means of vibratory waves emitted and collected by piezoelectric transducers (Hu et al. 2008), by means of highly directive, high power microwaves (Glaser 1973), and use of rectenna for conversion of microwave energy to DC electrical energy with a conversion energy exceeding 95% (Akkermans et al. 2005, Ali et al. 2005, Shams et al. 2007, Basset et al. 2007), (d) by means of resonant coupling effect between the inductive coils of two LC circuits, (Karalis et al. 2008, Kurs 2007, Gao 2007, Low et al. 2009), etc.

The method of wireless electrical energy transfer by means of resonant coupling of two coils capa-ble of transferring energy within a very short distance in order of few centimeters. But this method involves some important advantages such as (a) it can work in low frequencies which are harmless to the living beings, (b) it is the lowest cost method, (c) at resonant frequency, maximum absorption rate is guaranteed at the receiver coil. When two objects have the same resonant frequency, they can be easily coupled in a resonant way which causes transfer of energy from one object to another in an efficient way. This simple principle has to be exploited to transfer electrical energy from one point to another point by means of electromagnetic field. The resonant coupling can be broadly classified into two broad categories such as (a) inductive coupling (Mansor et al. 2008), (b) self-resonance coupling. Inductive coupling is a form of resonant coupling between two LC circuits having the same resonance frequency. The electrical energy is transferred from source to sink coil in a very efficient manner provided that both of them are separated by a very short distance. This phenomenon can easily be explained by the Faraday's law of electromagnetic induction (transformer emf equation: $\overline{\nabla} \times \overline{E} = -\partial \overline{B}/\partial t$; where $\overline{E}$ and $\overline{B}$ are the time varying electric and magnetic field vectors respectively). However the self-resonance coupling takes place between two coils due to the presence of very small parasitic capacitance distributed within those. Two coils having the same inductive values (L), having same structural parameters (l = length, A = cross-sectional area, N = number of turns) and same core material (having same constitutive parameters: σ = conductivity, ε = permittivity and μ = permeability) constitute the same parasitic capacitance value (C_p) theoretically. Thus the self-resonance frequency associated with the coils is $f_{rs} = 1/\sqrt{LC_p}$; where f_{rs} is much higher than the resonance frequency of LC circuit for inductive coupling ($f_r = 1/\sqrt{LC_l}$), since the value of the parasitic capacitance (C_p) of the coil is too low as compared to the lumped capacitance (C_l) used in the inductive coupling ($C_p \ll C_l$). In the year 2008, Karalis et al. (Karalis et al. 2008) reported around 40% efficiency of Wireless Power Transfer (WPT) between two self-resonance coils provided that $r \ll d \ll \lambda_r$; where r, d and λ_r are the radius of the coils, separation between them and wavelength corresponding to the resonance frequency respectively.

4 A BRIEF SUMMERY OF THE PRESENT WORK

In the present work, a comprehensive experimental investigation on Wireless Power Transfer (WPT) have been carried out to study the near-field electromagnetic coupling between transmitting and

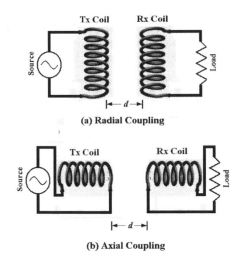

(a) Radial Coupling

(b) Axial Coupling

Figure 2. Pictorial view of (a) radial and (b) axial coupling between transmitter (Tx) and receiver (Rx) coils.

receiving electrically-small loop antennas as well as the self-resonance coupling between them. For self-resonance coupling mechanism, the frequency of the transmitted signal is kept same as the self-resonance frequency (f_{rs}) of the transmitting and receiving antenna system; however, for the near-field electromagnetic coupling mechanism, the frequency of the transmitted signal is kept well above the f_{rs} subject to obtain high radiation efficiency of the transmitting antenna. Experiments have been carried out to study the Power Transfer Efficiency (PTE) of the antenna system coupled in axial and radial orientations as shown in Figures 2 (a) and (b) respectively, for both the abovementioned coupling mechanisms. Results show that the better PTE is achieved in axial coupling mode as compared to the radial coupling mode when the self-resonance coupling mechanism is used; but the usable frequency range (i.e. the bandwidth) is found to be very small less than 1 MHz for this coupling mechanism. The observation is reversed (i.e. PTE associated with radial coupling orientation is found to be far better than that in axial coupling orientation) when near-field electromagnetic coupling mechanism is used and the bandwidth for WPT for this mechanism is also found to much broader ~50 MHz.

5 CONCLUSION

Authors have briefly described the historical background of the wireless power transmission in this part of the entire work. This initial part of the paper may be regarded as the introductory part of authors' experimental studies on wireless power transfer via near-field coupling of electrically-small transmitting and receiving loop antennas which have been recently carried out at the Antenna and Propagation Laboratory, SKFGI, Mankundu. Further, the theoretical modeling of the transmitting and receiving loop antennas as the wireless power transfer module and the corresponding experimental studies have been discussed in the successive parts later on.

REFERENCES

Akkermans, J. A. G., van Beurden, M. C., Doodeman, G. J. N., and Visser, H. J. (2005). Analytical models for low-power rectenna design," *IEEE Antennas and Wireless Propagation Letters*, 4, 187–190.

Ali, M., Yang, G., and Dougal, R. (2005). A new circularly polarized rectenna for wireless power transmission and data communication," *IEEE Antennas and Wireless Propagation Letters*, 4.

Basset, P., Andreas Kaiser, B. L., Collard, D., and Buchaillot, L. (2007). Complete system for wireless powering and remote control of electrostatic actuators by inductive coupling. *IEEE/ASME Transactions on Mechatronics*, 12(1), 23–31.

Bondyopadhyay, P. K. (1998). Sir J. C. Bose's Diode Detector Received Marconi's First Transatlantic Wireless Signal of December 1901 (The "Italian Navy Coherer" Scandal Revisited). *Proc. IEEE*, 86(1).

Gao, J. (2007). Traveling magnetic field for homogeneous wireless power transmission. *IEEE Transactions on Power Delivery*, 22(1), 507–514.

Glaser, P. E. (1973). Method and apparatus for converting solar radiation to electrical power. U.S. Patent 3 781 647.

Hertz, H. (1888). Ueber die Ausbreitungsgeschwindigkeit der elektrodynamischen Wirkungen. *Annalen der Physik*, 270(7), 551–569.

Hu, H., Hu, Y., Chen, C., and Wang, J. (2008). A system of two piezoelectric transducers and a storage circuit for wireless energy transmission through a thin metal wall. *IEEE Transactions on Ultrasonics, Ferroelectrics, and Frequency Control*, 55(10), 2312–2319.

Karalis, A., Joannopoulos, J., Soljacic, M. (2008). Efficient wireless non-radiative mid-range energy transfer. *Elsevier Annals of Physics*, 323, 34–48.

Kurs, A. (2007). Power transfer through strongly coupled resonances. *Massachusetts Institute of Technology, Master of Science in Physics Thesis*.

Low, Z. N., Chinga, R. A., Tseng, R., and Lin, J. (2009). Design and test of a high-power high-efficiency loosely coupled planar wireless power transfer system. *IEEE Transactions on Industrial Electronics*, 56(5), 1801–1812.

Mansor, H., Halim, M., Mashor, M., and Rahim, M. (2008). Application on wireless power transmission for biomedical implantable organ. *Springer-Verlag Biomed. Proceedings*, 21, 40–43.

Maxwell, J. C. (1864). A dynamical theory of the electromagnetic field. *Proceedings of Royal Society London*, XIII, 531.

NASA (2003). Beamed laser power for UAVs. *Dryden Flight Research Center.*

Poynting, J. H. (1884). On the Transfer of Energy in the Electromagnetic Field. *Philosophical Transactions of the Royal Society of London,* 175, 343–361.

Shams, K. M. Z., and Ali, M. (2007). Wireless power transmission to a buried sensor in concrete. *IEEE Sensors Journal*, 7(12), 1573–1577.

Tesla, N. (1891). Experiments With Alternate Currents of Very High Frequency and Their Application to Methods of Artificial Illumination. New York, NY, USA: AIEE.

Tesla, N. (1904). The Transmission of Electrical Energy without Wires. *Electrical World and Engineer.*

Tesla, N. (1914). Apparatus for transmitting electrical energy. U.S. Patent 1119732.

Frontiers in Computer, Communication and Electrical Engineering – Acharyya (Ed.)
© 2016 Taylor & Francis Group, London, ISBN: 978-1-138-02877-7

Wireless power transmission—part II: Theoretical modeling of transmitting and receiving electrically-small loop antennas

Nirvik Patra, Debodyuti Banerjee, Subhashri Chatterjee, T.K. Sengupta,
Subhendu Chakraborty & Aritra Acharyya
Supreme Knowledge Foundation Group of Institutions, Mankundu, Hooghly, West Bengal, India

ABSTRACT: In this part of the paper, the theoretical modeling of electrically-small loop antennas as transmitting and receiving modules for wireless power transmission have been discussed in detail. The proposed experimental setup for the measurements has also been described in brief at the end.

1 THEORY OF ELECTRICALLY-SMALL LOOP ANTENNAS

Loop antennas may be classified into two categories such as (a) electrically-small and (b) electrically-large. If the overall length, i.e. the circumference (C) of a loop antenna is less than one-tenth of the wavelength of the transmitting/receiving signal (i.e. $C < \lambda/10$), then it is called electrically-small. However, if the circumference of a loop antenna is comparable to the free-space wavelength (i.e. $C \sim \lambda$) then it is considered as electrically-large. In the present paper, the frequency range under study is 100 KHz–60 MHz and the corresponding wavelength range is 5–3000 m. But the diameter of the transmitter (Tx) and receiver (Rx) coils are $2a = 8$ cm. therefore the overall length of both the coils are $C = 2\pi a \approx 0.50$ m $<< \lambda$; therefore these can be considered as electrically-small.

In this section, at first the theory of electrically-small loop antenna is presented in brief. And finally the expression for power transfer efficiency has been established for WPT between two near-field coupled Tx and Rx electrically-small loop antennas. The near-field properties of the electrically-small loop antennas are taken into consideration since those are used for the WPT within a very short distance (i.e. $d \sim 2a << \lambda$). A single thin wire loop is assumed to be positioned symmetrically on the xy plane, at $z = 0$, as shown in Figure 2. Since the overall length of the loop is very small as compared to the free-space wavelength ($C << \lambda$), thus constant current distribution $I_\phi = I_0$ (where I_0 is a constant) throughout the loop is a valid approximation. Now, by following the standard procedure adopted by C. A. Balanis (Balanis 2010), the near-field electric and magnetic field components can easily be obtained. The near-field electric field components are given by (Balanis 2010).

$$E_r = E_\theta = 0, \tag{1}$$

$$E_\phi = \eta_0 \frac{(\beta a)^2 I_0 \sin\theta}{4r}\left[1 + \frac{1}{j\beta r}\right]e^{-j\beta r}. \tag{2}$$

And the near-field magnetic field components are given by (Balanis 2010)

$$H_r = j\frac{\beta a^2 I_0 \cos\theta}{2r^2}\left[1 + \frac{1}{j\beta r}\right]e^{-j\beta r}, \tag{3}$$

$$H_\theta = -\frac{(\beta a)^2 I_0 \sin\theta}{4r}\left[1 + \frac{1}{j\beta r} - \frac{1}{(\beta r)^2}\right]e^{-j\beta r}, \tag{4}$$

$$H_\phi = 0. \tag{5}$$

In equations (1)–(5), E_r, E_θ, E_ϕ and H_r, H_θ, H_ϕ are the r, θ, ϕ components of electric and magnetic fields in spherical coordinate system respectively (where $\vec{E}(r,\theta,\phi) = E_r \hat{a}_r + E_\theta \hat{a}_\theta + E_\phi \hat{a}_\phi$ and $\vec{H}(r,\theta,\phi) = H_r \hat{a}_r + H_\theta \hat{a}_\theta + H_\phi \hat{a}_\phi$), $\eta_0 = \sqrt{\mu_0/\varepsilon_0} = 120\pi$ is the intrinsic impedance of free-space, $\beta = 2\pi/\lambda$ is the phase constant, a is the radius of the loop and r is the distance of the near-field observation point (P) from the origin (outside the loop) as shown in Figure 1.

If the electrically-small loop antenna has N number of turns, then the near-field electric and magnetic field components can be obtained by simply multiplying N with those for single turn loop. Therefore the near-field electric and magnetic field components of a multiple turn loop can be obtained as

$$E_r = E_\theta = 0, \tag{6}$$

$$E_\phi = \eta_0 \frac{(\beta a)^2 NI_0 \sin\theta}{4r}\left[1 + \frac{1}{j\beta r}\right]e^{-j\beta r}, \tag{7}$$

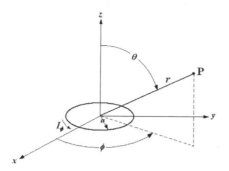

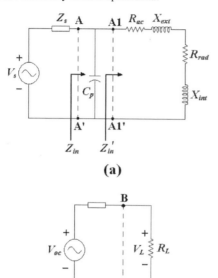

Figure 1. Geometrical arrangement for analyzing a single turn electrically small loop antenna.

(a)

(b)

Figure 2. Equivalent circuits of the electrically-small loop antenna in both (a) transmitting and (b) receiving modes.

$$H_r = j\frac{\beta a^2 N I_0 \cos\theta}{2r^2}\left[1 + \frac{1}{j\beta r}\right]e^{-j\beta r}, \qquad (8)$$

$$H_\theta = -\frac{(\beta a)^2 N I_0 \sin\theta}{4r}\left[1 + \frac{1}{j\beta r} - \frac{1}{(\beta r)^2}\right]e^{-j\beta r}, \qquad (9)$$

$$H_\phi = 0. \qquad (10)$$

Now the complex power density at any point outside the multiple turn electrically-small loop can be calculated as

$$\overline{W} = \frac{1}{2}\left(\vec{E} \times \vec{H}^*\right) = \frac{1}{2}\left(-E_\phi H_\theta^* \, \hat{a}_r + E_\phi H_r^* \, \hat{a}_\theta\right). \qquad (11)$$

The r- and θ-components of $\overline{W}$ is given by

$$W_r = \eta_0 \frac{(\beta a)^4}{32} N^2 |I_0|^2 \left(\frac{\sin\theta}{r}\right)^2 \left[1 + \frac{j}{(\beta r)^3}\right], \qquad (12)$$

$$W_\theta = -j\eta_0 \frac{\beta^3 a^4}{16 r^3} N^2 |I_0|^2 \sin\theta \cos\theta \left[1 + \frac{1}{(\beta r)^2}\right], \qquad (13)$$

respectively. Therefore the radiated power can be calculated a (using equations (12) and (13))

$$
\begin{aligned}
P_{rad} &= \operatorname{Re}\left[\oiint_S \overline{W}.\overline{dS}\right] \\
&= \operatorname{Re}\left[\oiint_S \left(W_r \, \hat{a}_r + W_\theta \, \hat{a}_\theta\right).\overline{dS}\right] \\
&= \eta_0\left(\frac{\pi}{12}\right)(\beta a)^4 N^2 |I_0|^2.
\end{aligned}
\qquad (14)
$$

Moreover, the radiated power may be also written as $P_{rad} = (1/2)|I_0|^2 R_{rad}$ (where R_{rad} is the radiation resistance), which can be equated to equation (14) and from that equality, the value of R_{rad} can be calculated as

$$
\left.
\begin{aligned}
R_{rad} &= \eta_0\left(\frac{\pi}{6}\right)(\beta a)^4 N^2 \\
&= \eta_0\left(\frac{2\pi}{3}\right)\left(\frac{\beta S}{\lambda}\right)^2 N^2 \\
&= 20\pi^2\left(\frac{C}{\lambda}\right)^2 N^2
\end{aligned}
\right\},
\qquad (15)
$$

where $S = \pi a^2$ and $C = 2\pi a$ are the cross-sectional area and circumference of the coil having circular cross-section. Since R_{rad} is the function of λ, therefore it is a function of frequency ($R_{rad}(f)$).

2 EQUIVALENT CIRCUITS OF TRANSMITTING AND RECEIVING LOOP ANTENNAS

The equivalent circuit of the Tx loop antenna (electrically-small) is shown in Figure 2(a). The input impedance Z_{in} across **A1A1′** terminals is given by

$$
\begin{aligned}
Z_{in} &= R_{in} + jX_{in} \\
&= (R_{rad} + R_{ac}) + j(X_{ext} + X_{int}),
\end{aligned}
\qquad (16)
$$

where R_{rad} is the radiation resistance given by equation (15), R_{ac} is the AC loss resistance of the loop conductor, $X_{ext} = 2\pi f L_{ext}$ is the external inductive reactance, $X_{int} = 2\pi f L_{int}$ is the internal inductive reactance of the loop antenna, C_p is the distributed stray or parasitic capacitance responsible for self-resonance of the loop antenna, V_s is the source voltage phasor and Z_s is the internal impedance of the source.

The AC loss resistance of the loop conductance can be calculated from the Dowell's equation (Dowell 1966) given by

$$R_{ac}(f) = \left(\frac{l}{\delta\sigma t_e}\right)mN\left[\Xi_1 + n_p^2\left(\frac{2}{3}\right)(m^2-1)\Xi_2\right], \quad (17)$$

where l is the length of the coil, $\delta = \sqrt{1/\pi f\sigma\mu}$ is the skin depth of the conductor at frequency f, σ is the conductivity of the conductor, m is the number of layers, Ξ_1 is the skin effect factor given by

$$\Xi_1 = \frac{\sinh(2\Delta') + \sin(2\Delta')}{\cosh(2\Delta') - \cos(2\Delta')}, \quad (18)$$

and Ξ_2 is the proximity effect factor given by

$$\Xi_2 = \frac{\sinh(2\Delta') - \sin(2\Delta')}{\cosh(2\Delta') + \cos(2\Delta')}, \quad (19)$$

where Δ' represents the modified penetration depth $\Delta' = \sqrt{n_p}\Delta$, Δ represents the penetration ration given by $\Delta = t_e/\delta$, t_e is the equivalent diameter of the wire having circular cross-section given by $t_e = \sqrt{\pi/4}\,t$, t is the actual diameter of the wire, $n_p = t_e N/2a$ is the porosity factor. Thus equation (17) takes into account both the skin and proximity effects.

In the present work, both the Tx and Rx coils have a single layer, i.e. $m = 1$ and other dimensions are given by (i) $t = 4.877$ mm (6 SWG wire), (ii) $2a = 8$ cm, (iii) $l = 8$ cm, (iv) $N = 30$. The conductor of the wire is copper; therefore the values of $\sigma = 5.96 \times 10^7$ S m^{-1}, $\mu = 1.256629 \times 10^{-6}$ H m^{-1} which are taken from the reference (Technical Data for Copper 2015). Thus the value of R_{ac} can easily be calculated from equation (17) as a function of frequency using the above mentioned parameter values.

Moreover, the external and internal inductances of the coils can be calculated from (Kraus 1992)

$$L_{ext} = \mu_0 aN\left[\ln\left(\frac{16a}{t}\right) - 2\right], \quad (20)$$

$$L_{int}(f) = \left(\frac{aN}{\pi ft}\right)\sqrt{\frac{\pi f\mu_0}{\sigma}}, \quad (21)$$

where $\mu_0 = 4\pi \times 10^{-7}$ H m^{-1} is the permeability of free-space.

If the self-resonance frequency (f_{rs}) of the coils can be measured experimentally, then from the knowledge of f_{rs}, L_{ext} and L_{int}, the value of the parasitic capacitance associated with the coils can easily be determined. The input admittance of the coil can be written as

$$Y_{in} = G_{in} + jB_{in} = \frac{1}{Z_{in}} = \frac{1}{R_{in} + jX_{in}}, \quad (22)$$

where $G_{in} = R_{in}/(R_{in}^2 + X_{in}^2)$ and $B_{in} = -X_{in}/(R_{in}^2 + X_{in}^2)$ are the conductance and susceptance. Now, at the self-resonance frequency (f_{rs}), the magnitude of the susceptance of C_p ($|B_p| = 1/2\pi f_{rs}C_p$) must be equal to the magnitude of the susceptance of Y_{in} (i.e. $|B_p|_{f=f_{rs}} = |B_{in}|_{f=f_{rs}}$). Thus at the self-resonance frequency

$$C_p =$$
$$\frac{\left(R_{rad}(f_{rs}) + R_{ac}(f_{rs})\right)^2 + 4\pi^2 f_{rs}^2\left(L_{ext} + L_{int}(f_{rs})\right)^2}{4\pi^2 f_{rs}^2\left(L_{ext} + L_{int}(f_{rs})\right)}.$$
$$(23)$$

The equivalent circuit of the Rx coil in receiving mode is shown in Figure 2(b); where V_{oc} is the open circuit voltage induced across it, Z_{in} is the input impedance of it given by the equation (16), V_L is the voltage across the R_L. Thus the V_L can be obtained as

$$V_L = V_{oc}\frac{R_L}{R_L + Z_{in}}. \quad (24)$$

3 CALCULATION OF POWER TRANSFER EFFICIENCY

Now the input power to the Tx coil is given by

$$P_{in} = \left(\frac{V_{in(p-p)}}{2\sqrt{2}}\right)^2\left(\frac{1}{|Z'_{in}|}\right)\cos\varphi_{in}, \quad (25)$$

where $Z'_{in} = Z_{in} \| (1/2j\pi fC_p)$ is the input impedance of the Tx coil across AA' terminals, $\varphi_{in} = \tan^{-1}(\text{Im}(Z'_{in})/\text{Re}(Z'_{in}))$ and $V_{in(p-p)}$ is the peak-to-peak signal amplitude measured across AA' terminals. Similarly, the power absorbed by the pure resistive load RL at the receiving side is given by

$$P_{out}(f,d) = \left(\frac{V_{L(p-p)}(f,d)}{2\sqrt{2}}\right)^2\left(\frac{1}{R_L}\right), \quad (26)$$

where $V_{L(p-p)}(f,d)$ is the peak-to-peak signal amplitude measured across BB' terminals which is obviously the functions of both frequency (f) of the transmitted signal and separation (d) between the Tx and Rx coils. Therefore the Power Transfer Efficiency (PTE) can be calculated as

$$PTE(f,d) = \left(\frac{P_{out}(f,d)}{P_{in}} \right). \qquad (27)$$

4 EXPERIMENTAL SETUP

The block diagram of the experimental setup is shown in Figure 3. The AC signal is amplified through a Class-B power amplifier and then fed into the transmitting coil. A DC regulated ±15 V bipolar power supply is used to provide the appropriate bias to the Class-B power amplifier. A Digital Storage Oscilloscope (DSO-1) is connected across the transmitting coil to obtain the direct readout of the peak-to-peak amplitude ($V_{in(p-p)}$) and frequency (f) of the transmitted signal. In the receiver side, a pure resistive load $R_L = 1 \ \Omega$ is connected across the receiver coil. Another Digital Storage Oscilloscope (DSO-2) is connected across the load (R_L) to obtain the read out of the peak-to-peak amplitude

Figure 4. Transmitter (Tx) and receiver (Rx) coils coupled in (a) axial and (b) radial orientations. Snap-shots of the arrangement for power pattern measurement of the WPT system coupled in (c) axial and (d) radial orientations.

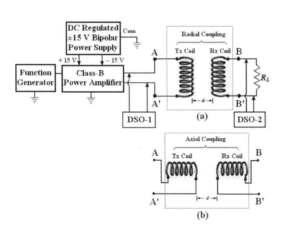

Figure 3. Experimental setup to study the WPT system for both (a) radial and (b) axial coupling of Tx and Rx coils.

($V_{L(p-p)}(f, d)$) and frequency (f) of the RF voltage appeared across R_L. Figures 4(a)–(d) show the snaps of the Tx and Rx coils coupled in axial and radial orientations along with the arrangement for power pattern measurement of the WPT system in both the said orientations.

REFERENCES

Balanis, C.A. (2010). Antenna Theory: Analysis and Design. 3rd Edition, *Wiley-India Edition, India*, 231–246.
Dowell, P. (1966). Effect of eddy currents in transformers windings. *Proceedings of the Institution of Electrical Engineers*, 113(8), 1387–1394.
Technical Data for Copper. (2015). Retrieved from http://periodictable.com/Elements/029/data.html.
Kraus, J.D. (1992). Electromagnetics. 4th *Edition, McGraw-Hill Book Co., New York, USA*.

Frontiers in Computer, Communication and Electrical Engineering – Acharyya (Ed.)
© *2016 Taylor & Francis Group, London, ISBN: 978-1-138-02877-7*

Wireless power transmission—part III: Experimental study

Nirvik Patra, Debodyuti Banerjee, Subhashri Chatterjee, T.K. Sengupta,
Subhendu Chakraborty & Aritra Acharyya
Supreme Knowledge Foundation Group of Institutions, Mankundu, Hooghly, West Bengal, India

ABSTRACT: The authors have carried out a comprehensive experimental study on Wireless Power Transfer (WPT) by using both the self-resonance coupling and near-field electromagnetic coupling between transmitting and receiving electrically-small loop antennas. For the first coupling mechanism, the frequency of the transmitted signal is kept same as the self-resonance frequency of the transmitting and receiving antenna system; however, for the second coupling mechanism, the frequency of the transmitted signal is kept well above the self-resonance frequency subject to obtain high radiation efficiency of the transmitting antenna. Experiments have been carried out to study the Power Transfer Efficiency (PTE) of the antenna system coupled in axial and radial orientations for both the abovementioned coupling mechanisms. Results show that the better PTE is achieved in axial coupling mode as compared to the radial coupling mode when the self-resonance coupling mechanism is used; but the usable frequency range (i.e. the bandwidth) is found to be very small less than 1 MHz for this coupling mechanism. The observation is reversed when near-field electromagnetic coupling mechanism is used and the bandwidth for WPT for this mechanism is also found to much broader ~50 MHz.

1 MEASUREMENT OF FREQUENCY RESPONSE

The aim of the present work as well as the proposed setup for the experimental measurements have been already described in the earlier parts of the paper respectively (Patra *et al.* 2016a,b). Initially the frequency response of the WPT system have been studied for both radial and axial coupling (R-coupling and A-coupling) of Tx and Rx coils by keeping the separation between them $d = 0$ cm and varying the frequency of the transmitted signal (generated from function generator) from 100 KHz – 100 MHz. The power transfer efficiency in both the above mentioned coupling modes (PTE_r and PTE_a) have been calculated by using the formula given in equation (27) of Part II (Patra *et al.* 2016b) and from the peak-to-peak amplitudes of the transmitted and received signals ($V_{in(p-p)}$ and $V_{L(p-p)}$ ($f,d = 0$)) readout from DSO-1 and DSO-2 respectively for each frequency. The plots of PTE_r and PTE_a as functions of frequency (within the range of $f = 100$ KHz – 100 MHz) are shown in Figure 1. Now the total inductance ($L_T(f) = L_{ext} + L_{int}(f)$) of Tx and Rx coils can be calculated as functions of frequency from equations (20) and (21) of Part II. It is observed from Figure 1 that the self-resonance frequency (f_{rs}) of both the coils (where both PTE_r and PTE_a sharply attain their respective peaks) is found to be $f_{rs} = 1.4$ MHz. From equations (20)

and (21) of Part II, the value of L_T at resonance frequency (1.4 MHz) is found to be $L_T = 5.86$ μH. Thus the parasitic capacitance (C_p) of the coil may be obtained from equation (23) of Part II and its value is $C_p = 2.20$ nF at 1.4 MHz. It is interesting to observe from Figure 1 that the PTE_a is much greater

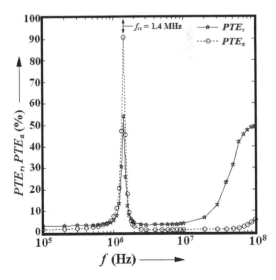

Figure 1. Frequency response of WPT system in both radial and axial coupling of Tx and Rx coils for zero centimeter separation between them.

than PTE_r at the resonance frequency $f_{rs} = 1.4$ MHz $(PTE_a|_{(f=f_r)} = 90.67\%$ and $PTE_r|_{(f=f_r)} = 53.91\%)$. Stronger magnetic field coupling in A-coupling orientation at resonance frequency as compare to the near-field electromagnetic in R-coupling orientation is the cause of higher value of PTE_a than PTE_r at f_{rs}. Very small value of radiation resistance $R_{rad} = 82.07$ µΩ and comparatively larger value of loss resistance $R_{ac} = 211.27$ nΩ cause a very small radiation efficiency $\epsilon_{rad} = (R_{rad}/(R_{rad} + R_{ac})) \times 100 = 3.88 \times 10^{-4}\%$ in R-coupling orientation at 1.4 MHz. Consequently, the near-field electromagnetic coupling in R-coupling orientation is very weak at 1.4 MHz. However, at the resonance frequency $f_{rs} = 1.4$ MHz, the impedance of both the coils become minimum and purely resistive; which causes maximum flow of AC current through then and leads to maximum magnetic field coupling in axial orientation (i.e. in A-coupling orientation).

Further, it is very interesting to observe from Figure 1 that at the higher frequencies especially $f > 50$ MHz, the values of PTE_r significantly predominate over the values of PTE_a. This fact can be explained from Figure 2. The loss resistance (R_{ac}) and radiation resistance (R_{rad}) of the Tx coil have been calculated as functions of frequency and plots of R_{ac}, R_{rad} versus frequency are shown in Figure 2. It is observed from Figure 2 that the values of R_{rad} are much smaller and almost negligible as compared to the values of R_{ac} at the lower frequencies. But the rate of increase of R_{rad} with respect to frequency (i.e. dR_{ac}/df) is much larger than that of R_{ac} (i.e. dR_{ac}/df). It is interesting to note that the values of R_{rad} become almost comparable to the values of R_{ac}, when the frequency increases beyond 10 MHz. At $f = 54.1$ MHz, both of them

become equal, $R_{rad} = R_{ac} = 1.83$ ohm and corresponding radiation efficiency becomes $\epsilon_{rad} = 50\%$. The values of R_{rad} over take those of R_{ac} while the frequency increases above 54.1 MHz; which leads to sharp increase of radiation efficiency of the Tx coil. Thus the near-field electromagnetic coupling in R-coupling orientation improves sharply for the higher frequencies $f > 10$ MHz and due to the significant increase of ϵ_{rad} beyond 54.1 MHz, PTE_r increase significantly. But at the same time, it may be noted that these frequencies ($f > 10$ MHz) are much larger than the self-resonance frequency of the coils. Thus at higher frequencies (10 MHz ≤ f ≤ 60 MHz), the magnitude the input impedance of the coils are much greater than that of its value at resonance. Thus the AC current flowing in both the coils is very small, causing very small magnetic coupling. As a result of small amount of magnetic coupling, the PTE_a values are very small at the said frequency range (10 MHz ≤ f ≤ 60 MHz).

2 MEASUREMENT OF POWER TRANSFER EFFICIENCY VERSUS DISTANCE CHARACTERISTICS

Power transfer efficiency versus distance characteristics of the WPT system have been also investigated by varying the separation between T_x and R_x coils coupled in axial (i.e. magnetic field coupling) and radial (i.e. near-field electromagnetic coupling) directions. The above mentioned characteristics are shown in Figures 3 and 4 for 1.4 MHz and 60 MHz frequencies respectively.

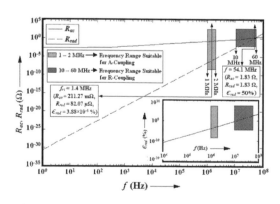

Figure 2. Variations of radiation resistance and loss resistance of the coils with frequency. Inset of the figure shows the variation of radiation efficiency of the transmitting loop antenna with frequency. In all the graphs the frequency ranges suitable for A-coupling (1–2 MHz) and R-coupling (10–60 MHz) are highlighted by the light gray and deep gray boxes respectively.

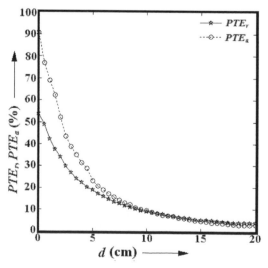

Figure 3. Power transfer efficiency versus distance plots of WPT system in radial and axial coupling orientations at self-resonance frequency ($f_{rs} = 1.4$ MHz).

It is observed from Figure 3 that the value of PTE_a is larger than the values of PTE_r when the separation of the coils are small ($d \le 8$ cm) at the self-resonance frequency ($f_{rs} = 1.4$ MHz). However, both the PTE_a and $PTEr$ become comparable for higher separation between the coils ($d > 8$ cm) at 1.4 MHz. It suggests that the axial coupling must be preferred over the radial coupling to obtain better power transfer efficiency, especially for smaller separation or very small distance WPT at self-resonance frequency. Further, from Figure 4 it can be concluded that at 60 MHz (i.e. $f \gg f_{rs}$), radial coupling orientation is highly suitable for WPT and its performance is even better than the axial coupling orientation at self-resonance frequency for larger separation between the Tx and Rx coils.

3 MEASUREMENT OF RADIATION PATTERNS

Normalized power patterns of the R_x coil in yz plane for both R- and A-coupling orientations at self-resonance frequency $f_{rs} = 1.4$ MHz for $d = 5$ cm are shown in Figure 5. The same patterns for 60 MHz transmitted frequency for the same separation ($d = 5$ cm) have been shown in Figure 6. At self-resonance frequency, half power beam width for A-coupling orientation is found to be larger ($HPBW_a = 108.11°$) than that for R-coupling ($HPBW_r = 84.13°$). The same nature is also observed in the normalized power patterns

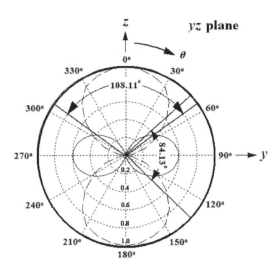

Figure 5. Normalized power patterns of the receiving coil in yz plane for both radial and axial coupling orientations at self-resonance frequency ($f_{rs} = 1.4$ MHz) for 5 cm separation between transmitter and receiver coils.

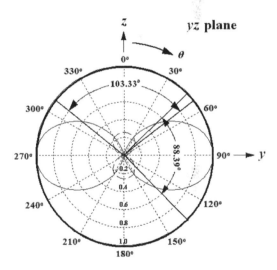

Figure 6. Normalized power patterns of the receiving coil in yz plane for both radial and axial coupling orientations at 60 MHz for 5 cm separation between transmitter and receiver coils.

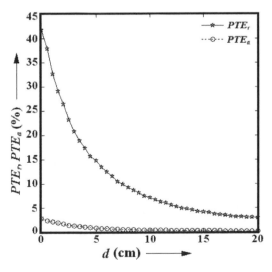

Figure 4. Power transfer efficiency versus distance plots of WPT systems in radial and axial coupling orientations at $f = 60$ MHz.

at 60 MHz (i.e. $HPBW_a = 103.33° > HPBW_r = 88.39°$). The smaller beam widths of R-coupling orientation for both the cases suggest better directivity of that orientation as compared to the A-coupling orientation. Better directivity naturally improves the received power density at larger distance from the transmitter. Therefore, R-coupling is always the preferable orientation for longer distance WPT.

4 CONCLUSION

A comprehensive experimental study have been carried out on wireless power transfer by using both the self-resonance coupling and near-field electromagnetic coupling between transmitting and receiving electrically-small loop antennas. Experiments have been carried out to study the frequency response, Power Transfer Efficiency (PTE) versus distance characteristics and power pattern characteristics of the antenna system coupled in axial and radial orientations for both the abovementioned coupling mechanisms. It may be concluded from the experimental results that the near-field electromagnetic coupling mechanism is the better choice for WPT between transmitting and receiving loop antennas aligned in radial direction as compared to the self-resonance coupling mechanism between transmitting and receiving loop antennas aligned in axial direction especially when the separation between the antennas is larger. The results are highly encouraging to make larger prototypes of the WPT system for transmission of high power to confirm the better suitability of radial coupling orientation for usable amount of far-field power transmission.

REFERENCES

Patra, N., Banerjee, D., Chatterjee, S., Sengupta, T. K., and Acharyya, A. (2016a). Wireless Power Transmission—Part I: A Brief History. In Proceedings of C2E2 2016, SKFGI, Mankundu, WB, India, (pp. 5–7).
Patra, N., Banerjee, D., Chatterjee, S., Sengupta, T. K., and Acharyya, A. (2016b). Wireless Power Transmission—Part II: Theoretical Modeling of Transmitting and Receiving Electrically-Small Loop Antennas. In Proceedings of C2E2 2016, SKFGI, Mankundu, WB, India, (pp. 8–11).

Evaluation of ionization rates of charge carriers in a semiconductor via a generalized analytical model based on multistage scattering phenomena—part I: Wurtzite-GaN

Apala Banerjee, Subhashri Chatterjee, Adrija Das, Subhendu Chakraborty & Aritra Acharyya
Supreme Knowledge Foundation Group of Institutions, Mankundu, Hooghly, West Bengal, India

ABSTRACT: A comprehensive analytical model based on multistage scattering phenomena has been used to evaluate the impact ionization rates of electrons and holes in Wz-GaN within the field range of 4.0×10^7–2.0×10^8 V m^{-1}. The numerical results obtained from the proposed analytical model within the field range under consideration have been compared with the ionization rate values calculated by using the empirical relations fitted from the experimentally measured data. The calculated values of impact ionization rates of electrons and holes in Wz-GaN are found to be in close agreement with the experimental results especially for the electron and hole concentrations of 10^{23} m^{-3} which are same as those taken into account in the experiment.

1 INTRODUCTION

Avalanche multiplication by means of impact ionization of charge carriers within the active regions reversed biased *p-n* junctions plays a significant role in various microwave, millimeter-wave and optoelectronic semiconductor devices. Several theoretical models to describe and formulate the impact ionization rates of charge carriers in a semiconductor material had been proposed by different researchers since last sixty years. Out of those, most acceptable analytical models were proposed by Wolff (Wolff 1954) in the year 1954 and Shockley (Shockley 1961) in the year 1961. In the year 1961, Mole and Meyer (Mole *et al*. 1961) modified the Shockley's theory by considering different possible ways through which electron can cause ionization by acquiring energy equal and greater than the ionization threshold energy. Two years later in 1963, Mole and Overstraeten (Mole *et al*. 1963) proved that the theories of Wolff (Wolff 1954) and Shockley (Shockley 1961) are applicable for low and high field conditions respectively. One year earlier of that, i.e. in the year 1962, Baraff (Baraff 1962) had plotted the impact ionization rates of charge carriers as function of electric field from the numerical simulation technique without any low or high field approximations. In the year 1975, Ghosh and Roy (Ghosh *et al*. 1975) shown that the higher carrier density in semiconductors enhance the energy loss due to electron-electron collisions which causes significant deterioration in ionization rate of charge carriers. They considered the energy loss occurred due to electron-electron interactions

in addition to the usual scattering events like optical phonon scattering ionization collisions in their analysis. Ten years later in 1985, Singh and Pal (Singh *et al*. 1985) adopted the similar approach and evaluated the ionization rates of electrons and holes in <100> oriented in GaAs. The calculated results were in close agreement with the experimental results of Ito *et al*. (Ito *et al*. 1978) and Pearsall *et al*. (Pearsall *et al*. 1978). In addition to that the theory proposed by Singh and Pal (Singh *et al*. 1985) was successful to explain the behavior of holes in GaAs as spin-orbit splitoff holes and heavy holes at lower and higher field ranges respectively.

Generation of Electron Hole Pairs (EHP) by impact ionization process basically a multistage phenomena. Several combinations of optical phonon scattering as well as carrier-carrier collision events may take place prior to the generation of an EHP via impact ionization. This microscopic view of an EHP generation by means of impact ionization has been first taken into account by Acharyya *et al*. (Acharyya *et al*. 2014) in the year 2014. They considered the all possible combinations of optical phonon scattering and carrier-carrier collision events prior to the impact ionization and thereby generation of an EHP. Finally the probability of impact ionization has been obtained from a trinomial distribution function which describes the above mentioned multistage scattering phenomena. Using the probability of impact ionization, finally the analytical expression of ionization rate of charge carriers in a semiconductor material has been developed and using that the ionization rates of electrons and holes in 4H-SiC have been calculated as functions of electric field. The calculated results show better agreement with the experimental data (Konstantinov et al. 1997) as compared to those calculated from the analytical expressions of Ghosh et al. (Ghosh et al. 1975) and Singh et al. (Singh et al. 1985).

In the present paper, similar approach (as in ref. Acharyya et al. 2014) has been taken into account to calculate impact ionization rates of both electrons and holes in Wurtzite-GaN (Wz-GaN). The calculated results have been compared with the values of electrons and holes ionization rates of Wz-GaN obtained from the empirical relation with the experimentally measured data of Kunihiro et al. (Kunihiro et al. 1999) within the electrical field range under consideration. The calculated results are found to be in good agreement with the experimental results.

2 THEORETICAL MODEL

Acharyya et al. (Acharyya et al. 2014) considered the multistage scattering phenomena by taking into account a w-stage process; where $w = 1, 2, 3, 4, \ldots, N, N+1$. During the first N stages, all possible combinations of optical phonon scattering as well as carrier-carrier scattering events are assumed to be occurred. At the final $(N+1)^{th}$ stage, the impact ionization is assumed to be occurred. They have formulated the above mentioned w-stage process by means of a trinomial distribution function. And finally by considering all possible values of the number of stages, i.e. values of w varying from 1 to ∞, the total probability of multistage impact ionization initiated by electrons or holes can be expressed as ($P_{Tc} \equiv P_{Te}$ for electrons and $P_{Tc} \equiv P_{Th}$ for holes)

$$P_{Tc}(\xi) = \frac{p_{i(e,h)c} \exp\left(-\dfrac{E_{i(e,h)c}}{q\xi l_{rc}}\right)}{\left[1 - \left\{p_{rc}\exp\left(-\dfrac{E_r}{q\xi l_{ic}}\right) + p_{cc}\exp\left(-\dfrac{E_{cc}}{q\xi l_{irc}}\right)\right\}\right]},$$ (1)

where $p_{i(e,h)c}$, p_{rc}, p_{cc} are the probabilities of occurrence of impact ionization, optical phonon generation, carrier-carrier interaction after reaching the respective energies ($p_{i(e,h)c} \equiv p_{i(e,h)e}$, $p_{rc} \equiv p_{re}$ and $p_{cc} \equiv p_{ee}$ for electrons, $p_{i(e,h)c} \equiv p_{i(e,h)h}$, $p_{rc} \equiv p_{rh}$ and $p_{cc} \equiv p_{hh}$ for holes), E_r is the energy of optical phonons, E_{cc} is the average energy loss due to carrier-carrier interation ($E_{cc} \equiv E_{ee}$ for electron-electron and $E_{cc} \equiv E_{hh}$ for hole-hole interaction), $E_{i(e,h)c}$ is the ionization threshold energy ($E_{i(e,h)c} \equiv E_{i(e,h)e}$ for electrons

and $E_{i(e,h)c} \equiv E_{i(e,h)h}$ for holes), $l_{rc} = \left(l_r^{-1} + l_{cc}^{-1}\right)^{-1}$, $l_{ic} = \left(l_{i(e,h)c}^{-1} + l_{cc}^{-1}\right)^{-1}$, $l_{irc} = \left(l_{i(e,h)c}^{-1} + l_r^{-1}\right)^{-1}$; where l_r, l_{cc} and $l_{i(e,h)c}$ are the mean free paths associated with optical phonon scattering, carrier-carrier collisions and impact ionization respectively ($l_{cc} \equiv l_{ee}$, $l_{i(e,h)c} \equiv l_{i(e,h)e}$ for electrons and $l_{cc} \equiv l_{hh}$, $l_{i(e,h)c} \equiv l_{i(e,h)h}$ for holes), q is the electronics charge ($q = 1.6 \times 10^{-19}$ C) and ξ is the applied electric field.

Now considering the energy balance equation, i.e. applied energy per unit length equals to energy loss due to optical phonon collision, carrier-carrier interaction and impact ionization, along with the relative probabilities of impact ionization of charge carriers, impact ionization rates of charges carriers may be obtained as

$$\alpha_e(\xi) = \frac{P_{Te}(\xi)\left(1 + \dfrac{l_r}{l_{ee}}\right)\left(q\xi - \left\langle\dfrac{dE_{ee}}{dx}\right\rangle\right)}{\left(1 - P_{Te}(\xi)\right)E_r + P_{Te}(\xi)\left(1 + \dfrac{l_r}{l_{ee}}\right)E_{i(e,h)e}},$$ (2)

$$\alpha_h(\xi) = \frac{P_{Th}(\xi)\left(1 + \dfrac{l_r}{l_{hh}}\right)\left(q\xi - \left\langle\dfrac{dE_{hh}}{dx}\right\rangle\right)}{\left(1 - P_{Th}(\xi)\right)E_r + P_{Th}(\xi)\left(1 + \dfrac{l_r}{l_{hh}}\right)E_{i(e,h)h}},$$ (3)

where $\langle dE_{ee}/dx\rangle$ and $\langle dE_{hh}/dx\rangle$ are the average energy loss per unit length due to electron-electron and hole-hole collisions respectively (Acharyya et al. 2014). The effect of carrier density on $\alpha_{e,h}(\xi)$ has been incorporated by the dependence of l_{ee} and l_{hh} on the electron and hole concentrations (i.e. n and p) respectively. Those are taken to be $l_{ee} = (A_e)^c n^{-1/3}$ and $l_{hh} = (A_h)^d p^{-1/3}$ (Singh et al. 1985), where A_e, A_h, c and d are dimensionless fitting parameters which may be adjusted in numerical calculations for obtaining the best fit of the experimental data.

3 NUMERICAL RESULTS AND DISCUSSION

Numerical calculations have been carried out to obtain the ionization rate of electrons and holes in Wz-GaN as functions of applied electric field within the field range $4.0 \times 10^7 - 2.0 \times 10^8$ V m⁻¹ using the analytical expressions given in equations (2) and (3) in which the multistage scattering phenomena have been taken into account for obtaining the total probability of impact ionization of the charge carriers. The calculated values of $\alpha_{e,h}$ are compared with the empirical relations fitted from the experimental data measured by Kunihiro et al. (Kunihiro et al. 1999) given by

$$\alpha_{e,h}\left(\xi\right)=A_{e,h}\exp\left[-\left(\frac{B_{e,h}}{\xi}\right)^{m}\right], \qquad (4)$$

where the ionization coefficients A_e, B_e, A_e, B_h are 138.00×10^8 m^{-1}, 14.28×10^8 V m^{-1}, 6.867×10^8 m^{-1}, 8.72×10^8 V m^{-1} respectively for $\xi \leq 10^8$ V m^{-1} and 122.70×10^8 m^{-1}, 13.63×10^8 V m^{-1}, 3.84×10^8 m^{-1}, 7.95×10^8 V m^{-1} respectively for $\xi > 10^8$ V m^{-1}, while $m = 1$ for the entire field range under consideration (Kunihiro et al. 1999). Other material parameters of Wz-GaN such as bandgap ($E_g = 3.39$ eV), ionization threshold energy of electrons ($E_{i(e,h)\,e} = 3.6612$ eV) and holes ($E_{i(e,h)h} = 3.5934$ eV) and corresponding mean free paths ($l_{i(e,h)e} = 840$ Å and $l_{i(e,h)h} = 1360$ Å respectively), optical phonon energy ($E_r = 91.2$ meV) and corresponding mean free path ($l_r = 42$ Å), permittivity ($\varepsilon_r = 8.9$), etc. at 300 K are taken from the experimental reports (Electronic Archive 2015).

Variations of ionization rate of electrons and holes in Wz-GaN obtained from the analytical model presented in this paper with inverse of the electric field for different electron and hole concentrations respectively are shown in Figures 1 and 2. The same variations obtained from the empirical relations given in equations (4) (Kunihiro et al. 1999) are also shown in Figures 1 and 2. The parameters A_e, c, A_h and d associated with the mean free path of electron-electron and hole-hole

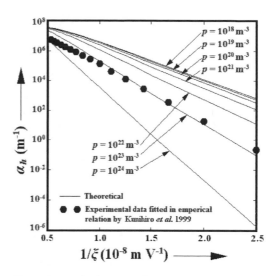

Figure 2. Ionization rate of holes in Wz-GaN versus inverse of applied electric field. Points (•) represent the plot of empirical relation given by equation (4) fitted from experimental data of Kunihiro et al. (Kunihiro et al. 1999) (hole concentration in (Kunihiro et al. 1999) is $p = 10^{23}$ m^{-3}).

collisions (l_{ee} and l_{hh}) are adjusted in numerical calculations for the best fit of the experimental data. In the present analytical model the best fittings for electron and hole ionization rates described by equations (2) and (3) with the empirical relations describing the field variations of the same parameters, i.e. equations (4) are obtained from $l_{ee} = (0.75)^{3.12}\,n^{-1/3}$ and $l_{hh} = (0.77)^{3.18}\,p^{-1/3}$ m respectively. It can be observed from Figures 1 and 2 that the analytical expressions of ionization rates considering the multistage scattering phenomena presented in this paper are in close agreement with respect to the experimental data (Kunihiro et al. 1999) at any electric field especially for the electron and hole concentrations of 10^{23} m^{-3} which are same those taken for the experiment (Kunihiro et al. 1999). It is also noteworthy from Figures 1 and 2 that the impact ionization rate of charge carriers decreased significantly when the carrier concentrations are increased. This decrement of impact ionization rates are more pronounced for the carrier concentrations of 10^{21} m^{-3} and above of it. Degradation of impact ionization probabilities for both electrons and holes as a consequence of increase of energy loss per unit length due to increased amount of electro-electron and hole-hole collisions is responsible for decrement of impact ionization rate of charge carriers at higher carrier concentrations. Moreover the degradations impact ionization rate of both electrons and holes are found to be more severe at lower electric field

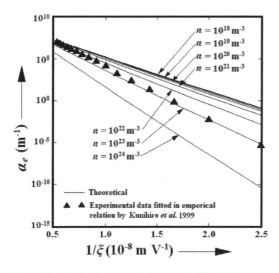

Figure 1. Ionization rate of electrons in Wz-GaN versus inverse of applied electric field. Points (▲) represent the plot of empirical relation given by equation (4) fitted from experimental data of Kunihiro et al. (Kunihiro et al. 1999) (electron concentration in (Kunihiro et al. 1999) is $n = 10^{23}$ m^{-3}).

values especially below 1.0×10^8 V m^{-1} for all carrier densities. At lower electric fields the supplied energy per unit length ($q\xi$) is smaller and comparable to the energy loss per unit length due to the carrier-carrier interactions which leads to more degradation in impact ionization probabilities at those electric fields as compared to at higher fields. Thus the values of ionization rates are more affected by energy loss due to carrier-carrier collisions at lower electric fields.

4 CONCLUSION

In this paper, a comprehensive analytical model based on multistage scattering phenomena has been used to evaluate the impact ionization rates of electrons and holes in Wz-GaN within the field range of 4.0×10^7–2.0×10^8 V m^{-1}. The numerical results obtained from the proposed analytical model within the field range under consideration have been compared with the ionization rate values calculated by using the empirical relations fitted from the experimentally measured data. The calculated values of impact ionization rates of electrons and holes in Wz-GaN are found to be in close agreement with the experimental results especially for the electron and hole concentrations of 10^{23} m^{-3} which are same as those taken into account in the experiment. The effect of varying charge density and corresponding variations of energy loss due to carrier-carrier collisions on the impact ionization rate of charge carriers in the base semiconductor can be taken into account for the analysis or simulation of several microwave, millimeter-wave and optoelectronic devices by using the analytical expressions developed in this paper.

REFERENCES

Acharyya, A., Banerjee, J. P. (2014). A generalized analytical model based on multistage scattering phenomena for estimating the impact ionization rate of charge carriers in semiconductors. *Journal of Computational Electronics*, 13, 917–924.

Baraff, G. A. (1962). Distribution functions and ionization rates for hot electrons in semiconductors. *Phys. Rev.* 128, 2507.

Electronic Archive: New Semiconductor Materials, Characteristics and Properties (2015). Retrieved from http://www.ioffe.ru/SVA/NSM/Semicond/GaN/ index.html.

Ghosh, R., and Roy, S. K. (1975). Effect of electron-electron interactions on the ionization rate of charge carriers in semiconductors. *Solid-State Electronics*, 18, 945–948.

Ito, M., Kagawa, S., Kaneda, T., and Yamaoka, T. (1978). Ionization rates for electrons and holes in GaAs. *Journal of Applied Physics*, 49, 4607.

Konstantinov, A. O., Wahab, Q., Nordell, N., and Lindefelt, U. (1997). Ionization rates and critical fields in 4H-Silicon Carbide. *Applied Physics Letters*, 71, 90–92.

Kunihiro, K., Kasahara, K., Takahashi, Y., Ohno, Y. (1999). Experimental evaluation of impact ionization coefficients in GaN. *IEEE Electron Device Letter*, 20, 608–610.

Moll, J. L., and Meyer, N. I. (1961). Secondary multiplication in silicon. *Solid-State Electronics*, 3, 155–158.

Moll, J. L., and Overstraeten, R. V. (1963). Charge multiplication in silicon p-n junctions. *Solid-State Electronics* 6, 147–157.

Pearsall, T. P., Capasso, F., Nahory, R. E., Pollack, M. A., and Chelikowsky, J. R. (1978). The band structure dependence of impact ionization by hot carriers in semiconductors: GaAs. *Solid-State Electronics*, 21, 297–302.

Shotckey, W. (1961). Problems related to *p-n* junctions in silicon. *Solid-State Electronics*, 2, 35–67.

Singh, S. R., and Pal, B. B. (1985). Ionization rates of electrons and holes in GaAs considering electron-electron and hole-hole interactions. *IEEE Trans. on Electron Devices*, 32(3), 599–604.

Wolff, P. A. (1954). Theory of electron multiplication silicon and germanium. *Phys. Rev.* 95, 1415.

Evaluation of ionization rates of charge carriers in a semiconductor via a generalized analytical model based on multistage scattering phenomena—part II: Type-IIb diamond and 6H-SiC

Apala Banerjee, Adrija Das, Subhashri Chatterjee, Subhendu Chakraborty & Aritra Acharyya
Supreme Knowledge Foundation Group of Institutions, Mankundu, Hooghly, West Bengal, India

ABSTRACT: The analytical model based on multistage scattering phenomena as discussed in the earlier part of the paper has been used to evaluate the impact ionization rates of charge carriers in type-IIb diamond and 6H-SiC within the field ranges of 4.0×10^7–1.0×10^8 V m^{-1} and 5.0×10^7–5.0×10^8 V m^{-1} respectively. The calculated results have been compared with the ionization rate values calculated by using the empirical relations fitted from the experimentally measured data. The calculated values of impact ionization rates of charge carriers in both type-IIb diamond and 6H-SiC are found to be in close agreement with the experimental results especially for the electron and hole concentrations of 10^{22} m^{-3} which are same as those taken into account in the experiments.

1 INTRODUCTION

The generalized analytical model based on multistage scattering phenomena proposed by Acharyya *et al.* (Acharyya *et al.* 2014) has been used to evaluate the impact ionization rates of charge carriers in type-IIb diamond and 6H-SiC within the field ranges of 4.0×10^7–1.0×10^8 V m^{-1} and 5.0×10^7–5.0×10^8 V m^{-1} respectively. The detail of the theoretical model has already been discussed in the earlier part of the paper (Chatterjee *et al.* 2016). The numerical results obtained from the proposed analytical model within the field ranges under consideration have been compared with the ionization rate values calculated by using the empirical relations fitted from the experimentally measured data (Konorova *et al.* 1983, Dmitriev *et al.* 1983). The calculated values of impact ionization rates of charge carriers in both type-IIb diamond and 6H-SiC are found to be in close agreement with the experimental results especially for the electron and hole concentrations of 10^{22} m^{-3} which are same as those taken into account in the experiments (Konorova *et al.* 1983, Dmitriev *et al.* 1983).

2 CALCULATION OF IMPACT IONIZATION RATES OF CHARGE CARRIERS IN TYPE-IIB DIAMOND

Numerical calculations have been carried out to obtain the ionization rate of electrons and holes in type-IIb diamond as functions of applied electric field within the field range 4.0×10^7–1.0×10^8 V m^{-1} using the analytical expressions given in equations (2) and (3) of the Part I of this paper (Chatterjee *et al.* 2016) in which the multistage scattering phenomena have been taken into account for obtaining the total probability of impact ionization of the charge carriers. The calculated values of $\alpha_{e,h}$ are compared with the empirical relations fitted from the experimental data measured by Konorova *et al.* (Konorova *et al.* 1983) given by

$$\alpha_{e,h}(\xi) = A_{e,h} \exp\left[-\left(\frac{B_{e,h}}{\xi}\right)^m\right], \qquad (1)$$

where the ionization coefficients A_e, B_e, A_e, B_h are 193.50×10^8 m^{-1}, 7.749×10^8 V m^{-1}, 193.50×10^8 m^{-1}, 7.749×10^8 V m^{-1} respectively, while $m = 1$ for the entire field range under consideration (Konorova *et al.* 1983). It is observed from the experimental measurement of Konorova *et al.* (Konorova *et al.* 1983), the ionization rates of both electrons and holes are same (i.e. $\alpha_e = \alpha_h$) in type-IIb diamond for the entire field range under consideration. Other material parameters of type-IIb diamond such as bandgap ($E_g = 5.48$ eV), ionization threshold energy of electrons ($E_{i(e,h)e} = 6.1376$ eV) and holes ($E_{i(e,h)h} = 6.1376$ eV) and corresponding mean free paths ($l_{i(e,h)e} = 3400$ Å and $l_{i(e,h)h} = 3400$ Å respectively), optical phonon energy ($E_r = 160.0$ meV) and corresponding mean free path ($l_r = 170$ Å), permittivity ($\varepsilon_r = 5.7$), etc. at room temperature (i.e. $T = 300$ K) are taken from the experimental reports (Electronic Archive 2015).

Variations of ionization rate of electrons and holes in type-IIb diamond obtained from the analytical model presented in the Part I of this paper with inverse of the electric field for different electron and hole concentrations are shown in Figure 1. The same variation obtained from the empirical relation given in equation (1) (Konorova *et al.* 1983) is also shown in Figure 1. The parameters A_e, c, A_h and d associated with the mean free path of electron-electron and hole-hole collisions (l_{ee} and l_{hh}) are adjusted in numerical calculations for the best fit of the experimental data. In the present analytical model the best fittings for electron and hole ionization rates described by equations (2) and (3) of the Part I of this paper with the empirical relations describing the field variations of the same parameters, i.e. equation (1) are obtained from $l_{ee} = (0.03)^{0.54} n^{-1/3}$ and $l_{hh} = (0.03)^{0.54} p^{-1/3}$ m respectively. It can be observed from Figure 1 that the analytical expressions of ionization rates considering the multistage scattering phenomena presented in this paper are in close agreement with respect to the experimental data (Konorova *et al.* 1983) at any electric field especially for the electron and hole concentrations of 10^{22} m^{-3} which are same those taken for the experiment (Konorova *et al.* 1983). It is also noteworthy from Figure 1 that the impact ionization rate of charge carriers decreased significantly when the carrier concentrations are increased. This decrement of impact ionization rates are more pronounced for the carrier concentrations of 10^{21} m^{-3}

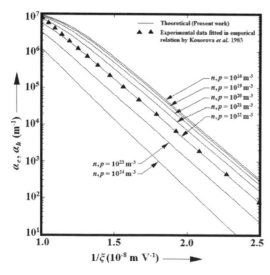

Figure 1. Ionization rates of electrons and holes in type-IIb diamond versus inverse of applied electric field. Points (▲) represent the plot of empirical relation given by equation (1) fitted from experimental data of Konorova *et al.* (Konorova *et al.* 1983) (electron and hole concentrations in (Konorova *et al.* 1983) is $n, p = 10^{22}$ m^{-3}).

and above of it. Degradation of impact ionization probabilities for both electrons and holes as a consequent of increase of energy loss per unit length due to increased amount of electro-electron and hole-hole collisions is responsible for decrement of impact ionization rate of charge carriers at higher carrier concentrations. Moreover the degradations impact ionization rate of both electrons and holes are found to be more severe at lower electric field values especially below 7.0×10^7 V m^{-1} for all carrier densities. At lower electric fields the supplied energy per unit length ($q\xi$) is smaller and comparable to the energy loss per unit length due to the carrier-carrier interactions which leads to more degradation in impact ionization probabilities at those electric fields as compared to at higher fields. Thus the values of ionization rates are more affected by energy loss due to carrier-carrier collisions at lower electric fields.

3 CALCULATION OF IMPACT IONIZATION RATES OF CHARGE CARRIERS IN 6H-SILICON CARBIDE

Numerical calculations have also been carried out to obtain the ionization rates of both types of charge carriers in 6H-SiC as functions of applied electric field within the field range 5.0×10^7–5.0×10^8 V m^{-1} using the analytical expressions given in equations (2) and (3) of the Part I of this paper (Chatterjee *et al.* 2016) in which the multistage scattering phenomena have been taken into account for obtaining the total probability of impact ionization of the charge carriers. The calculated values of $\alpha_{e,h}$ are compared with the empirical relations fitted from the experimental data measured by Dmitriev *et al.* (Dmitriev *et al.* 1983) given by the equations (1). The ionization coefficients A_e, B_e, A_e, B_h are 4.65×10^6 m^{-1}, 12.00×10^8 V m^{-1}, 4.65×10^8 m^{-1}, 12.00×10^8 V m^{-1} respectively, while $m = 1$ for the entire field range under consideration (Dmitriev *et al.* 1983). It is observed from the experimental measurement of Dmitriev *et al.* (Dmitriev *et al.* 1983), the ionization rate of holes is 100 times greater than that of electrons (i.e. $\alpha_e = \alpha_h/100$) in 6H-SiC for the entire field range under consideration. Other material parameters of 6H-SiC such as bandgap ($E_g = 2.86$ eV), ionization threshold energy of electrons ($E_{i(e,h)e} = 5.2052$ eV) and holes ($E_{i(e,h)h} = 4.4330$ eV) and corresponding mean free paths ($l_{i(e,h)e} = 1600$ Å and $l_{i(e,h)h} = 600$ Å respectively), optical phonon energy ($E_r = 104.2$ meV) and corresponding mean free path ($l_r = 40$ Å), permittivity ($\varepsilon_r = 9.66$), etc. at 300 K are taken from the experimental reports (Electronic Archive 2015).

Variations of ionization rate of electrons and holes in 6H-SiC obtained from the analytical model presented in earlier part of this paper with inverse of the electric field for different electron and hole concentrations respectively are shown in Figures 2 and 3. The same variations obtained from the empirical relations given in equations (1) (Dmitriev *et al.* 1983) are also shown in Figures 2 and 3. The parameters A_e, c, A_h and d associated with the mean free path of electron-electron and hole-hole collisions (l_{ee} and l_{hh}) are adjusted in numerical calculations for the best fit of the experimental data. In the present analytical model the best fittings for electron and hole ionization rates described by equations (2) and (3) of Part I of this paper with the empirical relations describing the field variations of the same parameters, i.e. equations (1) are obtained from $l_{ee} = (3.75)^{3.86}$ $n^{-1/3}$ and $l_{hh} = (2.65)^{2.05}$ $p^{-1/3}$ m respectively. It can be observed from Figures 2 and 3 that the analytical expressions of ionization rates considering the multistage scattering phenomena presented in this paper are in close agreement with respect to the experimental data (Dmitriev *et al.* 1983) at any electric field especially for the electron and hole concentrations of 10^{22} m^{-3} which are same those taken for the experiment (Dmitriev *et al.* 1983).

It is also noteworthy from Figures 2 and 3 that the impact ionization rate of charge carriers decreased significantly when the carrier concentrations are increased. This decrement of impact ionization rates are more pronounced for the carrier concentrations of 10^{21} m^{-3} and above of it. Degradation of impact ionization probabilities for both

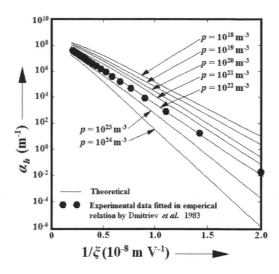

Figure 3. Ionization rate of holes in 6H-SiC versus inverse of applied electric field. Points (▲) represent the plot of empirical relation given by equation (1) fitted from experimental data of Dmitriev *et al.* (Dmitriev *et al.* 1983) (hole concentration in (Dmitriev *et al.* 1983) is $p = 10^{22}$ m^{-3}).

electrons and holes as a consequence of increase of energy loss per unit length due to increased amount of electro-electron and hole-hole collisions is responsible for decrement of impact ionization rate of charge carriers at higher carrier concentrations. Moreover the degradations impact ionization rate of both electrons and holes are found to be more severe at lower electric field values especially below 1.0×10^8 V m^{-1} for all carrier densities. At lower electric fields the supplied energy per unit length ($q\xi$) is smaller and comparable to the energy loss per unit length due to the carrier-carrier interactions which leads to more degradation in impact ionization probabilities at those electric fields as compared to at higher fields. Thus the values of ionization rates are more affected by energy loss due to carrier-carrier collisions at lower electric fields.

4 CONCLUSION

In this part of the paper, the analytical model based on multistage scattering phenomena as discussed in the earlier part has been used to evaluate the impact ionization rates of charge carriers in type-IIb diamond and 6H-SiC within the field ranges of 4.0×10^7–1.0×10^8 V m^{-1} and 5.0×10^7–5.0×10^8 V m^{-1} respectively. The calculated results have been compared with the ionization rate values calculated by using the empirical relations fitted from the

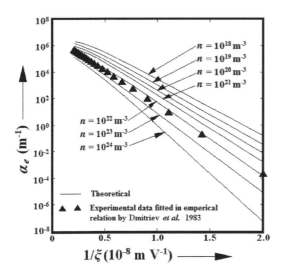

Figure 2. Ionization rate of electrons in 6H-SiC versus inverse of applied electric field. Points (▲) represent the plot of empirical relation given by equation (1) fitted from experimental data of Dmitriev *et al.* (Dmitriev *et al.* 1983) (electron concentration in (Dmitriev *et al.* 1983) is $n = 10^{22}$ m^{-3}).

experimentally measured data. The calculated values of impact ionization rates of charge carriers in both type-IIb diamond and 6H-SiC are found to be in close agreement with the experimental results especially for the electron and hole concentrations of 10^{22} m^{-3} which are same as those taken into account in the experiments. The effect of varying charge density and corresponding variations of energy loss due to carrier-carrier collisions on the impact ionization rate of charge carriers in the base semiconductor can be taken into account for the analysis or simulation of several microwave, millimeter-wave and optoelectronic devices by using the analytical expressions developed in this paper.

REFERENCES

Acharyya, A., Banerjee, J.P. (2014). A generalized analytical model based on multistage scattering phenomena for estimating the impact ionization rate of charge carriers in semiconductors. *Journal of Computational Electronics*, 13, 917–924.

Chatterjee, S., Das, A., and Acharyya, A. (2016). Evaluation of Ionization Rates of Charge Carriers in a Semiconductor via a Generalized Analytical Model Based on Multistage Scattering Phenomena—Part I: Wurtzite-GaN. in Proceedings of 3rd International Conference C2E2 2016, SKFGI, Mankundu, WB, India.

Dmitriev, A.P., Kanstantinov, A.O., Litvin, D.P., and Sankin, V.I. (1983). Impact ionization and supperlattice in 6H-SiC. *Sov. Phys.–Semicond*, 17, 686–689.

Electronic Archive: New Semiconductor Materials, Characteristics and Properties (2015). Retrieved from http://www.ioffe.ru/SVA/NSM/Semicond/GaN/index.html.

Konorova, E.A., Kuznetsov, Y.A., Sergienko, V.F., Tkachenko, S.D., Tsikunov, A.K., Spitsyn, A.V., and Danyushevski, Y.Z. (1983). Impact ionization in semiconductor structures made of ion-implanted diamond. *Sov. Phys.–Semicond*, 17, 146–149.

Frontiers in Computer, Communication and Electrical Engineering – Acharyya (Ed.)
© 2016 Taylor & Francis Group, London, ISBN: 978-1-138-02877-7

Design and development of smart traffic lighting

Suman Roy, Prasenjit Rakshit, Soumyadeep Nandy, Pranabesh Chakraborty,
Sourav Mukhopadhyay & Vishwanath Gupta
Department of Electrical Engineering, Supreme Knowledge Foundation Group of Institutions, Mankundu, India

ABSTRACT: In the proposed work, a fuzzy logic based inference system based waiting time predictor system for traffic control has been designed and the hardware model for the traffic lighting system has been developed using ARDUINO micro-controller kit. The fuzzy logic inference system has been designed using raw traffic data collected manually at a particular road intersection and the predicted waiting time has been found nearly equal to the observed waiting time. The hardware model has been designed with a special feature incorporated into it to prevent excess fuel wastage and also prevent accidents. The results that have been found are satisfactory.

1 INTRODUCTION

In recent years, there has been significant development in the use of automation while designing traffic lighting systems. Automation of the traffic lights has been instrumental in reducing the fuel wastage and smooth running of the traffic. Various soft computing techniques have come to the fore where the automation of the traffic lights have been concerned (Tahilyani, Darbari & Shukla 2013). In this proposed work, fuzzy logic along with Arduino micro-controller has been used to design and develop a smart traffic lighting system for a particular road intersection.

2 SMART TRAFFIC LIGHTING

Smart traffic lighting system differs from the traditional traffic lighting system, which is a combination of advanced signaling devices positioned at pedestrian crossings, road intersections and other places, to ascertain smooth movement of traffic. Our proposed system has combined the existing technology with fuzzy logic inference system to create a traffic light system that ascertains smooth moving of traffic and saving of fuel.

3 PREDICTION OF WAITING TIME USING FUZZY LOGIC INFERENCE SYSTEM

The designed fuzzy logic inference system is a 2-input (Road Condition and number of vehicles), 1-Output (Waiting time) and 9 Rule-Based system. The value of "Road condition" is kept in the range of 0 to 1, the value of the number of vehicles is kept between 0 and 100. These values are chosen based on the heuristic knowledge. The rules are formed based on the heuristic knowledge of the required waiting time by manually collecting data at different times of the day at Dunlop More, South of 24 parganas. The manually collected data is shown in Table 1. The output has a range of 0 to

Table 1. Manually collected raw traffic data.

Time of the day	Average number of vehicles per 2 minutes	Waiting time in seconds
12 Midnight–6 AM	8	25
6 AM–8 AM	20	40
8 AM–11 AM	42	70
11 AM–4PM	34	50
4PM–8PM	50	75
8PM–12 Midnight	25	50

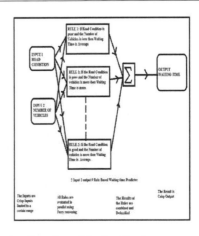

Figure 1.   Overview of the fuzzification process.

100 seconds. The developed Rule-Base is based on the Mamdani type interface which is a collection of the IF-THEN rules. The defuzzyfication technique is used in the centroid method. The overview of the fuzzification process is shown in Figure 1. The designed FIS editor window and the designed Rule-Base are shown in Figure 2 and Figure 3, respectively.

4 DEVELOPMENT OF AUTOMATIC TRAFFIC LIGHT USING ARDUINO MICRO-CONTROLLER KIT

After the fuzzy logic inference system has predicted the waiting time, the hardware for the proposed automatic traffic lighting system was developed using Arduino Micro-Controller Kit. The devel-

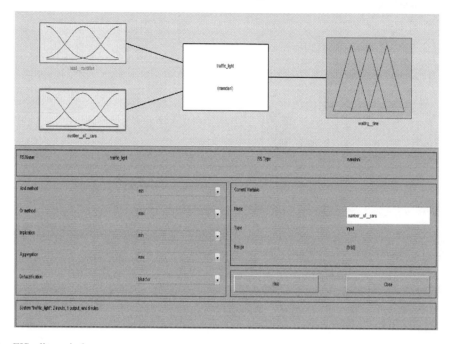

Figure 2.   FIS editor window.

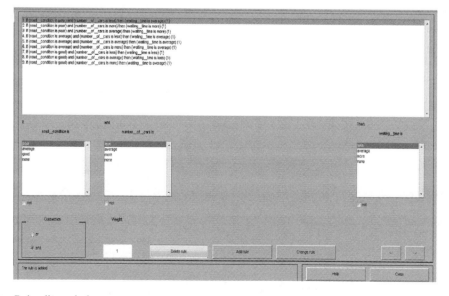

Figure 3.   Rule editor window.

oped hardware model has taken into consideration four traffic light units viz. Unit 1, Unit 2, Unit 3, and Unit 4. Unit 1 and Unit 2 have the same light pattern at any particular time as they have been placed on the opposite crosswise intersections while Unit 3 and Unit 4 have the same light patterns between themselves but opposite light pattern with respect to Unit 1 and Unit 2 and have been placed on the remaining two crosswise intersections as shown in Figure 4. In the proposed work, a slight modifica-

tion has been done in the conventional traffic lighting system, which is that the amber light starts to blink along with the red and green light 5 seconds before the end of waiting period and free movement period so as to indicate the driver to get ready to either start the engine or to slow down, respectively. This modification prevents unnecessary fuel usage during the waiting period and the accident prevention during the free movement period.

5 RESULTS OBTAINED

The fuzzy logic inference system predicts the waiting time for the vehicles satisfactorily as is shown in Table 2. One of the results of the predicted waiting time by the designed Fuzzy Logic Inference System is shown in Figure 5. The Road condition for a particular location is constant for a considerable time period so it has been taken as a constant during the waiting time prediction and the designed Fuzzy Logic Inference System that predicts the waiting time based on the number of vehicles for a particular road.

The developed hardware model works satisfactorily for a single LED as a single traffic light. The developed hardware model in the working condition is shown in Figure 6. Due to unavailability of the amber colored LED, White LED has been used in place of Amber. The signal phase and cycle length for each light of the smart traffic lighting system is given in Table 3.

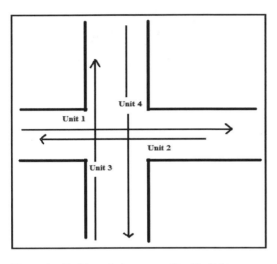

Figure 4.   Position of placement of traffic lights.

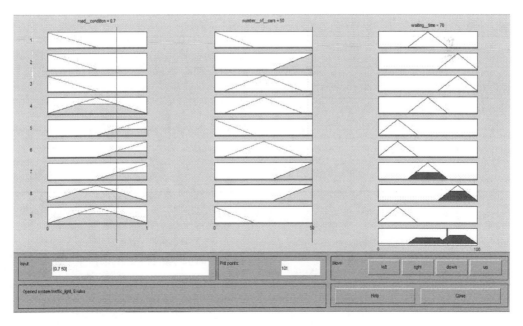

Figure 5.   Predicted waiting time when road condition is 0.7 and number of vehicles is 50.

Table 2. Predicted waiting time vs. observed waiting time.

Average number of vehicles per 2 minutes	Waiting time predicted by fuzzy logic inference system (seconds)	Waiting time observed manually (seconds)
8	24	25
20	40	40
34	45	50
42	64	70
50	70	75
25	48	50

Figure 6. Smart traffic light in operation.

Table 3. Signal phase and cycle length of each light at any one unit.

		Cycle length						
	Average number of vehicles per 2 minutes	Unit 1 and Unit 2			Unit 3 and Unit 4			Total cycle length
Time		Red	Green	Amber (blinks)	Red	Green	Amber (blinks)	
12 Midnight–6 am	08	OFF	OFF	Blinks	OFF	OFF	Blinks	Not Applicable
6 am–8 am	20	OFF	0 to 40 secs	35 to 40 secs	0 to 40 secs	OFF	35 to 40 secs	40 secs
8 am–11 am	42	OFF	0 to 64 secs	59 to 64 secs	0 to 64 secs	OFF	59 to 64 secs	64 secs
11 am–4 pm	34	OFF	0 to 45 secs	40 to 45 secs	0 to 45 secs	OFF	40 to 45 secs	45 s
4 pm–8 pm	50	OFF	0 to 70 secs	65 to 70 secs	0 to 70 secs	OFF	65 to 70 secs	70 s
8 pm–12 midnight	25	OFF	0 to 48 secs	43 to 48 secs	0 to 48 secs	OFF	43 to 48 secs	48 s

6 CONCLUSION

Both the designed fuzzy logic waiting time predictor and the developed hardware traffic light model work satisfactorily for a particular considered case. However, more than one traffic lighting system has been positioned at different intersections, and can also be controlled simultaneously. The modification made in the lighting pattern helps to save fuel and prevent accidents.

7 FUTURE SCOPE OF THE WORK

The interfacing between the designed Fuzzy Logic Waiting time predictor system and the developed hardware model can be done to make a coherent system. The Fuzzy Rule base and membership functions can be further modified. The hardware model can be further developed to a drive having more number of LEDs.

REFERENCES

Ross T.J. (2008), "Fuzzy Logic with Engineering Applications" 2nd Edition, John Wiley

Wen W. (2008), "A dynamic and automatic traffic light control expert system for solving the road congestion problem", *Expert Systems with Applications, Volume 34, Issue 4,* pp 2370–2381.

Tahilyani S., Darbari M. & Shukla P.K. (2013), "Soft Computing Approaches in Traffic Control Systems: A Review", *AASRI Conference on Intelligent Systems and Control, Volume 4,* pp 206–211.

Mehan S., Sharma V. (2011), "Development of traffic light control system based on fuzzy logic", ACAI'11 Proceedings of the International Conference on Advances in Computing and Artificial Intelligence, pp 162–165.

Frontiers in Computer, Communication and Electrical Engineering – Acharyya (Ed.)
© 2016 Taylor & Francis Group, London, ISBN: 978-1-138-02877-7

Influence of band-to-band tunneling induced shift of ATT phase delay on millimeter-wave properties of DDR IMPATTs— part I: Theoretical modeling

Partha Banerjee
Department of Electronics and Communication, Techno India, Salt Lake, Kolkata, India

Prasit Kumar Bandyopadhyay
Department of Electronics and Communication, Dumkal Institute of Engineering and Technology, Murshidabad, West Bengal, India

Subhendu Chakraborty & Aritra Acharyya
Supreme Knowledge Foundation Group of Institutions, Mankundu, Hooghly, West Bengal, India

ABSTRACT: The effect of tunneling on the high frequency properties of Double-Drift Region (DDR) Impact Avalanche Transit Time (IMPATT) diodes based on different semiconductors such as Si, InP, GaAs and 4H-SiC designed to operate at 94 GHz window frequency has been investigated by using a generalized double—iterative computer simulation method. Drift-diffusion model is used for DC and high frequency analysis of DDR IMPATTs operating in Mixed Tunneling Avalanche Transit Time (MITATT) mode. The theoretical modeling and simulation methodology have been discussed in this part of the paper.

1 INTRODUCTION

The rapid developments in the process technology of IMPATT device have made possible the fabrication of this device having narrow depletion layer width. IMPATT devices are finding useful application as powerful and efficient sources in various communication systems operating in millimetre-wave and sub-millimeter-wave bands of frequency. The advantages of millimeter-wave and sub-millimeter-wave frequencies are many fold such as increased resolution, higher penetrating power of these signals through cloud, dust, fog etc, low power supply voltage and reduced system size (Midford *et al.* 1979). Several atmospheric window frequencies are available in the millimetre-wave frequency range (30–300 GHz) such as 35, 94, 140, 220 GHz. This has attracted the attention of researchers working in this area to design and develop IMPATT diodes which can deliver appreciable amount of millimeter wave power of the order of watts at the chosen window frequency. Further the decrease of DC to RF conversion efficiency of the device at higher mm-wave frequencies of operation can be compensated by using impurity bumps in the depletion layer of flat profile structures leading to quasi Read hi-lo and lo-hi-lo devices (Mukherjee *et al.* 2009). The incorporation of impurity bump in the depletion layer of IMPATTs leads to constriction of avalanche zone and simultaneous increase of the breakdown field to very high values in the range of 6×10^7 to 10×10^7 V m^{-1} (Dash *et al.* 1992). Higher RF power output can be obtained from IMPATTs based on wider bandgap materials such as SiC, GaN and InP (Mukherjee *et al.* 2010). Some theoretical studies on the RF performance of heterojunction IMPATTs based on III-V and IV-IV semiconductors have been reported in the literature (Bailey 1992) which show that these devices are suitable for generation of high millimeter wave power along with high conversion efficiency. The thinner depletion layer ($W < 0.80\,\mu$m) of mm-wave IMPATT diode and the higher peak electric field ($E_m > 6 \times 10^7$ V m^{-1}) near the junction are the favourable conditions for tunnel generation of Electron Hole Pair (EHP) provided vacant states are available in the conduction band opposite to the filled states of the valance band. Both the tunnel generated and the impact ionization generated carriers move in the drift region to produce the necessary transit time delay for IMPATT action. This mode of operation of IMPATT diode is known as mixed tunneling avalanche transit time mode or MITATT mode.

W.T. Read (Read 1958) followed by other researchers (Kwok *et al.* 1972, Chive *et al.* 1975) carried out the analysis of IMPATT diode based on the simplifying assumption of equal ionization rates of electrons and holes in the respective semiconductor.

Elta and Haddad (Elta et al. 1978) used an effective ionization rate of charge carriers in their analysis by introducing the concept of dead space. Luy et al. (Luy et al. 1989) considered a time dependent tunnel current and observed that the DC to RF conversion efficiency of the device decreases due to tunneling induced phase distortion. A detailed numerical simulation of the DC and high frequency properties of DDR MITATT device based on Si, InP, GaAs and 4H-SiC operating at W-Band is carried out and the results are presented and compared in the next part of this paper. A pure field dependent tunnel generation rate for electrons is taken from which the tunnel generation rate for holes is calculated from the energy band diagram. The present method provides not only the integrated terminal properties of MITATTs such as negative resistance, negative conductance of the device but also the microscopic negative resistance distribution profile in the depletion layer of the device. The effect of tunnel current on RF performance of the device has been determined separately in avalanche and drift zone to identify the region of the depletion layer where the effect is more predominant. In this part of the paper the method of analysis incorporating the effect of tunneling in reverse biased DDR p^+pnn^+ IMPATT device is described. Also the simulation technique of MITATT mode of operation has been discussed in brief.

2 BAND-TO-BAND TUNNELING

One-dimensional model of a reverse biased p^+pnn^+ DDR IMPATT device is shown in Figure 1. The DC and RF performance of MITATT device are obtained at 94 GHz under mixed tunneling and avalanche breakdown condition. In this analysis the tunneling generation rate for electrons [$g_{Tn}(x)$] is obtained from quantum mechanical considerations as reported in (Dash et al. 1992, Kane 1961). Thus

$$g_{Tn}(x) = A_T E^2(x) \exp\left(-\frac{B_T}{E(x)}\right), \quad (1)$$

where the coefficients A_T and B_T are given by

$$A_T = \frac{q^2}{8\pi^3\hbar^2}\left(\frac{2m^*}{E_g}\right)^{\frac{1}{2}}, \quad (2)$$

$$B_T = \frac{1}{2q\hbar}\left(\frac{m^* E_g^3}{2}\right)^{\frac{1}{2}}. \quad (3)$$

The symbols used in Equations (1)–(3) carry their usual significance. The tunneling generation rate for holes can be obtained from Figure 1. The phenomenon of tunneling is instantaneous and the tunnel generation rate for holes is related with that for electrons i.e. $g_{Tp}(x) = g_{Tn}(x')$. The tunnel generation of an electron at x' is simultaneously associated with the generation of a hole at x, where $(x–x')$ is the spatial separation between the edge of conduction band and valence band at the same energy. Taking F as the measure of energy from the bottom of the conduction band on the n-side and making use of the idea that the vertical difference between x and x' is E_g, x' can be easily obtained by referring to Figure 1 as,

$$x' = x\left(1 - \frac{E_g}{F}\right)^{\frac{1}{2}} \text{ for } 0 \le x \le x_0, \quad (4)$$

$$x' = W - (W - x)\left(1 + \frac{E_g}{F_B - F}\right)^{\frac{1}{2}} \text{ for } x_0 \le x \le W, \quad (5)$$

In Figure 1 x_a/x_b is the position where the electron energy in the valence/conduction band equals the energy corresponding to the bottom/top of the conduction/valance band at the n-side/p-side edge of the depletion layer. W is the total depletion width of the device and x_0 is the position of the junction.

The hole generation rate due to tunneling is zero between $x = 0$ and $x = x_a$ (Figure 1) as within this region electrons in the valance band have no available states in the conduction band for tunneling. Similarly, non-availability of states in the conduction band for tunneling to take place between

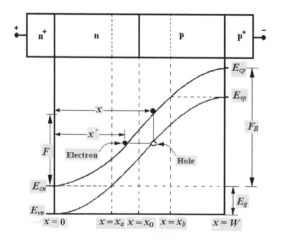

Figure 1. Structure and energy band diagram of a reverse biased DDR IMPATT Diode showing the locations of tunnel generated electrons and holes.

$x = x_b$ and $x = W$ (Figure 1) makes no contribution of tunnel generated electrons in this region (Dash *et al.* 1992).

3 DESIGN METHODOLOGY AND MATERIAL PARAMETERS

The frequency of operation of an IMPATT diode essentially depends on the transit time of charge carriers to cross the depletion layer of the diode. IMPATT diodes having symmetrical Double drift p^+pnn^+ structure are first designed for operation at 94 GHz window by using computer simulation technique and the transit time formula of Sze and Ryder (Sze *et al.* 1971) given by $W_{n,p} = 0.37v_{sn,sp}/ f_d$; where $W_{n,p}$, $v_{sn,sp}$ and f_d are the total depletion layer width (*n*- or *p*-side), saturation velocity of electrons/holes and design frequency respectively. Here n^+- and p^+-layers are highly doped substrates whose doping concentrations are taken to be 10^{26} m^{-3}. The background doping concentrations of *n* and *p* – depletion regions are initially chosen according to the design frequency. The electric field profile using the above doping profile is obtained from the simulation. The input doping profile is adjusted so that the electric field just punches through the depletion layers ($W_{n,p}$) for a particular value of f_d and a particular biasing current density J_0. Small-signal computer simulation based on Gummel-Blue approach (Gummel *et al.* 1967) is then carried out to obtain the admittance characteristics of the device. The optimum frequency (f_{opt}) for peak negative conductance is determined from the admittance characteristics. If the magnitude of f_{opt} differs very much from f_d, the value of J_0 is varied and the computer simulation program is run till the value of f_{opt} is nearly equal to the value of f_d. The bias current density is thus fixed for the particular design frequency. The material parameters like ionization coefficients, saturation drift velocities, mobilities of charge carriers, dielectric constants, bandgaps, effective masses of the base materials (Si, InP, GaAs and 4H-SiC) used for the analysis are taken from (Electronic Archive 2015).

4 SIMULATION TECHNIQUE

The computer simulation technique is used to study the DC and small-signal properties of DDR IMPATT diode, one dimensional model of the *p-n* junction is considered in the present simulation method.

The DC analysis of DDR IMPATT described in details elsewhere (Acharyya *et al.* 2014) is first carried out by using a double iterative field maximum computer method. In this method the computation starts from the location of field maximum near the metallurgical junction. The spatial distributions of both DC electric field and carrier current densities in the depletion layer are obtained by using the above method, which involves iteration over the magnitude of field maximum (E_m) and its location in the depletion layer. A software package has been developed for simultaneous numerical solution of Poisson's equation, combined continuity equations and the space charge equation subject to appropriate boundary conditions by taking into account the effects of mobile space charge, carrier diffusion and tunneling. Thus the electric field and carrier current density profiles are obtained for DDR device operating in both IMPATT and MITATT modes. The boundary conditions for the electric field at the depletion layer edges are given by (Figure 2)

$$E(0) = 0 \text{ and } E(W) = 0, \quad (6)$$

where, $x = 0$ and W define the p^+- and n^+-edges of the depletion layer. The boundary conditions for normalized current density $P(x) = (J_p - J_n)/ J_0$ (where, J_p = hole current density, J_n = electron current density) at the edges are given by

$$P(0) = \left(\frac{2}{M_p} - 1\right) \text{ and } P(W) = \left(1 - \frac{2}{M_n}\right), \quad (7)$$

where, $M_p = J_0/J_{ps}$ = hole current multiplication factor and $M_n = J_0/J_{ns}$ = electron current multiplication factor. The necessary device equations are simultaneously solved satisfying the appropriate boundary conditions given in (6) and (7). The field dependence of electron and hole ionization rates (α_n and α_p) and saturated drift velocities of electron

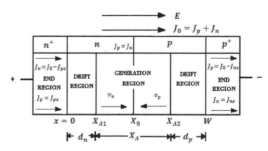

Figure 2. The different active regions of DDR IMPATT diode.

(v_{sn}) and holes (v_{sp}) at 500 K are incorporated in the simulation of DC electric field and carrier currents density profiles. The conversion efficiency is calculated from the semi quantitative formula

$$\eta(\%) = \frac{2m}{\pi} \times \frac{V_D}{V_B}, \qquad (8)$$

where, V_D = Voltage drop across the drift region, V_B = Breakdown voltage and m is the RF voltage modulation index. Avalanche breakdown takes place when the electric field at the junction is large enough such that the charge multiplication factors (M_n, M_p) become infinitely large $(\approx 10^6)$. The breakdown voltage is obtained by integrating the electric field profile over the total depletion layer width, i.e.

$$V_B = \int_0^W E(x)dx. \qquad (9)$$

The high-frequency analysis of DDR IMPATT diode is carried out to obtain distribution of negative resistivity in the depletion layer which provides considerable insight into high frequency performance and the possible measure to improve the same. The range of frequency for which the device exhibits negative conductance is obtained from the admittance plot. The DC parameters are fed as input data for the small signal analysis. The depletion layer edges of the device, are fixed from the DC analysis and taken as the starting and end points for the small signal analysis. Two second order differential equations are framed by resolving the diode impedance $Z(x, \omega)$ into its real part $R(x, \omega)$ and imaginary part $X(x, \omega)$ given by

$$\frac{\partial^2 R}{\partial x^2} + \left[\alpha_n(x) - \alpha_p(x)\right]\frac{\partial R}{\partial x}$$
$$- 2r_n\left(\frac{\omega}{\bar{v}}\right)\frac{\partial X}{\partial x} + \left[\begin{array}{c}\left(\frac{\omega^2}{\bar{v}^2}\right) - H(x) \\ -\frac{qr_p}{\bar{v}\varepsilon}\left(g_T'(x) + g_T'(x')\right)\end{array}\right]R, \quad (10)$$
$$- 2\bar{\alpha}\left(\frac{\omega}{\bar{v}}\right)X - 2\left(\frac{\bar{\alpha}}{\bar{v}\varepsilon}\right) = 0$$

$$\frac{\partial^2 X}{\partial x^2} + \left[\alpha_n(x) - \alpha_p(x)\right]\frac{\partial X}{\partial x}$$
$$- 2r_n\left(\frac{\omega}{\bar{v}}\right)\frac{\partial R}{\partial x} + \left[\begin{array}{c}\left(\frac{\omega^2}{\bar{v}^2}\right) - H(x) \\ -\frac{qr_p}{\bar{v}\varepsilon}\left(g_T'(x) + g_T'(x')\right)\end{array}\right]X. \quad (11)$$
$$+ 2\bar{\alpha}\left(\frac{\omega}{\bar{v}}\right)R + 2\left(\frac{\omega}{\bar{v}^2\varepsilon}\right) = 0$$

The boundary conditions for R and X are given by (n side and p side respectively)

$$\frac{\partial R}{\partial x} + \frac{\omega X}{v_{ns}} = -\left(\frac{1}{v_{ns}\varepsilon}\right) \text{ and } \frac{\partial X}{\partial x} - \frac{\omega R}{v_{ns}} = 0 \text{ at } x = 0,$$
$$(12)$$

$$\frac{\partial R}{\partial x} - \frac{\omega X}{v_{ps}} = \left(\frac{1}{v_{ps}\varepsilon}\right) \text{ and } \frac{\partial X}{\partial x} + \frac{\omega R}{v_{ps}} = 0 \text{ at } x = W,$$
$$(13)$$

where $Z(x, \omega) = R(x, \omega) + i X(x, \omega)$. A double-iterative simulation scheme is used to solve those equations simultaneously by satisfying the boundary conditions. The device negative resistance (Z_R) and reactance (Z_x) are computed through the numerical integration of the $R(x)$ and $X(x)$ profiles over the space-charge layer width, W. Thus

$$Z_R = \int_0^W R\,dx \text{ and } Z_X = \int_0^W X\,dx \qquad (14)$$

The negative conductance $(-G)$, Susceptance (B) and the quality factor (Q_P) of the device are evaluated by using the following relations

$$|-G(\omega)| = \frac{Z_R}{\left(Z_R^2 + Z_X^2\right)}, \qquad (15)$$

$$|B(\omega)| = \frac{-Z_X}{\left(Z_R^2 + Z_X^2\right)}, \qquad (16)$$

$$Q_p = -\left(B/G\right)_{f=f_{opt}}. \qquad (17)$$

It may be noted that both $-G$ and B are normalized to the area of the diode. The avalanche frequency (f_a) is the frequency at which the susceptance (B) changes its nature from inductive to capacitive. Again it is the lowest frequency at which the real part (G) of admittance becomes negative and oscillation starts to build up in the circuit. At the resonant frequency of oscillation, the maximum RF power output (P_{RF}) from the device is calculated by using the following expression

$$P_{RF} = \frac{1}{2} \cdot V_{RF}^2 \cdot |G_p| \cdot A, \qquad (18)$$

where V_{RF} is the amplitude of the RF swing ($V_{RF} = V_B/2$, assuming 50% modulation of the breakdown voltage V_B), G_p is the diode negative conductance at the operating frequency and A is the junction area of the diode.

REFERENCES

Acharyya, A., Mukherjee, M., Banerjee, J. P. (2014). Effects of Tunnelling Current on mm-wave IMPATT Devices. *International Journal of Electronics*, published online, DOI: http://dx.doi.org/10.1080/002072 17.2014.982211.

Bailey, M. J. (1992). Heterojunction IMPATT diodes. IEEE Transactions on Electron Devices, 39(8), 1829–1834.

Chive, M., Constant, E., Lefebvre, M., and Pribetich, J. (1975). Effect of tunneling on high efficiency IMPATT avalanche diode. *Proc. IEEE (Lett.)*, 63, 824–826.

Dash, G. N., and Pati, S. P. (1992). A generalized simulation method for MITATT-mode operation and studies on the influence of tunnel current on IMPATT properties. *Semicond. Sci. Technology*, 7, 222–230.

Electronic Archive: New Semiconductor Materials, Characteristics and Properties (2015). Retrieved from http://www.ioffe.ru/SVA/NSM/Semicond/index.html.

Elta, M. E., and Haddad, G. I. (1978). Mixed tunneling and avalanche mechanism in p-n junctions and their effects on microwave transit time devices. *IEEE Trans. Electron Devices*, 25, 694–702.

Gummel, H. K., and Blue, J. L. (1967). A small-signal theory of avalanche noise in IMPATT diodes. *IEEE Trans. on Electron Devices*, 14(9), 569–580.

Kane, E. O. (1961). Theory of tunneling. *J. Appl. Phys.*, 32, 83–91.

Kwok, S. P., and Haddad, G. I. (1972). Effect of tunneling on an Impatt oscillator. *J. Appl. Phys.*, 43, 3824–3830.

Luy, J. F., and Kuehnf, R. (1989). Tunneling assisted Impatt operation. *IEEE Trans. Electron Devices*, 36, 589–595.

Midford, T. A., and Bernick, R. L. (1979). Millimeter Wave CW IMPATT diodes and Oscillators. *IEEE Trans. Microwave Theory Tech.*, 27, 483–492.

Mukherjee, M., Banerjee, S., and Banerjee, J. P. (2010). Dynamic characteristics of iii-v and iv-iv semiconductor based transit time devices in the terahertz regime: a comparative analysis. *Terahertz Sci. and Tech.*, 3, 97–109.

Mukherjee, M., and Mazumder, N. (2009). Effect of charge-bump on high-frequency characteristics of α-SiC based double drift ATT diodes at MM-wave window frequencies. *IETE J. of Research*, 55, 118–127.

Read, W. T. (1958). A proposed high-frequency negative-resistance diode. *Bell Syst. Tech. J.*, 37, 401–466.

Sze, S. M., and Ryder, R. M. (1971). Microwave Avalanche Diodes. *Proc. of IEEE, Special Issue on Microwave Semiconductor Devices*, 59, 1140–1154.

Frontiers in Computer, Communication and Electrical Engineering – Acharyya (Ed.)
© 2016 Taylor & Francis Group, London, ISBN: 978-1-138-02877-7

Influence of band-to-band tunneling induced shift of ATT phase delay on millimeter-wave properties of DDR IMPATTs—part II: Simulation results

Partha Banerjee
Department of Electronics and Communication, Techno India, Salt Lake, Kolkata, India

Prasit Kumar Bandyopadhyay
Department of Electronics and Communication, Dumkal Institute of Engineering and Technology, Murshidabad, West Bengal, India

Subhendu Chakraborty & Aritra Acharyya
Supreme Knowledge Foundation Group of Institutions, Mankundu, Hooghly, West Bengal, India

ABSTRACT: The simulations have been carried out based on the technique described in the earlier part of the paper to study the millimeter-wave performance of DDR IMPATTs based on different semiconductors such as Si, InP, GaAs and 4H-SiC. The simulation results show that the millimeter-wave performance of the device based on Si, InP, GaAs and 4H-SiC as regards RF power delivery and conversion efficiency degrades when tunneling is incorporated in the analysis. But the deterioration of device performance of MITATTs based on relatively wider bandgap 4H-SiC material is less pronounced than those based on other materials.

1 STRUCTURAL AND DOPING PARAMETERS

Double-Drift Region (DDR) IMPATT diodes based on different materials such as Si, InP, GaAs and 4H-SiC are designed and optimized for CW operation at W-Band (near 94 GHz). The method of analysis presented in the earlier part of the paper is used to simulate the DC and RF properties of DDR IMPATT diodes designed for optimum performance. The design parameters of the devices are given in Table 1. The junction diameter of the device is taken to be $D = 35$ μm (for CW mode of operation at W-Band) (Luy *et al.* 1987).

2 BAND-TO-BAND TUNNELING

Figure 1 shows that the tunneling and avalanche generation rates increase sharply with electric field in all the devices except in case of 4H-SiC based DDR device where the increase of avalanche generation rate is much sharper than the increase of tunneling generation rate with electric field. This can be understood from equations (1) – (3) of Part I of this paper. The values of both E_g and m^* are higher for 4H-SiC compared to those for other materials. Thus the value of the coefficient B_T in equation (1) is higher for 4H-SiC which causes a slower rate of increase of tunneling generation rate in comparison of that of avalanche generation rate.

Table 1. Structural and doping parameters of 94 GHz DDR IMPATT diodes.

Base Material	W_n (μm)	W_p (μm)	N_D ($\times 10^{23}$ m^{-3})	N_A ($\times 10^{23}$ m^{-3})	N_{n+}, N_{p+} ($\times 10^{26}$ m^{-3})	J_0 ($\times 10^8$ A m^{-2})
Si	0.390	0.380	1.20	1.25	1.00	3.4
InP	0.350	0.352	1.60	1.60	1.00	1.4
GaAs	0.320	0.320	1.60	1.60	1.00	6.0
4H-SiC	0.550	0.550	3.50	3.50	1.00	5.0

Figure 2 shows the spatial variations of both avalanche generation rate and tunneling generation rate in the depletion layer of DDR devices based on different materials. These generation rates exhibit peaks near the junction plane followed by sharp falls on either side of the junction. The g_{Tpeak}/g_{Apeak} – ratio is found to be maximum in InP DDR (1.59) followed by GaAs DDR (0.76), Si DDR (0.31) respectively and minimum in 4H-SiC DDR (0.21). Figure 2 shows that the avalanche generation rate predominates over tunneling generation rate in the space charge layer for DDR IMPATTs based on Si, GaAs and 4H-SiC while the converse is true for DDR IMPATTs based on InP.

3 INFLUENCE OF TUNNELING ON STATIC AND DYNAMIC CHARACTERISTICS

The simulated has been carried out to study the DC and small-signal parameters of DDR devices based on all four semiconductor materials in both IMPATT (without considering tunneling) and MITATT (considering tunneling) modes. It is observed that the DC and RF performances of MITATT devices degrade with respect to those of IMPATT devices due to the effect of tunneling. It is interesting to note that this degradation is strongly related to the ratio of the peak electron tunneling generation rate to the peak avalanche generation rate (g_{Tpeak}/g_{Apeak}) of the corresponding device. Higher value of g_{Teak}/g_{Apeak} – ratio leads to more degradation of the RF performance of the device. The results show that the decrease of peak DC electric fields are 8.5%, 2.4%, 1.53%, 1.2%, breakdown voltages are 7.31%, 3.7%, 2.5%, 0.97% and DC to RF conversion efficiencies are 9.8%, 6.5%, 4.5%, 0.82% for InP, GaAs, Si and 4H-SiC based DDR IMPATTs respectively due to the effect of tunneling. Thus the degradation of the performance of the devices is pronounced in InP DDR device.

The admittance characteristics i.e. Conductance-Susceptance Plots of the devices in both IMPATT and MITTAT modes are shown in Figure 5. It is observed from Figure 3 that the decrease of the magnitude of negative conductance is most pronounced (6.4%) in InP DDR. The optimum frequency shifts to higher values for all DDRs and this shift is found to be maximum in InP DDR (38 GHz) as compared to all DDR devices based on GaAs (27 GHz), Si (22 GHz) and 4H-SiC (9 GHz). Q-factor of the device is an indicator of the growth rate and stability of oscillation. Lower Q-factor close to unity leads to better RF performance of the device. It is observed that the Q-factors degrade for all the DDRs in MITATT mode.

The negative resistivity profiles of DDR devices in both IMPATT and MITATT shown in Figures 4–6, which exhibit two peaks in the two drift regions separated by a minimum near the junction in the avalanche region for both IMPATT and MITATT mode operation. The magnitude of negative resistivity peak of the device is higher in n-side than p-side for Si DDR (Figure 4) while the reverse is observed to be true in InP, GaAs and 4H-SiC DDRs (Figures 5, 6). Ratio of the negative resistivity peaks on n-side and p-side is indirectly

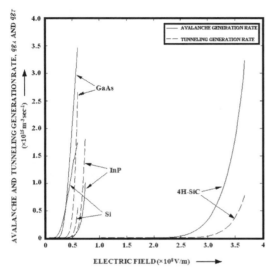

Figure 1. Avalanche and electron tunneling generation rates (qg_A and qg_T) versus electric field.

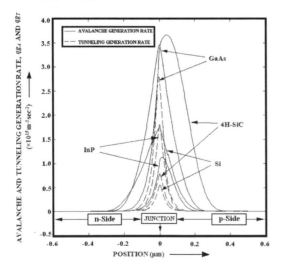

Figure 2. Spatial variation of avalanche and electron tunneling generation rates (qg_A and qg_T) in the active layers of the DDR IMPATT/MITATT diodes operating at W-Band.

connected to the ratio of electron to hole ionization rates of respective materials (Banerjee *et al.* 1989). It is well known that in Si, α_n is greater than α_p, while the α_p is greater than α_n in InP, GaAs and 4H-SiC. That is why the magnitude of negative resistivity peak is found to be higher in *p*-side of the device compared to *n*-side of it for InP, GaAs and 4H-SiC DDRs. It is interesting to observe that the magnitude of negative resistance falls appreci-

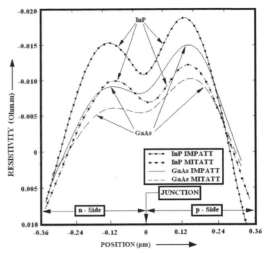

Figure 5. Negative resistivity profiles of the InP and GaAs flat-DDR diodes in both IMPATT and MITATT modes.

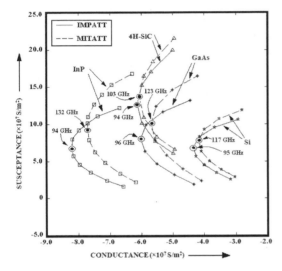

Figure 3. Small-signal conductance versus susceptance plots of the diodes in both IMPATT and MITATT modes.

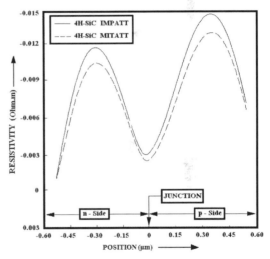

Figure 6. Negative resistivity profile of the 4H-SiC flat-DDR diode in both IMPATT and MITATT modes.

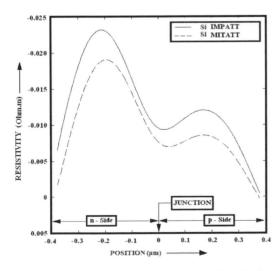

Figure 4. Negative resistivity profiles of the Si flat-DDR diode in both IMPATT and MITATT modes.

ably due to the effect of tunneling in all the devices based on Si, GaAs, InP and 4H-SiC DDRs.

It may further be noted that the decrease in the magnitude of the negative resistance peak due to tunneling is more appreciable on *p*-side of the device than on the *n*-side. This can be understood by considering that the tunnel generation rate due to electrons at any point x is a function of the electric field at the same point x while that due to holes at x depends on the electric field at some other point x', given in equations (4) and

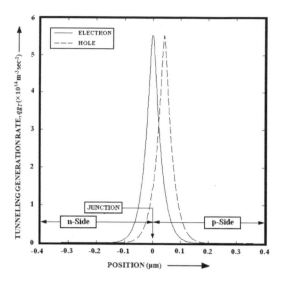

Figure 7. Spatial variation of tunneling generation rates of electrons and holes (qg_{Tn} and qg_{Tp}) in the active layers of the Si flat-DDR.

4 CONCLUSION

A generalized method based on double-iterative numerical analysis for Mixed Tunneling And Avalanche Transit Time (MITATT) mode of operation of a reverse biased *p-n* junction is presented in this paper. Application of this method for flat-DDR diode structures based on different semiconductor base materials such as Si, GaAs, InP, 4H-SiC operating at W-Band indicate that the effect of tunneling degrades the device performance in MITATT mode due to phase distortion caused by tunneling. The results indicate that the effect of tunneling is found to be more pronounced in InP DDR device and least pronounced in 4H-SiC DDR device. Therefore it is expected that the DDR IMPATTs based on 4H-SiC should provide superior RF performance in MITATT mode at higher frequency bands. Further the design data, presented in this paper, are expected to be helpful for realization of IMPATT oscillators for communication systems in millimeter-wave and sub-millimeter-wave frequencies.

(5) of the Part I of this paper. Thus the tunneling generation rate of electrons attains its peak value close to the junction while that of holes attains the peak value slightly away from the junction on the *p*-side (Figure 7) due to the effect of tunneling. The effective tunneling generation rate defined as, $q(g_T(x)+g_T(x'))$ is found to be predominant on *p*-side of the avalanche zone.

REFERENCES

Banerjee, J.P., Pati, S.P. and Roy, S.K. (1989). High frequency characterisation of double drift region InP and GaAs diodes. *Appl. Phys. A*, 48(5), 437–443.

Luy, J.F., Casel, A., Behr, W. and Kasper, E. (1987). A 90-GHz double-drift IMPATT diode made with Si MBE. *IEEE Trans. Electron Devices*, 34, 1084–1089.

Influence of band-to-band tunneling induced shift of ATT phase delay on millimeter-wave properties of DDR IMPATTs— part III: Calculation of shift of ATT phase delay due to tunneling

Partha Banerjee
Department of Electronics and Communication, Techno India, Salt Lake, Kolkata, India

Prasit Kumar Bandyopadhyay
Department of Electronics and Communication, Dumkal Institute of Engineering and Technology, Murshidabad, West Bengal, India

Subhendu Chakraborty & Aritra Acharyya
Supreme Knowledge Foundation Group of Institutions, Mankundu, Hooghly, West Bengal, India

ABSTRACT: A novel method is proposed by the author to calculate the tunneling induced shift of Avalanche Transit Time Phase (ATT) delay in MITATT mode of the device from the spatial shift of the simulated negative resistivity profiles due to tunneling. The method is used to calculate the shift of ATT phase delays due to tunneling in DDR IMPATTs based on Si, InP, GaAs and 4H-SiC, designed to operate at 94 GHz.

1   PROCEDURE OF CALCULATING THE TUNNELING INDUCED SHIFT OF ATT PHASE DELAY

The Avalanche Transit Time (ATT) phase delays of the 94 GHz diodes can be calculated from the corresponding $R(x)$-profiles for the following two cases from which the shift of ATT phase delay due to tunneling current can be computed: (i) without considering the tunneling current; which gives ATT phase delay in pure avalanche mode and (ii) considering the tunneling current; which gives ATT phase delay in mixed mode.

The $R(x)$-profiles of the diodes for the above two cases exhibit negative specific resistance peaks in the drift layers, but the magnitudes and locations of these maxima change when tunneling effect is taken into account in MITATT devices. The spatial shift of the negative specific resistance maxima for a particular base material determines the shift of ATT phase delay of MITATT due to tunneling current. If the distances of the peaks from the junction on the p-side of ATT device in the avalanche and mixed mode are x_{pa} and x_{pm} and the corresponding optimum frequencies f_a and f_m respectively, then the phase delays on the p-side of the device at x_{pa} and x_{pm} for cases (i) and (ii), are obtained from the following relations

$$\varphi_{pa} = \frac{2\pi x_{pa} f_a}{v_{ps}}, \quad \varphi_{pm} = \frac{2\pi x_{pm} f_m}{v_{ps}}. \tag{1}$$

Similarly the phase delays on the n-side of the device can be obtained from the following relations

$$\varphi_{na} = \frac{2\pi x_{na} f_a}{v_{ns}}, \quad \varphi_{nm} = \frac{2\pi x_{nm} f_m}{v_{ns}}, \tag{2}$$

where x_{na} and x_{nm} are the distances of the $R(x)$-peaks from the junction on the n-side of the ATT device in the pure avalanche and mixed mode respectively. The condition for obtaining maximum power is that the total phase lag should satisfy the following relations at x_{pa} and x_{na} given by, $\varphi_{pa} + \varphi_{pt} = \varphi_{na} + \varphi_{nt} = \pi$; where φ_{pa} and φ_{na} are the avalanche phase delays and φ_{pt} and φ_{nt} are the transit time phase delays.

The shifts of ATT phase delays (in MITATT device) due to the effect of tunneling are determined from the following relations:

$$\delta_p = \varphi_{pm} - \varphi_{pa} \quad \text{at } p\text{-side}, \tag{3}$$

$$\delta_n = \varphi_{nm} - \varphi_{na} \quad \text{at } n\text{-side}. \tag{4}$$

Table 1. Simulated DC and small-signal parameters.

Device	Si Flat-DDR		InP Flat-DDR		GaAs Flat-DDR		4H-SiC Flat-DDR	
	IMPATT	MITATT	IMPATT	MITATT	IMPATT	MITATT	IMPATT	IMPATT
Bias current density, J_0 ($\times 10^8$ A m^{-2})	3.4	3.4	1.4	1.4	6.0	6.0	5.0	5.0
Peak electric field, E_m ($\times 10^7$ V m^{-1})	6.0125	5.9404	6.7499	6.1762	6.0192	5.8747	36.1247	35.5720
Breakdown voltage, V_B (V)	24.25	23.64	28.82	26.71	22.88	22.03	212.41	210.35
Efficiency, η (%)	10.62	10.14	15.90	14.34	9.89	9.24	16.40	16.27
Peak operating frequency, f_p (GHz)	95	117	94	132	96	123	94	103
Peak conductance, G_P ($\times 10^7$ S m^{-2})	−4.3419	−4.1676	−8.2456	−7.7174	−6.0287	−5.6971	−6.2250	−6.0687
Peak susceptance, B_P ($\times 10^7$ S m^{-2})	6.6244	7.7904	6.8436	7.8599	8.1159	10.1450	12.7510	13.7700
Quality factor, $Q_P = (-B_P/G_P)$	1.53	1.86	0.83	1.19	1.35	1.78	2.05	2.27
Negative resistance, Z_R ($\times 10^{-8}$ Ohm.m^2)	−0.6921	−0.5340	−0.7181	−0.6360	−0.5898	−0.4208	−0.3091	−0.2680
RF power output, P_{RF} (W)	0.3192	0.2911	0.8563	0.6882	0.3946	0.3456	35.1070	33.5653

The results presented in the next section show that the effect of tunneling causes an effective shift of the $R(x)$-profile towards the p-side of the device. That is why on the p-side of the device, $\varphi_{pm} > \varphi_{pa}$; which means δ_p is positive. But on the n-side of the device $\varphi_{nm} < \varphi_{na}$; which means δ_n may be negative or positive. Generally p-side of the device is predominantly affected by the tunneling than the n-side of that, which is explained in the following section. Thus the value of δ_p is greater than the value of δ_n (i.e. $\delta_p / \delta_n > 1$). The overall ATT phase delay shift in MITATT device is obtained by averaging over δ_p and $|\delta_n|$ and given by,

$$\delta = \frac{\delta_p + |\delta_n|}{2}. \tag{5}$$

The simulated values of DC and small-signal parameters of DDR devices based on all four semiconductor materials in both IMPATT (without considering tunneling) and MITATT (considering tunneling) modes are given in Table 1.

It can be observed from Table 1 that the RF power output of the devices based on different semiconductors decreases in MITATT mode as compared to IMPATT mode. Maximum power output can be obtained from an IMPATT device when the phase difference between total current and AC voltage is 180°. In IMPATT mode of operation, total current gets 180° phase difference with RF voltage due to both avalanche build-up and transit time of charge carriers, while in MITATT mode tunnel current develops the phase delay with RF voltage due to only the transit time. This fact causes phase distortion between the total current and AC voltage leading to lower RF power output of the device (Acharyya et al. 2014).

2 CALCULATION OF THE TUNNELING INDUCED SHIFT OF ATT PHASE DELAY

The authors have involved a novel method of calculating the tunneling induced shift of ATT phase delay from a study of the shift of negative resistivity peaks due to tunneling in the depletion layers of the device. The principle of the calculation of this is outlined in Section 1 and the results are given below. The ATT phase delay shifts in MITATT mode of the devices are obtained from the $R(x)$-profiles (Figures 4 – 6 of Part II of this paper) of the diodes. The magnitude of shift of ATT phase delay for 94 GHz Si, InP, GaAs and 4H-SiC DDR devices are given in Table 2. This is found to be highest in InP DDR and lowest in 4H-SiC DDR. Thus higher $g_{Tpeak}/$

Table 2. ATT phase delay shifts in MITATT mode.

Base Material and Device Structure	ATT Phase Delay Shift δ ($\times 10^{-3} \pi$ radian)
Si Flat-DDR	130.2
InP Flat-DDR	257.6
GaAs Flat-DDR	169.4
4H-SiC Flat-DDR	41.7

g_{Apeak} – ratio corresponds to higher ATT phase delay shift; which in turn leads to greater deterioration of device performance due to tunneling.

3 CONCLUSION

A novel method of calculating the tunneling induced shift of ATT phase delay is proposed by the authors. Using this method the tunneling induced shift of ATT phase delay is calculated for DDR IMPATTs based on Si, InP, GaAs and 4H-SiC, designed to operate at 94 GHz.

REFERENCE

Acharyya, A., Mukherjee, M. & Banerjee, J.P. (2014). Effects of Tunnelling Current on mm-wave IMPATT Devices. *International Journal of Electronics*, published online, DOI: http://dx.doi.org/10.1080/002072 17.2014.982211.

Frontiers in Computer, Communication and Electrical Engineering – Acharyya (Ed.)
© 2016 Taylor & Francis Group, London, ISBN: 978-1-138-02877-7

Effect of gate voltage and structural parameters on the Subthreshold Swing and the DIBL of Si-SiO$_2$ GAA quantum wire transistor

Arpan Deyasi

Department of Electronics and Communication Engineering, RCC Institute of Information Technology, Kolkata, India

N.R. Das

Institute of Radio Physics and Electronics, University of Calcutta, Kolkata, India

ABSTRACT: Subthreshold Swing (SS) and Drain-Induced Barrier Lowering (DIBL) of gate-all-around quantum wire transistor is computed using a coupled-mode space approach. Schrödinger and Poisson equations are self-consistently solved involving a Non-Equilibrium Green's Function (NEGF) technique under ballistic limit. Gate voltage, channel width and an insulating layer thickness are varied to observe the effect on SS and DIBL considering the occupancy of carriers at a few lower sub-bands only. The result plays an important role in designing the quantum wire transistor based on an integrated circuit involving quasi-ballistic electron transport, and also for switching applications.

1 INTRODUCTION

For modeling of transistor with a low power consumption and minimum leakage, a subthreshold swing and drain-induced barrier lowering are the two crucial parameters, which play a pivotal role in improving the performance of a QWT-based integrated circuit. With the increase of a short-channel effect owing to the downsizing of a transistor, the performance of conventional transistors degraded due to higher leakage current and high power requirement (Lu et al., 2008). The quantum wire transistor is the best possible alternative (Gilbert et al., 2005, Clement et al., 2013) in this condition, which remarkably improves the performance of the transistor with an effective gate control (Zhang et al., 2010). Electronic and transport properties of the nanowire transistors are significantly different than the conventional bulk transistors (Lundstorm et al., 2002), which can be realized through atomistic simulations. Extensive works have already been carried out for computing current flow and significant electrical properties in a nanowire transistor (Chowdhury et al., 2014, Jin et al., 2008, Lundstorm 2000) and are reported for different geometries (Sajjad et al., 2008, Fiori et al., 2007) involving various numerical methods (Fitriwan et al., 2008, Vashaee et al., 2006). Theoretical analysis of resonant tunneling diode (Do et al., 2006), DG MOSFET (Sabry et al., 2011), and CNTFET (Xinghui et al., 2011) is already reported in various literatures due to their unique electrical properties which are beneficial for realizing quantum integrated circuit.

Among the various numerical methods used so far by researchers, Non-Equilibrium Green's function formalism is one of the best techniques as it provides the near accurate calculation for different devices with various geometries (Ren et al., 2000, Datta 2000). Real space approach under NEGF method is popular due to less complexity (Venugopal et al., 2002), but the mode space approach may be used for those devices where the expansion of the characteristic modes of the Hamiltonian is required. In this technique, the Schrödinger equation and the Poisson equation are self-consistently solved to compute electric potential, based on which channel current is computed.

In this paper, subthreshold swing and DIBL of gate-all-around quantum wire transistors are computed using NEGF formalism for Si-SiO$_2$ material composition excluding the effect of scattering. Channel width and thickness of the insulating layer are varied along with the vertical electric field to study the change in the electrical parameters. Dissipative effects are considered at both the source and drain ends in terms of self-energy. Results are important in designing the QWT-based integrated circuit.

2 THEORETICAL FOUNDATION

The Schrödinger equation in 3D domain can be written as:

$$\left[\begin{matrix} V(x,y,z) - \dfrac{\hbar^2}{2m_x*(y,z)}\dfrac{\partial^2}{\partial x^2} - \\[2mm] \dfrac{\hbar^2}{2}\dfrac{\partial}{\partial y}\left(\dfrac{1}{m_y*(y,z)}\dfrac{\partial}{\partial y}\right) - \\[2mm] \dfrac{\hbar^2}{2}\dfrac{\partial}{\partial z}\left(\dfrac{1}{m_z*(y,z)}\dfrac{\partial}{\partial z}\right) \end{matrix}\right]\psi(x,y,z) = E\psi(x,y,z)$$

(1)

where V(x, y, z) is the conduction band edge potential in the active device, and we assume the parabolic energy band with diagonal effective mass tensor due to the variation along y and z directions because of the transition between the channel and the insulating layer.

Considering 'N' number of nodes in the coupled mode space and 'M' number of sub-bands are occupied, the Hamiltonian of the device may be written as:

$$H\begin{bmatrix} \phi^1(x) \\ \phi^2(x) \\ ... \\ ... \\ \phi^M(x) \end{bmatrix} = E\begin{bmatrix} \phi^1(x) \\ \phi^2(x) \\ ... \\ ... \\ \phi^M(x) \end{bmatrix}$$

$$= \begin{bmatrix} h_{11} & h_{12} & ... & ... & h_{1M} \\ h_{21} & h_{22} & ... & ... & h_{2M} \\ ... & ... & ... & ... & ... \\ ... & ... & ... & ... & ... \\ h_{M1} & h_{M2} & ... & ... & h_{MM} \end{bmatrix}\begin{bmatrix} \phi^1(x) \\ \phi^2(x) \\ ... \\ ... \\ \phi^M(x) \end{bmatrix}$$

(2)

where

$$h_{mn} = \delta_{m,n}\left[-\frac{\hbar^2}{2}a_{mn}(x)\frac{\partial^2}{\partial x^2} + E^m_{sub}(x)\right]$$
$$-\frac{\hbar^2}{2}c_{mn}(x) - \hbar^2 b_{mn}(x)\frac{\partial}{\partial x}$$

(3)

All the symbols have the usual meanings.

Introducing dissipative effects in the retarded green's function:

$$G(E) = [ES - H - \sum_S(E) - \sum_D(E)]^{-1}$$

(4)

After computing drain current, subthreshold swing may be put into the form:

$$S = \left[\frac{1}{k_B T \log_e(10)}\frac{d(\mu_s - \mu_0)}{dV_G}\right]^{-1}$$

(5)

Considering is channel-gate capacitance and CP is channel-substrate capacitance, Eq. (5) may be modified as:

$$S = \left[\frac{1}{k_B T \log_e(10)}\frac{C_G q}{C_G + C_P}\right]^{-1}$$

(6)

DIBL is measured by:

$$D = -\frac{V_{th}^{DD} - V_{th}^{low}}{V^{DD} - V_D^{low}}$$

(7)

where V_{th}^{DD} is the threshold voltage measured for higher drain voltage, and V_{th}^{low} is the threshold voltage measured for lower drain voltage, V_D^{low} is the lower drain voltage at the linear region of the static characteristics.

3 RESULTS AND DISCUSSION

Using Eq. (6) and Eq. (7) derived in the section II of this paper; subthreshold swing and drain-induced barrier lowering are computed for the device as a function of channel width, insulating layer thickness and gate voltage. Figure 1 shows the variation of subthreshold swing with channel width for 1V V_{GS} for different insulating layer thicknesses. From the plot, it is seen that with increasing channel dimensions, the subthreshold swing reduces. This is due to the fact that larger channel width increases ON current due to enhanced gate-channel capacitive coupling; simultaneously, OFF current is reduced due to charge sharing.

The overall effect is represented by a monotonic reduction of subthreshold swing. But, with the

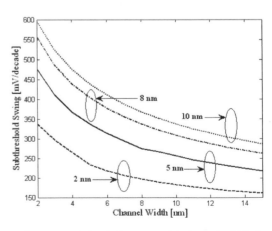

Figure 1. Subthreshold swing with channel width for different insulating layer thickness at $V_{GS} = 1$V.

increase of the insulator thickness for a given channel width, the swing increases. This is because the higher thickness of the insulator restricts the charge sharing, which enhances the off current, and results in increase of the subthreshold swing. This effect is shown in Fig. 2, where subthreshold swing is plotted as a function of insulating the layer thickness for a different channel width. One interesting feature may be noted down in this context that is rate of the decrease of the swing is higher when channel dimensions have been enhanced from 5 nm to 10 nm, but the rate reduces when the width is increased from 10 nm to 15 nm. Hence, the effect of channel width on the subthreshold swing width is predominant for lower magnitude, and also for a higher insulator thickness.

Fig. 3 shows the variation of DIBL with channel width, for different insulating layer thick-nesses with 1V V_{GS}. From the plot, it is observed that DIBL monotonically decreases with channel width, and it increases for higher insulator thickness. Higher channel width allows higher rate of carrier transport inside the channel, which increases ON current. This lowers the potential barrier more, which requires less effective voltage to overcome the channel resistance. Hence DIBL reduces. With the increase of insulating layer thickness, effective potential barrier in the channel increases, which requires a greater horizontal field to make carrier transport. This, in turn, increases DIBL, as depicted in Fig. 4.

Effect of gate voltage on subthreshold swing and DIBL is represented from Fig. 5 to Fig. 8. In Fig. 5 and Fig. 6, subthreshold swing is calculated for the three different values of V_{GS}. From both the

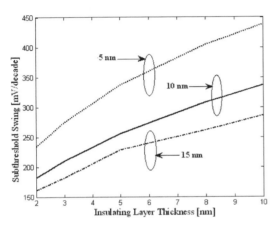

Figure 2. Subthreshold swing with insulating layer thickness for different channel width at $V_{GS} = 1$ V.

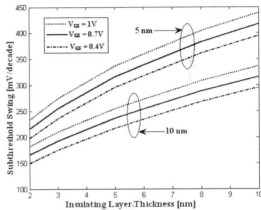

Figure 4. Drain-induced barrier lowering with insulating layer thicknesses for different channel widths at $V_{GS} = 1$ V.

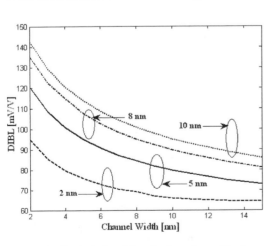

Figure 3. Drain-induced barrier lowering with channel width for different insulating layer thicknesses at $V_{GS} = 1$ V.

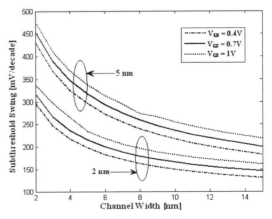

Figure 5. Subthreshold swing with channel width for different V_{GS} for two insulating layer thicknesses.

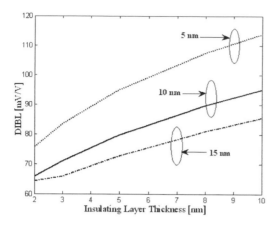

Figure 6. Subthreshold swing with insulating layer thickness for different V_{GS} for two channel widths.

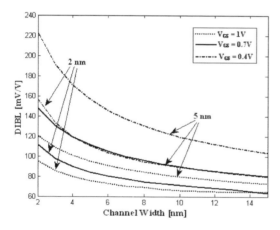

Figure 7. Drain-induced barrier lowering with channel width for different V_{GS} for two insulating layer thicknesses.

figures, it may be concluded that with the increase of V_{GS}, subthreshold swing increases when other structural parameters remain constant. The reason behind the variation may be stated as: higher V_{GS} increases ON current, simultaneously the charge sharing between channel and substrate also increases towards the drain end. This enhances OFF current. The overall effect is the increase of subthreshold swing.

Fig. 7 and Fig. 8 show the effect of V_{GS} on DIBL. Since the increase of V_{GS}, the channel current increases; hence, channel resistance decreases. Thus, the effective potential barrier against carrier transport is reduced, which makes a lower magnitude of DIBL. However, it may be noted that the rate of change of DIBL for lower to moderate value of gate voltage (0.4 v to 0.7 V) is signifi-

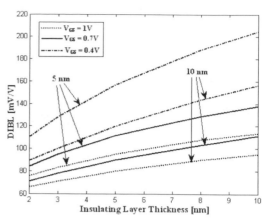

Figure 8. Drain-induced barrier lowering with insulating layer thickness for different V_{GS} for two channel widths.

cantly compared to the rate when V_{GS} is changed from moderate to higher magnitude (0.7 V to 1 V). Thus, the precise estimation of DIBL is important for low to moderate bias application.

4 CONCLUSION

Subthreshold swing and DIBL of GAA quantum wire transistors are computed using the self-consistent solutions of Schrödinger equation and Poisson equation. Solution is obtained through NEGF technique under ballistic limit. Dissipative effect is considered at both the ends of the channel. Si is used as material for channel with SiO_2 as a dielectric. Structural parameters and gate voltage are varied to study the behavior of the computed properties. Result will play a crucial role in designing quantum wire transistor based integrated circuit for nanoelectronic applications.

REFERENCES

Chowdhury. B.N. & Chattopadhyay. S., 2014. Investigating The Impact of Source/Drain Doping Dependent Effective Masses on The Transport Characteristics of Ballistic Si-Nanowire Field-Effect-Transistors, Journal of Applied Physics, 115, 124502.

Clément. N., Han. X.L. & Larrieu. G., 2013. Electronic transport mechanisms in scaled gate-all-around silicon nanowire transistor arrays, Applied Physics Letters, 103, 263504.

Datta. S., 2000, Nanoscale device modeling: Green's function method. Superlattices and Microstructures, 28, 253–278.

Do. V.N., Dollfus. P., & Nguyen. V.L., 2006. Transport and noise in resonant tunneling diode using self-consistent Green's function calculation, Journal of Applied Physics, 100, 093705.

Fiori. G., & Iannaccone. G., 2007. Three-Dimensional Simulation of One-Dimensional Transport in Silicon Nanowire Transistors, IEEE Trans. Nanotechnology, 6, 524–529.

Fitriwan. H., Ogawa. M., Souma. S., & Miyoshi. T., 2008. Fullband Simulation of Nano-Scale MOSFETs based on A Non-Equilibrium Green's Function Method, IEICE Trans. Electronics, E91-C, 105–109.

Gilbert. M.J. & Ferry. D.K., 2005. Quantum Interference in Fully Depleted Tri-Gate Quantum-Wire Transistors—The Role of Inelastic Scattering, IEEE Transactions on Nanotechnology, 4 (5), 599–604.

Jin. S., Amherst. M.A., Fischetti. M.V., & Tang. T.W., 2008. Theoretical Study of Carrier Transport in Silicon Nanowire Transistors Based on the Multisubband Boltzmann Transport Equation, IEEE Transactions on Electron Devices, 55, 2886–2897.

Lu. W., Xie. P. & Lieber. C.M., 2008. Nanowire Transistor Performance Limits and Applications, IEEE Transactions on Electron Devices, 55(11), 2859–2876.

Lundstorm. M., 2000. Fundamentals of carrier transport, 2nd Ed, Cambridge, U.K, Cambridge University Press.

Lundstrom. M. & Ren. Z., 2002. Essential Physics of Carrier Transport in nanoscale MOSFETs, IEEE Transactions on Electron Devices, 49, 133–141.

Ren. Z., Venugopal. R., Datta. S., Lundstrom. M., Jovanovic. D., & Fossum. J.G., 2000. The Ballistic Nanotransistor: A Simulation Study, IEDM Tech. Digest, 715–718.

Sabry. Y.M., Abdolkader. T.M., & Farouk. W.M., 2011. Simulation of quantum transport in double-gate MOSFETs using the non-equilibrium Green's function formalism in real-space: A comparison of four methods, International Journal of Numerical Modeling: Electronic Networks, Devices and Fields, 24, 322–334.

Sajjad. R.N., Bhowmick. S., & Khosru. Q.D.M., 2008. Cross-sectional Shape Effects on the Electronic Properties of Silicon Nanowires, IEEE International Conference on Electron Devices and Solid-State Circuits, 1–4.

Vashaee. D., Shakouri. A., Goldberger. J., Kuykendall. T., Pauzauskie. P., & Yang. P., 2006. Electrostatics of nanowire transistors with triangular cross-sections, Journal of Applied Physics, 99, 054310.

Venugopal. R., Ren. Z., Datta. S. & Lundstrom. M., 2002, Simulating Quantum Transport in Nanoscale Transistors: Real versus Mode-Space Approaches. Journal of Applied Physics, 92, 3730–3739.

Xinghui. L., Junsong. Z., Zhong. Q., Fanguang. Z., Jiwei. W., & Chunhua. G., 2011. Study on transport characteristics of CNTFET based on NEGF theory, International Conference on Electron Devices and Solid-State Circuits, 1–2.

Zhang. X.G., Chen. K.J., Fang. Z.H. & Jun. X., 2010. Dual gate controlled single electron effect in silicon nanowire transistors, 10th IEEE International Conference on Solid-State and Integrated Circuit Technology, 923–925.

Frontiers in Computer, Communication and Electrical Engineering – Acharyya (Ed.)
© 2016 Taylor & Francis Group, London, ISBN: 978-1-138-02877-7

50 Hz cascaded twin-tee notch filter for removal of power line interference from human electrocardiogram—part I: Circuit design

Subhendu Chakraborty, Shuvajit Roy, Subhashri Chatterjee, Adrija Das,
Monisha Ghosh & Aritra Acharyya
Supreme Knowledge Foundation Group of Institutions, Mankundu, Hooghly, West Bengal, India

ABSTRACT: A cascaded twin-tee active notch filter topology has been proposed in this paper to remove the 50 Hz power line interference from the human electrocardiogram signal. Two second order twin-tee active notch filters having different Q-factors have been cascaded to obtain the proposed topology of the fourth order. The rejection bandwidth and the amount of attenuation given to the center frequency of the cascaded filter have been optimized by adjusting the Q-factor of the second stage. Detail of the design methodology has been discussed in this part of the paper.

1 INTRODUCTION

The removal of the power line interference from the recorded biopotential signals from the human body is one of the major fields of study for the biomedical engineers. The strong electric field that originated from 50 Hz power lines induces relatively large potentials over the human body, which corrupts the recorded biopotential signals such as Electrocardiogram (ECG), Electroencephalogram (EEG), Electromyogram (EMG), etc. (Huhta et al. 1973, Chimeno et al. 2000). Some well known techniques such as guard-shield and driven-right-leg (Spinelli et al. 1999, Van Rijn et al. 1990) are used in the biopotential amplifiers in order to reduce the coupling of power line interferences on the recorded biopotential signals. However, enhancement of 'voltage divider effect' due to the large variability of electrode impedances converts the common-mode interference over the body into differential-mode interference, which cannot be rejected, even by using the best quality biopotential amplifiers (Gruetzmann et al. 2007).

Notch filtering is one of the easiest ways of rejecting power line interference from recorded biopotential signals especially in case of human ECG. Both analog and digital notch filters may be used for the above mentioned purpose. Second order analog 50 Hz notch filters such as improved twin-tee (Stout 1976), state-variable topology (Kerwin et al. 1967), and active synthetic inductor RLC (Franco 1988) filters are very popular for the said purpose. Single notch fundamental IIR, single notch fundamental FIR, and multiple notch fundamental and harmonics FIR digital notch filters are also used to reject the power line interference

from biopotential signals. However, due to simplicity in circuit design and implementation, analog notch filtering methods are more convenient as compared to their digital counterparts.

In the present paper, the authors have proposed a fourth order cascaded improved twin-tee notch filter topology for obtaining a very high attenuation at center frequency without significant broadening of rejection bandwidth as compared to its single stage second order counterpart. The Q-factor of the proposed filter is kept well below 10 ($Q < 10$). The values of resistance and capacitance associated with the twin-tee topology are calculated for obtaining the center frequency very near to 50 Hz and those are kept the same for both the stages in the cascaded topology. The performance of the proposed filter has been examined in both time and frequency domains and has been compared with the single stage second order 50 Hz improved notch filter to establish the superiority of the proposed topology.

2 CIRCUIT DESIGN

The circuit diagrams of the single stage second order improved twin-tee notch filter as well as the proposed fourth order cascaded notch filter are shown in Figures 1 (a) and (b), respectively. The center frequency (f_0) for both the cases can be obtained from

$$f_0 = \frac{1}{2\pi RC}, \tag{1}$$

where R and C are the resistance and capacitance values of the twin-tee network. If the value of the capacitance is assumed to be $C = 10\,\mu F$, then from

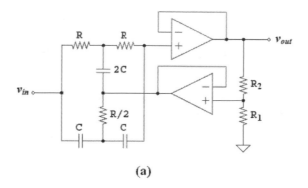

(a)

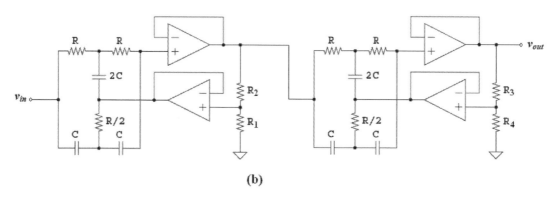

(b)

Figure 1.   Circuit diagrams of (a) single stage second order improved twin-tee notch filter and (b) proposed fourth order cascaded notch filter.

equation (1), the values of the resistance obtained is $R = 318.3099 \ \Omega$. Now practically the branches having resistance values R can be implemented by $R = 1 \ \mathrm{K}\Omega \ \| \ 470 \ \Omega \ \| \ 56 \ \mathrm{K}\Omega$, i.e. the parallel connections of 1 KΩ, 470 Ω, and 56 KΩ resistances. The value of the practical R becomes 317.9128 Ω, which is very near to the ideal one. The corresponding $f_0 = 50.0625$ Hz is very close to the 50 Hz. The branches having resistance values $R/2$ can be simply implemented by a parallel combination of two practical R (i.e. (1 KΩ ‖ 470 Ω ‖ 56 KΩ) ‖ (1 KΩ ‖ 470 Ω ‖ 56 KΩ)) and the 2C can be implemented by using the parallel combination of the two 10 µF capacitances.

The magnitude of the transfer function of the first and second stages of the cascaded filter are given by (Stout 1976):

$$|H_1(j\omega)| = \frac{(\omega^2 - \omega_0^2)}{\sqrt{(\omega^2 - \omega_0^2)^2 + 16(1 - K_1)^2 \omega^2 \omega_0^2}}, \qquad (2)$$

$$|H_2(j\omega)| = \frac{(\omega^2 - \omega_0^2)}{\sqrt{(\omega^2 - \omega_0^2)^2 + 16(1 - K_2)^2 \omega^2 \omega_0^2}}, \qquad (3)$$

respectively; where $K_1 = R_2/(R_1 + R_2)$, $K_2 = R_3/(R_3 + R_4)$, and $\omega_0 = 2\pi f_0$. The Q-factors of the individual stages are given by $Q_1 = 1/4(1 - K_1)$ and $Q_2 = 1/4(1 - K_2)$ and corresponding bandwidths are given by $BW_1 = 4(1 - K_1)f_0$ and $BW_2 = 4(1 - K_2)f_0$.

The overall transfer function (magnitude) of the cascaded filter is given by

$$|H(j\omega)| = |H_1(j\omega)| \times |H_2(j\omega)|. \qquad (4)$$

Now, the value of R_3 keeps varying from 1 to 40 KΩ. The Q_2 varies from 1–10, keeping the $R_1 = R_4 = 1$ K and $R_2 = 40$ KΩ (i.e. $Q_1 = 10.25$) and the corresponding magnitudes of the overall transfer function 20 $\log_{10}|H(j\omega)|$ (in dB) as well as overall bandwidth (BW in Hz) of the cascaded filter have been calculated. The 3-dB lower and upper cutoff frequencies (f_l and f_u) have been determined by equating $|H(j\omega)| = 1/\sqrt{2}$ and determining the corresponding frequencies numerically. Thus, the overall bandwidth is $BW = (f_u - f_l)$.

The 20 $\log_{10}|H(j\omega)|_{f = 50 \ \mathrm{Hz}}$ vs. Q_2 as well as BW vs. Q_2 plots are shown in Figure 2. It is observed from the Figure 2 that the bandwidth decreases sharply with the increase of the Q-factor of the second stage, whereas the amount of attenuation 50 Hz frequency also decreases rapidly with the increase of the same. Therefore, a trade-off between the bandwidth and attenuation has to be done. Thus, one particular value of Q_2 has to be chosen for which the overall rejection bandwidth is not that high as well as the attenuation at 50 Hz is not that low. Keeping this fact in mind, the value of Q_2 is chosen to be 8.0, for which the value of $R_3 = 31$ KΩ.

The frequency response of the designed fourth order cascaded filter and the frequency responses of the individual stages having Q-factors 10.25 and 8.0, respectively, are shown in Figure 3.

The bandwidths of the first, second stand alone stages and the overall cascaded notch filters are

found to be 4.88 Hz, 6.25 Hz, and 8.68 Hz, respectively. The amount of attenuation provided by the above mentioned filters at 50 Hz frequency are found to be −48.75, −50.90, and −99.65 dB, respectively. The overall Q-factor of the fourth order cascaded filter is 5.76 which is lower than that of the individual stages.

3 CONCLUSION

A cascaded twin-tee active notch filter topology has been proposed in this paper to remove the 50 Hz power line interference from the human electrocardiogram signal. Two second order twin-tee active notch filters having different Q-factors have been cascaded to obtain the proposed topology of the fourth order. The rejection bandwidth and the amount of attenuation given to the center frequency of the cascaded filter have been optimized by adjusting the Q-factor of the second stage. The performance of the proposed filter has been evaluated by means of both time and frequency domain analysis in the next part of the paper.

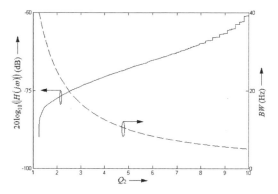

Figure 2. Magnitude of overall transfer function (in dB) at 50 Hz frequency as well as bandwidth (in Hz) versus Q-factor of the second stage.

REFERENCES

Chimeno, M., and Pallas-Areny, R. (2000) A comprehensive model for power line interference in biopotential measurements. *IEEE Trans. Instrum. Meas.*, 49, 535–540.
Franco, S. (1988) Design with Operational Amplifiers and Analog Integrated Circuits. New York: McGraw-Hill, pp. 162–174.
Gruetzmann, A., Hansen, S., and Muller, J. (2007) Novel dry electrodes for ECG monitoring. *Physiol. Meas.*, 28, 1375–1390.
Huhta, J.C., and Webster, J.G. (1973) 60-Hz interference in electrocardiography. *IEEE Trans. Biomed. Eng.*, BME-20, 91–101.
Kerwin, W.J., Huelsman, L.P., and Newcomb, R.W. (1967) State-variable synthesis for insensitive integrated circuit transfer functions. *IEEE J. Solid-State Circuits*, 2, 87–92.
Spinelli, E., Martinez, N., and Mayosky, M. (1999) A transconductance driven-right-leg circuit. *IEEE Trans. Biomed. Eng.*, 46, 1466–1470.
Stout, D.F. (1976) Handbook of Operational Amplifier Circuit Design. New York: McGraw-Hill, pp. 12.1–12.
Van Rijn, A.C.M., Peper, A., and Grimbergen, C.A. (1990) High-quality recording of bioelectric events. *Med. Biol. Eng. Comput.*, 28, 389–397.

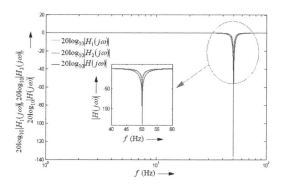

Figure 3. Frequency responses of the first, second stand alone stages as well as the overall cascaded notch filter.

Frontiers in Computer, Communication and Electrical Engineering – Acharyya (Ed.)
© 2016 Taylor & Francis Group, London, ISBN: 978-1-138-02877-7

Research on the 50 Hz cascaded twin-tee notch filter for the removal of power line interference from human electrocardiogram—part II: Simulation study

Subhendu Chakraborty, Shuvajit Roy, Subhashri Chatterjee, Adrija Das,
Monisha Ghosh & Aritra Acharyya
Supreme Knowledge Foundation Group of Institutions, Mankundu, Hooghly, West Bengal, India

ABSTRACT: The simulation study has been carried out to establish the superiority of the proposed fourth order cascaded twin-tee active notch filter topology to compare the results withits single stage second order counterpart. Noiseless human Electrocardiogram (ECG) signal has been added with different levels of interfering 50 Hz power line signal, which are used as inputs of the filters and the corresponding outputs have, thus, been analyzed. Both the time domain and frequency domain analyses of the filtered (by both second single stage order and fourth order cascaded twin-tee notch filters) and unfiltered noisy ECG signals have been presented in this part of the paper for a better inference.

1   INTRODUCTION

In this section of the paper, the performance evaluation of the 50 Hz cascaded fourth order twin-tee notch filter has been carried out and compared with its single stage second order counterpart. Human ECG signals that have been retrieved from the MIT-BIH Arrhythmia Database available at Physionet are of an interference-free record (Goldberger *et al*. 2000). The first six cycles of a typical interference-free ECG signal ($v_{ecg}(t)$) has been shown in Figure 1.

The 50 Hz sinusoidal signal ($v_{infr}(t)$) is added with the signal $v_{ecg}(t)$ in order to introduce the power line interference in it. The corrupted ECG signal has been given by:

$$v_{ecg}^{uf}(t) = v_{ecg}(t) + v_{infr}(t), \qquad (1)$$

where $v_{infr}(t) = V_{infr}\,sin(100\pi t)$, V_{infr} is the amplitude of the interfering signal. Now, the corrupted EGC signal has been Fourier transformed by using the Fast Fourier Transform (FFT) algorithm in MAT-LAB (The MathWorks™) in order to obtain the frequency domain information ($V_{ecg}^{uf}(j\omega)$) of it. Sampling frequency and the number of the frequency points of the discrete Fourier transform have been taken to be $f_s = 1000$ Hz and $N = 5000$, respectively. Finally, $V_{ecg}^{uf}(j\omega)$ is multiplied by the transfer function of the first stage of the second order notch filter ($H_1(j\omega)$) as well as the transfer function of the overall cascaded fourth order notch filter ($H(j\omega)$) in order to obtain the frequency domain of the reconstructed signals ($V_{ecg}^{f}(j\omega)$)

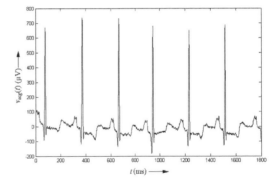

Figure 1.   First 6 cycles of a typical interference-free ECG signal.

(Chakraborty *et al*. 2016). Inverse FFT has been carried out on $V_{ecg}^{f}(j\omega)$ to obtain the time domain of the reconstructed signals. The amplitude of the 50 Hz interfering signal (V_{infr}) has been varied within the range of 1–500 μV to study the effect of the power line interference on the performance of the filters. In order to access the amount of distortion in both the corrupted and reconstructed signals, a relative residue energy parameter has been defined, which is given by:

$$\chi_{uf/f}^{Order} = \frac{\sum_{i=1}^{M}\left(u_{sig}[i] - v_{ecg}[i]\right)^2}{\sum_{i=1}^{M}\left(v_{ecg}[i]\right)^2}, \qquad (2)$$

53

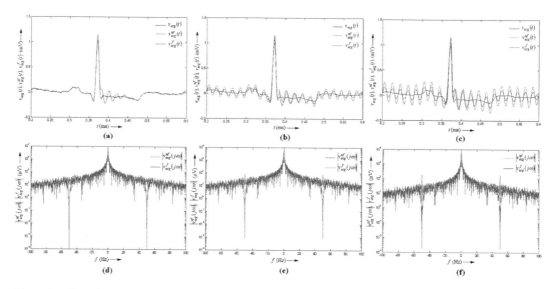

Figure 2. Time domain of the original, corrupted, and reconstructed EGC signals are shown in (a) – (b) for the amplitudes of the interfering signal of 1, 10, and 100 µV. Reconstruction has been done via second order single stage (first stage) notch filter (Chakraborty *et al.* 2015). Frequency domain of the corrupted and reconstructed EGC signals are shown in figures (d) – (f) for the same values of the amplitudes of the interfering signal.

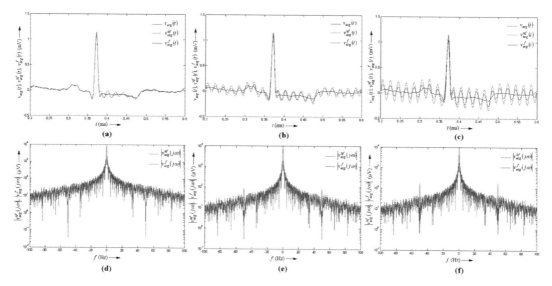

Figure 3. Time domain of the original, corrupted, and reconstructed EGC signals are shown in figures (a) – (b) for the amplitudes of the interfering signal of 1, 10, and 100 µV. Reconstruction has been done via fourth order cascaded notch filter (Chakraborty *et al.* 2015). Frequency domain of the corrupted and reconstructed EGC signals are shown in figures (d) – (f) for the same values of the amplitudes of the interfering signal.

where $\chi_{uflf}^{Order} \equiv \chi_{uf}^{0}$ is used for the unfiltered corrupted ECG signal, $\chi_{uflf}^{Order} \equiv \chi_{f}^{2}$ is used for the reconstructed ECG signal filtered by second order single stage notch filter and $\chi_{uflf}^{Order} \equiv \chi_{f}^{4}$ is used for the reconstructed ECG signal filtered by fourth order cascaded notch filter,

$u_{sig}[i] \equiv v_{ecg}^{uf}[i]$ is the unfiltered corrupted ECG signal, $u_{sig}[i] \equiv v_{ecg}^{f(2)}[i]$ is the reconstructed ECG signal filtered by second order single stage notch filter, and $u_{sig}[i] \equiv v_{ecg}^{f(4)}[i]$ is the reconstructed ECG signal filtered by fourth order cascaded

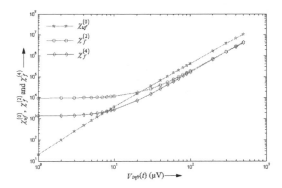

Figure 4. Relative residue energy parameter of unfiltered corrupted signal, reconstructed signal filtered by second order single stage and fourth order cascaded notch filters vs. amplitude of the interfering signal.

notch filter, and M is the total number of samples that are under consideration.

It is obvious that the amount of distortion in the unfiltered signal increases linearly with the increase of the amplitude of the 50 Hz interfering signal. But in filtered signals, the amount of distortion has reduced significantly as observed in Figures 2 and 3. It is observed from Figure 4 that the relative residue parameter of the signal filtered by the fourth order cascaded notch filter is significantly lesser than that of the unfiltered signal within the range of 10–500 μV of the amplitude of 50 Hz interfering signal. However, the same range is found to be of the range of 30–500 μV in the signal filtered by the second order single stage notch filter, which is smaller than that of its fourth order counterpart. But below these specified ranges, the relative residue parameters of both the types of filtered signals remain at constant values (significantly lesser in the signal filtered by fourth order notch) due to the inherent low frequency oscillations produced by the filters. Removal of some harmonics of the clean ECG signals near 50 Hz by the filters cause those low frequency oscillations to occur, which are irremovable.

2 NOISE PERFORMANCE

Signal to interference ratio at the input side of a filter can be expressed as:

$$SIR_{in} = \frac{\sum_{i=1}^{M}\left(\left|v_{ecg}\left[i\right]\right|\right)^2}{\sum_{i=1}^{M}\left(\left|v_{\inf r}\left[i\right]\right|\right)^2}. \tag{3}$$

Similarly, the signal to interference ratio at the output of a filter may be defined as:

$$SIR_{out}^{Order} = \frac{\sum_{i=1}^{M}\left(\left|u_{sig}\left[i\right]\right| - v_{\inf r}\left[i\right]\right)^2}{\sum_{i=1}^{M}\left(\left|u_{sig}\left[i\right]\right| - v_{ecg}\left[i\right]\right)^2}. \tag{4}$$

Here, the interfering signal may be considered as the noise to the original ECG signal. Thus, the noise figure of a filter may be written as:

$$NF = 10\log_{10}\left(\frac{SIR_{in}}{SIR_{out}^{Order}}\right). \tag{5}$$

Equivalent noise temperature of a filtered system may be calculated from:

$$T_e = T_0\left(10^{\frac{NF}{10}} - 1\right), \tag{6}$$

where $T_0 = 300$ K (room temperature).

Figure 5 shows the plots of the noise figure of the second order single stage notch filters, that is the stand-alone first stage (NF_1) and second stage (NF_2) and the overall fourth order cascaded notch filter (NF) vs. the amplitude of the interfering signal. It is observed from Figure 5 that the overall fourth order cascaded notch filter possesses a lower noise figure as compared to its second order single stage counterpart for the entire range of interfering signal amplitude under investigation. Moreover, the stand-alone second stage having lower Q-factor possesses a slightly better noise figure as compared to the stand-alone first stage having a higher Q-factor. Magnitudes of the noise

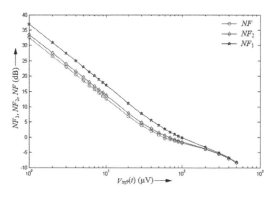

Figure 5. Noise figure of the second order single stage notch filters (stand-alone first stage and second stage) and fourth order cascaded notch filter vs. amplitude of the interfering signal.

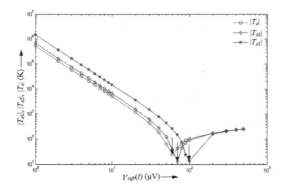

Figure 6. Magnitudes of the noise temperature of the second order single stage notch filters (stand-alone first stage, and second stage) and fourth order cascaded notch filters vs. amplitude of the interfering signal. Right sides of the vertical arrows contain negative values of the noise temperature.

temperature of the abovementioned filters vs. The amplitude of the interfering signal plots has been shown in Figure 6. It is obvious that the smaller noise figure of the fourth order cascaded notch filter leads to a smaller equivalent noise temperature of the same second order stand-alone filters.

3 CONCLUSION

A detailed simulation study has been done to investigate the 50 Hz power line interference rejection capability of the proposed fourth order cascaded twin-tee notch filter topology. The noise performance of the proposed topology has been compared with the conventional second order modified twin-tee notch filter topology. The proposed cascaded topology shows a better noise figure throughout a large range of interference and shows a significantly better 50 Hz power line interference rejection capability.

REFERENCES

Chakraborty, S., Roy, S., Chatterjee, S., Das, A., Ghosh, M. and Acharyya, A. (2015). 50 Hz Cascaded Twin-Tee Notch Filter for Removal of Power Line Interference from Human Electrocardiogram—Part I: Circuit Design. In Proceedings of C2E2 2016, SKFGI, Mankundu, WB, India, CRC Press, Taylor & Francis Group.

Goldberger, A.L., Amaral, L.A.N., Glass, L., Hausdorff, J.M., Ivanov, P.C., Mark, R.G., Mietus, J.E., Moody, G.B., Peng, C.K. and Stanley, H.E. (2000). PhysioBank, PhysioToolkit, and PhysioNet: components of a new research resource for complex physiologic signals. *Circulation*, 101, e215–220.

Digital Phase Lock Loop based on Discrete Energy Separation Algorithm

S. Sarkar
Department of Information Technology, GCELT, Kolkata, West Bengal, India

B.N. Biswas
Chairman Education Division, SKF Group of Institute, Mankundu, West Bengal, India

U. Maulik
Department of Computer Science and Engineering, Jadavpur University, Kolkata, West Bengal, India

ABSTRACT: This paper introduces a new Digital Phase Lock loop (DPLL) in the presence of the three discrete energy separation algorithms, DESA-1a, DESA1, and DESA2 modules, with the incorporation of the Teager Energy Operator (TEO). An additional phase modulation input, along with its frequency modulation input, is present in the digitally controlled oscillator of the loop. The incorporation of DESA module and phase modulation input in the DPLL is an entirely new proposal in the design of the DPLL. DPLL with DESAs are simulated in MATLAB SIMULINK environment. The new model is capable of demodulating the Frequency Modulated (FM) signal with proper perfection and has achieved a faster accusation time for the frequency step signal. Superior output Signal to the Noise Ratio (SNR) is also achievable in the modified DPLL.

1 INTRODUCTION

One of the most common modulation techniques in many naturally occurring signals is the Frequency Modulation (FM). The Digital Phase Lock Loop (DPLL) has been extensively used for demodulation of the FM signal. In recent years, a variety of designs have emerged for DPLL (Biswas 1985) with the advancement in digital circuit technology. In order to obtain faster acquisition and proper demodulation of the signal, a Digitally Controlled Oscillator (DCO) with an appropriate filtering technique is used (Sarkar, Chatterjee, Maulik, & Biswas 2012, Sarkar, Maulik, & Biswas 2013) in the design of the digital phase lock loop. Again, the inclusion of the Discrete Energy Separation Algorithm (DESA) (Maragos, Quatieri, & Kaiser 1993) module in the DPLL gives better results in terms of demodulation, acquisition of frequency step signal, and the output signal to the noise ratio (SNR).

2 TEAGER ENERGY OPERATOR (TEO) AND DISCRETE ENERGY SEPARATION ALGORITHM (DESA)

2.1 Teager Energy Operator (TEO)

The Teager Energy Operator (TEO) (Maragos, Quatieri, & Kaiser 1993, Pal, Chatterjee, & Biswas 2012, Bouchikhi & Boudraa 2012) is a powerful

operator, capable of extracting the signal energy based on mechanical and physical considerations. It has been successfully used in various speeches and biomedical applications (Semmaoui, Drolet, Lakhssassi & Sawan 2012).

To derive the TEO expression, let us consider the unmodulated signal $x(t) = A\cos(\omega_c t + \theta)$, where ω_c is the carrier frequency, A is the amplitude, and θ is the phase of the signal. According to Kaiser (Maragos, Quatieri, & Kaiser 1993), the following differential equations can be considered as a starting point for the operator. The differential equation of the system that is capable of generating such a waveform is the Equation (1).

$$\frac{d^2x}{dt^2} + \omega_c^2 x = 0 \qquad (1)$$

This is the system equation of an undriven linear undamped linear oscillator. It could be a mechanical system consisting of a mass attached to a spring of constant. The periodic oscillation or displacement of the spring can be expressed as ω_c which is also the same as an unmodulated signal.

2.1.1 The continuous teager energy operator

To find out the continuous time TEO considering $A\cos\omega t$ and according to (Maragos, Quatieri, & Kaiser 1993), it yields:

$$\left(\frac{d(x)}{dt}\right)^2 - x(t)\frac{d^2x}{dt^2} = (A\omega)^2 \qquad (2)$$

2.1.2 *The discrete teager energy operator*

Consider a digital signal $x(n) = A\cos(\Omega n + \phi)$, where $\Omega = 2pif / F_s$ is the digital frequency, f is the analog frequency, and F_s is the sampling frequency. The arbitrary phase is ϕ.

Using simple mathematics, the discrete Teager Energy Operator can be written as:

$$x^2(n) - x(n-1)x(n+1) = A^2\sin^2\phi \qquad (3)$$

Thus, the following basic definition of the discrete Teager Energy Operator from the Equation (3) can be written as the Equation (4)

$$\Psi[x(n)] = x^2(n) - x(n+1)x(n-1) \qquad (4)$$

So, for a given signal $x(n)$, the estimation of the instantaneous energy of the signal can be done by substituting $x(n)$ in the formula above. From the context of differential energy operators, the discrete time TEO can be defined (Maragos, Quatieri, & Kaiser 1993) using the Lie bracket as Equation (5):

$$[x,x] = [\dot{x}]^2 - [x.\ddot{x}] = \Psi(x) \qquad (5)$$

The TEO block is implemented in Figure 1.

2.2 *Discrete Energy Separation Algorithm*

The Energy Separation Algorithm (ESA) is the method to separate the amplitude and the frequency components using energy operators from a signal. The ESA can be broadly classified into two major categories as Continuous Energy Separation Algorithm (CESA) and Discrete Time, Energy, Separation Algorithms (DESA). The DESA are again of three major types such as DESA 1a, DESA 2, and DESA 1. In this paper we are considering only discrete energy separation algorithm for our proposed design.

2.2.1 *DESA-1a*

In 1993, Maragos, Kaiser, and Qatari proposed the DESA algorithms (Maragos, Quatieri, & Kaiser 1993) to estimate the amplitude and instantaneous frequency of discrete time AM-FM signals. Now if we carry on as (Maragos, Quatieri, & Kaiser 1993) to derive the three DESAs algorithm, we can get the frequency (Ω_c) and the amplitude (A) for the DESA-1a.

$$\Omega_c = \arccos(1 - \frac{\Psi[x(n) - x(n-1)]}{2\Psi[x(n)]}) \qquad (6)$$

$$A = \sqrt{\frac{\Psi(x[n])}{1 - (1 - \frac{\Psi[x(n)-x(n-1)]}{2\Psi[x(n)]})^2}} \qquad (7)$$

2.2.2 *DESA1*

The DESA-1 can be easily formulated using the result of the DESA-1a as the asymmetric difference for approximating the derivative of x in DESA-1a, a further improvement, may be done by averaging the term $\psi[x(n) - x(n-1)]$ with the $\psi[x(n+1) - x(n)]$. This is done in the case of DESA-1. Repeating the analysis of DESA-1a using forward differences $z(n) = \psi[x(n+1) - x(n)] = y(n+1)$, we get the Equations (8) and (9),

$$\Omega_c = \arccos\left(1 - \frac{\Psi[y(n)] + \Psi[y(n+1)]}{4\Psi[x(n)]}\right) \qquad (8)$$

$$A = \sqrt{\frac{\Psi(x[n])}{1 - (1 - \frac{\Psi[y(n)]+\Psi[y(n+1)]}{4\Psi[x(n)]})^2}} \qquad (9)$$

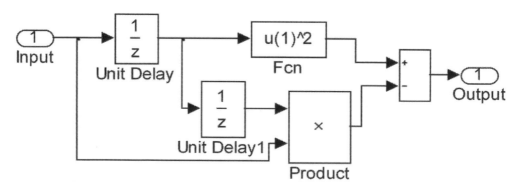

Figure 1. Discrete Teager Energy Operator.

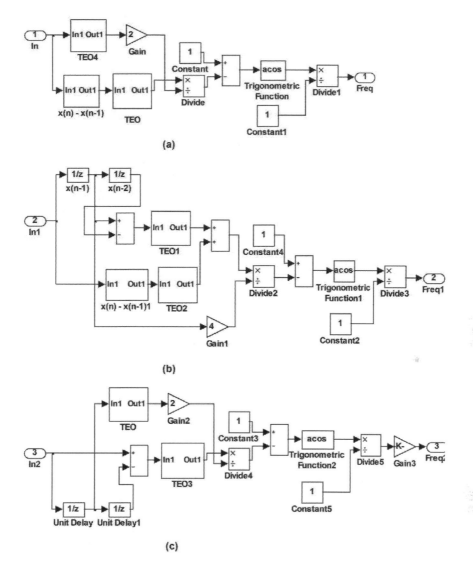

(a)

(b)

(c)

Figure 2.   MATLAB SIMULINK model of DESA-1a (a), DESA1 (b), and DESA2 (c).

2.2.3   *DESA2*

Instead of using two sample derivatives in DESA-1 if we use three sample derivatives, or three sample symmetric differences, then we obtain the DESA-2 algorithm. The new expressions for amplitude and frequency are given as:

$$\Omega_c = 1/2\arccos\left(1 - \frac{\Psi[x(n+1) - x(n-1)]}{2\Psi[x(n)]}\right) \qquad (10)$$

$$A = \frac{\Psi(x[n])}{\sqrt{\Psi[x(n+1) - x(n-1)]}} \qquad (11)$$

The MATLAB SIMULINK implementation of three DESAs are depicted in Figure 2.

3   DPLL USING DESA-1A, DESA1, AND DESA2

Here, we have proposed a new DPLL of a second order where the output signal is fed into the frequency modulation and the phase modulation input of a digitally controlled oscillator (DCO). For the conventional DPLL, the Phase detector output is fed into the frequency modulation input of a DCO through an integrating filter and an amplifier of gain G_{FM}. In our publications (Sarkar, Chatterjee, Maulik, & Biswas 2012, Sarkar, Maulik, & Biswas 2013) we have introduced the phase modulation input in the proposed DPLL to enhance its performances. On the other hand, in our proposed model the phase detector output is fed into the input of the DESA modules (Figure 3) and the subsequent frequency outputs of the DESA blocks are fed into the frequency modulation input of DCO through the amplifier of gains $G_{DESA-1a}$, G_{DESA1}, and G_{DESA2} of the respective DESA blocks. Once more the phase detector output is fed into the phase modulation input of the DCO through an amplifier of gain G_{PM} (Figure 3).

4 NUMERICAL ANALYSIS OF THE DPLL USING DESA

The equivalent open loop transfer function of the DPLL with the phase modulation input can be written from Figure 3. So the equivalent open loop transfer function can be written as:

$$G(z) = \frac{(G_{FM}+G_{PM})z - G_{PM}}{z^2 - 2z + 1} \quad (12)$$

From (12), the closed loop transfer function can be found out. Hence Equation (13) is given as:

$$H_{DPLL_{PM}}(z) = \frac{(G_{FM}+G_{PM})z - G_{PM}}{z^2 - z(G_{FM}+G_{PM}-2) + (1-G_{PM})} \quad (13)$$

The terms are explained above.

The introduction of the DESA modules gives similar results as Equation (13). Here, the frequency modulation gain, G_{FM}, input is absent, but the gain G_{DESA} and the frequency output of DESA module Ω_c are introduced.

$$H_{DPLL\_DESA}(Z) = \frac{z(G_{PM}+G_{DESA}\Omega_c) - G_{PM}}{z^2 + z(G_{PM}+G_{DESA}\Omega_c - 2) +}$$
$$(1-G_{PM}) \quad (14)$$

Again the one in the side loop bandwidth B_L for a digital PLL is defined (Lindsey & Chie 1981) by:

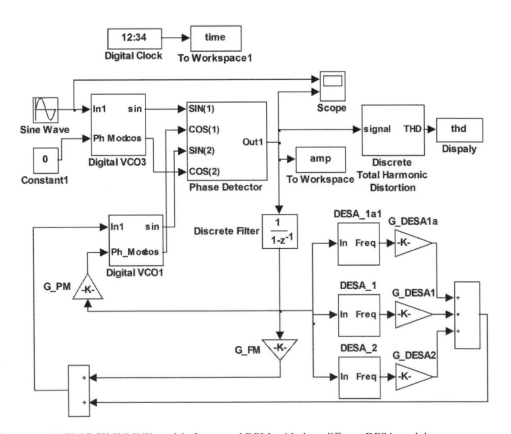

Figure 3.   MATLAB SIMULINK model of proposed DPLL with three different DESA modules.

$$\frac{B_L}{B_i/2} = \frac{1}{2\pi j} \oint H(z)H(z^{-1})z^{-1}dz \qquad (15)$$

The noise bandwidth can be calculated using Equation (14) and Equation (15) as:

$$\frac{B_L}{B_i/2} = \frac{[(G_{PM}+G_{DESA}\Omega_c)^2 + G_{PM}^2](2-G_{PM}) + 2G_{PM}[(G_{PM}+G_{DESA}\Omega_C)(G_{PM}+G_{DESA}\Omega_C-2)]}{G_{PM}G_{DESA}\Omega_C(4-2G_{PM}-G_{DESA}\Omega_C)} \qquad (16)$$

Again we know that:

$$SNR_{out} = \frac{SNR_{input}}{\frac{2B_L}{B_i}} \qquad (17)$$

Using Equation (16) and Equation (17), it is observed that the improvement of an output Signal to the Noise Ratio (SNR) in the presence of the DESA module in the proposed DPLL is over the earlier one. Increase in SNR promises further improvement in the proposed design as shown in Figure 4.

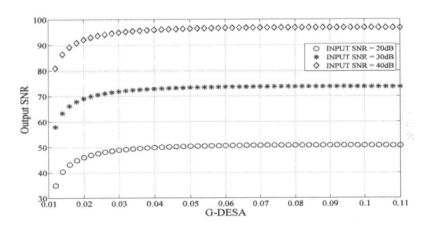

Figure 4. Output SNR versus G_{DESA} with Input SNR as a parameter.

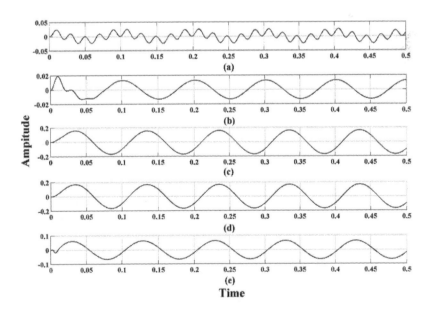

Figure 5. MATLAB Simulation response of FM demodulated signal in conventional DPLL (a), using control of phase modulation (b), using control of phase modulation and DESA-1a (c), using phase modulation and DESA1 (d), and using phase control and DESA2 (e).

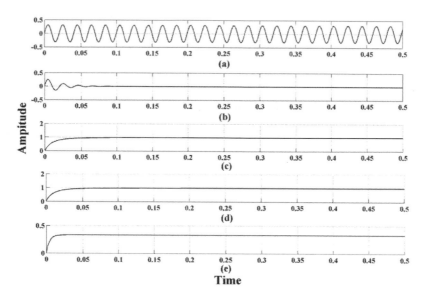

Figure 6. Acquisition response of the frequency step signal in conventional DPLL (a), using control of phase modulation (b), using control of phase modulation and DESA-1a (c), using phase modulation DESA1 (d) and using phase control and DESA2 (e).

5 MATLAB SIMULATION RESPONSES

The FM demodulated signal tracking behavior and the signal acquisition of the frequency step signal have been observed from the Figures 5 and 6. It is evident from those figures that there is a significant improvement in the performance of DPLL in the presence of the DESA modules. For acquisition of the frequency step signal, the maximum overshoot is lower with the use of DESA2 compared to DESA-1a and DESA1 modules in our proposed DPLL. Again from Table 1 of (Maragos, Quatieri, & Kaiser 1993), we get that the mean absolute and the RMS error is smallest in DESA2 compared to other two DESAs. Maragos, Quatieri, and Kaiser (1993) realized that DESA2 is very much easier to implement and it is fastest of all three DESAs. Therefore, use of the DESA2 module is preferable in the proposed model.

6 CONCLUSION

This document suggests an original variety of DPLL design methodologies in the presence of a discrete energy separation algorithm and phase modulation input in the DCO for distortionless demodulation of the FM signal. From the simulation, it is observed that the signal tracking time and signal to noise ratio improved extremely in the presence of DESA. From the above, it is evident

that the presence of all the three DESA modules give a performance of the DPLL.

REFERENCES

Biswas, B.N. (1985). *Phase Lock Theories and Applications.* New Delhi: Oxford and IBH.

Bouchikhi, A. & A. Boudraa (2012). Multicomponent am-fm signal analysis based on emd-b splines esa. *Signal Processing, Elsevier. 92*, 2214–2228.

Lindsey, W. & C. Chie (1981). A survey of digital phase locked loops. *Proc. IEEE. 69*, 410–431.

Maragos, P., T. Quatieri, & J. F. Kaiser (1993). Energy separation in signal modulation with application to speech analysis. *IEEE Trans. Signal Processing. 41*, 3024–3051.

Pal, S., S. Chatterjee, & B.N. Biswas (2012). Energy based discriminator: Some issues. *International Journal of Engineering Research and Applications. 2*, 1735–1741.

Sarkar, S., B. Chatterjee, U. Maulik, & B. N. Biswas (2012). Elimination of truncation and round off error and enhancement of stability using a new split loop dpll. *Int. J. Communication System. 25*, 1059–1067.

Sarkar, S., U. Maulik, & B.N. Biswas (2013). Performance of the new split loop dpll in additive wide band gaussian noise. *International Journal of Adaptive Control and Signal Processing.557*, 1–27.

Semmaoui, H., J. Drolet, A. Lakhssassi, & M. Sawan (2012). Setting adaptive spike detection threshold for smoothed teo based on robust statistics theory. *IEEE Transactions on Biomedical Engineering 59*, 1596–1599.

Frontiers in Computer, Communication and Electrical Engineering – Acharyya (Ed.)
© 2016 Taylor & Francis Group, London, ISBN: 978-1-138-02877-7

Vibrational signal analysis for bearing fault detection in mechanical systems

Naynjyoti Boro, Hrishikesh Das, Abhi Ghosh & Ganesh Roy
Instrumentation Engineering Department, BTAD, Central Institute of Technology, Kokrajhar, Assam, India

ABSTRACT: Rotating machine plays a vital role in the plant. Bearing is one of the critical components in the rotating machinery. Various maintenance techniques such as breakdown maintenance, preventive maintenance are performed for the protection of a machine. Vibration analysis of bearing is the predictive maintenance technique used for fault diagnosis of the machine bearing. Fault in the bearing causes changes in the vibration of the bearing. The main objective of this project is to detect the bearing faults (outer race) in the mechanical system using the vibration signal from the bearing. The vibrational signal is sensed by a piezoelectric transducer, which is analyzed using LabVIEW software. For avoiding the difficulties in the time domain analysis, FFT is employed for a quick comparison between a healthy and defective bearing.

1 INTRODUCTION

The electrical machine plays a vital role in the plant. Vibration signal analysis has an important tool in the area of fault diagnosis (Sukhjeet Singh et al. 2014). Vibration based diagnosis has always been considered as a reliable and easy factor to use. Vibration occurs in the machine in different plants. The risk of machine faults can remarkably create serious danger in the day to day life and productive activities of the people. Vibration can produce the noise as well as energy, which are very much harmful to human health. It can also affect the equipment life and operation stability of the machine. It causes great losses to the plant. So maintenance is a very important factor for a machine, which is essential (Heta S. Shah et al. 2013, Ioan Liță, Daniel et al. 2010). The effect of vibration can be eliminated or reduced from the machine by using vibrational signal monitoring and analysis. Vibration is considered as a good indicator of machine monitoring system. In this paper, the monitoring and analysis of the vibration signals are taken care of using LabVIEW software. The monitoring and analysis of the vibration signal are concerned for implementing the predictive maintenance and fault detection (Asan Gani et al. 2002). Various information about the states change in the vibration signal and fault features of the machines are displayed in the vibration signal. So the measurement and analysis of the vibrational signals are necessary for reducing the effect of the vibration. Sufficient training is required to familiarize with the different parameters. So that the given analysis

method can be effectively applied for eliminating the effect of the vibration. The performance of the rotating machine can also be determined by the measurement and analysis of the vibration signal. The vibration fault simulation system is developed to achieve a better understanding about the faults of the machine (Zhao Hua. 2011). In this paper a LabVIEW based data acquisition system is demonstrated in a very simple way. The obtained results are carefully analyzed using the graphical tools. This simple and quick technique for detecting bearing faults using software control has not been reported previously as far as the knowledge of the authors goes.

2 HARDWARE SETUP

The basic block diagram of the system is illustrated in Figure 1. Vibration monitoring and Analysis

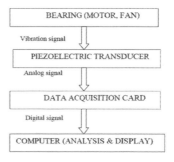

Figure 1. Block diagram of the system.

system consists of the components like piezoelectric transducer, data acquisition card, and computer. The bearing is fitted with the shaft of an ac induction motor.

2.1 Motor (split phase ac induction motor)

One of the important equipment to convert electrical energy to mechanical energy in the modern production is the motor shown in Figure 2, and it plays an important role in the recent industrial plants. Typical components of the machine like shafts, bearing, gears, rotors, drive belts are monitored. The common problem associated with the machine components are Imbalance, misalignment, ball bearing, looseness, bent shaft, journal bearing, gear problem, impeller blade problem, and motor problem. When a machine runs, it creates vibration; this vibration gives a vibration signal, which is sensed by an acceleration transducer. Acceleration transducer is mounted on the machine. Specifications of the analyzed motor are given below.

2.2 Bearing

A bearing (Figure 3) is a machine element that constrains relative motion to only the desired motion, and reduces the friction between the moving parts. The design of the bearing may provide for free linear movements of the moving part or for free rotation around a fixed axis. Also, it may prevent a motion by controlling the vectors of normal forces that bear on the moving parts. Many bearings also facilitate the desired motion as much as possible,

such as by minimizing friction. Bearings are classified broadly according to the type of operation, the motions allowed, or to the directions of the loads (forces) applied to the parts. The tested bearing specifications are given in the bellow.

a. Number of balls (Nb) = 8
b. Pitch diameter (D) = 25 mm
c. Contact angel of bearing = 0°
d. Ball Diameter = 7 mm
e. Shaft Speed = 2800 rpm

2.3 Piezoelectric transducer

Mostly the Piezoelectric transducer shown in Figure 4 is a widely used transducer for vibration monitoring and analysis system. The piezoelectric transducer contains a piezoelectric crystal element. It is preloaded by a mass of certain value. The whole assembly is enclosed in a strong protective covering. An electrical output (low voltage or charge) is generated by the piezoelectric crystal, when the crystal is physically stressed by the vibration of the machine. The crystal is stressed due to the variable inertial force of the mass and produces an electrical signal proportional to the acceleration of that mass. For the acceleration measurement, this small acceleration signal can be amplified to a standard unit. This acceleration signal is converted (electronically integrated) into a velocity or displacement signal within the sensor. This is commonly considered as the ICP (Integrated Circuit Piezoelectric) type transducer or sensor. It has an extensive frequency range. It can be performed well in accurate phase measurements and also in wider temperature range. It resists damage due to severe vibrations and shocks. Now a day's internal amplifier is included in most of the PE transducer. It has some advantages. These are: it has provision for providing a relative immunity to the effects of poor cable insulation and high output to weight ratio. Specifications of the PE transducer are given below.

i. Flat frequency range (HZ): 20–1500HZ or 20–10000 Hz.
ii. Temperature limitation (°C): −50 to +120°C.
iii. Sensitivity range (mV/g): 100 mV/g.

Figure 2.   Split phase ac induction motor.

Figure 3.   Bearing.

Figure 4.   Piezoelectric transducer.

2.4 *Data acquisition*

A data acquisition system is defined as the process in which physical signals are transformed into electrical signals, which are then measured and converted to digital signal for processing, collection, and storage by a computer. In this work the NI myRIO 1900 is used, which is shown in Figure 5. This hardware module has the maximum voltage of +10 v. The maximum voltage range is –10 v to + 10 v. The ADC resolution of this module is of a 16 bit.

3 SETUP AND PROCEDURE

The experimental setup consists of a split phase asynchronous 15 kW induction motor coupled with mild steel shaft of 15 mm diameter mounted on one deep groove ball bearing fixed in one Plummer block. The experimental setup is shown in Figure 6. A voltage regulator has been attached with the motor in order to adjust the motor speed which can be varied 0 to 2800 rpm. A piezoelectric transducer is coupled to the bearing with the help of the adhesive. The motor bearings are healthy. The faults in the bearings that are measured in the mechanical system are installed artificially. The output from the accelerometer given to the data acquisition system of (NI myRIO 1900) attached with a PC having NI LabVIEW software is installed properly. Also, the rotational speed of the shaft is measured with a tachometer. Here, two sets of experiment are conducted, first with the healthy and second with the defective bearing installed in the

Figure 5.   Data acquisition card NI myRIO 1900.

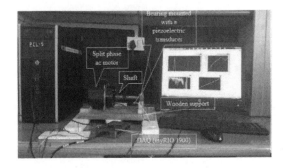

Figure 6.   System setup.

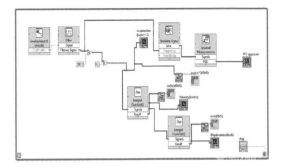

Figure 7.   Block diagram of LabVIEW program.

mechanical system. If there is a fault in the inner/ outer raceway, the balls will pass over the defected area, which produces a vibration while rotating on the machine with a fixed frequency. This frequency is known as a characteristic frequency (f_{bng}), which is associated with a different type of bearing faults. The mathematical calculation for finding out the f_{bng} is given in equation (1).

4 VIBRATION SIGNAL ANALYSIS & PROGRAMMING

The input signal is coming from the piezoelectric sensor with a high frequency is processed with a Butterworth band pass filter in LabVIEW program. As a result the vibration signal is recovered. These data is in 3U (acceleration, velocity, displacement). For further analysis, data must be converted in DC RMS and at the same time, the system can analyze the filtered signals in both the time domain and the frequency domain. Here, Fast Fourier Transformation is applied for identifying the frequency at which the fault came. Which means that it informs how much amplitude of each frequency component in the signal is used. The block diagram of the simulated system is illustrated in Figure 7.

5 MATHEMATICAL CALCULATION

5.1 *Calculation for acceleration, velocity and displacement*

All waveform graph indicators of "Y" scale provide the amplitude value for each signal or frequency. It is the real time signal but overall measurements using the Root Mean Square (RMS) are the most common vibration measurements used. It is important to measure the true RMS not the mean.

$$peak = \frac{peak - peak}{2} \ \& \ peak \times 0.707 = RMS$$

The acceleration is obtained in inch/(sec)² unit from mv/g as per following calculation.

$$1g = 32.2 \, feet \, / \, (\sec)^2 = 386.1 \, inch \, / \, (\sec)^2$$

It can be transferable in any terms like displacement or velocity by following very easy steps.

5.2 *Bearing frequency calculation*

Two sets of experiments are conducted for this work. The first one is with the healthy bearing and second is with the defective bearings installed in the mechanical system. If there is fault on the inner/outer raceway, the rolling element passes over the defect point, which produces an impulse while rotating on the machine at a given frequency. This frequency is known as a characteristics frequency, f_{bng} is associated with different types of bearing faults. Due to the periodicity occurrence of the abnormal physical phenomenon related to the existence of the fault. Characteristics frequencies are functions of the bearing geometry and the mechanical rotor frequency, the outer race fault frequency for vibration f_{bng} is given as:

$$f_{bng} = N_b \times \omega_{inner} \left(\frac{1 - \dfrac{d}{D} \cos \alpha}{2} \right) \qquad (1)$$

where, N_b = Number of balls, ω_{inner} = Shaft speed, d = Ball diameter, D = Pitch diameter, α = Contact angle of the bearing.

In the experiment it is found that:

$N_b = 8$, $\omega_{inner} = 2800$, rpm = 46.6 Hz, d = 6 mm, D = 25 mm, $\alpha = 0°$

Now, $f_{bng} = 8 \times 46.6(\dfrac{1 - \frac{6}{25} \cos 0^0}{2}) = 141.6 \, Hz$

Therefore, the f_{bng} of a healthy bearing frequency must not be greater than 141.6 Hz.

6 RESULT AND DISCUSSION

Two bearings are tested for the real time experiment. Out of the two bearings the condition of one bearing is good and the other one is bad. The result shows the clear difference in their vibration frequency. Figures 8, 9, and 10 shows the vibration signal in the form of acceleration, velocity, and displacement, respectively.

From the acquired FFT spectrum shown in Figure 11 of the healthy bearing it is seen that the bearing frequency approximately around 60 Hz, which is less than the healthy bearing maximum

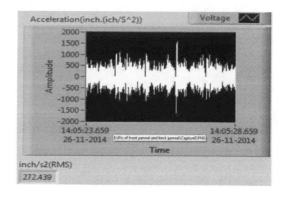

Figure 8.   Vibration Signal in the form of acceleration.

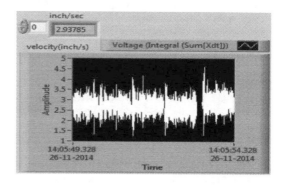

Figure 9.   Vibration Signal in the form of velocity.

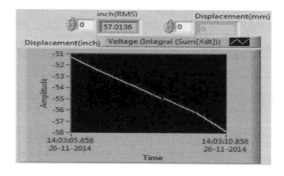

Figure 10.   Vibration Signal in the form of displacement.

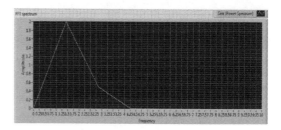

Figure 11.   FFT Spectrum for the Healthy Bearing.

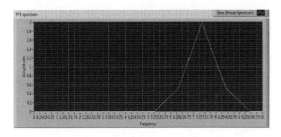

Figure 12.   FFT Spectrum for the Faulty Bearing.

frequency 141.6 Hz. Hence, it is clear that the bearing is healthy.

From the Figure 12 it is observed that the obtained FFT spectrum of the faulty bearing has the bearing frequency of approximately 340 Hz. Therefore, the frequency is greater than the healthy bearing maximum frequency 141.6 Hz. The scale is used as along x axis 1 small division = 10 Hz for the two FFT plots.

7   CONCLUSION

The project approaches systematically to measure the vibrational signal in the real time system. This measurement is automated with one button click. The computer acquires the data from the vibration sensor, and mathematical analysis is performed in LabVIEW15 software. The information can be obtained about the velocity, acceleration, and displacement of vibration signal as well as its RMS values, which are very important in the fault diagnosis of the mechanical system. Users can also get frequency domain analysis (FFT). This feature is incorporated with the versatility of the LabVIEW vibration measurement tool box. Users can also diagnose the faulty bearing in comparison with the healthy bearing with the help of FFT spectrum generated from the vibrational signal of the bearing. Most important thing is that the acquired data allows for the simultaneous solution of the governing equation that describes the theoretical model. This fundamental research helps in building a real-time vibrational measurement system. Later on such a system may be upgraded to measure real time vibrational signals generated by industrial machinery, which will be useful in making key decisions for the maintenance of the machine.

REFERENCES

Asan Gani and M.J.E. Salami (2002). A LabVIEW based Data Acquisition System for Vibration Monitoring and Analysis. 2002 Student Conference on Research and Development Proceedings, Shah Alam, Malaysia.

Heta S. Shah, Pujaben N. Patel, Shashank P. Shah, and Manish T. Thakker. (2013). Channel Vibration Monitoring and Analyzing SystemUsing LabVIEW. (NUiCONE). http://www.ni.com/.

Ioan Liţă, Daniel Alexandru Vişan, and Ion Bogdan Cioc (2010). Virtual Instrumentation Application for Vibration Analysis in Electrical Equipments Testing. 978-1-4244-7850-7/2010/$26.00 ©2010 IEEE, 33rd Int. Spring Seminar on Electronics Technology.

Sukhjeet Singh, Amit Kumar, and Navin Kumar (2014). Motor Current Signature Analysis for Bearing Fault Detection in Mechanical Systems. 3rd (ICMPC 2014), Procedia Materials Science 6. 171–177.

Zhao Hua. (2011). Application of LabVIEW in the Design of Data Acquisition and Signal Processing System of Mechanical Vibration. International Conference on Mechatronic Science, Electric Engineering and Computer. Jilin, China.

Frontiers in Computer, Communication and Electrical Engineering – Acharyya (Ed.)
© 2016 Taylor & Francis Group, London, ISBN: 978-1-138-02877-7

Differential Biogeography Based Optimization applied to Load Frequency Control problem

Dipayan Guha
Dr. B.C. Roy Engineering College, Durgapur, West Bengal, India

Provas Kumar Roy
Jalpaiguri Government Engineering College, Jalpaiguri, West Bengal, India

Subrata Banerjee
National Institute of Technology-Durgapur, West Bengal, India

ABSTRACT: This present paper proposes a new stochastic optimization technique called Differential Biogeography Based Optimization (DBBO) for optimal and effective solution of Load Frequency Control (LFC) problem. DBBO is a hybridization of relatively new optimization algorithms namely Differential Evolution (DE) and Biogeography Based Optimization (BBO). To show the effectiveness, DBBO algorithm is applied to the two-area nonlinear interconnected non-reheat thermal power system and dynamic stability of concerned power system is investigated considering 10% step load perturbation in area-1. Secondary controller gains are simultaneously optimized using DBBO employing Integral Time Absolute Error (ITAE) based fitness function. The performance of proposed controller is compared with original BBO, DE and other similar population based techniques. The simulation results show that DBBO can significantly improve the dynamic performance of concerned power system compared to BBO, DE and other optimization techniques.

1 INTRODUCTION

Modern interconnected power system network is made up with several control areas and these areas are interconnected by tie-line. For satisfactory operation of the power system, there should be a balance between real power generation and load demand. It is well known that power system is always subjected to load variations, which may result in deviation of area frequency and tie-line interchange power from their scheduled values. In this aspect, Load Frequency Control (LFC) plays a vital role to maintain frequency and tie-line power flow at their nominal values. The function of LFC can be viewed as a regulatory control function which tries to match the generation trend within an area to the trend of the randomly changing load, so as to keep balance between total power generation and load demand under normal and disturbed conditions.

Literature survey reveals that LFC has received the great attention of the researchers over the last few decades in order to improve dynamic stability of power system. Different control strategies like conventional [1–4], robust [5], fractional order controller [6], fuzzy logic controller [7], artificial neural network [8] etc. have been proposed for this pur-

pose. The growth in size and complexity of power systems have necessitated the use of an intelligent system that combines the knowledge, techniques and methodologies from various sources for real time control of power system. In this context, various meta-heuristic stochastic population based optimization techniques like hybrid bacteria foraging optimization algorithm—particle swarm optimization (BFOA-PSO) [3], BBO [4], craziness based PSO [9], Differential Evolution (DE) [10], BFOA [11], oppositional BBO [12] etc. have been proposed in LFC system. These algorithms have emerged as efficient computational tools for solving LFC problem in the power system.

The said optimization techniques pertaining to the tuning of LFC parameters suffer from the heavy computational burden and premature convergence, which results in degradation of search capability and computational ability. Further, in the line of '*no-free-lunch*' theorem, there are no meta-heuristic optimization techniques well suited for all optimization problems. In other words, the existing algorithms may give satisfactory results in solving some problems but not all.

Having knowledge of the aforesaid discussion, an attempt has been made in this article to design an effective and optimal solution of LFC problem employing a novel hybrid optimization method, called Differential Biogeography Based Optimization (DBBO). In this paper, DE method is included in original BBO to improve the computational efficacy and convergence performance of original BBO. To demonstrate the ability of DBBO algorithm, an extensively used two-area non-reheat thermal power system with Generation Rate Constraint (GRC) is designed and 10% step load perturbation is given in area-1 to investigate the dynamic performance.

2 MATERIAL AND METHODS

2.1 System under study

A widely employed two-area non-reheat thermal power plant [2, 3] as depicted in Figure 1 is considered for design and analysis purpose. Each area has rating of 2000 MW with nominal loading of 1000 MW. The relevant system parameters are available in [3].

The appropriate value of GRC of the steam turbine is included in the system modeling. In practical power system situation, power generations can only change at a specified maximum limit and therefore, GRC is always considered with steam turbine, otherwise system will experience large momentary disturbances that may cause instability in power system network. The limiting value of GRC in the thermal power plant is 2–5% [2].

2.2 Controller structure and objective function

To control the frequency and tie-line power flow, an optimal Proportional-Integral (PI) controller is

designed using DBBO and provided in each area. The error inputs to the controller are the respective Area Control Errors (ACE) given by:

$$ACE_1 = B_1 \Delta f_1 + \Delta P_{tie,12}$$
$$ACE_2 = B_2 \Delta f_2 - \Delta P_{tie,12}$$
(1)

The controlled inputs (u_1, u_2) to the plant with PI-controller structure is defined as follows:

$$u_1 = K_{p1} ACE_1 + K_{i1} \int ACE_1 dt$$
$$u_2 = K_{p2} ACE_2 + K_{i2} \int ACE_2 dt$$
(2)

where K_p and K_i are proportional and integral gains of PI-controller, respectively. The proposed DBBO algorithm is employed to search optimal settings of said controller through minimization of Integral Time Multiplied Absolute Error (ITAE) based fitness function, which is defined as follows:

$$J_{ITAE} = \int_{t=0}^{T_{final}} \left(|\Delta f_1| + |\Delta f_2| + |\Delta P_{tie,12}| \right) * t * dt$$
(3)

The design problem can be formulated as an optimization problem which is controller parameters bounds. Thus, the optimization problem with DBBO tuned PI-controller is defined as follows:

Minimize J:

Subjected to: $K_{p,min} \leq K_p \leq K_{p,max}$; $K_{i,min} \leq K_i \leq K_{i,max}$

For the present study, PI-controller gains are optimally selected between [−1, 1].

3 DIFFERENTIAL BIOGEOGRAPHY BASED OPTIMIZATION

BBO, proposed by Simon in 2008 [13], is a relatively new optimization technique entered in the domain EA's based on the science of biogeography. Biogeography can easily be defined as the study of distribution of species in nature. The BBO algorithm simulates the emigration and immigration of species between islands, where each island represents a candidate solution for the optimization problems. The main feature of BBO is that the original population is maintained from one generation to another generation. The original BBO suffers from the lack of diversity, slow convergence rate, premature convergence and easily get trapped in local minima. One possible way of improving convergence speed and computational ability of BBO is the hybridization with another algorithm. In this case, DE algorithm can be effectively employed since it is simple, fast, easy to use and ease of comprehensibility [14]. In this hybrid method, both DE and BBO share a

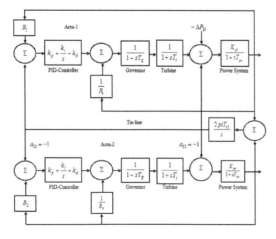

Figure 1. Transfer function model of concerned two-area thermal power system.

single population. The process will start by activating DE algorithm which updates the solution at first and then transfer to the control of BBO. BBO performs migration and mutation operation on the generated population and selects the fittest solution based on their fitness value. The newly generated solution competes to its parent solution in order to enhance the convergence rate of BBO. Additionally, the fittest solutions will be copied to the next generation to replace the worst ones. The candidate solutions provided by BBO are then used by DE for the next generation. For more details regarding DBBO, readers are referred to [14].

4 DBBO APPLIED TO LFC SYSTEM

In this article, a hybrid optimization technique called Differential Biogeography-Based Optimization (DBBO) has been proposed for the first time to solve LFC problem in the power system. In this section, different algorithmic steps of DBBO technique applied to LFC system are enumerated as follows:

Step 1 Randomly generate initial population within the search space.

Step 2 Compute the fitness value of individual population using (3) and sort the population from best to worst.

Step 3 Update current population (X_i) according to DE updating strategy as described in [14].

Step 4 Map fitness value to the number of species.

Step 5 Calculate emigration and immigration rate for individual population using BBO.

Step 6 Perform migration and mutation on individual population as described in [13].

Step 7 Calculate fitness value using (3) of the newly updated population M_i.

Step 8 If, $f(M_i) < f(X_i)$; then select M_i be the current population. Otherwise, go to step 3 until the termination criterion is met.

5 RESULTS AND DISCUSSION

To show the ability of proposed DBBO algorithm to cope with nonlinear interconnected power system with different PI-coefficients, an extensively used [2, 3] two-area non-reheat thermal power plant is considered for investigation. Initially, the linear model of test system, as shown in Figure 1, is studied excluding GRC of steam turbine and then the steam turbine nonlinearity is included in the system modeling. The simulations were performed on an Intel core (TM) i3 processor 2.4 GHz and 2 GB memory computer in the MATLAB 7.8.0 (R2009a) environment. For the present analysis, 50 population size and maximum 50 iterations are considered. The comparative convergence performance of DBBO, BBO, and DE is shown in Figure 2. It is clearly viewed from Figure 2 that proposed DBBO takes 35–40 iterations to reach the optimal point, which justifying the choice of maximum iterations of 50 for the present study.

To demonstrate the superiority of proposed DBBO algorithm, an extensive comparative analysis have been performed with original BBO, DE, hBFOA-PSO, BFOA, PSO, and GA for the similar test system with identical controller structure. The optimum controller gains, minimum fitness value and settling time of frequency and tie-line power oscillations with and without GRC are presented in Tables 1–2. The dynamic performances of con-

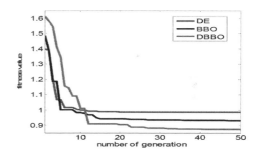

Figure 2. Convergence characteristic of different algorithms.

Table 1. Controller gains, fitness value and settling time of frequency and tie-line power excluding GRC.

EA's	Controller gains (–ve)				ITAE value	Settling time		
	K_{p1}	K_{i1}	K_{p2}	K_{i2}		Δf_1	Δf_2	ΔP_{tie}
:PI	0.2790	0.5406	0.7502	0.0008	**0.8714**	**4.60**	**5.28**	**5.89**
BBO:PI	0.2746	0.5944	0.1497	0.0109	0.8968	5.72	6.57	6.15
DE:PI	0.2884	0.6370	0.7173	0.0015	0.9324	5.47	5.38	6.05
hBFOA-PSO	0.3317	0.4741	0.3317	0.4741	1.1865	7.39	7.65	5.73
BFOA	0.3317	0.4741	0.3317	0.4741	1.8379	5.52	7.09	6.35
PSO	0.3597	0.4756	0.3597	0.4756	1.2142	7.37	7.82	5.0
GA	0.2346	0.2662	0.2346	0.2662	2.7475	10.6	11.4	9.37

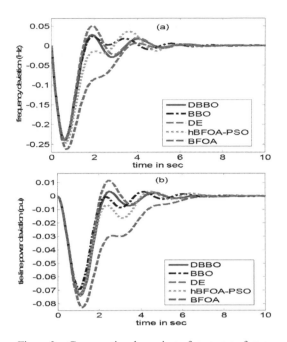

Figure 3. Comparative dynamic performances of concerned power system after 10% SLP in area-1.

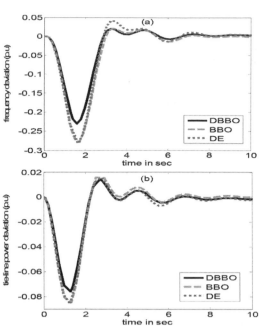

Figure 4. Comparative dynamic performances of concerned power system after 10% SLP in area-1 in the presence of GRC.

Table 2. Controller gains, fitness value and settling time of frequency and tie-line power including GRC.

EA's	Controller gains (–ve)				ITAE value	Settling time		
	K_{p1}	K_{i1}	K_{p2}	K_{i2}		Δf_1	Δf_2	ΔP_{tie}
DBBO:PI	0.2846	0.6413	0.6739	0.0019	**0.8731**	**5.79**	**6.57**	**6.97**
BBO:PI	0.2805	0.6580	0.6277	0.0013	0.9526	5.80	6.61	7.13
DE:PI	0.2788	0.7028	0.5253	0.0010	0.9970	8.62	7.72	12.4

cerned power system without GRC are presented in Figure 3. For better comparison, the system dynamics with original BBO, DE, hBFOA-PSO [3], and BFOA [3] are also given in Figure 3. Critical observation of Figure 3 and Table 1 shows that proposed DBBO tuned PI-controller yields greater system performance in terms of settling time, fitness value.

In the second phase of study, GRC of steam turbine is included in the system and its impact on the dynamic stability of power system has been studied. The dynamic responses of concerned power system with GRC are depicted in Figure 4 and setting time of frequency and tie-line power oscillations are given in Table 2. Critical analysis of Table 2 shows that proposed DBBO tuned PI-controller outperforms original BBO and DE.

6 CONCLUSION

In this paper, a hybrid optimization technique called differential biogeography based optimization is designed and implemented in the power system to solve LFC problem. The proposed DBBO algorithm is used to search optimal settings of PI-controller employing ITAE based fitness function. To establish the superiority of DBBO, the static and dynamic responses are compared with other similar population based optimization methods. The result reveals that proposed DBBO provide better transient as well as steady state response and increased stability margin when compared with other similar population based optimization techniques.

REFERENCES

[1] Nanda, J., Mishra, S. & Saikia L.C. 2009. Maiden Application of Bacterial Foraging-Based Optimization Technique in Multiarea Automatic Generation Control. *IEEE Trans Power Sys* 24(2): 602–609.

[2] Sahu, R.K., Panda, S. & Padhan, S. 2015. A hybrid firefly algorithm and pattern search technique for automatic generation control of multi-area power systems. *Int. J Elect Power Energy Sys* 64: 9–23.

[3] Panda, S., Mohanty, B. & Hota, P.K. 2013. Hybrid BFOA–PSO algorithm for automatic generation control of linear and nonlinear interconnected power systems. *Applied Soft Comput*13: 4718–30.

[4] Guha, D., Roy, P.K. & Banerjee, S. 2014. Optimal Design of Superconducting Magnetic Energy Storage Based Multi-Area Hydro-Thermal System Using Biogeography Based Optimization. *In Proc. Int. Conf of Emerging Appl of Inform Tech* ISI Kolkata, 52–57.

[5] Hosseini, H., Tousi, B., Razmjooy, N., & Khalilpour, M. 2013. Design Robust Controller for Automatic Generation Control in Restructured Power System by Imperialist Competitive Algorithm. *IETE J of Res* 59(6): 745–52.

[6] Pan. I. & Das, S. 2015. Fractional-order load-frequency control of interconnected powersystems using chaotic multi-objective optimization. *Appl Soft Comp* 29: 328–44.

[7] Prakash, S. & Sinha, S.K. 2015. Neuro-Fuzzy Computational Technique to Control Load Frequency in Hydro-Thermal Interconnected Power System. *J. Inst. Eng. India Ser. B*96(3):273–282

[8] Saxena S.C., Kumar, V. & Waghmare, L.M. 2002. Cascade Control of Interconnected System Using Neural Network. *IETE J of Res* 48(6): 461–69.

[9] Gozde, H. & Taplamacioglu, M.C. 2011. Automatic generation control application with craziness based particle swarm optimization in a thermal power system. *Int. J Elect Power Energy Sys* 33: 8–16.

[10] Padhan, S., Sahu, R.K. & Panda, S. 2014. Automatic generation control with thyristor controlled series compensator including superconducting magnetic energy storage units. *Ain Shams Engg J* 5: 759–774.

[11] Guha, D., Roy, P.K. & Banerjee, S. 2015. Oppositional Biogeography Based Optimization Applied to SMES and TCSC Based Load Frequency Control with Generation Rate Constraints and Time delay. *Int. J Power Energy Conversion (in press)*.

[12] Ali, E.S. & Abd-Elazim, S.M. 2013. BFOA based design of PID controller for two-area Load Frequency Control with nonlinearities. *Elect Power Energy Sys* 51: 224–31.

[13] Saimon, D. 2008. Biogeography based optimization. *IEEE Trans Evolutionary Computation* 12(6): 702–713.

[14] Boussaid, I., Chatterjee, A., Siarry, P. & Ahmed-Nacer, M. 2011. Two-stage update biogeography-based optimization using differential evolution algorithm (DBBO). *Computers Operation Res* 38: 1188–98.

Frontiers in Computer, Communication and Electrical Engineering – Acharyya (Ed.)
© 2016 Taylor & Francis Group, London, ISBN: 978-1-138-02877-7

Wide beam microstrip patches with grounded E-shaped edges to improve the polarization purity

Subhradeep Chakraborty
Department of ECE, Siliguri Institute of Technology, Siliguri, West Bengal, India
Sir J.C. Bose School of Engineering, Department of ECE, Mankundu, West Bengal, India

Rakshapada Poddar & Sudipta Chattopadhyay
Department of ECE, Siliguri Institute of Technology, Siliguri, West Bengal, India

Raktim Guha
Sir J.C. Bose School of Engineering, Department of ECE, Mankundu, West Bengal, India

ABSTRACT: Rectangular microstrip patch with grounded E shaped edges are proposed for wide radiation beam along with an improved polarization performance. Around 64° and 98° of the 3 dB beam width is obtained in E and H plane, respectively. Moreover, the proposed structure exhibits excellent cross polarization isolation compared to peak co polarization gain. The measured results are theoretically justified with simulations and excellent agreement between them has been revealed.

1 INTRODUCTION

In the present state of the art research of wireless communication technologies, Rectangular Microstrip Patch (RMP) is one of the most explored type of planar antennas. A Conventional RMP (CRMP) exhibits a number of attractive features like small size, light weight, compatibility with MMICs, conformability to the curved surface along with moderate radiation performance (Garg et al. 2001). However, a CRMA often suffers from its poor polarization purity (less co-polarization to cross polarization ratio or CP-XP ratio) of the radiated field at the far-field, which limits its use in various practical applications where wide coverage is required over a wide elevation angle (Guha et al. 2011). The XP radiation is very much pronounced in the H plane than in the E plane, as found in the open literature (Garg et al. 2001). It affects the polarization purity (CP-XP ratio) significantly. Thus, the improvement of CP-XP ratio of RMAs is of a very great importance and it is probably one of the decisive factors for the use of RMAs.

Some investigations on the modified structure of a conventional patch antenna were reported by (Loffler et al. 1999, Granholm et al. 2001) for the XP radiation reduction. Around 13 to 20 dB of CP-XP isolation is obtained from those. A single microstrip antenna with an air substrate with a complex feeding mechanism was presented by (Chiou et al. 2002), where the structure produced a dual polarization, along with 25 dB and 23 dB

suppression of XP radiation compared to the maximum CP gain for excitation at different ports. However, all of these structures were quite bulky, and complex to design.

Recently, Defected Ground Structures (DGS) has been used increasingly in the RMAs for the suppression of XP radiation as suggested by (Kumar et al 2014, Ghosh et al. 2015). Using linear and bracketed DGS, around 15–25 dB of CP-XP ratio is found. However, the use of DGS influences an increase in the back radiation in terms of XP radiation from the RMA.

Some works (Lee et al. 2000, Wong et al. 2005) show completely new configurations of the microstrip patches, where the patch element was shortened with the ground plane. But those do not deal with the polarization purity of RMP. A recent paper (Ghosh et al. 2014) presents a CRMA with grounded non-radiating edges intended to achieve around 32–34 dB CP-XP ratio in the H plane with a nearly equal XP suppression in E-plane. Nevertheless, in case of the proposed structure by (Ghosh et al. 2014) the resonant frequency is shifted to the higher side of the spectrum by around 30%. This increases the antenna size and seems to be difficult to integrate it with the modern miniaturized wireless devices. Moreover, the proposed antenna (Ghosh et al. 2014) shows poor bandwidth of around 5%. Following Ghosh et al. 2014, another very recent report (Poddar et al.) shows the conventional structure with three shorts at non radiating edges for low XP with a broad impedance bandwidth. However, the CP radiation beams are not improved in that work. The H plane beam width is only 50° with a distorted E plane beam, which is obtained from the paper (Poddar et al.).

Therefore, in order to circumvent the shortcomings, a RMP with grounded E shaped non radiating edges is proposed. Around 30–40 dB of CP-XP isolation over a wide angular range along with the broad H plane beam width is evident from the proposed structure.

2   PROPOSED STRUCTURE

The proposed configuration of the RMP with grounded E shaped non radiating edges is shown in Figure 1, and the value of the different parameters of the proposed RMP is presented in Table 1. It is a fact that in a thin CRMP some higher order orthogonal modes resonate along the width of the patch, which contribute significantly in XP radiations. In order to perturb the field distribution of these modes the two slots (dimensions $l_c \times w_c$) are cut on the non-radiating edges of the patch. It ruptures the higher order mode distribution beneath the patch without affecting the dominant mode field pattern. Now, the non radiating edges are shorted with very thin strips of metal (copper) (dimensions $l_s \times 0.5$ mm) to the ground plane. It completely mitigates the higher order mode field distribution.

3   RESULTS AND DISCUSSIONS

A Pair of prototypes (CRMP and proposed grounded E-shaped RMP) with the dimensions mentioned in the Table 2. have been utilized for the

Table 2.   Simulated and Measured half power beam width of the conventional and proposed patch.

| Structure | 3 dB beam width | | | |
| | E plane | | H plane | |
	Simulated	Measured	Simulated	Measured
Conventional	61°	66°	63°	64°
Present	64°	68°	91°	94°

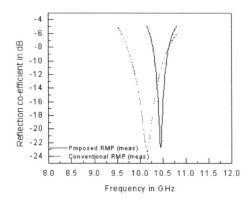

Figure 2.   Measured reflection coefficient profile for conventional and proposed RMP.

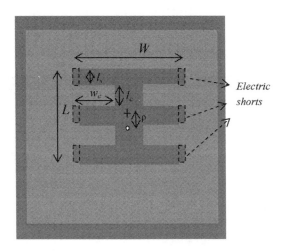

Figure 1.   Schematic representation of the proposed microstrip patch (top view).

Table 1.   Detail parameters of the proposed RMA on 60 mm × 60 mm ground plane. Substrate thickness $h = 1.575$ mm, $\varepsilon_r = 2.33$. (All dimensions are in mm).

L	W	l_s	l_c	w_c	ρ
8.5	12.75	0.5	3.2	5	0.8

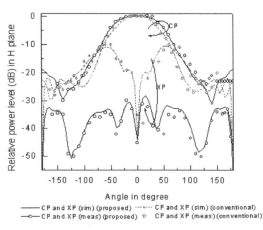

Figure 3a.   Measured and Simulated H plane radiation pattern of conventional and proposed RMP.

present investigation. The simulated and measured results are documented in the same plot for comparison. From the reflection co-efficient profile, it is found that both the conventional and the proposed structures resonate near 10 GHz (Figure 2). The complete radiation patterns of the CRMP and the proposed antenna at the resonating frequency are presented in Figure 3. For CRMP, H plane XP level is appreciable near the broadside direction with the peak XP level of −10 dB near ±50° (Figure 3a). On the contrary, when the non-radiating edges of the CRMP are shaped like 'E', the fields resonating along the width of the patch are perturbed significantly. Moreover, shorting the E-shaped edges completely nullifies the corresponding field distribution and mitigates the radiation from those edges.

Consequently, the improvement of the XP performance due to the modification of a conventional patch structure is clear from the Figure 3a. The H plane XP pattern is suppressed by around 20 dB compared to the CRMP as is observed from Figure 3a. High CP-XP isolation of more than 30 dB for the present antenna is found for whole ±180° angular range (Figure 3a) Excellent agreement between simulated and measured results are apparent from the figure. In E plane, significant suppression in the XP radiation is observed with the proposed structure compared to CRMP Figure 3b. However, in both the cases, E plane XP level is below −30 dB. Nevertheless, Figure 3 depicts improved polarization purity in both of the principle planes for proposed RMP.

It is also evident from Figure 3a that the radiated beam becomes broader in the H plane, compared to the E plane for the proposed RMP. Wide beam in the H plane is obviously beneficial for a wide angular coverage in wireless communication. From

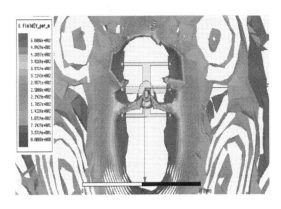

Figure 4. Substrate field distribution for proposed patch at the dominant mode.

Table 2, it is evident that the 3dB beam width of the proposed RMP is 47% more wider than that of a CRMP. The zoomed view of a substrate field distribution has been explored in Figure 4. The close inspection of the figure shows that there is a significant asymmetry in the field distribution in the H plane, while in the E plane appreciable asymmetry is not seen. This may be attributed for the wide beam in the H plane.

4 CONCLUSION

A simple and new grounded E shaped patch is proposed for the wide beam and improved cross polarization performance. It finds potential applications in the field of wireless communication where wide beam and improved polarization purity is required. However, more investigations are required with such structures for more improved radiations and bandwidth performances.

REFERENCES

Chiou, T.W., Wong, K-L. 2002 Broad-Band Dual-Polarized Single Microstrip Patch Antenna With High Isolation and Low Cross Polarization *IEEE Transactions on Antennas And Propagation* 50(3): 399–401.

Garg, R., Bhartia, P., Bhal, I. and Ittipibun, A. 2001 *Microstrip Antenna Design Handbook,* Norwood, Artech House.

Ghosh, A., Ghosh, D., Chattopadhyay, S., Singh, L. L. K. 2015 Rectangular Microstrip Antenna on Slot Type Defected Ground for Reduced Cross Polarized Radiation *IEEE Antennas and Wireless Propagation Letters* 14: 321–324.

Ghosh, D., Ghosh, S.K., Chattopadhyay, S., Nandi, S., Chakraborty, D., Anand, R., Raj, R. and Ghosh, A. 2014 Physical and Quantitative Analysis of Compact Rectangular Microstrip Antenna with Shorted

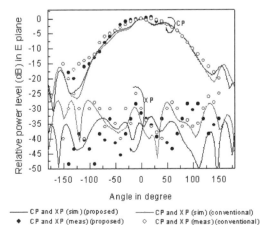

CP and XP (sim) (proposed) — CP and XP (sim) (conventional)
♦ CP and XP (meas) (proposed) ◇ CP and XP (meas) (conventional)

Figure 3b. Measured and Simulated H plane radiation pattern of conventional and proposed RMP.

Non-Radiating Edges for Reduced Cross-Polarized Radiation Using Modified Cavity Model *IEEE Antennas and Propagation Magazine* 56(4): 61–72.

Granholm, J., Woelders, K. 2001 Dual Polarization Stacked Microstrip Patch Antenna Array With Very Low Cross-Polarization *IEEE Transactions on Antennas And Propagation* 49(10): 1393–1402.

Guha, D., Antar, Y. M. M. (Eds) 2011 *Microstrip and Printed Antennas—New Trends, Techniques and Applications*, U.K, John Wiley.

High Frequency Structure Simulator, v 14.

Kumar, C. and Guha, D. 2014 DGS integrated Rectangular microstrip patch for improved polarization purity with wide impedance bandwidth *IET Microwave, Antennas & Propagation* 8(8): 589–596.

Lee, K.F., Guo, Y.X., Hawkins, J.A., Chair, R. and Luk, K.M. 2000 Theory and experiment on microstrip patch antennas with shorting walls *IEE Proceedings Microwaves, Antennas and Propagation* 147(6): 521–525.

Loffler, D., Wiesbeck, W. 1999 Low-cost X-polarised broadband PCS antenna with low cross-polarisation *Electronics Letters* 35(20): 1689–1691.

Poddar, RP., Chakraborty, S., Chattopadhyay, S. Improved Cross Polarization and Broad Impedance Bandwidth from Simple Single Element Shorted Rectangular Microstrip Patch: Theory and Experiment, *Frequenz* (In press).

Wong, K. L. 2005 Internal Shorted Patch Antenna for a UMTS Folder-Type Mobile Phone, *IEEE Transactions on Antennas And Propagation* 53(10): 3391–3394.

Frontiers in Computer, Communication and Electrical Engineering – Acharyya (Ed.)
© 2016 Taylor & Francis Group, London, ISBN: 978-1-138-02877-7

Smooth sliding mode control of a nonlinear CSTR using an inverse hyperbolic function-based law

Abhinav Sinha
TATA Consultancy Services Limited, Pune, Maharashtra, India

Rajiv Kumar Mishra
School of Electronics Engineering, Kalinga Institute of Industrial Technology (KIIT University), Bhubaneswar, Odisha, India

ABSTRACT: A Continuous Stirred Tank Reactor (CSTR) is a typical example of an industrial equipment for chemical processes that exhibit dynamics of a second order nonlinear system. Nonlinear and coupled nature of CSTR poses a challenge in the design of the robust control with a large operating region. Industrial processes require a good state estimation and disturbance rejection. Under parameter variations and fast changing dynamics, a sliding mode controller is presented in this work to provide robustness to the system in a very short time. Unlike traditional controllers with a fixed gain, this controller is based on an inverse hyperbolic function whose gain is a nonlinear and variable depending on the value of the function. The stability of the controller has been proved using the Lyapunov method. Robustness has been confirmed using numerical simulations.

1 INTRODUCTION

A Continuous Stirred Tank Reactor (CSTR) is a benchmark equipment in many industries that requires a continuous addition and withdrawal of reactants and products. A CSTR may be assumed to be somewhat opposite of an idealized well-stirred batch and tubular plug-flow reactors. CSTRs are incorporated to achieve an optimal productivity of a chemical process by maintaining high conversion rates and thus maximize economy. This chemical process exhibits a nonlinear dynamic behavior during a process start-up, shutdown, and during upsets in steady state conditions. These dynamics are unfavorable and need to be modified by a control action that provides a robustness to it. This process is nonlinear in nature and hence dynamical modelling, control, simulation, and controller tuning requires substantial effort. Earlier control strategies like PID (Ray 1981) and feedback linearization (Yayong Zhai 2012) lack the versatility to compensate the complexities of the dynamic process. Several linear methods of control have been proposed in literature to maintain stability and achieve robustness but they fail to provide performance of a high degree. Operation of these controllers are restricted to a linear regime and yield suboptimal results when they are operated outside their linear limits.

Kravaris and Palanki (1998) and Morari and Zafiriou (1989) developed a linear controller using Taylor's linearization for the system whose uncertainties are assumed to be bounded. Alvarez- Ramirez (1994) showed that global stability can't be achieved when approximation is made using local linearization. Other methods include complete or partial feedback linearization of the system dynamics. Input-output feedback linearization was a widely used approach earlier, but it requires full measurement of states (Colantonio, Desages, Romagnoli, and Palazoglu 1995) by input-output feedback. This is practically not possible owing to the difficulties in the measurement of the concentration of the reactants directly online. A possible control strategy, which is nonlinear in nature, was proposed by Kravaris and Arkun (1991). The approach involves modifying the dynamics of the system to make it a pole assignment problem. This method is known as a linearization of the dynamics (Colantonio, Desages, Romagnoli, and Palazoglu 1995) and incorporates state coordinate transformation for a linear input-state behavior. The problem of the unmeasured states was tackled using the design of observers (Wu, 2000, Chen and Peng, 2004, Pan et al., 2007, Graichen et al., 2009, Di Ciccio et al., 2011, Hoang et al., 2012 and Antonelli and Astolfi, 2003). Although, guaranteed asymptotic stability seems promising but the effect of disturbance remains overlooked. Advanced industrial chemical processes like alkylation of benzene with ethylene, requires a faster response (Zhao, Zhu, and Dubbeldam 2014) and robustness of a high degree. Disturbance creeping is strictly undesirable in the mentioned scenario. Under continuous perturbations, observer design approach to control such a nonlinear process ceases to yield optimal results. Providing a high gain controller may achieve faster response but it can also lead to an unwanted control effort saturation (Zhao, Zhu, and Dubbeldam 2014), especially, when the initial track error is very large. This also lacks safety in such an environment and is not permissible. The choice of system outputs is considered in practical applications posing limitations on the controller design using these methods. In many cases, the actual nonlinear system fails to get linearized completely and there remains a part of the transformed system, which is nonlinear (Kravaris and Arkun, 1991). This part has now some zero dynamics as its dynamic properties and it can't be ignored in the design of the controller and can preclude the satisfactory process response (Colantonio, Desages, Romagnoli, and Palazoglu 1995). Under continuous perturbations, the above strategies fail to provide robustness and high performances. Hence, the system calls for a nonlinear controller that can provide a high degree of robustness. Adaptive and fuzzy techniques can also be applied to such a problem but they have to be combined with other control techniques to yield optimal performances. The main reason behind this is inspite of being a non-model based control and providing a good approximation to hard nonlinearities, they are not able to cope up very fast with the changing dynamics. Hence, we propose a Sliding Mode Control (SMC) to achieve a control scheme, which fully accounts for model uncertainties and matched parameter variations. SMC is a variable structure control composed of independent structures and a switching logic (Zak 2003). The advantage of using this control is that we can tailor the system of dynamical behavior by a particular choice of the sliding function. However, high frequency oscillations are also referred to as chattering (K. David Young, Vadim I. Utkin and Umit Ozguner 1999), which comes into picture when this discontinuous control (K. David Young, Vadim I. Utkin and Umit Ozguner 1999) is used. This can excite unmodeled parameters and can cause serious damage to the actuators and plant dynamics. This arises due to the use of signum function in the control law and its discontinuous nature. Hence, chattering reduction (Levant 1993) is also a problem to be considered in the design of the controller. In this work, we have used a reaching law based on the Inverse Hyperbolic Function (IHF) to formulate our control law.

2 REACTOR DESCRIPTION

The mathematical model of the system is nonlinear due to the presence of exponential terms. The model (Yayong Zhai 2012), as obtained from the theory of thermodynamics and chemical kinetics, is presented here.

$$\dot{C}_a = \frac{Q}{V}(C_{a_0} - C_a(t)) - k_0 C_a(t) e^{-\frac{E}{RT(t)}} \quad (1)$$

$$\dot{T} = \frac{Q}{V}(T_0 - T(t)) - k_1 C_a(t) e^{-\frac{E}{RT}}$$
$$+ k_2 Q_c(t)[1 - e^{-\frac{k_3}{Q_c(t)}}](T_{c_0} - T(t)) \quad (2)$$

The above model is based on the complex nonlinear chemical reaction system of CSTR. Under the assumption of complete mixing, the reactor is getting cooled continuously and the volume of the output chemical product (B) is equal to the volume of the input reactant (A). Also, this reactor is supposed to exhibit an irreversible exothermic reaction A→B.

In the above model, the symbols hold the following significance.

C_a: Concentration of resultant B
$T(t)$: Reactor temperature
C_{a_0}: Feed concentration
T_0: Feed temperature
Q: Feed flow rate
$Q_c(t)$: Coolant flow rate
T_{c_0}: Coolant temperature

However, for our computational purpose, we have adopted the following state space model, as found in (Colantonio, Desages, Romagnoli, and Palazoglu 1995). This model is based on dimensionless modeling for an exothermic first order reaction. Symbols have their usual meanings as provided in (W.H Ray 1991).

$$\dot{x}_1 = -x_1 + Da(1 - x_1)exp\left(\frac{x_2}{1 + x_2/\gamma}\right) - d_2 \quad (3)$$

Dimensionless Parameters for CSTR

activation energy	$\gamma = E / RT_0$
adiabatic temperature rise	$B = \frac{(-\Delta H)c_{A_{f_0}}}{Qc_p T_{f_0}}$
Damkohler number	$Da = k_0 exp(-\gamma V) / F_0$
heat transfer coefficient	$\beta = hA / Qc_p F_0$
dimensionless time	$t = t' / (F_o / V)$
dimensionless composition	$x_1 = (c_{A_{f_0}} - c_A) / c_{A_{f_0}}$
dimensionless temperature	$x_2 = (T - T_{f_0}) / T_{f_0}$
dimensionless control input	$u = (T - T_{c_0}) / T_{f_0}$
feed temperature disturbance	$d_1 = (T_f - T_{f_0}) / T_{f_0}$
feed composition disturbance	$d_2 = (c_{A_f} - c_{A_{f_0}}) / c_{A_{f_0}}$

$$\dot{x}_2 = -x_2 + BDa(1 - x_1)exp\left(\frac{x_2}{1 + x_2 / \gamma}\right)$$
$$- \beta(x_2 - x_{2_c}) + \beta u + d_2 \qquad (4)$$

The dimensionless parameters for the above model are provided in the table below.

d_1 and d_2 are disturbances in the inlet temperature and concentrations, respectively.

3 PROBLEM FORMULATION

The control objective is to develop a controller with accurate set point tracking and disturbance rejection in the process, such that the controlled output is satisfactory and acceptable. Also, the response is required to be quick.

The error candidate is defined as:

$$e = x_2 - x_{2_{ref}} \qquad (5)$$

Our primary objective in achieving the stated is to stabilize the origin to the tracking error (Abhinav Sinha and Rajiv Kumar Mishra 2015). It is a prime requirement in the design process that the error approaches some small vicinity of zero after a transient of acceptable duration.

4 CONTROLLER DESIGN

The design of controller using sliding modes requires the choice of a manifold and a driving control to push the states onto the manifold (Zak 2003). The manifold under this study is a simple linear one, given by:

$$\sigma(x) = c_1 x_1 + c_2 x_2 \qquad (6)$$

c_i are the weighing parameters affecting system trajectory, and the performance of the system; whose values we have selected by minimizing the quadratic index. Only relative weights are of significance (Abhinav Sinha, Pikesh Prasoon, Prashant K. Bharadwaj and Anuradha C. Ranasinghe 2015), and hence any arbitrary choice of weight matrix Q can be taken as per design consideration.

The control effort u(t) is given by:

$$u(t) = \beta^{-1}[x_2 - BDa(1 - x_1)exp\left(\frac{x_2}{1 + x_2 / \gamma}\right)$$
$$+ \beta(x_2 - x_{2_c}) - d_2 - \mu sinh^{-1}(m + w \mid \sigma \mid)] \qquad (7)$$

This is a novel reaching law that reduces the reaching time as well as reduces the high frequency

oscillations. This function is odd and monotonous, hence, it acts as a good switching candidate. This law causes the gain of the system to vary nonlinearly. The gain is high when the states are away from the manifold and low at the time of the sliding motion. It has been chosen in a way to ensure sliding motion. It should be noted that parameters $m \geq 0$, $\mu > 0$, and $w > 0$.

5 STABILITY ANALYSIS

Stability is the most important factor to be considered in any design problem. Analysis of the stability has been carried out considering a Lyapunov candidate of $\frac{1}{2}\sigma^2(x)$. Negative definiteness of this candidate ensures stability in Lyapunov sense. By a simple mathematical calculation, it can be easily shown that $\dot{V}(x) = \sigma\dot{\sigma} < 0$ for all $\mu > 0$. Also, $m, w > 0 \Rightarrow sinh^{-1}(m + w(\mid \sigma \mid) > 0$

Hence, the control law is stable in Lyapunov sense.

6 SIMULATION RESULTS

The plant model has been simulated in the ideal conditions using Mathworks MATLAB and Simulink and the disturbance is randomly generated. The variable x_{2_c} is taken to be as 0. Following parametric values have been incorporated (Leonid Poslavsky and Jeffrey C. Kantor 1991).

Clearly, the response is quick as desired, as well as chattering is completely eliminated. The temperature reference has been set to 300 K and 350 K in two tests. In both runs, the reference is tracked equally well. Steady state is reached in a very short span and the response does not deviate, thereby, guaranteeing robustness. The response has been shown in the Figure 1. The proposed controller provides advantages over the conventional PID and other linear controllers by reducing the time to reach steady state. For a controller using feedback linearization, e.g., (Yayong Zhai 2012), the response is shown in Figure 2. It can be seen that the response takes more than 0.5 sec to enter steady state. However, the system attains the same in 0.1 sec with the proposed controller.

Numerical values of parameters used in simulation

heat transfer coefficient	β	1.5
activation energy	γ	20
adiabatic temperature rise	B	11
Damkohler number	Da	0.135

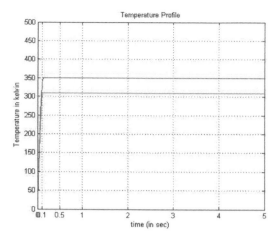

Figure 1. Temperature profile of the plant with proposed controller.

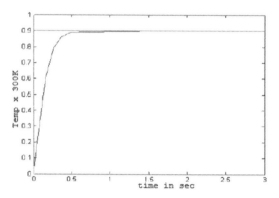

Figure 2. Temperature profile of the plant with the controller based on feedback linearization.

7 CONCLUSION

A novel robust and nonlinear controller has been designed for a continuous stirred tank reactor. The model has been kept intact in its nonlinear form and no attempt of linearization was carried out. The controller has been designed based on the sliding mode control with the reaching law based on an inverse hyperbolic function. This reaching law has the advantage of reducing chattering as well as the reaching phase by varying the gain in a nonlinear fashion. The control law has been analyzed for the stability and results that were presented to demonstrate high performance of the controller. This controller is highly efficient and there is no need to increase the order of the sliding mode to achieve the same efficiency. Thus, physical complexity in the design has been simplified.

REFERENCES

Abhinav Sinha and Rajiv Kumar Mishra (2015, March). Robust altitude tracking of a miniature helicopter uav based on sliding mode. In *IEEE 2nd International Conference on Innovations in Information, Embedded and Communication Systems*, Coimbatore, India.

Abhinav Sinha, Pikesh Prasoon, Prashant K.Bharadwaj and Anuradha C. Ranasinghe (2015, Jan). Nonlinear autonomous control of a two-wheeled inverted pendulum mobile robot based on sliding mode. *In IEEE International Conference on Computational Intelligence and Networks*, Bhubaneswar, India.

Colantonio, M.C., A.C. Desages, J.K. Romagnoli, and A. Palazoglu (1995). Nonlinear control of a cstr: Disturbance rejection using sliding mode control. *Industrial and Engineering Chemistry Research 34*(7), 2383–2392.

David Young, K., Vadim I. Utkin and Umit Ozguner (1999, May). A control engineer's guide to sliding mode control. *IEEE transactions on Control Systems Technology 7*(3), 328–342.

Leonid Poslavsky and Jeffrey C. Kantor (1991, June). Sliding mode control of an exothermic continuous stirred tank reactor. *In American Control Conference*, Boston, MA, USA.

Levant, A. (1993). Sliding order and sliding accuracy in sliding mode control. *International Journal of Control 58*(6), 1247–1263.

Ray, W.H. (1981). *Advanced Process Control*. Mc-Graw Hill, New York.

Ray, W.H., Kravaris C., A.Y. (1991). *Geometric Nonlinear Control- An Overview, CPC IV*. AIChE Publications, New York.

Yayong Zhai, S.M. (2012, July). Controller design for continuous stirred tank reactor based on feedback linearization. In *2012 Third International Conference on Intelligent Control and Information Processing*, Dalian, China.

Zak, S.H. (2003). *Systems and Control*. 198 Madison Avenue, New York, New York, 10016: Oxford University Press.

Zhao, D., Q. Zhu, and J. Dubbeldam (2014).Terminal sliding mode control for continuous stirred tank reactor. *Chemical Engineering Research and Design*.

Frontiers in Computer, Communication and Electrical Engineering – Acharyya (Ed.)
© 2016 Taylor & Francis Group, London, ISBN: 978-1-138-02877-7

A unified FDTD approach in electromagnetics metamaterials

B. Mandal, S.K. Singh & A. Biswas
Faculty of Engineering and Technology, NSHM Knowledge Campus, Durgapur, West Bengal, India

Aritra Acharyya
Supreme Knowledge Foundation Group of Institutions, Mankundu, Hooghly, West Bengal, India

A. Ghosal
Institute of Radio Physics and Electronics, Calcutta University, West Bengal, India

A.K. Bhattacherjee
NIT Durgapur, West Bengal, India

D.P. Chakraborty
Techno India University, Saltlake City, Kolkata, India

ABSTRACT: This Paper presents the implementation of Finite Difference Time Domain (FDTD) technique for metamaterial structures with absorbing boundary condition. Finite Difference Time Domain (FDTD) technique can be used to model metamaterial by treating them as dispersive material. FDTD simulations will use the Convolutional Perfectly Matched Layer (CPML) to model the absorbing boundary condition. CPML is used to absorb the outgoing EM waves in order to model free space calculation domain. The contribution of this work is toward the development of a powerful FDTD engine for modern metamaterial analysis. Our implementation could be used to improve the analysis of a number of electromagnetic problems.

1 INTRODUCTION

Metamaterial is periodic structures created by many identical scattering objects which are stationary and small compared to the wavelength of electromagnetic wave applied to it so that when combined with different elements, these materials have the potential to be coupled to the applied electromagnetic wave without modifying the structure. They are known by a number of names including negative index materials (because the refractive index is negative), DNG (double negative index materials as opposed to DPS double positive index materials), left-handed materials (the E and H fields and the wave vector k form a left-hand triad and as a consequence k and the Poynting vector S point in opposite directions in an isotropic material). Due to their unusual properties that are not readily available in nature, metamaterial has been drawing significant attentions in many research areas, including theoretical, experimental as well as numerical investigations. Victor Veselago theoretically analyzed the propagation of plane wave in a material with both negative permittivity and negative permeability in his original research,

which is the first study in this area and key to the advancement of recent researches in Electromagnetic (EM) metamaterials (V.G. Veselago 1968). Metamaterial gain its properties from its structure rather than directly from its composition. For conventional materials in nature, the basic unit is the constitutive molecules, which are of the order of the angstrom (0.1 nm) (K.Yee 1966). There are many ways to generate different effective permittivity and permeability for a metamaterial. For example, changing host materials; varying the size, shape, and composition of the inclusions; changing the density, arrangement, and alignment of these inclusions (C. Argyropoulos et al. 2009). Materials with negative parameters are innately dispersive and their permittivity and permeability will alter with respect to frequency (K. Yee 1966). At the meanwhile, many new applications also use nonlinear and gain materials to tune the properties of the metamaterials. These degrees of freedom make the metamaterial tunable in its properties and thus provide us a variety of applications for metamaterial. Plasmonic metamaterials are negative index metamaterials and can be used to couple the incident light with the metal dielectric material to create self-sustaining, propagating electromagnetic waves known as surface plasmon polaritons (J.A. Roden et al. 2000). Metamaterials can be integrated with nonlinear media to provide the option to change the power of the incident wave, which overcomes the limitation of many common optical materials that have weak nonlinearity.

2 METHOD

The major computational electromagnetic modeling method, Finite-difference time-domain technique tackles problems by providing a full wave solution. Since it naturally treats dispersive and nonlinear material, and covers a wide range frequencies in one single run, FDTD has been a primary means to numerically model scientific and engineering problems, like antennas, photonic crystals, transmission lines, circuit simulations and so on J.A. Roden et al. 1998). Thus all the classical electromagnetic as well as the computational electromagnetic methods are based on this set of equations. The equation 1 and 2 are the differential form of Ampere's Law and Faraday equations, respectively. These two equations are used to solve numerical calculation in FDTD method.

$$\nabla \times H = \epsilon \frac{\partial E}{\partial t} + \sigma E \qquad (1)$$

$$\nabla \times E = -\mu \frac{\partial H}{\partial t} \qquad (2)$$

where E and H are the electric filed and magnetic field, respectively; ε and μ are the electrical permittivity and magnetic permeability, respectively; σ is the electric conductivity loss; t is the time. The well-known Yee algorithm comes from Kane Yee's publication in 1966 (P. H. Harms et al. 1997), which described the basis of the modern FDTD model to solve Maxwell's curl equations in time domain. The special grid cell that Yee originally used is called Yee cell and it discreteness the three dimensional space into cubes.

As we can be seen from the Figure 1, electric fields are along the edges of the cube while the magnetic fields are located at the center of the cube's 6 surfaces. Each unit cell has 12 electric fields and 6 magnetic fields. Since each electric field is shared by the adjacent 4 unit cells, each unit cell possesses 3 electric fields, namely Ex, Ey and Ez. Similar rule applies to magnetic fields and each unit cell possesses 3 magnetic fields, Hx, Hy and Hz. In a 3D point of view, each electric field E is surrounded by 4 circulating H fields, which interprets the Ampere's Law. Similarly, every magnetic field H is surrounded by 4 circulating E fields, representing Faraday equation.

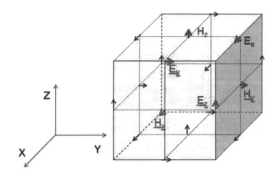

Figure 1. FDTD grid unit.

In three-dimensional case, each of these two vector equations Equation 1 and Equation 2 can be converted into 3 scalar equations, forming 6 scalar equations.

$$\left. \begin{aligned} \frac{\partial E_x}{\partial t} &= \frac{1}{\varepsilon}\left(\frac{\partial H_z}{\partial y} - \frac{\partial H_y}{\partial z} - \sigma E_x \right) \\ \frac{\partial E_y}{\partial t} &= \frac{1}{\varepsilon}\left(\frac{\partial H_x}{\partial z} - \frac{\partial H_z}{\partial x} - \sigma E_y \right) \\ \frac{\partial E_z}{\partial t} &= \frac{1}{\varepsilon}\left(\frac{\partial H_y}{\partial x} - \frac{\partial H_x}{\partial y} - \sigma E_z \right) \end{aligned} \right\} \qquad (3)$$

$$\left. \begin{aligned} \frac{\partial H_x}{\partial t} &= -\frac{1}{\mu}\left(\frac{\partial E_z}{\partial y} - \frac{\partial E_y}{\partial z} + \sigma_m H_x \right) \\ \frac{\partial H_y}{\partial t} &= -\frac{1}{\mu}\left(\frac{\partial E_x}{\partial z} - \frac{\partial E_z}{\partial x} + \sigma_m H_y \right) \\ \frac{\partial H_z}{\partial t} &= -\frac{1}{\mu}\left(\frac{\partial E_y}{\partial x} - \frac{\partial E_x}{\partial y} - \sigma_m H_z \right) \end{aligned} \right\} \qquad (4)$$

For a function f of space and time, such as

$$f(x,y,z,t) = f(i\Delta x, j\Delta y, k\Delta z, n\Delta t) = fn(i,j,k) \qquad (5)$$

where Δt is the time increment, Δx, Δy, Δz are the space increments; n is the time mark of current calculation, i, j and k denote the space location of current calculation, the differential form in time and space can be written as

$$\left. \frac{\partial f(x,y,z,t)}{\partial t} \right|_{t=n\Delta t} \approx \frac{f^{n+\frac{1}{2}}(i,j,k) - f^{n-\frac{1}{2}}(i,j,k)}{\Delta t} \qquad (6)$$

$$\left. \frac{\partial f(x,y,z,t)}{\partial x} \right|_{x=i\Delta x} \approx \frac{f^n(i+\frac{1}{2},j,k) - f^n(i-\frac{1}{2},j,k)}{\Delta x} \qquad (7)$$

To make the time domain update for the above Maxwell's equations of differential form, similar manipulations are applied. Those scalar equations from Equation 3 and Equation 4, taking Ex and Hx for example, can be written as

$$E_x^{n+1}(i-\frac{1}{2},j,k)=CA(m)\cdot E_x^{n}(i-\frac{1}{2},j,k)+CB(m)$$

$$-\left(\frac{H_z^{n+\frac{1}{2}}(i-\frac{1}{2},j+\frac{1}{2},k)-H_z^{n+\frac{1}{2}}(i-\frac{1}{2},j-\frac{1}{2},k)}{\Delta y}\right.$$

$$\left.\frac{H_y^{n+\frac{1}{2}}(i-\frac{1}{2},j,k+\frac{1}{2})-H_y^{n+\frac{1}{2}}(i-\frac{1}{2},j,k-\frac{1}{2})}{\Delta z}\right)$$

$$\times E_y^{n+1}(i,j-\frac{1}{2},k)=$$

$$\cdots\cdots\cdots$$

$$E_z^{n+1}(i,j,k-\frac{1}{2})=\cdots\cdots\cdots$$

(8)

$$H_x^{n+\frac{1}{2}}(i,j-\frac{1}{2},k-\frac{1}{2})$$

$$=CC(m)\cdot H_x^{n-\frac{1}{2}}(i,j-\frac{1}{2},k-\frac{1}{2})$$

$$+CD(m)\cdot\left(\frac{E_z^{n}(i,j,+\frac{1}{2},k-\frac{1}{2})-E_z^{n}(i,j-\frac{1}{2}k-\frac{1}{2})}{\Delta y}\right.$$

$$\left.-\frac{E_y^{n}(i,j-\frac{1}{2},k+\frac{1}{2})-E_y^{n}(i,j-\frac{1}{2},k-\frac{1}{2})}{\Delta z}\right)$$

$$H_y^{n+\frac{1}{2}}(i-\frac{1}{2},j,k-\frac{1}{2})=\cdots\cdots$$

$$H_z^{n+\frac{1}{2}}(i-\frac{1}{2},j-\frac{1}{2},k)=\cdots\cdots$$

(9)

where CA, CB, CC and CD are coefficients, m is the number of material. After discretization of the space, as well as time derivatives, scalar form of Equation 8 and Equation 9 are used for the electric and magnetic fields update at different time steps, respectively. Thus, one by another, the electric and

$$E=0\to\underline{H\to E\to H}\cdots\cdots\underline{H\to E}\quad time$$
$$\underline{}_{\Delta t}$$

$t = 0 \qquad\qquad\qquad t = end$

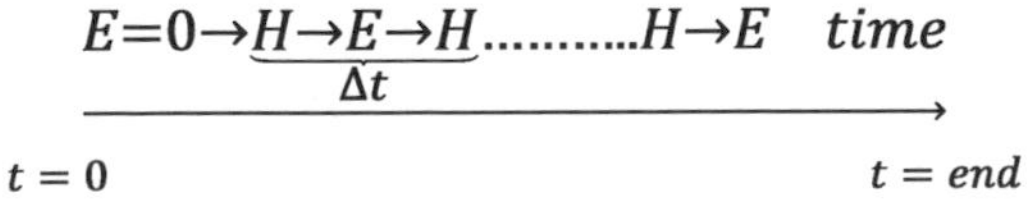

Figure 2.   General FDTD field update procedure.

magnetic fields are calculated as time goes on, as shown in Figure 2.

3   BOUNDARY CONDITION

In the numerical calculations, all the data, including cell discretization, electric/magnetic fields, material parameters, are stored in the computer memory. However, any computer has a memory limit. As a result, our FDTD model must have a boundary. Usually, this outer boundary is Perfect Electric Conductor (PEC) which has tangential electric fields as zero. PEC will reflect electromagnetic waves for sure. Therefore, absorbing boundary condition will be needed in FDTD calculation. In this dissertation, FDTD simulations will use the Convolution Perfectly Matched Layer (CPML) to model the absorbing boundary condition. As shown in Figure 3, CPML is used to absorb the outgoing EM waves in order to model free space calculation domain.

4   TOTAL FIELD AND SCATTERED FIELD MODEL

To model the effect of scattering of objects from plane wave incidence, Total Field/ Scattered Field (TF/SF) technique is applied in FDTD algorithm. As illustrated in Figure 4 (a), the calculation domain is divided into 3 regions. The CPML region absorbs the outgoing wave to model an infinite free space. The scattered field region only simulate the scattered electric/magnetic fields that caused by the plane wave incidence. Finally, the total field region, as shown inside the red box, presents the total field for the wave interaction with scattering objects. Figure 4 (b) shows a Matlab simulation for a PEC object. In this two dimensional simulation, the incident wave is a plane wave with Gaussian signature going through from lower left to upper

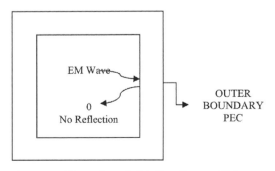

Figure 3.   Illustration of CPML effect on EM wave reflection.

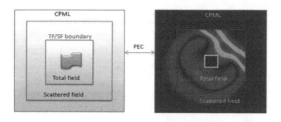

Figure 4.   (a) Total field/scattered field Technique effects on EM wave scattering; (b) Matlab Simulation for PEC.

right. The scattering object is denoted by the white square and has no field inside because its boundary is PEC. In the total field region, both incident wave and scattered field present. In the scattered field region, only the field scattered by the PEC square is observed. Beyond the scattered region is the CPML region where the EM fields are always decaying towards the outer PEC boundary and causes no reflection.

5   CONCLUSION

It can be seen that general procedure may be derived from those cases   so that any material could be incorporated into FDTD as long as certain field relation is provided. For the implementation of sub gridding system, higher order of spatial interpolation is expected to improve the calculation accuracy and reduce the reflection from the coarse grids/sub grids boundary and higher order of sub grids could also be considered for even smaller structures.

REFERENCES

Argyropoulos, C., Y. Zhao, Y. Hao (2009), "A Radially dependent dispersive finite difference time domain method for the evaluation of electromagnetic cloaks," IEEE Transactions on Antennas and Propagation, vol. 57, pp. 1432–1441.

Harms, P.H., J.A. Roden, J.G. Maloney and M.P. Ke (1997), "Numerical analysis of periodic structures using the split-field algorithm," in Proc. 13th Annual Review of Progress in Applied Computational Electromagnetics.

Kao, Y.-C.A., and R.G. Atkins (1996), "A finite difference-time domain approach for frequency selective surfaces at oblique incidence," in Antennas and Propagation Society International Symposium, 1996. APS. Digest.

Roden, J.A., S.D. Gedney, M.P. Kesle, J.G. Maloney and P.H. Harms (1998), "Time-domain analysis of periodic structures at oblique incidence: Orthogonal and non orthogonal FDTD implementations," Microwave Theory and Techniques, IEEE Transactions, vol. 46, no. 4, pp. 420–427.

Roden, J.A., and S.D. Gedney (2000), "Convolutional PML (CPML): An efficient FDTD implementation of the CFS-PML for arbitrary media," Microwave and optical technology letters, vol. 27, no. 5, pp. 334–338.

Taflove, A., and S.C. Hagness (2005), Computational Electrodynamics: The Finite Difference Time Domain Method, 3rd ed., Norwood, MA: Artech House.

Veselago, V.G. (1968), "The electrodynamics of substances with simultaneously negative values of ε and μ," Physics-Uspekhi, vol. 10, no. 4, pp. 509–514, 1968.

Yee, K., (1966), "Numerical solution of initial boundary value problems involving Maxwell's equations in isotropic media," Antennas and Propagation, IEEE Transactions on, vol. 14.3, pp. 302–307.

Yee, K., (1966), "Numerical solution of initial boundary value problems involving maxwell's equation sinisotropic media," IEEE Transactionson Antennasand Propagation, vol. 14, pp. 302–307.

Frontiers in Computer, Communication and Electrical Engineering – Acharyya (Ed.)
© 2016 Taylor & Francis Group, London, ISBN: 978-1-138-02877-7

Development of a low-cost field detector unit for safety of operating personnel in Low Tension line

T.S. Biswas, A. Baug, R. Ghosh, B. Chatterjee & S. Dalai
Electrical Engineering Department, Jadavpur University, Kolkata, West Bengal, India

ABSTRACT: This paper proposes a simple and reliable low-cost field detector unit for the safety of maintenance personnel attending to Low Tension (LT) lines. The device is a small and light-weight non-contact detector that can indicate the presence of electric field by variation in glow of an LED. The need for safety devices amongst maintenance personnel attending to LT lines has been highlighted. The local field enhancement around a maintenance personnel attending to a line has been simulated using finite element method. The knowledge of the enhanced field has been used to design the maximum distance at which the detector unit will operate. It is expected that a low cost field detector will encourage utilities to promote the use of safety devices amongst maintenance personnel.

1 INTRODUCTION

The continuous and reliable availability of electrical power has become a critical requirement in our daily life. Resources all over the world are heavily dependent on the continuity of electric supply. With rapid modernization and industrial development, the demand for electricity has been steadily going up in recent times. The high demand for electricity, coupled with the demand for increased reliability has made the task of the utilities challenging. Though most power utility companies around the world have largely been able to meet the demands, system outages and failures cannot be eradicated. Hence, the need for regular, timely and quick maintenance of power distribution systems has become important.

In the context of developing countries like India, where the safety requirements for maintenance personnel are not stringent, fatal accidents are common. Hundreds of maintenance personnel meet with serious accidents while attending to live lines every year. One of the major causes for this is the inability of the power utilities to allocate funds for state-of-the-art safety devices for the operating personnel. With thousands of personnel working on power distribution lines every day, providing each of them with expensive safety devices is not a viable solution to the utilities. To encourage power utilities to take proper safety measures, one solution could be the development of a low cost field detector designed to detect appropriate electric fields. For such a design to be effective it is necessary to first know the potential distribution around the tower and line on which the maintenance personnel will work (Milutinov *et al.* 2009). This will help in designing the field detector by incorporating the knowledge of safe expose limit for a human body.

In this paper, a study has been done to determine how the stress increases if one stands under the low voltage distribution line. Based on this study, a low-cost non-contact field detector unit has been developed that is capable of indicating the presence of electric fields from a distance. The detector device is a small light weight unit that may be conveniently put inside the pockets of the personnel (McNulty *et al.* 2011). As and when the personnel get too close to the line, an indicator will indicate the proximity of the personnel to the power line, thereby saving them from accidents.

2 ELECTRIC FIELD NEAR A HUMAN BODY UNDER AN OVERHEAD LT LINE

The human body is composed of some biological materials with a large amount of water. The relative permittivity of human body thus becomes very high along with high conductivity as compared to solid non conducting materials like bones. In addition, human body contains free electric charges due to the presence of highly organized electron transport system in the body. These charges move in response to forces exerted by electric field in nearby power lines and as a consequence, local enhancement of field occurs in human body (King *et al.* 2004). Electric field under a LT line increases as one travels from the vertically from the ground towards the line. In the presence of a human body beneath the line conductors, the electric field at the ground level is distorted due to the accumulated surface charges on the body. The field strength may increase by a factor that may reach five times of the magnitude for low voltage line. This local enhancement of electric field, in turn, aids in increase of potential, local to the human body.

The system studied in this paper consists of a model of a human body standing under an overhead line of 415 V. Due to the large conductivity (about 0.1 S/m) and very high equivalent dielectric constant of the human body), the external power frequency electric fields near the human body are perpendicular to the body surface As a result it is possible to treat the human body as conducting. It needs to be emphasized that the axial current flowing through the human body depends primarily on the height of the body and is independent of the cross-section. Hence, the human body may be approximated as a cylinder, having the same length and mean cross sectional area as that of an average human body (King *et al.* 2004). Height of the model is taken as 1.8 m and diameter of waist and legs are taken as 400 mm and 200 mm respectively. Horizontal field exposure is not considered since exposure to electric fields of horizontal orientations results in small induced currents (El-Makkaway *et al.* 2007).

In order to investigate the effect of the overhead line on a human standing directly beneath the line, a model of the human body under a 3-phase 415 V line has been simulated in COMSOL. Figure 1 clearly shows that the disturbance in the electric field is negligible near the line, but the disturbance is significant near to the human body. From Table 1, it can be seen that field in presence of a human body is near about 5 times higher than field in absence of body at 1.8 m height from ground level.

Table 1. Comparison of field intensity at different heightsdue to presence of a human body (perturbed field).

Height from ground level (in m)	Electric field intensity (in V/m)		
	Unperturbed field (E_1)	Perturbed field (E_2)	Ratio (E_2/E_1)
5.00	13.02	13.21	1.02
4.00	4.07	4.18	1.03
3.50	2.43	2.52	1.04
3.00	1.69	1.91	1.13
2.50	1.23	1.60	1.30
2.20	1.07	1.65	1.54
1.85	0.98	1.99	2.03
1.80	0.89	4.06	4.56

According to electrostatic field theory, in presence of electric field the human body gets polarized and bound charges of opposite nature are induced inside the surface of the body depending upon the strength of the field. Presence of this bound charge in the body attracts more charge from the environment which in turn becomes a cyclic process that leads to more deformation in the electric field around the human body (El-Makkawy *et al.* 2007, Yildrim *et al.* 2007). The body surface with lower radius of curvature, which has higher charge density, has higher electric field intensity. As the human body has high permittivity, it has very low electric field intensity inside the body and the field is more uniform in nature, whereas, the intensity at the body surface becomes high. It may also be concluded that the nearer a person is to a line, the higher the field he is exposed to. This is particularly true in case of maintenance personnel who are responsible for line maintenance.

3 DEVELOPMENT OF A FIELD DETECTOR CIRCUIT

Exposure of the human body to a 50 Hz electric field due to an overhead high voltage power transmission line or distribution line can lead to two types of potential hazards. First of these is the steady state current that is induced in the body on exposure to the electric field for a prolonged period of time. The second is the hazard due to direct electric shock when current passes through the body on contact with live parts associated with a power line (Deno *et al.* 1977). The main aim of the present work is to develop a simple low-cost device which can detect the presence of electric field and can indicate when this field intensity crosses a threshold value. This device will be a non-contact voltage detector. The advantage of this type of device is

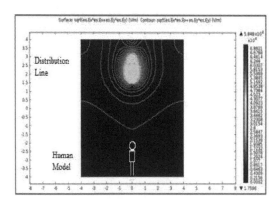

Figure 1. Field distribution due to presence of human body beneath a 3-phase 415 V distribution line.

that it is able to measure weak field from a distance, especially in the midst of several objects where one or more of them are at a hazardous potential irrespective of their relative position.

Electric field detectors can be usefully divided into two types: (1) <u>antennas</u>, which typically have a size of 1/10 of a wavelength to a full wavelength and pick up high frequency electromagnetic radiation far from the source; and (2) <u>sensors</u>, which have sizes much smaller than a wavelength and can pick up low frequency fields in either the near zone or far zone (Williams *et al.* 2003).

It is already established in the literature that the electric field due to any overhead transmission line or live object is responsible for induced surface charge in nearby metallic electrode or body present within that field. From studies it is understood that the amount of charge that will be induced in this electrode will be very small. An operational amplifier with JFET input stage is used in this scheme. The opamp is driven as a half wave active rectifier to detect the presence of alternating field. A Light-Emitting Diode (LED) at the output stage is used as a simple indicator, the intensity of which varies with the intensity of electric field in which the sensor electrode is placed. A circular metallic disc is used to detect the presence of field with diameter 10.9 cm. Here an op-amp, along with silicon junction diode 1 N4148 has been used as an active half wave rectifier. The presence of the electric field will act as the input to the non-inverting terminal of the op-amp. Since the system has unity gain, the output terminal will show the potential to be same as the potential at non-inverting terminal. A LED with current limiting resistor is used as indicator.

In the experimental setup, two conducting plates, named live plate and capture plate as depicted in Figure 3, are taken which have 10.9 cm diameter. Live plate is connected to the variable voltage source and the capture plate is the antenna of the detector circuit. Electric field in the capture plate was varied by two methods- (1) the potential of the live plate

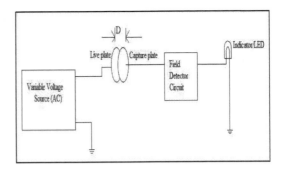

Figure 3. Experimental setup for field detection.

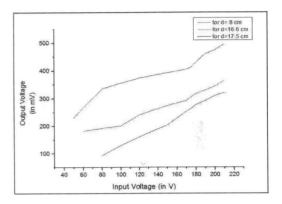

Figure 4. Variation of output voltage with different input voltage applied across the detector plates keeping the distance fixed.

was varied, keeping the distance same between them and (2) the distance (d) between the plates was varied, keeping the voltage same on the live plate.

The output of the detector circuit is an LED. This scheme is done for a layman to visually perceive the presence of hazardous field and at the same time, to make the detector unit small, portable and rugged with minimal battery consumption for long run. But using a LED at output, quantitative analysis cannot be carried out. Hence, the output of the detector has been replaced with a resistor and a true r.m.s. reading Multimeter. Figure 4 shows the output voltages from the detector for different input voltages in the live plate, keeping the distance between the plates as a constant. Figure 5 shows the output voltages from the detector for different distance *d* keeping the voltage same on the live plate.

The other important consideration for the performance evaluation of the device is the detection capability of this device when the output is replaced by the LED, corresponding to the scheme presented in Figure 2. The advantage of LED is that, it can detect the peak value and has a very fast turn on time (in the order of μ sec). Table 2 shows

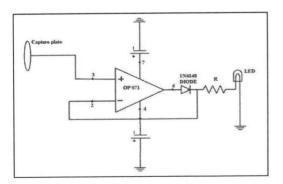

Figure 2. Electric field detector circuit.

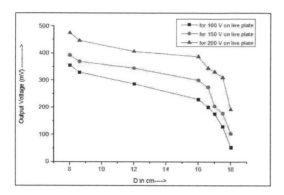

Figure 5. Variation of output voltage with distance between plates keeping the input voltage fixed.

Table 2. Maximum distance at which the LED can detect the presence of field corresponding to different voltages applied to the live plate.

Input voltage on live plate (V)	Distance between plates (m)
60	0.18
80	0.32
120	0.44
160	0.58
180	0.68
200	0.75
220	0.83

the maximum distances from which the detector circuit is able to detect the presence of different voltages on the live plate to produce a perceptible glow in the output LED.

4 CONCLUSIONS

In this paper the development of a low-cost electric field detection device has been reported for maintenance personnel working on LT lines. Based on the finite element method, change in electric field near a grounded human body standing beneath a LT line is computed. It has been shown that in the presence of a human body under an overhead line, the electric field around the line has negligible changes, but the local electric field around the human body is considerably enhanced. The knowledge of this enhancement in electric field has been utilized to design the maximum distance at which the detector unit will operate. The main contribution of the paper is the development of a simple and low-cost field detector circuit that can be used for the safety of the personnel attending to line maintenance. The detector device is a small light

weight unit that may be carried inside the pockets of the maintenance personnel working on LT lines. As and when the personnel get too close to a live line, an indicator will indicate the proximity of the personnel to the power line, thereby saving them from accidents. It is expected that the low-cost detection unit reported in this paper will help power utilities to encourage the use of safety devices among the maintenance personnel.

REFERENCES

Abdel-Salam, M. & Abdallah H. M. (1995). "Transmission-Line Electric Field Induction in Humans Using Charge Simulation Method", IEEE Transactions on Biomedical, Engineering, Vol. 42, No. 11, pp. 1105–1109, November 1995.

Ahmadi H., Mohseni, S., Akmal, A. A. S. (2010). "Electro-magnetic Fields Near Transmission Lines—Problems And Solutions" Iran. J. Environ. Health. Sci. Eng., Vol. 7, No. 2, pp. 181–188, 25 February 2010.

Deno, D W. (1977). "Currents Induced in the Human Body by High Voltage Transmission Line Electric Field—Measurement and Calculation of Distribution and Dose", IEEE Transactions on Power Apparatus and Systems, Vol. PAS-96, No.57, pp. 1518–1528. September/October 1977.

El-Makkawy, S. M. (2007). "Numerical Determination of electric field induced currents on human body Standing under a high voltage transmission line".

Hossam-Eldin, A. A. (2001). "Effect of Electromagnetic Fields from Power Lines on Living Organisms", IEEE 7th International Conference on Solid Dielectrics, June 25–29, 2001, Eindhoven, the Netherlands, pp. 438–441.

King, R. W. P. (2004). "A Review of Analytically Determined Electric Fields and Currents Induced in the Human Body When Exposed to 50–60-Hz Electromagnetic Fields", IEEE Transactions on Antennas and Propagation, Vol. 52, No. 5, May 2004, pp. 1186–1192.

McNulty, W. (2011). "Voltage detection and indication by electric field measurement", Transmission and Distribution Construction, Operation and Live-Line Maintenance (ESMO), IEEE PES 12th International Conference on 16–19 May 2011 Publisher: IEEE.

Milutinov M., Juhas A., Prsa M. (2009) "Electromagnetic field underneath overhead high voltage power line", 4th International Conference on Engineering Technologies—ICET 2009, Novi Sad, April 28–29, 2009.

Williams, K. R., De Bruyker, D. P. H., Limb, S. J., Amendt, E.J., Overland, D. A. (2003). "Vacuum Steered-Electron Electric-Field Sensor" Submitted to IEEE/ASME Journal of Microelectromechanical Systems-JMEMS-2013–0021.

Yildirim H. & Kalenderli, O. (2003). "Computation of electric field induced currents on biological bodies near High Voltage transmission lines", Xth International Symposium on High Voltage Engineering, Netherlands, Smit(ed), pp. 1–4.

Zipse, D. W. (1993). "Health Effects of Extremely Low-Frequency (50- and 60-Hz) Electric and Magnetic Fields", IEEE Transactions on Industry Applications, Vol. 29, No. 2, March/April 1993, pp. 447–458

Frontiers in Computer, Communication and Electrical Engineering – Acharyya (Ed.)
© 2016 Taylor & Francis Group, London, ISBN: 978-1-138-02877-7

Congestion control in Cognitive Radio networks using fractional order rate reaching law based sliding modes

Tirtha Majumder, Rajiv Kumar Mishra & Sudhansu Sekhar Singh
School of Electronics Engineering, Kalinga Institute of Industrial Technology (KIIT University), Bhubaneswar, Odisha, India

Abhinav Sinha
TATA Consultancy Services Limited, Pune, Maharashtra, India

Prasanna Kumar Sahu
National Institute of Technology, Rourkela, Odisha, India

ABSTRACT: Limited bandwidth and exponentially rising user demand calls for an exigent need of throughput maximization. In order to maintain Quality of Service (QoS) in any communication network requires accurate means to control congestion of the network. The focus of this work is the onerous task of designing an effective congestion controller to taper off packet losses to zero in Cognitive Radio (CR) networks. Huge number of requests vitiate the performance due to limited buffer capacity and bandwidth. A controller has been designed using sliding mode, which is a variable structure control technique manifested for its robustness and disturbance rejection capabilities. Exertions are made to reduce design complexity by incorporating fractional order rate reaching law based sliding mode. An optimal design strategy has been used in the synthesis of the controller. The efficacy of the controller has been confirmed by numerical simulations.

1 INTRODUCTION

Compelling developments in communication and networks have ushered heavy increase in user demand and traffic. This acute rise in traffic has resulted in congestion in communication networks which degrades Quality of Service (QoS). Cognitive Radio Network (CRN) is an intelligent system that promises effective utlization of wireless spectrum by dynamically modifying the transmission parameters to improve the bandwidth usage. One of the common problems in most wireless spectrums is their underutilization in spite of their capability to serve increased user demands and traffic. Various technologies like Adaptive Radio and Software Defined Radio are used by CRN to steer the traffic from occupied and congested channels to the vacant ones for smooth concurrent communicaton. This technique felicitates the secondary (low priority) users to use the unused licensed band of primary (high priority) users temporarily, thereby significantly improving spectrum efficiency by opportunistic spectrum utilization.

There are two types of traffic queues in CRN on which the QoS are categorized, *viz.*, high priority or premium service and low priority or ordinary service. Former provides bandwidth to the delay

and loss sensitive user while latter regulates the input flow and delay is not given much concern here. The leftover capacity of former is used for the latter. Best effort traffic is given the least significance where consumed bandwidth is left over from both services. Varying needs and data communication rate of different users may result in congestion of the network. Controlling this serious problem is a challenging task and is addressed in this work.

Several control strategies have been proposed to tackle this problem. Most techniques involved linearizing the nonlinear model that simplifies the design problem but limits the region of operation. Other control techniques such as Adaptive and Fuzzy algorithms (A. Pitsillides and J. Lambert 1997), (Y.C. Liu and C. Douligeris 1997), (A. Pitsillides and A. Sekercioglu 1999) have also been used but they are unable to cope up with fast changing scenario as they are non-model based strategies. They provide well adaptation and good approximation to hard nonlinearities but have to be combined with other strategies to yield high performance. Random Early Detection (RED) (C. Chrysostomou, A. Pitsilliides, L. Rossides, M. Polycarpou and A. Sekerciouglu 2003) and its variants are also proposed in literature but their parameters are very sensitive to network load (Ming Yan, Tatjana D. Kolemisevska-Gugolovska, Yuanwei Jing and Georgi Dimirovski 2007). Other methods such as Active Queue Management (Ming Yan, Tatjana D. Kolemisevska-Gugolovska, Yuanwei Jing and Georgi Dimirovski 2007) are also available. Sliding mode control is a state dependent feedback controller with model based strategy that is known to be very effective in controlling uncertain dynamical systems. It provides advantages over other controls by eliminating the need for exact modeling and providing very good disturbance rejection. Using this control, the dynamic behavior of the system can be tailored using a particular choice of sliding function. The controller is designed to drive and then constrain the states of the system to lie in the vicinity of switching function. Under uncertainties such as link failures and in both steady state and nonstationary conditions, this controller delivers a high performance. However, the hardware used in realizing the controller may introduce infinite switching, known as chattering and is undesirable. In this work, we present Fractional Order Rate reaching law based sliding mode control to tackle the congestion control problem in CR networks.

2 NETWORKS DYNAMICAL MODEL

The CR network is analytical modelled using *queuing model* or *Fluid Flow Model* (J. Filipiak 1988), (Bouyoucef and Khorasani 2007). The model used here is simpler than probabilistic model and allows distributed control and reduces computational time in performance evaluations. Let us consider an arbitrary channel $\bar{C}$ with maximum capacity of C_{max} (David Tipper and Malur K. Sundareshan 1990), $N(t)$ is the number of packets in the system, i. e., *queue+server* at time t and $x(t)$ as the state variable that represents the average number of packets in the system at any given time t. $x(t)$ is also the *ensemble average* of the number of packets in the system at time t, i. e., $x(t) = E\{N(t)\}$. If $d(t)$ and $a(t)$ represent the flow out and into the system respectively, then we can define f_{in} and f_{out} as ensemble average of the flow in and out of the queue respectively, i. e., $f_{out}(t) = E\{d(t)\}$ and $f_{in}(t) = E\{a(t)\}$. Thus, by flow conservation law, we have

$$\dot{x}(t) = f_{in}(t) - f_{out}(t) \tag{1}$$

It is assumed that the storage capacity of the queue is unlimited and the users are arriving at the queue according to a nonstationary Poisson process with rate $\lambda(t)$, the model for M/G/1 queue is

$$\dot{x}(t) = -C(t)\frac{x(t)}{1 + x(t)} + \lambda(t) \tag{2}$$

Here, $x(t)$ is the queue length of the buffer, C is the assigned to be link capacity and the nonlinear term $\lambda(t)$ denotes incoming traffic rate.

Premium services require guaranteed delivery within given loss bound and delay but rate regulation is not allowed beyond the delay bound (Tirtha Majumder, P.K. Sahu, S.S. Singh, Abhinav Sinha and Rajiv Kumar Mishra 2015). The buffer state is controlled to be close to a reference value such that maximum allowable delay and packet loss bound is guaranteed in premium service (A. Pitsillides, P. Ioannou, M. Lestas and L. Rossides 2005). Assignment of maximum bandwidth C_{max} is done dynamically to the premium service such that $0 \le C_p(t) \le C_{max}$. Ordinary services can tolerate queuing delay and allow the regulation of flow but packet losses are not allowed. The bandwidth for ordinary service can be given as $C_s(t) = C_{max} - C_p(t)$ where $C_s(t) > 0$.

By suitable change of variable, the model can also be rewritten in the observable canonical form, which will further aid us in optimization and development of controller. Hence, we have

$$\dot{x}_1 = x_2$$
$$\dot{x}_2 = -C(t)\frac{x_2(t)}{(1 + x_1(t))^2} + \dot{\lambda}(t) \tag{3}$$

We can also write in general

$$\dot{x}_1 = f_1(x)$$
$$\dot{x}_2 = f_2(x) + b(x)u + \varphi \tag{4}$$

with u and φ are control effort and matched uncertainties respectively.

3 PROBLEM FORMULATION

The control target is to maintain desired queue length in the buffer, provided the buffer length of the router has been defined in priori. The error variable is required to be stabilized to the origin for accurate tracking, and is defined as

$$e_i = x_i - x_{iref} \tag{5}$$

4 CONTROLLER DESIGN

The ontogenesis of sliding mode controller has been carried out in two steps *scilicet*, development of a stable hyper surface that is the geometrical locus consisting of boundaries, and formation of the control law that is composed of equivalent and corrective control.

Let us define the surface variable as

$$\sigma(x) = \sum_{i=1}^{n} c_i x_i = c_1 x_1 + c_2 x_2 = c^T X \tag{6}$$

with c_i that are weighting parameters affecting the system trajectory and states (Abhinav Sinha and Rajiv Kumar Mishra 2013) and in turn, its performance.

Choice of c_i has been done by minimizing the quadratic cost function based on optimal integral rule. This requires second order sufficiency, so the model from equation (3) has been taken and put into regular form.

$$\dot{Z} = \begin{bmatrix} \mathbf{0} & \mathbf{I} \\ \mathbf{0} & g \end{bmatrix} Z + \begin{bmatrix} \mathbf{0} \\ b \end{bmatrix} u + \begin{bmatrix} \mathbf{0} \\ \bar{\varphi} \end{bmatrix} \quad (7)$$

The surface variable can be rewritten as

$$\sigma(Z) = p_1 z_1 + p_2 z_2 = p^T Z \quad (8)$$

From equation (7), we have

$$\dot{Z} = f^* + b^* u + \phi^* \quad (9)$$

The design of control law requires $\dot{\sigma}(Z) = 0$. We propose a control law of the form

$$u = \frac{-p^T f^*}{p^T b^*} - \frac{p^T \phi^*_{max} + \mu |\sigma|^\alpha}{p^T b^*} \quad (10)$$

μ is a design parameter in the above law and $\alpha \in (0,1)$. The formulated law is based on fractional power reaching law which is continuous, bounded and robust to perturbations.

5 STABILITY ANALYSIS

Stability is the prime implication to be taken in account in any design problem. Analysis of stability has been carried out by considering a Lyapunov candidate of $\frac{1}{2}\sigma^2(x)$. Negative definiteness of this candidate ensures stability in Lyapunov sense. By simple mathematical calculations, it can be easily shown that $\dot{V}(x) = \sigma\dot{\sigma} < 0$ for all $\mu > 0$. Hence, the control law is stable in Lyapunov sense and also ensures finite time reachability. The weighting parameters govern the dynamics of the system during sliding. The solution for the differential equation of equation (8) in $z_1(t)$ yields the following solutions

$$z_1(t) = \exp(-\frac{p_1}{p_2}) z_1(0) \quad (11)$$

and $z_2(t) = -\frac{p_1}{p_2} \exp(-\frac{p_1}{p_2}) z_1(0) \quad (12)$

As long as $p_1 p_2 > 0$, the state Z will show exponential convergence to zero, whatever initial conditions $Z(0)$ may be. Alternatively we may say that exponential convergence to zero will occur in finite time

when all the roots of the polynomial $\gamma(s) = p_1 + p_2 s$ are essentially present in the negative half plane.

6 SIMULATION RESULTS

The simulation has been performed using Mathworks MATLAB™ and Simulink™ and the disturbance is randomly generated. The network parameters that have been considered here are overall server capacity as 1000 packets/sec, disturbance input rate as 80 packets/sec and desired queue length as 50 packets (R. Barzamini and M. Shafiee 2013). Other design and surface parameters have been selectively tuned. The results of using higher order sliding mode control, as used in (R. Barzamini and M. Shafiee 2013) et al. are shown here. From Figures 1 and 2, it is clear that desired queue length has been achieved in around 0.5 seconds. Since, higher order sliding modes have been used here (R. Barzamini and M. Shafiee 2013) to counteract chattering (K. David Young, Vadim I. Utkin and Umit Ozguner 1999), (Levant 1993), the profile is smooth. However, increasing the order also increases the complexity of the design. We have used first order sliding mode instead. From Figures 3

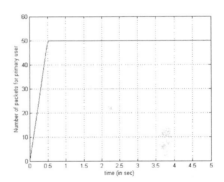

Figure 1. Queue for primary user using higher order sliding mode.

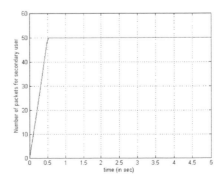

Figure 2. Queue for secondary user using higher order sliding mode.

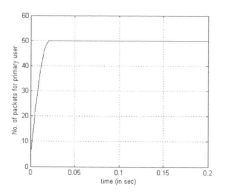

Figure 3. Queue for primary user using proposed law.

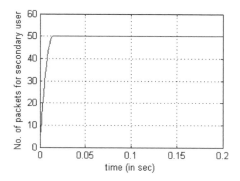

Figure 4. Queue for secondary user using proposed law.

and 4, it is clear that the proposed law can provide better results under same conditions without increasing design complexity, given selective tuning of the design parameters. The results are self explanatory.

7 CONCLUSION

A sturdy controller has been synthesized to tackle the problem of congestion in cognitive radio networks. A simple and low order fluid flow model of the network has eased the synthesis of the controller based on sliding mode control. The control algorithm has been derived analytically considering the original nonlinear model of the network. This resulted in an increase in the range of operation. The contemplated controller is easier to design and insensitive to network anomalies. Asymptotic stability has been guaranteed in Lyapunov sense. Notable improvements have been observed using the proposed law and packet loss has been nullified. The proposed controller also knocked out the need of a higher order sliding mode controller, as the problem of chattering has been tackled, provided all the design parameters be carefully tuned.

REFERENCES

Abhinav Sinha and Rajiv Kumar Mishra (2013, December). Smooth sliding mode controller design for robotic arm. In *Proc. IEEE International Conference on Control, Automation, Robotics and Embedded Systems(CARE '13)*, Jabalpur, India.

Barzamini, Y.C. and M. Shafiee (2013). Congestion control of differentiated services networks by sliding mode control. *ELECTRONIKA IR ELEKTRO-TECHNIKA 19*(1).

Bouyoucef, K. and K. Khorasani (2007, July). A sliding mode-based congestion control for time delayed differentiated-services networks. In *Proc. 15th Mediterranean Conference on Control and Automation*, Athens, Greece.

Chrysostomou, C., A. Pitsilliides, L. Rossides, M. Polycarpou and A. Sekerciouglu (2003, Mar). Congestion Control in Differentiated services networks using Fuzzy- RED. *Control Engineering Practice*, 1153–1170.

David Tipper and Malur K. Sundareshan (1990, Dec). Numerical methods for modelling computer networks under non stationary conditions. *IEEE Journal of Selected Areas in Communication 8*(9), 1682–1695.

David Young, K., Vadim I. Utkin and Umit Ozguner (1999, May). A control engineer's guide to sliding mode control. *IEEE transactions on Control Systems Technology 7*(3), 328–342.

Filipiak, J. (1988, Jan). Modelling and control of dynamic flows in communication networks. *IEEE Journal of Selected Areas in Communication*, 202.

Levant, A. (1993). Sliding order and sliding accuracy in sliding mode control. *International Journal of Control 58*(6), 1247–1263.

Liu, Y.C. and C. Douligeris (1997, Feb). Rate Regulation with Feedback Controller in ATM Networks- A Neural Network Approach. *IEEE Journal of Special Areas in Communication 15*(2), 200–208.

Ming Yan, Tatjana D. Kolemisevska- Gugolovska, Yuanwei Jing and Georgi Dimirovski (2007, September). Robust discrete-time sliding mode control algorithm for tcp networks congestion control. In *Proc. 8th IEEE International Conference on Telecommunications in Modern Satellite, Cable and Broadcasting Services (TELSIKS '07)*, Serbia.

Pitsillides, A., P. Ioannou, M. Lestas and L. Rossides (2005, Feb). Adaptive nonlinear congestion controller for a differentiated-services framework. *IEEE/ACM Transactions on Networking 13*(1), 94–107.

Pitsillides, A., and A. Sekercioglu (1999, May). Fuzzy logic based effective congestion controller. *TCD workshop on Applications of Computational Intelligence to Telecommunications*.

Pitsillides, A., and J. Lambert (1997). Adaptive Congestion Control in ATM based networks. *Journal of Computer Communication* (20), 1239–1258.

Tirtha Majumder, P.K. Sahu, S.S. Singh, Abhinav Sinha and Rajiv Kumar Mishra (2015, March). Robust nonlinear congestion controller for cognitive radio based wireless network. In *Proc. 2nd IEEE-DRDO International Conference on Innovations in Information, Embedded and Communication Sytems (ICIIECS '15)*, Coimbatore, India.

Frontiers in Computer, Communication and Electrical Engineering – Acharyya (Ed.)
© 2016 Taylor & Francis Group, London, ISBN: 978-1-138-02877-7

The dynamic compensation of the reactive power for the integration of wind power in a weak distribution network

Ritam Misra
Midcontinent Independent System Operator, Eagan, Minnesota, USA

ABSTRACT: In the near future, wind energy will be the most cost effective source of electrical power, but it will also bring some integration-related power-quality issues like reactive power compensation and voltage regulation when it is integrated with a power system network, especially in a weak network. Induction generators that we use in wind power generators draw a reactive power from the connected system. Thus, the power system engineers are very much concerned about the integration of wind power to a weak power system network. We are mainly concerned about the voltage regulation and reactive power compensation for the integration of wind power in a weak network. A centralized STATCOM (Static Synchronous Compensator) is proposed for the dynamic reactive power compensation. The STATCOM is connected at a point of interconnection between the wind farm and the network, and, thus, the system will absorb the wind power while maintaining its voltage level.

1 INTRODUCTION

Wind power is a great way to generate clean renewable energy, and the innovations in the wind technology being pursued over the last decade or so is a reminder that with the right tools, we can turn the movement of the air into fuel for our energy-hungry lifestyles. The technology is growing exponentially with the current power crisis. Wind turbine consists of a conventional squirrel cage induction generator, which has the following characteristics: firstly, it consumes the redundant reactive power in a large turbine and weak distribution network system, secondly, this type of generator slows down the voltage restoration after a voltage collapse and leads to the voltage and rotor speed instability. Generator consumes the reactive power when the voltage restores, but when the voltage doesn't restore quickly, the generator consumes an even huge amount of reactive power, which results in the voltage and rotor speed instability. So to prevent this, the shunt capacitors are connected at the generator terminals. We perform the dynamic compensation of the reactive power in order to reduce the exchange of the reactive power between wind farm and the distribution network. For that purpose, we use the FACTS devices like STATCOM (Static Synchronous Compensator) and SVC (Static Var Compensator).

The main problem that we are trying to solve is the voltage regulation and reactive power compensation that arises because of the integration of wind farm to the weak power system network.

The STATCOM produces more reactive power than SVC during the voltage collapse. Also STATCOM has a fast response time than the SVC because when STATCOM is used with Voltage Sourced Converter, there will be no time delay associated with the thyristor firing.

2 BACKGROUND

2.1 Literature review

The research presented in (Xu, L., Yao, L. & Sasse, C. 2006. Comparison) focuses on the system stability based on Fixed Speed Induction Generator (FSIG) and examines the use of SVC and STATCOM for the wind farm integration. The reactive power absorbed by the FSIG after the fault makes the system unstable. The use of SVC and STATCOM improves the system stability. The model is developed in PSCAD/EMTDC. A detailed examination is conducted on the impact of SVC/STATCOM on the system recovery after a fault. STATCOM provides a better dynamic performance and a better reactive power support as compared to SVC. The paper (Narimani, M. & Varma, R. K. 2010. Application ...) shows the dynamic reactive power compensation using Static Var Compensator. To improve the performance of an SVC, a fuzzy controller is used instead of a PI controller. This is shown by conducting a simulation using MATLAB/Simulink. This paper (Dongxu, Z. & Xiaoming, L. 2010. Research) mainly focusses on the voltage stability method based on dynamic reactive power compensation. Simulation is performed in MATLAB/Simulink and the results show that this method can offer dynamic voltage support, stabilize the bus voltage of wind turbine group in wind farms thereby improving the system's operational characteristics. This method can also be applied for the dynamic reactive power compensation and voltage stability control of the asynchronous generator groups and asynchronous motor generator groups. DFIG based wind farm, which has the advantage of a variable speed operation also faces voltage stability problems during the grid side disturbances. The STATCOM at the point of common coupling is connected to maintain stable voltage. As a dynamic reactive power compensator, it protects the DFIG-based wind farm interconnected to weak distribution network (Pokharel, B. & Wenzhong, G. 2010. Mitigation). SVC has also been used for the dynamic reactive power compensation instead of a fixed capacitor banks. The voltage stability problem has also been studied from the aspect of a large-capacity vicarious fan-speed wind power generator. Also, the relationship between the angular velocity of the wind turbine and terminal voltage of asynchronous machine can be found (Zhang, J., Zhongdong. Y., Xiao, X. & Di, Y. 2009. Enhancement).

3 PROPOSED APPROACH

3.1 Test system

The test system is shown in Figure 1 as a single line diagram. The source of the network is a132 kV, 60 Hz grid supply point, feeding a distribution system of 33kV. A step down transformer of 132/33 kV, 62.5 MVA is connected to the source. There are two loads in the system of 50 MW, a 0.9 lagging power factor and 6 MW, and a 0.9 lagging power factor at a distance of 50 km from the transformer. The transmission line is modeled as a Π line. The total MVA loading is 62.22 MVA considering the transmission and distribution losses in

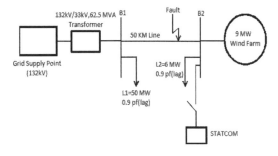

Figure 1.   Single line diagram of the system.

the system it is overloaded, thus, truly representing a weak distribution network. A 9 MW wind farm is connected with the transmission line. The wind farm consists of six 1.5 MW wind turbines. The STATCOM, which is connected at the wind farm connection provides a dynamic compensation of the reactive power. The variable pitch wind turbines are conventionally used in the wind farm and consists of squirrel cage induction generators. The turbines have its own protection scheme which monitors voltage, current, and machine speed (Shinde, M., Patil, K. D. & Gandhare, W. Z. 2009. Dynamic).

4 IMPLEMENTATION

4.1 System simulation

As most of the distribution systems are unbalanced, three phase models are used here for the study. The test system is modeled using MATLAB/SIMULINK. Phasor simulations have been mainly used to simulate the model. Variable step ode23tb solver is used for the simulation of the system. The simulation time is 5 seconds. For the simulation purposes there are mainly four cases which would be considered, which are:

1. Without wind farm and STATCOM
2. With wind farm and without STATCOM
3. With wind farm and STATCOM (without fault)
4. With wind farm and STATCOM (with fault)
5. Changing the source voltage

The active and reactive power at all the buses and in the wind farm has been measured. After the connection of STATCOM the reactive power provided by the STATCOM will also be presented. For the system studying the 50 MVA base power and 33 kV base voltages has been taken.

In each case, the voltage, active and reactive power at 33 kV Bus-1 and Bus-2 is taken. The active and reactive power of the wind farm is also measured. When the STATCOM is connected, the reactive power of the STATCOM is also measured (Shinde, M., Patil, K. D. & Gandhare, W. Z. 2009. Dynamic).

5 RESULTS

5.1 Case 1 – without wind farm and STATCOM

For this mode, the wind farm and STATCOM were skipped while running the simulation. Only the distribution system and two loads were kept in the model. In this mode only the voltages at 33 kV Bus-1 and Bus-2 are measured to ascertain that the test system is a weak system.

Figure 2 shows the voltages at 33 kV Bus-1 and Bus-2, it is clear from the figure that the voltage at 33 kV Bus-1 is 0.92 p.u. is the voltage at 33 kV Bus-2 is 0.88p.u. Thus, these voltages are below 0.95 p.u, the distribution network is really weak.

5.2 Case 2 – with wind farm and without STATCOM

For this mode, the wind farm is connected to the weak distribution network in the above mode. Here, the simulation is performed in order to integrate the 9 MW wind power in a weak distribution network, without using the STATCOM. Figure 3 shows the voltages at 33 kV Bus-1 and Bus-2.

Figure 4 shows the active power at 33 kV Bus-1 and Bus-2, it is clear from the figure that before tripping the wind turbines, they have supplied active power to the network. It can be seen that even after the connection of wind turbine, the network is still a weak network.

The reactive power at 33 kV Bus-1 and Bus-2 are shown in Figure 5. From this figure, it can be seen that before tripping the turbine generators have drawn reactive power from the network. Therefore, the voltages at 33 kV Bus-2 and Bus-1 are decreasing,

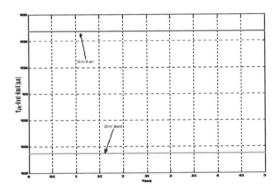

Figure 2.   Voltages at 33 kV Bus-1 and Bus-2.

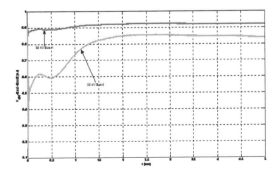

Figure 3.   Voltages at 33 kV Bus-1 and Bus-2.

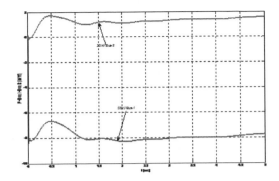

Figure 4.   Active power at 33 kV Bus-1 and Bus-2.

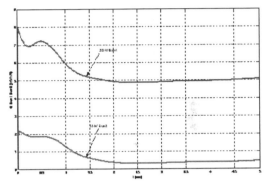

Figure 5.   Reactive power at 33 kV Bus-1 and Bus-2.

which causes the under voltage tripping of the wind turbine generators.

5.3 Case 3 – with wind farm and STATCOM (without fault)

In this mode, the wind farm with the dynamic compensation by STATCOM is connected to the weak distribution network in the above mode. The purpose of running a simulation in this mode is to integrate 9 MW wind power in a weak distribution network, with a dynamic compensation of reactive power using the STATCOM.

Figure 6 shows the voltages at 33 kV Bus-1 and Bus-2, it can be seen from the figure that initially the voltage at Bus-2 is less than that at Bus-1, but due to the reactive power injection by STATCOM, the voltage at Bus-2 goes beyond the voltage at Bus-1. Figure 7 shows the active power supplied by the wind turbine generators to the distribution network. It is seen from the figure that in this case, the wind turbine generators are not tripped. But they are supplying 9 MW power to the distribution network. Figure 8 shows the reactive power drawn by the induction generators from the network and it can be seen from the figure that initially wind tur-

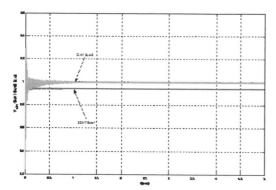

Figure 6.   Voltages at 33 kV Bus-1 and Bus-2.

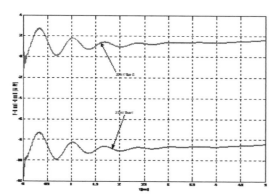

Figure 7.   Active power at 33 kV Bus-1 and Bus-2.

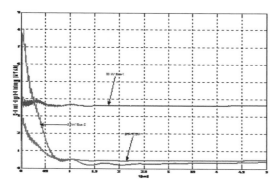

Figure 8.   Reactive power at 33 kV Bus-1 and Bus-2.

bine generators draw more reactive power, but later on the reactive power demand has been stabilized.

5.4   *Case 4 – with wind farm and STATCOM (with fault)*

This mode is the same as the above mode of simulation, but in this case a three phase fault for

2 to 2.15 seconds is made. The purpose of running the simulation in this mode is to verify the dynamic reactive power compensation capability of STATCOM during the event of a fault, while integrating the wind power in a weak distribution network.

Figure 9 shows the voltages at 33 kV Bus-1 and Bus-2, and it is seen from the figure that the voltage recovery after the fault is accelerated due to

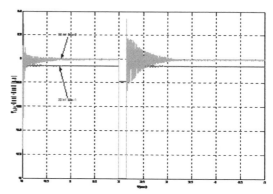

Figure 9.   Voltages at 33 kV Bus-1 and Bus-2.

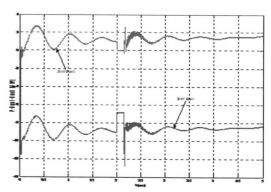

Figure 10.   Active power at 33 kV Bus-1 and Bus-2.

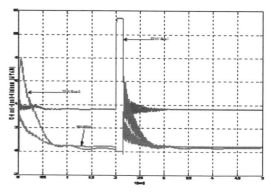

Figure 11.   Reactive power at 33 kV Bus-1 and Bus-2.

STATCOM and the system voltage restores before the initiation of the protection systems. Thus, the wind turbine generators do not trip in the event of a short duration fault. Figure 10 shows the active power at 33 kV Bus-1 and Bus-2. Figure 11 shows the reactive power at 33 kV Bus-1 and Bus-2 and STATCOM, it can be seen that the STATCOM is supplying the reactive power to the network even in the event of short duration fault at its point of interconnection.

6 CONCLUSION

Evaluation is done about the dynamic power compensation capability of the STATCOM for the integration of the wind power in a weak distribution network. The dynamic power compensation capability is also done during an external three phase fault. The evaluation reveals that the reactive power compensation by STATCOM makes it possible to integrate the wind farm in a weak distribution network. STATCOM prevents large deviations of the bus voltage due to the reactive power drawn by the wind turbine generators and also after the fault rapid recovery of the voltage has been resulted.

REFERENCES

Dongxu, Z. & Xiaoming, L. 2010. Research on the Voltage Stability Control for Wind Turbine Group Based on Dynamic Reactive Power Compensation. *Power and Energy Engineering Conference(APEEC), Asia Pacific. 28–31 March 2010.*
Narimani, M. & Varma, R.K. 2010. Application of static var compensator (SVC) with fuzzy controller for grid integration of wind farm. *Canadian Conference on Electrical and Computer Engineering.*
Pokharel, B. & Wenzhong, G. 2010. Mitigation of disturbances in DFIG-based wind farm connected to weak distribution system using STATCOM. *North American Power Symposium (NAPS).*
Shinde, M., Patil, K.D. & Gandhare, W.Z. 2009. Dynamic Compensation of Reactive Power for Integration of Wind Power in a Weak Distribution Network. *International Conference On Control, Automation, Communication and Energy Conservation. 4–6 June 2009.*
Xu, L., Yao, L. & Sasse, C. 2006. Comparison of Using SVC and STATCOM for Wind Farm Integration. *International Conference on Power System Technology, 22–26 Oct 2006.*
Zhang, J., Zhongdong. Y., Xiao, X. & Di, Y. 2009. Enhancement voltage stability of wind farm access to power grid by novel SVC," Industrial Electronics and Applications. *4th IEEE Conference on Industrial Electronics and Applications. 25–27 May 2009.*

Frontiers in Computer, Communication and Electrical Engineering – Acharyya (Ed.)
© 2016 Taylor & Francis Group, London, ISBN: 978-1-138-02877-7

The Oppositional Chemical Reaction Optimization algorithm for the optimal tuning of the Power System Stabilizer

S. Paul & A. Maji
Dr. B.C. Roy Engineering College, Durgapur, West Bengal, India

Provas Kumar Roy
Jalpaiguri Government Engineering College, Jalpaiguri, West Bengal, India

ABSTRACT: In this paper, the co-ordination scheme to improve the stability of a power system by the optimal design of the power system stabilizer has been proposed. This paper presents a meta-heuristic population-based algorithm named Oppositional Chemical Reaction Optimization (OCRO) algorithm of optimally tuning parameters of the Power System Stabilizers (PSSs), which have been added to the excitation system to dampen the low frequency oscillation that lies in the large power system. Eigen value based objective function incorporating the damping factor and damping ratio of the lightly damped electromechanical modes is to be considered in the PSSs design problem. The effectiveness of the proposed OCRO along with the conventional CRO is demonstrated on a single machine infinite bus from the Heffron-Phillips model. To show the superiority of the proposed method, the results are presented over a wide range of loading conditions. Simulated results of the proposed OCRO method are compared with those obtained by CRO, Gravitational Search Algorithm (GSA), and Oppositional GSA (OGSA) to demonstrate the feasibility of the proposed algorithm.

1 INTRODUCTION

The electric power system is operated close to its capacity limits, increasing the risk of the stability problems. One problem that faces the power systems is the low frequency oscillations arising due to the disturbances. These oscillations may sustain and grow to cause a system separation if no adequate damping is provided. Power System Stabilizer (PSS) is the most effective device for dampening the low frequency oscillations and increasing the stability margin of the power system. PSS is used to generate supplementary control signals for the excitation system in order to mitigate these types of oscillations. The stability of the power system is affected due to disturbances like a sudden change in load, loss of a generator, and widespread use of the high gain fast acting excitation system. The research of the PSS for improving the stability of the power system has been conducted since the late 1960s. In the last two decades, various types of PSSs were introduced. Many new intelligent methods such as neural networks (Segal et al. 2004) and fuzzy logic (Kocaarslan et al. 2005) were used in the PSS design. Modern control methods such as adaptive control was also used in PSS design (Shaoru et al. 2009). Basically, two main group of approaches have been investigated in literature: (a) methods based on linear models with several

variants (classic tuning methods, soft computing methods, and robust approaches) and (b) methods based on nonlinear dynamic models. Classical control techniques including self-tuning regulators, pole placement, robust control, and pole shifting were successfully investigated in the past for the design of power system stabilizers.

In the past decade, various population based optimization methods such as particle swarm optimization (PSO) (Shayeghi et al. 2010), Genetic Algorithm (GA) (Zhang et al. 2000), Honey Bee Mating Optimization (HBMO) (Niknam et al. 2011) have been efficiently implemented to find the optimum set of parameters to effectively design the PSS. Abd-Elazim et al. (2013) suggests that PSS tuning by hybrid PSO and Bacterial Foraging Algorithm (BFA) was performed on the three machine nine bus power system. Detailed analysis of the PSS tuning with non-linear model based algorithm is provided in reference to (Yee et al. 2006). Though, the above stated optimization techniques have successfully been applied in PSS design, they suffer from many demerits like GA requires a very long run time depending on the size of the system under study. Also, it suffers from the optimal settings of the algorithm parameters and gives rise to repeat the revisiting of the same suboptimal solutions. In case of PSO, the premature convergence nature may trap the algorithm into the local optimum, hence, reducing the chances of reaching the global optimal solutions.

In view of the above discussion, the main aim of the present study is to design and implement a new Evolutionary Algorithm (EA) named Oppositional Chemical Reaction Optimization (OCRO). In this article, the proposed OCRO based PSS design problem is applied in the single machine power systems from the Heffron-Phillips model. The computed results are compared with those of OGSA and GSA to verify the superiority of the proposed algorithm.

The rest of the paper is organized as follows: the problem of formulation is illustrated in section 2. A brief outline of CRO algorithm is discussed in section 3. Section 4 gives the outline of the Oppositional Based Learning (OBL) concept. Section 5 presents different algorithmic steps of OCRO. Section 6 reports the comparative simulation results. Finally, section 7 concludes this article.

2 PROBLEM FORMULATION

2.1 System model

The synchronous machine can be represented by a set of four first order linear differential equations (1–4). These equations represent a fourth order generator model. Higher machine's models are also proposed based on the varying degrees of complexity which provides better results.

$$\frac{d\delta}{dT} = \omega - \omega_0 \tag{1}$$

$$\frac{d\omega}{dT} = \frac{1}{2H}(-D\omega + T_m - T_e) \tag{2}$$

$$\frac{dV_q'}{dT} = \frac{1}{T_{do}'}(-V_q' + (x_d - x_d')i_d + V_{fd}) \tag{3}$$

$$\frac{dV_d'}{dT} = \frac{1}{T_{qo}'}(-V_d' + (x_q - x_q')i_q) \tag{4}$$

where, V_{fd} is the voltage proportional to field voltage; V_d', V_q' are the voltages proportional to damper winding and field flux; i_d, i_q are the d-q components of armature currents; T_{do}', T_{qo}' are the d-q axis transient time constants. The overall block diagram for Heffron-Phillips model that illustrates the linearized model of SMIB is shown in Figure 1. The constants K_1, K_2, K_3, K_4, K_5, K_6 are modeled in (Padiyar et al. 1996)

2.2 PSS controller structure

The model of the Conventional Power System Stabilizer (CPSS) structure is illustrated in Figure 2.

The CPSS structure under study consists of two phase-lead compensation blocks, a gain block, and a signal washout block. It can be shown that by appropriate selection of the CPSS parameters, a satisfactory performance can be obtained during the system disturbance.

The transfer function of the ith PSS is given by:

$$CPSS = K_{pss} \frac{T_w s(1+sT_1)(1+sT_3)}{(1+sT_w)(1+sT_2)(1+sT_4)} \tag{5}$$

2.3 Objective function

The stability of a power system can be evaluated based on its eigenvalues. The unstable poles or lightly damped stable poles result in power system oscillations and instability. Hence, displaying these poles to the left side of the s-plane (shown in Figure 3) can guarantee the stability of the power system with the corresponding improvement in the system performance.

This paper concentrates on the relocation of the unstable and lightly damped electromechanical modes of the oscillations. The parameters of the PSS may be chosen to minimize the following objective functions (Shayeghi et al. 2010).

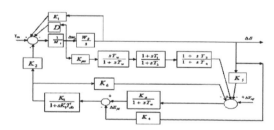

Figure 1. Block diagram of single machine with CPSS.

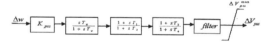

Figure 2. Conventional PSS.

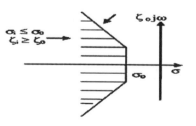

Figure 3. D-shaped sector in the negative half of s-plane.

$$f_1 = \min(abs(\sigma_0 - \sigma_i)); \quad i = 1, 2, \ldots, n \tag{6}$$

$$f_2 = \min(abs(\zeta_1 - \zeta_i)); \quad i = 1, 2, \ldots, n \tag{7}$$

where, σ_i and ζ_i are the real part and the damping ratios of the ith operating conditions; n is the number of states for which the optimization is carried out. The relative stability is determined by σ_0. The value of σ_0 is problem dependent. The design procedure can be formulated as the following constrained optimization problem, where the constraints are the PSS parameters that are bound (Shayeghi et al. 2010). In this paper, the value of σ_i and ζ_i are chosen to be −1.5 and 0.2, respectively, and the ranges of the PSS parameters are selected to be $1 \leq K_{pss} \leq 200$, $0.01 \leq T \leq 2$.

3 CHEMICAL REACTION OPTIMIZATION ALGORITHM

Chemical Reaction Optimization (CRO) algorithm (Lam et al. 2010) is a population-based metaheuristic optimization technique inspired by the nature of chemical reactions, mimicking the process of reactions where molecules collide with the walls of the container and with each other. In CRO, a candidate solution for a specific problem is encoded as a molecule. A molecule possesses two kinds of energies, i.e., Potential Energy (PE) and Kinetic Energy (KE). PE represents the objective function of a molecule while the KE of a molecule represents its ability of escaping from a local minimum.

Moreover, the chemical reaction is assumed to take place in a closed container, and there are four types of elementary reactions implemented in CRO, namely, (a) on-wall Ineffective Collision, (b) decomposition, (c) inter molecular ineffective collision, and (d)synthesis.

3.1 On-wall ineffective collision:

In this reaction, a single molecule is involved and it is allowed to collide on the wall of the container. If the energy criteria has been satisfied then the present molecular structure (ω) is changed to (ω') Needed to be satisfied for the above reaction is:

$$\sum_\omega PE_\omega + \sum_\omega KE_\omega \geq \sum_{\omega'} PE_{\omega'} \tag{8}$$

where ω is the existing molecular structure; ω' is the converted molecular structure.

Uni-molecular reactions involving a single molecule which hits on the wall of the container causing some of its energy to be transferred to a central energy buffer (Environment) and (KE) of the molecule with new structure can be calculated by the following equation:

$$KE_{\omega'} = (PE_\omega - PE_{\omega'} + KE_\omega) \times a \tag{9}$$

where a is a random number.

3.2 Decomposition:

This is also anuni-molecular reaction. Here, one particular molecule converts into two new molecules with different structures.

i.e. $\omega \rightarrow \omega_{1 +} \omega_2'$ \hfill (10)

This happens when

$$PE_\omega + KE_\omega + buffer \geq PE_{\omega_1'} + PE_{\omega_2'} \tag{11}$$

3.3 Inter-molecular ineffective collision:

This is a reaction involving two molecules interaction. When two molecules with structure ω_1 and ω_2 collide with each other, they are changed into two new molecules with structure ω'_1 and ω'_2, if the following energy criteria is satisfied:

$$PE_{\omega_1} + PE_{\omega_2} + KE_{\omega_1} + KE_{\omega_2} \geq PE_{\omega'} + PE_{\omega_2'} \tag{12}$$

E_{inter} denotes extra energy after the inter molecular collision.

$$E_{inter} = \left(PE_{\omega_1} PE_{\omega_2} KE_{\omega_1} + KE_{\omega_2} \right) - \left(PE_{\omega_1} + PE_{\omega_2'} \right) \tag{13}$$

Kinetic energy of ω_1' is given by:

$$KE_{\omega_1'} \rightarrow E_{inter} \times \varepsilon_1 \tag{14}$$

Kinetic energy of ω_2' is given by:

$$KE_{\omega_2'} \rightarrow E_{inter} \times (1 - \varepsilon_1) \tag{15}$$

3.4 Synthesis

This reaction is also an intermolecular reaction. Here, two molecules with a molecular structure ω_1 and ω_2 collide to form a single molecule with structure ω'. If the following energy criteria hold:

$$KE_{\omega'} = PE_{\omega_1} + KE_{\omega_1} + PE_{\omega_2} + KE_{\omega_2} - PE_{\omega'} \tag{16}$$

4 OPPOSITIONAL BASED LEARNING

Opposition-Based Learning (OBL) (Tizhoosh et al. 2005) is a new concept in computational intelligence. Generally speaking, evolutionary optimization methods start with some initial solutions and try to improve their performance toward some optimal solutions. We can improve our chance to start with a closer solution by checking the opposite solution simultaneously. By doing this, the closer one can be chosen as an initial solution. The same approach can be applied not only to the initial solutions but also continuously to each iteration in the current population. In the OBL method, it generates a second set of solutions, which is opposite to the original solution set so that our probability of choosing better solutions can increase. Thus, by comparing a number to its opposite, a smaller search space is needed.

4.1 Definition (opposite number)

while evaluating a solution to a given problem, simultaneously computing its opposite solution will provide another chance for finding a candidate solution, which is closer to the global optimum.

Let $z \in [m,n]$ be a real number, then opposite number z^* is defined as:

$$z^* = m + n - z \qquad (17)$$

4.2 Definition (opposite point)

Let $Z = (z_1, z_2, ..., z_d)$ be a point in a d-dimensional space, where $z_1, z_2, ..., z_d \in R$ and $z_i \in [m_i, n_i] i \in 1, 2, ..., d$. The opposite point $Z^* = (z_1^*, z_2^*, ..., z_d^*)$ is defined by:

$$z_i^* = m_i + n_i - z_i \qquad (18)$$

By employing an opposite point definition, opposition-based optimization can be defined as follows.

4.3 Opposition-Based Optimization

Let $Z = (z_1, z_2, ..., z_d)$ be a point in the d-dimensional space. Assume $f(Z)$ as a fitness function, which is used to measure the candidate's fitness. According to the definition of the opposite point, $Z^* = (z_1^*, z_2^*, ..., z_d^*)$ is the opposite of $Z = (z_1, z_2, ..., z_d)$. Now if $f(Z^*) \geq f(Z)$ then point Z can be replaced with Z^*; otherwise, keep the current point Z. Hence, the current point and its opposite point are evaluated simultaneously in order to continue with the faster one.

The definition can be extended to the d-dimensional search space as follows:

$$z_i^o = rand(z_i^*) \qquad (19)$$

5 OCRO ALGORITHM

Step 1: Randomly initialize the population (P) and generate the opposite population (OP).

Step 2: Calculate fitness function of P and OP and select n_p number of the fittest solution from P and OP. Each fitness value represents a Potential Energy (PE). An initial Kinetic Energy (KE) is assigned to all the molecules.

Step 3: Depending on the PE value a few best solutions are kept as elite solution.

Step 4: On-wall Ineffective Collision, decomposition, inter molecular ineffective collision, and synthesis are applied to enhance the search space using equations. (11–14).

Step 5: Generate opposite population from the current solution based on the jumping rate.

Step 6: The feasibility of each solution is checked by satisfying the operational constraints.

Step 7: Select n_p number of fittest solution from P and OP. Sort the solution from the best to the worst and replace the worst solution by the best elite solution.

Step 8: If the stop criterion does not satisfy stop, then go to step 4.

6 SIMULATION RESULTS AND DISCUSSION

Proposed OCRO and CRO algorithms have been applied on the two objective functions namely; eigen value minimization (Case I) and damping ratio minimization (Case II) and their results are compared with the results of OGSA and GSA for verifying its feasibility. The algorithm is tested over a wide range of loading conditions. Three different cases as light $(P = 0.65, Q = 0.55)$, nominal $(P = 1.25, Q = 0.50)$, and heavy $(P = 1.45, Q = 0.70)$ loading are considered as system uncertainties.

Maximum number of iteration of 100 is taken for a simulation study. Proposed OCRO are run for 50 population size. Programs are developed using Matlab 7.6 on a personal computer having 2.53 GHZ core i5 processor with 4GB RAM. Different trails are made for solving PSS problem. The values for the input parameters are deduced from the literature (Lam et al. 2010).

The calculated results for different loading conditions are demonstrated in Tables 1–3. The simulation results of the proposed algorithms have been compared with (Paul et al. 2014). It is clearly visible from these tables that the OCRO has a better dynamic performance in terms of stability with the corresponding improvement in the overshoot (O_s) and undershoot (U_s) than the CRO and other population based optimization techniques like OGSA and GSA for all operating conditions. Also, the computed results improve the damping of low frequency oscillations. Also, it should be pointed out that for light load in case I, the proposed OCRO attains a damping ratio of 0.2371, which is better than the previously reported 02379, 0.2287, obtained by CRO and OGSA. From the Table 3 it can also be clearly seen that there is a considerable improvement in the fitness function value of (0.0347) using OCRO in comparison to CRO, OGSA (Paul et al. 2014) and GSA (Paul et al. 2014) in case of the nominal load in Case II. These results clearly prove the superiority of the proposed algorithm.

The dynamic rotor deviation response obtained by OCRO under nominal load and heavy load for case I and light load for case II are depicted in Figures 4–6. It can be clearly seen that the overshoot, undershoot and settling time using OCRO are better than OGSA and GSA. With changing operating conditions from light to heavy load, while the performance of OGSA and GSA become meager, the OCRO stabilizer have stable and robust performances. These Figs. clearly demonstrates the superiority of the OCRO based PSS.

Table 1. Comparative results of the electromechanical modes for different algorithms for Case I under light load.

Methods	Eigen Value	O_s ($\times 10^{-4}$)	U_s ($\times 10^{-4}$)	ω_n	ξ	Fitness value ($\times 10^{-4}$)
OCRO	**−1.5000 ± 3.1122i**	**3.3140**	**9.4729**	**3.1122**	**0.4342**	**0.0043**
CRO	−1.5000 ± 3.1854i	3.3138	9.4726	3.1122	0.4329	0.0056
OGSA	−1.5000 ± 3.1602i	3.3131	9.4865	3.1602	0.4288	0.0074
GSA	−1.5000 ± 3.1067i	3.3131	9.4713	3.1067	0.4348	0.0381

Table 2. Comparative results of the electromechanical modes for different algorithms for Case I under heavy load.

Methods	Eigen Value	O_s ($\times 10^{-4}$)	U_s ($\times 10^{-4}$)	ω_n	ξ	Fitness value ($\times 10^{-4}$)
OCRO	**−1.3819 ± 6.7698i**	**9.3682**	**16.0000**	**6.7698**	**0.2000**	**0.0347**
CRO	−1.3798 ± 6.7923i	9.3518	15.9821	6.7912	0.2000	0.0631
OGSA	−1.3892 ± 6.8057i	9.3425	15.6839	6.8057	0.2000	0.0727
GSA	−1.3672 ± 6.6995i	9.4051	16.1162	6.6995	0.1999	0.5264

Table 3. Comparative results of the electromechanical modes for the different algorithms for Case II under nominal load.

Methods	Eigen Value	O_s ($\times 10^{-4}$)	U_s ($\times 10^{-4}$)	ω_n	ξ	Fitness value ($\times 10^{-4}$)
OCRO	**−1.5000 ± 6.1432i**	**7.0782**	**13.000**	**6.1432**	**0.2372**	**0.0106**
CRO	−1.5000 ± 6.2341i	6.9065	13.000	6.4512	0.2379	0.0298
OGSA	−1.5000 ± 6.3855i	4.6761	13.0352	6.3855	0.2287	0.0319
GSA	−1.5000 ± 6.7786i	4.8443	13.4608	6.7786	0.2161	0.0859

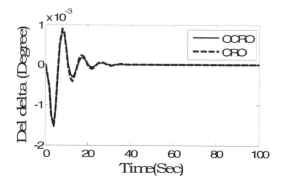

Figure 4. Generator response for nominal load for Case I.

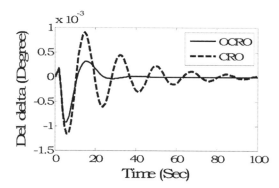

Figure 5. Generator response for heavy load for Case I.

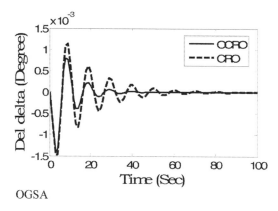

OGSA

Figure 6. Generator response for light load for Case II.

7 CONCLUSION

In this article, a novel approach named oppositional based CRO along with conventional CRO are developed and successfully applied on a Single Machine Infinite Bus (SMIB) system to find the optimal design parameters of the PSS controllers.

The viability of the proposed method is applied under various loadings. The computed results obtained are compared with other optimization techniques like OGSA and GSA. OCRO gives better results for both the two objective functions taken in this study. It may be concluded that the new proposed algorithm shows more potential than the other.

REFERENCES

Abd-Elazim, S.M. & Ali, E.S. 2013. A hybrid particle swarmoptimization and bacterial foraging for optimal power system stabilizer design. *Electr Power Energy Syst*; 46: 334–41.

Kocaarslan, I. & Cam, E. 2005. Fuzzy logic controller in interconnected electrical power systems for load-frequency control. *Int J Electr Power Energy Syst*;27: 542–9.

Lam, A.Y.S., & Li, V.O.K. 2010."Chemical Reaction inspiredmetaheuristic for optimization", *IEEE Transactionson on Evolutionary Computation*,14(3): 381–399.

Niknam, T., Mojarrad, H.D., Meymand, H.Z. & Firouzi, B.B. 2011. A new honey bee mating optimization algorithm for non-smooth economic dispatch. *Electr Power Energy Syst*;36: 896–908.

Padiyar, K.R. 1996. "Power system dynamics: stability and control", Bangalore: *Interline Publishing Private Ltd.*

Paul, S. & Roy P.K. 2014." Optimal Design of Power System Stabilizer Using Oppositional Gravitational Search Algorithm", *IEEE Int. Conf. on Non Conventional Energy*, 360–65.

Segal, R., Sharma, A. & Kothari, M. 2004. A self-tuning power system stabilizer based on artificial neural network. *Int J Electr Power Energy Syst*;26: 423–30.

Shaoru, Z. & Fang Lin, L. (2009). An improved simple adaptive control applied to power system stabilizer. *IEEE Trans Power Electron*;24: 369–75.

Shayeghi, H., Shayanfar, H.A., Safari, A. & Aghmashah, R.A. 2010."Robust PSSs design using PSO in a multimachine environment" *Energy Convers Manage*, 51: 696–02.

Tizhoosh, H.R. 2005.Opposition-based learning: a new scheme for machine intelligence, *Proceedings of International Conference on Computational Intelligence for Modeling Control and Automation: 695–701.*

Yee, S.K, & Milanovi, c.J.V. 2006. Nonlinear time-response optimization method for tuning power system stabilizers. *IEE Proc.—Gen. Transm. Distrib.* 153 (3): 269–275.

Zhang, P., & Coonick, A.H. 2000.Coordinated synthesis of PSS parameters in multimachine power systems using themethod of inequalities applied to genetic algorithms. *IEEE Trans Power Syst*;15: 811–6.

Frontiers in Computer, Communication and Electrical Engineering – Acharyya (Ed.)
© 2016 Taylor & Francis Group, London, ISBN: 978-1-138-02877-7

Neural network based multi objective optimization—a new algorithm

Debasish Roy
Techno India University, Salt Lake, Kolkata, India

ABSTRACT: The difference in Single objective and multi-objective optimizations is primarily in the cardinality in the optimal set; from a practical standpoint a user needs only one solution, no matter whether the associated optimization problem is single-objective or multi-objective. A multi-objective optimization problem can be converted into a single-objective optimization problem by the method of scalarizing an objective vector into a single composite objective function. This procedure of handling multi-objective optimization problems called Predilection Based MOO and is much simpler, yet still being more subjective than the any other procedure. In the evolutionary approaches the most popular methods are by evolutionary or genetic algorithms. This paper has proposed a new method for deriving Pareto Optimal Front by use of Neural Network. The algorithm has been presented here. A numerical experiment has also been performed to validate the algorithm.

1 INTRODUCTION

In principal, the search space in the context of multiple objectives can be divided into two non-overlapping regions, namely one which is non-optimal and one which is optimal. It is palpably true for two-objective problem and this can be generalized to problems with more than two objectives. In the case of incongruous objectives, usually the set of optimal solutions contains more than one solution. Thus, it can be conjectured that there are two raison d'être in a multi-objective optimization:

1. A set of solutions as close as possible to the Pareto-optimal front.
2. A set of solutions as diverse as possible.

Artificial neural network are nonlinear information or signal processing devices, which are built from interconnected elementary processing devices called neurons. An Artificial Neural Network (ANN) is an information-processing paradigm that is inspired by the way biological nervous systems, such as the brain, that process information. The key element of this paradigm is the novel structure of the information processing system. It is composed of a large number of highly interconnected processing elements called neurons, working in union to solve specific problems. ANNs, like people, learn by example. It resembles that brain in two respects, firstly the knowledge is acquired by the network through a learning process, and, secondly inter-neuron connection strengths known as synaptic weights are used to store the knowledge.

In this paper ANN is used to estimate Pareto Optimal front in case of MOOP.

2 LITERATURE REVIEW:

Efficient solutions in the decision set were also proved too frequently to map onto the same solution of the outcome set (Benson 1995), leading to redundant solutions in the decision set. There exists number of algorithms that solve in the decision space but are hard to apply practically as computational demands increase substantially as problem size increases (Steuer, 1986), (Armand & Malivert, 1991), (Aramand, 1993), (Sayin, 1996).There are other methods for solving problems like use of homotopy to find efficient solutions for convex multi-objective programming (Lin, Zhu, & Sheng, 2003).

Some of the most popular methods are narrated in following paragraphs.

The ε-constraint method chooses a single-objective to be optimized while every other objective is treated as a constraint (Haimes, Lasdon, & Wismer, 1971).

The two phase method (Ulungu & Teghem, 1995) involves finding all supported non-dominated solutions by solving the weighted sum problem in the first phase and finding all other non-dominated solutions in the second phase, usually using enumerative techniques which develop search spaces from the supported non-dominated solutions found in the first phase. The outer approximation algorithm (Benson H.P, 1998) which works on the outcome space has the advantage of no need for backtracking or bookkeeping which is needed when solving in the decision space (Benson, 1997). Hybrid vector maximization approach (Benson, 1998) though was introduced in 1951 (Kuhn & Tucker, 1951) was implemented in 1997. Simple partitioning technique came into being as an extension of outer approximation algorithm (Ban, 1983), (Horst & Tuy, 1996), but (Benson B. L., 1998) the algorithm also generated weakly efficient points in the outcome space. A feasible basis for the linear program can be decomposed (Benson & Sun, 2000) into a finite union of subsets with a one-to-one correspondence between the weights and efficient extreme solutions in the outcome space and the result prompted development of a weight set decomposition algorithm (Benson & Sun, 2002). The network optimization problems used an algorithm evolved from scalarisation theorem and single-objective duality theory (Pureto & Nickel, 2005). Two distinctive approaches to duality are one based on the duality relationship between minimal and maximal elements of a set and its complement, and another using polarity between convex polyhedral sets and the epigraph of its support function (Luc, 2011). Improvements to existing duality relations have also been explored. The outer approximation algorithm has a dual variant using geometric duality theory to derive a dual variant of the algorithm (Ehrgott, Lohne, & Shao, 2012).

Sequentially efficient points and rays can be generated using extreme ray generation method (Ida, 2011). The inequality constraints can be added to the polyhedral feasible region. Adjacency between efficient extreme points can be utilized to generate maximal efficient faces (Krichen, Masri, & Guitouni, 2012). The algorithm explores efficient extreme points and uses simplex pivots to find adjacent vertices of the current extreme point.

Adaptive—constraint method can be used and shown that problems of higher dimensions require a bound of number of objectives (Laumanns, Thiele, & Zitzler, 2006).Tri-criteria problems can be solved by finding solutions of sub-problems and thereby an algorithm emerged (Klamroth & Dachert, 2015).

3 MULTI OBJECTIVE OPTIMISATION

The concept of domination is used by most multi-objective optimization algorithms. Let there be M objective functions. In order to cover minimization and maximization of objective functions, the operator « between two solutions p and q is used and p«q is used to denote that solution p is better than solution q on all objective functions. Likewise, p»q for a particular objective implies that solution p is worse than solution q on the objective space.

Minimize/Maximize: $\Phi m(x)$, $m = 1, 2... M$;

Subject to: $\theta_j(x) \leq 0$, $j = 1, 2... J$;
$\psi_k(x) = 0$, $k = 1, 2... K$;
$x_i^{(L)} \leq x_i \leq x_i^{(U)}$, $i = 1, 2,........, n$.

A solution x is a vector of n decision variable x = $(x_1, x_2, ... ,x_n)^T$. The variable bounds constitute the last set of constraints, restricting each decision variable x_i to take a value within a lower $x_i^{(L)}$ and an upper $x_i^{(U)}$ bound and also comprise the decision variable space or simply the decision space.

In the problem formulation there are J inequality and K equality constraints. The constraint functions are denoted by the terms $\theta_j(x)$ and $\psi_k(x)$. The inequality constraints are treated as lesser-than-equal-to types, although a greater-than-equal-to type inequality constraint is also taken care of in the above formulation. The type of problem this paper is trying to solve has following format:

$$Minimise\, Z = (\frac{\sum c_i x_i}{\sum d_j y_j}, \frac{\sum e_i x_i}{\sum f_j y_j}, ...)$$

Such that: $\dfrac{\sum c_i x_i}{\sum d_j y_j} \leq 1$

$$\frac{\sum e_i x_i}{\sum f_j y_j} \leq 1$$

$$\sum d_j y_j \neq 0$$

$$\sum f_j y_j \neq 0$$

$$x_i, y_i \in R^+$$

4 NEURAL NETWORK BASED MULTI OBJECTIVE OPTIMISATION

The proposed New Algorithm based on Neural Network is presented below:

Step 1: Define n m-dimensional random points.
Step 2: Define q fractional functions dependent on m-dimensional points.
Step 3: Evaluate n q-dimensional points.
Step 4: Find n/k non-dominant points from n points evaluated in Step 3.
Step 5: Partition the output in non-dominant and dominated sets.
Step 6: Define a feed forward network m inputs and 1 output.
Step 7: Train the network output for non-dominated points be set to 1 and dominated points be set to 0.

Step 8: Train a network with set of inputs as output of Step 3 where combined partitioned output is input and output is as defined in Step 7.

Step 9: Simulate the output, if the error values are not within the range, retrain the network or increase number of layers in feedforward back propagation network.

Step 10: Repeat step7 till the results are satisfactory.

Step 11: Take another set of m random inputs. Obtain output from the network. Test if the results corroborate, that is error is within the margin of definition.

Step 12: If error is outside the margin, increase number of layers and repeat step 5–7.

Step 13: Run the network with more set of random inputs and generate the Pareto front.

5 THE EXPERIMENT

5.1 Setup

Minimise: $y_1 = \dfrac{3x_1 + 4x_2 + 1}{5x_1 + 7x_2 + 5}, y_2 = \dfrac{4x_1 + 5x_2 + 2}{7x_1 + 9x_2 + 3}$

Subject to: $5x_1 + 7x_2 + 5 \neq 0,\ 7x_1 + 9x_2 + 3 \neq 0$

$$\frac{3x_1 + 4x_2 + 1}{5x_1 + 7x_2 + 5} \leq 1, \frac{4x_1 + 5x_2 + 2}{7x_1 + 9x_2 + 3} \leq 1$$

5.2 Operation

The algorithm has been implemented in MATLAB. A part of the code is presented below:

```
function [] = NNMOOP2()
[xy_nd xy_d xy_inp] = MOOP4(100);
xy_out = ones(3,100); [m n] = size(xy_nd);
xy_out(1:2,1:n) = xy_nd;
xy_out(1:2,n+1:100) = xy_d;
xy_out(3,n+1:100) = 0.0;xyi = xy_out(1:2,:);
xyo = xy_out(3,:);net = newff(xyi,xyo,30);
xySim = sim(net,xyi);net.trainParam.
epochs = 100;
net = train(net,xyi,xyo);xyTrainO = sim(net,xyi);
[xy_nd1 xy_d1 xy_inp1] = MOOP4(100);
xy_out1 = ones(3,100);[m1 n1] = size(xy_nd1);
xy_out1(1:2,1:n1) = xy_nd1;
xy_out1(1:2,n1+1:100) = xy_d1;
xy_out1(3,n1+1:100) = 0.0;
xyi1 = xy_out1(1:2,:);
xySim1 = sim(net,xyi1);
disp([xy_out1(3,:)' xySim1' ]);
figure;
plot(1:100,xy_out1(3,:),'*B',1:100,xySim1,'oR');
```

Table 1. Sample of 100 point output. F1 = First Function Value, F2 = Second Function Value. NDM/DM = Non Dominated and Dominated Points, NDM = 1, DM = 0 = Non Dominated points are marked 1 and dominated points are marked 0.

F1	F2	NDM = 1, DM = 0	EF1	EF2	NDM/DM
0.5439916	0.5446757	1	0.5489116	0.5602151	1
0.5411736	0.5652664	1	0.5362089	0.5664375	0
0.5376216	0.5589858	1	0.5157171	0.5679379	1
0.5376216	0.5589858	1	0.5296044	0.5619226	1
0.5411736	0.5652664	1	0.4991266	0.5702679	1
0.5376216	0.5589858	1	0.5287474	0.5702679	1
0.5437886	0.5693342	1	0.5106052	0.568453	0
0.5411771	0.5652719	1	0.5496189	0.5607274	1
0.5376216	0.5589858	1	0.533112	0.5616756	1
0.5376216	0.5589858	1	0.5220115	0.57649	1
0.5376216	0.5589858	1	0.5281739	0.562853	1

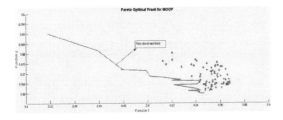

Figure 1. Pareto optimal front.

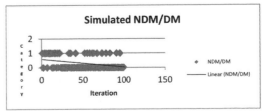

Figure 2. The simulated output of NN.

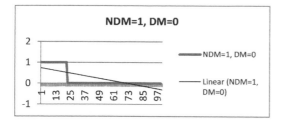

Figure 3.   The actual output classification.

```
title('Second Network with new data');
xy_err1 = xySim1-xy_out(3,:);
disp('New error');
disp(std(xy_err1));
```

5.3   Results

The results have been presented in Table 1 and Figures 1–3. Std Deviation of error between Actual output and Simulated Output: 0.1290.

5.4   Conclusion

As can be seen from output of errors the Neural Network has reasonably traced the non-dominated front. The standard deviation of error is also within acceptable range. However, in case of further refinement, the experiment may be conducted with larger layers of Neural Network or with more training epochs.

ACKNOWLEDGEMENTS

I thank my guides Dr. Anjan Sarkar and Dr. Abhijit Mitra for inspiring me to write the paper and guiding me in this direction of research.

REFERENCES

Aramand, P. (1993). Finding all maximal efficient faces in multiobjective linear programming. *Mathematical Programming, 61*(1), 357–375.

Armand, P., & Malivert, C. (1991). Determination of the efficient set in multiobjective optimisation. *Journal of Optimisation Theory and Application, 70*(3), 467–490.

Ban, V.T. (1983). A finite algorithm for minimising a concave function under linear constraint and its application. *In Proceedings of IFIP Working Conference on recent advances on system modelling and optimisation.*

Benson, H.P. (1995). A geometrical analysis of efficient outcome set in multiple objective convex programs with linear criteria functions. *Global Optimisation, 6*(3), 231–251.

Benson, H.P. (1997). Hybrid Approach for Solving Multi-objective linear programs:the bicriteria case. *Acta Mathemetica Vietnamica, 22,* 29–51.

Benson, B.L. (1998). *Evolution of Commercial Laws.* London: Macmillan Press.

Benson, H.P. (1998). Hybrid Approach for solving mltiobjective linear programs in outcome space. *Journal of Optimisation Theory and Applications, 98,* 17–35.

Benson, H.P., & Sun, E. (2000). Pivoting in Outcome Polyhedron. *Journal of Global Optimisation, 16,* 301–323.

Benson, H.P., & Sun, E. (2002). A weight set decomposition algorithm for finding all efficient extereme points in the outcome set of multiple objective linear program. *European Journal of Operation Research, 139*(1), 26–41.

Ehrgott, M., Lohne, A., & Shao, L. (2012). A dual variant of Benson's outer approximation algorithm for multiple objective linear programming. *Journal of Global Optimisation, 52*(4), 757–778.

Haimes, Y.Y., Lasdon, L.S., & Wismer, D.A. (1971). On a bricriteria formulation of the problems of the integrated system identification and system optimisation. *IEEE transaction on Systems and Man and Cybernetics, 1*(3), 296–297.

Horst, R., & Tuy, H. (1996). *Global Optimsation: Deterministic Approaches.* Berlin: Springer.

Ida, M. (2011). Efficient solution generation for multiobjective linear programming basedon extreme ray generation method. *European Journal of Operation Research, 160*(1), 242–251.

Klamroth, K., & Dachert, K. (2015). A linear bound on the number of scalarizations needed to solve discrete tricriteria optimization problems. *Journal of Global Optimisation, 61,* 643–676.

Krichen, S., Masri, H., & Guitouni, A. (2012). Adjacency based method for generating maximal efficient faces in multiobjective linear programming. *Applied Mathematical Modelling, 36*(12), 6301–6311.

Kuhn, H.W., & Tucker, A.W. (1951). Nonlinear Programming. *Second Barkeley Symposium, 2,* 481–492.

Laumanns, M., Thiele, L., & Zitzler, E. (2006). An efficient, adaptive parameter variation scheme for metaheuristics based on the epsilon-constraint method. *European Journal of Operation Research, 169*(3), 932–942.

Lin, Z.H., Zhu, D.L., & Sheng, Z.P. (2003). Finding a minimum efficient solution of a convex multiobjective program. *Journal of Optimisation Theory and Application, 118*(I), 587–600.

Luc, D. (2011). On Duality in Multiple Objective Linear Programming. *European Journal of Operation Research, 118*(3), 158–168.

Pureto, J., & Nickel, S. (2005). *Location Theory: A Unified Approach.* Springer.

Sayin, S. (1996). An algorithm based on facial decomposition for finding efficient set in multiple objective linear programming. *Operation Research Letters, 19*(2), 87–94.

Steuer, R.E. (1986). *Multiple Criteria Optimisation: Theory Computation and Application.* New York: John Wiliey.

Ulungu, E.L., & Teghem, J. (1995). The two phases method: An efficient process to solve bi-objective combinatorial optimisation problems. *Foundaions of Computing and Decision Sciences, 20*(2), 149–165.

Frontiers in Computer, Communication and Electrical Engineering – Acharyya (Ed.)
© 2016 Taylor & Francis Group, London, ISBN: 978-1-138-02877-7

Available Transfer Capacity evaluation through BBO and GWO algorithms

Kingsuk Majumdar
Dr. B.C. Roy Engineering College, Durgapur, India

Provas Kumar Roy
Jalpaiguri Government Engineering College, Jalpaiguri, India

Subrata Banerjee
National Institute of Technology, Durgapur, India

ABSTRACT: Information of the Available Transfer Capacity (ATC) is very important in the Deregulated Environment (DE). In this paper, ATC calculation is emphasized in view of the Optimal Power Flow (OPF) technique through some soft computational methods viz. Biogeography-Based Optimization (BBO) and Gray Wolf Optimization (GWO). The ATC is the key factor on the basis of the effect of transaction on transmission to allow or disallow bilateral transmission transection. The OPF has many objectives in the DE with open market conditions and these calculations aid to Independent System Operators (ISO) to handle the congestion threat on the transmission lines and assure system security and reliability. The two methods of soft computing, i.e., BBO and GWO are applied to find out the aforesaid criteria in the ATC calculation through the OPF, a corrective tool, which hints a new generation schedule to resist congestion and gives clues to the values of controlled parameters (e.g. active power generations, tap setting, reactive power injection) to avoid the violation of power system constrains (bus voltage limit and reactive power injection limit). The proposed methods are tested on WSCC 3 machine 9-bus and IEEE 30 bus test systems to evaluate ATC and their results, which are compared and it is observed that the GWO results are better than those obtain by BBO in both cases.

1 INTRODUCTION

The Available Transfer Capacity (ATC) is a vital area of the power system, which helps the system to operate securely and avoid the generation deficiency that may occur during a power transfer. The transmission congestion appears when the demand for power transmission exceeds the system's capability limit. The condition of congestion can be controlled by the proper planning and designing of the power network, i.e., by increasing the network's capacity to fulfill its need. Nowadays, the aforesaid criteria requires a serious concern by the transmission grid, system planner and needs an optimized engineering study about it, which is one of the most challenging aspects in a multi-buyer's competitive wholesale system [1]. It can be controlled by a Regional or Zonal congestion transmission charges [2]. As this simple concept is a dynamics and real time manner, hence, it is difficult to implement the views of the congestion management, economics of energy market and commitment of deregulation of power toward the society. Thus, ATC is defined as a measure of

transfer capability or available room in the physical transmission network, for transfers of power for further commercial activity, over and above already committed uses without violating the power system constrains, which is a key requirement for fostering the generation competition and customer choices and it should be available on a publicly accessible Open Access Same time Information System (OASIS). ATC information is very vital for independent system operators (ISO), planners and operators of the bulk power markets [3], and it also gives strategy to handle the congestion. The present computations [4–5] are usually oversimplified or time consuming. Without a fast computation algorithm, the center computer of the ISO would not calculate at a faster speed as it is required by the market. As the dc model is proposed in [4], it has many assumptions, which leads to erroneous results, hence, an ac model with a fast algorithm is required to fulfil the need regarding ATC, which involves the determination of Total Transfer Capability (TTC) and two margins – Transmission Reliability Margin (TRM) and Capacity Benefit Margin (CBM) [6–7]. TTC is the largest flow in the selected interface for which there are no thermal overloads, voltage limit violations, and voltage collapse and/or any other system security problems. ATC can also be defined as the TTC minus the base case flow. Hence, the determination of the TTC is a key factor for computing ATC. In this paper, transient stability or oscillations are not addressed, which can be approximated by flow limits. So far ATC is mainly calculated by the classical methods like dc linear sensitivity analysis continuation power flow (CPF), decomposition, repetitive power flow (RPF), and linear available transfer capability (LATC) methods [8–10]. However, these approaches possess some drawbacks. In most of these methods, the reactive power and voltage constraints are neglected and the transmission line thermal limits are the only effective restraints and they are unable to give an optimum value. Later on, OPF is adopted with classical methods to enhance the feasibility of this calculation. But the objective functions of the OPF problem are generally non-linear and non-convex in nature, which needs the special care to get the global optimal solutions. The traditional optimization approaches such as non-linear programming (NLP) [11], quadratic programming (QP) [12], linear programming (LP) [13], and interior point method (IPM) [13] are used successfully to solve the OPF problem. The particle swarm optimization (PSO) was first introduced by Kennedy and Eberhart [15], which is a flexible, robust, and population-based stochastic search/optimization algorithm with inherent parallelism and in recent years this method has earned the popularity over its competitors and is increasingly gaining acceptance for solving many power system problems like economic dispatch [17], and OPF [19]. So far optimization methods like BBO [16] or GWO [17] is not reported prominently in this field though these are very powerful tools to evaluate many power system problems. In this paper, BBO introduced by Simon *et al.* [16] is utilized for the solution of ATC problems. Moreover, gray wolf optimization (GWO) introduced by Mirjalili *et al.* [17] is also implemented for the same. The effectiveness and application of the proposed BBO and GWO algorithms are demonstrated by implementing these in WSCC 3 machine 9 bus and standard IEEE 30-bus systems and their performances are compared with each other's.

The rest of the article is composed as follows: At the very beginning of this paper, a brief overview of the ATC and its solution methods are presented. In the next section, mathematical formulation of the objective function in view of OPF is evaluated. In the third section, a general overview of BBO technique to evaluate proposed problem is discussed. Afterwards, the GWO method is applied to handle the proposed problem. The simulation results

and a comparative study of the above two methods are discussed in Section 5. In the tail end section, a conclusion is drawn to summarize the concept of this paper and future scope of the works for the same.

2 MATHEMATICAL MODELING

In this section, OPF tool is used to deal with TTC challenges. The objective function consists of the transfer of possible maximum power from a specific set of generators in a source area to load in a sink area through some particular tie-lines without the violation of load flow constrains and system operation limits. The sum of real power loads in the sink area at the maximum power transaction is defined as the TTC value. The objective function for the OPF-based TTC calculation is expressed mathematically as shown in (1) [20].

$$F_T = \Phi_{TL}F_{TL} + F_{TLA} + \Phi_{GP}F_{GP} + \Phi_V F_{Vmis} + \Phi_{GQ}F_{GQmis} \tag{1}$$

where, $\Phi_{TL}, \Phi_{GP}, \Phi_V, \Phi_{GQ}$ are different positive scalar values for the objective function.

The equality constrains of load flow can be defined as:

$$P_{G_i} - P_{D_i} - \sum_{j=1}^{NB} V_i Y_{ij} V_j \cos\left(\theta_{ij} + \delta_j - \delta_i\right) = 0 \tag{2}$$

$$Q_{G_i} - Q_{D_i} + \sum_{j=1}^{NB} V_i Y_{ij} V_j \sin\left(\theta_{ij} + \delta_j - \delta_i\right) = 0 \tag{3}$$

The generators constraints are as below:

$$P_{Gi}^{\min} \leq P_{G_i} \leq P_{Gi}^{\max}; i = 1, ..., NG \tag{4}$$

$$Q_{Gi}^{\min} \leq Q_{Gi} \leq Q_{Gi}^{\max}; i = 1, ..., NG \tag{5}$$

$$V_i^{\min} \leq V_i \leq V_i^{\max}; i = 1, ..., NB \tag{6}$$

where, $P_{Gi}^{\min}, Q_{Gi}^{\min}, V_i^{\min}, P_{Gi}^{\max}, Q_{Gi}^{\max}$ and $V_i^{\max}$ are minimum and maximum active, reactive generated power and voltage limits, respectively, and *NG, NB* denotes number of generators and total bus numbers, respectively, in the system.

Tap setting transformers are shown as below:

$$T_i^{\min} \leq T_i \leq T_i^{\max}; i = 1, ..., NT \tag{7}$$

where, $T_i^{\min}, T_i^{\max}$ are tap setting limit and NT represents total number of tap setting transformers.

Similarly, shunt reactor compensations can be presented as:

$$Q_{SHi}^{\min} \le Q_{SHi} \le Q_{SHi}^{\max}; i = 1, ..., NSH \qquad (8)$$

where, $Q_{SHi}^{\min}, Q_{SHi}^{\max}$ are shunt reactors limit and NSH represents number of shunt reactors.

The objective function can be divided into several parts as:

For a more practical and accurate model of the cost function of generators, multiple valve steam turbines are incorporated for flexible operational facilities. Total cost of generating units with valve point loading is given by [21]:

$$F_{GP} = \sum_{i=1}^{NG} \left(a_i P_{Gi}^2 + b_i P_{Gi} + c_i + \left| e_i \sin\left(f_i \left(P_i^{\min} - P_i \right) \right) \right| \right) \qquad (9)$$

where, a_i, b_i, c_i, e_i and f_i are cost co-efficient of the ith unit with valve point effect. To evaluate ATC or TTC for a system [6], power flow should be enhanced through some particular lines and can be represents as:

$$F_{TLA} = \left| \left(\frac{1}{S_{TLi}} \right) \right| \qquad (10)$$

where, S_{TLi} is power flow through the ith line.

To restrict the overloading of the transmission line cost function which can be formulated as below:

$$F_{TL} = \sum_{i=1}^{NTL} \left(S_{TL\max_i} - S_{TL_i} \right)^2 \qquad (11)$$

where, $S_{TL\max i}$ is the maximum thermal loading, i.e., maximum capacity of the ith transmission line.

To restrict the voltage missmatch and generate the reactive power violations of the proposed model in the following equations, which can be defined as:

$$F_{Vmis} = \sum_{i=1}^{NL} \left(V_i^{\lim} - V_i \right)^2 \qquad (12)$$

$$F_{GQmis} = \sum_{i=1}^{NG} \left(Q_{Gi}^{\lim} - Q_{Gi} \right)^2 \qquad (13)$$

The aforesaid objective function of the propose problem (1) can be minimized by the proposed algorithms, i.e., BBO and GWO without violating the power system constrains.

3 BIOGEOGRAPHY-BASED OPTIMIZATION

Biogeography deals with how new species arise, how species become extinct and how species migrate from one island to another. A habitat is any island (area), which is geographically isolated from other islands. Areas that are well suited as residences for biological species are supposed to have a high habitat suitability index (HSI). The variables that characterize habitability are called as suitability index variables (SIYs). SIYs can be treated as the independent variables of the habitat, and HSI can be calculated using these variables. The basic BBO algorithms are applied to any optimization problem consisting of a migration operator followed by a mutation.

3.1 Migration operation

In the migration stage of BBO, the next iteration solutions are generated from the present iteration by immigrating solution features to other islands and receiving solution features emigration from other islands. The immigration and emigration rate of kth island can be formulated as described in (14) and (15)

$$\lambda_i = I_m \left(1 - \frac{S_i}{S_{\max}} \right) \qquad (14)$$

$$\mu_i = \frac{E_m S_i}{S_{\max}} \qquad (15)$$

where λ_i, μ_i are immigration and emigration rates, respectively, for the kth individual; $I_m, E_{\max}$ are the maximum possible immigration rate and emigration rate, respectively; S_i is the number of species of the ith individual and $S_{\max}$ is the maximum number of species.

3.2 Mutation operation

The mutation process of BBO is used to enhance the density of the population to achieve good solutions. In the mutation process, SIV of a habitat is randomly modified based on the mutation rate as follows [22]:

$$m_i = m_{\max} \left(\frac{1 - P_i}{P_{\max}} \right) \qquad (16)$$

where m_i is the mutation rate for the ith habitat; $m_{\max}$ is the maximum mutation rate; $P_{\max}$ is the maximum species count probability and P_i is the species count probability for the ith habitat and is described as [22]:

$$P_i = \begin{cases} -(\lambda_i + \mu_i)P_i + \mu_{(i+1)}P_{(i+1)} & S = 0 \\ -(\lambda_i + \mu_i)P_i + \lambda_{(i-1)}P_{(i-1)} + \mu_{(i+1)}P_{(i+1)} & 1 \leq S \leq S_{max} - 1 \\ -(\lambda_i + \mu_i)P_i + \lambda_{(i-1)}P_{(i-1)} & S = S_{max} \end{cases}$$

$$(17)$$

4 GREY WOLF OPTIMIZATION

Grey Wolf Optimization (GWO) is another new meta-heuristic algorithm inspired by grey wolves and it mimics the leadership hierarchy and hunting mechanism of them. Three main steps of hunting such as searching for prey, encircling prey, and attacking prey are applied for optimizing a specified problem using GWO. The group size is between 5 and12 on an average and there are four types of wolves that are considered as alpha, beta, delta, and omega. The alpha wolves have the highest priority and are mostly responsible for making decisions about hunting, sleeping place, time to wake, and so on. The next level in the hierarchy of grey wolves is beta. The betas are subordinate wolves which help the alpha in decision-making. After that there is delta wolf category and the lowest rank is the omega that plays the role of a scapegoat [17].

4.1 *Encircling prey*

As mentioned above, grey wolves encircle the prey during the hunt and it can be mathematically modelled encircling the behaviour of the following equations which are proposed

$$\vec{E} = \left| \vec{D}.\vec{X}_P(t) - \vec{X}(t) \right| \tag{17}$$

$$\vec{X}(t+1) = \vec{X}_p(t) - \vec{B}.\vec{E} \tag{18}$$

where $\vec{X}_p, \vec{X}$ are the positions of the prey and grey wolf, respectively. Here $\vec{B}, \vec{D}$ are as follows

$$\vec{B} = 2\vec{b}.\vec{r}_1 - \vec{b} \tag{19}$$

$$\vec{D} = 2\vec{r}_2 \tag{20}$$

where $\vec{b}$ linearly decreases from 2 to 0 over the course of iterations and $\vec{r}_1, \vec{r}_2$ are the random numbers between [0 1].

4.2 *Hunting*

Grey wolves have the inherent ability to recognize the location of the prey and encircles them and this is usually guided by the alpha. In order to mathematically represent the hunting mechanism, the following equations can be presented:

$$\vec{E}_\alpha = \left| \vec{D}_1.\vec{X}_\alpha - \vec{X} \right|, \vec{E}_\beta = \left| \vec{D}_2.\vec{X}_\beta - \vec{X} \right|, \vec{E}_\delta = \left| \vec{D}_3.\vec{X}_\alpha - \vec{X} \right|$$

$$(21)$$

$$\vec{X}_1 = \vec{X}_\alpha - B_1.(\vec{E}_\alpha), \vec{X}_2 = \vec{X}_\beta - B_2.(\vec{E}_\beta), \vec{X}_3 = \vec{X}_\delta - B_3.(\vec{E}_\delta)$$

$$(22)$$

$$\vec{X}(t+1) = \frac{\vec{X}_1(t) + \vec{X}_2(t) + \vec{X}_1(t)}{3} \tag{23}$$

4.3 *Attacking prey (exploitation)*

The hunt process has been ended by attacking the prey when it stops moving. In order to mathematically model the approaching prey, we decrease the value of $\vec{b}$ from 2 to 0 over the course of iterations. Consequently, random values of $\vec{B}$ are in [−1,1], the next position of a search agent can be in any position between its current position and the position of the prey. Hence, |B|<1 forces the wolves to attack towards the prey.

4.4 *Search for prey (exploration)*

Grey wolves mostly search according to the position of the alpha, beta, and delta. The value of |B|>1 forces the grey wolves to diverge from the prey to hopefully find a fitter prey. The vector $\vec{D}$ contains random values in [0, 2]. This component provides random weights for the prey in order to stochastically emphasize (*D>1*) or deemphasize (*D<1*) the effect of prey in defining the distance in (17).

The pseudo code for GWO is shown in Figure 1.

5 RESULT AND DISCUSSION

The applicability and validity of the BBO and GWO techniques is tested on WSCC 9 bus and IEEE 30-bus system to solve the ATC problems. For WSCC, 9 bus system data are taken from MATPOWER 4.1. The load buses are taken at 5, 6, and 8 and ATC is calculated for line number is 7, i.e., in between bus 2 and 8. The IEEE 30-bus system contains 6 generating units connected at the bus number 1,2,13,22,23, and 27, twenty four loads, four regulating transformers are connected between the line numbers 6–9, 6–10, 4–12 and 27–28; forty one transmission lines and nine reactive VAr compensators at the bus numbers 10, 12, 15, 17, 20, 21, 23, 24, and 29. The full system

Initialize the grey wolf population X_i (i = 1, 2, ..., n)
Initialize b, B, and D
Calculate the fitness of each search agent
Xα=the best search agent
Xβ=the second best search agent
Xδ=the third best search agent
while *(t < Max number of iterations)*
for *each search agent*
Update the position of the current search agent by (24)
end for
Update b, B, and D
Calculate the fitness of all search agents
Update Xα, Xβ, and Xδ
t=t+1
end while

Figure 1. The pseudo code of Grey Wolf Optimization.

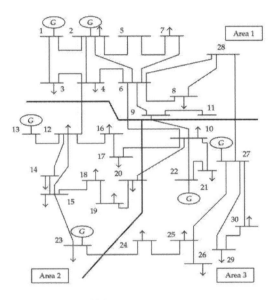

Figure 2. IEEE 30 bus test system.

data are taken form [23] and it can be divided into 3 areas say area-1, area-2 and area-3 shown in Figure 2. In this proposed test system, TTC is calculated from area-1 (seller) to area-2 (buyer) by the tie-line 4-12 i.e. line no 15 and loads of the area-2 at bus numbers 12,16,17,14,18, and 15, respectively, are increased by a minimum of ten percent to a maximum fifty percent randomly without changing its load power factor as that of base load case. The program is written in MATLAB-7 software (32 bits) and executed on a 1.7 GHz core i3 processor with 4-GB RAM. In this simulation study, $\Phi_{TL}, \Phi_{GP}, \Phi_V, \Phi_{GQ}$ are taken as 25, 0.00001, 10, and 10, respectively. Owing to the unpredictability of optimization algorithms, several trials should be

made for selecting the appropriate values of input parameters. After several trials, following parameters are found to be the best for successful implementation of BBO and GWO algorithms:

For BBO: habitat modification probability = 1, mutation probability = 0.005, maximum immigration rate, $I_{max} = 1$, maximum emigration rate, $E_{max} = 1$, step size used for numerical integration of probabilities = 1, elitism parameter = 15% of total habitat e.g. total population.

For GWO: No such initial input parameters are required which is a great advantage of this algorithm.

The population size and the maximum number of the generations are 50 and 100 both cases. It is observed from Table 1 that in both cases GWO is superior to that of the BBO. In case of IEEE 30 bus test system there are 6.45% and 10.39% line overflow at line numbers 10 and 26, respectively, when they are tested by BBO. But GWO there is only 6.41% at the line number 10. The optimized control parameters are given in Tables 2 and 3.

Table 1. Objective function values for BBO and GWO.

Parameters	BBO IEEE30	GWO IEEE30	BBO WSCC9	GWO WSCC9
F_{TL} (p.u.)	0.0539	0.0205	0	0
F_{LTA} (p.u.)$^{-1}$	2.3949	3.1963	0.6244	0.5654
$S_{TL(line\,flow)}$ (p.u.)	0.4175	0.3129	1.6014	1.7685
F_{GP} mu/hr	1125.990	841.5889	6814.4993	8060.4653
F_{Vmis} (p.u.)	0	0	0	0
F_{QGmis} (p.u.)	0	0	0	0
F_T	3.7537	3.7181	0.6926	0.6460

Table 2. Best control variables result for BBO and GWO for IEEE 30 bus test system.

Parameters	BBO	GWO	Parameters	BBO	GWO
P_{G1} (p.u.)	0.5578	1.1454	Q_{SH1} (p.u.)	0.0449	0.0101
P_{G2} (p.u.)	1.2609	0.3654	Q_{SH2} (p.u.)	0.0014	0.0231
P_{G3}(p.u.)	0.3147	0.1530	Q_{SH3} (p.u.)	0.0462	0.0385
P_{G4} (p.u.)	0.3314	0.2366	Q_{SH4} (p.u.)	0.0442	0.0104
P_{G5} (p.u.)	0.1575	0.1662	Q_{SH5} (p.u.)	0.0096	0.0466
P_{G6} (p.u.)	0.1367	0.1842	Q_{SH6} (p.u.)	0.0165	0.0108
V_1 (p.u.)	1.0204	1.1000	Q_{SH7} (p.u.)	0.0215	0.0167
V_2 (p.u.)	1.0195	1.0869	Q_{SH8} (p.u.)	0.0263	0.0023
V_3 (p.u.)	1.0371	1.0841	Q_{SH9} (p.u.)	0.0331	0.0325
V_4 (p.u.)	0.9956	1.0349	L_{PD1} (p.u.)	0.3880	0.1482
V_5 (p.u.)	1.0167	1.0341	L_{PD2} (p.u.)	0.1110	0.1168
V_6 (p.u.)	1.0210	1.0434	L_{PD3} (p.u.)	0.3277	0.1437
T_1	0.9285	1.0272	L_{PD4} (p.u.)	0.1535	0.1021
T_2	1.0925	0.9128	L_{PD5} (p.u.)	0.1023	0.0863
T_3	0.9231	1.0398	L_{PD6} (p.u.)	0.1268	0.1256
T_4	1.0313	1.0368			

115

Table 3. Best control variables result for BBO and GWO for WSCC 9 bus test system.

Parameters	BBO	GWO
P_{G1} (p.u.)	1.0364	1.1868
P_{G2} (p.u.)	1.6013	1.7559
P_{G3} (p.u.)	1.1234	1.2375
V_1 (p.u.)	1.0832	1.0446
V_2 (p.u.)	1.0986	1.0547
V_3 (p.u.)	1.0976	1.0273
L_{PD1} (p.u.)	1.2395	1.0171
L_{PD2} (p.u.)	0.1084	0.2524
L_{PD3} (p.u.)	0.1180	0.6265

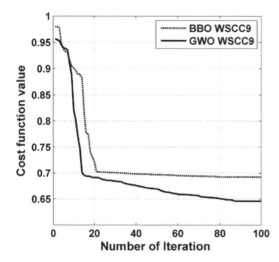

Figure 3. Convergence curves for WSCC 9 Bus.

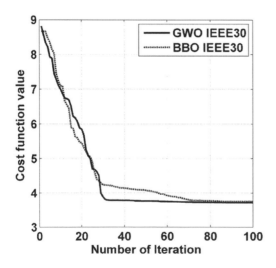

Figure 4. Convergence curves for IEEE 30 Bus.

The cost function convergence curves are shown in Figures 3 and 4 for WSCC 9 bus and IEEE 30 bus test system, respectively.

6 CONCLUSION

In this paper, theeffectiveness of the BBO and GWO optimizations are studied to achieve maximum possible ATC for IEEE 30 bus and WSCC 9 bus test systems. ATC programs along with OPF concept is implemented as a tool for solving the transmission management problem. By comparing both the results, it can be concluded that in case of GWO optimization, the power transfer from area-1 to area-2, i.e., from 4 to 12 tie-line is enhanced. But active power generation cost in later cases are higher than that of the former can be ignored. As the OPF is incorporated with these techniques, hence, these data can be shown online so that the buyers or sellers will be aware that such a transaction is possible. In case of IEEE 30 bus test system there are some line overflows, which can be mitigated by the Flexible AC Transmission System (FACTS) and can be extended in the future research work.

REFERENCES

[1] F. P. Sener and E. R. Greene, "Top three technical problems -a survey report," IEEE Transactions on Power Systems, vol.12, no.1, pp. 230–244, 1997.
[2] L. Philipson and H. Lee Willis, "Understanding electric utilities and deregulation," Marcel Dekker Inc., New York, 1998.
[3] Ian et.al, "Electric Power Transfer Capability: Concepts, Applications, Sensitivity and Uncertainty," www.pserc.cornell.edu/tcc/tutorial/TCC_Tutorial.pdf
[4] R. D. Christie, B. F. Wollenberg, I. Wangensteen, "Transmission Management in Deregulated Environment," Proceedings of the IEEE, vol.88, no.2, pp. 449–451, 2000.
[5] M. Shaaban, Y. Ni, F. F. Wu, "Transfer Capability computations in Deregulated Power Systems," Proceedings of the 33rd Hawaii International Conference on System Sciences-2000.
[6] G.C.Ejebe, J.Tong, J.G.Waight, J.G.Frame, X.Wang, W.F.Tinney, "Available Transfer Capability Calculations," IEEE Transactions on Power Systems, vol.13, no.4, pp. 1521–1527, 1998.
[7] H. Farahmand, M. Rashidinejad, A. Mousavi, and A. A. Gharaveisi, et all, "Hybrid Mutation Particle Swarm Optimisation method for Available Transfer Capability enhancement," International Journal of Electrical Power & Energy Systems, vol. 42, pp. 240–249, 2012.
[8] R.D. Christie, B.F, Wollenberg, and I. Wangensteen, "Transmission management in the deregulated

environment', IEEE Proceedings, vol. 88, no. 2, pp. 170–195, 2000.

[9] G.C. Ejebe, J.G. Waight, M. Santos-Nieto and W.F. Tinney, "Fast calculation of linear available transfer capability", Proceedings the Power Industry Computer Applications Conference, pp. 255–260, 1999.

[10] M. H. Gravener, C. Nwankpa, "Available Transfer Capability and first order sensitivity," IEEE Transaction on Power Systems, vol.14, no.2, pp. 512–518, 1999.

[11] H. Wei, H. Sasaki, J. Kubokawa, and R. Yokoyama,,"An interior point nonlinear programming for optimal power flow problems with a novel data structure,"EEE Power & Energy Society, vol. 13, pp. 870–877, 1998.

[12] R.S. Wibowo, Nursidi, H. Satriyadi, D.F.P. Uman, A.Soeprijanto, O.Penangsang, "Dynamic DC optimal power flow using quadratic programming," Information Technology and Electrical Engineering (ICITEE), International Conference, pp. 360–364, 2013.

[13] S. K. Mukherjee, A.Recio, C. Douligeris, "Optimal power flow by linear programming based optimization," Southeastcon '92, Proceedings., IEEE, vol. 2, pp. 527–529, 1992.

[14] W. Hua, H. Sasaki, J.Kubokawa,"Interior point method for hydro-thermal optimal power flow," Energy Management and Power Delivery, 1995. Proceedings of EMPD '95., 1995 International Conference, vol. 2, pp.607–612, 1995.

[15] J. Kennedy, R. Eberhart, Particle swarm optimization, in: Proceedings of the IEEE Conference on Neural Networks (ICNN'95), vol. 6, Perth, Australia, (1995), pp. 1942–1948.

[16] Dan Simon, "Biogeography-Based Optimization," IEEE transactions on Evolutionary computation, vol. 12, No. 6, pp. 702–713, 2008.

[17] S. Mirjalili, S. M. Mirjalili, A. Lewis, "Grey Wolf Optimizer", Advances in Engineering Software, vol. 69, pp. 46–61,2014

[18] D. Gong, Y. Zhang, C. Qi, "Environmental/economic power dispatch using a hybrid multi-objective optimization algorithm," Electrical Power and Energy Systems, vol. 32, pp.607–614, 2010.

[19] A. Srinivasan, P. Venkatesh, B. Dineshkumar, "Optimal generation share based dynamic available transfer capability improvement in deregulated electricity market," IEEE Transactions on Power Systems, vol.54, pp. 226–234, Nov2014.

Optimal location of capacitor in radial distribution network using Chemical Reaction Optimization algorithm

S. Sultana & S. Roy
Dr. B.C. Roy Engineering College, Durgapur, India

P.K. Roy
Jalpaiguri Government Engineering College, Jalpaiguri, India

ABSTRACT: Presently, the consolidation of the capacitor banks on the radial distribution network has a direct impact on the development of technology and the energy disasters that the world is encountering. Distribution system provides the final link between high voltage transmission system and the low voltage end users, so it's very essential to keep the system healthy with minimum losses. Studies have also shown that the non-optimal location of the capacitor banks increase the power loss, so that the optimal placement of the capacitor bank is very necessary in the radial distribution system. In this paper, a new and efficient Chemical Reaction Optimization (CRO) method is proposed to find the optimal and simultaneous placement for the capacitor in the radial distribution systems with an objective of the reduction of power losses. To verify the effectiveness and efficiency of the proposed method, it is tested on the 34-bus and 69-bus radial distribution systems. The simulation results of the proposed methods are compared with those of other population based optimizations like the Plant Growth Simulation Algorithm (PGSA), Teaching Learning Based Optimization (TLBO), Direct Search Algorithm (DSA), Particle Swarm Optimization (PSO), and Genetic Algorithm (GA) in order to demonstrate that the proposed approach is more useful than other techniques.

1 INTRODUCTION

A capacitor can be defined as a two terminal passive electrical component, which is used to store electric charge. Electricity distribution is a final stage in the delivery of electricity to end users. A distribution network carries electricity from the transmission system and delivers it to the consumers. It is the most visible part of the supply chain. Presently, due to their simplicity most of the distribution system is radial in nature, because of high R/X ratio, the I^2R loss in the radial distribution system is also high compared to the high voltage transmission system. The loss minimization in the distribution systems is assumed to be of a greater significance. Since, the trend towards distribution automation is to improve their reliability, efficiency, and service quality, the most efficient operating scenario for economic viability variations is required. Among many of their merits, line loss and Total Harmonic Distortion (THD) reduction, voltage profile improvement, and reactive power compensation can be the salient specifications of the capacitors. Therefore, it is necessary to find the optimum location and size of the capacitor to minimize the power loss in the radial distribution system.

Presently, capacitor placement problem is a very attractive and well researched topic. So, many researchers focused on the analytical approach to determine the optimum location and size of the capacitor in the radial distribution system (Salama *et al.* 2000, Baran *et al.* 1989). An analytical algorithm and fuzzy real coded GA is used to optimize the capacitor for enhancing voltage stability in the radial distribution system (Wafa *et al.* 2014). Besides the analytical approach, many researchers developed a classical based optimizing technique for the purpose of the placement of the capacitor in the radial distribution system. A Mixed Integer Nonlinear Programming approach (MINLP) was developed for the purpose of optimal capacitor placement (Nojavan *et al.* 2014). Jabr proposed a conic and Mixed Integer Linear Programming approach (MILP) to optimize the capacitor for minimizing the costs associated with the capacitor banks, peak power, and energy losses (Jabr *et al.* 2008). A Clustering Based Optimization (CBO) was implemented to minimize the sum of costs for power/energy losses and capacitor costs in the radial distribution system (Todorovski *et al.* 2014). A plant growth simulation algorithm (PGSA) was presented to solve the optimal placement of the capacitor to minimize the power loss and improve the voltage profile (Rao *et al.* 2011). An immune multi-objective algorithm was developed for the purpose of the optimal placement of the capacitor in the radial distribution system (Jiang *et al.* 2008).

Besides the classical based optimization approach, several population based optimizations are developed by many authors. Particle Swarm Optimization (PSO) was proposed for finding the optimal location and size of the capacitor (S.P. Sing *et al.* 2012). Genetic Algorithm (GA) based on the mechanism of natural selection (Pahwa *et al.* 1994) was used to determine the optimal placement of capacitors. Bacterial Foraging Optimization (BFO) based fuzzy logic was developed for the optimum location of the capacitor in order to minimize the cost of the peak power, reducing energy loss, and improving voltage profile (Vahidi *et al.* 2011).

The literature survey shows that the aforesaid techniques successfully solved the capacitor placement problem. However, the slow convergence toward optimal solution is the main drawback for most of these techniques. Furthermore, these techniques often converge to the local optimal solution instead of the global optimal solution.

In this article, a recently developed heuristic algorithm called Chemical Reaction Optimization (CRO) algorithm introduced by Lam and Li (Li *et al.* 2010) was utilized for the solution of the HTS problems. The effectiveness of the proposed CRO algorithm has been demonstrated by implementing it in two standard radial distribution systems and its performance has been compared with other algorithms available in the literature.

The rest of this paper has been organized as follows: Problem formulation is given in Section 2. The key points of the proposed CRO technique are described in Section 3. In Section 4, the proposed techniques applied to the capacitor allocation problem in the radial distribution system have been illustrated. Two cases based on medium and large scale power systems are studied and the simulation results are illustrated in Section 5. Section 6 summarizes the conclusion.

2 PROBLEM FORMULATION

2.1 *Power loss minimization*

Due to large reactance to resistance ratio of the radial distribution system, the power loss is significantly high. So, to run the radial system in the most economical way, the goal is to reduce the system power loss. The real power loss in a radial distribution is as:

$$P_{RLOSS} = \sum_{i=2}^{n_N} \begin{pmatrix} P_{gzi} - P_{dzi} - v_{yi} v_{zi} Y_{yzi} * \\ \cos(\partial_{yi} - \partial_{zi} + \theta_{yzi}) \end{pmatrix} \quad (1)$$

Here, P_{RLoss} is the real power loss; P_{gzi} is the real power generation at bus zi; P_{dzi} is active power demand at bus zi; v_{zi} is the voltage magnitude at bus zi. Y_{yzi} is the magnitude of the bus admittance matrix between yi and zi; ∂_{yi} is voltage angle of bus yi; ∂_{zi} is voltage angle of bus zi; θ_{yzi} is the phase angle between yi and qi of the bus admittance matrix.

2.2 *Constraints*

2.2.1 *Load balance constraint*
For each bus, the following balance constraint equation should be fulfilled:

$$P_{gzi} - P_{dzi} - v_{zi} \sum_{j=1}^{M} v_{zj} Y_{zj} \cos(\delta_{zi} - \delta_{zj} + \theta_{zj}) = 0 \quad (2)$$

$$Q_{gzi} - Q_{dzi} - v_{zi} \sum_{j=1}^{M} v_{zj} Y_{zj} \cos\left(\delta_{zi} - \delta_{zj} + \theta_{zj}\right) = 0 \quad (3)$$

where $zi = 1, 2 .... n_N$

2.2.2 *Voltage limit*
The voltage must be kept within the standard limits of each bus.

$$v_{zi}^{\min} < v_{zi} < v_{zi}^{\max} \quad (4)$$

2.2.3 *Capacitor power constraint*
A DG power capacity depends on the energy resources of any given location; therefore, it is essential to keep the DG power capacity within its minimum and maximum levels.

$$P_{gzi}^{\min} \le P_{gzi} \le P_{gzi}^{\max}. \quad (5)$$

3 CHEMICAL REACTION OPTIMIZATION (CRO)

Chemical Reaction Optimization (CRO) is a recently developed efficient optimization technique proposed by Lam and Li (Li *et al.* 2010). CRO, a population-based meta-heuristic technique, it is successfully used for solving various optimization problems in the recent past. CRO mimics the interactions of the molecules in the chemical reactions. A chemical system undergoes a chemical reaction when it is unstable, that is when it possesses excessive energy. It manipulates itself to release the excessive energy in order to stabilize itself and this manipulation is called as chemical reactions. Generally, in a chemical reaction molecules interact with each other aiming to reach the minimum state of free energy. Through a sequence of intermediate reactions, the resultant molecules (i.e. the products in a chemical reaction) tend to stay at the most stable state with the lowest free energy. A molecule possesses two kinds of energies such as Potential Energy (PE) and Kinetic Energy (KE). There are four types of reactions taken into considerations, which are elaborated below:

3.1 On-wall ineffective collision

This type of collision is considered as a uni-molecular collision where a single molecule is allowed to collide with the wall of a container. In this case, the molecular structure ω and PE of the molecule that hits the wall undergoes subtle changes. Therefore, the present molecular structure (β) is changed and it becomes a new molecule (β'). When a molecule hits a wall, a portion of its KE will be lost; the lost energy is then stored in the form of a central energy buffer (the initial energy buffer size of the container). Thus, the KE alters its magnitude and KE for the new molecule (β') which is as follows:

$$KE_{\beta'} = (PE_\beta + KE_\beta - PE_{\beta'}) \times R \qquad (6)$$

Here, '(R)' is the random number that lies in between $[KE_{lossrate}, 1]$, where $KE_{lossrate}$ is a parameter of CRO.

3.2 Decomposition

Similar to that of the on-wall collision, decomposition is also a uni-molecular reaction, when the molecule β hits a wall of the container, it breaks up into two molecules β_1' and β_2 with different structures.

i.e. $\beta \rightarrow \beta_1' + \beta_2$

In decomposition, two steps are considered: (i) the molecule has sufficient energy to complete the decomposition; (ii) the molecule should get energy from the energy center.

Step 1: The KE of the resultant molecules is shown as follows:

$$KE_{\beta_1'} = (PE_\beta + KE_\beta - PE_{\beta_1'} - PE_{\beta_2}) \times R \qquad (7)$$

$$KE_{\beta_2} = (PE_\beta + KE_\beta - PE_{\beta_1'} - PE_{\beta_2}) \times (1 - R) \qquad (8)$$

Here, R is a random number generated from the interval [0, 1] considering the constraint as follows:

$$PE_\beta + KE_\beta \geq PE_{\beta_1'} + PE_{\beta_2} \qquad (9)$$

Step 2: Here, the KE of the resultant molecules is shown as follows:

$$KE_{\beta_1'} = (PE_\beta + KE_\beta - PE_{\beta_1'} - PE_{\beta_2} + buffer) \times m1 \times m2 \qquad (10)$$

$$KE_{\beta_2} = (PE_\beta + KE_\beta - PE_{\beta_1'} - PE_{\beta_2} + buffer) \times m3 \times m4 \qquad (11)$$

Here, $m1$, $m2$, $m3$, and $m4$ are random numbers uniformly generated from the interval [0,1], and the constraint is as follows:

$$PE_\beta + KE_\beta + buffer \geq PE_{\beta_1'} + PE_{\beta_2} \qquad (12)$$

3.3 Inter-molecular ineffective collision

This reaction occurs when two molecules collide with each other and bounce back. After collision, the participating molecules change into two new molecules if the following energy criterion has been satisfied:

$$PE_{\beta_1} + KE_{\beta_1} + PE_{\beta_2} + KE_{\beta_2} \geq PE_{\beta_1'} + PE_{\beta_2} \qquad (13)$$

Then KE of the two new molecules are as follows:

$$KE_{\beta_1'} = (PE_{\beta_1} + KE_{\beta_1} + PE_{\beta_2} + KE_{\beta_2} - PE_{\beta_1'} - PE_{\beta_2}) \times s \qquad (14)$$

$$KE_{\beta_2} = (PE_{\beta_1} + KE_{\beta_1} + PE_{\beta_2} + KE_{\beta_2} - PE_{\beta_1'} - PE_{\beta_2}) \times (1 - s) \qquad (15)$$

Here, s is an arbitrary number generated from the interval [0, 1].

3.4 Synthesis

In this type of reactions, two molecules collide with each other and combine to form one new single molecule. Unlike the on-wall collision, synthesis reactants undergo a vigorous change in their molecular structures. Suppose two molecules α_1 and α_2 collide to form a single molecule with structure α'. The condition is as follows:

$$KE_{\beta'} = PE_{\beta_1} + KE_{\beta_1} + PE_{\beta_2} + KE_{\beta_2} - PE_{\beta'} \qquad (16)$$

4 CRO ALGORITHM APPLIED CAPACITOR ALLOCATION PROBLEM IN RADIAL DISTRIBUTION SYSTEM

The following steps must be taken to apply the CRO:

Step 1: The preliminary structure of each molecule should be randomly selected while satisfying different inequality constraints of the control variables. In the proposed area, capacitor size and locations are considered as the control variables.

Step 2: Power losses can be found by running the load flow problem. In this paper, a direct load flow algorithm based on the BIBC (Bus-Injection to Branch-Current) matrix and the BCBV (Branch-Current to Bus-Voltage) matrix (Teng *et al.* 2003) is used.

Step 3: Calculate the potential energy (fitness value 0 for each molecule) using (1)

Step 4: Update the capacitor size and its position using on-wall ineffective collision, decomposition, inter-molecular ineffective collision and synthesis process.

Step 5: Check whether the independent variables (rating of capacitor) violate the operating limits or not. The non-feasible solutions are replaced by randomly generated feasible solutions.

Step 6: Go to Step 3 until the current iteration number reaches the pre-specified maximum iteration number.

5 TEST SYSTEMS AND RESULTS

In this segment, two different test cases are discussed to show the effectiveness and efficiency of the proposed CRO method. For establishment, the suggested algorithm is implemented on 12.66 KV 34-bus and 69-bus radial distribution systems to calculate the actual position as well as the rating of various capacitors to reduce the total power losses. The proposed method for optimal placement of the capacitor is implemented on a personal laptop having core i-5, 2.5 GHz processor and 4 GB of RAM with the help of MATLAB software.

5.1 Test case 1 (For 34 bus radial distribution system)

The first test case of this proposed method is a 34-bus radial distribution system. This system has 34 buses and 33 branches, which are shown in Figure 1. The rated real and reactive power loads of this type of systems are 3.715 MW and 2.3 MVAR, respectively. The line and load data are taken from the literature (Sahoo *et al.* 2006). The simulation results along with the ratings of the capacitor obtained by the suggested CRO and the other population based optimization methods like PGSA, which is taken from the literature is shown in Table 1. From Table 1, it is observed

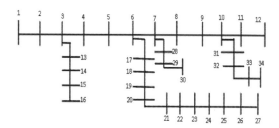

Figure 1. Single line diagram of the 34-bus radial distribution system.

Table 1. Optimal location and settings of the multiple capacitor and corresponding loss using CRO for 34-bus system.

Algorithm	CRO			PGSA (Rao *et al.* 2011)		
Optimal capacitor placement	10	18	24	19	20	22
Optimal size of capacitor (MVAR)	0.6914	0.9410	0.8310	1.200	0.200	0.639
Power loss(MW)	0.16023			0.16107		
CT time (Sec)	1.67			NA		
Chemical reaction optimization (CRO)						
Active power loss (MW)	0.16024					
Reactive power loss (MVAR)	0.04469475					
Voltage profile	0.97419					
Voltage deviation (VD)	0.034012					
Voltage stability index (VSI)	0.81555					
Minimum voltage	0.950304					

Table 2. Optimal location and settings of the multiple capacitor and corresponding loss using CRO for 69-bus system.

Algorithm	Optimal location of capacitor	Optimal size of capacitor (MVAR)	Power Loss (MW)	CT time (sec)
CRO	11	0.3919	0.14509	1.70
	18	0.2515		
	61	1.2334		
TLBO (Sultana *et al.* 2014)	12	0.600	0.14635	15.76
	61	1.050		
	64	0.150		
DSA (Raju *et al.* 2012)	15	0.450	0.1470	NA
	60	0.450		
	61	0.900		
PSO (Prakash *et al.* 2007)	46	0.241	0.15248	NA
	47	0.365		
	50	1.015		
GA (Sydulu *et al.* 2007)	59	0.100	0.15662	NA
	61	0.700		
	64	0.800		
Chemical reaction optimization (CRO)				
Active power loss (MW)	0.14509			
Reactive power loss (MVAR)	0.0676			
Voltage profile	1.3784			
Voltage deviation (VD)	0.0559			
Voltage stability index (VSI)	0.7527			
Minimum voltage	0.9314			

that the proposed CRO method incurs a power loss of 0.16023 MW, which is significantly smaller as compared to the loss incurred by the PGSA that is 0.16107 MW. So it may be concluded that the power loss incurred by the proposed CRO method is much better than that of the PGSA. The convergence characteristic of the loss for multiple capacitors using CRO is shown in Figure 3.

5.2 Test case 2 (For 69 bus radial distribution system)

The second test case for implementing the proposed method is a 69-bus radial distribution system. This system has 69 buses and 68 branches which are shown in Figure 2. The rated real power load is 3.715 MW while the rated reactive power load is 2.3 MVAR. The line and load data are taken from (Sahoo *et al.* 2006). The result of the proposed CRO method is compared with the other population based optimization techniques like TLBO, DSA, PSO, and GA, which are listed in Table 2. From Table 2, it is shown that the power loss incurred by the suggested method is 0.14509 MW, which is significantly smaller than the power loss incurred by TLBO (0.14635 MW), DSA (0.1470 MW), PSO (0.15248 MW), and GA (0.15662 MW). So, it is clear that the effectiveness of the suggested CRO

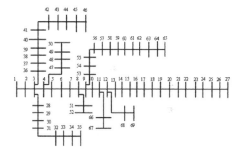

Figure 2. Single line diagram of the 69-bus radial distribution system.

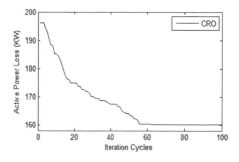

Figure 3. Power loss convergence graphs using CRO algorithm of the 34-bus system for multiple capacitor.

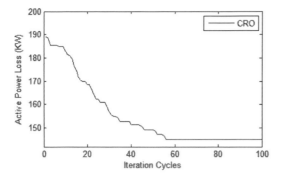

Figure 4. Power loss convergence graphs using CRO algorithm of the 69-bus system for multiple capacitor.

method is much better than TLBO, DSA, PSO, and GA. The convergence characteristic of the loss for multiple capacitors using CRO is shown in Figure 4.

6 CONCLUSION

In this paper, a new and efficient CRO technique is suggested for the purpose of optimal sizing and sitting of the capacitor in a radial distribution system. The proposed method is tested and validated on the 34-bus and 69-bus radial distribution systems. From the simulation results, it is observed that the proposed method successfully reduces the line losses. The proposed algorithm is compared with the other population based optimization techniques like PGSA, TLBO, DSA, GA, and PSO. The results show that the effectiveness of the proposed techniques on the minimization of the system power loss is better than the other techniques. So, from this study, it can be concluded that the proposed technique is very effective in finding the optimal sizing and sitting of the capacitor.

REFERENCES

Abul'Wafa, A.R. (2014). Optimal capacitor placement for enhancing voltage stability in distribution systems using analytical algorithm and Fuzzy-Real Coded GA. Elect Power Syst Res, 55, 246–252.

Baran, M.E., & Wu, F.F. (1989). Optimal Sizing of Capacitors Placed on a Radial Distribution System. IEEE Trans. Power Delivery, 1, 1105–1117.

Huang, T.L., Hsiao, Y.T., Chang, C.H., & Jiang, J.A. (2008). Optimal placement of capacitors in distribution systems using an immune multi-objective algorithm, Int. J. Electr. Power Energy Syst., 30, 184–192.

Jabr, R.A. (2008). Optimal placement of capacitors in a radial network using conic and mixed integer linear programming, Elect Power Syst Res, 78, 941–948.

Lam, A., & Li, V. (2010). Chemical-reaction-inspired metaheuristic for optimization. IEEE Trans. Evol. Comp., 14, 381–399.

Nojavan, S., Jalali, M., & Zare, K. (2014). Optimal allocation of capacitors in radial/mesh distribution systemsusing mixed integer nonlinear programming approach, Elect Power Syst Res, 107, 119–124.

Ng, H.N., Salama, M.M.A., & Chikhani, A.Y. (2000) Classification of capacitor allocation techniques. IEEE Tran. On Power Delivery, 15, 387–392.

Prakash, K., & Sydulu, M. (2007). Particle swarm optimization based capacitor placement on radial distribution systems, In: IEEE power engineering society general meeting, 1–5, 24th–28th June.

Raju, M.R., Murthy, K.V.S.R., & Avindra. (2012). Direct search algorithm for capacitive compensation in radial distribution systems, Int. J. Elect. Power Energy Syst., 42(1), 24–30.

Rao, R.S., Narasimham, S.V.L., & Ramalingaraju, M. (2011) Optimal capacitor placement in radial distribution system using plant growth simulation algorithm, Int. J. Electr. Power Energy Syst., 33(5), 1133–1139.

Sultana, S, & Roy, P.K. (2014). Optimal capacitor placement in radial distribution systems using teaching learning based optimization, 54, 387–398.

Singh, S.P., & Rao, A.R. (2012). Optimal allocation of capacitors in distribution systems using particle swarm optimization, Int. J. Electr. Power Energy Syst., 43, 1267–1275.

Sydulu, M, & Reddy, V.V.K. (2007). Index and GA based optimal location and sizing of distribution system capacitors, In: IEEE power engineering society general meeting, 1–4, 24th–28th June.

Sahoo, N.C., & Prasad, (2006). K.. A fuzzy genetic approach for network reconfiguration to enhance voltage stability in radial distribution systems, Int. J. Energy Convers and Manag. Energy Syst., 47, 3288–3306.

Sundharajan, & Pahwa, A. (1994). Optimal selection of capacitors for radial distribution systems using genetic algorithm, IEEE Trans. Power Systems, 9(3), 1499–1507.

Tabatabaei, S.M., & Vahidi, B. (2011). Bacterial foraging solution based fuzzy logic decision for optimal capacitor allocation in radial distribution system, Elect Power Syst Res, 81, 1045–1050.

Teng, J.H. (2003). A direct approach for distribution system load flow solutions, IEEE Trans Power Deliver., 18(3), 882–887.

Vuletic, J., & Todorovski, M. (2014). Optimal capacitor placement in radial distribution systems using clustering based optimization, Elect Power Syst Res, 62, 229–236.

Frontiers in Computer, Communication and Electrical Engineering – Acharyya (Ed.)
© 2016 Taylor & Francis Group, London, ISBN: 978-1-138-02877-7

Application of Improved Particle Swarm Optimization technique for thinning of Elliptical Array antenna

Rajesh Bera & Durbadal Mandal
Department of Electronics and Communication Engineering, National Institute of Technology Durgapur, West Bengal, India

Sakti Prasad Ghoshal
Department of Electrical Engineering, National Institute of Technology Durgapur, West Bengal, India

Rajib Kar
Department of Electronics and Communication Engineering, National Institute of Technology Durgapur, West Bengal, India

ABSTRACT: This paper describes optimal thinning of an Elliptical Antenna Arrays (EA) of uniformly excited isotropic antennas which can generate directive beam with minimum relative Side Lobe Level (SLL). The Improved Particle Swarm Optimization (IPSO) method, which represents a new approach for optimization problems in electromagnetic, is used in the optimization process. The IPSO is used to determine an optimal set of 'ON-OFF' elements that provide a radiation pattern with maximum SLL reduction. Optimization is done without prefixing the value of First Null Beam Width (FNBW). The variation of SLL with eccentricity of thinned array is also reported. Simulation results show that the number of array elements can be reduced by more than 40% of the total number of elements in the array with a simultaneous reduction in SLL significantly.

1 INTRODUCTION

Usually the radiation pattern of a single antenna element is relatively wide and each element provides low directivity. Antenna arrays increase the directivity without enlarging the size of single element. Generally, the overall array properties as directivity and gain, direction of maximum directivity, First Null Beam Width (FNBW), SLL, Half Power Beam-Width (HPBW) etc. can be controlled and optimized by adjusting the number of elements, the inter-element spacing, their excitation coefficients, their relative phases, the geometrical arrangement of the overall array (linear, circular, elliptical etc.) and the relative pattern of the individual elements (Ballanis 2005).

A linear array has excellent directivity and it can form the narrowest main-lobe in a given direction, but it does not work well in all azimuth directions. Since a circular array does not have edge elements, directional patterns synthesized with a circular array can be electronically rotated in the surface of the array without a significant change of the beam shape (Ioannides & Ballanis 2005). Circular array pattern has no nulls in azimuth plane (Ballanis 2005). In smart antenna (Chryssomallis

2000) applications to reject SNOI the array pattern should have several nulls in the azimuth plane. This can be implemented by the use of elliptical arrays instead of circular arrays (Lotfi *et al.* 2008).

The circular array antenna is of high side-lobe geometry. If the inter-element distance of array elements is decreased to reduce the side lobes, the mutual coupling influence becomes more significant. For mitigating high SLLs, concentric arrays are utilized in (Mahmoud *et al.* 2007). Also concentric circular array antennas have several advantages including the flexibility in array pattern synthesis and design both in narrowband and broadband beam forming applications (Dessouky *et al.* 2006).

Thinning an array means switching off some elements in a uniformly spaced or periodic array to generate a pattern with low SLL. In this proposed method, the locations of the elements are fixed and all the elements have two states either "on" or "off" (Similar to Logic "1" and "0" in digital domain), depending on whether the element is connected to the feed network or not. In the "off" state, either the element is passively terminated to a matched load or open circuited. If there is no matching between the elements, it is equivalent to removing "off" element from the array. There are many published articles (Razavi & Forooraghi 2008, Haupt 1994) dealing with the synthesis of thinned array. Element behavior in a thinned array is described in (Achwartzman 1967).

Classical optimization methods have several disadvantages such as: i) highly sensitive to starting points when the number of solution variables and hence the size of the solution space increase, ii) requirement of continuous and differentiable objective functions etc. But there are various evolutionary optimization tools for thinning such as Genetic Algorithm (GA), Particle Swarm Optimization (PSO) (Mandal *et al.* 2010) etc, which do not suffer from the above disadvantages. The PSO algorithm has proven to be a better alternative to other evolutionary algorithms such as Genetic Algorithms (GA), Ant Colony Optimization (ACO) etc. for optimal design of antenna array. In this work, for the optimization of complex, highly non-linear, discontinuous and non-differentiable array factors of EA design, IPSO is adopted for improving the global search ability.

2 DESIGN EQUATION

In this section, the expression for the array factor of EA as shown in Figure 1 is derived through the analysis of coordinate system.

If the center of an ellipse is located at the origin on the x-y plane, then the parametric equation of the ellipse in rectangular coordinate system is given by,

$$x = a\cos\varphi$$
$$y = b\sin\varphi \quad 0 \le \varphi < 2\pi \tag{1}$$

where a, b are semi-major axis and semi-minor axis, respectively, and φ is the angle between positive section of x axis and a point (x, y) of the ellipse in x-y plane.

Also, the ellipse eccentricity e can be defined as:-

$$e = \frac{c}{a} = \sqrt{1 - \frac{b^2}{a^2}} \tag{2}$$

where $c = \sqrt{a^2 + b^2}$ is the half of the distance between the two foci. Thus, for an elliptical N-element array with its center in origin of x-y plane, the following expressions are obtained.

I have:-

$$R_n = a\cos\varphi_n a_x + b\sin\varphi_n a_y$$
$$a_r = \sin\theta\cos\varphi a_x + \sin\theta\sin\varphi a_y + \cos\theta a_z \tag{3}$$

where $\varphi_n = 2\pi(n-1)/N$ is the angle in the x-y plane between the × axis and the n-th element.

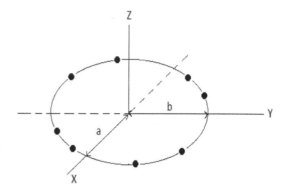

Figure1. Elliptical Array (EA) Structure.

The general form of array factor of an array can be written as [1]

$$AF(\theta,\varphi) = \sum_{n=1}^{N} A_n e^{j(\alpha_n + kR_n \cdot a_r)} \tag{4}$$

Thus, the array factor of an elliptical array can be derived by substituting (3) in (4)

$$AF(\theta,\varphi) = \sum_{n=1}^{N} A_n e^{jk\sin\theta(a\cos\varphi_n\cos\varphi + b\sin\varphi_n\sin\varphi)} \tag{5}$$

Normalized power pattern in dB can be expressed as follows:

$$P(\theta,\varphi) = 10\log_{10}\left[\frac{|AF(\theta,\varphi)|}{|AF(\theta,\varphi)|_{\max}}\right]^2$$
$$= 20\log_{10}\left[\frac{|AF(\theta,\varphi)|}{|AF(\theta,\varphi)|_{\max}}\right] \tag{6}$$

3 IMPROVED PARTICLE SWARM OPTIMIZATION

Improved Particle Swarm Optimization (IPSO), modified version of PSO, is applied to compute the distributions of the turned on and turned off elements in EA. The detail of PSO is not given; some references on PSO are given in (Bera & Roy 2013, Jin & Rahmat-Samii 2007, Everhart & Shi 2001, Kennedy & Eberhard 1995).

In PSO, mathematically, velocities of the particles are modified according to the following equation:

$$V_i^{k+1} = w * V_i^k + C_1 * rand_1 * \left(pbest_i - S_i^k\right)$$
$$+ C_2 * rand_2 * \left(gbest - S_i^k\right) \tag{7}$$

The searching point in the solution space may be modified by the following equation.

$$S_i^{k+1} = S_i^k + V_i^{k+1} \qquad (8)$$

The global search ability of PSO is enhanced with the help of the following modifications. This modified PSO is termed as IPSO (Mangoud & Elragal 2009).

i) The two random parameters $rand_1$ and $rand_2$ of (7) are self-governing. If both the parameters are large, both the personal and social experiences are over used and the particle is driven too far away from the local best promising. If both are small, both the personal and social experiences are not used fully and the convergence speed of the procedure is reduced. So, instead of taking fully independent $rand_1$ and $rand_2$, one single random parameter r_1 is chosen so that when r_1 is large, $(1-r_1)$ is small and vice versa. Moreover, to control the balance of global and local searches, another random parameter r_2 is introduced. For birds flocking for food, there could be some uncommon cases that after the position of the particle is changed according to (8), a bird may not, due to inertia, fly toward a region at which it thinks is the most hopeful for food. So, in the step that follows, the direction of the bird's velocity should be inverted in order for it to fly back into the promising region. $sign(r_3)$ is introduced for this purpose. Both cognitive and social parts are modified consequently.

Finally, the modified velocity of jth component of ith particle is expressed as follows:

$$V_i^{(k+1)} = r_2 * sign(r_3) * V_i^k + (1-r_2) * C_1 * r_1$$
$$* \left\{ pbest_i^k - S_i^k \right\} + (1-r_2) * C_2 * (1-r_1) * \left\{ gbest^k - S_i^k \right\}$$
$$(9)$$

where r_1, r_2 and r_3 are the random numbers between 0 and 1 and $sign(r_3)$ is a function defined as:

$$sign(r_3) = -1 \text{ when } r_3 \leq 0.05, = 1 \text{ when } r_3 > 0.05. \qquad (10)$$

After defining the far-field radiation pattern, the next step in the design process is to create the objective function that is to be minimized. The objective function is defined using the array factor in such a way that the purpose of the optimization is satisfied. For the optimization problem of the side lobe reduction, the array factor values at the side lobe peaks must be less than the reference pattern. To satisfy this, the objective function "cost function" (CF) to be minimized with the IPSO is

$$CF = C_1[(SLL_c - SLL_d)/SLL_d]$$
$$+ C_2[(FNBW_c - FNBW_d)/FNBW_d] \qquad (11)$$

Where, SLL_d & SLL_c are desired and computed value of SLL and $FNBW_d$ & $FNBW_c$ are desired and computed value of FNBW respectively.

The first term in (11) is used to reduce the SLL to a desired level. The second term in (11) is introduced to keep FNBW of the optimized pattern to a desired level which is at most 125% of reference pattern in this particular case.

C1 and C2 are weighting coefficients to control the relative Importance of each term in (11). Because the primary aim is to achieve a minimum SLL, the value of C1 is higher than the value of C2.

4 COMPUTATIONAL RESULTS

The paper describes the application of IPSO algorithm for the reduction of the maximum SLL without prefixing the value of First Null Beam Width (FNBW) of thinned Elliptical Array (EA) of isotropic elements.

In the numerical analysis, two cases are considered where antenna elements in elliptical arrays are 15 and 20. First the SLLs of fully populated arrays are computed then IPSO technique is applied to obtain maximum SLL by changing combinations of 'ON' and 'OFF' states of elements in each case. The eccentricity of the antenna array, in each case, is varied. In the both cases, inter-element spacing is kept fixed at d = 0.5λ.

Performances of optimized thinned array and corresponding fully populated array antennas with variation of eccentricity of the EAs are tabulated in Table 1. The value of semi-major axis 'a' and semi-minor axis 'b' shows in Table 1 can be calculated from the equations given below

$$circumference = \pi \sqrt{2(a^2 + b^2) - 0.5(a-b)^2} \qquad (12)$$

$$e = \sqrt{1 - \frac{b^2}{a^2}} \qquad (13)$$

Table 1. Performances of thinned elliptical arrays.

No. of elements	Eccentricity e = 0	Eccentricity e = 0.2	Eccentricity e = 0.4	Eccentricity e = 0.6
N = 15	a = 1.194	a = 1.2064	a = 1.2457	a = 1.323
	b = 1.194	b = 1.1823	b = 1.1417	b = 1.058
	$SLL_{max} =$ −18.48dB	$SLL_{max} =$ −20.94dB	$SLL_{max} =$ −21.3dB	$SLL_{max} =$ −23.8dB
N = 20	a = 1.59	a = 1.607	a = 1.66	a = 1.763
	b = 1.59	b = 1.575	b = 1.52	b = 1.41
	$SLL_{max} =$ −17.44dB	$SLL_{max} =$ −15.68dB	$SLL_{max} =$ −18.44dB	$SLL_{max} =$ −18.02dB

Table 2. Excitation amplitude distribution of 15 element thinned array for various values of eccentricity.

Eccentricity	Distribution of ON and OFF elements	Number of 'ON' elements	SLL_{max} (dB)
0	1 1 0 0 1 0 1 1 1 0 0 1 0 1 1	9	−18.48
0.2	1 1 0 1 0 0 1 1 1 0 0 0 0 1 1	8	−20.94
0.4	1 1 1 0 0 0 0 1 1 1 0 1 0 0 1	8	−21.30
0.6	1 1 0 0 1 0 1 1 1 1 0 1 0 0 1	9	−23.80

Table 3. Excitation amplitude distribution of 20 element thinned array for various values of eccentricity.

Eccentricity	Distribution of ON and OFF elements	Number of 'ON' elements	SLL_{max} (dB)
0	1 1 0 0 0 0 1 0 1 1 1 1 1 0 1 0 0 1 1 1	12	−17.44
0.2	1 1 1 0 1 1 0 0 1 1 1 1 1 0 0 1 0 0 1 0	12	−15.68
0.4	1 1 1 0 0 0 1 0 0 1 1 1 1 1 0 1 0 0 1 1	12	−18.44
0.6	1 1 1 0 1 0 0 0 1 1 1 1 1 0 0 1 0 0 1 1	12	−18.02

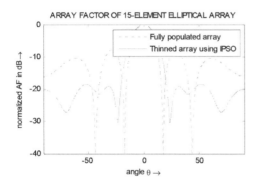

Figure 2. Normalized absolute power patterns in dB of the 15 element optimized thinned array for e = 0 (Circular case).

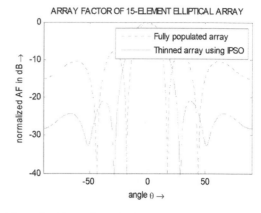

Figure 3. Normalized absolute power patterns in dB of the 15 element optimized thinned array for e = 0.2.

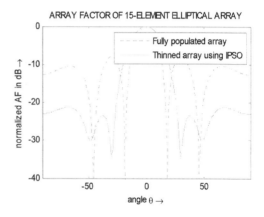

Figure 4. Normalized absolute power patterns in dB of the 15 element optimized thinned array for e = 0.4.

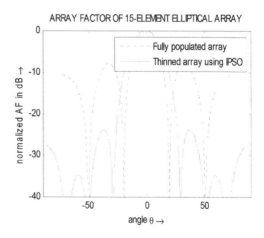

Figure 5. Normalized absolute power patterns in dB of the 15 element optimized thinned array for e = 0.6.

The optimal excitation amplitudes of thinned EA using IPSO are shown in Tables 2 and 3 for N = 15 & 20 respectively. IPSO algorithm is applied to compute the distributions of the 'ON' and 'OFF' elements. Thinning is the ratio of the number of 'OFF' elements to the total number of elements in the array.

Figures 2–5 represents the optimized pattern of 20 elements EA for various values of eccentricity. Also Figures 6–9 represents the optimized pattern of 20 elements EA for various values of eccentricity. Patterns for corresponding fully populated arrays are also present by dotted line in above mentioned figures.

Table 2 shows that for eccentricity e = 0 (circular array), maximum SLL is equal to −18.48 dB and for the value of eccentricity e = 0.6, maximum SLL is equal to −23.80. So, by the variation of eccentricity of elliptical array may give much better SLL

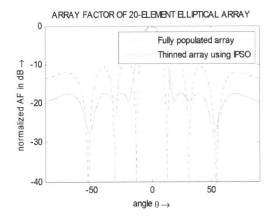

Figure 6. Normalized absolute power patterns in dB of the 20 element optimized thinned array for e = 0 (Circular case).

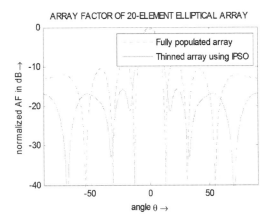

Figure 7. Normalized absolute power patterns in dB of the 20 element optimized thinned array for e = 0.2.

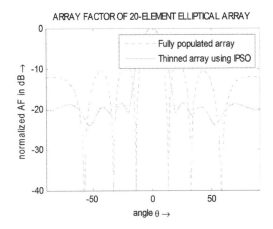

Figure 8. Normalized absolute power patterns in dB of the 20 element optimized thinned array for e = 0.4.

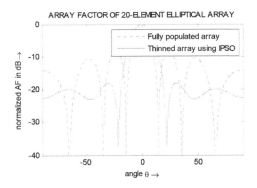

Figure 9. Normalized absolute power patterns in dB of the 20 element fully populated array and the corresponding optimized thinned array for e = 0.6.

Table 4. Performances of 15 element thinned EAs for various eccentricity.

Eccentricity	Filled Ratio (%)	Maximum SLL (dB)	First Null Beam Width (FNBW) in degree
e = 0	60.00	−18.48	47.2
e = 0.2	53.33	−20.94	56.8
e = 0.4	53.33	−21.30	59.6
e = 0.6	60.00	−23.80	58.4

Table 5. Performances of 20 element thinned EAs for various eccentricity.

Eccentricity	Filled Ratio (%)	Maximum SLL (dB)	First Null Beam Width (FNBW) in degree
e = 0	60.00	−17.44	38.4
e = 0.2	60.00	−15.68	34.0
e = 0.4	60.00	−18.44	40.4
e = 0.6	60.00	−18.02	43.6

than the circular array although the number of 'ON' elements is same in both these cases. Also we can see the similar thing in Table 3.

Tables 4 and 5 respectively shows the overall performances of 15 and 20 element elliptical array for various eccentricities. The filled ratios of the arrays are also calculated which are less than or equal to 60% for the cases. Filled ratio is defined as the ratio of the number of 'ON' elements to the total number of elements in the array. Performance of our proposed 15 element thinned array with respect to maximum SLLs is compared with the referred paper [13] shown in Fig. 10, which shows that our proposed method with different cost function gives better results for various eccentricity of elliptical array.

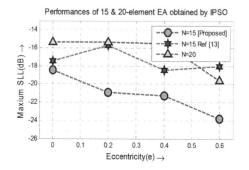

Figure 10. Variation of SLL_{max} for different values of eccentricity with $d = 0.5\lambda$.

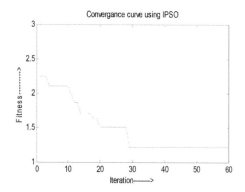

Figure 11. Convergence curve corresponding to the optimized thinned array pattern shown in Fig. 4 using IPSO.

The population size using IPSO, by which the antenna array is optimized, is 40 and the maximum number of iteration cycles is 60. The minimum CF values are recorded against number of iteration cycles to get the convergence profile of the algorithm, which is shown in Figure 11.

All computations were done in MATLAB 7.5 on core (TM) 2 duo processor, 3.00 GHz with 4 GB RAM.

5 CONCLUSIONS

This paper illustrates Improved Particle Swarm Optimization (IPSO) for thinning of 15 & 20 element Elliptical Array Antennas of isotropic elements. The simulation results show that the number of antenna array elements can be brought down more than 40% of total element in the array using IPSO with simultaneous reduction in SLL less than –23dB. The first null beam width of the synthesized array pattern with fixed inter-element spacing is close to that of the fully populated array of same shape and size. Simulation results are also compared with other published article to show the

effectiveness of our proposed method. Thus, IPSO algorithm can efficiently handle the problem of thinning of EA, showing a significant improvement for SLLs with a significant reduction in the number of elements, which will reduce the cost of designing the antenna array considerably.

REFERENCES

Balanis C.A. 2005. Antenna Theory and Design, *3rd Edition, John Wiley & Sons.*

Bera R. & Roy J.S. 2013. Thinning of elliptical & concentric elliptical antanna arrays using particle swarm optimization. *Microwave Review, Vol. 19, No. 1,* pp. 2–7.

Chryssomallis M. 2000. Smart antennas. *IEEE Antennas and Propagation Magazine, vol. 42, no. 3,* pp. 129–136.

Dessouky M., Sharshar H., & Albagory Y. 2006. Efficient sidelobe reduction technique for small—sized concentric circular arrays. *Progress in Electromagnetics Research, PIER, vol. 65,* pp. 187–200.

Eberhart R.C., & Shi Y. 2001. Particle swarm optimization: Developments, applications & resources. Proc. Congr. *Evolutionary Computation, Vol. 1,* pp. 81–86.

Haupt R.L. 1994. Thinned arrays using genetic algorithms. *IEEE Trans. Antennas Propag.,* Vol. 42, No. 7, pp. 993–999.

Ioannides P. & Balanis C.A. 2005. Uniform circular arrays for smart antennas. IEEE *Antennas and Propagation Magazine,* vol. 47, no. 4, pp. 192–206.

Jin N. & Rahmat-Samii Y. 2007. Advances in particle swarm optimization for antenna designs: Real-number, binary, single—objective & multiobjective implementations. *IEEE Trans. Antennas Propag., Vol. 55, No. 3,* pp. 556–567.

Kennedy J. & Eberhard R.C. 1995. Particle Swarm Optimization. *Proc. of IEEE Int'l Conf. on Neural Networks, Piscataway, NJ, USA,* pp. 1942–1948.

Lotfi A.A., Ghiamy M., Moghaddasi M.N. & Sadeghzadeh R A. 2008. An investigation of hybrid elliptical antenna arrays. *IET Microw. Antennas Propag., vol. 2, no. 1,* pp. 28–34.

Mahmoud K.R., El-Adway M., Ibrahem S.M.M., Basnel R., Mahmoud R., & Zainud-Deen S.H. 2007. A comparition between circular and hexagonal array geometries for smart antenna systems using particle swarm algorithm. *Progress in Electromagnetics Research, PIER, vol. 72,* pp. 75–90.

Mandal D., Ghoshal S.P. & Bhattacharjee A.K. 2010. Design of Concentric Circular Antenna Array With Central Element Feeding Using Particle Swarm Optimization With Constriction Factor & Inertia Weight Approach & Evolutionary Programing Technique. *Journal of Infrared Milli Terahz Waves, vol. 31 (6),* pp. 667–680.

Mangoud M.A. & Elragal H.M. 2009. Antenna array pattern synthesis & wide null control using enhanced particle swarm Optimization. *Progress In Electromagnetics Research B, vol. 17,* pp. 1–14.

Razavi, A. & Forooraghi K. 2008. Thinned arrays using pattern search algorithms. *Progress In Electromagnetics Research, PIER 78,* pp. 61–71.

Schwartzman L. 1967. Element behavior in a thinned array. *IEEE Trans. Antennas Propag., Vol. 15, No. 7,* pp. 571–572.

Frontiers in Computer, Communication and Electrical Engineering – Acharyya (Ed.)
© 2016 Taylor & Francis Group, London, ISBN: 978-1-138-02877-7

Linear phase FIR bandstop filter design using Colliding Bodies Optimization technique

S.K. Saha
Department of ETC, NIT Raipur, India

ABSTRACT: In this paper, Colliding Bodies Optimization (CBO) technique is applied for finding the most promising optimal set of linear phase FIR filter coefficients. CBO is a class of meta-heuristic search technique which is free from any kind of internal parameter and governed by the conservation law of momentum of colliding bodies. In order to establish the superiority of CBO over other optimization techniques, performance of CBO for FIR filter design is compared with the other techniques namely, Real Coded Genetic Algorithm (RGA), Particle Swarm Optimization (PSO) and Differential Evolution (DE). 20th order Bandstop (BS) FIR filter is designed with the proposed CBO and other afore-mentioned algorithms individually for the comparison of efficiency in performance of optimization techniques. A comparative study of simulation results reveals the optimization efficacy of the CBO over the other optimization techniques for the solution of the multimodal, non-differentiable, non-linear, and constrained FIR filter design problem.

1 INTRODUCTION

Digital Filter find its application in different fields of signal processing. It usually comes in two categories: Finite Impulse Response (FIR) and Infinite Impulse Response (IIR). FIR filter is an attractive choice because of the ease in design and assured stability. By designing the filter taps to be symmetrical about the centre tap position, a FIR filter can be guaranteed to have linear phase response (Litwin 2000, Parks & Burrus 1987).

Remez Exchange algorithm proposed by Parks and McClellan is used for the design of exact linear phase weighted Chebyshev FIR filter (Parks & McClellan 1972). Further a computer program is developed for the design of FIR digital filter by (McClellan et al. 1973). The program is to be iterated many times in order to meet the filter specifications in terms of stop band attenuation, cut-off frequency and filter length (Rabiner 1973).

The objective function for the design of optimal digital filters involves accurate control of various parameters of frequency spectrum and is thus highly non-uniform, non-linear, non-differentiable and multimodal in nature. Classical optimization methods cannot optimize such objective functions and cannot converge to the global minimum solution. So, evolutionary optimization methods have been implemented for FIR filter design.

Different meta-heuristics and stochastic optimization methods such as Genetic Algorithm (GA) (Saha et al. 2013, Ahmad & Antoniou 2006,

Mastorakis et al. 2003, Lu & Tzeng 2000), simulated annealing (Chen 2000), Tabu Search (Karaboga et al. 1997), differential evolution (Liu et al. 2010) and artificial bee colony optimization (Karaboga 2009) etc are developed to solve various engineering problems. GA proves itself to be more efficient in terms of obtaining local optimum while maintaining its moderate computational complexity but they are not very successful in determining the global minima in terms of convergence speed and solution quality (Ababneh & Bataineh 2008).

Particle Swarm Optimization is an evolutionary optimization technique developed by (Kennedy & Eberhart 1995). Several works have already been done in order to explore the flexibility of FIR filter design provided by PSO (Ababneh & Bataineh 2008, Najjarazadeh & Ayatollahi 2008). LMS and Minimax strategies are also applied in (Najjarazadeh & Ayatollahi 2008).

Several modifications of the conventional PSO technique have been made to increase its efficiency. PSO is used with the differential evolution (Luitel & Venayagamoorthy 2008) to design optimal filter. Quantum-behaved Particle Swarm Optimization (QPSO) proposed by (Sun et al. 2006 & 2004) is a novel algorithm based on the PSO and quantum model. Quantum infused PSO is also utilized for the design of digital filters (Luitel & Venayagamoorthy 2008). DEPSO and PSO-QI have been used for FIR filter design problem in Sarangi et al. 2011). Recently, particle swarm optimization with constriction factor and inertia weight approach with wavelet mutation (PSOCFIWAWM) has been applied for FIR filter design problem in (Saha et al. 2014).

Most of the above algorithms show the problems of fixing algorithm's control parameters, premature convergence, stagnation and revisiting of the same solution over and again (Biswal et al. 2009, Ling et al. 2008). In order to overcome these problems, in this paper, the Colliding Bodies Optimization (CBO) technique (Kaveh & Mahdavi 2014) and a novel fitness function are employed for the design of FIR bandstop (BS) filter.

The rest of the paper is arranged as follows. In section II, the FIR BS filter design problem is formulated. Section III discusses the CBO technique. Section IV describes the simulation results obtained for FIR BS filter using PM algorithm, RGA, PSO, DE and the proposed CBO based approach. Finally, section V concludes the paper.

2 PROBLEM FORMULATION

A digital FIR filter is expressed as follows:

$$H(z) = \sum_{n=0}^{N} h(n)z^{-n}, n = 0,1, ..., N \qquad (1)$$

where N is the filter's order with (N + 1) number of filter's coefficients, h(n). In this design approach, the FIR filter is positive even symmetric. So, N is even number and (N/2 + 1) number of h(n) coefficients are actually optimized, which are finally concatenated to find the required (N + 1) number of filter coefficients.

Now for (1), the particle, i.e., the coefficient vector $\{h_0, h_1, ..., h_{N/2}\}$, which is optimized, is represented in (N/2 + 1) dimension.

The frequency response of the FIR digital filter can be calculated as,

$$H(e^{j\omega_k}) = \sum_{n=0}^{N} h(n) e^{-j\omega_k n}; \qquad (2)$$

where $\omega_k = \frac{2\pi k}{N}$; $H(e^{j\omega_k})$ is the Fourier transform complex vector. This is the FIR filter's frequency response. The frequency is sampled in $[0, \pi]$ with S points. An error function given by (3) is the approximate error used in PM algorithm for filter design [3].

$$E(\omega) = G(\omega) \left[H_d(e^{j\omega}) - H_i(e^{j\omega}) \right] \qquad (3)$$

where $H_d(e^{j\omega})$ is the frequency response of the designed approximate filter; $H_i(e^{j\omega})$ is the

frequency response of the ideal filter; $G(\omega)$ is the weighting function used to provide different weights for the approximate errors in different frequency bands. For ideal BS filter, $H_i(e^{j\omega})$ is given as,

$$H_i(e^{j\omega}) = 0 \quad \textbf{for} \quad \omega_{sl} \leq \omega \leq \omega_{sh};$$
$$= 1 \quad \textbf{otherwise} \qquad (4)$$

where ω_{sl} and ω_{sh} are the lower and upper cut-off frequencies in stopband. The major drawback of PM algorithm is that the ratio of δ_p/δ_s is fixed. To improve the flexibility in the error function to be minimized, so that the desired levels of δ_p and δ_s may be specified, the error function given in (5) is considered as fitness function in many literatures (Ababneh & Bataineh 2008, Sarangi et al. 2011). The error fitness to be minimized using the evolutionary algorithms, is defined as:

$$J_1 = \max_{\omega \leq \omega_p} \left(|E(\omega)| - \delta_p \right) + \max_{\omega \geq \omega_s} \left(|E(\omega)| - \delta_s \right) \qquad (5)$$

It is found that the proposed filter deign approach as in (6) results in considerable improvement over the PM and other optimization techniques in terms of higher stop band attenuation and better transition width.

$$J_2 = \sum abs \left[abs \left(|H_d(\omega)| - 1 \right) - \delta_p \right]$$
$$+ \sum \left[abs \left(|H_d(\omega)| - \delta_s \right) \right] \qquad (6)$$

For the first term of (6), $\omega \in$ passband, including a portion of the transition band and for the second term of (6), $\omega \in$ stopband, including the rest portion of the transition band. The portions of the transition band chosen depend on passband edge and stopband edge frequencies.

Unlike other error fitness functions (Ababneh & Bataineh 2008, Sarangi et al. 2011) which consider only the maximum errors, J_2 involves summation of all absolute errors for the whole frequency band, and hence, minimization of J_2 yields much higher stopband attenuation and lesser stopband ripples. Transition width is also kept reduced.

3 OPTIMIZATION TECHNIQUES EMPLOYED

Evolutionary algorithms have the potential to adapt to their ever changing dynamic environment through the previously acquired knowledge. Descriptions of fundamental algorithms such as RGA, PSO and DE are presented in (Saha et al. 2014). In this section only CBO is described.

3.1 The CBO algorithm

3.1.1 Theory

The main objective of the present study is to formulate a new simple and efficient meta-heuristic algorithm which is called Colliding Bodies Optimization (CBO) (Kaveh & Mahdavi 2014). In CBO, each solution candidate X_i containing a number of variables (i.e. $X_i = \{X_{i,j}\}$) is considered as a Colliding Body (CB). The massed objects are composed of two main equal groups; i.e. stationary and moving objects, where the moving objects move to follow stationary objects and a collision occurs between pairs of objects. This is done for two purposes: (i) to improve the positions of moving objects and (ii) to push stationary objects towards better positions. After the collision, new positions of colliding bodies are updated based on new velocity by using the collision laws.

The CBO procedure can briefly be outlined as follows:

1. The initial positions of CBs are determined with random initialization of a population of individuals in the search space:

$$x_i^0 = x_{\min} + rand(x_{\max} - x_{\min}), \ i = 1, 2, \dots, n. \quad (7)$$

where x_i^0 determines the initial value vector of the ith CB. x_{min} and x_{max} are the minimum and the maximum allowable values vectors of variables; $rand$ is a random number in the interval $[0, 1]$; and n is the number of CBs.

2. The magnitude of the body mass for each CB is defined as:

$$m_k = \frac{1/fit(k)}{\sum_{i=1}^{n} 1/fit(i)} \quad k = 1, 2, \dots, n \quad (8)$$

where $fit(i)$ represents the objective function value of the agent i; n is the population size. It seems that a CB with good values exerts a larger mass than the bad ones.

3. The arrangement of the CBs objective function values is performed in ascending order as shown in Figure 1(a). The sorted CBs are equally divided into two groups:
   The lower half of CBs (stationary CBs): These CBs are good agents which are stationary and the velocity of these bodies before collision is zero. Thus:

$$v_i = 0, \ i = 1, 2, \dots, n/2 \quad (9)$$

CBs move toward the lower half. Then, according to Figure 1(b) the better and worse CBs, i.e. agents with upper fitness value, of each group

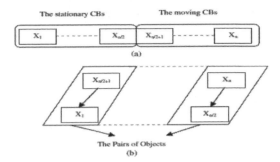

Figure 1. (a) CBs sorted in increasing order and (b) colliding object pairs.

will collide together. The change of the body position represents the velocity of these bodies before collision as:

$$v_i = x_i - x_{i-n/2}, \ i = n/2 + 1, \dots, n \quad (10)$$

where, v_i and x_i are the velocity and position vector of the ith CB in this group, respectively; $x_{i-n/2}$ is the ith CB pair position of x_i in the previous group.

4. After the collision, the velocities of the colliding bodies in each group are evaluated. The velocity of each moving CBs after the collision is obtained by:

$$v_i' = \frac{(m_i - \varepsilon m_{i-n/2})v_i}{m_i + m_{i-n/2}}, \ i = n/2 + 1, \dots, n \quad (11)$$

where v_i and v_i' are the velocity of the ith moving CB before and after the collision, respectively; m_i is mass of the ith CB; $m_{i-n/2}$ is mass of the ith CB pair. Also, the velocity of each stationary CB after the collision is:

$$v_i' = \frac{(m_{i+n/2} + \varepsilon m_{i+n/2})v_{i+n/2}}{m_i + m_{i+n/2}}, \ i = 1, \dots, n/2 \quad (12)$$

where $v_{i+n/2}$ and v_i' are the velocity of the ith moving CB pair before and the ith stationary CB after the collision, respectively; m_i is mass of the ith CB; $m_{i+n/2}$ is mass of the ith moving CB pair; ε is the value of the COR parameter whose law of variation will be discussed in the next section.

5. New positions of CBs are evaluated using the generated velocities after the collision in position of stationary CBs. The new positions of each moving CB is:

$$x_i^{new} = x_{i-n/2} + rand \circ v_i', \ i = n/2 + 1, \dots, n \quad (13)$$

where x_i^{new} and v_i' are the new position and the velocity after the collision of the ith moving

CB, respectively; $x_{i-n/2}$ is the old position of ith stationary CB pair. Also, the new positions of stationary CBs are obtained by:

$$x_i^{new} = x_i + rand \circ v_i', \quad i = 1,...,n/2 \qquad (14)$$

Where x_i^{new}, x_i and v_i' are the new position, old position and the velocity after the collision of the ith stationary CB, respectively. *rand* is a random vector uniformly distributed in the range $(-1, 1)$ and the sign "o" denotes an element-by-element multiplication.

6. The optimization is repeated from Step 2 until a termination criterion, such as maximum iteration number, is satisfied for the present paper. It should be noted that, a body's status (stationary or moving body) and its numbering are changed in two subsequent iterations.

Apart from the efficiency of the CBO algorithm, parameter independency is an important feature that makes CBO superior over other meta-heuristic algorithms. Also, the formulation of CBO algorithm does not use the memory which saves the best-so-far solution (i.e. the best position of agents from the previous iterations).

3.1.2 *The Coefficient of Restitution (COR)*

The meta-heuristic algorithms have two phases: exploration of the search space and exploitation of the best solutions found. In the meta-heuristic algorithm it is very important to have a suitable balance between the exploration and exploitation (Kaveh & Talatahari 2010). In the optimization process, the exploration should be decreased gradually while simultaneously exploitation should be increased.

In this paper, an index is introduced in terms of the coefficient of restitution (COR) to control exploration and exploitation rate. In fact, this index is defined as the ratio of the separation velocity of two agents after collision to approach velocity of two agents before collision. Three variants of COR values are considered and presented below.

1. The perfectly elastic collision: In this case, COR is set equal to unity. It can be seen that in the final iterations, the CBs investigate the entire search space to discover a favorite space (global search).
2. The hypothetical collision: In this case, COR is set equal to zero. In this case, the movements of the CBs are limited to very small space in order to provide exploitation (local search). Consequently, the CBs are gathered in a small region of the search space.
3. The inelastic collision: In this case, COR decreases linearly to zero and ε is defined as:

$$\varepsilon = 1 - iter/iter_{max} \qquad (15)$$

where *iter* is the actual iteration number and $iter_{max}$ is the maximum number of iterations. It can be seen that the CBs get closer by increasing iteration. In this way a good balance between the global and local search is achieved. Therefore, in the optimization process COR is considered such as the above equation.

4 RESULTS AND DISCUSSIONS

4.1 *Analysis of magnitude response of FIR BS filter*

This section presents the simulation study for the design of FIR BS filter. The filter order (N) is taken as 20, which results in the number of coefficients as 21. The sampling frequency is taken to be $f_s = 2$ Hz. The number of frequency samples (S) is 128. Each algorithm is run for fifty times to obtain best results.

The parameters of the filter to be designed using RGA, PSO, DE and CBO are as follows: Pass band ripple (δ_p) = 0.1, stop band ripple (δ_p) = 0.01. The lower passband (normalized) edge frequency (ω_{pl}) = 0.25; lower stopband (normalized) edge frequency (ω_{sl}) = 0.35; upper stopband (normalized) edge frequency (ω_{sh}) = 0.65; upper passband (normalized) edge frequency (ω_{ph}) = 0.75; transition width = 0.1. Table 1 shows the optimized filter coefficients obtained for FIR BS filter using RGA, PSO, DE and CBO algorithm individually.

Table 2 shows the comparative results of performance parameters in terms of maximum and average stopband ripple (normalized), transition width (normalized) and execution time for 100 iteration cycles for BS filter using PM, RGA, PSO, DE and the CBO, respectively. It is noticed that with small transition width CBO attains the lowest value of stopband ripple.

Table 3 also summarizes maximum, mean, variance and standard deviation for passband ripple (normalized) and stopband attenuation in dB for the designed BS filter using PM, RGA, PSO, DE and CBO, respectively. It is observed from Table 3 that the CBO achieves the best stop band attenuation (35.50 dB), as compared to others.

Figs. 2–5 show the magnitude responses of the BS filter using PM, RGA, PSO, DE and CBO. The magnitude responses in dB are plotted in Fig. 2 for the BS filter. The normalized magnitude response is shown in Fig. 3. Fig. 4 shows the normalized passband ripple and normalized stopband ripple is shown in Fig. 5. From the above figures and tables, it is observed that CBO based designed filter achieves well results in magnitude response (dB) with highest stopband attenuation for BS filter, as compared to others.

Table 1. Optimized coefficients of the FIR BS filter of order 20.

h(N)	RGA	PSO	DE	CBO
h(1) = h(21)	0.04087486099888	0.03803153039471	0.04589309650152	0.03875790077795
h(2) = h(20)	−0.00014306726313	0.00978872795414	−0.00054101197105	−0.00838892043043
h(3) = h(19)	0.06892412877542	0.05961759428540	0.06770857664214	0.05299909537482
h(4) = h(18)	0.00050761523950	0.00922601970397	0.00082276653114	−0.01630251507250
h(5) = h(17)	−0.06763806404466	−0.07826643279934	−0.07412324124248	−0.07802955626120
h(6) = h(16)	0.00022874372402	0.01158486435957	0.00389104365616	−0.01171433324790
h(7) = h(15)	−0.07359273266571	−0.08970179312418	−0.07621979186568	−0.08352692295001
h(8) = h(14)	0.00185475134293	0.01081379721712	−0.00053119770204	−0.00695868596627
h(9) = h(13)	0.30409887625951	0.29495264657382	0.30164753233390	0.29602844956797
h(10) = h(12)	−0.00325409665580	0.00207583340139	−0.00187372146712	−0.00802236950223
h(11)	0.58550996882805	0.57949794018090	0.58319314795976	0.57530336379387

Table 2. Other comparative results of performance parameters of all algorithms for the FIR BS filter of order 20.

| | FIR BS filter of order 20 | | |
Algorithm	Maximum, average Stop band ripple (normalized)	Transition width (normalized)	Execution time for 100 cycles
PM	0.07627, 0.07616	0.0875	−
RGA	0.03186, 0.02480	0.0933	6.7303
PSO	0.02902, 0.02176	0.0923	4.1629
DE	0.02756, 0.02093	0.0910	4.6253
CBO	0.0167, 0.00719	0.1004	5.2943

Table 3. Statistical parameters of passband ripple and stop band attenuation for different algorithms for the FIR BS filter.

| | FIR BS filter of order 20 | | | | | | | |
| | Passband ripple (normalized) | | | | Stopband Attenuation (dB) | | | |
Algorithm	Maximum	Mean	Variance	Standard Deviation	Maximum	Mean	Variance	Standard Deviation
PM	0.0763	0.0761	2e-8	1.41e-4	22.35	22.36333	8.89e-05	0.00943
RGA	0.135	0.12733	7.76e-5	8.807e-3	29.91	32.26333	2.81416	1.67755
PSO	0.170	0.1205	2.16e-3	4.6477e-2	30.74	33.51333	4.63776	2.15355
DE	0.161	0.1326	4.62e-4	2.1499e-2	31.19	33.85333	4.86269	2.20515
CBO	0.157	0.12117	1.174e-3	3.4264e-2	35.50	46.3575	68.88722	8.29983

4.2 Convergence profiles of RGA, PSO, DE and CBO

Convergence profile presents the algorithms' efficiency in finding the optimal solution in multidimensional search space in terms of the error fitness value with iteration cycle. In Fig. 6 convergence profiles for all concerned algorithms are shown. From Fig 6, it is that CBO converges to much lower error fitness value of 0.8653 as compared to RGA, PSO and DE of 3.28, 2.564 and 1.869 which yield suboptimal higher error fitness values. The execution time for CBO is also moderately low and may be verified from Table 2. With a view to the above fact, it may finally be inferred that the performance of CBO is the best among all the algorithms considered in this paper. All optimization programs are run in MATLAB 7.5 version on core (TM) 2 duo processor, 3.00 GHz with 2 GB RAM.

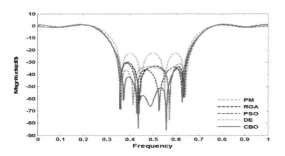

Figure 2. dB plots for the FIR BS filter of order 20.

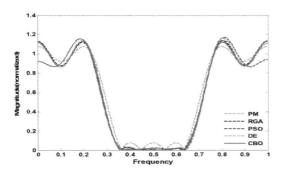

Figure 3. Normalized plots for the FIR BS filter of order 20.

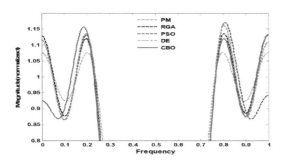

Figure 4. Normalized passband ripple plots for the FIR BS filter of order 20.

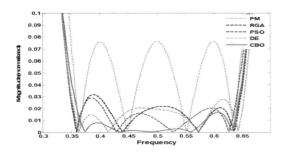

Figure 5. Normalized stopband ripple plots for the FIR BS filter of order 20.

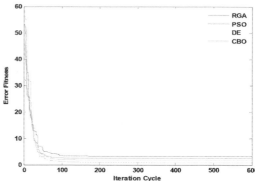

Figure 6. Convergence profiles for RGA, PSO, DE and CBO in case of FIR BS Filter of order 20.

5 CONCLUSIONS

In this paper, a recently proposed Colliding Bodies Optimization (CBO) technique is applied to the solution of FIR bandstop filter design with optimal filter coefficients. Comparison of the results of PM, RGA, PSO, DE and CBO algorithms have been made. It is revealed that CBO has the ability to converge to the best quality near optimal solution and possesses the best convergence characteristics in much less execution time among the algorithms.

The simulation results clearly indicate that the CBO demonstrates the best performance in terms of magnitude response, minimum stop band ripple and maximum stop band attenuation with the small transition width and execution time. Thus, the CBO may be used as a good optimization tool for obtaining the filter coefficients in any practical digital filter design problem of digital signal processing systems.

REFERENCES

Ababneh J.I. & Bataineh M.H. 2008. Linear phase FIR filter design using particle swarm optimization and genetic algorithms. *Digital Signal Processing, vol. 18,* pp. 657–668.

Ahmad S.U. & Antoniou A. 2006. A genetic algorithm approach for fractional delay FIR filters. *IEEE International Symposium on Circuits and Systems,* pp. 2517–2520.

Biswal B., Dash P.K. & Panigrahi B.K. 2009. Power quality disturbance classification using fuzzy C-means algorithm and adaptive particle swarm optimization. *IEEE Trans. Ind. Electron., vol. 56, no.1,* pp. 212–220.

Chen S. 2000. IIR Model Identification Using Batch-Recursive Adaptive Simulated Annealing Algorithm.

6th Annual Chinese Automation and Computer Science Conference, pp.151–155.

Fang Wei, Sun Jun, Xu Wenbo & Liu Jing 2006. FIR Digital Filters Design Based on Quantum-behaved Particle Swarm Optimization. *First International Conference on Innovative Computing, Information and Control*, vol 1, pp. 615–619.

Karaboga D., Horrocks D.H., Karaboga N. & Kalinli A. 1997. Designing digital FIR filters using Tabu search algorithm. *IEEE International Symposium on Circuits and Systems*, vol.4, pp. 2236–2239.

Karaboga N. 2009. A new design method based on artificial bee colony algorithm for digital IIR filters. *Journal of the Franklin Institute, vol. 346, no.4*, pp. 328–348.

Kaveh A. & Mahdavi V.R. 2014. Colliding bodies optimization: A novel meta-heuristic method. *An International Journal of Computers and Structures, Elsevier, vol. 139*, pp. 18–27.

Kaveh A. & Talatahari S. 2010. A novel heuristic optimization method: charged system search. *Acta Mechanica, vol. 213, no. 3–4*, pp. 267–289.

Kennedy J. & Eberhart R. 1995. Particle swarm optimization. *Proceedings of the IEEE International Conference on Neural Networks, vol. 4*, pp. 1942–1948.

Ling S.H., Iu H.H.C., Leung F.H.F. & Chan K.Y. 2008. Improved hybrid particle swarm optimized wavelet neural network for modelling the development of fluid dispensing for electronic packaging. *IEEE Trans. Ind. Electron., vol. 55, no. 9*, pp. 3447–3460.

Litwin L. 2000. FIR and IIR digital filters. *IEEE Potentials*, pp. 28–31.

Liu G., Li Y.X., & He G. 2010. Design of Digital FIR Filters Using Differential Evolution Algorithm Based on Reserved Gene. *IEEE Congress on Evolutionary Computation*, pp. 1–7.

Lu H.-C. & Tzeng S.T. 2000. Design of arbitrary FIR log filters by genetic algorithm approach. *Signal Processing, 80*, pp. 497–505.

Luitel B. & Venayagamoorthy G.K. 2008. Differential Evolution Particle Swarm Optimization for Digital Filter Design. *IEEE Congress on Evolutionary Computation*, pp. 3954–3961.

Luitel B. & Venayagamoorthy G.K. 2008. Particle Swarm Optimization with Quantum Infusion for the Design of Digital Filters. *Swarm Intelligence Symposium*, pp. 1–8.

Mastorakis N.E., Gonos I.F. & Swamy M.N.S. 2003. Design of Two Dimensional Recursive Filters Using Genetic Algorithms. *IEEE Transaction on Circuits and Systems-I. Fundamental Theory and Applications, vol. 50*, pp. 634–639.

McClellan J.H., Parks T.W. & Rabiner L.R. 1973. A computer program for designing optimum FIR linear phase digital filters. *IEEE Trans. Audio Electro acoust., AU-21*, pp. 506–526.

Najjarzadeh M. & Ayatollahi A. 2008. FIR Digital Filters Design: Particle Swarm Optimization Utilizing LMS and Minimax Strategies. *International Symp. on Signal Processing and Information Technology*, pp. 129–132.

Parks T.W. & Burrus C.S. 1987. Digital Filter Design. *Wiley*, New York.

Parks T.W. & McClellan J.H. 1972. Chebyshev approximation for non recursive digital filters with linear phase. *IEEE Trans. Circuits Theory, CT-19*, pp. 189–194.

Rabiner L.R. 1973. Approximate design relationships for High-pass FIR digital filters. *IEEE Trans. Audio Electro acoust., AU-21*, pp. 456–460.

Saha S.K., Ghoshal S.P., Kar R. & Mandal D. 2013. A novel firefly algorithm for optimal linear phase FIR filter design. *International Journal of Swarm Intelligence Research, IGI Global, vol. 4, no. 2*, pp. 29–48.

Saha S.K., Kar R., Mandal D. & Ghoshal S.P. 2014. Harmony search algorithm for infinite impulse response system identification. *Computers and Electrical Engineering, Elsevier, vol. 40, no. 4*, pp. 1265–1285.

Saha S.K., Kar R., Mandal D. & Ghoshal S.P. 2014. Optimal linear phase FIR filter design using particle swarm optimization with constriction factor and inertia weight approach with wavelet mutation. *International Journal of Hybrid Intelligent Systems, IOS Press, vol. 11, no. 2*, pp. 81–96.

Sarangi A., Mahapatra R.K. & Panigrahi S.P. 2011. DEPSO and PSO-QI in digital filter design. *Expert Systems with Applications, vol. 38, No.9*, pp. 10966–10973.

Sun J., Feng B & Xu W.B. 2004. Particle Swarm Optimization with Particles Having Quantum Behaviour. *Proc. Congress on Evolutionary Computation*, pp. 325–331.

Frontiers in Computer, Communication and Electrical Engineering – Acharyya (Ed.)
© *2016 Taylor & Francis Group, London, ISBN: 978-1-138-02877-7*

An intelligent controller for the enhancement of voltage stability and power oscillation damping of an isolated micro grid

Asit Mohanty, Meera Viswavandya & Sthita Pragyan
CET Bhubaneswar, India

ABSTRACT: This paper focuses on the stability improvement in a standalone hybrid power system through a reactive power compensation and also discusses the oscillation damping in the system. A sliding mode based on the controller for the DFIG based wind turbine is proposed. For an improved stability and reactive power compensation a Linearized small signal model of STATCOM is considered. The Compensation has been carried out with the STATCOM Controller for different loadings and with several wind inputs.

1 INTRODUCTION

DFIG based Wind Turbines have been preferred and have got much attention as one of the preferred technologies for the wind energy conversion systems. The DFIG based wind turbine offers many advantages in comparison to others such as reduction of the inverter cost, better potential to control torque and having a better efficiency in wind energy extraction. Role of the DFIG based wind turbine in Isolated/Standalone hybrid power systems is very crucial and effective. Standalone systems are small power systems located at remote places to cater the local power demands of the places situated far away from the main grid. Generally, two or more renewable sources are combined to form a hybrid power system where shortage due to one source is compensated by the other. In a typical standalone configuration of Wind Diesel Micro hydro hybrid systems, diesel generators work as a backup with a wind turbine to provide power to remote loads. Normally Synchronous generators are preferred as Diesel Generator and SCIG/DFIG/PMIG are preferred to Wind Turbine for a better performance and for their rugged characteristics. In a typical Wind-diesel-Micro hydro hybrid system, wind is unpredictable and the system faces variable input loads. This disturbs the stability of the hybrid power system and needs capacitor banks and FACTS for enhanced stability and better reactive power compensation [1]–[3]. The challenges of the power quality issues like voltage instability and reactive power compensation are generally met by the use of FACTS (Flexible AC Transmission System) devices [4]–[6].

The advantage of the sliding mode control is that it is insensitive to the system parameters change and load disturbance. It shows good robustness, quick response, and an easy realization. Sliding Mode Control (SMC) has been proposed here to control DFIG based hybrid power system. For the sliding mode controller, the Lyaponov stability method has been applied to keep the nonlinear system under control SMC provides a fast and accurate dynamic response [7]–[8].

2 MATHEMATICAL MODELING

A detailed small signal model of the hybrid system is shown in Figure 1 and the reactive power balance equation of the system is mentioned below. The small change in the load ΔQ_L, the system terminal voltage varies affecting the reactive power. The reactive balance equation is $\Delta Q_{SG} + \Delta Q_{SVC} - \Delta Q_L - \Delta Q_{IG} - \Delta Q_{IGH} = 0$. The System Model Equation is through the transfer function equation, which is given below [14].

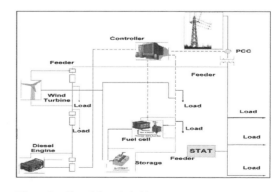

Figure 1.   Standalone hybrid system.

$$Q_{SG} + \Delta Q_{STAT} = \Delta Q_L + \Delta Q_{IG} + \Delta Q_{IGH} \quad (1)$$

$$\Delta V(S) = [\Delta Q_{SG}(S) + \Delta Q_{COM}(S) - \Delta Q_L(S) - \Delta Q_{IG}(S) - \Delta Q_{IGH}(S)]$$

$$R_{IG} = \frac{R_y}{R_Y^2 + X_{eq}^2} V^2 \text{ and } Q_{IG} = \left[\frac{X_{eq}}{R_Y^2 + X_{eq}^2} + \frac{1}{X_M}\right] V^2 .$$

$\dfrac{V^2}{X_m}$ term contributes towards

Electromagnetic energy storage equation for the modelling, the reactive power absorbed by the Induction generator is:

$Q_{IG} = \left\{\dfrac{X_{eq}}{R_Y^2 + X_{eq}^2}\right\} V^2$. The modified synchro-

nous generator equation is $Q_{SG} = \dfrac{(E'_q V \cos\delta - V^2)}{X'_d}$ for incremental change:

$$\Delta Q_{SG} = \frac{V \cos\delta}{X'_d \Delta E'_q} + \frac{E'_q q \cos\delta - 2V}{X'_d \Delta v} \quad (3)$$

The state space representation of the system is $\dot{X} = AX + Bu + Cw$.

The state, control, and disturbance vectors are x, u, and w of the wind diesel hybrid system and A, B, C are the matrices of the appropriate dimensions.

$$\underline{X} = [\Delta I_{dr}^{ref}, \Delta V, \Delta\delta, \Delta E_{fd}, \Delta V_a, \Delta V_f, \Delta E'_q]^T$$

$$\underline{u} = [\Delta V_{ref}] \quad \underline{w} = [\Delta Q_L]$$

$$\Delta Q_{SG}(s) = K_a \Delta E'q(s) + K_b \Delta V(s) \quad \text{Where}$$

$$K_a = \frac{V \cos\delta}{X'd} \quad K_b = \frac{E'q \cos\delta - 2V}{X'd} \quad (4)$$

The reactive power injected by the system bus is:

$$Q + V^2 B - KV_{dc} VB\cos(\delta - \alpha) + KV_{dc} VG\sin(\delta - \alpha) = 0$$

$$(5)$$

$$\Delta Q_{STATCOM}(S) = K_J \Delta\alpha(S) + K_K \Delta V(S) \quad (6)$$

$$K_K = KV_{dc} B_{\sin\alpha} K_L = KV_{dc} B_{\cos\alpha} \quad (7)$$

3 SLIDING MODE CONTROLLER DESIGN

The sliding mode control method works in three stages. (i) Sliding Surfaces (ii) Conditions of convergence, and (iii) Controller Design. The detailed algorithm is mentioned below. In order to have a stable operation and to enable independent control through

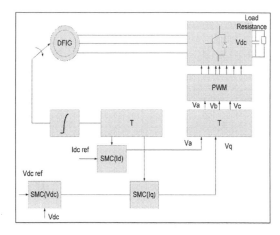

Figure 2. Sliding mode controller based DFIG Turbine.

smc based DFIG has been used for the smooth operation. ol active and reactive power of the DFIG The PI and SMC controllers have been formulated with the dynamic model equations and the results are compared. A schematic of sliding mode controller based DFI turbine ia shown in Figure 2.

$s(x,t) = (\frac{d}{dt} + \lambda)^{n-1} e(t)$ $e(t)$ is the error in the output state:

$e(t) = x_{ref}(t) - x(t)$ λ is a positive coefficient.

The convergence condition is defined by the equation of Lyapunov:

$s.\dot{s} \prec 0$. Consequently, the structure of a controller consists of two parts; a first one concerning the exact linearization and the second being that of stabilizing.

The equations $u(t) = u_{eq}(t) + u_n$, $u_{eq}(t)$ corresponds to the equivalent control, it is calculated on the basis of the system behaviour along the sliding mode described by: $\dot{s}(x,t) = 0$

$u_n = -K \text{sgn}(s)$ Where $K \rangle 0$
K is the control gain

$$\text{sgn}(s) = \begin{cases} 1 & s \rangle 0 \\ 0 & s \rangle 0 \\ -1 & s \langle 0 \end{cases}$$

4 SIMULATION RESULTS

From the MATLAB based simulation, which has been carried out taking a sliding mode based STATCOM Controller for the wind diesel micro hydro hybrid power system, the settling points and peak overshoots of the different parameters of the hybrid systems are noted. The reactive power and voltage stability of the hybrid power system have been optimized with a step load change of 5%. Variation of all the system parameters under system uncertainties as shown in Figure 3(a–d) are noticed using traditional PI Controller and sliding mode based STATCOM

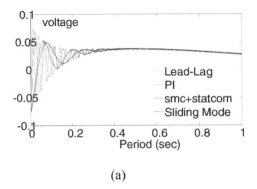

(a)

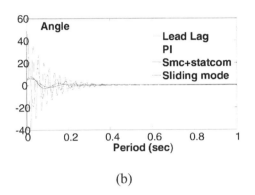

(b)

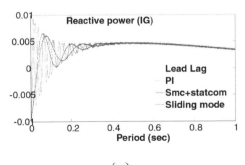

(c)

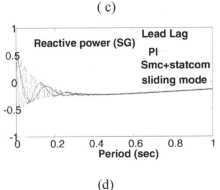

(d)

Figure 3. (a–d) Simulation results of the different parameters with the change in load and wind power.

controller. In the sliding mode approach the components I_d and I_q are regulated making I_d is zero; the controller is designed in the total system including switching devices. A sliding mode control method has been proposed and used for the control of the DFIG based hybrid power system. For reactive power compensation and stability analysis a STAT-COM has been used. Simulation results for Figure 3(a–d) show good performances with proposed sliding mode control and shows a good choice of the parameters control.

5 CONCLUSION

The reactive power compensation and voltage control of the hybrid system show better stability results with the sliding mode controller. With STATCOM and sliding mode control, the simulation results show a faster response without overshoot and the robust performance to parametric variation and disturbances in all the system.

REFERENCES

[1] Hingorani N.G, Gyugyi L. 2000.Understanding FACTS, Concepts and Technology of Flexible AC TransmissionSystem, *IEEE Power Engineering Society*: New York.
[2] Kaldellis J. et al. 1999. Autonomous energy system for remote island based on renewable energy sources, *in proceeding of EWEC*: Nice.
[3] Padiyar K.R, Verma R.K. 1991. Damping Torque Analysis of Static VAR System controller, *IEEE Transaction of Power System*: vol 6. No 2, PP458–465, May.
[4] Padiyar K.R, 2008. FACTS Controllering in Power Transmission system and Distribution, *New Age International Publishers*.
[5] Bansal R.C, Bhatti T.S, Kumar V. 2007. Reactive Power Control of Autonomous Wind Diesel hybrid Power System using ANN, *Proceeding of the IEEE Power Engineering Conference.*
[6] Mohanty Asit, Viswavandya Meera, Ray Prakash K, Patra Sandipan.2014.Stability analysis and reactive power compensation issue in a microgrid with a DFIG based WECS, *Electrical Power and Energy System.*
[7] Massoum A, Fellah M-K, Meroufe A.2005. Sliding mode control for a permanent magnet synchronous machine fed by three levels inverter using a singular perturbation decoupling. *Journal Of Electrical & Electronics Engineering*: Vol 5. Istanbul University.
[8] Louze L, Nemmour A, Khezzar A, Hacil M.E, Boucherma M.2002. Cascade sliding mode controller for self-excited induction generator. *Revue des Energies Renouvelables*: Vol. 12 N°4 (2009) 617–626 Ohm.

Frontiers in Computer, Communication and Electrical Engineering – Acharyya (Ed.)
© *2016 Taylor & Francis Group, London, ISBN: 978-1-138-02877-7*

Identification of lyapunov function for testing stability of nonlinear systems using BFO

C.M. Banerjee & A. Baral
Department of Electrical Engineering, Indian School of Mines, Dhanbad, Jharkhand, India

ABSTRACT: This paper proposes a methodology to access system stability for any nonlinear autonomous/non autonomous system. Here, Lyapunov Stability Theorem has been utilized to find that domain of attraction. It is known that obtaining suitable Energy Function that satisfies Lyapunov Stability criteria is difficult to formulate for a given nonlinear system. Hence the present paper is aimed at identifying the Energy Function for a given nonlinear system using an optimization technique. The proposed methodology, based on Bacteria Foraging Optimization (BFO) Algorithm is capable of identifying the domain of attraction along with the Energy Function which satisfies Lyapunov stability criteria. The discussed technique is first applied on Synchronous machine dynamics and thereafter, on a non autonomous system to show its practical effectiveness.

1 INTRODUCTION

The aim of this paper is to find suitable Energy Function and at the same time maximize domain of attraction for stability analysis of any nonlinear system. Available literature shows that an Energy Function, which does not satisfy the properties of stability, does not guarantee the instability of a given nonlinear system. The Lyapunov Energy Function for a given system being non-unique, the system under consideration can be said to be stable if any other Energy Function can be identified which satisfies stability conditions. Furthermore, it is advantageous to have the area of such region as large as possible (Panikhom et al., 2012).

Information regarding stability as obtained from the Eigen value (Nagrath, 2011) does not give satisfactory result for a nonlinear system. It is understood that for a given system, the definition of the Energy Function $V(x) = X^T P X$ can be altered by changing symmetric matrix P only. Here, X is the state vector of the system considered. The dimension of matrix P depends on number of state variables of the system. The discussed method not only identifies the Energy Function corresponding to a symmetric matrix P but also ensures that the Energy Function satisfies the stability conditions on its corresponding domain of attraction. It can be observed that identification of such an Energy Function is difficult especially for a nonlinear system. Therefore, in the present paper an optimization technique is used for addressing the above issue.

In the present paper, Bacteria Foraging Optimization is used to find suitable Lyapunov Function and maximize domain of attraction. Related analysis presented in this paper shows that compared to other optimization technique like, Adaptive Tabu Search (ATS) (Panikhom et al., 2012), application of BFO leads to identification of a much larger domain of attraction.

2 LYAPUNOV STABILITY: BRIEF THEORY

It is imperative that a description about Lyapunov Stability Theorem is provided before the actual problem formulation is presented. The following section briefly describes Lyapunov Stability theorem and stability criteria.

2.1 *Lyapunov Stability Theorem (Nagrath, 2011)*
• Stability *Property*:
For a system whose dynamics can be described by the equation $\dot{x} = f(x); f(0) = 0,$ *t*here exists a scalar Energy Function $V(x)$ which satisfies the following properties for all x in the region $\|x\| \leq \alpha$ where $\alpha > 0$.

   a. $V(x) > 0; x \neq 0$ and $V(0) = 0$ i.e. $V(x)$ is positive definite scalar function.
   b. $V(x)$ has continuous partial derivatives for all x.
   c. $\dot{V} \leq 0$ i.e. derivative of the Energy Function must be at least negative semi definite function.

Then it can be concluded that the system is stable at origin.

• *Asymptotically Stability Property*:
If the property (a), (b) of *Stability Property* remain valid and property (c) is substituted by $\dot{V} < 0$, $x \neq 0$ i.e. the function is negative definite function then the system is said to be asymptotic stable. The generalized quadratic form of Energy Function that has gained popularity among researchers is given by equation (1)

$$V(x) = X^T P X \qquad (1)$$

2.2 Domain of Attraction (Vidyasagar, 1993)

Information related to stability region R_{st} holds important information to design any system. In a stable domain of attraction, the Energy Function should satisfy the following properties

1. $V(x)$ should be positive definite in region R_{st}.
2. $\dot{V}(x)$ should be negative definite in region R_{st}.
3. The domain of attraction should contain equilibrium points.

If the above conditions are satisfied and the initial operating point of the system lies within R_{st} then as $t \rightarrow \infty$ the system operating point will converge to the equilibrium point. If, $V(x)$ denotes the Energy Function in a domain then the domain of attraction can be defined by equation (2)

$$\psi = \{x \in R^N \mid V(x) \leq d\}, d > 0 \qquad (2)$$

It is understood that the contour of $V(x)$ can be varied by changing real symmetric matrix P. The aim of this paper is to find suitable Energy Function $V(x)$ for a particular nonlinear system and maximize domain of attraction within which $V(x)$ satisfies stability conditions. To satisfy all above mentioned condition, Bacteria Foraging optimization is used in the present work.

3 BACTERIA FORAGING OPTIMIZATION

Foraging animal always try to maximize energy consumption per unit time during foraging period. BFO algorithm is modeled by keeping in mind the foraging behavior of Escherichia coli (E.Coli) bacteria. It is worth mentioning here that the movement of E.Coli bacteria during foraging is also known as chemo-tactic behavior (Passino et al., 2002).

In a search field nutrient may be present in patches. In any foraging technique the predator present in the search space locates these prey items (food) and attacks them. There are distinct advantages of social and intelligent foraging. For which communication between agents is required. Due to this, locating nutrients become easier and success rate is also increased. In BFO, bacteria attract or,

repel according to their position in nutrient or noxious substance.

E.Coli bacteria use its flagella for locomotion. E.coli bacteria swim and tumble using flagella. Within the search space E.Coli bacteria always try to move in positive nutrient direction. Bacteria's movement is controlled by attraction and repel (Das et al., 2009). The dynamics of bacteria movement can be summarized in following three steps (Passino et al., 2002).

• In neutral substances it tumbles and run in random direction.
• In homogeneous nutrient concentration i.e. zero nutrient gradient, bacteria mean run length, mean speed increase and mean tumble time decrease.
• In nutrient gradient bacteria spent more time in swimming less in tumbling. They move towards positive nutrient direction up to a certain period.

E.coli bacteria senses nutrient gradient from the difference between nutrient concentrations on the two ends of bacteria. After maximum chemotactic step of all bacteria's lifetime are over, bacteria with lower health die and those with better health split for reproduction (Chen et al., 2008). It should be kept in mind that during this process the total population bacteria are kept constant. This is called *Reproduction Event*. After maximum reproduction step, some bacteria are eliminated and disperse near nutrient location according to probability of elimination. This is called *Elimination and dispersal event*.

4 IMPLEMENTATION PROCEDURE

It is understood that the Energy Function $V(x)$ can be changed by changing real symmetric matrix P. To generate suitable $V(x)$, that satisfies Lyapunov Stability criteria, a proper real symmetric matrix P needs to be generated. If a system has two state variable x_1 and x_2 then Energy Function $V(x)$ can be represented by equation (3)

$$V(x) = \left(P(1,1) \times x_1^2\right) + \left(2 \times P(2,1) \times x_1 \times x_2\right) + \left(P(2,2) \times x_2^2\right) \qquad (3)$$

$\dot{V}(x)$ can be calculated from the system state variable x_1 and x_2 and their derivative $\dot{x}_1$ and $\dot{x}_2$ as illustrated in equation (4)

$$\dot{V}(x) = \left(2 \times P(1,1) \times x_1 \times \dot{x}_1\right) + \left(2 \times P(2,1) \times \{[x_1 \times \dot{x}_2] + [\dot{x}_1 \times x_2]\}\right) + \left(2 \times P(2,2) \times x_2 \times \dot{x}_2\right) \qquad (4)$$

The matrix P predicted by BFO at particular iteration 'i' is used to compute $V(x)$ according to equation (3). Once a closed path 'R_{st}' having area (A_i) for function $V(x)$ is identified, the following condition are checked. These conditions must be satisfied in order to classify $V(x)$ as a suitable candidate of Lyapunov Function.

- $V(x)$ must be positive definite for all points within the region
- $\dot{V}(x)$ must be negative definite for all points within the region

If both the above conditions are not satisfied the *cost_function* (J) is assigned to a large number (denoting a noxious substance or invalid state) else it is updated according to equation (5)

$$J_i = 1/A_1 \qquad (5)$$

It is understood that in order to test the above referred conditions, coordinates within the region 'R_{st}' must be identified. It is observed that the use of '*inpolygon*' function (available in MATLAB) to detect the suitable points is too time consuming. So a more efficient algorithm having less computational burden is devised for the discussed method.

The '*GRADE*' (Grosman et al., 2005) to which stability conditions on the detected domain of attraction are satisfied, can be checked by taking a small region whose radius is 10^{-3} and center is at the equilibrium point using equation (6).

$$\text{GRADE } G_i = -r_i \times 2^{-r_i*(D_+ + V_- - 1)} \qquad (6)$$

Where, r_i is the radius of the region and D_+ is number of points where $\dot{V}(x)$ is positive(Grosman et al., 2005) and V_- is number of points where $V(x)$ is negative.

5 RESULT AND DISCUSSION

The applicability of Bacteria Foraging Optimization in identifying P is illustrated by considering two examples.

Example 1: The dynamics of a synchronous machine is considered to be the first example. The system equation that govern the dynamics of synchronous machine (Genesio et al., 1984) is given by

- $\dot{x}_1 = x_2$
- $\dot{x}_2 = -D \times x_2 - \sin(x_1) + \sin(\delta)$

Where the power angle is indicated by x_1 and the corresponding speed deviation is indicated by x_2. In the present work it is assumed that the values of D and δ are 0.5 and 0.412 respectively. $x_2 = 0$, $x_1 = 0.412$ is the equilibrium point of interest.

The other critical points for the system described by above are $x_2 = 0$, $x_1 = (0.412 + 2\pi)$ and $x_2 = 0$, $x_1 = (-0.412 + \pi.)$.

The expression of $V(x)$, obtained using P matrix (corresponding to the minimum value of *cost_function* (J)) is given in equation (7)

$$V(x) = \left(0.2280 \times x_1^2\right) + \left(2 \times 0.1927 \times x_1 \times x_2\right)$$
$$+ \left(0.7720 \times x_2^2\right) + 7 \qquad (7)$$

It can be observed from Figure 1 that BFO converges after approx 1200 number of iterations for Example 1. It should be mentioned here that Figure 1 does not show the value of J for which $V(x)$ does not form a closed path as this condition denotes an highly noxious substance. The area of stability region (shown in Figure 2) is observed to much larger than the area found using Adaptive Tabu Search (Panikhom et al., 2012). By considering 10^{-3} radius circle (Grosman et al., 2005) with equilibrium point as its center, the GRADE (Grosman et al., 2005) at which the stability condition satisfied is 99.72%. This implies $V(x)$ is sufficient to shift the operating point of system into the equilibrium point of interest.

The surface plot of $\dot{V}(x)$ corresponding to the Energy Function given in equation (8) is shown in Figure 3. It can be observed from Figure 2 and Figure 3 that the identified Energy Function satisfies the Lyapunov stability criteria satisfactorily. Table 1 summarizes the performance of BFO over ATS (Panikhom et al., 2012) for the nonlinear system considered as Example 1.

Example 2: A non autonomous system means that the system is time varying. The system equation for such a non autonomous system (Panikhom et al., 2012), considered as Example 2, is given as

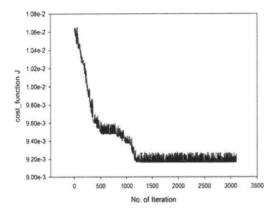

Figure 1. Variation of *cost_function* J for Example 1 with iteration count.

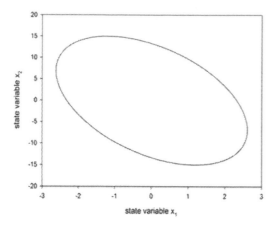

Figure 2. Domain of attraction identified for Example 1 with area 109.224.

tion (8). The range of t selected is an exemplary one. The proposed methodology is equally acceptable for any range of t.

$$V(x) = \left(0.8858 \times x_1^{\,2}\right) + \left(\left(2 \times -0.1295\right) \times x_1 \times x_2\right)$$
$$+ \left(0.1142 \times x_2^{\,2}\right) + 0.7857$$

(8)

Figure 4 shows the variation of *cost_function J* with number of iteration for the non autonomous system considered as Example 2. Like Example 1, Figure 4 also does not show the value of J for which $V(x)$ does not form a closed path in the solution space. The stability region identified for Example 2 is shown in Figure 5. Like Example 1, the area of the obtained stability region is observed to be much larger than the area found using Adaptive Tabu Search (Panikhom et al.,

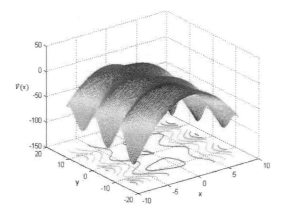

Figure 3. Surface plot of $\dot{V}(x)$.

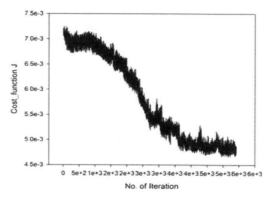

Figure 4. Variation of *cost_function J* for Example 2 with iteration count.

Table 1. Performance of BFO over ATS.

Method name	P(1,1)	P(2,1)	P(2,2)	AREA	GRADE (G_i)
ATS	0.2308	0.0632	0.322	11.2693	99.72%
BFO	0.2280	0.1927	0.772	109.224	99.72%

- $\dot{x}_1 = x_2$
- $\dot{x}_2 = 0.2 \times f(t) - 0.8 \times x_1 - 1.4 \times x_2$

where the $f(t)$ is a time varying function. For the present paper $f(t)$ is considered to be $sin(t)$ (Panikhom et al., 2012).

The expression of the Energy Function $(V(x))$ for minimum *cost_function J* using the P matrix (identified by BFO) for $2 \le t \le 5$ is given by equa-

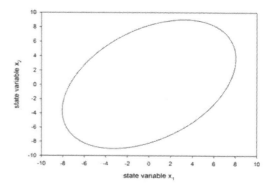

Figure 5. Domain of attraction identified for Example 2 with area 201.10.

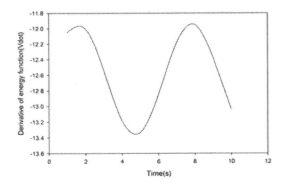

Figure 6. Variation of $\dot{V}(x)$ *with* time at $(x_1, x_2) = (2.8514, 4.3271)$.

2012). Figure 6 shows the variation of $\dot{V}(x)$ at a typical coordinate ($x_1 = 2.8514$ and $x_2 = 4.3271$) for the range of $1 \leq t \leq 10$. It is worth mentioning here that similar results are observed for other points within the search space. It is observed from Figure 6 that $\dot{V}(x)$ remains negative for the entire selected range of t selected. Hence, it can be reasoned that the P matrix identified using Bacteria Foraging Optimization is appropriate.

6 CONCLUSION

This paper discusses a Bacteria Foraging Algorithm based procedure for identifying suitable Lyapunov Energy Function for nonlinear system. The proposed method is capable of identifying the maximum domain of attraction for given system whose state equations are known. In the present work the capability of the discussed method is demonstrated by considering two examples namely, a synchronous machine and a non autonomous system. Analysis presented in this paper show that BFO based method is capable of identifying much larger domain of attraction compared to Adaptive Tabu Search.

REFERENCES

Chen, H., Zhu, Y., and Hu, K., "Self-adaptation in Bacterial Foraging Optimization algorithm", pp1026–1031, 2008.

Das, S., Biswas, A., Dasgupta, S., and Abraham A., "Foundations of Computational Intelligence", Vol. 3, Springer Berlin Heidelberg 2009.

Genesio, R., and Vicino, A., "New techniques for constructing asymptotic stability regions for nonlinear systems." IEEE Trans. Circuits Syst., vol. 31, No. 6, pp. 574–581, 1984.

Grosman, B. and Lewin, D. R., "Automatic generation of lyapunov functions using genetic programming", Vol. 16, No. 1, pp. 872–872, 2005.

Nagrath, I.J. and Gopal, M., "Control System Engineering", 5th Edition, 2011.

Panikhom, S. and Sujitjorn, S., "Numerical Approach to Construction of Lyapunov Function for Nonlinear Stability Analysis", Vol. 4, No. 17 pp. 2915–2919, 2012.

Passino, K.M., "Biomimicry of Bacterial Foraging for Distributed Optimization and Control," IEEE Control Systems Magazine, Vol. 22, No. 3, pp. 52–67, June 2002.

Vidyasagar, M., "Nonlinear Systems Analysis", Second Edition, Prentice Hall, Inc. Englewood Cliffs, New Jersey 1993.

Frontiers in Computer, Communication and Electrical Engineering – Acharyya (Ed.)
© *2016 Taylor & Francis Group, London, ISBN: 978-1-138-02877-7*

Development of a cross correlation based induction motor stator winding inter-turn fault severity indicator

Partha Mishra
EE Department, College of Engineering and Management, Kolaghat, India

Santanu Das
EE Department, Jalpaiguri Government Engineering College, Jalpaiguri, India

ABSTRACT: Stator winding inter-turn faults contribute to a major percentage of induction motor failure in the industries. A cross correlation technique based induction motor stator winding fault identification scheme has a significantly low computational burden that has been proposed in this paper. Presented scheme has been found to identify the inter turn fault conditions involving minor number of turns in three phase induction motor stator winding. Experimentally obtained, the three phase currents of the induction motor under healthy and faulty conditions were analyzed employing a suitably designed FIR digital filter and cross correlation technique. Captured motor current signals of the faulty phase were fed to the designed fault identification algorithm implemented in the form of a MATLAB program. The entire scheme has been found to identify stator winding inter-turn fault conditions quite efficiently and, thus, responses of cross-correlation between filtered motor currents under healthy and faulty conditions that have been proposed as an effective indicator for identifying such faults of varying severity.

1 INTRODUCTION

Induction Motors (IMs) are complex electromechanical devices utilized in most industrial applications for the conversion of electrical to mechanical form of energy. Most of the mechanisms in the industry are driven by three-phase induction motors which are energized either by constant frequency sinusoidal power supply or by ac drive. Induction motor is a widely accepted component in the industry due to its low cost, ruggedness, small size, low maintenance, and competent to work in any environmental condition such as heat, vibration, dust, moisture etc. Though IM is considered as a robust machine, it produces several types of electrical faults. Among all types of faults in IM, stator winding fault occurs mainly due to the extra voltage stress and high frequency current components are introduced by ac drives. Regarding the industrial perspective and other electrical appliances, IM fault monitoring and diagnosis is important to prevent motor failure to run the plant continuously for the extended hours. Any form of unexpected shutdown may lead to the wastage of raw materials, production time and hence economic losses. To reduce the effective maintenance cost, a contentious and efficient monitoring system capable of providing an indication about the inception of the fault is necessary to prevent premature failure of such machines.

According to IEEE and Electric Power Research Institute motor reliability study stator fault are responsible for 37% of the IM failures (Kliman *et al.* 1996, Thomson *et al.* 2001). The stator faults are due to turn- to- turn, phase to phase or winding to earth short circuit. This paper mainly describes the detection of inter-turn short-circuit introduced due to the different stresses mainly thermal and electrical in stator winding of IM. The stressful condition of the stator winding may lead to a breakdown of winding insulation. Insulation failure may occur in between turn-to-turn or phase-to- phase on the stator winding. This phenomenon causes a destructive effect on the stator winding.

During the last two decades, significant research works have been carried out for detection and diagnosis of inter-turn fault in IMs (Thomson *et al.* 2001, Gandhi *et al.* 2011). A number of online and offline techniques were proposed to find out the inter turn fault. For an online monitoring, Motor Current Signature Analysis (MCSA) has been proposed as the most prudent approach (Gandhi *et al.* 2011, Das *et al.* 2011). Other significant techniques are turn-to-turn capacitance calculation, online Partial Discharge (PD) analysis, and axially transmitted leakage flux. Few more popular offline techniques are Insulation Resistance (IR) measurement, Polarization Index (PI) measurement, dc and ac Hipot test, dc conducting test, capacitance impedance test, stator-current and voltage harmonic analysis, recording the negative-sequence impedance estimation. Other proposed methods include temperature analysis, magnetic field analysis, vibration analysis, and acoustic noise measurement (Kliman *et al.* 1996, Gandhi *et al.* 2011, Das *et al.* 2011, Das *et al.* 2014). Some modern knowledge based techniques such as artificial intelligence, neuro-fuzy technique; genetic algorithm, neural network, and Bayesian classifier have also been proposed as efficient (Gandhi *et al.* 2011, Das *et al.* 2011). But, most of the techniques suffer from either lack of high accuracy in fault detection and/or high computational burden.

It is still an open area of research to find out and propose a more efficient fault identification method with relatively low computational burden. This paper presents a simple yet efficient method with less computational burden to differentiate healthy from faulty conditions, which developed due to insulation degradation between turns in induction motor's stator winding. In this work, a few basic and popular signal conditioning and processing tools such as digital FIR band-pass filter and cross-correlation technique have been used. Motor phase currents are filtered to discard frequency components around fundamental and also very high frequency components. After a suitable digital filtering of faulty phase current, the extracted signal containing a specific frequency band has been found to be sensitive to fault. Output signal of the digital filter corresponding to a healthy motor current has been used as a reference signal for further analyses. Signature resulting from cross correlation between filtered signals corresponding to healthy motor phase current and faulty phase current has been found to be an effective indicator to identifying such faults and also their severity. Experimentally obtained comprehensive data sets corresponding to healthy and faulty conditions of the stator winding were used to verify and validate the proposed scheme.

2 EXPERIMENTAL DETAILS

In the present work, experimental studies have been conducted on a 1/2 HP, 400V, 4 pole, 3-phase induction motor with star connected stator winding, having 8 coils per phase and 70 turns per coil. To carry out the desired test, the following apparatus and arrangements have been used.

- A 3-phase auto-transformer, which has the capability to provide the variable voltage ranges from 0% to 125% of the primary side rated supply voltage. It has been used to maintain throughout the rated voltage to motor in spite of utility end voltage.

- A data acquisition system with **YOKOGAWA** 3-phase power meter [Model WT 230] to capture stator current signals under different experimental conditions.
- A customized stator winding where all stator coil ends and several tappings from different points of windings were brought out of the machine to a patch-board. This arrangement allowed several faulty conditions to be created by means of appropriate short-circuiting links. Exact positions of tappings in Y-phase winding have been presented in the Table 1 and also shown in Figure 1.

Photograph of the experimental setup is shown in Figure 2.

In this experimental study, induction motor has been tested under a no load condition. At first, a rated three phase supply from the auto-transformer is fed to the motor under healthy conditions and corresponding three line currents of the motor has been captured at 20 kHz sampling frequency by employing digital power meter interfaced with PC. Then, the inter-turn fault conditions were artificially created one-by-one involving 0.2% (1 turn) to 2.5% (14 turns) turns in Y-phase winding by short circuiting two external taps at a time as listed in Table 1. Motor current signals were captured in all fault cases for further analysis. Such type of faults change the stator symmetrical current to the asymmetrical. Exemplary plots of captured three phase motor current signals under healthy and few faulty cases are shown in Figure 3.

Stator inter-turn insulation failures and resulting short-circuiting of turns lead to impedance unbalance among different phase windings. From

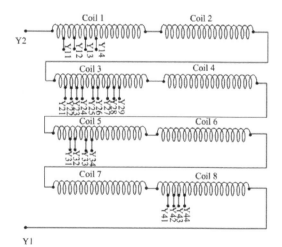

Figure 1.   Positions of the taps in Y phase winding.

Table 1. Fault conditions in Y-phase winding.

Sl. No.	Shorting between taps	No. of turns involved	% of stator winding turns/phase shorted
1	Y25-Y26	1	0.18
2	Y11-Y12	2	0.36
3	Y21-Y22	3	0.53
4	Y12-Y13	4	0.71
5	Y22-Y23	5	0.89
6	Y13-Y14	6	1.07
7	Y23-Y24	7	1.25
8	Y31-Y32	8	1.43
9	Y42-Y41	9	1.61
10	Y2-Y11	10	1.78
11	Y32-Y33	11	1.96
12	Y42-Y43	12	2.14
13	Y33-Y34	13	2.32
14	Y44-Y43	14	2.5

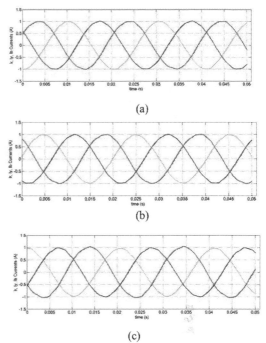

(a)

(b)

(c)

Figure 3. Captured motor current signals under (a) healthy, (b) 3 turns short, (c) 14 turns short in Y-phase winding.

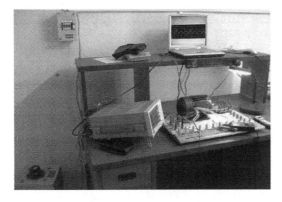

Figure 2. Photograph of the experimental setup.

the Figure 3, it is clear that such fault cases involving minor number of turns hardly introduce any visually recognizable unbalanced in the current signals. Therefore, it is necessary to implement a prudent technique to extract visually recognizable fault indicator, which would enable an easy interpretation as per an identification and also discrimination of faulty conditions from healthy motor is concerned. Since the objective of the work was to find out a suitable fault identification tool, the analyses have been carried out taking faulty phase motor currents only.

3 DATA ANALYSIS

Any kind of internal faults in the induction motor introduces an imbalance in the three phase motor currents, which may not be always identified through a mere visual inspection specially when the developed fault level is very low, i.e., in case of minor faults. But the current Concordia pattern and the orientation of the major axis of the corresponding pattern has been used to identify the faulty phase (Das et al. TENCON 2011). Therefore, in this analysis, the current in the faulty phase, which is of a Y-phase in present study has only been analyzed. In the first step of analysis, an aim has been set to mask the frequency components in the current profiles, which are not much sensitive to the variations of the fault levels. To achieve a Finite Impulse Response (FIR) type digital band-pass filter has been designed and found to be performing satisfactorily in this application as compared to other popular digital filters. This band-pass filter keeps a band of frequency components in the current signal of the faulty phase while discards low frequency and very high frequency components from the same. It has been done so considering the fact that fundamental components in the motor current does not play much role in detection of fault or abnormal conditions (Das et al. 2011). The magnitude of the fundamental components in the current signal varies significantly only with load. On the other side, very high frequency components can be treated as noise and those normally get introduced in motor currents while acquiring current signals at a high sampling frequency.

3.1 Brief theory of FIR digital filter

Digital filter is a discrete system, and it does a series of mathematic operations to the input signal, to obtain the desired response. Details of Finite Impulse Response (FIR) type digital filter design can be found in (Peng 2013). Summary of the important steps of the digital filter design is presented here for ready reference.

The transfer function for a linear, time-invariant, digital filter is usually expressed as (Peng 2013):

$$H(z) = \frac{\Sigma_j^M b j z^{-j}}{1 + \Sigma_{i=1}^N a_i z^{-i}} \tag{1}$$

where, a_i and b_j are coefficients of the filter in the Z-transform. There are different types of digital filters, and also different ways to classify them. Similar to the other filters, FIR filters can be classified into four categories, which are low-pass, high-pass, band-pass, and band-stop filter. According to the impulse response, there are usually two types of digital filters, which are Finite Impulse Response (FIR) filters and Infinite Impulse Response (IIR) filters. According to the formula in equation (1), if a_i is zero, then it is a FIR filter, otherwise, if there is at least one none-zero a_i, then it is an IIR filter. Usually, three basic arithmetic units are needed to design a digital filter; which are the adder, the delay, and the multiplier blocks.

Finite Impulse Response (FIR) filter is one of the most basic elements in digital signal processing and it can guarantee a strict linear phase frequency characteristic with any kind of amplitude frequency characteristic. Unit impulse response of FIR filter is finite; so, FIR filters are a stable system. FIR filter has a broad application in many fields, such as telecommunication, image processing, and so on.

The system function of FIR filter is:

$$H(z) = \sum_{n=0}^{L-1} h[n] z^{-n} \tag{2}$$

where, L is the length of the filter, and $h[n]$ is the impulse response.

At the time of designing a filter, it is necessary to specify a pass-bands, stop-bands, and transition bands. In pass-band, frequency components are needed to be passed without attenuation and in stop-band, frequency components need to be passed attenuated. Transition band contains frequencies, which are lying between the pass-band and stop-band. Therefore, the entire frequency range is split into one or even more pass-bands, stop-bands, and transition bands. Practically, the magnitude is not necessary to be constant in the pass-band of a filter. A small amount of ripple is usually allowed in the pass-band. Similarly, the filter response will not be zero in the stop-band. A small, nonzero value is also tolerable in the stop-band.

FIR filter is designed using least-squares error minimization has been employed in the present analysis. Cut off frequencies of FIR band-pass filter are chosen to be 80 Hz and 800 Hz. Figure 4 shows few samples of filtered output of faulty phase (Y-phase) motor current signals. Employing the Fast Fourier Transformation (FFT) tool the frequency responses of the filtered signals are verified as presented in Figure 5.

Frequency responses of the selected frequency band have been found to be sensitive to fault level and that may be observed in Figure 5 also.

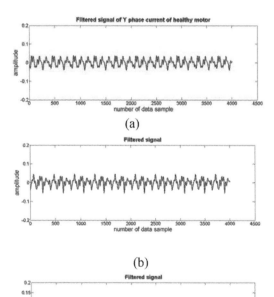

(a)

(b)

(c)

Figure 4. Filtered signals of Y-phase motor current under (a) healthy, (b) 2 turns short in Y-phase winding, and (c) 6 turns short in Y-phase winding.

152

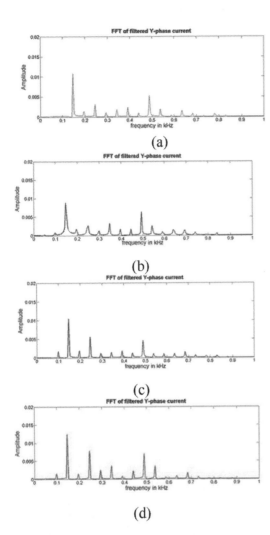

(a)

(b)

(c)

(d)

Figure 5. FFT profiles of filtered Y-phase motor current under (a) healthy, (b) 2 turns short in Y-phase winding, (c) 8 turns short in Y-phase winding, and (d) 12 turns short in Y- phase winding.

3.2 Brief theory of cross correlation

In signal and system analysis, the concept and implementation of autocorrelation and cross correlation plays an important role (Taghizadeh 2000). Correlation is a mathematical tool, which is used to analyze functions or series of data or value or time series data. Correlation is the mutual similarity or any other relationship between two or more signals. Autocorrelation is the correlation of the signal itself whereas cross correlation is the correlation of two different signals. In autocorrelation, generally, lagged tendency of a signal is observed through a long time series of data in the past and future values. The cross correlation function measures the

dependency of the values of one signal on another signal. It is the simplest tool used to compare similarity or dissimilarity between two signals.

The autocorrelation function of a random signal describes the general dependence of the values of the samples at one time on the values of the samples at another time. Consider a random process $x(t)$ (i.e. continuous-time), its autocorrelation function is written as:

$$R_{xx}(\tau) = \lim_{T \to \infty} \frac{1}{2T} \int_{-T}^{T} x(t)\, x(t + \tau)dt \qquad (3)$$

where, T is the period of observation.

$R_{xx}(\tau)$ is really valued and an even function with a maximum value at $\tau = 0$.

For a sampled signal, the autocorrelation is defined as either biased or unbiased as follows:

$$R_{xx}(m) = \frac{1}{N-|m|} \sum_{n=1}^{N-m+1} x(n)\, x(n+m-1) \qquad (4)$$

[Biased Autocorrelation]

$$R_{xx}(m) = \frac{1}{N} \sum_{n=1}^{N-m+1} x(n)\, x(n+m-1) \qquad (5)$$

[Unbiased Autocorrelation]

For $m = 1, 2, \ldots M + 1$, where M is the number of lags.

For cross correlation, two WSS (Wide Sense Stationary) process $x(t)$ and $y(t)$ it is defined by:

$$R_{xy}(\tau) = \lim_{T \to \infty} \frac{1}{T} \int_{-T}^{T} x(t)y(t + \tau)dt \qquad (6)$$

Or,

$$R_{yx}(\tau) = \lim_{T \to \infty} \frac{1}{T} \int_{-T}^{T} y(t)x(t + \tau)dt \qquad (7)$$

Where T is the observation time. For sampled signal it is defined as:

$$R_{yx}(m) = \frac{1}{N} \sum_{n=1}^{N-m+1} y(n)\, x(n+m-1) \qquad (8)$$

Where, $m = 1, 2, 3, \ldots N + 1$ and N is the record length (i.e. number of sample).

Properties of cross correlation function are as follows:

- $R_{xy}(m)$ is always a real valued function which may be positive or negative.
- $R_{xy}(m)$ may not necessarily have a maximum at $m = 0$ nor $R_{xy}(m)$ an even function.

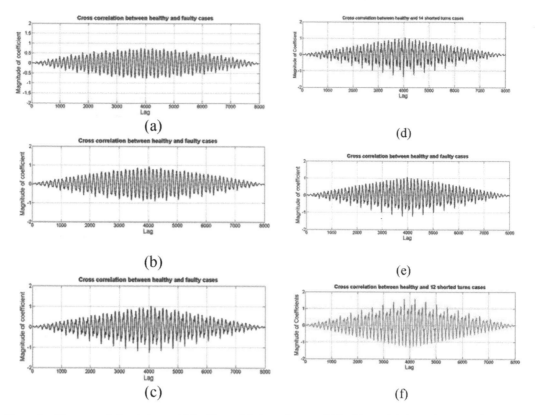

Figure 6. Cross correlation between (a) healthy and 2 shorted turns, (b) healthy and 5 shorted turns,(c) healthy and 6 shorted turns, (d) healthy and 14 shorted turns, (e) healthy and 10 shorted turns, (f) healthy and 12 shorted turns.

- $R_{xy}(-m) = R_{yx}(m)$
- $|R_{xy}(m)|^2 \leq R_{xx}(0)\, R_{yy}(0)$
- $|R_{xy}(m)| \leq \frac{1}{2}[R_{xx}(0) + R_{yy}(0)]$

When $R_{xy}(m) = 0$, $x(n)$ and $y(n)$ are said to be 'uncorrelated' or they are said to be statistically independent (assuming they have zeros mean.)

4 RESULTS AND DISCUSSIONS

In this investigation, filtered stator current signal of faulty phase has been found to be most sensitive to the faults in the respective phase winding. It has been verified through the FFT responses of the filtered signals under healthy and faulty cases as presented in Figure 5. Filtered stator current signals under healthy conditions of winding must have some differences from that of the faulty conditions of the stator winding. Since the level of stator winding inter-turn faults being studied is at a very low or minor level, the difference between healthy and faulty signals may not be significant enough. So to get a significant decimating information about the fault, a sensitive and highly accurate tool needs to

be employed. In view of that, cross correlation as a statistical signal processing tool has been adopted for the analyses. Lag verses coefficient plots obtained from cross correlation analyses have been found to provide encouraging information, which can be used to assess degree of deviation of motor fault conditions from its healthy state. As shown in Figure 6, the lag vs. coefficient magnitude plots of the cross correlation analysis between healthy and faulty signals evolves a typical pattern. Envelops of the patterns are found to increase monotonically in the vertical directions with increasing fault severity. That may be observed in presented Figure 6. Figure 6 (a), (b), (c), (d), (e), and (f) are responses of cross correlation between healthy and 2 shorted turns, healthy and 5 shorted turns, healthy and 6 shorted turns, healthy and 10 shorted turns, healthy and 12 shorted turns, and healthy and 14 shorted turns, respectively.

5 CONCLUSION

A new type of stator winding inter-turn fault indicator has been proposed in this paper. Digital filtering followed by cross correlation tools have been employed to develop this scheme, which is found to be accurately and unambiguously identifying short circuit faults in stator winding involving minor number of turns. Such minor faults may not be easily discernible at its early stage and may lead to a catastrophic failure of some critical industrial process driven by three phase induction motor. The entire analysis has been carried out on a faulty phase current of the motor. Proposed scheme has been validated on the motor developing inter-turn winding fault under a no-load operating condition only. Therefore, the effectiveness of the scheme still needs to be verified on motors running at load and also under supply voltage unbalance conditions. But, the outcome of this work has established the effectiveness of the digital filtering and cross correlation tools in detection of such minor faults.

REFERENCES

Das S., Koley C., Purkait P., & Chakravorti S. (2014) "Performance of a Load-Immune Classifier for Robust Identification of Minor Faults in Induction Motor Stator Winding", *IEEE Transactions on Dielectrics and Electrical Insulation*,. 21(1). 33–44.

Das S., Purkait P., Dey D., & Chakravorti S (2011). Monitoring of Inter-Turn Insulation Failure in Induction Motor using Advanced Signal and Data Processing Tools. *IEEE Transactions on Dielectrics and Electrical Insulation. 18(5)*. 1599–1608.

Das S., Purkait P.,& Chakravorti S. (2011). Relating Stator Current Concordia Patterns to Induction Motor Operational Abnormalities. *in Proc. IEEE TENCON, Bali, Indonesia*.

Gandhi A., Corrigan T., & Parsa L. (2011). Recent advances in modeling and online detection of stator interturn faults in electrical motors. *IEEE Trans. Ind. Electron*., 58(5).1564–1575.

Kliman G.B., Premerlani W.J., Koegl R.A., & Hoeweler D.(1996). A new approach to on-line turn fault detection in ac motors. in Proc. 31st IEEE IAS Annual Meeting, 1996.

Peng Su. (2013). Design and analysis of FIR digital filter based on Matlab.

Taghizadeh S.R. (2000). Digital Signal Processing Case Study.

Thomson W.T. & Fenger M.(2001). Current Signature Analysis to Detect Induction Motor Faults. *IEEE Ind. Appl. Magazine, pp. 26–34*.

Frontiers in Computer, Communication and Electrical Engineering – Acharyya (Ed.)
© 2016 Taylor & Francis Group, London, ISBN: 978-1-138-02877-7

Webpage prediction using latest substring association rule mining

R.P. Chatterjee
Department of Computer Science and Engineering, Meghnad Saha Institute of Technology, Kolkata, India

M. Ghosh, M.K. Das & R. Bag
Department of Computer Science and Engineering, Supreme Knowledge Foundation Group of Institutions, Mankundu, West Bengal, India

ABSTRACT: Web page prediction plays an important role by predicting and fetching probable web pages of the next request in advance, resulting in reducing the user latency. This paper proposes a web page prediction model giving significant importance to the user's interest using the clustering techniques and the navigational behavior of the user through latest substring association Rule. This method achieves a better precision compared to the recent methods in the web usage mining.

1 INTRODUCTION

The users surf the internet either by entering the URL or search for some topic or through the link of the same topic. For searching and for link prediction, clustering plays an important role. Website designers want to increase the number of visitors and the time that these visitors spend on their website. In order to accomplish that, they have to supply an attractive content. Also, to make their content attractive, website designers and content providers need to know what their potential visitors want, in order to organize their content according to their visitors needs, and, if possible, according to individual preferences. Researchers use different techniques like Markov Model [Jin X. *et al.* 2003], association rule mining, clustering [Dutta R. *et al.* 2011] and so on. Web usage mining [Barsagade *et al.* 2003] is the application of data mining techniques to extract knowledge from web data, where at least one of the structure or usage data is used in the mining process. Web usage mining has various application areas such as web pre-fetching, link prediction, site reorganization and web personalization. Most important phases of web usage mining are the reconstruction of user sessions by using heuristic techniques and discovering useful patterns from these sessions by using pattern discovery techniques like association rule mining, Apriori etc. We propose an integrated system for applying data mining techniques such as association rules on access log files.

1.1 *Related work*

A web page prediction model gives us significant importance to the user's interest using the clustering [Dutta R. *et al.* 2011] techniques and the navigational behavior of the users through Markov model. The clustering technique is used for the accumulation of the similar web pages. Similar web pages of the same type reside in the same cluster; the cluster containing web pages have the similarity with respect to topic of the session. The clustering algorithms [Dutta R. *et al.* 2011] considered are K-pages that are stored in the form of cellular automata to make the system more memory efficient. Sequential classifiers [6] from association rules obtained through data mining on a large web log data have been proposed by [Yang Q. *et al.*] and by significant statistical correlations the next likely web page to be predicted. Another web page prediction method is web pre-fetching [Jin X., Xu H. 2003], which predicts the next request for web pages based on the current request of users that are analyzing the server log, and fetching them in advance and loading them into the server cache. It reduces the perceived access delay to some extent and improves the service quality of web server.

2 PROPOSED METHOD

Given a web log, the first step is to clean the raw data. We filter out documents that are not requested directly by users. These are image requests in the log that are retrieved automatically after accessing requests to a document containing links to these files. We consider web log data as a sequence of distinct web pages, where user sessions can be observed by unusually long gaps between consecutive requests.

We have created a unique ID for each web page link that exists in the web log. After the binary context corresponding to that unique ID to count how many times a particular link of a web page has been visited by users for a particular session are created. Next Apriori algorithm (Agrawal *et al.* 1993) to find out frequent web pages from all the previously user visited web pages has been used.

2.1 *Apriori algorithm*

This algorithm is used for mining frequent item sets for Boolean rules where frequent subsets have extended one item at a time (a step known as candidate generation (Agrawal *et al.* 1993), and groups of candidates are tested against the data. The algorithm terminates when no further successful extensions are found. It first finds all frequent 1-itemsets (Agrawal *et al.* 1993), and then discovers 2-itemsets and continues by finding increasingly larger frequent item sets.

Key Concepts:

- Frequent Item sets: The sets of the item, which has minimum support (sup).
- Apriori Property: Any subset of the frequent item set must be frequent.
- Joint Operation: To find Lk, a set of candidate k-item sets are generated by joining Lk-1with itself.

For each rule of the form LHS→RHS, we define the support and confidence as follows:

$$sup = count(LHS, RHS)/count(Table) \qquad (1)$$
$$conf = count(LHS, RHS)/count(LHS) \qquad (2)$$

In the equations above, the function count (Table) returns the number of records in the log table, and the count (LHS) returns the number of records that match the left-hand-side LHS of a rule. In the beginning, each page with sufficient support forms a (length–1) support pattern. Then, in the main step, for each k value greater than 1 and up to the maximum reconstructed session length, supported patterns (patterns satisfying the support condition) with length k + 1 are constructed by using the supported patterns with length k and length 1 as follows. If the last page of the length-k

Table 1. Latest substring association rule.

W1	W2	Substring rules
A,B,C	D	{C} ->D
B,A,C	D	

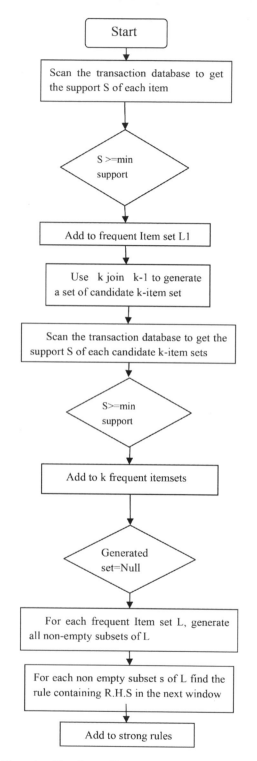

Figure 1. Flowchart of latest substring association rule mining

158

pattern has a link to the page of the (length–1) pattern, then by appending that page length-k+1 candidate pattern is generated. At some k value, if no new supported pattern is constructed, the iteration halts.

2.2 *Latest substring association rule*

After finding out the frequent web pages according to apriori algorithm we use "Latest Substring Association Rule" to generate the rules as shown in Table 1.

The "latest-substrings" are in fact the suffixes of the strings in W1 window. These rules not only take into account the order and adjacency information, but also how recent the information about the LHS string is. In this representation, only the substring ending in the current time (which corresponds to the end of the window W1) qualifies to be of the LHS rule.

In our example, we can easily observed that in window W1 the first session contains the sequence {A,B,C} and the second session contains {B,A,C}. But the suffixes 'C' contains both the sessions at the end and 'D' is the only predicted page presented in the window W2. Hence, only one rule{C->D} can be generated. In this way, we use the "Latest Substring Association Rule" in the web page ranking. The proposed approach has been depicted by the flow chart diagram in Figure 1.

3 RESULTS AND DISCUSSION

For a test case that consists of a sequence of web page visits, the prediction for the next page visit is correct if the RHS of the selected rule occurs in window W2. For N different test cases, let C be the number of correct predictions. Then the precision is defined as:

$$precision = C/N \qquad (3)$$

First, we measured support of our input file, which contains user access path. Results have been presented in Table 2 and Figure 2. Among them we find mostly frequent access path by using Apriori algorithm (Agrawal *et al.* 1993). From the frequent user access path we made the Association rule (Agrawal *et al.* 1994) and there corresponding Confidence in the different steps of the user access path. If the rule contains minimum confidence threshold then we have treated the corresponding test case as correct. We have made the experiment on 10 different test cases for r = 4, r = 6, r = 8, r = 10, and r = 12 and after all calculate the precision by using the above mentioned formula for each value of r and plot them into the graph above.

Table 2. User access paths steps and corresponding precision value.

User Access Path steps (r)	Precision
4	50%
6	60%
8	60%
10	60%
12	60%

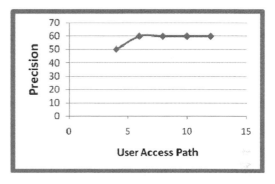

Figure 2. User access path with respect to precision value has been shown.

4 CONCLUSION

In this paper, we surveyed the association rule mining techniques using Apriori Algorithm and the latest substring association rules, which has been experimented and having a good result. The previously used algorithms have too many parameters for somebody who is a non-expert in data mining and the obtained rules are far too many, most of them are non-interesting and with low comprehensibility .We have overcome this problem in our methods but there are some acute problems which we had observed during our experiment. The larger the set of frequent itemsets, the more the number of rules are presented to the user, many of which are redundant. We have also faced a problem of dynamic itemset counting, i.e, the web pages are frequent or not it should be decided dynamically.

REFERENCES

Agrawal, R., Imielinski, T. and Swami, A.N. (1993). Mining association rules between sets of items in large databases. *In Proceedings of the 1993 ACM SIGMOD International Conference on Management of Data*, 207–216.

Agrawal, R. and Srikant, R. 1994. Fast algorithms for mining association rules. *In Proc. 20th Int. Conf. Very Large Data Bases*, 487–499.

Baralis, E. and Psaila, G., Designing templates for mining association rules. *Journal of Intelligent Information Systems*, 9(1): 7–32, July 1997.

Barsagade, N., Web Usage Mining and Pattern Discovery CSE 8331, December 8, 2003.

Dutta, R. and Kundu, A., Mukhopadhyay D. Clustering Based Web Page Prediction, 2011.

Jin, X. and Xu, H. 2003. An Approach to Intelligent Web Pre-fetching Based on Hidden Markov Model, Dec 12, 2003.

Yang, Q., Li, T. and Wang, K., School of Computing Science, Simon Fraser University, Burnaby, BC, Canada V5A 1S6, Building Association-Rule Based Sequential Classifiers for Web-Document Prediction.

Frontiers in Computer, Communication and Electrical Engineering – Acharyya (Ed.)
© 2016 Taylor & Francis Group, London, ISBN: 978-1-138-02877-7

Development of a compact, portable setup for demonstration of corona phenomenon

S. Mukherjee, R. Ghosh, B. Chatterjee & S. Dalai
Electrical Engineering Department, Jadavpur University, Kolkata, West Bengal, India

ABSTRACT: The development of a low-cost, compact and portable corona setup for classroom demonstration has been reported in this paper. Since an in-depth understanding of corona is difficult based solely on the theoretical aspects, a portable corona setup for classroom demonstration is expected to be more beneficial to students. The developed setup may be used to demonstrate visual corona at both positive and negative electrodes. It is a basic model that efficiently demonstrates the fundamental stages of corona such as inception of visual corona, appearance of streamers and then final breakdown. The students were made to note down their observations based on the demonstration, and the observations were correlated with the physics behind corona effect during the class. It was found that the use of the corona setup helped the students have a better understanding of the theory behind corona formation.

1 INTRODUCTION

Engineering education all over the world has predominantly been a combination of theory sessions and hands-on laboratory sessions. Traditionally, the students begin by attending theory sessions where they gain a theoretical understanding of different engineering phenomena. Subsequently, they perform laboratory experiments on some of the more important phenomena they have studied. Since most engineering curriculums give priority to theory classes rather than practical demonstrations, only some of the more common phenomena are demonstrated to the students through hands-on experiments due to time limitations.

The normal practice while teaching engineering students is to first organize a lecture where the basic theory related to a particular phenomenon is explained to the students. This is true even in various areas of electrical engineering, where it is difficult for the students to grasp the fundamental idea without looking at the phenomenon itself. Later, a laboratory demonstration may be conducted wherein the students are expected to look at the phenomenon and correlate their observations with what they have learnt in the theory sessions. The problem with this approach is that the students were expected to understand and correlate a practical engineering problem based purely on theoretical knowledge. This can often lead to the development of misconceptions regarding engineering phenomena among students (Goris *et al.* 2012, Trotskovsky *et al.* 2013).

This is where the importance of models in engineering comes in. Models are a simple, cost effective and safer alternative to demonstrate real-world phenomena (Rothenberg *et al.* 1989). Nowadays, models have been widely accepted to provide a more authentic understanding of science and engineering among students (Trotskovsky *et al.* 2015). Models can also help in avoiding the complexity and hazards involved with certain engineering phenomena. However, models have not been fully utilized as yet to improve the quality of imparting engineering education to students. Consequently, a little modification in the way certain complex phenomena are taught in the classes can go a long way in improving student satisfaction.

One such area where the use of models can significantly improve the student understanding is the corona effect in electrical engineering. Corona is a partial discharge brought about by the ionization of air surrounding a conductor that is electrically charged. To have an in-depth knowledge on corona, understanding the theory behind its occurrence is often insufficient. Instead, a demonstration of the corona effect in the classroom through a model, and correlating the observations with the theory behind corona formation, is expected to be more beneficial to the students.

With this in mind, the authors have developed a corona setup that is simple, cost-effective and easily reproducible. The entire setup can be housed inside a box of size 30 cm × 15 cm × 15 cm and may be carried easily into the classrooms. The setup gives a voltage output of 25 kV_p dc with the distance between the electrodes varying from 12 mm to 40 mm. Using this setup, the entire phenomenon starting from the inception of visual corona to breakdown may be demonstrated. It was found that the classroom demonstration helped the students have a better understanding of the theory behind corona formation and to make various observations which have been enumerated later in the paper.

2 DESCRIPTION OF THE CORONA SETUP

2.1 *Construction*

Figure 1 shows a schematic representation of the corona setup developed for classroom demonstration. A low duty cycle oscillator stage at 50 Hz is constructed with the help of a widely available 555 timer IC. The output from the 555 timer is fed to a MOSFET driver stage for current amplification. The output of the MOSFET is fed to the primary of a fly-back transformer which is designed to generate high voltage saw-tooth signals at a relatively high frequency. Commercially available EHT fly-back transformers have in-built rectifier circuits to produce dc output. Hence, the output of the circuit is half wave rectified dc with 25 kVp. This 25 kVp output is fed to the corona pointed electrode arrangement. The developed corona setup has been shown in Figure 2.

2.2 *Working*

As stated earlier, the corona setup is housed in a box which can be conveniently carried into classrooms. The setup is plugged into an ordinary 230 V, 50 Hz power socket and the phenomenon is demonstrated to students. The following paragraph summarizes the overall working of the setup:

As the power supply is switched on and the distance between electrodes is at 40 mm nothing is seen. The distance is gradually decreased using the slider. As the distance is reduced to 38 mm, a faint violet glow appears at the negative electrode;

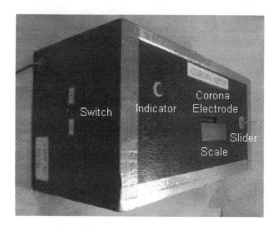

Figure 2. Photograph of the developed corona setup.

but nothing can be seen at the positive electrode. As the distance is further decreased, the intensity and size of the glow will increase. At a distance of 34 mm a faint bluish glow appears at the positive electrode. With a further decrease in the distance, the visual corona at both electrodes increases in size and finally at about 24 mm transient arcing starts to appear. As the separation is reduced to about 12 mm, complete breakdown takes place.

3 OBSERVATIONS

The demonstration of the corona setup in the laboratory was found to be fruitful for the students' understanding of the phenomenon. The students expressed their profound interest in the demonstration session by actively participating in the learning process. The students were asked to note down their observations in detail. The teachers then correlated the observations made by the students with the physics involved with the observations. The observations from the corona setup and the corresponding explanations have been enumerated below.

Observation 1: Visual corona is observed at the negative electrode while nothing can be seen at the positive electrode when the distance between the two electrodes is about 38 mm. Visual corona begins to appear at the positive electrode when the distance is about 34 mm. This implies that the corona inception voltage in a non-uniform field is lower with the negative polarity of voltage than that with the positive polarity.

Explanation: On application of voltage, the avalanche is first formed and then the transition takes place and a streamer develops. The corona inception is mainly controlled by the space charges before the formation of streamer. On the

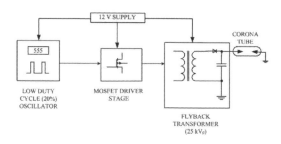

Figure 1. Block diagram of corona setup.

other hand breakdown is the result of growth of streamer.

In the case of negative point electrode to positive plate electrode, the avalanche moves towards the plate with the electrons at its head, as shown in Figure 3. This will result in the separation of charges: the electrons with higher velocity move faster and the positive ions, though sluggish, move in the direction of higher field which results in higher drift velocity. The position of the space charges is shown in Figure 4. Presence of positive ion space charge in the vicinity of the negative point results in the increased field strength at the cathode. This may exceed the corona inception voltage of air and corona may start around the negative point.

In the case of positive point to negative plate electrode configuration, the avalanche formed in a region of high field moves towards the anode. The electrons with higher velocity and moving towards a high field region are swept into the anode leaving behind the positive space charge drifting towards the low field region. This is elaborated in Figure 5 and Figure 6. The presence of positive ion space charge before the anode substantially reduces the

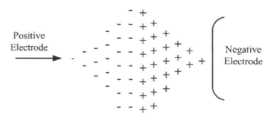

Figure 5. Electron and positive ion avalanche in a positive point to negative plate electrode arrangement.

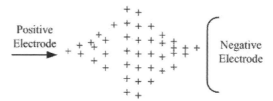

Figure 6. Space charge distribution in a positive point to negative plate electrode arrangement.

Figure 7. Bluish glow at the positive electrode (left) and violet glow at the negative electrode(right).

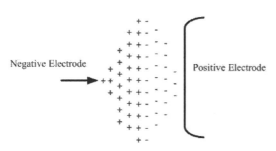

Figure 3. Electron and positive ion avalanche in a negative point to positive plate electrode arrangement.

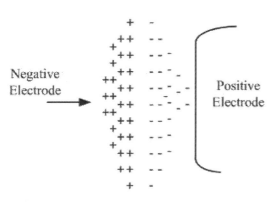

Figure 4. Space charge distribution in a negative point to positive plate electrode arrangement.

field near the anode and to reach the critical ionization gradient, applied voltage must be increased.

Observation 2: In case of negative corona, there is violet glow whereas positive corona has a bluish glow. This may be seen from Figure 7.

Explanation: The emission of light during the corona is caused by the recombination of electrons and positive ions leading to the formation of neutral atoms. As the electrons fall back to their previous energy level, photons of light are emitted. Near the negative electrode, the electrons are repelled and the ions are attracted, whereas the opposite happens near the positive electrode: the electrons are attracted and ions are repelled. As a result, electrons have a higher drift velocity near the negative electrode. This results in higher energy emission due to recombination. This explains why negative corona has a violet glow and positive corona has a bluish glow.

Observation 3: Positive corona appears to be a little smaller than the corresponding negative corona.

Explanation: Negative corona has a non-ionizing plasma region between the inner and outer regions which is absent in positive corona. A negative corona can be divided into three regions around the electrode. In the inner region, inelastic collisions occur between high-energy electrons and neutral atoms, which lead to avalanches. This inner region is known as the ionizing plasma region. In the intermediate region, electrons do not have sufficient energy to cause avalanche ionization. They collide with neutral atoms to form negative ions. This non-ionising region, however, remains a part of the plasma region and has the ability of taking part in characteristic plasma reactions as different polarity species are present. This region is called the non-ionizing plasma. The outer electrons (usually of much lower energy than inner electrons) combine with neutral atoms to produce negative ions. These ions flow toward the positive electrode. Additionally, a flow of free electrons also takes place toward the positive electrode. The outer region is known as the unipolar region.

The positive corona can be divided into two regions. The inner region contains ionizing electrons and positive ions which collide with neutral atoms to create an avalanche of ion/electron pairs. This forms the plasma region. In the outer region, positive ions slowly migrate away from the positive electrode. Very little electrons are present in this region. In the boundary of the two regions, photons strike neutral atoms liberating secondary electrons. These secondary electrons are re-accelerated into the plasma. Only a small percentage of the secondary electrons remain in the outer region. The outer region is known as the unipolar region.

4 CONCLUSIONS

The corona setup is a basic point-to-point electrode setup to demonstrate visual corona at both positive and negative electrodes. It is a very basic model that efficiently demonstrates inception of visual corona, appearance of streamers and then final breakdown. The power consumption of the experimental set up is very low, typically around 2.4 W. The model described here is a low cost, simple yet effective for corona investigations. Using this setup, a demonstration of the corona effect in the classroom was given. It was found that the students were able to observe some salient features of corona formation and were able to correlate them with the theory behind corona formation.

REFERENCES

Goris T.V. and Dyrenfurth, M.J. "Concepts and misconceptions in engineering, technology and science. Overview of research literature," Proc. Amer. Soc. Eng. Educ. IL/IN Sectional Conf., Valparaiso, IN, USA, Mar. 17, 2012.

Rothenberg, J. "The nature of modelling," in Artificial Intelligence, Simulation, Modeling, L. E. Widman, K. A. Lopara, and N. R. Nielsen, Eds. New York, NY, USA: Wiley, 1989, pp. 75–92.

Trotskovsky, E., Sabag N. and Waks, S. "Students' Achievements and Misunderstandings When Solving Problems Using Electronics Models—A Case Study," IEEE Trans. Educ., vol. 58, no. 2, pp. 104–109, 2015.

Trotskovsky, E., Waks, S., Sabag, N. and Hazzan, O. "Students' misunderstandings and misconceptions in engineering thinking," Int. J. Eng. Educ., vol. 29, no. 1, pp. 1–12, 2013.

Design of the bipolar, floating HVDC source for the insulation diagnostics

A. Kumar, N. Haque, R. Ghosh, B. Chatterjee & S. Dalai
Department of Electrical Engineering, Jadavpur University, Jadavpur, Kolkata, India

ABSTRACT: This paper describes the design and development of a bipolar, floating 5 kV DC source. The paper explores a cost effective alternative to the high voltage power supply units and comprises of easily available components that are used in low voltage systems. The generated voltage is controllable throughout a wide range (from 1 kV to 5 kV) by a closed loop control system. Experimental results demonstrated that the developed source is capable of providing a stable voltage output for low loads, typically suitable for the insulation diagnostic purpose where the current required is in the order of microamperes. A simple yet effective metering scheme is implemented to get a stable DC output voltage even under floating conditions. It is believed that the developed high voltage source can be very useful for insulation testing and scientific research purposes.

1 INTRODUCTION

High Voltage DC power supplies are quite frequently used in areas of testing and research in the field of dielectric response analysis, applied physics research, and other industrial applications that require testing of insulation having low leakage current. Especially, in the field of electrical engineering, high voltage sources have very prominent roles. Due to these reasons, the development of low cost, yet compact high voltage DC source has attracted researchers and engineers for quite a long time. Various methods have been utilized to develop high voltage DC sources (Law *et al.* 2010, Jin *et al.* 1997, Benwell *et al.* 2008, *Wai et al.* 2007). A good high voltage power source should have a wide adjustable range, stable operation over the entire selectable range along with a full short circuit protection and proper safety measures (Kuffel *et al.* 2000). For laboratory purposes, it is intended that the source should be lightweight, portable, and should have the facility of interchanging the polarity and earthing of any of the terminals for unground, as well as grounded sample under test. Keeping all things in mind, a low cost, portable, high voltage DC power supply source is developed in the laboratory. The developed source has an adjustable range of 1–5 kV with floating, stable output at 1 mA. Any of the terminals can be earthed for diagnostic purpose of different types of dielectric samples, both grounded and ungrounded. An effective metering scheme that can indicate the output voltage under the earthing of any of its output terminals is also integrated into the source.

2 OPERATING PRINCIPLE

The developed scheme, by virtue of its operating principle, can be divided into few sub-sections that are explained subsequently.

2.1 Flyback Transformer

Flyback Transformer (FBT), also known as Line Output Transformer (LOPT) or Extra High Tension Transformer (EHT), is a major equipment to generate low power high voltage DC in Cathode Ray Tubes (CRT) for several years. Basically, it generates high voltage from horizontal line output pulses at a high frequency, which are rectified using a half-wave rectifier circuit. The primary winding of the flyback transformer is connected to a series of electronic switching arrangement from a low voltage (around 12 volt) DC supply. Usually, the switch comprises of a MOSFET or power transistor. In these types of transformers, the primary winding has a few number of turns while the secondary has had a very large number of turns. When the switch is closed, the current in the primary builds up within milliseconds. Thus, energy is stored in the core. When the switch is suddenly switched on, a very high voltage is induced in the primary circuit due to high inductance and the high rate in the change of current. This overvoltage is sufficient to destroy the electronic switching element. This voltage should be limited by using a fast operating diode with a breakdown voltage lower than that of the switching elements. According to the turn's ratio of the transformer, the secondary output becomes very high; the rectifier diode conducts and delivers the current to the load by releasing energy from the core. The same process is repeated in each cycle. The output voltage is made stable by using a bank of ballast capacitors. The frequency of switching the primary is normally kept around 10–15 kHz to match the operating characteristics of the ferrite core transformer used in this work.

2.2 Basic scheme

The heart of the circuit is based on a commonly available 555 timer IC with an unstable mode of frequency that is 10–15 kHz, and a duty cycle of 20%–40%.

The switching element used in the circuit is an n channel MOSFET. A fast acting Zener diode is used parallel to the MOSFET to limit the maximum primary voltage at an appropriate voltage level.

A low voltage, isolated tertiary winding of 24 V in the EHT is used for the feedback loop. This tertiary voltage is rectified and filtered to give a smooth DC output. The schematic representation of the experimental set-up is shown in Figure 1.

The scheme of feedback loop is simple. A series of silicon diode is used at the base of an NPN silicon transistor. As soon as the voltage at the anode of the diode exceeds 1.4 V, the transistor conducts and puts the IC in reset mode. Thus, according to the setting of variable resistance, the output of the tertiary winding must vary to keep a constant voltage of 1.4 V at its wiper. Since, the high voltage winding, the tertiary winding and the main primary winding are all magnetically coupled the final high voltage output varies by varying the control resistance. This scheme produces a stable high voltage output from the developed source while keeping both output terminals floating from the ground. Either of the terminals can be earthed according to the user's choice. The terminals can also be reversed for a reverse polarity.

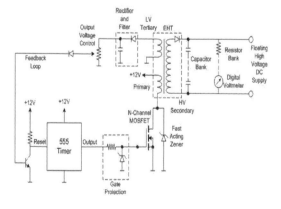

Figure 1. Schematic of the experimental set-up.

2.3 Metering scheme

The output voltage of the developed source is too high to be measured with an ordinary voltmeter. Therefore, a proper metering scheme is needed so that the output can be measured with a reasonable accuracy. In this work, a resistor bank (shown in Figure 1) is utilized for an indirect output voltage measurement. The basic principle of the metering scheme is to feed the digital voltmeter a very small fraction of the actual output current. If the impedance of the meter and series resistance is known, then the output can be easily calculated from the digital reading.

Since the source current is itself the order of a few milliamperes, the series resistance used in the metering scheme is in the order of 99 MΩ for a meter impedance of 1 MΩ so that a straight multiplication factor of 100 can be used with a load current of on 50 μA at 5 kV. But in practice, it is very difficult to achieve this value of a series resistance in a single package with low surface leakage and breakdown voltage greater than 5 kV. Hence, in this scheme, several carbon film resistors are used. Carbon film resistors were used because they show fewer variations with little temperature change. The surface of these resistors is covered with liquid polyurethane. This decreases the surface leakage considerably so that a higher accuracy is obtained in the reading. This has been verified using an electrostatic voltmeter.

3 RESULTS

The developed DC source was able to produce high voltage DC upto 5 kV. The DC output was smooth and absolutely stable when the load was less than 1 milliampere. The control circuitry discussed in the previous section enables the device to control the output voltage from 1 kV to 5 kV. By connecting any of the output terminals to ground, floating polarity output was easily achieved. As discussed before, two knobs were utilized for a smoother control one for the coarse and an another one for

Table 1. Output voltage at different knob configurations.

Coarse knob Position	Output voltage	
	Minimum voltage (kV)	Maximum voltage (kV)
1	1	1.1
2	1.9	2.2
3	2.5	3
4	3.8	5.2

fine tuning. Once the coarse knob is fixed, the output was further tuned using the second knob. The output voltage at different knob configurations is shown in Table 1. The output obtained through the metering scheme was further compared with Electrostatic Voltmeter (ESV) present in the laboratory. It was observed, that for load current of 100 µA, the meter output was very much similar to that of the ESV reading. However, with an increase of load current, some deviations were found. The comparison of the obtained metering output and ESV reading of the developed high voltage DC source is shown in Figure 2. This comparison is also tabulated in Table 2.

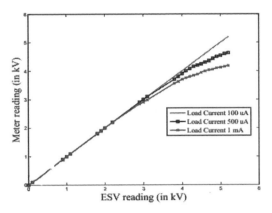

Figure 2. Comparison of obtained meter reading and ESV reading.

Table 2. Comparison of meter reading with ESV reading.

ESV	Meter reading		
	Load current (100 uA)	Load Current (500 uA)	Load Current (1 mA)
0	0	0	0
0.1	0.1	0.1	0.1
0.9	0.9	0.9	0.9
1	1	1	1
2.0	2.0	2.0	1.9
2.2	2.2	2.2	2
2.9	2.9	2.9	2.2
3.8	3.8	3.8	2.6
3.9	3.9	3.84	2.7
4.0	4.0	4.0	2.75
4.2	4.2	4.15	2.9
4.4	4.4	4.24	3.56
4.6	4.6	4.35	3.75
4.8	4.8	4.5	3.84
5.0	5.0	4.6	3.94

4 CONLUSION

In the present work, a cost effective high voltage DC source is developed using very low cost and easily available components. The developed source demonstrates stable and controllable output throughout a wide range for especially low loads (less than 1 mA). The output voltage ranges from 1 kV to 5 kV and it can be adjusted using two knobs. Additionally, the developed source is lightweight, portable, and has the facility of reversible polarity. Such a source can be extremely useful for the insulation diagnostic purpose as insulation diagnostic tests generally require high voltages but very low currents. It is believed that the voltage source developed can be used in the laboratory research on solid insulation materials, i.e., polymers, dry type insulation. It can be also used in space charge investigations.

REFERENCES

Benwell, A., Kovaleski, S., & Kemp, M. (2008, May). A resonantly driven piezoelectric transformer for high voltage generation. In IEEE International Power Modulators and High Voltage Conference, Proceedings of the 2008 (pp. 113–116). IEEE.

Jin, J.X., Dou, S.X., Liu, H.K., & Grantham, C. (1997). High voltage generation with a high T/sub c/superconducting resonant circuit. Applied Superconductivity, IEEE Transactions on, 7(2), 881–884.

Kuffel, J., Kuffel, E., & Zaengl, W.S. (2000). High voltage engineering fundamentals. Newnes.

Law, M.K., & Bermak, A. (2010). High-voltage generation with stacked photodiodes in standard CMOS process. Electron Device Letters, IEEE, 31(12), 1425–1427.

Wai, R.J., Lin, C.Y., Duan, R.Y., & Chang, Y.R. (2007). High-efficiency DC-DC converter with high voltage gain and reduced switch stress. Industrial Electronics, IEEE Transactions on, 54(1), 354–364.

Frontiers in Computer, Communication and Electrical Engineering – Acharyya (Ed.)
© *2016 Taylor & Francis Group, London, ISBN: 978-1-138-02877-7*

Minimization of return loss using minimum steps coaxial coupler for ka-band helix TWT

R. Guha
Sir J.C. Bose School of Engineering, Mankundu, Hooghly, West Bengal, India

N. Purushothaman & S.K. Ghosh
Council of Scientific and Industrial Research, Central Electronics Engineering Research Institute, Pilani, Rajasthan, India

ABSTRACT: In this paper the impedance matching between the characteristic impedance of coaxial coupler and helix and the return loss improvement has been done using step by step modeling of the input and output coaxial coupler with minimum number of steps in the centre conductor for ka-band helix travelling-wave tube. Applying single port excitation and proper boundary conditions in CST Microwave Studio, for both input and output section in a helix travelling-wave tube, the return loss has been investigated for single step as well as multiple steps uniform, quarter-wavelength long sections, coaxial coupler along with the sever and the attenuator coating in the T-shaped dielectric support rods. Using this technique more than −15 dB return loss, throughout the frequency band, and −46 dB return loss, at the design frequency, has been achieved.

1 INTRODUCTION

Coaxial coupler is one of the crucial parts of the helix travelling-wave tubes (TWTs). It acts as an impedance transformer to convert the characteristic impedance of the helix slow-wave structure to the standard 50 Ω TNC connector. It feeds the radio-frequency signal into the slow-wave structure and extracts the amplified signal from the slow-wave structure after suitable beam-wave interaction. That is why the design and optimization of this coaxial coupler sections is very difficult. In a helix TWT attenuator coating is applied to the three dielectric support rods (each separated azimuthally by 120°) to stabilize the device against oscillations.

Some research works about the designing and optimization of coaxial coupler has been reported earlier. Some techniques are proposed by the different authors to make the design of the coaxial coupler simple and time efficient for the simulation purpose (Sinha *et al.* 2000, Ghosh *et al.* 2007, Agrawal *et al.* 2011, Rao *et al.* 2012). In order to reduce the reflected wave, generated from the impedance mismatch, multi-section impedance transformer with arbitrary length was used (Sinha *et al.* 2000). But this technique suffers from a huge computational time because of its arbitrary length sections. In a complete helix travelling wave tube,

attenuator coating on the dielectric support rods are applying to get a considerable reduction of the reflected wave from the sever end. To make a prediction about the reflection due to impedance mismatch and coaxial-to-helix discontinuity at the input and output coupler ports one should have consider the attenuator coating at the sever end (Rao *et al.* 2009).

In above all cases, for modeling simplicity, authors directly connect the coaxial coupler with the helix, considering the diameter of the first section of the coupler equal to the helix width. That consideration limits the diameter of the coaxial coupler sections. Practically the coaxial coupler is connected with the helix with the help of platinum tape, of high conductivity, through laser welding.

In this paper, the authors are presenting a technique that not only reduce the simulation effort but also avoid the hit-and-try or iterative method by not applying the impedance transformation technique directly on the load impedance. This technique is also useful for understanding the nature of the reflected wave from different discontinuities. The full simulation of the modeled coaxial coupler along with this platinum tape, sever and attenuator coated dielectric support rods and ceramic window has been done using CST Microwave Studio, which gives a practical insight.

2 THEORETICAL BACKGROUND AND DESIGN OF COAXIAL COUPLER

In any coaxial coupler, for helix TWT, the reflection at the coupler port is mainly contributed by the five discontinuities, namely, i) dielectric discontinuity between window ceramic (alumina, $\varepsilon_r = 9.4$) and the vacuum, inside the metal barrel, ii) coaxial step discontinuity, between the conductors of the coupler, iii) coupler to platinum tape discontinuity, iv) platinum tape to helix discontinuity and v) attenuator discontinuity at the sever end. When the RF signal is fed into the coupler port, the RF wave reflected back from these discontinuities and setup a standing wave pattern in the coupler arm, and it is not easy to analyze this reflection characteristic from the each discontinuity separately from the superimposed reflected wave at the coupler port. But using this proposed method one can characterize the effect of, more or less, each discontinuity separately. Initially, the design of the coupler has been carried out by making the characteristic impedance of a single section coaxial coupler, of one quarter-wavelength ($\lambda/4$) long, equal to the characteristic impedance of the slow-wave structure, calculated from the equivalent circuit analysis (B.N. Basu, 1996). Slow-wave structure parameters used for analysis and simulation, are taken from a tube under development at CSIR-CEERI. From theoretical point of view, due to the prefect impedance matching, the return loss for the single section impedance transformer will be minimum. But the reflected wave still exists with a considerable value, due to the both platinum tape discontinuities and attenuator discontinuity at the sever end.

2.1 Transmission line concept for coaxial coupler design

From transmission line concept, one can say that the load impedance of the transmission line is the helix characteristic impedance (Z_h) and the coaxial coupler is an impedance transformer with characteristic impedance Z_0 that transform the load impedance into an input impedance (Z_{in}). So, one can write the relation of Z_{in} and Z_h for a loss-less line as (Peter A. Rizzi, 1986)

$$Z_{in} = Z_0 \frac{Z_h Cos\beta d + jZ_0 Sin\beta d}{Z_0 Sin\beta d + jZ_h Cos\beta d} \qquad (1)$$

where β is the phase constant ($= 2\pi/\lambda$) and d is the length of the line. For $\lambda/4$ long lines βd will be $\pi/2$. Using this present method Z_0 is equal to the Z_h. So, in this case, equation (1) is reduced to,

$$Z_{in} = Z_h \qquad (2)$$

The characteristic impedance of the coaxial coupler has been evaluated from,

$$Z_0 = \frac{60}{\sqrt{\varepsilon_r}} \ln \frac{b}{a} \qquad (3)$$

where ε_r is the relative permittivity of the medium between inner and outer conductor, b is radius of the outer conductor and a is the radius of the inner conductor.

2.2 Attenuator coating on the dielectric support rods

The attenuator coating on the support rods has been done using thin conducting carbon coating by the process of pyrolytic cracking of hydrocarbons such that coating thickness is increased toward the sever end. Usually, the carbon-coated rods are quality controlled by measuring the loss profile through insertion loss (S_{21}) in a reduced-height rectangular waveguide, excited in the dominant TE_{10} mode, measurement setup (Naidu et al. 2009). Now the measured insertion loss values are used to find the equivalent bulk conductivity using the following equations given in (Naidu et al. 2009)

$$\sigma = \frac{b^2 \omega \mu_0}{4\omega_{rod}^2 Z_{VI}^2} \left(\frac{1}{|S_{21}|} - 1 \right)^2 \qquad (4)$$

the characteristic impedance of the waveguide (Z_{VI}) is the voltage–current characteristic impedance defined as

$$Z_{VI} = \left(\frac{\pi b}{2a} \right) Z_{TE} \qquad (5)$$

with

$$Z_{TE} = \frac{k_0}{\beta wg} \eta_0 \qquad (6)$$

where σ is the equivalent bulk conductivity of the lossy dielectric rod and w_{rod} is the equivalent width of the support rod of a square-shaped cross-sectional geometry having an area equal to its actual cross-sectional area, b is the distance between the broad walls of the reduced-height waveguide, Z_{TE} is the TE-mode wave impedance of the waveguide with k_0 as the free-space propagation constant, β_{wg} is the propagation constant of the waveguide and η_0 as the free-space intrinsic impedance (377 Ω), $|S_{21}|$ is the magnitude of voltage-scattering parameter corresponding to the insertion loss.

Fig. 1 shows the tip loss profile and pitch tapering for the output section of the ka-band helix TWT. Fig. 2 shows the CST-MWS studio model of the single section coaxial coupler, with ceramic window, along with the attenuator coated dielectric T-shaped support rods.

3 RESULTS AND DISCUSSION

From the equivalent circuit analysis (B.N. Basu, 1996), we got the characteristic impedance of the helix at the center frequency is 84.94 Ω. Fig. 3 rep-

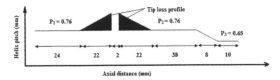

Figure 1. Tip loss and tapering pitch profile of a two section ka-band helix traveling wave tube.

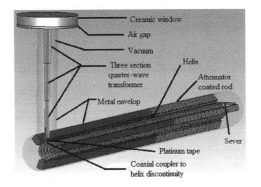

Figure 2. CST-Microwave Studio model of three quarter-wave sections coaxial coupler with sever, attenuator and ceramic window.

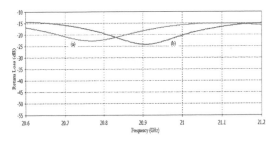

Figure 3. Return loss comparison of a single section quarter-wave impedance transformer with characteristic impedance equal to helix characteristic impedance: (a) from analysis it is 84.94Ω, (b) after optimization it becomes 85.63Ω.

resents the comparative study of the return loss of single section quarter-wave impedance transformer using analytical and optimized characteristic impedance value of the helix using CST-Microwave Studio. The design parameters of coaxial line coupler has been listed in Table 1.

The coating length of each support rod is taken approximately 22 mm, from the sever end, which give the total loss around 25–30 dB in the structure at the sever end by providing a tapered profile loss on the individual rods, the bulk conductivity variation with the distance from the sever end is shown in Fig. 4. To apply this equivalent bulk conductivity value to the modeled rods, the authors divide the dielectric rods into small discrete rods and then assign the conductivity value to each small discrete part as per the distance from the sever end, as shown in Fig. 2 and 4.

In Fig. 4, the value of the equivalent bulk conductivity goes down near to zero at 10 mm from the sever end. So, the same value of conductivity, at 10 mm from the sever end, can be imposed on remaining part of the coating length.

After finalizing the diameter, 0.48 mm, of the first part of the coaxial coupler, we use the binomial multi-section impedance transformation technique to design the entire coupler along with ceramic window assembly with improved impedance matching, as shown in Fig. 2. Using this three section coaxial coupler a very good return loss has been achieved for the input coupler, i.e. −15dB loss throughout the frequency band and −46dB loss at the design frequency, as shown in Fig. 5.

The same coupler has been used for output section, where the tapering pitch profile has been

Table 1. Coaxial coupler design parameters.

Parameters	Dimensions
1st section radius (mm)	0.240
2nd section radius (mm)	0.290
3rd section radius (mm)	0.385

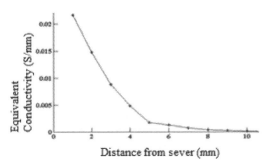

Figure 4. Equivalent bulk conductivity variation depend upon distance from sever end.

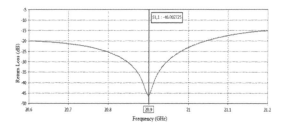

Figure 5. Return loss versus frequency response for a three quarter wave sections input coaxial coupler.

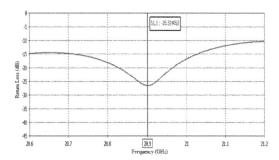

Figure 6. Return loss versus frequency response for a three quarter-wave sections output coaxial coupler.

considered for increasing the efficiency of the helix traveling wave tube and −10 dB loss has been achieved throughout the frequency band and −26 dB loss has been get at the design frequency, as shown in Fig. 6.

The input and output coaxial coupler consists of RF window assembly and an impedance RF transformer assembly. Coaxial ceramic window of Alumina has been preferred over Beryllia because of its low RF loss, high mechanical strength and non-toxicity. The thickness of the Alumina window was chosen as minimum as possible (0.9 mm) for low RF loss and its diameter was chosen to make the input impedance nearly 50Ω, same as RF connector.

4 CONCLUSION

In this paper. the authors have proposed a technique that is very useful to determine the helix characteristic impedance by minimum reflection method and also gives a insight about the reflection from the different discontinuities. This three quarter-wave sections coaxial axial coupler, with platinum tape, attenuator coating, sever and ceramic window, gives a insight of the practical coaxial coupler with significant reduction of return loss. This is a very general technique and this technique can be applied to all types of coupler and along with the different frequency bands.

REFERENCES

Agrawal, A.K., Raina, Sushil, Kumar, Lalit. (2011). A Simple Method for the Design of a Coupler for Helix TWTs. IEEE.

Basu, B.N. (1996). *Electromagnetic Theory and Applications in Beam-Wave Electronics.* (World Scientific, Singapore).

Ghosh, T.K. *et al.* (2007). Optimization of coaxial couplers. *IEEE Trans. Electron Devices*, vol. 54, no.8, pp. 1753–1759.

Naidu, V.B., Datta, S.K., Rao, P.R.R., Agrawal, A.K., Reddy, S.U.M., Kumar, L., and Basu, B.N. (2009). Three-dimensional electromagnetic analysis of attenuator coated helix-support rods of a traveling-wave tube, *IEEE Transaction on Electron Devices.*

Rao, K.V., Naidu, V.B., Rao, P.R.R.; Datta, S.K. (2009). Simulation of RF coupler of a multi-section TWT with matched sever-loss. *Proc. IEEE International Vacuum Electronics Conference*, pp. 455–456.

Rao, P. Raja Ramana., Datta, Subrata Kumar. and Kumar, Lalit. (2012). Optimization of couplers of TWT using TDR method. IEEE.

Rizzi, Peter A. (1986). Microwave engineering, passive circuits.

Sinha, A.K., Singh, V.V.P., Srivastava, V. and Joshi, S.N. (2000). On the design of coaxial coupler having multi-section short transformer for compact sized power helix traveling wave tubes. *Proc. Int. Vac. Electron. Conf.*, Monterey, CA, pp. p2.25–p2.26.

Frontiers in Computer, Communication and Electrical Engineering – Acharyya (Ed.)
© 2016 Taylor & Francis Group, London, ISBN: 978-1-138-02877-7

Human identification by gait using wavelet transform and the analysis of variance

M. Ghosh & S. Chatterjee
Department of Computer Science Engineering, Supreme Knowledge Foundation Group of Institutions, Mankundu, West Bengal, India

D. Bhattacharjee
Department of Computer Science Engineering, Jadavpur University, Kolkata, West Bengal, India

ABSTRACT: In this work a novel feature extraction technique has been discussed, which uses a discrete wavelet decomposition and Daubechies4 filters up to four levels of decomposition on gait energy image. Statistical parameter variance has been considered for feature extraction. RBF network has been applied for the recognition of probe gait sequence with reference to gallery gait sequences. The experimental results show a very good recognition rate for different degrees of movement with respect to the surveillance camera.

1 INTRODUCTION

Gait recognition is the term used in the human identification domain to represent the automatic extraction of visual indication that describes the activity of a walking person under a surveillance camera. Gait recognition is rather promising among those biometrics that are used to evaluate and compute biometrical data.

It is a soft biometric for the identification of an individual by way in which they walk. The classification of people at a distance can be accomplished via an unobtrusive biometric without any interface or cooperation from the subject.

The biometric methodologies like face, iris, and fingerprints have their own limitations, i.e., most of the face recognition techniques are used to recognize the frontal or side faces or based with some particular turn, but if only the back side of the head is seen then it is of no use. Other biometrics such as fingerprint and iris cannot be applied when the person unexpectedly appears in the surveillance.

1.1 Literature review

(Li *et al.* 2010) have presented the use of lower leg and ankle for the extraction of a gait feature where the silhouette was divided into three consecutive parts and Discrete Cosine Transform was used to transform the amplitude angle sequences.

Moreover, Ghosh *et al.* (2015) have proposed a novel idea of extracting features from the gait signature by the Fourier descriptor and anatomical landmarks. A fractal scale descriptor, which is based on discrete wavelet analysis, was proposed by (Shen *et al.* 1999), with the idea of a Generalized Multi-Resolution Analysis (GMRA). Using a time-varying sequence, it describes the self-similarity of gait appearance. Combined features of fractal scale and wavelet moments improve the results of gait recognition. The method is simple and improves the flexibility of the wavelet moments, but the recognition rate can be improved.

2 PROPOSED METHOD

The classification of the human gait in this paper consists of five parts, Silhouette Extraction, Creation of GEI of different angles, DWT with Daubechies4, Feature Extraction using Analysis of Variance, Matching of the probe image sequence with the Gallery of image sequences. Figure 1 shows the complete overview of the proposed human GEI classifications. The sample database of a person at different angles of movement with respect to a fixed surveillance camera has been shown in Figure 2.

2.1 Silhouette extraction

To recognize the people using gait depends upon the silhouette shape of an image sequence, which is based on the individual changes during the time.

Considering the silhouette images the information related to the silhouette needs to be extracted by applying silhouette extraction method (Ghosh *et al.* 2012).

```
┌─────────────────────────────┐
│     Silhouette Extraction   │
└─────────────────────────────┘
              │
              ▼
┌─────────────────────────────┐
│   Creation of GEI of dif-   │
│       ferent angles         │
└─────────────────────────────┘
              │
              ▼
┌─────────────────────────────┐
│   DWT            with       │
│   Daubechies4               │
└─────────────────────────────┘
              │
              ▼
┌─────────────────────────────┐
│   Feature Extraction us-    │
│   ing analysis of Variance  │
└─────────────────────────────┘
              │
              ▼
┌─────────────────────────────┐
│   Matching of the probe     │
│   sequence with the Gallery │
│   sequences                 │
└─────────────────────────────┘
```

Figure 1.   Flowchart describing GEI (Gait Energy Image).

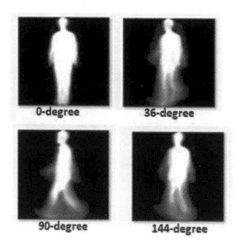

Figure 2.   Sample Database of a person at different angles of movement with respect to a fixed surveillance camera.

2.2   Gait Energy Image (GEI)

Gait Energy Image (GEI) can be defined as:

$$M(x,y) = 1/N \sum_{k=1}^{P} I_k(x,y,k) \qquad (1)$$

where, $I_k(x,y,k)$ represents the silhouette image at time k in a sequence and N is the number of frames in the complete gait cycles. GEI saves the storage space and computation time, and it is time-normalized accumulative energy image.

2.3   Wavelet transform

In a Discrete Wavelet Transform (DWT), filters of different cutoff frequencies are used to evaluate the signal at different scales (Saeid et al. 2008). The signal is passed through a series of high pass filters to evaluate the high frequencies and is passed through a series of low pass filters to evaluate the low frequencies. Wavelet decomposition includes waveforms, i.e., the wavelet function and the scaling function. The wavelet function represents high frequencies related to the detailed parts of an image. The scaling function represents the low frequencies or smooth parts of an image. The scaled and translated basis functions in 2D DWT (Nandini et al. 2011) are defined in Eq. (1) and Eq. (2).

$$\phi_{i,a,b}(m,n) = 2^{i/2}\phi(2^i m - a, 2^i n - b) \qquad (2)$$

$$\psi^j_{i,a,b}(m,n) = 2^{i/2}\psi^j(2^i m - a, 2^i n - b) \qquad (3)$$

The discrete Wavelet Transform can decompose a given signal into other signals known as the approximation and detail coefficients. The given function, f(t), can be represented by the following equation:

$$p(t) = \sum_{j=2}^{m} \sum_{n=-\alpha}^{\alpha} d(j,n)\phi(2^{-1}t - n)$$
$$+ \sum_{n=-\alpha}^{\alpha} b(m,n)\theta(2t - n) \qquad (4)$$

where: $\phi(t)$ is the mother wavelet and $\theta(t)$ is the scaling function. The variable $b(m,n)$ is called as the approximation coefficient at scale m and $d(j,n)$ is called as the detail coefficients at scale j. The approximation and detail coefficients can be expressed as:

$$b(m,n) = 1/\sqrt{2^m} \int_{\alpha}^{\alpha} p(t)\theta(2^{-m}t - n)dt \qquad (5)$$

$$w(j,n) = 1/\sqrt{2^1} \int_{-\alpha}^{\alpha} p(t)\phi(2^{-1}t - n)dt \qquad (6)$$

Based on the choice of the mother wavelet φ(t) and scaling function θ(t), different families of wavelets can be constructed (Gilbert et al. 1997).

2.3.1 *Daubechies wavelet*

Daubechies wavelets known as orthogonal wavelets and can be written as dbN, where N is the order, db is the family name of the wavelet. These wavelets use overlapping windows, for this high frequency coefficient spectrum reflects all high frequency changes. (Harbo *et al.* 2001).

The two functions are used to define Scaling and Wavelet functions.

2.3.1.1 Scaling function

The following conditions are necessary for {h (m)} to produce a valid scaling function

$$p(t) = 1/\sqrt{2} \sum\nolimits_m h(m)p(t-m) \qquad (7)$$

2.3.1.2 Wavelet function

The wavelet function can be defined in terms of the scaling function

$$q(t) = \sqrt{2} \sum\nolimits_{h=0}^{n-1} v_k p(2t-m) \qquad (8)$$

The Daubechies D4 transform has four wavelet and scaling function coefficients. The scaling function coefficients are:

$$p1 = 1 + \sqrt{3}/4\sqrt{2},$$
$$p2 = 3 + \sqrt{3}/4\sqrt{2} \quad p3 = 3 - \sqrt{3}/4\sqrt{2},$$
$$p4 = 1 - \sqrt{3}/4\sqrt{2}$$

The wavelet function coefficient values are:

$$a0 = b_3 \quad a1 = -b_2 \quad a2 = b_1 \quad a3 = -b_0$$

2.4 *Variance*

The variance is one of the several descriptors of a probability distribution. The variance of a random variable X is its second central moment, the expected value of the squared deviation from the mean

$$\mu = E(Y) \qquad (9)$$

$$Var(Y) = E[(Y - \mu)^2] \qquad (10)$$

This definition includes random variables that are discrete, continuous, neither, or mixed. The variance can be described as the covariance of a random variable itself.

2.5 *RBF Network*

The basic idea of the Radial Basis Function (RBF) Networks is that it is obtained from the theory of function approximation. The methods were developed for the exact interpolation of a set of data points in a multidimensional space (Rahul *et al.* 2010). The purpose is to project every input vector a_i, onto the corresponding target b_i, such that the function can be denoted as:

$$b_i = f(a_i) \qquad (11)$$

where $i = 1 \ldots, n$. n = number of objects.

According to the radial basis function, exact mapping can be obtained using a set of n basis functions and can be defined as:

$$f(a_i) = \sum_{j=1:n} p_j \phi(\|a_i - a_j\|) \qquad (12)$$

where p_j denotes weights and a_i and a_j denotes input object and the center of the basis function, respectively. The function $|a_i - a_j|$ denotes the distances between a_i and a_j.

3 RESULTS

This paper focuses on the idea of using variance as a statistical parameter where a small variance indicates the data points to be very close to the expected value while high variance points to the data points are to be spread out around the mean. The approximated value has been used to calculate the variance. Using the CASIA GEI database of a sample person as seen for the four different angles where the variance of a person varies very little for three different sequences for the same angle. However, for different angles, variance varies. So the threshold value (T) is needed for each degree.

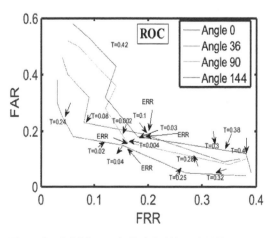

Figure 3.   ROC for angle 0°, 36°, 90°, and 144°.

Table 1. Table contains the approximate data with different sequences.

Degree	Sequence number	Variance			
0°	1	0.8540	8538.4	0.8525	0.8451
	2	0.8450	8448.8	0.8436	0.8365
	3	0.8650	0.8648	0.8635	0.8573
36°	1	0.6387	0.6385	0.6375	0.6329
	2	0.6423	0.6421	0.6411	0.6371
	3	0.7081	0.7080	0.7070	0.7070
90°	1	0.5976	0.5973	0.5957	0.5906
	2	0.5975	0.5971	0.5954	0.5903
	3	0.5959	0.5955	0.5940	0.5887
144°	1	0.6529	0.6527	0.6510	0.6466
	2	0.6473	0.6471	0.6453	0.6405
	3	0.6354	0.6352	0.6334	0.6285

Table 2. Different recognition parameters of different degree.

Degree	Correct Recognitions	Correct rejection rate	False acceptance rate	False rejection rate
0°	85%	15%	15%	15%
36°	81%	19%	19%	19%
90°	84%	16%	16%	16%
144°	82%	18%	18%	18%

So $var^d = var^d \pm T$ (13)

where var is the variance and d represents the degree.

Here for 0° angle T has been chosen experimentally as ± 0.04, for 36° it is 0.1, for 90° it is ± 0.004, for 144° it is ± 0.03. Results are presented in Tables 1 and 2.

Figure 3 shows the receiver Operating Characteristics Curve (ROC) for different angle of movement. For 0°, 36°, 90°, and 144° angle the Equal Error Rate (ERR) is 0.15, 0.19, 0.16, and 0.18, respectively.

4 CONCLUSION

This paper considers four different angles of a walk with respect to the surveillance camera. This work can be extended to different angles to have a good recognition rate since one statistical parameter has been considered for the feature extraction considering more feature extraction parameters can make the recognition rate at a high level.

REFERENCES

Ghosh, M., Bhattacharjee D. (2012). Human Identification by Gait Using Corner Points. *I.J. Image, Graphics and Signal Processing*, pp 30–36.

Ghosh, M., Bhattacharjee, D. (2015). Gait recognition for Human Identification Using Fourier Descriptor and Anatomical Landmarks. *I.J. Image, Graphics and Signal Processing*, pp 30–38.

Gilbert, Strang, and Truong, Nguen. (1997). Wave lets and Filter Banks. *Wellesley-Cambridge Press, MA*, pp. 174–220, 365–382.

La Cour Harbo, Jensen. (2001). Ripples in mathematics Berlin. *Springer*, pp 157–160.

Li, Yi-Bo., and Yang, Qin. (2010). Gait extraction and recognition based on lower leg and ankle, *International Conference on Intelligent Computation Technology and Automation*.

Nandini, C., K, Sindhu., and N.C, Ravi Kumar. (2011). Gait Recognition by Combining Wavelets and Geometrical Features, *IEEE Second International Conference on Intelligent Agent and Multi-Agent Systems*, Chennai.

Rahul, K. (2010). Evolutionary Radial Basis Function Network For Classificatory Problems International Journal of Computer Science and Applicaions. *Techno mathematics Research Foundation* Vol. 7 No. 4, pp. 34–49.

Saeid, R., and Moravejian., Reihaneh., & Farhad Mohamad, K,. (2008). Gait Recognition Using z Wavelet Transform, *Fifth International Conference on Information Technology: New Generations*.

Shen, D, & Ip, H.H.S. (1999). Discriminative Wave let Shape Descriptors for Recognition of 2-D Patterns, Pattern Recognition 32, pp. 151–165.

Frontiers in Computer, Communication and Electrical Engineering – Acharyya (Ed.)
© 2016 Taylor & Francis Group, London, ISBN: 978-1-138-02877-7

A fiber optic sensor for the detection of Partial Discharge within the High Voltage power transformer

B. Sarkar, C. Koley, N.K. Roy & P. Kumbhakar
National Institute of Technology, Durgapur, West Bengal, India

ABSTRACT: High Voltage (HV) power transformers are the most costly and critical equipment in any power system. Failures of such an apparatus lead to catastrophic losses and unwanted power interruptions. Majority of these failures are due to the weakness in insulation. Therefore, for economic and reliable operations, it is essential to monitor the insulation conditions of the HV transformers continuously. Partial Discharges (PDs) are reported as the main reason for the degradation of insulation quality and, therefore, the measurement of PD can provide an early indication of the failure. There are many existing methods for sensing these partial discharges, among them very few methods can be applied for continuous monitoring in online system. In the present work, a Fiber Optic sensor has been developed, with the help of Fiber Bragg Grating (FBG) as a primary sensing element, by detecting the acoustic wave generated from the PDs inside the high voltage transformer tank. Due to the inherent dielectric properties of the FBG and optical fiber the proposed system can be easily placed inside the transformer tank, without considering any separate insulation system for the proposed PD monitoring system. On the other hand, FBG is immune to the Electromagnetic Interference (EMI), and the optical fiber has very low transmission loss compared to the electrical conductor, therefore, the measured signal can be easily transmitted to the monitoring station.

1 INTRODUCTION

Partial Discharges (PDs) are the localized electrical discharges that only partially bridge the insulation between conductors and which can or cannot occur adjacent to the conductor [1]. The previous studies concluded that the Partial Discharge (PD) is a prime responsible phenomenon for the degradation of insulation of the HV power transformers [2–3], among others such as accumulation of mechanical, thermal, and electrical stresses occurring during long service period. Each discrete PD is the result of an electrical breakdown due to the unwanted stress for thermal, electrical, environmental effect within the insulation in the area of the homogeneities [4]. Measurable physical effects of the PDs are emissions of electromagnetic waves (non-visible and visible UV light, radio frequency wave), acoustics wave, and sudden abrupt changes in the voltage or current [5].

Conventional and standard techniques [1] use a coupling capacitor connected across the test system to measure the sudden short current pulse, which flows through the insulation defect where the discharge occurs. This is accurate but requires an electrical connection with the test system.

As the sudden electrical discharge due to PD generates acoustic waves, mainly high frequency ultrasonic waves, which can propagate through multiple paths with different media and different propagation velocities [6–7], thus, various researchers have proposed PD detection and localization system by placing single or multiple ultrasonic sensors inside or outside the equipment [8–11]. The main problem with such acoustic sensors is that the requirement of the electrical wiring around HV terminals, and the EMI noise, which limits its online applications for continuous monitoring. Moreover, the intensity of the PD generated acoustic signals is quite low and the output of piezoelectric sensors employed for the purpose is in the order of few micro-volts, which require a high gain and high input impedance amplifier to be placed in the proximity of the sensor [12–13].

In this context, an optical fiber based PD sensor can be a promising method, as the optical fiber is dielectric material so they can be easily placed inside the equipment without any insulation arrangements. On the other hand, it has very low transmission losses; therefore, the measured signal can be transmitted to longer distances to the substation control room without the use of any electrical amplification system and is free from EMI [14–15]. These features of optical fibers are the main motivation behind the development of optical fiber based sensor.

2 BASIC THEORY OF THE FBG BASED PARTIAL DISCHARGE (PD) SENSOR

An FBG is a periodic perturbation of the refractive index written into a segment of Ge-doped single mode fiber along the fiber length, which is formed by the exposure to a spatial pattern of the ultraviolet light in the region of around 244–248 nm and was first demonstrated by Hill et al. (1978). However, it was not until the transverse holographic fabrication method was developed by Meltz et al. (1989).

When the FBG is illuminated by a broadband light source, a set of beams reflected from the set of partially reflecting planes formed by the periodic core index modulation interfere with each other. For conservation of momentum, wavevector of the reflected wave (k_r) must be equal to the subtraction of the incident wavevector (k_i) and the grating wavevector (K).

$$k_r = k_i - K \tag{1}$$

When the Bragg condition is satisfied, $k_r = -k_i$ and Equation 1, can be written as:

$$-\frac{2\pi}{\lambda_B} n_{eff} = \frac{2\pi}{\lambda_B} n_{eff} - \frac{2\pi}{\Lambda} \tag{2}$$

The simplified form of the Equation 2 is related to the Bragg wavelength and the effective refractive index and grating period is:

$$\lambda_B = 2n_{eff} \Lambda \tag{3}$$

where, λ_B is the Bragg wavelength, Λ is the grating period and n_{eff} is the effective refractive index. In the most general case, the index perturbation $\delta n(z)$ takes the form of a phase and amplitude-modulated periodic waveform:

$$\delta n(z) = \delta n_0(z)\left[1 + m\cos\left(\frac{2\pi z}{\Lambda} + \phi\right)\right] \tag{4}$$

where, $\delta n_0 z$ is the amplitude of the photo induced index change. The local reflectivity is the complex ratio of the forward and backward going wave amplitudes. It is related to a multiplicative phase factor. The grating can then be represented by adding this index change described in Equation 4, to the original refractive index of the core as follows:

$$n(z) = n_{core} + \delta n(z) \tag{5}$$

Variation of the reflectivity of a grating is given by:

$$R = \frac{\sinh^2(l\sqrt{k^2 - \hat{\sigma}^2})}{\cosh^2(l\sqrt{k^2 - \hat{\sigma}^2}) - \dfrac{\hat{\sigma}^2}{k^2}} \tag{6}$$

where l provides the length of the grating and k and $\hat{\sigma}$ are the coupling coefficients. Values of k and $\hat{\sigma}$ can be calculated with the following equations:

$$k = \frac{\pi \delta n_{eff}}{\lambda} \tag{7}$$

$$\hat{\sigma} = \delta + \sigma \tag{8}$$

where,

$$\sigma = \frac{2\pi n_{eff}}{\lambda} \tag{9}$$

$$\delta = 2\pi n_{eff}\left(\frac{1}{\lambda} - \frac{1}{\lambda_B}\right) \tag{10}$$

where, $\lambda_B = 2n_{eff}\Lambda$ for an infinitesimally grating. The maximum reflectivity at the Bragg wavelength can be written as:

$$R_{max} = \tanh^2(kl) = \tanh^2(\Omega)$$

$$= \frac{\sinh^2[\eta(V)\delta n_{eff}]\sqrt{1 - \Gamma^2 \dfrac{N\Lambda}{\lambda}}}{\cosh^2[\eta(V)\delta n_{eff}]\sqrt{1 - \Gamma^2 \dfrac{N\Lambda}{\lambda}} - \Gamma^2}$$

$$= \tanh^2\left[\frac{N\eta(V)\delta n_{eff}}{n}\right] \tag{11}$$

where, N is the number of periodic variations and Ω is the angular frequency that can be found out with the help of following equation

$$\Omega = \pi n_{eff}\left(\frac{l}{\lambda_B}\right)\left(\frac{\Delta n}{n}\right)\eta(V) \tag{12}$$

$$\Gamma(\lambda_B) = \frac{1}{\eta(V)\delta n_{eff}}\left[\frac{\lambda}{\lambda_B} - 1\right] \tag{13}$$

The factor $\eta(V) \approx 1 - 1/V^2$ and $V \geq 2.4$ is the fraction of the integrated fundamental mode intensity contained in the core (V is the normalized frequency of the fiber). It is seen that R is directly proportional to the grating length l and the index perturbation ($\Delta n/n$), which is normally determined by the equation exposure power and time of the UV radiation for a specified fiber. The Full Width Half Maximum (FWHM) can be calculated as

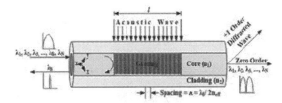

Figure 1. Basic working principle of FBG.

$$\Delta\lambda_B = \lambda_B S \sqrt{\left(\frac{\Delta n_{eff}}{2n}\right)^2 + \left(\frac{1}{N}\right)^2} \qquad (14)$$

Due to the axial strain there will be a change of radial and axial length in the fiber. As the radial change is very small as compared to the axial change, neglecting the radial change it can be written as:

$$\frac{\Delta l}{l} = \varepsilon = -\frac{(1-2v)}{E}\Delta P \qquad (15)$$

FBG shown in Figure 1 is a periodic structure written in a small section on the core of a single mode fiber and its transmission characteristics can be used as an in-coupler reflection band-pass filter as shown by [8]. The reflected peak Bragg wavelength is highly sensitive to the change in geometry due to the strain (ε) caused either by the mechanical stress or temperature change (ΔT). PD generated acoustic signal induces ultrasonic pressure wave having an impulsive nature, which propagates through the lossless homogenous liquid dielectric medium at a speed of 1390 m/s as given by [9]. Finally, the strain produced by the PD event, neglecting the temperature effect, the Bragg wavelength shift can be expressed as:

$$\frac{\Delta\lambda_B}{\lambda_B} = (1-\rho_c)(\varepsilon + \varepsilon_A) \qquad (16)$$

where, ε_A is the strain produced by the high frequency acoustic wave and ρ_c is the effective strain optic coefficient.

3 EXPERIMENTAL PROCEDURE

3.1 Sensor assembly

GE-doped FBG has a core diameter of 8.2 μm and a cladding diameter of 125 μm manufactured by Central Glass and Ceramic Research Institute (CGCRI), Kolkata. It also has the standard characteristics values: $n_1 = 1.4494$, $n_2 = 1.444$, $n_{eff} = 1.4485$, $v = 0.16$, $E = 70$ GPa, $p_{11} = 0.121$, $p_{12} = 0.270$,

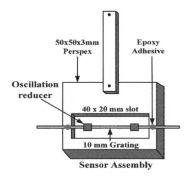

Figure 2. Sensor assembly.

$\Lambda = 535$ nm, $l = 10$ mm, strain sensitivity -1.0 pm $\mu\varepsilon^{-1}$, and temperature sensitivity -10 pm °C^{-1} at 1550 nm. A waveguide is needed to fix the fiber to make it convenient for placing it inside the PD source. The waveguide must be able to withstand the PD environment and will not create any abnormalities on the test setup. A 3 mm thick, hard acrylic sheet chosen as a waveguide material on which the FBG has been fixed is shown in Figure 2.

3.2 Test system

A 60 kV, AC, Power frequency portable transformer oil test set having cell size of $100 \times 75 \times 100$ mm is made by 10 mm thick acrylic sheet and capacity of 600 ml has been taken as a PD source. The cell contains a needle and a flat electrode combination as an HV electrode and ground electrode. Optical Spectrum Analyzer (Yokogawa, Ando AQ6317B) of wavelength range: 600–1700 nm, resolution: 10 pm (min) has been used to analyze the optical signals. A 50 nm BW, 20 mW SLED (Dense light) Source having central peak at: 1550 nm and Santec (TSL-510), type 'C', wave length-1260–1630 nm, SMF, 5 pm resolution, 13 dBm peak output power used as Broad ban and tunable laser source. A power meter (Newport Model-2936-C, Sl. No.: 10282) and photo receiver (Newport IR, Model-1811) of range DC-125 MHz is used to detect PD. Teraxian makes a compact unit containing the circulator and connectors, Agilent DSO6054 A, 500 MHz, 4 channel, 4 Gs/S, and a laptop is taken as a part of the experiment for recording and analyzing s. Schematic diagram and actual photograph of the setup is illustrated in Figure 3.

4 RESULT AND DISCUSSION

The recorded PD signals from the detector at two different applied voltages have been presented in Figure 4(a) and (b), during applied voltage of

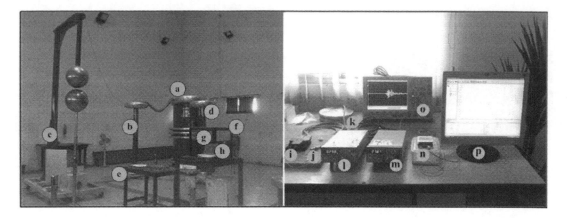

Figure 3. Experimental setup at the laboratory for the measurement of PD, using optical and electrical methods. a: 300 kV, AC power frequency source, b: Capacitive voltage divider (CVD), c: Sphere-sphere gap setup, d: Coupling capacitor (C_k), e: Quadripole (AVK-D), f: Oil filled prototype test transformer, g: PD sensor, h: Optical fiber coupler, i: Laser source, j: FC/ PC connector with mating sleeve, k: Single mode optical fiber, l: Current controller, m: Temperature controller, n: Fiber Bragg grating analyzer (FBGA), o: Digital storage oscilloscope (DSO), p: Computer.

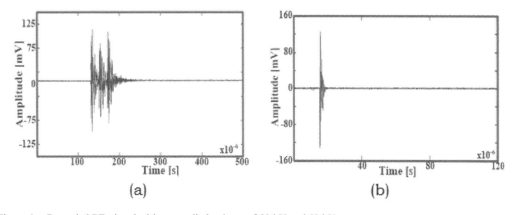

Figure 4. Recorded PD signal with an applied voltage of 30 kV and 50 kV.

30 and 50 kV. From the Figure it can be observed the proposed PD sensor is able to capture the acoustic wave as generated the occurrence of PD. It can be further noticed that the amplitude of the captured waveform also increases due to applied voltage, which is due to the fact that at higher voltages the magnitude of the PDs also increase.

5 CONCLUSION

The presented system is an FBG based acoustic sensor, which is capable of surviving in the harsh environment of the transformer interiors without compromising the transformer's normal functionality, which is the prime criteria for the online PD detection as well as to find out its location.

Experimental results show that the presented system has enough potential to capture acoustic signals emitted by PD sources in the oil transformer in both the domains, i.e., time and frequency domain. By keeping one end free and by putting extra load, the FBG acts as an oscillation reducer. Finally, it can be concluded that the FBG based fiber-optic sensor can be used to detect and localize the PDs within high voltage transformer.

REFERENCES

[1] IEC International Standard 60270, "High-Voltage Test Techniques – Partial Discharge Measurements", International Electrotechnical Commission (IEC), Geneva, Switzerland, 3rd Edition 2000.

[2] R. Bartnikas, "Partial Discharges, Their Mechanism, Detection and Measurement", IEEE Transactions on Dielectrics and Electrical Insulation Vol. 9 No. 5, October 2002.

[3] F.H. Kreuger, Industrial High Voltage, Delft University Press, 1991.

[4] C.L. Wadhwa, "High Voltage Engineering (second edition)" 2007, New Age International (P) Limited, publishers.

[5] E. Kuffel, W.S. Zaengl, J. Kuffel, "High Voltage Engineering Fundamentals, Second edition" Reprinted 1986, Newnes.

[6] IEEE Guide for the Measurement of Partial Discharges in AC Electric Machinery, IEEE P1434/D1.1, 4 October 2010.

[7] Duval, M.: 'A review of faults detectable by gas-in-oil analysis in transformers', IEEE Electr. Insul. Mag., 2002, 18, (3), pp. 8–17.

[8] W. Sikorski, W. Ziomek, "Detection, Recognition and Location of Partial Discharge Sources Using Acoustic Emission Method" available online at www.intechopen.com.

[9] S.E.U. Lima, O. Frazão, R.G. Farias, F.M. Araújo, L.A. Ferreira, V. Mirandac J.L. Santos, "Acoustic Source Location of Partial Discharges in Transformers", Proc. of SPIE Vol. 7653 76532N-1.

[10] S. Biswas, C. Koley, B. Chatterjee, S. Chakravorti, "A Methodology for Identification and Localization of Partial Discharge Sources using Optical Sensors", IEEE Transactions on Dielectrics and Electrical Insulation Vol. 19, No. 1; February 2012, pp. 18–28.

[11] Peter Kung, Lutang Wang, Maria I. Comanici, Lawrence R. Chen, "Detection and location of PD activities using an array of fiber laser sensors", IEEE International Symposium on Electrical Insulation (ISEI) Conference: 2012, pp. 511–515.

[12] Mm Yaacob, Ma Alsaedi, Jr Rashed, Am Dakhil, and Sf Atyah, "Review on Partial Discharge Detection Techniques Related to High Voltage Power Equipment Using Different Sensors", Photonic Sensors / Vol. 4, No. 4, 2014: 325–337.

[13] Kee-Joe Lim, Seong-Hwa Kang, Kang-Won Lee, Sung-Hee Park1 & Jong-Sub Lee, "Partial Discharge Signal Detection by Piezoelectric Ceramic Sensor and The Signal Processing", Journal of Electroceramics, 13, 487–492, 2004.

[14] S. Karmakar, N.K. Roy, P. Kumbhakar, "Monitoring of high voltage power transformer using direct optical partial discharge detection technique", J Opt 38 (4): pp. 207–215.

[15] Xiaodong Wang, Baoqing Li, Harry T. Roman, Onofrio L. Russo, Ken Chin, and Kenneth R. Farmer, "Acousto-optical PD Detection for Transformers", Ieee Transactions On Power Delivery, Vol. 21, No. 3, July 2006, pp. 1068–1073.

Frontiers in Computer, Communication and Electrical Engineering – Acharyya (Ed.)
© 2016 Taylor & Francis Group, London, ISBN: 978-1-138-02877-7

2D-thermal model for estimation of heat-dissipation in SiC based p-i-n switches used for RF-communication

Jhuma Kundu
NSHM Knowledge Campus, Durgapur, West Bengal, India

Abhijit Kundu
West Bengal University of Technology, Kolkata, India

Maitreyi Ray Kanjilal
Narula Institute of Technology, West Bengal University of Technology, Kolkata, India

Moumita Mukherjee
CMSDS, DRDO-Kolkata, (Ministry of Defence, Government of India), University of Calcutta, Kolkata, India

ABSTRACT: Operation of p-i-n switch under pulse mode condition is presented in this paper. The junction temperature is increased by 357°K with thermal heat flux 1.5×10^{10} Watt/m^2. The p-i-n switch model is done for two poly types of SiC-, α-SiC, and β-SiC. The analysis revels that this β-SiC is better than α-SiC as far as thermal model is concerned. It offers high break down voltage for which it can be used at high temperature and the device is capable to handle the high power for signal routing between instruments and Devices Under Test (DUT). Incorporating a switch into a switch matrix system enables one to route signals from multiple instruments to single or multiple DUTs. This allows multiple tests to be performed with the same setup, eliminating the need for frequent connects and disconnects. For these properties p-i-n diode is suitable as a switch at millimeter wave frequency.

1 INTRODUCTION

P-i-n diode is a solid state device that operates as a variable resistor at millimeter wave frequency. The impedance estimation of p-i-n switch is required to determine the power dissipation [1–3] under forward bias condition and the impedance of p-i-n diode is controlled by stored charge (Q) in the intrinsic-region [1, 4]. The power dissipation in a device or circuit is an important factor which has to be considered when the device will be used to generate high current. And heat is generated by the dissipation of power in the junction of the diode which absorbs the small amount of power and converts it in the heat. So heat sink is required without any distortion of the signal (Figure 1). Therefore during the fabrication of the semiconductor device applicable to high voltage and high frequency, precaution should be taken to control the temperature increasing within the device especially at the junction. It is required to simulate and design the

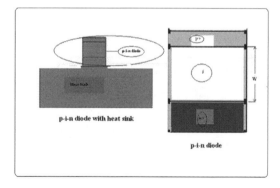

Figure 1. Geometry structure of p-i-n switch with heat sink.

device prior to fabrication in the development of the system [5] for calculation of optimum efficiency at low cost. The property of the device varies from material to material. The switching property of the device is controlled by the carrier mobility (μ), diffusion coefficient etc. This high frequency switching device is used in wireless and millimeter wave systems [5–6].

2 THEORY

The application of the device at high frequency requires higher breakdown voltage to handle high power. The WBG semiconductor shows high breakdown voltage [5, 7]. The breakdown voltage of p-i-n diode is given by—

$$V_b = \int_0^W E(z)dz \tag{1}$$

where electric field E(z) is constant and z represents the width of i region. So it can be written as—

$$V_b = E \cdot W \tag{2}$$

Applying Guss law—

$$\nabla \cdot E = \rho / \varepsilon \tag{3}$$

where ρ is the charge density and ε is the relative permittivity of GaN. If the charge is distributed by ionized donors (N_i) in the lightly doped n region or i region then—

$$\rho(z) = N_i(z) \tag{4}$$

It is also constant in the depletion region and zero in the diffusion region. Using the Gauss's law the electric field is given by—

$$E(z) = zN_i q / \varepsilon \tag{5}$$

The p-i-n diode is a 2-terminal device which finds application as a variable resistance at forward bias as a function of frequency [2, 8]. Intrinsic region resistance (R_i) of p-i-n diode is a function of width (W) of the i region (Figure 1), effective carrier mobility (μ_{eff}), carrier life time (τ) and forward bias current (I_f) and it can be written as—

$$R_i = \left(W^2 / 2I_f \mu_{eff} \tau \right) \tag{7}$$

At forward bias condition the stored charge ($Q = I_0\tau$) depends on carrier life time (τ) and forward bias current (I_0). The junction resistance (R_j) varies with frequency and as a function of frequency it can be expressed—

$$R_j(f) = (kT/qI_0)\,\beta\tanh(W/2L)\cos(\phi - \theta/2) \tag{8}$$

The carrier diffusion length (L) has an important role in controlling the junction impedance of the switch. The diffusion length can be expressed as—

$$L = \sqrt{D_{eff}\,\tau} \tag{9}$$

The total series resistance of p-i-n switch can be written as—

$$R_s = 2R_j + Ri \tag{10}$$

The resistance has an important role for power dissipation [9, 10]. In p-i-n switch the power dissipation is mainly controlled by series resistance R_s, characteristic impedance Z_0 and available power P_{av} of p-i-n diode. It can be expressed as—

$$P_d = \frac{4R_T Z_0}{(Z_0 + 2R_T)^2} \tag{11}$$

For dissipation of power the heat is produced at the junction of the switch [11]. And the heat capacity also depends on geometry structure of p-i-n diode and considering the volume of the diode as V, it can be written as—

$$H_c = C_p \rho V \tag{12}$$

Heat is also converted into temperature. At ambient temperature (T_a) for a given power dissipation (P_d), the junction temperature (T_j) can be expressed as—

$$T_j = T_a + \theta_j P_d \tag{13}$$

The thermal impedance (θ_j) of heat flow can be determined by the following equation—

$$\theta_j = \int_0^l dl / AK \tag{14}$$

where K is the conductivity of the medium, A is the cross-section area and l is the length of heat flow path.

3 RESULT AND ANALYSIS

Through the simulation model, and using equations (1)–(5) the breakdown voltage of the p-i-n switch has been estimated first at reverse biased condition. The widths of the i region of the simulated devices are optimized through this modelling technique. The break down voltages of 3C-SiC, 4H-SIC, and 6H-SiC based p-i-n switches are then compared for optimization of the switch and the results are shown in Figure 2(a) Among these three materials 3C-SiC offers highest breakdown voltage than its other counterparts for the same width of the i region. The width of the intrinsic region is optimized through this modelling technique for high power and high frequency operation. Figure 2(b) shows that 3C-SiC offers less thermal power dissipation than other two materials. So 3C-SiC is preferred for p-i-n RF-switch at millimetre wave communication. Power dissipation is occurred by absorbing the small amount of power at the junction of p-i-n switch and converting it into heat.

Analysis of high frequency (35 GHz) p-i-n switch under continuous wave thermal operation is done through COMSOL Multiphysics software. Figure 3 shows the steady state thermal image of p-i-n switch under high power condition and Figure 4 shows the thermal power dissipation pattern of a p-i-n switch which is mounted on a Cu substrate. It is interesting to observe that even under a high power operation, the junction temperature rise is not so significant; this is due to the good thermal conductivity property of SiC and this is the importance of using such wide band gap semiconductor for RF switch fabrication.

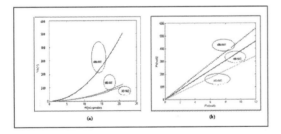

Figure 2. (a) Breakdown voltage of p-i-n switch (b) Power dissipation of p-i-n switch.

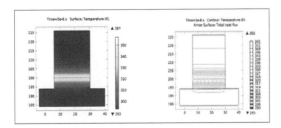

Figure 3. The junction temperature in Kelvin (K) of SiC based p-i-n switch.

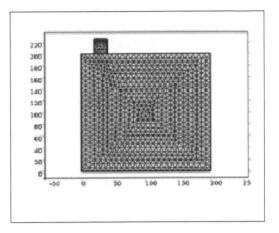

Figure 4. Thermal diffusion pattern of p-i-n switch.

REFERENCES

Caverly, R.H. & Hiller, G. (1990) The small signal ac impedance of gallium arsenide and silicon p-i-n diode, solid state electron, Vol.33, no.10, (PP.1255–1263).

Ghosh, D. & Ray (Kanjilal), M. & Kundu, A. (2013) Electrical Response on MESFET using WBG Semiconductor as Potential Substrate, IEE Conference on Computation and Communication Advancement (IC3 A).

Gated, E., & et al. (2007) An improved physics- Based Formulation of the Microwave p-i-n diode impedance, IEEE microwave and Wireless components Letters, Vol.17. No.3.

Hiller, C.R. (1987) Microwave resistance of Gallium Arsenide and Silicon p-i-n Diode, IEEE MTT-S Digest.

Hinojosa A.I. & Resendize, L.M. & Torchynska, T.V (2010) Numerical Analysis of the Performance of p-i-n Diode Microwave Switches Based on Different Semiconductor Materials, Int. J. Pure Appl. Sci. Tech.

Kundu, A. & Ray (Kanjilal) & Mukherjee, M. & Ghosh, D. (2013) Switching Characteristics of p-i-n Diode Using Different Semiconductor Materials, International Journal of Advanced Technology and Engineering Research (IJATER) (ISSN 2250–3536),(pp-19), Volume 3, Issue 1.

Konczakowska, A. & Cichosk, J. & Dokupil, D. & Flisikowski, P. (2011) The Low Frequency Noise Behavior of SiC MESFETs, 21st International Conference on Noise and Fluctuation, IEEE.

Leenov, D. (1964) The Silicon p-i-n diode as a microwave radar protector at megawatt levels, IEEE Trans. Electron Devices, Vol. ED-11, no.2 (PP.53–61).

Milligan, J.W. & Henning, J. & Allen, S.T. & Ward, A. & Parikh, P &.Smith, R.P. Transition of SiC MESFET Technology", Cree, Inc., 4600 Silicon Drive, Durham, NC27703.

Size, S.M. (1989) "Physics of Semiconductor Devices", John Willy and Sons, New York.

Frontiers in Computer, Communication and Electrical Engineering – Acharyya (Ed.)
© 2016 Taylor & Francis Group, London, ISBN: 978-1-138-02877-7

Load balancing in cloud computing using a local search technique—Tabu Search

Brototi Mondal, Madhuchanda Das, Chayanti Mukherjee & Oishika Das
Supreme Knowledge Foundation Group of Institutions, Mankundu, Hooghly, West Bengal, India

ABSTRACT: Cloud computing offers distributed, virtualized, and elastic resources as utilities to end users, which has the potential to support full realization of "computing as a utility" in the near future. Cloud computing uses Virtual Machine (VM) instead of physical machine to host, store, and interconnect the different components. Load balancing is being used to assign massive amount of requests to different VMs in such a way that none of the nodes can get loaded heavily or lightly. Thus, it can be considered as an optimization problem. In this paper, a load balancing strategy based on Tabu Search has been proposed which balances the load of the cloud infrastructure. The proposed algorithm is tested and evaluated on universal datasets using an existing simulator known as CloudAnalyst. The overall performance of the proposed algorithm is compared with the existing approaches like First Come First Serve (FCFS), Round Robing (RR), and a local search algorithm i.e. Stochastic Hill Climbing (SHC).

1 INTRODUCTION

Cloud computing is an area that is experiencing a rapid advancement in both academia and industry. This distributing computing mechanism utilizes the high speed of the internet to move jobs from private PC to the remote computer clusters (big data centers owned by the cloud service providers) for data processing. Load balancing in cloud refers to distributing client requests across multiple application servers that are running in a cloud environment.

As load balancing is one of the hot topics among the researchers, there have been various research efforts on the approaches of load balancing, like Minimum Execution Time (MET) scheduling, Min-Min scheduling algorithms, Minimum Completion Time (MCT) etc., are reported in literature. Also a few soft computing techniques are used to optimize the work load like Ant Colony (Mishra et al. 2012), Stochastic Hill Climbing (Mondal et al. 2012), and Simulated Annealing (Mondal et al. 2015). Load balancing using Genetic Algorithm (GA) improves the response time and the resource utilization effectively (Dasgupta et al. 2013).

In this paper, we propose the use of Tabu Search (TS) to find efficient solutions to the cloud load balancing problem which improves the overall performance by reducing the average Response Time (RT) by a significant amount. A comparison on the improved performance is also investigated.

The rest of paper is organized as follows.

In section 2, we describe the development of proposed TS algorithm for load balancing which is followed by the simulation results and analysis with an overview of CloudAnalyst (Wickremasinghe et al. 2010) that is described in section 3. And finally, section 4 concludes this paper.

2 LOAD BALANCING USING TABU SEARCH

To handle a massive amount of requests by equally spreading the work load on each computing node, to maximize their utilization and minimize the response time, efficient load balancing algorithms are needed.

2.1 Tabu Search

Tabu search is a "meta-heuristic" as well as "local-search" algorithm for finding the optimized result to a wide variety of classical and practical problems. One main ingredient of Tabu Search (TS) is the use of *adaptive memory* to guide problem solving.

For a given problem to be solved, using tabu search, three things need to be defined: a set V of feasible solutions (for the problem), a *neighborhood structure* $N(s)$ for a given solution $s \in V$, and a *tabu list*, *TL*. An initial solution $s\kappa$ is chosen from the set of feasible solution V. This initial solution is usually chosen randomly. Once an initial solution is chosen, the algorithm goes into a loop that terminates when one or more of the stopping conditions are met. After that, the next solution s_{i+1} is selected from the neighbors of the current solution

s_i, $s_{i+1} \in \leq N(s_i)$. In a basic tabu search, all possible solutions $s \in \leq N(s_i)$ is considered, and the best solution is chosen as the next solution s_{i+1}.

2.2 Load balancing based on Tabu Search

The parameters used in our proposed algorithm are described as follows:

Let us assume s_0 contains the initial candidate VM chosen randomly and s contains the intermediate best candidate VM. The tabuList is used to remember the last k solutions that have been visited, and to avoid these solutions sCandidate and bestCandidate represent each candidate and best candidate of sNeighborhood respectively.

To calculate the fitness value, we formulate the problem of load balancing as follows:

Let's assume n is the number of VMs that is present in the cloud at any particular instance of time and m is the number of jobs submitted by cloud users which have to be allocated. Each VM will have a Virtual Machine Vector (*VMV*) indicating the current status of VM's utilization. *MIPS* indicate how many million instructions can be executed by that machine per second. β and *DL* is the cost of execution of instruction and delay cost respectively. The delay cost is an estimate of penalty, which the cloud service provider needs to pay to the customer in the event of job being finished in the actual time rather than being more than the deadline advertised by the service provider,

$$VMV = g \, (MIPS, \beta, DL) \qquad (1)$$

Similarly each request submitted by cloud user can be represented by a request unit vector (*RUV*). The type of service that is required by a request are, Software as a Service (SAAS), Infrastructure as a Service (IAAS) and Platform-as-a-Service (PAAS) is represented by *T*. *NIC* indicates the count of instruction in the request determined by the processor. Request Arrival Time (*RAT*) is the clock time of arrival of request in the system and the worst case completion time (*wc*) is the minimum time required to complete the request by a processing unit. Thus Equation 2 gives the different attributes of requests.

$$RUV = g \, (T, NIC, RAT, wc) \qquad (2)$$

The cloud service provider needs to allocate these N jobs among M number of processors such that C cost function as indicated in equation 3 is minimized.

$$C = w1 * \beta \, (NIC \div MIPS) + w2 * DL \qquad (3)$$

The proposed algorithm is given below:

Proposed Algorithm:

1. s ← s$_0$
2. sBest ← s
3. tabuList ← NULL
4. while either maximum number of iteration is exceeded or optimum solution is found do
5. bestCandidate ← NULL
6. for all sCandidate in sNeighborhood do
7. if fitness(sCandidate)> fitness(bestCandidate) then
8. bestCandidate ← sCandidate
9. end if
10. end for
11. s ← bestCandidate
12. if fitness(bestCandidate) >fitness(sBest)then
13. sBest ← bestCandidate
14. end if
15. tabuList ← tabuList U bestCandidate
16. if tabuList size exceeds maximum size then
17. Remove first candidate from tabuList
18. end if
19. end while
20. Return sBest

3 EXPERIMENTAL ANALYSIS

This section presents our experimental study; simulation environment, and performance evaluation of our proposed algorithm.

3.1 Simulation environment

We have conducted this research with CloudAnalyst simulation toolkit which is modified with our proposed algorithm.

The CloudAnalyst also enables a modeler to repeatedly execute a series of simulation experiments with slight parameters variations in a quick and easy manner. The simulation configuration is provided in Table 1. The world is divided into six regions, modeling a group of users, which represents the six major continents of the world. It is assumed out of the total registered users 5% are online simultaneously during the peak time and only one tenth is online during the off-peak. Size of VMs used to host applications in the experiment is 100MB. VMs have 1GB of RAM memory and have 10MB of available bandwidth. Simulated hosts have x86 architecture. Each simulated data center hosts a particular amount of VMs dedicated for the application. Machines have 4 GB of RAM and 100GB of storage. Each machine has 4 CPUs, and each CPU has a capacity power of 10000 MIPS.

3.2 Result

We consider several scenarios for experimentation. Experiment is started by taking one Data

Table 1. Simulation configurations.

S. no	User base	Region	Simultaneous online users during peak hrs	Simultaneous online users during off-peak hrs
1	UB1	0–N. America	5,70,000	57,000
2	UB2	1–S. America	7,00,000	70,000
3	UB3	2–Europe	4,50,000	45,000
4	UB4	3–Asia	9,00,000	90,000
5	UB5	4–Africa	1,25,000	12500
6	UB6	5–Oceania	1,50,000	30,500

Table 2. Simulation scenario and calculated overall average Response Time (RT) in (MS) using one data center.

Sl. no	Cloud configuration	DC specification	RT using TS	RT using SHC	RT using RR	RT using FCFS
1	CC1	Each with 25 VMs	327.34	329.02	330	330.11
2	CC2	Each with 50 VMs	326.11	329.01	329.42	329.42
3	CC3	Each with 75 VMs	326.66	329.34	329.67	329.44

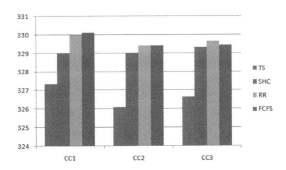

Figure 1. Performance analysis of proposed TS with SHC, FCFS and RR using One Data Center.

Center (DC) having 25, 50, and 75 VMs to process all the requests around the world. Table 2 describes this simulation setup with the calculated overall average Response Time (RT) in ms for TS, SHC, RR, and FCFS. The performance analysis graph for it is depicted in Figure 1 where cloud configuration is along x-axis and response time in ms is along y-axis. Subsequently two, three, four, five DCs are considered with combination 25, 50, and 75 VMs for each CCs as given in Tables 3, 4, 5, and 6. The corresponding performance analysis graphs are displayed beside them in Figures 2, 3, 4, and 5.

Table 3. Simulation scenario and calculated overall average Response Time (RT) in (MS) using two data centers.

Sl. no	Cloud configuration	DC specification	RT using TS	RT using SHC	RT using RR	RT using FCFS
1	CC1	Each with 25 VMs	362.12	370.44	376.27	381.34
2	CC2	Each with 50 VMs	357.47	365.15	372.49	377.52
3	CC3	Each with 75 VMs	356.21	364.73	369.78	375.56
4	CC4	Two DCs with 25, 50 VMs each	353.33	361.72	367.91	373.87
5	CC5	Two DCs with 25, 75 VMs each	354.24	362.23	369.45	372.23
6	CC6	Two DCs with 75, 50 VMs each	349.54	357.04	356.01	361.61

Table 4. Simulation scenario and calculated overall average Response Time (RT) in (MS) using three data centers.

Sl. no	Cloud configuration	DC specification	RT using TS	RT using SHC	RT using RR	RT using FCFS
1	CC1	Each with 25 VMs	351.34	361.82	366.17	368.34
2	CC2	Each with 50 VMs	350.11	358.25	363.52	367.52
3	CC3	Each with 75 VMs	347.66	355.73	360.18	366.56
4	CC4	Each with 25, 50 and 75 VMs	351.34	359.01	361.21	367.87

Table 5. Simulation scenario and calculated overall average Response Time (RT) in (MS) using four data centers.

Sl. no	Cloud configuration	DC specification	RT using TS	RT using SHC	RT using RR	RT using FCFS
1	CC1	Each with 25 VMs	346.64	354.35	359.35	360.95
2	CC2	Each with 50 VMs	342.16	350.71	356.93	359.97
3	CC3	Each with 75 VMs	338.36	346.46	352.09	358.44
4	CC4	Each with 25, 50 and 75 VMs	336.74	344.31	351	355.94

Table 6. Simulation scenario and calculated overall average Response Time (RT) in (MS) using five data centers.

Sl. no	Cloud configu- ration	DC specification	RT using TS	RT using SHC	RT using RR	RT using FCFS
1	CC1	Each with 25 VMs	337.24	342.86	348.57	352.05
2	CC2	Each with 50 VMs	326.56	332.84	339.76	345.44
3	CC3	Each with 75 VMs	324.32	329.46	335.88	342.79
4	CC4	Each with 25, 50 and 75 VMs	321.67	326.64	334.01	338.01

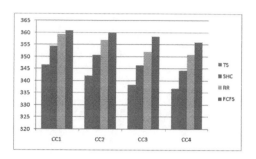

Figure 5. Performance analysis of proposed TS with SHC, FCFS and RR using Five Data Center.

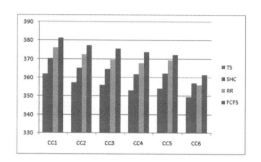

Figure 2. Performance analysis of proposed TS with SHC, FCFS and RR using Two Data Center.

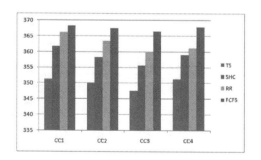

Figure 3. Performance analysis of proposed TS with SHC, FCFS and RR using Three Data Center.

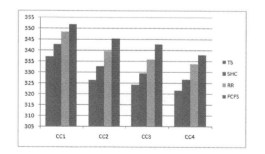

Figure 4. Performance analysis of proposed TS with SHC, FCFS and RR using Four Data Center.

4 CONCLUSION

In this paper, we have proposed an efficient algorithm to distribute massive work load in cloud computing, based on a well-known optimization method, Tabu Search (TS). The performance analysis in terms of response time (ms) among the proposed TS algorithm and some existing algorithms shows that the proposed method not only outperforms but also guarantees the QoS requirement of customer requests. Though TS algorithm based on load balancing strategy balances the incoming requests among the available VM efficiently, we shall strive to find more and more efficient methods using soft computing approaches for load balancing to obtain better results.

REFERENCES

Dasgupta, K., Mandal, B., Dutta, P., Mondal, J. K. & Dam, S (2013). A Genetic Algorithm (GA) based Load Balancing Strategy for Cloud Computing. CIMTA-2013, Procedia Technology 10 (2013) 340–347.

Mishra, R., & Jaiswal, A. (2012). Ant colony Optimization: A Solution of Load balancing in Cloud. *International Journal of Web & Semantic Technology (IJWesT)*, 2(3), pp. 33–50.

Mondal, B., Dasgupta, K., and Dutta, P. (2012). Load Balanc ing in Cloud Computing using Stochastic Hill Climbing-A Soft Computing Approach. Proc. of *C3IT2012*, Elsevier, Procedia Technology 4(2012), pp.783–789.

Mondal, B., Choudhury, A. (2015). Simulated Annealing (SA) based Load Balancing Strategy for Cloud Computing. (IJCSIT) International Journal of Computer Science and Information Technologies, Vol. 6 (4), 2015, 3307–3312

Wickremasinghe,B., Calheiros, R. N., and Buyya,R. (2010). CloudAnalyst: A CloudSim-based Visual Modeller for Analysing Cloud Computing Environments and Applications. Proceedings of the 24th International Conference on Advanced Information Networking and Applications (AINA 2010), Perth, Australia.

Frontiers in Computer, Communication and Electrical Engineering – Acharyya (Ed.)
© 2016 Taylor & Francis Group, London, ISBN: 978-1-138-02877-7

High density salt and pepper noise removal by selective mean filter

A. Bandyopadhyay, K. Chakraborty & R. Bag
Department of Computer Science Engineering, Supreme Knowledge Foundation Group of Institutions, Mankundu, West Bengal, India

A. Das
Department of Computer Science Engineering, Netaji Subhash Engineering College, Garia, Kolkata, West Bengal, India

ABSTRACT: In this paper, a selective mean filter is proposed. A two stage method has been implemented, comprising of Noisy pixel detection followed by restoration. Primarily the noisy pixels are determined by a detection procedure and a binary flag image is generated. Then the selective mean operation is applied on the each of the noisy pixel and replacement is done by the approximated value. A prefixed threshold value is used to re-approximate the recently replaced pixels. Those pixels are further smoothed by a conditional mean operation. Finally, the four border pixels are considered and checked for noisiness which is then removed by neighborhood pixel replacement. This filter provides improbable noise removal result on corrupted images; it clearly outperform the existing filters with respect to MSE and PSNR comparison. It also shows to be robust to very high levels of noise, retrieving meaningful detail at noise levels as high as 90%.

1   INTRODUCTION

Noise is a random variation of intensities in an image which is visible as grains in the image. In other words, the pixels in the image portray varied intensity values instead of true (noise-free) pixel values. To remove or reduce the noise in images certain noise removal filters are deployed. Those filters help in enhancing the visibility of the image by smoothing the entire image, but they do produce a vague result in cases of fine, low contrast details. One of the types of noise in images is the Impulse Noise (Bandyopadhyay et al. 2015), also called as spike noise or independent noise. The visibility of this noise appears on an image as random black and white dots, hence the name salt and pepper noise (Banerjee et al. 2014). Sharp and sudden changes of image signal results in the production of this type of noise. Dust particles in the image acquisition source or over heated faulty components may also cause this type of noise. Here the noisy pixels take either maximum or minimum gray levels i.e. '255' or '0', (Bandyopadhyay et al. 2015) respectively; producing white dot for the maximum value and black dot for the minimum value (Chan et al. 2005). Now, to remove these noises from images, ample image de-noising filters are available, which removes the noise from the image, while preserving the details. The Mean Filter (Kundu et al. 1984) was the first filter used by A. Kundu (in 1984) for noise removal.

The filter used a 3×3 window for smoothing, but it did not consider whether the pixel was corrupted or not, hence replacing the non-noisy pixel in the process. Then the Standard Median Filter (Tukey. 1971) was introduced and revised by Tukey, which had the disadvantage of poor performance if the number of corrupted pixels surpassed the number of non-corrupted pixels in the 3×3 window. To overcome this shortfall, the Decision Based Algorithm (DBA) (Srinivasan et al. 2007) came through; this replaced only the pixel that was detected as corrupt. But this method also had its share of disadvantage by producing a streaking effect due to consecutive replacement of the neighborhood pixels. Ultimately the Decision Based Un-symmetric Trimmed Median Filter (DBUTMF) (Aiswarya et al. 2010) and Modified Decision Based Un-symmetric Trimmed Median Filter (MDBUTMF) (Esakkirajan et al. 2011) was introduced which had the advantage that the actual value could be regained from the mean if all the pixels were noisy in the 3×3 window, though this filter didn't perform good in the noise density 70%–90%. Further a filter (Banerjee et al. 2015) was designed by S. Banerjee which performed better at moderate noise densities. Then a Noise Adaptive Fuzzy Switching Median Filter (NAFSMF) (Toh et al. 2010) was proposed which applied fuzzy reasoning to handle the noise present. But effective noise removal at high noise density was still un-achievable.

In this paper a selective mean filter is proposed where arithmetic mean is used in different stages and noisy pixels are approximated by checking with a prefixed threshold value. The proposed filter outperforms the above discussed filters in high noise densities (70%–90%). The whole noise detection and removal procedure has been divided into sections where section 2 illustrates the proposed methodology, section 3 represents Result and Discussion and section 4 depicts the conclusion.

2 PROPOSED METHOD

Here we consider $x_{i,j}$ for $(i,j) \in A \equiv \{1,2,3, \ldots M\} \times \{1,2,3, \ldots\ldots N\}$ be the gray intensity level at pixel location (i,j) of a true $M \times N$ image G. A salt and pepper noisy grayscale image G_1 $(M \times N)$ is taken where a pixel having intensity value '0–4' or '251–255' will be considered as a noisy one. The proposed method consist of two segments: 2.1 noise detection and 2.2 restoration.

2.1 Noise detection

The image G_1 $(M \times N)$ is taken and every pixel is considered for checking. Concurrently a same size binary flag image F_1 $(M \times N)$ is generated where $f_{i,j}$ is considered a pixel value at the location (i,j).

If $x_{i,j} = 0$ or $x_{i,j} = 1$ or $x_{i,j} = 2$ or $x_{i,j} = 3$ or $x_{i,j} = 4$ or
$\qquad x_{i,j} = 255$ or $x_{i,j} = 254$ or $x_{i,j} = 253$ or
$\qquad x_{i,j} = 252$ or $x_{i,j} = 251$

then
$\qquad f_{i,j} = 0$
else
$\qquad f_{i,j} = x_{i,j}$

Repeating the above procedure a same size image G_2 $(M \times N)$ is generated.

2.2 Restoration

The noisy image has to be restored after being affected by salt & pepper noise. We perform the Restoration using the Selective Mean Operation.

STEP 2.2.1 Replacement by selective mean operation

Consider G_2 $(M \times N)$ for (i,j) starting from $(2,2)$ to $(M–1,N–1)$.
If $x_{i,j} = 0$ and $(x_{i-1,j}, x_{i+1,j}, x_{i,j-1}, x_{i,j+1}) \neq 0$ then

Replace $x_{i,j} = (\sum_{k=i-1}^{i+1} \sum_{r=j-1}^{j+1} x_{k,r})/n$

where n is the no of non noisy pixels in the said matrix. Repeating the above procedure image G_3 $(M \times N)$ is generated.

STEP 2.2.2 Neighborhood mean operation

The image G_3 $(M \times N)$ is taken. Consider G_3 $(M \times N)$ for (i,j) starting from $(2,2)$ to $(M–1, N–1)$.
Taking $x_{i,j}$ as the center we create a (3×3) matrix. Calculate for each pixel in the said matrix

$A1 = (x_{i,j-1} + x_{i,j} + x_{i,j+1})/3$
if $(x_{i,j}) > A1$
then
$\qquad A2 = (x_{i,j}) – A1$
else
$\qquad A2 = A1 – (x_{i,j})$

After completing this process an image G_4 $(M \times N)$ is generated.

STEP 2.2.3 Threshold comparison & flag image re-creation

If A2 is greater than the pre-fixed threshold value = 10, then,

$\qquad f_{i,j} = 1$
else

$\qquad f_{i,j} = 0$

Repeating the above procedure a Flag image F_2 $(M \times N)$ is generated.

STEP 2.2.4 Flag Image converted to the noisy image

In F2 If $(f_{i,j} = 1)$ Replace $f_{i,j} = 0$.
Examine G_4 $(M \times N)$ for (i,j) starting from $(2,2)$ to $(M–1,N–1)$.pixels Repeat STEP 2.2.1 until all the pixels are traversed.

The above procedure is performed so that the transformed pixels that exceed the threshold value are once again denoted as noisy and hence the Selective Mean is re-applied to ensure ultimate noise removal.

STEP 2.2.5 Border Operation

- Upper Border Mean Calculation
  Examine F_2 $(M \times N)$ for (i,j) starting from $(1,2)$ to $(1,N–1)$.
  if $x_{i,j} = 0$ then
  $\qquad$ replace $x_{i,j} = (x_{i,j-1}+x_{i,j+1})/2$
  If adjacent pixels corresponding to $x_{i,j}$ are '0' then consider the next neighborhood pixels (both in the left and right direction) for row wise replacement.

- Left Border Mean Calculation
  Examine F_2 $(M \times N)$ for (i,j) starting from $(2,1)$ to $(M–1,1)$.
  if $x_{i,j} = 0$ then
  $\qquad$ replace $x_{i,j} = (x_{i-1,j} + x_{i+1,j})/2$
  If adjacent pixels corresponding to $x_{i,j}$ are '0' then consider the next neighborhood pixels

(both above and below the column) for column wise replacement.

- **Right Border Mean Calculation**
  Examine F_2 (M × N) for (i,j) starting from (2,N) to (M–1,N).
  if $x_{i,j} = 0$ then
  $$\text{replace } x_{i,j} = (x_{i-1,j} + x_{i+1,j})/2$$
  If adjacent pixels corresponding to $x_{i,j}$ are '0' then consider the next neighborhood pixels (both above and below the column) for column wise replacement.

- **Lower Border Mean Calculation**
  Examine F_2 (M × N) for (i,j) starting from (M,2) to (M, N–1).
  if $x_{i,j} = 0$ then
  $$\text{replace } x_{i,j} = (x_{i,j-1} + x_{i,j+1})/2$$
  If adjacent pixels corresponding to $x_{i,j}$ are '0' then consider the next neighborhood pixels (both in the left and right direction) for row wise replacement.
  The four corner pixels, i.e. (1,1), (M,1), (1,N), (M,N) are replaced by the mean of the adjacent neighborhood uncorrupted pixels.

3 RESULTS AND DISCUSSION

The proposed method is assessed on the base of Mean Square Error (MSE), Peak Signal-to-Noise Ratio (PSNR). The visual performances of proposed work are tested step-wise and the results are shown in Figure.3 Quantitative performances of the de-noising techniques are measured by Mean Square Error (MSE), Peak Signal-to-Noise Ratio (PSNR) as defined in equation (I) and (II) respectively.

$$\text{MSE} = \left(\sum_{M,N} \frac{\left(I(m,n) - \hat{I}(m,n) \right)^2}{M \times N} \right) \quad \text{(I)}$$

MSE is the mean square error between original image (I) and de-noised image ($\hat{I}$). M and N are the number of rows and columns in the input image, respectively.

$$PSNR = 10 \log_{10} \left(\frac{255^2}{MSE} \right) \quad \text{(II)}$$

Table 1 shows the PSNR comparison with respect to Lena image for different existing filters with the proposed filter at variable noise density (50%–90%).

Figure 1 and Figure 2 shows the visual result of proposed filter after application on Barbara and Cameraman image respectively. Figure 3 shows

Table 1. PSNR for different filters for Lena image at different noise densities.

Image Filters		50%	60%	70%	80%	90%
Lena	SMF	15.42	11.13	9.93	8.70	6.60
	DBA	26.42	24.81	22.62	20.37	17.11
	DBUTMF	27.08	25.52	23.41	20.93	17.92
	MDBUTMF	28.18	26.40	24.30	21.70	18.40
	S. BANERJEE	30.02	28.73	27.20	25.37	22.52
	NAFSMF	29.87	28.75	27.58	26.08	22.72
	Proposed	**30.20**	**29.52**	**27.86**	**26.61**	**24.92**

Figure 1. Proposed filter on Barbara image: a) Original image b) 70% noisy image c) Output image.

Figure 2. Proposed filter on Cameraman image: a) Original image b) 70% noisy image c) Output image.

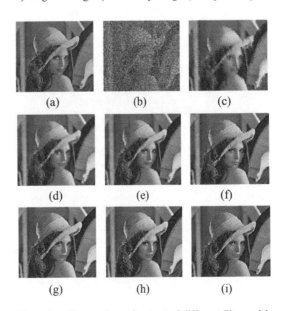

(a) (b) (c)

(d) (e) (f)

(g) (h) (i)

Figure 3. Comparison of output of different filters with the proposed filter: (a) Original Lena image (b) 60% noisy image (c) SMF (d) DBA (e) DBUTMF (f) MDBUTMF (g) S. BANERJEE (h) NAFSMF (i) Proposed.

Table 2. Mse for different filters for Lena image at different noise densities.

Image	Filters	60%	70%	80%	90%
Lena	SMF	5047.54	6608.15	8771.63	14225.91
	DBA	215.31	357.33	597.14	1267.88
	DBUTMF	183.26	297.22	524.90	1054.88
	MDBUTMF	148.96	241.59	439.62	939.89
	S. BANERJEE	87.70	123.90	191.92	365.66
	NAFSMF	87.71	115.63	163.33	349.20
	Proposed	72.95	107.91	142.25	210.41

the quality of the reconstructed image for different filters compared to the proposed filter at 60% noise density. Table 2 illustrates the comparison of MSE between different filters with the proposed at varied noise density (50%–90%). The qualitative and quantitative result shows that the proposed filter outperforms the above stated filters in all respect.

4 CONCLUSION

A varied Mean filter is proposed in this paper based on the selective arithmetic mean of the corrupted pixels. Further, in this method, the noise removal is significant even at very high noise density cases. Qualitative and quantitative evaluation of the proposed method proves that it outperforms other existing methods for all the considered images. Though this filter outdoes most of the conventional filters it still may be enhanced by introducing further modification in future.

REFERENCES

Aiswarya, K., Jayaraj, V., and Ebenezer, D. (2010). A new and efficient algorithm for the removal of high density salt and pepper noise in image and video. *Proc. of Second Int. Conf. Computer Modeling and Simulation.* 409–413.

Bandyopadhyay, A., Kumari, M., Pooja., Das, A., and Bag, R. (2015). Detection and Removal of High Density Random Valued Impulse Noise. *CiiT International Journal of Digital Image Processing*, 8(7), 242–246.

Bandyopadhyay, A., Banerjee, S., Das, A., and Bag, R. (2015). A Relook and Renovation over State-of-Art Salt and Pepper Noise Removal Techniques. *I.J. Image, Graphics and Signal Processing*, vol-9, 61–69.

Banerjee, S., Bandyopadhyay, A., Bag, R., and Das, A. (2015). Moderate Density Salt & Pepper Noise Removal. *IJECT* 10(6), Issue 1, Spl- 1 Jan–March.

Banerjee, S., Bandyopadhyay, A., Bag, R., and Das, A. (2014). Neighborhood Based Pixel Approximation for High Level Salt and Pepper Noise Removal. *CiiT International Journal of Digital Image Processing*, 8(6), 346–351.

Chan, H. R., Ho, W. C., and Nikolova, M. (2005). Salt-and-pepper noise removalby median-type noise detectors and detail-preserving regularization. *IEEE Trans. Image Process.*, 10(14), 1479–1485.

Esakkirajan, S., Veerakumar, T., Subramanyam, N. A., and PremChand, H. C. (2011). Removal of High Density Salt and Pepper Noise Through Modified Decision Based Un-symmetric Trimmed Median Filter. *IEEE Signal Process. Lett.*, 5(18).

Kundu, A., Mitra, S. K., and Vaidyanathan, P. P. (1984). Application of Two-Dimensional Generalized Mean Filtering for Removal of Impulse Noises from Images. *IEEE Transactions on Acoustics, Speech and Signal Processing*, 3(32).

Srinivasan, S. K., and Ebenezer, D. (2007). A new fast and efficient decision based algorithm for removal of high density impulse noise. *IEEE Signal Process. Lett.*, 3(14), 189–192.

Toh V. K. K., and Isa, M. A. N. (2010). Noise adaptive fuzzy switching median filter for salt-and-pepper noise reduction. *IEEE Signal Process. Lett.*, 3(17) 281–284.

Tukey, W. J. (1971). *Exploratory Data Analysis (preliminary ed.)*. Reading, MA: Addision-Wesley.

Frontiers in Computer, Communication and Electrical Engineering – Acharyya (Ed.)
© 2016 Taylor & Francis Group, London, ISBN: 978-1-138-02877-7

Comparison of electric and thermal stress distribution in underground cable for different insulating materials

Soumita Pal, Vishwanath Gupta & Abhijit Lahiri
Department of Electrical Engineering, Supreme Knowledge Foundation Group of Institutions, West Bengal, India

ABSTRACT: In this proposed work, a single phase underground cable is considered to study the electric stress and thermal stress distribution as a means to find out the location and value of maximum electric stress and thermal stress when operated at a system voltage of 1kV under power frequency condition. Three different insulating materials viz. Teflon, nylon and polyethylene are considered to study the effect of these stress distributions. The cable under consideration is simulated and the electric and thermal stress distributions are both obtained by using COMSOL Multiphysics version 4.3b software. This comparative study of the electric and thermal stress distribution, the location and value of maximum electric and thermal stress will help to assess the type of insulating material that is to be used in an underground cable in a particular application.

1 INTRODUCTION

The increased need for electric power combined with a competitive deregulated market environment have forced utilities to refocus their attention on reliability. This has created a significant need for diagnostic methods and technologies to assess the condition of the underground cable systems, commonly used in low voltage and medium voltage solutions (Kanikella 2014).

Insulating materials plays a vital role in the electrical and thermal design and performance of underground cables for both steady state and transient state conditions. The choice of insulating material for high voltage and medium voltage applications depends on the electrical and thermal properties of the materials (Mejia 2008). A good underground cable should have uniform electrical and thermal stress distribution.

The aim of the proposed work is to obtain the electric stress and thermal stress distribution in an underground cable which will be to locate the point of occurrence of maximum electric stress and thermal stress and the maximum values (Shaker, El-Hag, Patel U, Jayaram 2014).

2 MODEL DESCRIPTION

In this proposed work electric and thermal stress distribution in a XLPE insulated PVC sheathed 1kV (normalized voltage) single core underground cable operated at power frequency has been studied to observe their effects for different insulating materials keeping the material of the sheath same.

The cable under consideration is shown in Figure 1. In Figure 1, AB denotes the conductor-insulator boundary, CD represents the insulator-sheath boundary, EF rwepresents the sheath boundary in contact with outside environment. GH, IJ and KL are the horizontal sections along which the electric and thermal stress distributions are calculated. The dimensions of the cable are given in Table 1 and the

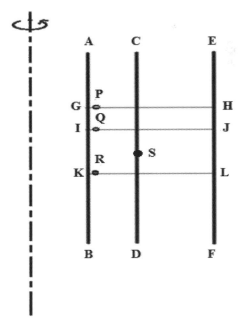

Figure 1. Schematic diagram of cable under consideration.

dielectric and thermal properties of the different insulating materials are provided in Table 2.

3 SIMULATION DETAILS

For obtaining electrical and thermal stress distribution corresponding differential equations are solved using proper boundary conditions. The differential equations related with electrical stress are as follows:

$$\nabla \cdot D = \rho v \tag{1}$$

$$E = -\nabla V \tag{2}$$

$$D = \epsilon E \tag{3}$$

Where
D = Electric flux in Coulombs/metre^2
E = Electric field in Volt/metre
V = Electric Potential in Volts
ρv = Volume charge in Coulombs/metre^3
ϵ = Permittivity in Farads/metre

The surface charge density is incorporated using equation 4.

$$n(D_1 - D_2) = \rho_s \tag{4}$$

The conductor insulator boundary is assigned with $\rho_s = 0$.

In the proposed work temperature is considered as a measure of the thermal stress. More is the temperature more is the thermal stress generated. The differential equations related with thermal stress are as follows:

$$\rho C_p \mu \cdot \nabla T = \nabla \cdot (K \nabla T) + Q \tag{5}$$

$$\nabla \cdot J = Q j \tag{6}$$

$$E = -\nabla V \tag{7}$$

$$J = \sigma E + Je \tag{8}$$

where,
ρ = Density of material in kg/metre^3
C_p = Heat capacity at constant pressure in Joule/(kg.K)
T = Temperature in Kelvin
K = Thermal conductivity in Watt/(metre. Kelvin)
J = Surface radiosity in Watt/metre^2
Q = Heat source in Watt/metre^3
E = Electric field in Volt/metre
V = Electric Potential in Volts

The differential equations for conductive heat flux is given by equation (9)

$$-n \cdot (-K \nabla T) = h \cdot (T_{ext} - T) \tag{9}$$

Where,
h = Heat transfer coefficient in Watt/(metre^2. Kelvin) and the differential equation for surface to ambient radiation is given by equation (10)

$$-n \cdot (-K \nabla T) = \epsilon \sigma \left(T_{amb}^4 - T^4 \right) \tag{10}$$

Where
σ = Electrical conductivity in Seimens/metre
ϵ = Surface emissivity

The solutions of the above differential equations are obtained by using COMSOL Multiphysics software that works on the principle of Finite Element Method (FEM).

4 RESULTS AND DISCUSSIONS

4.1 *Electric stress distribution*

The variations of electric stress along the radial directions GH, IJ and KL that are at 2.5 m, 2.0 m

Table 1. Cable under consideration.

Name	Conductor area (in mm^2)	Insulation thickness (in mm)	Sheath thickness (in mm)
XLPE insulated PVC sheathed single core cable	1.5	0.7	1.4

Table 2. Properties of insulating materials.

Materials used as insulating materials	Relative permittivity	Thermal conductivity (w/m-k)	Density (kg/m^2)	Specific heat at constant pressure (J/kg-k)	Emissivity	Break down strength (v/m)
Polyethylene	2.3	0.51	960	1250	0.9	160
Nylon	4	0.25	1150	1700	0.85	56
Teflon	2	0.25	2200	1172	0.92	173

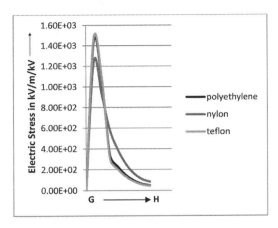

Figure 2. Variation of electric stress along GH using different insulating materials.

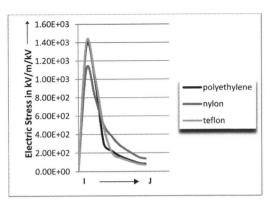

Figure 3. Variation of electric stress along IJ using different insulating materials.

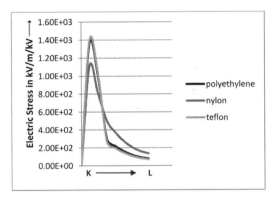

Figure 4. Variation of electric stress along KL using different insulating materials.

and 1.5 m from BDF are studied. Figure 2 presents such variations for different insulating materials along GH. Figure 3 and Figure 4 present the same along IJ and KL respectively. In each of these cases it may be observed that the maximum value of electric stress is maximum when Teflon is used as the insulating material while it is minimum when Nylon is used in the place of Teflon.

Moreover, the point of occurrence of the maximum stress in each of these cases is close to the conductor-dielectric boundary as indicated by the points P, Q and R at the three different heights respectively.

Table 3 presents the values of the maximum electric stresses along the radial direction of the cable at the three different heights that are considered for the three different insulating materials. From Table 3 it may be observed that the maximum electric stress varies from 1.44×10^3 kV/m/kV to 1.36×10^3 kV/m/kV when Polyethylene is used as the insulating material while it ranges from 1.24×10^3 kV/m/kV to 1.11×10^3 kV/m/kV in the case of Nylon used as the insulating material. For Teflon, the maximum electric stress is observed to vary from 1.47×10^3 kV/m/kV to 1.41×10^3 kV/m/kV.

The variation of electric stress for different insulating materials along the insulator-sheath boundary i.e. along the line CD has also been studied. The results are presented in Figure 5. From Figure 5, it is seen that the electric stress in each of the cases increases at the two ends while it decreases in the region between the terminals.

4.2 Thermal stress calculation

Similar to the case of electric stress, the variations of thermal stress due to different insulating

Table 3. Maximum electric stress at different cross-sections for different materials.

| Cross-section | Maximum electric stress in kV/m/kV | | |
| | Materials | | |
	Polyethylene	Nylon	Teflon
GH	1.44E+03	1.24E+03	1.47E+03
IJ	1.36E+03	1.11E+03	1.41E+03
KL	1.38E+03	1.14E+03	1.44E+03

materials along the radial directions along GH, IJ and KL are studied and are presented in Figure 6, Figure 7 and Figure 8 respectively.

From each of these figures it may be observed that the maximum value of the thermal stress is greatest at each of the three cross sections considered when Nylon insulation is used while it is least when Polyethylene is used as the insulating material.

197

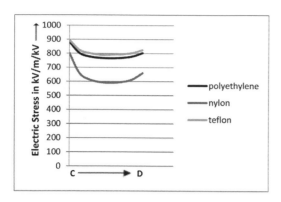

Figure 5. Variation of electric stress using different insulating materials along the insulator sheath boundary CD.

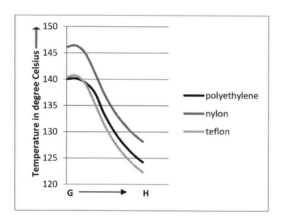

Figure 6. Variation of temperature along GH using different insulating materials.

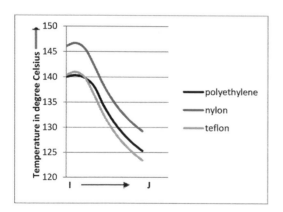

Figure 7. Variation of temperature along PQ using different insulating materials.

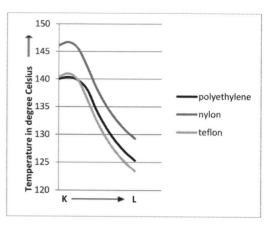

Figure 8. Variation of temperature along RS using different insulating materials.

Table 4. Maximum temperature at different cross-sections for different materials.

| | Maximum temperature in degree Celsius | | |
| | Materials | | |
Cross-section	Polyethylene	Nylon	Teflon
GH	140.15	146.38	140.68
IJ	140.33	146.67	140.98
KL	140.33	146.67	140.98

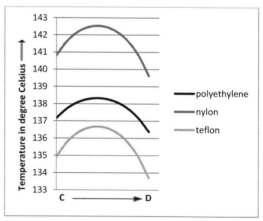

Figure 9. Variation of temperature along the insulator sheath boundary CD using different insulating materials.

The points of occurrence of the maximum temperature for each of the three insulating materials at the three different cross sections GH, IJ and KL are P, Q and R respectively. Table 4 presents the maximum temperature at different cross sections with different insulating materials. From Table 4 it may be observed that for a particular material, the value of the maximum temperature is fairly constant at three different cross sections.

The variations of temperature distribution along the insulation-sheath boundary CD due to different insulating materials are studied and are shown in Figure 9. From Figure 9 it may be noted that along CD, the value of the maximum temperature is maximum when Nylon is used while it is minimum with Teflon.

5 CONCLUSIONS

The proposed work gives an idea how different material properties affects the electric stress distribution, thermal stress distribution and the maximum value of electric stress and thermal stress a single core underground cable operating at power frequency. However, the different material properties do not affect the location of occurrence of maximum electric stress and thermal stress.

Also from the results obtained it may be said that the point of occurrences of the maximum electric stress and the maximum thermal stress are same at three different cross-sections of the cable when measured along its radial direction.

With respect to the electric stress, Teflon seems to be a better choice over Nylon and Polyethylene while with respect to thermal stress; Polyethylene seems to be superior compared to Nylon and Teflon.

From Figure 5 and Figure 9 it may be observed that the distribution of electric stress and thermal stress at the insulation-sheath boundary are different in nature. For each of the cases under study it is noted that from C to D electric stress has decreased uniformly first and after reaching point S on CD at a height of 1.5 m it again changes its curvature and starts increasing. When the thermal stress is computed along the same direction, it also increases uniformly and after reaching the same point S on CD, it changes its curvature and decreases uniformly. Moreover, if electric stress at the insulator-sheath boundary is considered then Nylon is the best choice among the three materials but if thermal stress is considered at the insulator-sheath boundary the Teflon proves to be a better choice over the other two materials.

REFERENCES

Kanikella S.K. (2013). Electric Field and Thermal properties of dry cable using FEM. *TELKOMNIKA, Vol.11, No.5, e-ISSN: 2087–278X*.

Kanikella S.K. (2014). Electric Field and Thermal properties of wet cable using FEM. *International journal of innovative research in Engineering and science, ISSN: 2319–5665, Issue 3 volume 4*.

Mejia J.C.H. (2008). Characterization of Real Power cable defects by diagnostic measurement. *Geargia Institute of Technology*.

Rajagopal K., Vittal K.P & Hemsingh L. (2012). Electric Field and Thermal properties of wet cable 10 KV and 15 KV. *TELKOMNIKA, Vol.10, No.7, e-ISSN: 2087–278X*.

Shaker Y.O., El-Hag A.H., Patel U. & Jayaram S.H. (2014). Thermal Modeling of medium Volatge Cable Terminations under square pulses. *IEEE Transactions on Dielectrics and Electrical Insulation Vol. 21, No. 3*.

Frontiers in Computer, Communication and Electrical Engineering – Acharyya (Ed.)
© 2016 Taylor & Francis Group, London, ISBN: 978-1-138-02877-7

An improved approach of cloud service brokerage model in multi-cloud environment

Bidisha Bhabani
Computer Science Engineering Department, Supreme Knowledge Foundation Group of Institutions, Mankundu, Hooghly, West Bengal, India

ABSTRACT: Cloud computing has emerged as one of the most promising and challenging technologies at present. Cloud Service Broker (CSB) creates a governed and secure cloud management platform to simplify the delivery of complex cloud services to cloud service customers. They enable customers to realize the full potential that cloud provider has to offer. CSB leads to creation of a system of highly distributed, task-oriented, modular, and collaborative cloud services managed by a broker. In order to effectively manage the complexity inherent in such systems, enterprises are anticipated to crucially depend upon CSB mechanisms. Here in this paper, an improved approach is proposed that serves secure brokerage in multi-cloud environment.

1 INTRODUCTION

According to NIST cloud computing is a model for enabling ubiquitous, convenient, on-demand network access to a shared pool of configurable computing resources (e.g., networks, servers, storage, applications, and services) that can be rapidly provisioned and released with minimal management effort or service provider interaction. (NIST sp: 500–292, 2011)

The five Essential Characteristics of cloud computing are a) *On-demand self-service* b) *Broad network access* c) *Resource pooling* d) *Rapid elasticity* and e) *Measured service*. The three Service Models are a) *Software as a Service* (*SaaS*) b) *Platform as a Service* (*PaaS*) and c) *Infrastructure as a Service* (*IaaS*). And the four Deployment Models are as follows: a) *Private cloud* b) *Community cloud* c) Public cloud and d) *Hybrid cloud*.

According to NIST Actors in Cloud are described in Table 1.

This paper is concerned about the Cloud Broker among all the actors in cloud computing.

The NIST Reference Architecture (Figure 1) defines a Cloud Broker as an entity that manages the use, performance, and delivery of cloud services, and negotiates relationships between Cloud Providers and Cloud Consumers. A Cloud Consumer may request cloud services from a Cloud Broker instead of directly contacting a Cloud Provider. Cloud Brokers provide a single point of entry for managing multiple cloud services. The key defining feature that distinguishes a Cloud Broker from a Cloud Service Provider is the ability

Table 1. Actors in cloud computing.

Actors	Definition
Cloud consumer	A person or organization that maintains a business relationship with, and uses service from, *cloud providers*.
Cloud provider	A person, organization, or entity responsible for making a service available to interested parties.
Cloud auditor	A party that can conduct independent assessment of cloud services, information system operations, performance and security of the cloud implementation.
Cloud broker	An entity that manages the use, performance and delivery of cloud services, and negotiates relationships between *cloud providers* and *cloud consumers*.
Cloud carrier	An intermediary that provides connectivity and transport of cloud services from *cloud providers* to *cloud consumers*.

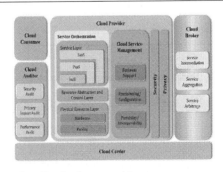

Figure 1. Cloud reference architecture.

to provide a single consistent interface to multiple differing providers, whether the interface is for business or technical purposes. In general, Cloud Brokers provide services in three categories:

a. *Intermediation*—A Cloud Broker enhances a given service by improving some specific capability and providing value-added services to Cloud Consumers by managing access to cloud services, identity management, performance reporting, enhanced security, etc.
b. *Aggregation*—A Cloud Broker combines and integrates multiple services into one or more new services and provides data & service integration and ensures the secure data movement between the Cloud Consumer and multiple Cloud Providers.
c. *Arbitrage*—Service arbitrage is similar to service aggregation except that the services being combined/consolidated are not fixed. Service arbitrage means a Broker has the flexibility to choose services from multiple service Providers.

2 LITERATURE REVIEW

The concepts of cloud bursting and cloud brokerage models are discussed in (Nair et al. 2010). The open management and security issues associated with the two models are also discussed in this paper. A possible architectural framework capable of powering the brokerage based cloud services that had been considered in the scope of OPTIMIS (OPTIMIS) (an EU FP7 project) is being proposed.

A new cloud brokerage service that reserves a large pool of instances from cloud providers and serves users with price discounts is proposed in (Wang et al. 2013). The design of dynamic strategies for the broker to make instance reservations with the objective of minimizing its service cost has been proposed.

A broker framework for balancing costs against security risks is presented in (Goettelmann et al. 2014). This framework selects cloud offers and generates deployment-ready business processes in a multi-cloud environment.

A framework of the cloud components is proposed and the role of cloud service broker system is discussed in (Jeon et al. 2014) that supports the allocation and control and management of virtual system based on the brokering function between the Cloud Service Provider (CSP) and Cloud Service Customer (CSC) by integrating and managing cloud resources in multiple heterogeneous cloud environment.

A model based execution-ware is proposed in (Baur et al. 2015) that helps coping with the challenges like heterogeneous nature of cloud, selection of cloud provider and its divergent offerings etc., by allowing the deployment of applications in a multi-cloud environment, based on a high-level model created by the user.

After having a detailed study of all these works it is known that a CSB model is actually expected to do all the functionalities which a real life broker does. We generally deal with brokers to make our work done without having any headache to search for and deal with the providers/suppliers, their offerings etc., be it renting/buying house, buying lands or buying railway/air tickets anything. We just have to inform about our requirements to the broker and the broker comes up with the best deal. But at the same time we are concerned about the authenticity of the broker so as the broker does. Keeping this in mind I have considered the work in (Nair et al. 2010) in which security is incorporated. But this work has some loopholes as follows. If after matching the user requirements, any single CSP is not able to fulfill them then no solution is given for this situation. If more than one request is made at a time then how the requests will be served is not mentioned. In this paper, an improved approach has been proposed considering the same architecture as mentioned in (Nair et al.) that can solve the above mentioned issues and serve in multi-cloud environment.

3 CLOUD BROKERAGE MODEL

CSB can stand for three different models as follows:

a. *Use of multiple CSP*—in this model, a user makes use of services provided by various cloud providers to fulfil an internal process.
b. *Brokering multiple providers to provide a Service Level Agreement (SLA)-based tiered pricing model*—in this, a user approaches a cloud broker with a given set of functional and SLA-based requirements and the cloud broker then picks up the best match in terms of the functions as well as variables like pricing, SLA parameters and other non-functional requirements like compliance and certification capabilities.
c. *Cloud Aggregation Ecosystem (CAE)*—this scenario offers the potential to treat both IT and business functions as a series of interconnected cloud services.

4 CLOUD BROKER ARCHITECTURE

In this section a more detailed look has been taken at the architecture that can be used to implement

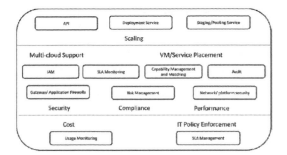

Figure 2. Cloud Broker Architecture.

the broker model involving the brokering of multiple providers to provide a SLA-based tiered pricing model to the customers of the broker.

The functional components (Figure 2), which are needed by the broker to provide these broker capabilities are specified below.

a. *The cloud API* is a mechanism through which consumers can interact with the cloud broker for performing cloud related actions including creating and managing cloud resources like compute and storage components.

b. *The deployment service unit* helps the broker to inform the cloud provider to start a Virtual Machine (VM) which consumer can use for compute purpose or to create a storage space for user's data.

c. *The staging/pooling service* ensures the ability of the cloud broker to provide a satisfactory performance for the various cloud actions requested by the customer. Since Cloud Services are offered on pay-as-you-go basis, and are highly scalable, service consumers can demand extra resources and the broker needs to ensure that a suitable cloud provider is in place to handle the request.

d. *Identity Access Management (IAM) unit* is a vital unit as it keeps record of all the service consumer required enterprise details, assigned cloud providers, type of service like storage or compute. Furthermore, the classification criteria's that broker decides with the service consumer is also stored in this unit.

e. *The SLA monitoring unit* constantly monitors SLAs negotiated by the SLA management unit. It checks to see if there are any impending SLA violations and if any, to take the specified preventive measure.

f. *The capability management and matching unit* keeps tab of all the capabilities provided by the various providers. Whenever a service consumer approaches the broker for cloud services, this unit matches the consumer's functional and

SLA requirements with the services each cloud provider offer and finds the most appropriate match.

g. *The audit unit* periodically audits broker's platform as much as possible using capabilities provided by the cloud provider. Audit is performed to ascertain the validity and reliability of information and to provide an assessment of internal controls.

h. *The risk management unit* identifies, assess and prioritize risks on the basis of effects of events. Its target is to minimize, monitor, and control the probability of such unfortunate events.

i. *Network/platform security unit* manages the overall security of the broker's platform. Provisions and policies have been adopted to prevent unauthorized access, misuse, modification, or denial of network or network accessible resources.

j. *The usage monitoring unit* monitors the usage of the services by the cloud consumer, and generates monthly bills for them.

k. *The SLA management unit* controls all the SLAs in place between the broker and consumer.

5 PROPOSED APPROACH

User needs to register itself with the Enterprise cloud portal to access cloud computing facility.

Authentication Phase:

Step 1: User asks for Digital Certificate (DC) of the cloud portal.

Step 2: Cloud portal provides its DC (acquired by a renowned Certificate Authority (CA)) to the user and asks for the user's DC as well.

Step 3: User sends its digital certificate (acquired by a renowned CA) while registering itself with the cloud portal.

Login Phase:

Step 1: User sends the request along with its user id and password to enterprise cloud portal.

Step 2: Enterprise cloud portal then sends this request to enterprise IAM. IAM authenticates the user on the basis of its user id and password and then the access user is entitled and grants access permission to the user.

Step 3: After authenticating the user if IAM responds positively, cloud portal converts the packet received in step 2 to an external token and encrypts it together with the compute request and forwards it to the broker. If the response of IAM is a negative, cloud portal informs the user about the denial of request.

Step 4: Broker decrypts the packet using its public key. On the basis of the request, broker contacts CSPs to start a VM and SLA negotiation.

4.1: The broker requests for initial proposals from all CSPs.

4.2: All providers propose initial offers based on their current capabilities and availability to fulfill brokers' requirements.

4.3: *Negotiation Process with providers:*

4.3.1: If there are multiple CSPs who can fulfill all the requirements, then the broker selects the best CSP.

4.3.2: If there is no CSP that can fulfill all the requirements, then the broker starts the negotiation process with CSPs.

4.3.2.1: Broker selects the best initial offer from all offers that are proposed by all CSPs according to the objective.

4.3.2.2: Broker adjusts its initial offer according to the offer selected in 4.3.2.1 to generate new counter offer and propose it to all CSPs.

4.3.2.3: A Provider evaluates broker's counter proposal.

4.3.2.4: If the counter offer proposed by Broker cannot be accepted, then a Provider proposes the counter offer.

4.3.2.5: *Termination of negotiation.* There are three termination conditions:

4.3.2.5.1: When negotiation deadline expires.

4.3.2.5.2: When the offer is mutually agreed by both the Customer and the Provider.

4.3.2.5.3: When a Broker is not able to accept any counter offer proposed by all the providers within the negotiation deadline.

Step 5: After a successful negotiation CSP(s) which is/are selected launches VM(s). CSP(s) replies back to the broker with the accepted proposal along with encrypted packet containing the information about the location of the VM; IP is the address of the VM and key through which a user can SSH (Secure Shell) (SSH) to the VM.

Step 6: Broker decrypts the packet received from the cloud provider using its secret key. It maps the ID with UID and forwards an encrypted packet containing UID, where IP is the address of the VM and key to the enterprise cloud portal.

Step 7: Enterprise cloud portal decrypts the packet received from the broker using its private key. It then forwards the decrypted packet to the respective user.

Step 8: User stores the UID for future reference and with the help of IP address and key, it SSH to the virtual machine. By this way a connection is established between the virtual machine and the user. The most important part is that the broker is successful in SLA negotiation between the Customer and the Provider.

6 CONCLUSION

Cloud computing is still evolving day by day with its new wings. For a detailed study of cloud computing and cloud broker one can refer the book (Buyya et al: Mastering Cloud computing 2013) and (Lucas-Simarro et al: A Cloud Broker Architecture for Multicloud Environments Chap-15 2014) respectively. Among various research areas in cloud computing CSB is a hot topic for researchers since CSB models are not standardized yet. There are still doubts regarding the necessity of cloud broker and its functions. In contribution to this research, this paper elaborates the functions of a cloud broker which further helps in an improved approach that has been proposed for a multi-cloud environment. There is always a scope for further improvement. Details of SLA negotiation with all its parameters can be taken care off in the future work.

REFERENCES

Baur, D., Wesner, S. and Domaschka, J. (2015). "Towards a Model-based Execution-Ware for Deploying Multi-Cloud Applications" *ESOCC 2014, CCIS 508, 124–138, SpringerInternational Publishing Switzerland 2015.*

Buyya, Dr. R., Vecchiolan Dr. C. and Selvi Dr. S.T. (2013). "Mastering Cloud Computing" Tata McGraw Hill.

Goettelmann, E., Dahman, K., Gateau, B. and Godart, C. (2014). "A Broker Framework for Secure and Cost-Effective Business Process Deployment on Multiple Clouds." *CAiSE (Forum/Doctoral Consortium)2014: 49–56.*

Jeon, H., Kim, J., Min, Y. and Seo, K.K. (2014). "A Framework of Cloud Service Broker System for Managing and Integrating Multiple Cloud Services" *Advanced Science and Technology Letters, 64, 93–96 http://dx.doi.org/10.14257/astl.2014.64.23.*

Lucas-Simarro, J.L., Aniceto, S.I., Moreno-Vozmediano, R., Montero, S.R. and M. Llorente, I. (2014). *A Cloud Broker Architecture for Multicloud Environments.* In Large Scale Network-Centric Distributed Systems, A.Y. Zomaya H. Sarbazi-Azad (ed.), Chap. 15, John Wiley & Sons, Inc.

Nair, S.K., Porwal, S., Dimitrakos, T., Ferrer, A.J., Tordsson, J., Sharif, T., Sheridan, C., Rajarajan, M.

and Khan, A.U. (2010). "Towards Secure Cloud Bursting, Brokerage and Aggregation" *Web Services (ECOWS), 2010 IEEE 8th European Conference, 189–196.*

NIST cloud computing reference architecture sp 500–292 (2011) http://www.nist.gov/customcf/get_pdf.cfm?pub_id = 909505.

OPTIMIS – Optimized Infrastructure Services, (2013). http://optimis-project.eu/

SSH http://support.suso.com/supki/SSH_Tutorial_for_Linux

Wang, W., Niu, D., Li, B. and Liang, B. (2013). "Dynamic cloud resource reservation via cloud brokerage," *Distributed Computing Systems (ICDCS), IEEE 33rd International Conference, 400–409, ISSN: 1063–6927.*

Frontiers in Computer, Communication and Electrical Engineering – Acharyya (Ed.)
© 2016 Taylor & Francis Group, London, ISBN: 978-1-138-02877-7

A study to find the most suitable set of prominent genes from microarray data for disease prediction

S. Dasgupta & G. Saha
Government College of Engineering and Leather Technology, Kolkata, India

R. Mondal
Government College of Engineering and Ceramic Technology, Kolkata, India

A. Chanda & R.K. Pal
Department of Computer Science and Engineering, University of Calcutta, Kolkata, India

ABSTRACT: Selection of prominent or significantly expressed genes from microarray data has become a very well-known area of research. There are different methods for attaining the purpose. It has been found that the gene lists returned by these methods are not always similar. In this work we have generated reduced gene lists as outputs from two different feature selection methods applied on two different microarray data sets. The feature selection methods are the random-forest method and a modified t-test based method. These reduced gene-sets are then used to train classifiers based on methods like Random forest, SVM and KNN Classifier. The classification accuracy of the classifiers give some idea about the most suitable list of candidate genes. These candidate gene sets are then further evaluated using biological tools or GO tools to find out the biological significance.

1 INTRODUCTION

The major set back in extracting and analyzing information from microarray data is the problem of high-dimensionality. Typically, such high dimensional data is characterized by thousands of genes in a few sample sizes. Studies on microarrays is an ever evolving area of research [1,2,3]. They help us to simultaneously measure the expression levels of a large (in the order tens of thousands) of genes and can be used extensively in applications related to biology and biomedicine. Microarrays used to pick out differentially expressed genes between two or more groups of patients can be used for many applications [4,5].

Clinically identifying a disease including cancer using gene expression data has two broad objectives: one is to obtain a genuine diagnosis for a patient with greatest accuracy and the second target is to identify the gene or the set of genes responsible for a particular type of disease [1,9].

In the current work we investigate two feature selection methods: the first being t-test along with False Detection Ratio and Fold-change value and the second being Random Forest (RF). The subset of genes selected by these methods is used to build classification models from training datasets which are samples with known class labels using standard classifiers like SVM, K-NN and RF. These models are then used to classify test datasets. The classification accuracies then evaluated in each case.

2 RELATED DEFINITIONS AND TERMS

2.1 *Microarray*

Microarrays are high dimensional data comprising of a large number of genes or features in contrast to the smaller or fewer number of samples or cases. A DNA microarray experiment can be used to simultaneously interpret the expression levels for thousands of genes [3,10]. The raw microarray data is transformed into gene expression matrices. Figure 1 shows an example. Usually, a row in the matrix represents a gene and a column represents a sample. The numeric value in each cell represents the expression level of a particular gene in a specific sample.

2.2 *t-test and p-values*

A statistical test measures the statistical significance of an experiment. A parametric test is usually conducted with an assumption that the underlying distribution of the data is known [2]. The outcome of any statistical test is measured by p-value [7]. In

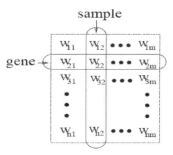

Figure 1. A typical microarray data.

gene selection problem, a large variance of distribution among different phenotypes means that the particular gene can strongly discriminate different phenotypes. Generally, the 2-sample t-statistic still can to some extent measure the difference in the distributions between the different groups. All statistical tests are represented in the form of a null hypothesis. The null hypothesis in this case is that the mean of two populations are equal to each other. A gene which opposes this hypothesis has a low p-value and is considered to be significantly differentially expressed. To test the null hypothesis, one needs to calculate the following values: x_1, x_2 (the means of the two samples), s_{12}, s_{22} (the variances of the two samples), and $n1$, n_2 (the sample sizes respectively of the two samples)

$$t = \frac{\bar{x}_1 - \bar{x}_2}{\sqrt{(s_1^2 / n_1 + s_2^2 / n_2)}}$$

2.3 FDR (False Detection Ratio)

In the process of generating differentially expressed genes from microarray data, as the number of genes that are tested gets large the probability of false error or type I error increases sharply. FDR is used as a control measure to lay stress on the ratio of errors among the set of differentially expressed genes [6].

2.4 Fold change

Fold change [8] does not give any assessment of statistical significance. The common t-statistic is used to pick up genes with small standard deviations; the fold-changes select genes with high degree of shift between control and treatment conditions. Since fold-change method produces gene lists that are more reproducible, it has recently been put as a suggestion that differentially-expressed genes a microarray experiment are best identified using fold-change.

2.5 Random forest for feature selection

The concept of Random Forest was first introduced by Leo Breiman in early 2001 [11]. Random Forest is an ensemble of classification trees. To classify a new object from an input vector, the same is put down each of the trees in the forest. Each tree gives a classification, and it is said that the tree "votes" for that class. Each tree is grown in the following way-

- From the original dataset N cases are sampled at random, with replacement to create a training set of N cases. This sampled set is used as the training set for growing the tree.
- Of the M input variables, a number m is chosen such that m is much smaller than M. Now at each node, m variables are selected at randomly from the M variables. The best split on these m variables is used to split the node. The value of m is fixed and does not change throughout the forest growing. Every tree is grown to the maximum size possible. There can be no pruning.

2.6 Classifiers

2.6.1 Support Vector Machine (SVM)

SVM is used to classify large data sets by introducing a linear or non-linear separating plane in the input space of a data set [14]. The separating surface depends only on a set of support vectors which is a subset of the original data. SVM builds a hyper plane or set of hyper planes in a high dimensional space, which used to achieve a good separation between training data points of any class and is called functional margin. The larger the functional margin the smaller the generalization error of the classifier. SVM models are built based on a kernel function [15,16]. This function transforms the input data into an n-dimensional space, the data is then partitioned with the construction of a hyper plane [12].

2.6.2 K-Nearest Neighbour (KNN)

The K-Nearest Neighbor (KNN) classier has been used extensively as a benchmark classifier [13]. The k-NN approach of classification, outputs a class membership value. The majority number votes that an object gets from its neighbors, is used to classify a particular object. The object is assigned to the class most common among its k nearest neighbors (k is a positive integer, typically small). When k = 1, then the object is put to the class of that single nearest neighbor.

2.6.3 Random-forest for classification

A tree with a low error rate is a considered as good classifier or the tree is said to be strong. Thus by increasing the strength of the individual trees the forest error rate is also lowered.

In the random-forest method there is a concept of OOB (out-of-bag) error. This itself is a measure of classification accuracy.

2.7 *Gene-Ontolology (GO)*

Gene ontology or GO, is a major bioinformatics tool used to relate gene and gene product attributes across all species [18]. GO is again used for GO annotations which relates sets of genes or gene-products with

3   METHODS AND FLOW CHARTS

In this work we have compared 2 different approaches of gene selection from microarray data on 2 different datasets. In the first method we have modified the existing 2 sample t-test by considering the false detection ratio as well as the log fold change in the gene expression and in the second method we have employed the random forest based gene selection method.

We have performed our work using the R software along with its Bioconductor Library. The datasets used for the experiment are given in

Table 1.   Datasets used.

Name	Reference	No. of genes	No. of samples
GDS3715	http://www.ncbi.nlm. nih.gov	12626	110
Golub Leukemia	http://www. broadinstitute.org	3015	38

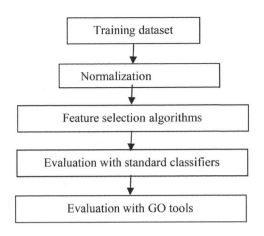

Figure 2.   Flowchart of the work.

Table 1. The flowchart of the work has been shown in Figure 2.

4   EXPERIMENTAL RESULTS

We have presented two different sets of results.

Classifier based results: The classification results are obtained from Tables 2 and 3.

GO results: In our experiment we have used the functional Annotation Tool of David [19] to determine the corresponding percentage of genes annotated along with their GO ID and the corresponding p-value. Such databases have also integrated many tools which can be used to measure the biological significance of a given gene list as shown in Table 4.

Table 2.   Classification accuracy with random forest as feature selection method.

Dataset Name	Size of Test set used	Classifiers			
		RF		SVM	KNN
		Importance Cut-off	Accuracy	Accuracy	Accuracy
GDS 3715	20	0.005	85	50	50
		0.01	80	50	60
		0.05	50	50	45
		0.1	55	45	65
Golub	15	0.005	66.7	46.7	100
		0.01	73.3	46.7	93.3
		0.05	86.7	86.7	80
		0.1	86.7	93.33	93.33

Table 3. Classifiction accuracy with modified t-test as feature selection method.

Dataset name	Size of test set used	Classifiers				
		RF		SVM		KNN
		Accuracy		Accuracy		Accuracy
GDS3715	20	100		75		100
Golub	15	93.3		100		100

Table 4. GO results using David's annotation tool.

List name	Category of GO	Total % of genes annotated	GO ID	P-value
GDS_corrected t-test	BP	92.0	GO:0051252 regulation of wRNA metabolic process	1.1E-6
	CC	88.6	GO:004424 intracellular part	1.3E-7
	MF	86.8	GO:0003700 sequence-specific DNA binding transcription factor activity	6.2E-9
Golub's corrected t-test	BP	95.8	GO:0002376 immune system process	4.2E-15
	CC	98.9	GO:0009986 Cell surface	3.6E-5
	MF	93.7	GO:0005515 protein binding	8.2E-9
GDS_RF (with cut-off <0.01)	BP	88.1	GO:0010557 positive regulation of micromolecule biosynthetic process	7.3E-8
	CC	87.0	GO:0005654 nucleoplasm	3.2E-3
	MF	91.0	GO:0030528 transcription regulator activity	6.3E-6
Golub's RF (with cut-off <0.1)	BP	95.1	GO:0051704 multiorganism prpcess	3.3E-6
	CC	98.8	GO:0005829 cytosol	3.1E-3
	MF	95.1	GO:0005515 protein binding	1.3E-8

5 DISCUSSION

We have used the Affymetrix probe_id of the genes as the gene signatures. It is to be noted that although the selected gene lists are different, yet most of the gene_ids are common to both the methods.

5.1 Analysis of classifier based results

The results with Random Forest as feature selection method measure accuracy at different importance cut off levels. These thresholds are based on permutation index of importance of the random forest algorithm. It is found that the accuracy values measured with cut off <0.01 with the GDS3715 dataset are over all moderate. It is this subset of list that has been chosen for verification with GO. The values measured with Golub et al [17] dataset gives good classification results with cut-off <0.1.

When corrected t-test is used as the feature selection algorithm, it gives very good classification results with both datasets.

5.2 Analysis of GO results

After verification with GO we find that the p-values obtained has the lowest degree of the order −15. There are also values which are not so significant. So we are yet to obtain better results using GO.

We have presented the results of the GO Annotations obtained using DAVID annotation Tool. The 4 lists of genes that have been generated using the 2 data sets and the 2 different approaches, are run using in the DAVID Tool. Each dataset is evaluated under the BP (Biological Process), CC (Cellular Condition) and MF (Molecular Function) category of GO as shown in column 2.

The number of genes annotated over the number of genes presented as the signature is represented in percentage form in column 3. The highest value being 98.9 for CC group of Golub's corrected t-test and the lowest being 86.8% for MF group of GDS corrected t-test. The lowest p-values for that particular group is reported. The highest being 3.2E-3 for CC group of GDS_RF signature and the lowest being 4.2E-15 for BP group of Golub's corrected t-test signature in column 5. The corresponding GO IDs and terms are reported in column 4.

6 CONCLUSION AND FUTURE WORK

Gene sets which have been verified by classifiers and give high accuracies may not always be biologically significant. Also biologically significant genes may not give high accuracies when evaluated through classifiers. We are trying to find a gene-set which will yield high accuracies both statistically and biologically. We will also explore the network that exists between a gene and a particular state. The final results are to be further verified in wet lab.

REFERENCES

[1] Debahuti Mishra, Barnali Sahu "Feature Selection for Cancer Classification: A Signal-to-noise Ratio Approach" International Journal of Scientific & Engineering Research, Volume 2, Issue 4, April-2011 1 ISSN 2229–5518.

[2] L. Ding, J. Pei, J. Ma and D.L. lee "A Rank Sum Test Method for Informative Gene Discovery" KDD '04 Seatle USA.

[3] Ian B Jeffery*1, Desmond G Higgins1 and Aedín C Culhane2J. DeRisi and L. Penland et al. Use of a cdna microarray to analyse gene expression patterns in human cancer. *Nature Genetics*, 14:457–460, 1996.

[4] Hsi-Che Liu, Pei-Chen Peng, Tzung-Chien Hsieh, Ting-Chi Yeh, Chih-Jen Lin, Chien-Yu Chen, Jen-Yin Hou, Lee-Yung Shih, Der-Cherng Liang, "Comparison of feature selection methods for cross-laboratory microarray analysis" IEEE/ACM Transaction on Computational Biology and Bioinformatics 2013.

[5] Herbert Pang, Stephen L. George, Ken Hui, and Tiejun Tong, "Gene Selection Using Iterative Feature Elimination Random Forests for Survival Outcomes." IEEE/ACM Transaction on Computational Biology and Bioinformatatics Vol. 9, NO. 5, September/October 2012.

[6] Anat Reiner, Daniel Yekutieli and Yoav Benjam "Identifying differentially expressed genes using False discovery rate controlling procedures" 19 no. 3 2003, pages 368–375 DOI: 10.1093 /bioinformatics/ btf87.

[7] Theo A. Knijnenburg, Lodewyk F.A. Wessels, Marcel J.T. Reinders and Ilya Shmulevich, "Fewer permutations, more accurate P-values" Bioinformatics Vol. 25 ISMB 2009, pages i161–i168 doi:10.109.

[8] Daniela M. Witten, Robert Tibshirani "A comparison of fold-change and the t-statistic for microarray data analysis" Stanford University, Stanford.

[9] Cheng-San Yang, Li-Yeh Chuang, Chao-Hsuan Ke, and Cheng-Hong Yang, "A Hybrid Feature Selection Method for Microarray Classification" IAENG International Journal of Computer Science IJGS 35_3_05.

[10] Guo, L. and Lobenhofer E.K. et al. (2006), "Rat toxicogenomic study reveals analytical consistency across microarray platforms", Nature Biotechnology 24, 1162–1169.

[11] Leo Breiman, "RANDOM FORESTS", 2001 available at https://www.stat.berkeley.edu/~breiman/randomforest2001.pdf

[12] Seeja, K.R., Shweta "Microarray Data Classification Using Support Vector Machine" International Journal of Biometrics and Bioinformatics (IJBB), Volume (5): Issue (1): 2011.

[13] Yang Song1, Jian Huang2, DingZhou1, Hongyuan Zha1,2, and C. Lee Giles1 "IKNN: Informative K-Nearest Neighbor Pattern Classification"

[14] Guyon, I., et al. (2002) "Gene selection for cancer classification using support vector machines." Mach. Learn., 46:389–422.

[15] Joachims, T., "Making large-scale SVM learning practical", Advances in Kernel Methods –Support Vector Learning, B. Schokopf et al. (ed.), MIT Press, 1999.

[16] Ben-Hur A, Ong CS, Sonnenburg S, Schölkopf B, Rätsch G, "Support Vector Machines and Kernels for Computational Biology." PLoS Comput Biol 4(10), 2008.

[17] T.R. Golub, D.K. Slonim, P. Tamayo, C. Huard, M. Gaasenbeek, J.P. Mesirov, H. Coller, M.L. Loh, J.R. Downing, M.A. Caligiuri, C.D. Bloomfield, E.S. Lander "Molecular Classification of Cancer: Class Discovery and Class Prediction by Gene Expression Monitoring" REPORTS www.sciencemag.org SCIENCE vol 286 15 October 1999.

[18] The Gene Ontology Consortium (January 2008). "The Gene Ontology project in 2008". Nucleic Acids Res. 36 (Database issue): D440–4. doi:10.1093/nar/ gkm883. PMC 2238979. PMID 17984083.

[19] DAVID's Bioinformatics Resources available at http://david.abcc.ncifcrf.gov.

Frontiers in Computer, Communication and Electrical Engineering – Acharyya (Ed.)
© 2016 Taylor & Francis Group, London, ISBN: 978-1-138-02877-7

Magnetic Field Tunable Avalanche Transit Time (MAGTATT) device

Partha Banerjee
Department of Electronics and Communication, Techno India, Kolkata, India

Aritra Acharyya
Department of Electronics and Communication, Supreme Knowledge Foundation Group of Institutions, Mankundu, Hooghly, West Bengal, India

Arindam Biswas
Department of Electronics and Communication, NSHM Knowledge Campus, Durgapur, West Bengal. India

A.K. Bhattacharjee
Department of Electronics and Communication, NIT, Durgapur, West Bengal, India

ABSTRACT: In this paper, a new device named Magnetic Field Tunable Avalanche Transit Time (MAGTATT) diode has been proposed. Besides the conventional electronic and optical tuning mechanisms, a third, new terminal frequency tuning technique of double-drift region Impact Avalanche Transit Time (IMPATT) has been formulated and studied by the authors. Here the static magnetic field is used for tuning the power output and operating the frequency of a 94 GHz DDR Si IMPATT source. A few more results have been presented in this paper; however the detail analysis and simulation results are reported elsewhere.

1 INTRODUCTION

Besides the electronic and optical tuning mechanisms, there is also another method available that can control the IMPATT properties externally, i.e. magnetic field tuning. The application of magnetic field in properly biased IMPATT diode causes a number of effects in its operation such as changes of the carrier motion, the carrier distributions and the electrostatic potential distribution. Only in certain limits the effect of the magnetic field on device properties can simply be visualized in terms of Hall voltage or carrier deflection. B. Glance was the first, who reported the frequency tuning of IMPATT oscillator by means of magnetic field, in the year 1973 (Glance 1973). Later, Hartnagel *et al.* reported the magnetic field tuning of frequency and power output of a 7.480 GHz Double-Drift Region (DDR) IMPATT oscillator (Hartnagel *et al.* 1975). But as far as authors' knowledge is concerned there is no theoretical analysis available in published literature, which studies the effect of magnetic field on the RF performance of IMPATT sources in detail.

The authors have developed a complete two-dimensional (2-D) large-signal model to study the above mentioned effect (Banerjee *et al.* 2015). The model is capable of analyzing the effects of the magnetic field applied along any direction on the RF performance of IMPATT sources. However, in this paper, a special case has been presented in which the magnetic field is applied along the transverse direction of the carrier flow, since it is the maximum influential direction that causes maximum carrier deflection leading to a highest modulation of RF properties of the device. The sensitivity of various static and large-signal parameters on externally applied transverse steady magnetic field of a DDR IMPATT diode, which are designed to operate at 94 GHz, have been calculated. The arrangement proposed in this paper as a whole may be regarded as Magnetic Field Tunable Avalanche Transit Time (MAGTATT) device.

2 A BRIEF VIEW ON THE DEVICE STRUCTURE, 2-D MODEL AND SIMULATION TECHNIQUE

For the simplicity in the arrangement of magnetic field coupling and 2-D devices modeling, the cross-sectional shape of the DDR IMPATT device is considered as rectangular instead of circular. This assumption ensures the use of Cartesian coordinate system which is far more convenient than the cylindrical coordinate system. However, the choice for the shape of the cross-section of the

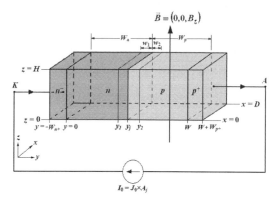

$$\vec{B} = (0, 0, B_z)$$

Figure 1. 3-D model of the n^+-n-p-p^+ structured DDR IMPATT diode ($w_1 = w_2 = 0.1$ μm).

diode hardly affects the simulation results since the cross-sectional area of the diode is more important than its shape.

The 3-D model of the aforementioned device structure is shown in Figure 1. However, in the modeling, z-direction need not be taken into account and 2-D model is enough for the analysis; the reason behind this is discussed later. In the present model, the cross-sectional shape of the metallurgical junction (i.e. the n-p junction) is taken to be rectangular for simplicity as mentioned earlier (Acharyya 2015). In Figure 1, the lengths of n^+, n, p, p^+-layers are denoted by W_{n+}, W_n, W_p and W_{p+} respectively. The doping concentrations of the above mentioned layers are denoted by N_{n+}, N_D, N_A and N_{p+} respectively. Width of the device along x-direction is D and height of the device along z-direction is H. Thus the cross-sectional area of the metallurgical junction is $A_j = DH$. In order to achieve high-quality RF performance of the device at 94 GHz and above, Molecular Beam Epitaxy (MBE) growth technique is the most favorable process to fabricate DDR Si IMPATT diode. However for the simplicity in mathematical formulation of the 2-D doping profile of the said device, simple formulation of diffusion technique has been adopted.

Optimizing some of the most important parameters, which is related to the diffusion technique, sharp, and the most favorable doping profile of the device must be achieved which almost matches with the MBE grown DDR structure, especially for the devices operating at or above 94 GHz. This consideration is justified from the prospective of realistic device simulation.

Due to the shortage of space, the details of the 2-D static and large-signal simulation techniques have not been included, in this paper. However, those detail formulations can be obtained from other source (Banerjee *et al.* 2015).

3 RESULTS AND DISCUSSION

A DDR IMPATT diode based on Si has been designed to operate at W-band (75–110 GHz) by choosing the optimum values of W_n, W_p, N_D, N_A, N_{n+}, and N_{p+} (Acharyya *et al.* 2013). As discussed the design parameters are given in Table 1. The width (along x-direction) and height (along z-direction) of the diode have been taken as $D = 10.0$ μm and $H = 96.211$ μm. Therefore the junction area of the diode is $A_j = DH = 9.6211 \times 10^{-10}$ m^2 which is same as the junction area of the diode having circular cross-section of radius 35.0 μm (Luy *et al.* 1987). The simulation procedure described in the earlier section has been used to simulate both the static and large-signal properties of the device for the bias current densities ranging from 2.6×10^8 to 3.8×10^8 A m^{-2} in the absence of the magnetic field (i.e. $B_z = 0$). After that the abovementioned simulations have been repeated by varying the magnetic field from 1.0 to 5.0 Tesla (T).

Variation of magnetic field sensitivity of breakdown voltage with magnetic field is shown in Figure 2 (a). It is observed that the value of breakdown voltage increases with the increase of mag-

Table 1. Structural and doping parameters.

Design parameters	Values
W_n (μm)	0.400
W_p (μm)	0.380
N_D ($\times 10^{23}$ m^{-3})	1.200
N_A ($\times 10^{23}$ m^{-3})	1.250
N_{n+} ($\times 10^{25}$ m^{-3})	5.000
N_{p+} ($\times 10^{25}$ m^{-3})	2.700
A_j (m^{-2})	9.6211×10^{-10}

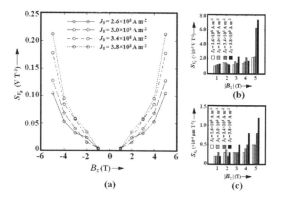

Figure 2. Graphs showing the sensitivities of (a) breakdown voltage, (b) avalanche region voltage drop and (c) avalanche region width with respect to the applied magnetic field for different bias current densities.

netic field. Application of magnetic filed in the perpendicular direction of the carrier movement (i.e. z- and y-directions respectively) consequences deflection of carries in $\pm x$ directions depending on

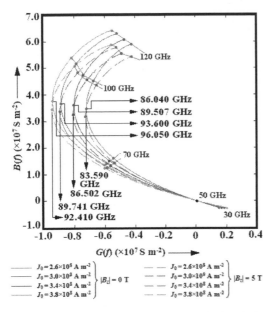

Figure 3. Admittance characteristics of the device under magnetic field intensities of 0 and ±5 T for different bias current densities.

their polarities. This, in turn, increases the effective depletion region width of the device for a given bias current density. Therefore, the breakdown occurs at higher voltage. In Figure 2 (a), its shown that the magnetic field sensitivity of breakdown voltage varies from 0.0025 to 0.1047 V T^{-1} due to the increase of magnetic field from ±1.0 to ±5.0 T for the bias current density of 2.6×10^8 A m^{-2}. It can also be observed that the avalanche region broadens due to the application of magnetic field which leads to increase in the avalanche region voltage drop. The magnetic field sensitivities of V_A and x_A are shown in the bar graphs of Figures 2 (b) and (c) respectively. Those are varying from 0.0010 to 0.0022 V T^{-1} and 0.2×10^{-4} to 0.5×10^{-4} μm T^{-1} for the same increase of magnetic field for the bias current density of 2.6×10^8 A m^{-2}. Due to the increase of both the V_A and x_A, the V_D/V_B and x_A/W increases with the increase of B_z. Increase of the ratio V_D/V_B indicates the increase in DC to RF conversion efficiency of the device (Scharfetter 1969) operating under magnetic field. Another noteworthy fact which can be observed from Figure 3 is that the sensitivities of abovementioned DC parameters are higher at higher bias current densities for the same magnetic field. Larger bias current increases the amount of carriers moving through the device under operational condition. Higher carrier densities show higher values of magnetic field sensitivity. The highest magnetic field sensitivities of V_B, V_A, and x_A are found to be 0.2128 V T^{-1}, 0.0074 V

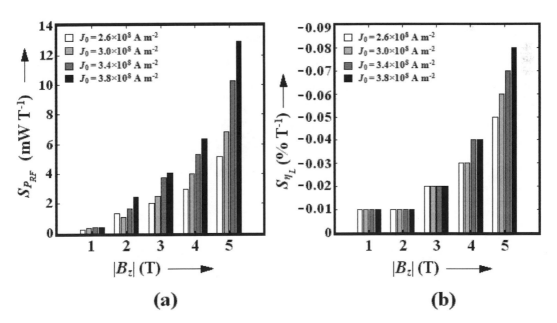

Figure 4. Bar graphs showing the sensitivities of (a) RF power output and (b) DC to RF conversion efficiency with respect to the applied magnetic field for different bias current densities.

T^{-1}, and 1.20×10^{-4} μm T^{-1} for $|B_z|$ varying from 4.0 to 5.0 T and for $J_0 = 3.8 \times 10^8$ A m^{-2}.

The admittance characteristics of the DDR Si IMPATT for different bias current densities are shown in Figure 3 in the absence of the magnetic field as well as in the presence of the magnetic field of ±5 T. It is noteworthy from Figure 3 that the magnitudes of the peak negative conductance slightly increase due to the presence of magnetic field while the values of the corresponding susceptance increase significantly. This fact leads to significant decrease in Q-factor which, in turn, improves both the oscillation growth rate and stability of oscillation due to the presence of magnetic field. It can be noted from the above study that both the breakdown voltage and magnitude of the peak negative conductance increases due to the application of external magnetic field, consequently the RF power output is increased. However the DC to RF conversion efficiency decreases in presence of the magnetic field due to the significant increase of input DC power as results of increased breakdown voltage.

The variations of the magnetic field sensitivity of P_{RF} and η_L with magnetic field are shown in the bar graphs in Figures 4 (a) and (b). These are monotonically increasing and decreasing functions of magnetic field respectively for a fixed bias current density. On the hand, these are also monotonically increasing and decreasing functions of bias current density respectively for a fixed value of magnetic field. The maximum values of magnetic field sensitivity of P_{RF} and η_L are obtained as 12.95 mW T^{-1} and 0.08% T^{-1} for $|B_z|$ varying from 4.0 to 5.0 T and for $J_0 = 3.8 \times 10^8$ A m^{-2}. The application of transverse static magnetic field causes lengthening of the transit distance of the charge carriers. This fact changes the transit time delay and hence affects the phase relationship between the terminal current and voltage waveforms. This leads to alteration in the admittance characteristics of the device and therefore causing the change in the large-signal properties.

4 CONCLUSION

In this paper, we have proposed the magnetic field sensitivities of various static and large-signal parameters of the device that have been designed to operate at W-band have been calculated from the large-signal simulation of MAGTATT device. Results show that the frequency and power tuning of the source of around 3.64 GHz and 26 mW are achievable with maximum sensitivities of about −1.59 GHz T^{-1} and 12.95 mW T^{-1} respectively for the application of transverse magnetic field varying from 4.0 to 5.0 T. The nature of both frequency and power tuning of the device due to application of transverse magnetic field is in well agreement with the experimental results carried out earlier.

REFERENCES

Acharyya, A. (2015). RF Performance of IMPATT Sources and Their Optical Control. *Lambert Academic Publishing, Germany*.

Acharyya, A., Banerjee, S., and Banerjee, J.P. (2013). Effect of Junction Temperature on the Large-Signal Properties of a 94 GHz Silicon Based Double-Drift Region Impact Avalanche Transit Time Device. *Journal of Semiconductors* 34, 024001–12.

Banerjee, P., Acharyya, A., Biswas, A., and Bhattacharjee, A.K. (2015). Effect of Magnetic Field on the RF Performance of Millimeter-Wave IMPATT Source, *Unpublished*.

Glance, B. (1973). A Magnetically Tunable Microstrip IMPATT Oscillator. *IEEE Transaction of Microwave Theory and Techniques*, 21, 425–426.

Hartnagel, H.L., Srivastava, G.P., Mathur, P.C., and Sharma, V. (1975). Effect of Transverse Magnetic Field on the Power Output and Frequency of IMPATT Oscillators. *physica status solidi (a)* 3,1 K147–K149.

Luy, J.F., Casel, A., Behr, W., and Kasper, E. (1987). A 90-GHz double-drift IMPATT diode made with Si MBE. *IEEE Trans. Electron Devices* 34, 1084–1089.

Scharfetter, D.L., and Gummel, H.K. (1969). Large-Signal Analysis of a Silicon Read Diode Oscillator. *IEEE Trans. on Electron Devices* 6, 64–77.

Frontiers in Computer, Communication and Electrical Engineering – Acharyya (Ed.)
© *2016 Taylor & Francis Group, London, ISBN: 978-1-138-02877-7*

A deviation based identification of random valued impulse noise towards image filtering using neighborhood approximation

S. Banerjee
Department of Computer Science Engineering, Narula Institute of Technology, Agarpara, Kolkata, West Bengal, India

A. Bandyopadhyay & R. Bag
Department of Computer Science Engineering, Supreme Knowledge Foundation Group of Institutions, Mankundu, West Bengal, India

A. Das
Department of Computer Science Engineering, Netaji Subhash Engineering College, Garia, Kolkata, West Bengal, India

ABSTRACT: In this paper a two phase scheme has been proposed for removal of random valued impulse noise from digital gray scale images. First phase detects the noisy pixels and second phase removes those by using updated neighbourhood based pixel approximation method. A same size binary flag image is also been generated using a variable threshold value which depends upon the standard deviation of the noisy image. The quantitative and qualitative result shows that the proposed filter outperform the existing with respect to MSE, PSNR and SSIM comparison even at 60% high noise density levels.

1 INTRODUCTION

Images are usually corrupted by impulse noise due to bit errors in transmission, circuitry of a digital camera, or signal acquisition (Banerjee *et al.* 2015). Generally impulse noise can be classified into two categorics namely Salt and Pepper Noise (SVN) and Random Valued Impulse Noise (RVIN). The intensity of a gray scale image is accumulated in a 8 bit integer giving 256 possible gray levels in the range '0' to '255' (Bandyopadhyay *et al.* 2015). The random valued impulse noise picks any value between the dynamic ranges (0–255) randomly (Dong *et al.* 2007, Bandyopadhyay *et al.* 2015). So, detection of noisy pixels takes a very important role to design a random valued impulse noise removal filter. There are several filters for Random Valued impulse noise removal. Among them median filter (Tukey 1971) is often used for its noise containment capability while preserving edges. But as all the pixels (both corrupted and uncorrupted) are changed by the median operation the median filter does not perform well at moderate or high noise densities. To remove these drawbacks different improvised median filters (Chen *et al.* 1999, Chen *et al.* 2001) were proposed. Those filters tried to remove the previous drawbacks but lagged in detail preservation.

A few years later Luo filter (Luo *et al.* 2007) was introduced which was able to preserve the minute details of images. But in this method the detection did not yield fruitfull result. The genetic programming (GP) filter (Petrovic *et al.* 2008) tried to solve this issue by a two cascaded detectors for detection and two corresponding estimators for reduction. This detection method was more successful than Luo filter. Then after a new PDE based method (Wu *et al.* 2011) was evolved for random valued impulse noise removal which followed anisotropic diffusion for filtering impulse noise. This gave a new dimension in detection and removal of Impulse noise. It was followed by robust outlyingness ratio (ROR) filter (Xiong *et al.* 2012) where all the pixels were divided into four clusters according to the ROR values. A two stage detection process was used for fine detection from each cluster. Based on ROR a robust direction based detector (RDD) (Prathiba *et al.* 2013) was proposed which used standard deviation of images for superior detection. In the recent years another filter named TDWM (Sarkar *et al.* 2014) was introduced. It has been tried to detect the noise by calculated threshold. Times to time many efforts were made to produce a satisfactory result over random valued impulse noise. But at high noise density effective detection is still an issue.

This paper takes a challenge to overcome the above said problems. In this paper standard deviation of the noisy image takes a vital role to assume the threshold value which is applied to announce the center pixel of a 3×3 matrix whether it is corrupted or uncorrupted. The whole noise detection and removal procedure has been divided into sections where section 2 presents the review of the filter used for noise removal, section 3 describes the proposed methodology, section 4 constitutes result and discussion and section 5 contains the conclusion.

2 NEIGHBOURHOOD BASED PIXEL APPROXIMATION

2.1 Review of Neighbourhood Based Pixel Approximation (NBPA) (Banerjee et al. 2014)

Firstly all the noisy pixels $x_{i,j}$ in the corrupted image are mapped by '0' (zero) then those are tried to de-noise by $x_{i-1,j},\ x_{i+1,j},\ x_{i,j-1},\ x_{i,j+1}$ pixels when all the functioning pixels are not '0' (zero). The row, column and clumping operations are introduced to decrease the noise density and to preserve the edges of the image. After that the remaining corrupted pixels are replaced by the arithmetic mean of non-zero pixels in the created 7×7 matrix centering the said pixel. The border operation is introduced to maintain the size of the image with the arithmetic mean of non-zero pixels in the created 6×6 matrix taking the said pixel at different corners for different side's border operations. The flowchart of NBPA is shown in Figure 1.

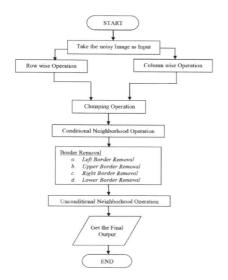

Figure 1. Flow chart of NBPA.

3 PROPOSED METHOD

Let $x_{i,j}$ for $(i,j) \in A \equiv \{1,2,3,...M\} \times \{1,2,3,......N\}$ be the gray intensity level at pixel location (i,j) of a true $M \times N$ image X and X_1 is the random valued noisy image of X. Now standard deviation of X_1 has been calculated with equation no (I).

$$\sigma = \sqrt{\frac{1}{N}\sum_{i=1}^{N}(x_i - \mu)^2} \qquad (I)$$

Simultaniously a same size binary flag image F_1 has been generated where $f_{i,j} = 0$ represents $x_{i,j}$ is uncorrupted and $f_{i,j} = 1$; a noisy one. Firstly it has been assumed that all the pixels in the image X_1 are uncorrupted i.e. $f_{i,j} = 0$.

Algorithm I

A, $W \times W$ ($W=5$) matrix is created centering the pixel $x_{i,j}$ in the image X_1. Elements $(w_1, w_2, w_3, ..., w_{25})$ in the matrix are sorted.

Consider $w = \frac{w_{12} + w_{13}}{2}$

if $\left(|x_{i,j} - w| < T\right)$ then $x_{i,j}$ is uncorrupted

else

corrupted and $f_{i,j} = 1$.

Where threshold value $T = \frac{\sigma}{2}$.

Algorithm II

---

if $f_{i,j} = 1$ then put $x_{i,j} = 0$.

An image X_2 is developed.

if $x_{i,j} \neq 0$ in X_2 create a 3×3 matrix taking $x_{i,j}$ as a center.

Let $A = \{x : x = \text{non zero pixel in the created matrix}\}$

and $B = \{y : y = |x_{i,j} - x| \leq 5\}$

if $n(B) \geq 2$ then replace $x_{i,j} = \frac{\sum y}{n(B)}$

After completion an image X_3 is generated from X_2.

The rest uncorrupted pixels in image X_3 are removed by NBPA (Banerjee *et al.* 2014) method.

4 RESULTS AND DISCUSSION

The proposed method is assessed on the base of Mean Square Error (MSE), Peak Signal-to-Noise

Ratio (PSNR), Structural Similarity Index Measurement (SSIM), and Mean Structural Similarity Index Measurement (MSSIM).Quantitative performances of the de-noising techniques are measured by Mean Square Error (MSE), Peak Signal-to-Noise Ratio (PSNR) as defined in equation (II) and (III) respectively. Edge preservation is shown by the Structural Similarity Index Measurement, and Mean Structural Similarity Index Measurement (MSSIM) as defined in the equation (IV) and (V) respectively.

$$MSE = \left(\sum_{M,N} \frac{\left(I(m,n) - \hat{I}(m,n) \right)^2}{M \times N} \right) \quad (II)$$

MSE is the mean square error between original image (I) and de-noised image ($\hat{I}$). M and N are the number of rows and columns in the input image, respectively.

$$PSNR = 10 \log_{10} \left(\frac{255^2}{MSE} \right) \quad (III)$$

$$SSIM(x,y) = \frac{(2\mu_x\mu_y + C_1)(2\sigma_{xy} + C_2)}{(\mu_x^2 + \mu_y^2 + C_1)(\sigma_x^2 + \sigma_y^2 + C_2)} \quad (IV)$$

$$MSSIM(x,y) = \frac{1}{M} \sum_{m=1}^{M} SSIM(x_m, y_m) \quad (V)$$

where μ_x is the mean of image x and σ_x is the standard deviation of image x, similarly μ_y is the mean of image y and σ_y is the standard deviation of image y, C_1, C_2 are the constants σ_{xy} is the covariance of x and y, given by:

$$\mu_x = \frac{1}{N} \sum_{i=1}^{N} x_i$$

$$\sigma_x = \left[\frac{1}{N-1} \sum_{i=i}^{N} (x_i - \mu_x)^2 \right]^{\frac{1}{2}}$$

$$\sigma_{xy} = \left(\frac{1}{N-1} \sum_{i=i}^{N} (x_i - \mu_x) \right)$$

Table 1 shows the PSNR comparison with respect to Lena image for different new filters with the proposed filter at variable noise density (30%–60%). The result shows that the quality of the reconstructed image for different filters.

So from the comparison it can be inferred that proposed filter produces better result than the existing filters reviewed in this paper.

Figure 2 shows the qualitative comparison of the restored output of different filters with the proposed. The visual result clarifies the superior image quality of the proposed with respect to the other filters discussed in this paper.

Figure 3 presents the SSIM and the Figure 4 presents the Mean Square Error graph respectively. The SSIM specifies the structural symmetry of an image after reconstruction. The index value shows that the proposed filter effectively preserves the edge details. The MSE graph in Figure 4 shows that the proposed method generates minimum error than the other filters.

Table 1. PSNR for different filters for Lena image at different noise densities.

Image	Filters	30%	40%	50%	60%
Lena	LUO	30.29	28.27	26.23	24.21
	GP	31.87	28.42	24.86	21.68
	NSDD	30.86	28.35	26.39	24.08
	ROR	24.93	21.51	18.69	15.61
	RDD	26.14	22.98	19.99	16.72
	TDWM	28.02	26.85	25.88	25.00
	Proposed	**30.68**	**29.41**	**27.41**	**25.10**

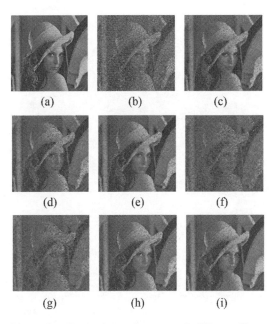

(a) (b) (c)

(d) (e) (f)

(g) (h) (i)

Figure 2. Comparison of output of different filters with the proposed filter: (a) Original Lena image (b) 50% noise corrupted image (c) Luo (d) GP (e) NSDD (f) ROR (g) RDD (h) A. Sarkar (i) Proposed.

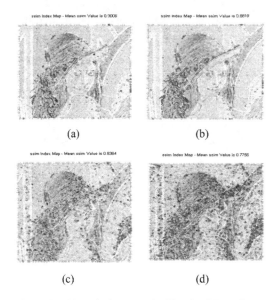

(a)   (b)

(c)   (d)

Figure 3. SSIM index map (MSSIM) of Lena image
at noise density: (a) 30% (MSSIM = 0.9006), (b) 40%
(MSSIM = 0.8819), (c) 50% (MSSIM = 0.8364), (d) 60%
(MSSIM = 0.7756).

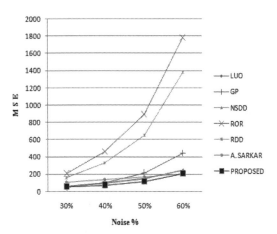

Figure 4. MSE vs noise density for Lena image ranges
from 30% to 60%.

5 CONCLUSIONS

The quality of a filter over random valued impulse
noise depends upon the capability of noise detection. If the detection accuracy is elevated, there are
so many existing filters for effective restoration.
Here a detection based random valued impulse
noise removal is introduced in this paper. Quantitative performance and the structural symmetry
are analyzed using the PSNR, MSE and SSIM
respectively. The analysis shows that the proposed
method outperforms the existing methods is all
respect. This method can produce better results
for other types of noise removal which is left for
further study.

REFERENCES

Banerjee, S., Bandyopadhyay, A., Bag, R., and Das, A.
(2014). Neighborhood Based Pixel Approximation
for High Level Salt and Pepper Noise Remov al. *CiiT
International Journal of Digital Image Processing*,
8(6), 346–351.
Banerjee, S., Bandyopadhyay, A., Bag, R., and Das, A.
(2015). Moderate Density Salt & Pepper Noise
Removal. *IJECT* 10(6), Issue 1, Spl–1 Jan–March.
Bandyopadhyay, A., Banerjee, S., Das, A., and Bag, R.
(2015). A Relook and Renovation over State-of-Art
Salt and Pepper Noise Removal Techniques. *I.J.
Image, Graphics and Signal Processing*, vol–9, 61–69.
Bandyopadhyay, A., Kumari, M., Pooja., Das, A., and
Bag, R. (2015). Detection and Removal of High Density
Random Valued Impulse Noise. *CiiT International
Journal of Digital Image Processing*, 8(7), 242–246.
Chen, T., Ma, K.K., and Chen, H.L. (1999). Tri-state
median-based filters in image denoising. *IEEE Trans.
Image Process.* 12(8), 1834–1838. (TSM).
Chen, T., and Wu, H. (2001). Adaptive impulse detection
using center-weighted median filters. *IEEE Signal
Process. Lett.*, 1(8), 1–3. (ACWM).
Dong, Y., Chan, H.R., and Xu, S. (2007). A detection
statistic for random valued impulse noise. *IEEE Trans.
Image Process.* 4(16), 1112–1120.
Luo, W. (2007). An efficient algorithm for the removal of
impulse noise from corrupted images. *Int. J. Electron.
Commun.* 8(61), 551–555. (Luo).
Prathiba, K., Rathi R., and Christopher C.S. (2013).
Random Valued Impulse Denoising using Robust
Direction based Detection. *Proceedings of 2013 IEEE
conference on Information and Communica tion Tech-
nologies (ICT 2013)*. (RDD).
Petrovic, N., and Crnojevic, V. (2008). Universal impulse
noise filter based on genetic programming. *IEEE
Trans. Image Process.*, 7(17), 1109–1120. (GP).
Sarkar, A., Changder, S., and Mandal, K.J. (2014).
A Threshold based Directional Weighted Median Filter
for Removal of Random Impulses in Thermal Images.
*IEEE 2nd International Conference on Business and
Information Management (ICBIM)*. (J K Mandal).
Tukey, W.J. (1971). *Exploratory Data Analysis (prelimi-
nary ed.)*. Reading, MA: Addision-Wesley.
Wu, J., and Tang, C. (2011). PDE-Based Random-Valued
Impulse Noise Removal Based on New Class of
Controlling Functions. *IEEE Transactions on Image
Processing*, 9(20). (NSDD).
Xiong., and Zhouping, Y. (2012). A Universal de noising
framework with a new impulse detector and nonlinear
means. (ROR).

Frontiers in Computer, Communication and Electrical Engineering – Acharyya (Ed.)
© 2016 Taylor & Francis Group, London, ISBN: 978-1-138-02877-7

A variable gain CMOS phase shifter for phased array beamformer applications

Dipankar Mitra, Alarka Sanyal, Palash Roy & Debasis Dawn
North Dakota State University, Fargo, ND, USA

ABSTRACT: In this paper, a variable gain phase shifter is developed using 180 nm CMOS process at S-band, which can control phase and gain independently. This has a potential application in a conformal phased array antenna, for beamforming within 360° range. The integrated phase shifter, along with the variable gain amplifier, helps in recovering the degraded radiation pattern precisely due to the conformal shaping of the antenna. The simulation results show state-of-the-art performances of 7dB conversion gain with variable gain and a continuous phase rotation of 360°. The fabricated chip, measuring an area of 1.125 mm², consumes a very low power of 17 mW.

1   INTRODUCTION

Phased array antenna systems can be used for wireless communication during space explorations where the antenna array can be placed on the suits of the astronauts. But due to movement of the astronauts, the radiation pattern of the antenna changes. A phase shifter can be used in this case to recover the degradation of the radiation pattern, by providing the required amount of phase shifts to each of the antenna array elements (Braaten, *et al.* 2013). There are several classifications of phase shifters, like passive phase shifters, active phase shifters (Vadivelu, Praveen Babu, *et al.* 2009), analog and digital phase shifters (Nagra, Amit S., *et al.* 1999), switched line phase shifters (Kang, Dong-Woo, *et al.* 2006), loaded-line phase shifters (Yu, Yikun, *et al.* 2008), and reflection type phase shifters (Malczewski, A., *et al.* 1999) among others.

Now a days, it is possible to integrate a high number of transistors into a single chip, with the transistors exhibiting both analog and digital functionality (Masaki, Ichiro, *et al.* 1995). The realization of System-On-Chip (SOC) solutions was possible using CMOS RFIC (Radio Frequency Integrated Circuits), where both analog and digital functionality can be integrated into a single process platform and where the digital circuitry can control analog radio with computer programs.

In our work, we have designed and fabricated a CMOS phase shifter, which is to be operated at the S frequency band. There are several circuits, which are integrated together in a single chip to create the variable gain phase shifter as shown in Fig. 1 are a passive hybrid, two active baluns, a double balanced Gilbert Cell, two Variable Gain Amplifiers

(VGAs) and two buffers. We used 180 nm CMOS process for the design. This paper has the following sections: 2. Phase Shifter Design, 3. Simulations Results, 4. Measurement Results and 5. Summary and Future Work.

2   PHASE SHIFTER DESIGN

The heart of the Phase shifter is a Double balanced Gilbert Cell (Fig. 2).

We have used a co-design methodology to optimize the design of the chip by taking into consideration the input/output impedances of each block in order to maximize the performance of the chip.

The input RF signal is fed into a passive hybrid (Fig. 3) to get two signals with phase difference of 90°: I (in phase) and Q (quadrature phase) signals. They are then passed through two active baluns (Fig. 4) to get 4 RF signals I⁺ (0°), I⁻ (180°), Q⁺ (90°), and Q⁻ (270°), which are fed to the Gilbert Cell of the phase shifter. The active baluns also compensate any amplitude loss to the input signal when it is passed through the passive hybrid. The

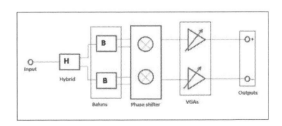

Figure 1.   Integrated CMOS phase shifter.

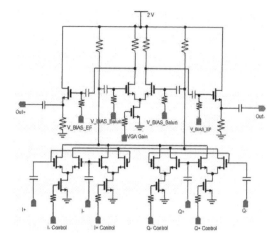

Figure 2. The schematic of the double balanced Gilbert cell with VGA and buffers.

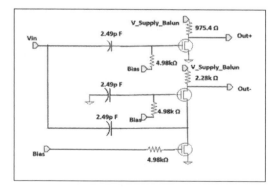

Figure 3. Passive hybrid.

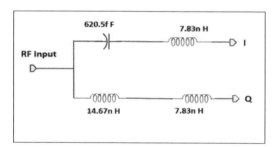

Figure 4. Active balun.

Gilbert Cell has 4 control voltage arms and the four controlling voltages are fed to the tail transistors of the Gilbert Cell. The output phase can be controlled by supplying control voltages to any of these two arms. This principle of operation is as depicted in Fig. 5.

For example, to get a 45° phase shift in the output signal, we can provide equal control voltages in the I+ Control and Q+ Control arms. The outputs of the Gilbert Cell are then passed through a two stage VGA in order to obtain an amplified output signal with variable gain capability. The output of the VGA is passed through two buffers where we get the final outputs of the phase shifter. All the blocks shown in Fig. 1, have been designed with a co-design methodology, taking into the considerations the input/output impedances matching of each block in order to maximize the performance of the chip.

3 SIMULATION RESULTS

We have used CADENCE® Spectre® (Cadence Design Systems, Inc.) simulator and layout is done using Virtuoso® tool. The input and the output of the phase shifter are matched to 50Ω for better RF performance. Table 1 shows the performance summary of the chip.

We have used a 2.2 GHz input RF signal with −30 dBm (10 mV peak) amplitude to carry out the simulations. The simulation results show an output signal of ~20 mV (peak) for single ended output and ~40 mV (peak) for differential output. Using the VGA, the output signal amplitude can be varied, as shown for single ended output in Fig. 6.

Fig. 7 shows the phase shifting property of the circuit in time domain over a range of 0° to 180° with 22.5° steps for better presentation, although the phase can be varied over the entire 0° to 360° range. Similarly, the frequency characteristics of the phase variation in the entire 360°degree is shown in Fig. 8.

The S parameter analysis S11 (input return loss), S22, S33 (output return loss), and S21 (conversion gain) are shown in Fig. 9. The circuit has an input return loss of better than −25 dB, and output

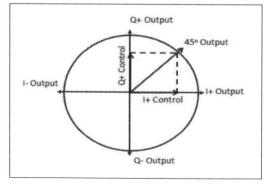

Figure 5. Phase control mechanism.

Table 1. Performance summary (Simulated).

Technology	0.18 um CMOS
Frequency	1 ~ 3 GHz
Phase Range	360°
Gain	+7 dB @ 2.2 GHz (variable)
Input Return Loss	<−25 dB @ 2.2 GHz
Output Return Loss	<−11 dB @ 2.2 GHz
Power Consumption	17 mW
Chip Area	1.5 mm × 0.75 mm (with pads)

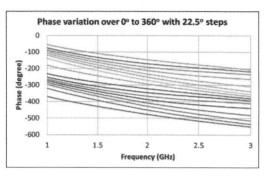

Figure 8. Phase shifting performance in the entire 360° range in frequency domain.

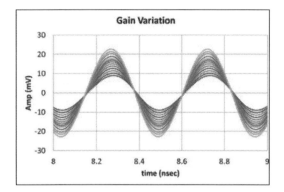

Figure 6. Gain variation of the circuit with a particular phase setting.

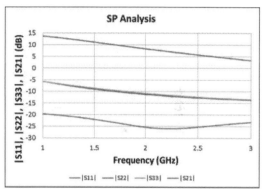

Figure 9. S parameter analysis.

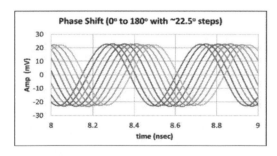

Figure 7. Phase shifting performance in time domain.

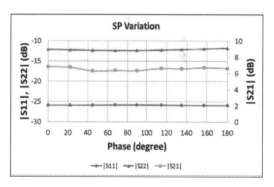

Figure 10. Stability of conversion gain and return loss with phase shifting.

return loss better than −11 dB and a conversion gain of close to 7 dB at 2.2 GHz frequency.

The uniqueness of the circuit is that the input/output return losses (S11, S22) and the conversion gain (S21) remains almost constant with the variation of the phase as shown in Fig. 10.

The large signal performances are shown in Fig. 11, where we can see that the 1 dB output compression power lies at around −15 dBm, which is sufficient to integrate the phase shifter in the LO path of the transmitter circuitry.

The microphotograph of the chip with a dimension of 1.5 mm × 0.75 mm with all the bond pads included is shown in Fig. 12.

4 MEASUREMENT

We used Cascade Microtech probe station for probing the chip and Agilent N5230 PNA-L Network Analyzer for measuring S-parameters performances. The measured insertion loss is 7 dB at 2.2 GHz and the input and output is well matched with input return loss is 20.04 dB and output return loss

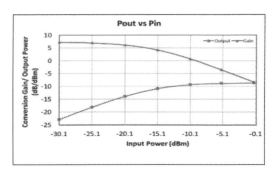

Figure 11. Large-signal performance of the phase shifter.

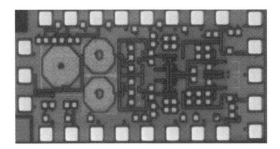

Figure 12. Microphotograph of the fabricated phase shifter die (1.5 mm × 0.75 mm).

is 10.05 dB @ 2.2 GHz. The measured phase rotation is 75° and we got measured 3.27 dB gain variation from 1 GHz to 4 GHz. These discrepancies with the simulation results are thought to be due to the active and passive device model mismatch. Table 2 summarizes the measured results.

5 CONCLUSION

A variable gain CMOS phase shifter is designed, fabricated and measured using 180 nm CMOS technology working at S-band frequency, which is to be integrated into a conformal phased array antenna for beamformer applications. The vector-modulator based phase shifter is integrated with an on-chip passive hybrid and active baluns capable of controlling each other, the phase and gain independently. This functionality of the phase shifter helps it to be integrated with conformal phased array beamformer to recover a degraded radiation pattern due to the conformal shape of the antenna array, which requires separate control of phase and amplitude of the input signal. The phase shifter exhibits state-of-the-art performances including a

Table 2. Summary of measured results.

Phase Range	75°
Insertion Loss	7 dB @ 2.2 GHz
Input Return loss	20.0 dB @ 2.2 GHz
Output Return Loss	10.0 dB @ 2.2 GHz
Gain Variation	3.27 dB
DC Power Consumption	17.48 mW

conversion gain of 7 dB and control of phase over the entire 360° range with a very low power consumption of 17 mW. The discrepancies between simulations and measured performances are thought due to the foundry supplied design model mismatch. As an extension of this work we are developing a phase shifter array with a closed loop digital control system for taking care of dynamical changes of surface curvature of a conformal phased array antenna.

ACKNOWLEDGEMENT

Authors would like to thank ND NASA EPSCoR for their support during this project under the agreement FAR0020852.

REFERENCES

Braaten, B.D., M.A. Aziz, S. Roy, S. Nariyal, I. Irfanullah, N.F. Chamberlain, M.T. Reich and D.E. Anagnostou. (2013). A Self-Adapting Flexible (SELFLEX) Antenna Array for Changing Conformal Surface Applications. *IEEE Transactions on Antennas and Propagation, vol. 61, no. 2*, (pp. 655–665).

Cadence spectre/virtuoso Simulator by Cadence Design Systems, Inc., 2655 Seely Avenue, San Jose, CA 95134.

Kang, Dong-Woo, et al. (2006). Ku-band MMIC phase shifter using a parallel resonator with 0.18-/spl mu/m CMOS technology. *Microwave Theory and Techniques, IEEE Transactions on* 54.1: 294–301.

Malczewski, A., et al. (1999). X-band RF MEMS phase shifters for phased array applications. *Microwave and Guided Wave Letters, IEEE* 9.12: 517–519.

Masaki, Ichiro, et al. (1995). New architecture paradigms for analog VLSI chips. *Vision Chips, Implementing Vision Algorithms Using Analog VLSI Circuits*: 353–375.

Nagra, Amit S., and Robert A. York. (1999). Distributed analog phase shifters with low insertion loss. *Microwave Theory and Techniques, IEEE Transactions on* 47.9: 1705–1711.

Vadivelu, Praveen Babu, et al. (2009). Integrated CMOS mm-wave phase shifters for single chip portable radar. *Microwave Symposium Digest. MTT'09. IEEE MTT-S International.*

Yu, Yikun, et al. (2008). A 60GHz digitally controlled phase shifter in CMOS. *Solid-State Circuits Conference. (ESSCIRC) 34th European. IEEE.*

Calculating absorption coefficient of Gaussian double quantum well structured with band nonparabolicity for photodetector in microwave spectra

Debasmita Sarkar & Arpan Deyasi

Department of Electronics and Communication Engineering, RCC Institute of Information Technology, Kolkata, India

ABSTRACT: Since the absorption coefficient of double Gaussian quantum well is analytically computed in the presence of electric field (applied along the direction of confinement) incorporating Lorentzian lineshape function for photodetector application in IR and Microwave region. The results are derived for the intraband transition between lowest three energy states, which are computed by solving time-independent Schrodinger equation by using the appropriate boundary conditions. First order band nonparabolicity of Kane type is used for near accurate estimation. Gaussian potential profile is considered due to its closest resemblance with ideal parabolic potential configuration. Structural parameters, material composition, and electric field are tailored to observe the effect on absorption coefficient. These results are important for the design of optical detector for microwave applications.

1 INTRODUCTION

Electronic and optoelectronic properties of a low-dimensional semiconductor structure has enticed interest among theoretical (Parashar *et al.*, 2011, Far *et. al.*, 2012) and experimental (Urban *et al.*, 2007, Kim *et al.*, 2012) researchers in the last decade, who were influenced by the advancements of numerical techniques (Ogawa *et al.*, 1998, Deyasi *et al.*, 2013) and fabrication methodologies (Wu *et al.*, 2009). Quantum well, wires, and dots are the choice of candidates among the nanostructures, where motion of carriers is confined along one, two, or three dimensions. Eigenenergy states of these structures are originated due to the quantization, geometrical shapes (Gangopadhyay *et al.*, 1997, Li *et al.*, 2001), and material compositions (Jia *et al.*, 2011, Balet *et al.*, 2004), which are the controllable parameters. These quantum structures are already used in designing optical transmitters (Zhang *et al.*, 2011, Raihan *et al.*, 2011), detectors (Deyasi *et al.*, 2014, Jha *et al.*, 2014), defense (Philips *et al.*, 2010), medical (Perera *et al.*, 2006), communication (Wei *et al.*, 2006), memory (Muller *et al.*, 2006), and applications. Photonic application of these devices, precisely for detector applications; can be studied from the knowledge of absorption coefficient, and hence its accurate estimation plays an important role for the device design.

In this paper, absorption coefficient of Gaussian double quantum well structure is theoretically calculated in presence of electric field for intraband transition between lowest three energy states in presence of conduction band nonparabolicity. The reason behind the consideration of Gaussian potential profile is that it is very close to the ideal parabolic potential profile and fabrication methodologies that are available till this date, is very much suitable for designing Gaussian variation of potential. Hence, computation using Gaussian potential well provides very close and accurate simulation results compared with theoretically ideal parabolic potential configuration. Structural parameters, material composition, and electric field are tuned to study the variation of absorbance. Lorentzian lineshape function is considered to include the effect of pulse broadening. The results are important for the application of these devices in photonic detection.

2 THEORETICAL FOUNDATION

Optical absorption coefficient for quantum well can be written as

$$\alpha(\omega) = \frac{q^2 m^*}{2\varepsilon_0 \varepsilon_r c n_r \hbar L m_0^2} \frac{P^2}{\hbar \omega}$$
$$\times \sum_{n,m} \left\langle g_v^{\ m} \middle| g_c^{\ n} \right\rangle \Theta(E_{mn} - \hbar\omega) \quad (1)$$

where the bracket vectors gives the overlap integral between z-dependent envelope functions of conduction band and valence band respectively. This provides the selection rule for transition between the subbands, and also for band-to-band transition. L is the width of quantum well, Θ is the Heaviside step function, n_r is the refractive index of the well material, N_w is the number of well, and P, the momentum matrix element is defined as

$$P = N \int_{cell} \psi_{k_c}^*(\vec{r}\,')p_A \psi_{k_v}(\vec{r}\,')d^3r' \qquad (2)$$

The overlap integral can be reformatted as

$$\left\langle g_v^m \middle| g_c^n \right\rangle = \delta_{k,k'} \int_{-L/2}^{L/2} \chi_m^h(z)\chi_n^e(z)dz \qquad (3)$$

where 'χ' denotes the wave functions. For quantum well,

$$\chi_n(z) = \sqrt{\frac{2}{L}} \cos\left(\frac{n\pi z}{L}\right) \qquad (4.1)$$

for odd 'n'

$$\chi_n(z) = \sqrt{\frac{2}{L}} \sin\left(\frac{n\pi z}{L}\right) \qquad (4.2)$$

for even 'n'

Hence, Eq. (1) may be modified as

$$\alpha(\omega) = \frac{2\pi q^2 \hbar}{\varepsilon_0 \varepsilon_r cn_r m^{*2} V} \frac{1}{\hbar\omega}$$
$$\times \sum_{i,j} \left|\langle i|p_z|j\rangle\right|^2 \delta(E_j - E_i - \hbar\omega)(f_{FD}^i - f_{FD}^j) \qquad (5.1)$$

where 'i' and 'j' are the initial and final states respectively, and f_{FD} is the Fermi-Dirac distribution function. The factor 2 is incorporated for spin degeneracy, and δ function is introduced to conserve the momentum.

For two consecutive states, Equation (5) may be written as

$$\alpha(\omega) = \frac{2\pi q^2 \hbar}{\varepsilon_0 \varepsilon_r cn_r m^{*2} V} \frac{1}{\hbar\omega} \left[\frac{8\hbar}{3L}\right]^2$$
$$\times \sum_{i,j} \delta(E_j - E_i - \hbar\omega)(f_{FD}^i - f_{FD}^j) \qquad (5.2)$$

We assume that the well is doped. We further consider that the Fermi energy level is above the ground state and first excited state is completely empty, then Equation (5) may be simplified as

$$\alpha(\omega) = \frac{n_s \pi q^2 \hbar}{\varepsilon_0 \varepsilon_r cn_r m^{*2} L} \frac{1}{\hbar\omega} \left[\frac{8\hbar}{3L}\right]^2 \delta(\Delta E - \hbar\omega) \qquad (6)$$

Where n_s is the charge density per unit area, other parameters has the usual significances. Defining oscillator strength between 1st state to 2nd state as

$$f_{21} = \frac{2}{m^* \hbar\omega} \left[\frac{8\hbar}{3L}\right]^2 \qquad (7)$$

Equation (7) can be put into the form

$$\alpha(\omega) = \frac{n_s \pi q^2 \hbar}{2\varepsilon_0 \varepsilon_r cn_r m^* L} f_{21} \delta(\Delta E - \hbar\omega) \qquad (8)$$

By incorporating Lorentzian lineshape function and replacing the ideal delta function in order to consider the broadening of peaks, final expression of absorption coefficient may be written as

$$\alpha(\omega) = \frac{n_s \pi q^2 \hbar}{2\varepsilon_0 \varepsilon_r cn_r m^* L} f_{21} \frac{\Gamma}{\pi\left[(\hbar\omega - \Delta E)^2 + \Gamma^2\right]} \qquad (9)$$

where Γ is half-width at half of the maximum.

3 RESULTS AND DISCUSSION

Using Equation (9), absorption coefficient of the proposed structure is computed and plotted as a function of wavelength for different structural parameters. Electric field is applied along the direction of quantum confinement, and first order nonparabolic dispersion relation is considered for simulation purpose. Fig. 1 shows the absorption coefficient with wavelength in presence of electric field for band nonparabolicity for Gaussian double quantum well and compared with overestimated parabolic band structure. Absorption coefficient is computed and plotted for the first three intraband transitions considering the first three eigenstates. It is observed that with increase of energy difference between the bands, i.e. intersubband transition energy increases. Band nonparabolicity consideration reduces the eigenenergy than that obtained for parabolic structure, and hence the separation between two consecutive states also reduces. This lowering of subband energy makes a shift of the peak of absorption coefficient profile higher towards the wavelength, and magnitude of the peak also increases.

In Fig. 2, absorption coefficient profile with wavelength for different material compositions in

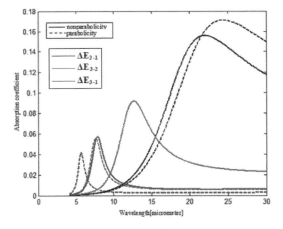

Figure 1. Absorption coefficient for Gaussian DQW for nonparbolic and parabolic dispersion relations.

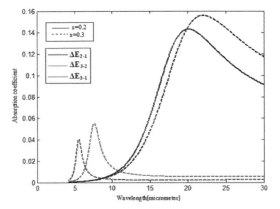

Figure 2. Absorption coefficient for Gaussian DQW for different material composition in presence of band nonparabolicity in presence of electric field.

barrier layer is plotted. The nonparabolic band structure is considered here for this simulation. It is observed that the peak value of absorption coefficient is increased with the increment of material composition value, i.e., with increasing the AlAs composition in barrier material, peak is redshifted. This is a fact that due to the increase of x, quantum confinement also increases. This enhances the relative magnitude of absorption coefficient, and it appears at a higher wavelength. Another important observation that can be made from the plot is that with lower percentage of x in $Al_xGa_{1-x}As$, only two eigenenergy levels exist which makes it possible for the absorption coefficient for $\Delta E_{2,1}$.

In the absence of electric field it is observed that electron confinement is possible only in ground states, there is no eigenenergies hence, no

intersubband transition energies exist. So, it may be concluded that only in the presence of external force this electron confinement is possible and absorption profile can be plotted. This is plotted in Fig. 3. It can also be observed that with the higher value of x, the magnitude of peaks also increases; hence, shift in absorption value is considerable.

Absorption coefficient is calculated as a function of operating wavelength for different well width, as shown in Fig. 4. This absorption profile depends on well width and difference of eigenenergies ($\Delta E_{j,i}$). For lower well dimension, it is found that the coefficient is more compared with that obtained for higher well width. This is due to the fact that with decrease of quantum well length, confinement decreases, but there is equal energy difference between the bands, i.e. intersubband

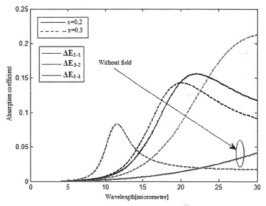

Figure 3. Absorption coefficient for Gaussian DQW for different material composition in presence of band nonparabolicity in presence & absence of electric field.

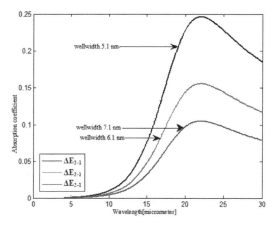

Figure 4. Absorption coefficient for Gaussian DQW for different well widths for band nonparabolicity in presence of electric field.

transition energy decreases. This reduces the peak of absorbance amplitude provided half-width at half maximum is kept constant throughout the simulation. Δ_E is kept same for all cases of well widths, so it is also constant. From the equation we can say that absorption coefficient is inversely proportional with total device length (well + barrier) so as the length of the well increases and the absorption coefficient decreases.

In Fig. 5, variation is calculated and plotted for different contact barrier width dimension. It is found that the coefficient is more compared with that obtained for higher contact barrier width. This can be explained as follows: As the barrier width increases the total length also increases, the electron confinement is less and absorption coefficient is inversely proportional to total length; absorption decreases with increase of length, here the energy difference between the bands remains the same, i.e., intersubband transition energy is invariant. This reduces the peak of absorbance amplitude.

In Fig. 6, comparative study is prepared for the absorption coefficient for different middle barrier widths without any external applied field with already estimated results of absorption profile in presence of the electric field. In the plot it is observed that without any external field applied the intersubband transition is not possible; to take the electron confinement form ground state to higher energy state electric field must be applied.

4 CONCLUSION

Absorption coefficient of Gaussian double quantum well structure is analytically computed in

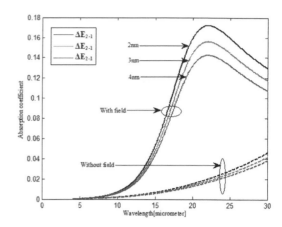

Figure 6. Absorption coefficient for Gaussian DQW for different middle barrier widths for band nonparabolicity in presence & absence of electric field.

presence of the electric field and conduction band nonparabolicity for intraband transition between lowest three quantum states of conduction band. Lorentzian lineshape function is considered to include the effect of pulse broadening, which makes the result very close to experimental findings. Gaussian potential profile is chosen to find the performance near accurately, as it is very close to the ideal parabolic potential configuration. Suitable tuning of structural parameters is made to observe the required blueshift/redshift, which essentially speaks in favor of using the device in microwave spectra.

REFERENCES

Balet. L.P., Ivanov. S.A., Priyatinski. A., Achermann. M., Kilmov. V.I., 2004, Inverted core/shell nanocrystals continuously tunable between Type-I and Type-II localization regimes, Nano Letters 4, 1485–1488.

Deyasi. A., Bhattacharyya. S., Das. N.R., 2013, Computation of intersubband transition energy in normal and inverted core-shell quantum dots using finite difference technique, Superlattices & Microstructures, 60, 414–425.

Deyasi. A., Das. N.R., 2014, Oscillator strength and absorption cross-section of core-shell triangular quantum wire for intersubband transition, Springer Proceedings in Physics: Advances in Optical Science and Engineering, 166, Chapter no: 78, pp. 629–635.

Far. S., Sterian. P., Fara. L., Iancu. M., Sterian. A., 2012, New Results in Optical Modeling of Quantum Well Solar Cells, International Journal of Photoenergy, p. 810801.

Gangopadhyay. S., Nag. B.R., 1997, Energy levels in finite barrier triangular and arrowhead-shaped quantum wires, Journal of Applied Physics 81, 7885–7889.

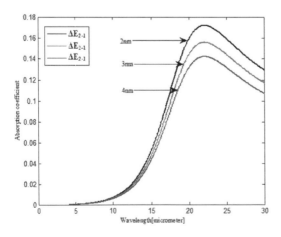

Figure 5. Absorption coefficient for Gaussian DQW for different contact barrier widths for band nonparabolicity in presence of electric field.

Jha. N., 2014, IR photodetector based on rectangular quantum wire in magnetic field, AIP Conference Proceedings, 1591, 434–436.

Jia. G., Wang. Y., Gong. L., Yao. J., 2011, Heterostructure type transformation of ternary $ZnTexSe_{1-x}$/ZnSe core-shell quantum dots, Digital Journal of Nanomaterials and Biostructures 6, 43–53.

Kim. D.H., You. J.H., Kim. J.H., Yoo. K.H., Kim. T.W., 2012, Electronic structures and carrier distributions of T-Shaped $Al_xGa_{1-x}As$/$Al_yGa_{1-y}As$ quantum wires fabricated by a Cleaved-Edge Overgrowth method, Journal of Nanoscience and Nanotechnology 12, 5687–5690.

Li. Y., Voskoboynikov. O., Lee. C.P., Sze. S.M., 2001, Electron energy state dependence on the shape and size of semiconductor quantum dots, Journal of Applied Physics, 90, 6416–6420.

Muller. C.R., Worschech. L., Forchel. A., 2006, Memory inhibition in quantum-wire transistors controlled by quantum dots, Physica Status Solidi (C), 3, 3794–3797.

Ogawa. M., Kunimasa. T., Ito. T., Miyoshi. T., 1998, Finite-Element analysis of quantum wires with arbitrary cross sections, Journal of Applied Physics 84, 3242–3249.

Parashar. T.K., Pal. R.K., 2011, Modeling of GaAs/$Al_{0.2}Ga_{0.8}$As quantum well gas detector for LWIR region, MIT International Journal of Electronics and Communication Engineering. 1, 97–100.

Perera. A.G.U., 2006, Quantum structures for multiband photon detection, Opto-Electronics Review, 14, 103–112.

Phillips. M.C., Taubman. M.S., Bernacki. B.E., 2010, Design and performance of a sensor system for detection of multiple chemicals using an external cavity quantum cascade laser, Proceedings of SPIE, 76080D.

Raihan. M.R., Li. Z., Liu. D., Hattori. H.T., Premaratne. M., 2011, Combining different in-plane photonic wire lasers and coupling the resulting field into a single-mode waveguide, Progress In Electromagnetics Research C, 21, 191–203.

Urban. D., Braun. M., König. J., 2007, Theory of a magnetically controlled quantum-dot spin transistor, Physical Review B, 76, 125306.

Wei. R., Deng. N., Wang. M., Zhang. S., Chen. P., Liu. L., Zhang. J., 2006, Study of self-assembled Ge quantum dot infrared photodetectors, 1st IEEE International Conference on Nano/Micro Engineered and Molecular Systems, 330–333.

Wu. Y. R., Lin. Y. Y., Huang. H. H., Singh. J., 2009, Electronic and optical properties of InGaN quantum dot based light emitters for solid state lighting, Journal of Applied Physics, 105, 013117.

Zhang. L., Yu. Z., Yao. W., Liu. Y., Feng. H., 2011, Optical properties of GaN/AlN quantum dots under intense laser field, Proceedings of SPIE, Optoelectronic Materials and Devices VI, 830813.

Frontiers in Computer, Communication and Electrical Engineering – Acharyya (Ed.)
© *2016 Taylor & Francis Group, London, ISBN: 978-1-138-02877-7*

Application of STATCOM in power quality improvement under different fault conditions in a power system network

Nabamita Roy

Electrical Engineering Department, MCKV Institute of Engineering, Liluah, Howrah, West Bengal, India

ABSTRACT: The present work proposes a model of STATCOM for power quality improvement under different fault conditions. A two bus double feed system has been considered for the study. Ten types of faults have been simulated for different locations on the transmission line. Total Harmonic Distortion (THD) of the voltage signal across the load has been computed in the MATLAB Simulink software. THD of all the voltage signals with and without the presence of STATCOM has been tabulated. Significant improvement has been obtained in the THD of the voltage signals in presence of STATCOM. All the simulations have been done in MATLAB software.

1 INTRODUCTION

Electric power systems providing energy to a large number of customers are spread over vast areas stretching from huge cities to remote location and rural places. This task is achieved through various power system components like generators, transformers, motors, and transmission and distribution lines which run thousands of miles, satisfying the customer needs through careful planning, design, installation and operation of such very complex networks. The network is subjected to constant disturbances, both natural and component generated, which may prove dangerous not only to the faulty component, but also to the neighboring equipment. It is therefore impervious that the damage caused by the disturbances is limited to a minimum by speedy isolation of the faulty section without hampering the functioning of the rest of the system.

Consumer awareness regarding reliable power supply has been growing day by day. Power quality is most common concern for power utilizes as well as for consumers. Today, the world needs increased amount of quality power for its growing population and industrial growth.

Development of compensation to enhance power quality has been an area of active interest for the past few decades (Ghosh et al. 2002, L. Gyugyi 1979, Y.H. Song et al. 1999, R. Arnold 2001, Olimpo et al. 2002, D.R. Patel et al. 2011). Passive compensators like shunt reactors and capacitors are uncontrolled devices and incapable of continuous variation in parameters. The emergence of custom power devices has led to development of new and fast compensators [R. Arnold, 2001]. The custom power devices include compensators like Distribution Static Compensator (DSTATCOM), Dynamic Voltage Restorer (DVR), Unified Power Quality Conditioner (UPQC), Battery Energy Storage System (BESS), and many more such controllers. These devices may be quite helpful in solving power quality problems. However, due to high cost, and for effective control they are to be optimally placed in the system.

A Fault is an unavoidable phenomenon in a power system. Power system faults are the major causes of deterioration of power quality causing interruption in power supply and voltage distortion. Detection of fault and its mitigation is a challenging task. THD of a voltage/current signal is an important measurable parameter of power quality. The present work involves determination of THD of the voltage profiles across a load in a 2-bus system under different fault conditions with and without STATCOM in the system.

The rest of this paper is organized as follows. Section 2 describes the simulated power system model in detail without STATCOM. The simulation of STATCOM in MATLAB has been discussed in section 3. The results of simulation followed by calculation of THD have been presented in section 4. Section 5 gives the conclusion of the present work.

2 POWER SYSTEM UNDER STUDY

A two bus double feed, 50 Hz, 3-phase power system network has been simulated using the Simpower Toolbox of MATLAB-7 and is shown in Figure 1. The length of the transmission line is 200 km. A 3-phase load is connected near the bus B2 of the transmission line. The system parameters have been provided in table 1.

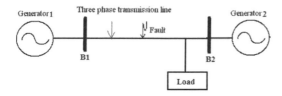

Figure 1. Single line diagram of three phase network without STATCOM.

Table 1. System parameters of Figure 1.

System components	Specifications
Generator 1 & Generator 2	18 KV, 50 Hz, X/R = 7
Transmission Line	3 phase, 50 Hz, 200 Km
Load	Three Phase RL load in parallel, P = 100 KW, Q = 100 MVAR

The sampling times of all the signals has been taken to be 78.28 μs and the time period of simulation in MATLAB has been taken up to 0.06 secs. The sampling frequency is 12.8 kHz. The following ten types of faults have been simulated in this system:

Single Line-Ground fault for phase A, B and C respectively, (i.e. AG, BG and CG).

Double Line fault (i.e. AB, BC and CA).

Double Line-Ground fault (i.e. ABG, BCG and CAG).

Three phase fault, i.e. LLL

All the faults have been initiated at 19 different locations starting from B1, each being 10 km apart. The fault resistance considered for the simulation is 0 Ω. The faults have been initiated after 20 ms (i.e. one cycle) and cleared at 30 ms. The fault inception angle is 0°. The total number of fault simulations made in this system are $19 \times 10 = 190$.

3 MODELLING OF STATCOM

The modeling of STATCOM with power system has been done in MATLAB SIMULINK environment. The STATCOM consists of a VSC, (Akil Ahemad et al. 2014, T. Devaraju et al. 2012, D.K. Tanti et al. 2011). The VSC in STATCOM has been modeled by connecting 3 arm IGBT bridge. Each IGBT is paralleled by a diode. The output of the converter should be a sine wave. In order to get the sine wave output at the converter end, a sine triangular PWM output is given to the IGBT's gates. When the IGBT is 'ON', the converter will act as an inverter and the DC capacitor

voltage is inverted to 3 phase AC. During the 'OFF' period of the IGBT, the converter will act as a full wave uncontrolled rectifier and the capacitor gets charged to the maximum value of the line voltage. The PWM generator is used to generate the gate pulses for the IGBTs of the VSC. Here, the sine and the triangular waveforms are generated. The sine wave which is of 50 Hz is compared with the triangular wave of 20 kHz. According to the comparison, the PWM pulses are produced. These pulses are given to gates of the IGBTs. By varying the modulation index, the magnitude of the converter output will vary and as well as by varying the phase angle of the modulating wave, the converter output voltage phase angle will also vary. Modulation index is defined as the ratio of the magnitude of modulating wave to carrier wave.

A SLD of the power system network with STATCOM is shown in Figure 2. Table 2 shows the system parameters.

4 RESULTS OF SIMULATION

The voltage profiles across the load in the present system for a specific LLL fault condition with and without STATCOM have been shown in Figure 3. The magnitudes of THD for a few particular fault conditions have been given in Table 3. It is observed that the proposed model of STATCOM has improved the THD of the load voltage signals in the present system during

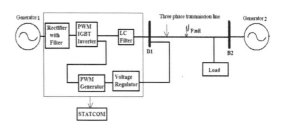

Figure 2. Single line diagram of the system with STATCOM.

Table 2. System parameters of figure 2.

System components	Specifications
Rectifier with Filter	Forward voltage 0.8 V Snubber Resistance = 100 Ω L = 200μ H, C = 5000 μ F
IGBT Inverter	Snubber Resistance = 5000 Ω
LC Filter	100 H, 50 MVAR
Voltage Regulator	Proportional Gain = 0.8, Integral gain = 500
PWM Generator	Carrier frequency = 2 kHz

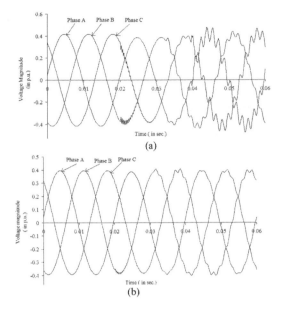

(a)

(b)

Figure 3.   (a) Voltage profiles across load without STAT-COM and (b) with STATCOM for LLL faults occurring at 100 km from B1.

fault conditions. The best performance of the STATCOM model has been obtained for the faults occurring at 10 km and 50 km from B1 in the present system.

5   CONCLUSION

The proposed model of STATCOM has improved the THD of the voltage signals in the present system. This study can be further extended for a multi-terminal system. STATCOM can be further tested by simulating fault conditions for differ-ent values of fault resistances and fault inception angles. For a multi-terminal system location of the STATCOM plays an important role in improve-ment of the power quality. The best location can be determined by application of Artificial Neural Network.

ACKNOWLEDGEMENT

The author is thankful to the 6 final year students (Suraj Yadav, Arijit Das, Diptesh Koley, Prabhat

Table 3.   Magnitudes of THD under different fault conditions with and without STATCOM.

Fault Condition	Magnitude of THD for each phase (without STATCOM)			Fault Condition	Magnitude of THD for each phase (with STATCOM)		
Fault Location = 100 Km from B1	A	B	C	Fault Location = 100 Km from B1	A	B	C
AG	0.68	0.61	0.53	AG	0.01	0.00	0.00
BG	0.03	0.05	0.03	BG	0.00	0.01	0.00
CG	0.02	0.02	0.04	CG	0.00	0.00	0.01
AB	0.22	0.22	0.00	AB	0.02	0.02	0.00
BC	0.00	0.06	0.06	BC	0.00	0.31	0.31
CA	0.12	0.00	0.12	CA	0.03	0.00	0.03
ABG	0.16	0.07	0.14	ABG	0.03	0.02	0.03
BCG	0.04	0.03	0.03	BCG	0.03	0.03	0.01
CAG	0.08	0.03	0.06	CAG	0.02	0.03	0.02
LLL	0.16	0.07	0.14	LLL	0.02	0.03	0.04
D = 10 KM, R_F = 0 Ω, θ = 0°	A	B	C	D = 10 KM, R_F = 0 Ω, θ = 0°	A	B	C
AG	0.45	0.94	0.39	AG	0.00	0.00	0.00
BG	0.16	0.31	0.16	BG	0.00	0.00	0.00
CG	0.03	0.03	0.06	CG	0.00	0.00	0.00
AB	0.06	0.06	0.00	AB	0.00	0.00	0.00
BC	0.00	0.05	0.05	BC	0.00	0.00	0.00
CA	0.09	0.00	0.09	CA	0.00	0.00	0.00
ABG	0.02	0.03	0.02	ABG	0.00	0.00	0.00
BCG	0.04	0.03	0.05	BCG	0.01	0.04	0.01
CAG	0.05	0.02	0.05	CAG	0.00	0.00	0.00
LLL	0.09	0.05	0.06	LLL	0.00	0.00	0.00

(*Continued*)

Table 3. (*Continued*)

Fault Condition	Magnitude of THD for each phase (without STATCOM)			Fault Condition	Magnitude of THD for each phase (with STATCOM)		
Fault Location = 150 Km from B1	A	B	C	Fault Location = 150 Km from B1	A	B	C
AG	0.72	0.80	0.75	AG	0.01	0.00	0.00
BG	0.10	0.20	0.10	BG	0.00	0.01	0.00
CG	0.11	0.11	0.19	CG	0.00	0.00	0.01
AB	0.29	0.29	0.00	AB	0.02	0.02	0.00
BC	0.00	0.09	0.09	BC	0.00	0.30	0.30
CA	0.18	0.00	0.18	CA	0.03	0.00	0.03
ABG	0.18	0.00	0.19	ABG	0.03	0.02	0.03
BCG	0.02	0.12	0.13	BCG	0.02	0.03	0.01
CAG	0.09	0.02	0.09	CAG	0.02	0.03	0.04
LLL	0.03	0.10	0.13	LLL	0.02	0.02	0.04
Fault Location = 190 Km from B1	A	B	C	Fault Location = 190 Km from B1	A	B	C
AG	0.07	0.36	0.24	AG	0.11	0.07	0.07
BG	0.17	0.17	0.32	BG	0.07	0.14	0.07
CG	0.13	0.13	0.23	CG	0.07	0.07	0.13
AB	0.43	0.43	0.00	AB	0.19	0.19	0.00
BC	0.00	0.25	0.25	BC	0.00	0.23	0.23
CA	0.47	0.60	0.58	CA	0.11	0.00	0.11
ABG	0.28	0.19	0.18	ABG	0.1	0.01	0.16
BCG	0.19	0.27	0.36	BCG	0.76	0.81	0.45
CAG	0.36	0.18	0.38	CAG	0.19	0.21	0.22
LLL	0.81	0.23	0.64	LLL	0.11	0.21	0.29
Fault Location = 50 Km from B1	A	B	C	Fault Location = 50 Km from B1	A	B	C
AG	0.07	0.36	0.24	AG	0.00	0.00	0.00
BG	0.17	0.17	0.32	BG	0.00	0.00	0.00
CG	0.13	0.13	0.23	CG	0.00	0.00	0.01
AB	0.43	0.43	0.00	AB	0.00	0.00	0.00
BC	0.00	0.25	0.25	BC	0.00	0.00	0.00
CA	0.47	0.60	0.58	CA	0.00	0.00	0.00
ABG	0.28	0.19	0.18	ABG	0.00	0.00	0.00
BCG	0.19	0.27	0.36	BCG	0.00	0.01	0.00
CAG	0.36	0.18	0.38	CAG	0.00	0.00	0.00
LLL	0.81	0.23	0.64	LLL	0.00	0.12	0.11

Kumar, Priyendu Ghosh and Monojit Das) of Electrical Engineering Department (2015 batch), MCKV Institute of Engineering, for their contribution to whom a part of this study was given as B.Tech project.

REFERENCES

Akil Ahemad, Sayyad Naimuddin, *Simulation Of D-Statcom In Power System*, IOSR Journal of Electrical and Electronics Engineering (IOSR-JEEE) e-ISSN: 2278–1676, p-ISSN: 2320–3331 PP 01–08.

Arnold, R. *Solutions to the Power quality problems*, Power Engineering Journal, Vol.–15, No.–2, pp. 65–73, April 2001.

Ghosh and G. Ledwich, *Power quality enhancement using custom power devices*, Kluwer Academic Publisher, London 2002.

Gyugyi, L. *Reactive power generation and control by thyristor circuits*, IEEE Transactions on Industry applications, Vol IA-15, No.–5, pp. 521–532, Sept/Oct 1979.

Olimpo Anaya-Lara and Acha, E. *Modeling and analysis of custom power systems by PSCAD/EMTDC*, IEEE Transactions on Power Delivery, Vol. 17, No. 1, pp. 266–272, January 2002.

Patil, D.R., & Komal K. Madhale, *Design And Simulation Studies Of D-Statcom For Voltage Sag, Swell Mitigation*, International Journal of Power System Operation and Energy Management ISSN (PRINT): 2231–4407, Volume–2, Issue–1, 2.

Song, Y.H., and Johns, A.T. *Flexible AC Transmission Systems (FACTs)*, IEE Power and Energy Series, London, UK, 1999.

Thangellamudi Devaraju, Akula Muni Sankar, Vyza. C. Veera Reddy, Mallapu Vijaya Kumar, *Understanding of voltage sag mitigation using PWM switched autotransformer through MATLAB Simulation*, World Journal of Modelling and Simulation, ISSN: 1746–7233, England, UK Vol. 8 (2012) No. 2, pp. 154–160.

Tanti, D.K., Brijesh Singh, Dr. Verma, M.K., Dr. Mehrotra, O.N. *An ANN Based Approach For Optimal Placement Of DSTATCOM For Voltage Sag Mitigation*, International Journal of Engineering Science and Technology (IJEST), ISSN: 0975–5462 Vol. 3 No. 2, pp.-827–835, Feb 2011.

Frontiers in Computer, Communication and Electrical Engineering – Acharyya (Ed.)
© 2016 Taylor & Francis Group, London, ISBN: 978-1-138-02877-7

Effects of close proximity and hybrid operation of HVAC & HVDC transmission lines under steady-state and fault conditions: A literature survey

Brajagopal Datta
National Institute of Technology, Yupia, Arunachal Pradesh, India

Kamaljyoti Gogoi & Saibal Chatterjee
NERIST, Nirjuli, Arunachal Pradesh, India

ABSTRACT: With the recent requirements of bulk power transfer, HVDC transmission lines has become more popular than the conventional HVAC transmission lines due to their ease of operation and certain benefits like more power transfer capability with same Right of Way, etc. One of the ways to incorporate the new HVDC transmission lines is to add them to the same mechanical poles of the existing HVAC networks leading to formation of a hybrid HVAC-HVDC transmission corridor. Also due to lack of geographical area, many a times there is a probability that the new HVDC transmission lines will be erected in close proximity of existing HVAC lines. When an AC line approaches a DC line, interferences would be generated upon the DC line due to the electromagnetic induction between the AC and DC lines. The interference includes the transverse voltage caused by capacitive coupling, the longitudinal electromotive force, and the fundamental frequency current on the DC line caused by inductive coupling. Thus elaborative study of this effects is required to understand them in both steady state and fault conditions. This becomes more important as many HVDC projects are operational or is going to be added soon to Indian Power Grid. The paper presents the work done in this area and scope of work that could be done for Indian Power Grid in this area. Focus of the study is to identify the prominent effects during steady state and fault conditions.

1 INTRODUCTION

In recent years, many HVDC power transmission systems has been introduced in our country, due its advantages over conventional HVAC transmission systems in long and bulk power transmission capabilities. Also, HVDC power transmission technology is an attractive solution for non-conventional power transmission applications in terms of increase in transfer capability with same right of way as is required by HVAC transmission system. HVDC connections can be justified from an economic and technical point of view for high transmitted power and/or long lines; in particular they are the adopted solutions for submarine cables and asynchronous interconnections. The higher initial costs of the conversion plants are compensated by the reduced losses during the line operations [1]. The operation of HVDC transmission lines in existing HVAC network can be done in two ways, hybrid transmission system where same mechanical structure can be used for both the lines and close proximity operation where the distance between two lines is usually between few hundred metres

[1]. The prior method is beneficial not only because of low economy but also due to no requirement of new right of way for new power transfer paths. However it has been observed that whenever an AC line approaches a DC line, interferences would be generated upon the DC line due to the electromagnetic induction between the AC and DC lines [2]. The interference includes the transverse voltage caused by capacitive coupling, the longitudinal electromotive force, and the fundamental frequency current on the DC line caused by inductive coupling. The transverse voltage induced upon the DC line would cause the insulation of the converting equipment connected to the DC transmission line breakdown. Moreover, the fundamental frequency current induced on the dc line would be converted to direct current component and harmonic current components after it flows into the converting equipment. The direct current component may cause saturation, noise and irregular operation of the converting transformer.

The complexity of the problem increases with different possibilities of abnormal operations of the lines due to both AC and DC faults of different types resulting into different kinds of responses obtained, as it has been seen from the literature survey. Limited work has been observed which addresses these issues of Indian Power Grid. The work therefore is important for the country because of its increasing HVDC networks for connecting the major generating region located in north-east region to the major loads in main lands with the limited right of way as available due to geographical distributions. So it is required to properly classify the major effects that are required to be studied on operation of these lines in both steady state and fault conditions with proper mathematical modelling of the system which may lead to modifications, if any.

In Section 2 induced overvoltage phenomena has been studied in detail for faults in both HVAC & HVDC transmission lines. Section 3 studies the effect of line length of HVDC transmission lines on transient stability. Section 4 describes work done in area of electromagnetic interference detection in close proximity operation of HVAC & HVDC transmission lines. Section 5 highlights the Indian Power Sector scenario where the aforesaid problem is expected to arise and the importance of the work.

2 INDUCED OVERVOLTAGES IN HVAC-HVDC HYBRID TRANSMISSION CORRIDOR

Adding a HVDC line to an existing HVAC transmission circuit or converting one circuit of a double circuit HVAC transmission system to HVDC has been proposed as an effective method of significantly increasing the transmission capacity of the power corridor [2]. Fig. 1 shows an example of how the existing tower of a Gulfport structure has been modified to include two dc conductors and two shield wires in addition to the three conductors of the 230 kV ac transmission circuit. This design was proposed by the Manitoba HVDC Research Centre [3], and takes into account factors such as structural integrity of the tower, electric field effects, insulation clearance and hot line maintenance. The HVDC circuit can be rated to carry up to 600 MW of power (bipolar operation). This would give a 200% increase in total power carrying capacity of the corridor, based on a maximum capacity of 300 MW for the HVAC line alone. The work has investigated the overvoltage on a hybrid HVDC-HVAC transmission system and suggests some design considerations which could be taken into account to reduce stresses on certain critical components which result from such an arrangement [3].

An ElectroMagnetic Transient simulation program (EMTDC) was used in the study. Fig. 2 shows a single line diagram of the study system. It includes a 230 kV, 300 MW ac circuit with a ±250 kV, 600 MW dc circuit connected into the same receiving and sending end ac systems. The study was carried out with an applied on the transmission lines at either the sending or receiving ends or at selected points along the 390 km right of way. The ac transmission line is rated at 230 kV, 300 MW and the HVDC scheme is a bipolar ±250 kV, 600 MW transmission system.

Sensitivity studies is performed at resistivity of 10 and 1000 Ω-m. The AC filters include tuned 11th, 13th filter, a high pass filter and, together with a fixed shunt capacitor, provide 300 MVAR of reactive power. The study system was adapted from an earlier investigation [4], and is based on a potential upgrade of an existing tie line between Manitoba Hydro and the Northern United States.

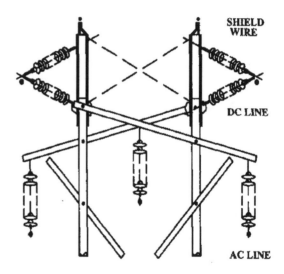

Figure 1. Gulfport structure: Modified top to accommodate HVDC transmission line for hybrid HVAC-HVDC transmission corridor [3].

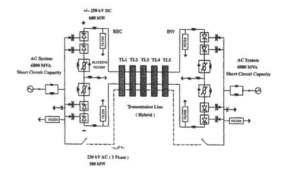

Figure 2. Simplified single line diagram of HVAC-HVDC hybrid transmission corridor [3].

238

The strong sending and receiving end systems (6800 MVA) are typical of this situation. The DC filters include a 12th harmonic filter and a high pass filter. A blocking filter (BF) consisting of a parallel L-C combination was included to limit the flow of 60 Hz current in the dc line. The electrodes and electrode lines were modelled as lumped R-L elements.

The results obtained from the study are shown in Fig. 3. The highest overvoltage occurs for a smoothing reactor inductance of 0.5 H which corresponds to a DC circuit resonance of approximately 120 Hz. Fig. 4 indicates that the magnitude of the overvoltage reduces rapidly when the dc circuit resonance is moved away from the second harmonics.

AC and DC Line overvoltage were calculated for the following types of faults:

- AC Single Line to Ground (LG)
- AC Line to Line to Ground (LLG)
- AC Line to Line (LL)
- AC 3Ø to Ground (LLLG)
- AC-DC Conductor (CF)

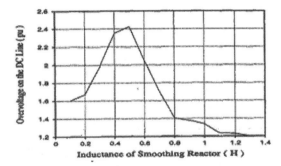

Figure 3. DC line overvoltage for single line to ground fault for different values of inductance [3].

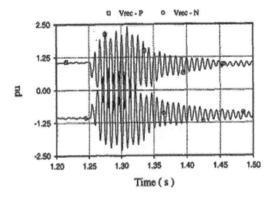

Figure 4. 2nd harmonic overvoltage on the DC line for an AC LG fault, L = 0.5 H [3].

- AC-DC Conductor to Ground (CFG)
- DC-Conductor to Ground (LGDC)

These faults were applied at either the sending or receiving ends or at any one of the five equally spaced locations in between. Each segment represents 65 km of line. The overvoltage were recorded at these locations for all faults at all locations. Faults were simulated at time intervals of 15o and the maximum overvoltage at any of the measuring locations was recorded. Different fault types for two different systems in terms of AC Short Circuit Capacity of 6800 and 1700 MVA, was performed. These correspond to short circuit ratios of 11.36 and 2.84 respectively.

As expected the HVDC line overvoltage are more severe for the weaker (1700 MVA) system. Note also that the HVAC-HVDC hybrid line considered here can also experience overvoltage due to the new fault types (CFG, LGDC, etc.) which would not be experienced by a HVDC line in isolation. The latter would experience only the LG DC type of fault.

The overvoltage are larger when the system is weaker for all faults involving only the HVAC conductors (LL, LLG, LLLG, and LG). However when the fault involves a dc conductor (CFG, CF, LGDC), the overvoltage is larger for the stronger system. This is due to the fact that the HVDC fault current is larger for the stronger system and thus there is a larger coupling into the healthy HVAC conductor.

The impact of not transposing the HVAC conductors was also studied for a contact fault between an HVAC and HVDC conductor not involving ground. The results indicated that significantly higher overvoltage can occur on the HVDC conductors if the HVAC conductors are not transposed.

3 IMPACT OF HVDC LINE LENGTH ON TRANSIENT STABILITY

Rotor angle stability is the ability of synchronous machines in a power system to remain in synchronism after been subjected to a disturbance. It depends on the ability of the machines to maintain equilibrium between the electromagnetic torque and the mechanical torque. The movement of the rotor is governed by the Newton's second law of motion as given by Equation 1.

$$\frac{2H}{\omega_o}\frac{d^2\delta}{dt^2} = P_m - P_e \qquad (1)$$

where, H is the per-unit inertia constant in sec
δ is the rotor angle in rad.

ω_o is the synchronous speed in rad/sec.
Pe is the electrical torque in per unit
Pm is the mechanical power in per unit.

When the power system is in steady state of equilibrium, the mechanical power and the electrical power of the generator in Equation 1 remain the same [5]. Hence, there is no accelerating power. This means that the rotor angle of the synchronous machines in the interconnected power system will be at a fixed angular position. If the systems is perturbed by a disturbance such as a three-phase AC fault, DC line fault or the loss of a generating unit, the equilibrium between Pe and Pm is lost and the speed of the generator changes. This change will lead to an imbalance in the active power generation by all the machines in the power system [5] [6].

The stability of such power systems containing HVDC transmission schemes are affected by DC line faults, converter faults, etc. In HVAC systems, relays and circuit breakers are used for fault detection and clearance. On the contrary, most of the faults in HVDC systems are self-clearing or are cleared through the action of converter controls [6]. In some cases, it may become necessary to take out a bridge or an entire pole out of service. The most common type of faults on HVDC lines is pole-to-ground faults. This fault blocks power transfer on the affected pole with the rest poles remaining intact.

During this fault, the short circuit causes the rectifier current to increase while the inverter current decrease. The rectifier current control acts to reduce the direct voltage and also reduce the current back to the current set point (normal operation current level). At the inverter side, the current level reduces below its current reference value. This causes the inverter to change from Constant Extinction Angle (CEA) control to Constant Current (CC) control. As a result, the voltage of the inverter reduces to zero and then reverses its polarity. Typically, the total time it takes for fault clearance and return to service is between 200 ms to 300 ms. The HVDC system response to HVAC fault is faster than that of the HVAC system. The HVDC system is capable of self-sustenance through the HVAC system fault or in severe cases, there will be a reduction in power or a complete shutdown till the HVAC system recovers from the fault.

When the HVAC fault is a distant 3-phase fault, there will be a reduction in the rectifier commutation voltage, which will lead to a reduction in the HVDC voltage and current. If the firing angle reaches its minimum value, the control measure will switch the rectifier from CC mode to Constant Ignition Angle (CIA) mode while the inverter changes to CC mode. Although in theory, DC power may be transmitted via the HVDC transmission line

when the rectifier voltage is very low, the resulting increase in reactive power consumption could damage the generators. This is mainly because the inverter would have to change from CEA to CC mode by lowering its voltage and increasing β [6]. DigSILENT power factory was used in the simulations. Fig. 5 shows the power system that was used in the simulations. There are four generators located in two areas; each has a capacity of 1800 MVA. In addition, area 2 has two distribution centres. The first distribution centre is at bus 7. This bus is connected to bus 5 via a 1600 MVA 500/11 kV step down transformer. The second distribution centre is at bus 8 which is connected to bus 5 via a 1,600 MVA 500/11 kV step down transformer. The load at bus 7 is 1,064 MW while the load at bus 8 is 1379 MW. Capacitor banks are connected to bus 5 and Bus 6 and contribute a total reactive power compensation of 1000 MVAR. In the following sub-sections, the impact of AC three-phase fault and DC line fault with respect to varying HVDC line length on the transient stability is investigated. First the DC fault is applied at the middle of the HVDC transmission line while the HVDC line length is varying. The same scenario is considered when there is a three-phase AC fault on the HVAC transmission line occurs in the system [7].

The study performed, used a 50 ms DC fault applied at 50% of the HVDC transmission line. The length of the HVAC line is fixed at 500 km, while the length of the HVDC line is varied between 100 km and 3,000 km. The pre-fault rotor angle of G2 decreases (i.e. the absolute value decreases) as the length of the HVDC transmission line increases. Also, the amplitude of the rotor angle oscillations decreased with the increase in the HVDC transmission line length. After the fault is cleared, it took on average about 20 seconds for the rotor angle to stabilize when the HVDC transmission line length is less than 3000 km. When the HVDC line length is 3000 km, it took about 16 seconds for the rotor angle to settle down. This is

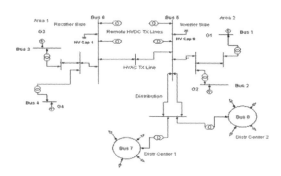

Figure 5. Power transmission network used for simulation [7].

expected because, as the length of the transmission line increases, the distance between the fault and G2 increases. Therefore, the effect of the fault on G2 reduces [7].

At steady state, G2 generated about 400 MW active power. At the time of the fault, the active power generation increased from 400 MW to about 1,300 MW to compensate for the reduction of power transmitted from area 1 due to the fault. After the fault was cleared, the active power oscillated between 500 MW and 300 MW before reaching its pre-fault value of 400 MW. The amplitude of the active power oscillations was the lowest when the HVDC line was at 3000 km. However, at this line length, the system experienced a dip in active power after the fault occurs. The time taken for the active power to reach its pre-fault value was the same for all the HVDC transmission line lengths [7].

Before the fault, the terminal voltage at G2 was 1.0 p.u. During the fault, the terminal voltage increased to 1.03 p.u. before it dipped to 0.98 p.u. Then the voltage oscillated between 1.03 p.u. and 0.98 p.u. before settling to 1.0 p.u. after the fault was cleared. The largest dip in terminal voltage was experienced at 100 km, while the largest post fault oscillations occur at 1,500 km line length [7].

At 3000 km, the system experiences the smallest terminal voltage oscillations. The time taken for the terminal voltage to settle was about 18 seconds, which is found to be the same across all transmission lines [7].

The amount of reactive power generated by G2 decreases as the length of the HVDC transmission line increases [7]. Similarly, the amplitude of the reactive power oscillation and the time taken to reach steady state also reduced as the length of the HVDC transmission line increased. For the 3,000 km HVDC transmission line, the reactive power generation settled to steady state at 15 seconds as compared to about 18 seconds for HVDC line length smaller than 3000 km [7].

The study further continued with introducing fault in HVAC line with fixed length while varying the length of HVDC line. A three phase fault was initiated at 50% of the HVAC transmission line for 50 ms. The HVAC transmission line is fixed at 500 km, while the length of the HVDC line is varied between 100 km and 3,000 km as before. The rotor angle decreased from −83.770 to −83.830 with the increase in the length of the HVDC transmission line. During the fault, the rotor angle of G2 oscillated between −83.610 and −83.850 across the four transmission lines before reaching steady state. The amplitude of the rotor angle oscillations and the time taken to reach steady state increased with the increase in length of the HVDC transmission line. The results of the terminal voltage indicated that the HVDC transmission line length

has no impact on the terminal voltage dip at G2 during the 3-phase to ground fault on the HVAC transmission line. Before the HVAC line fault, G2 was generating 400 MW. The highest increase in active power generation during the fault is with the 100 km HVDC transmission line. Although the time taken for the active power to stabilize using the four HVDC transmission line was the same, the amplitude of the active power oscillation increased with the increase in the length of the HVDC transmission line.

4 ELECTROMAGNETIC INTERFERENCE STUDIES IN CLOSE PROXIMITY OPERATION

When an HVAC line approaches a HVDC line, interferences would be generated upon the HVDC line due to the electromagnetic induction between the HVAC and HVDC lines. The interference includes the transverse voltage caused by capacitive coupling, the longitudinal ElectroMotive Force (EMF), and the fundamental frequency current on the dc line caused by inductive coupling [8]. The transverse voltage induced upon the HVDC line would cause the insulation of the converting equipment connected to the HVDC transmission line breakdown. Moreover, the fundamental frequency current induced on the HVDC line would be converted to direct current component and harmonic current components after it flows into the converting equipment [9]. The direct current component may cause saturation, noise and irregular operation of the converting transformer. Thus, it is necessary to analyze the electromagnetic interference in the HVDC line from the parallel ac line in close proximity.

The transverse voltage and the longitudinal EMF induced on a HVDC line by a parallel HVAC line in close proximity is calculated with the ElectroMagnetic Transient Program (EMTP) in this study [9]. In order to simulate the parallel HVAC-HVDC transmission lines system as shown in Fig. 6, a multi-conductor overhead transmission line model is established in EMTP. Three conductors marked A, B and C represent the phase conductors of HVAC line, two conductors marked and represent the positive and negative conductors of HVDC line and another four conductors marked N represent the ground wires of HVAC line and HVDC line. In order to obtain the electromagnetic interferences in different cases, the HVAC line is supposed to be operating in steady state; also the multi-conductor overhead transmission line model will be affected by a single-phase-to-ground fault.

To obtain the direct current component flowing into the converting transformer induced from the

HVAC line, the total impedance of HVDC line circuit should be computed firstly, which includes the pole conductor self-impedance of HVDC line, the impedances of smoothing reactors at both ends of the HVDC line and the internal impedance of converting equipment. Compared with the pole conductor self-impedance of HVDC line and the impedances of smoothing reactors, the internal impedance of converting equipment is very small and can be neglected.

Therefore, the total impedance of HVDC line circuit can be approximated to the summation of the pole conductor self-impedance of dc line and the impedances of smoothing reactors.

For overhead pole conductor of the HVDC line, the per-unit-length self-impedance can be calculated by Carson's model [10] accurately, but Carson's model is given by expressions with complex infinite integrals, which are very time-consuming to evaluate.

The study results showed that the transverse voltage induced on the HVDC line decreases rapidly with the separation distance between the HVAC and HVDC lines from 50 m to 150 m, but then reduces slowly when the separation distance is larger than 150 m. Taking the positive conductor as an example, it is observed that the induced transverse voltage decreases from 9.99 kV to 0.66 kV when the separation distance between the HVAC and HVDC lines increases from 50 m to 150 m. It

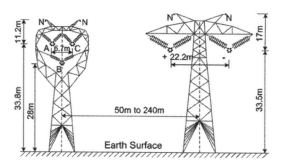

Figure 6. Cross section of parallel HVAC-HVDC transmission system [9].

means that the separation distance increases three times, the induced transverse voltage decreases to 1/15 of the original value. In addition, the transverse voltage induced on the positive conductor is much higher than that on the negative conductor when the separation distance between the HVAC and HVDC lines is small [10]. Two types of conductor arrangements were compared called as typical and compact arrangement where the phase conductors are arranged with separation of 11 m and 50 m respectively. The variation obtained in measurement of transverse and longitudinal EMF is as given in Table 1.

The study is further performed with single-line to ground fault in each of the HVAC conductors and then fault on HVDC conductors. The induced longitudinal EMF is much lower than the induced transverse voltage and the decreasing speed of the induced longitudinal EMF with the separation distance is slower than that of the induced transverse voltage [10]. However the obtained result showed that the characteristic of transverse and longitudinal EMF is very much dependent on type of fault and geometrical configuration of conductors.

5 CONCLUSIONS

From this literature survey, the complexity of the issue has been realized. The effects are not only dependent on types of fault but also on the geometrical configuration and system parameters. Overvoltage on the HVDC line increases with a decrease in system strength. The same is true for the HVAC lines for faults involving the HVAC conductors only. With HVAC-HVDC contact faults and HVDC line faults the HVAC overvoltage are smaller for the weaker system. Overvoltage for the DC line in a hybrid AC-DC environment is higher than for a DC line alone. However, transposition of the HVAC conductors reduces the overvoltage. Transient stability of the system after a HVDC transmission line fault is improved when longer HVDC transmission lines are used in parallel with a relatively shorter HVAC transmission lines. It can

Table 1. Effect of arrangement of HVAC phase conductors upon generated electromagnetic interference.

Seperation Distance (m)	Transverse Voltage (kV)		Longitudinal Voltage (kV)	
	COMAPCT	TYPICAL	COMAPCT	TYPICAL
50	9.99	18.34	1.38	5.18
100	1.8	2.49	0.57	2.16
200	0.337	0.287	0.309	0.92

Table 2. Possibilities of close proximity operation in India.

Name	Voltage kV	Length km	Power MW
Biswanath Chariali To Agra	800	1825	6000
Rihand To Dadri	500	814	1500
Talcher* To Kolar (East-South Connector)	500	1376	2000

*April 2006, work started for upgradation to 2500 MW.

also be mentioned that the generator terminal voltage dip is reduced with the increasing length of the HVDC transmission line. The effects developed in various fault conditions during a hybrid corridor are dependent on system configuration and design parameters. System behaviour is dependent on large number of factors. With new HVDC projects emerging in India, this kind of study on Indian power grid becomes very important as no such work has been done yet. The major HVDC projects of India where this study could be feasible are as listed in Table 2.

All this lines passes through the area where HVAC lines already exists and thus provides chances for study of the aforesaid problems. The obtained results from the study could also be used for design of Hybrid Transmission corridors for future projects of the country.

REFERENCES

Clerici, A., L. Paris, Per Danfors, "HVDC Conversion of HVAC to Provide Substantial Power Upgrading", IEEE Trans.onPowerDelivery,Vol.6, No. 1, January 1991, pp. 324–333.

Larsen, E.V., Walling, R.A. & Bridenbaugh, "Parallel ac/dc Transmission Lines Steady State Induction Issues", IEEE Transactions on Power Delivery, Vol. 4, No. 1, January 1989, pp. 67–673.

Nakra H.L., Bui L.X., Iyoda L., "System Consideration in Converting One Circuit of a Double Circuit ac Line to dc", IEEE Transaction on Power Apparatus and System, Vol. PAS-103, No.10, October 1984, pp. 3096–3103.

Oyedokun, D.T., K.A. Folly, A.V. Ubisse & L.C. Azimoh, "Interaction between HVAC- HVDC System: Impact of Line Length on Transient Stability", UPEC 2010, 45th International IEEE Conference, Aug 2010.

Oyedokun, D.T., K.A Folly, "Power Flow studies in HVAC and HVDC Transmision Lines", IASTED Africa PES 2008, Garborone, Botswana, 2008.

SIEMENS, "International Workshop for 800 kV High Voltage Direct Current (HVDC) Systems", New Delhi-India, 25th–26th Feb, 2005.

Szechtman, M., et. al., "CIGRE Benchmark Model for dc Controls", Electra, Vol. 135, April 1991, pp. 54–73.

Verdolin, R., A.M. Gole, E. Kuffel, N. Diseko & B. Bisewski, "Induced Overvoltages On AC-DC Hybrid Transmission System", IEEE Transactions on Power Delivery, Vol. 10, No. 3, July 1995.

Woodford, D.A. & Young, A.H., "Using dc to Increase Capacity of AC Transmission Circuits", IEEE Canadian Review, March 1989 pp. 15–18.

Woodford, D.A., "Secondary Arc Effect in ac/dc Hybrid Transmission", CEA report ST-3 12, February 1991.

Frontiers in Computer, Communication and Electrical Engineering – Acharyya (Ed.)
© 2016 Taylor & Francis Group, London, ISBN: 978-1-138-02877-7

Studies on frequency domain spectroscopy of transformer insulation considering distributed relaxation process

Sandip Kumar Ojha, Prithwiraj Purkait, Anchit Kumar, Asif Sultan, Indrajeet Kumar, Jayant Kumar Singh & S. Satyam
Department of Electrical Engineering, Haldia Institute of Technology, Haldia, India

ABSTRACT: In recent times, dielectric response measurements have gained immense attention for assessment of transformer oil-paper insulation condition. Time domain polarization measurements are one of the most popular dielectric testing techniques for the purpose. For accurate interpretation of test data various mathematical models of the dielectric response process have been proposed by various researchers over the years. Frequency domain spectroscopic results obtained from mathematical transformation of the time domain data has been found to be useful in analyzing dielectric test results. In complex dielectrics, it seems however, that the overall response be represented by a distribution of relaxation times (or frequencies), rather than a single "average" response characteristic. The present contribution provides in brief, the mathematical background for estimating the functions representing distribution of relaxation frequencies from dielectric response measurement data and further transforming it to frequency domain spectrum. Experimental results on laboratory scaled down models and also on real transformers have been presented to support the theoretical formulations.

1 INTRODUCTION

Information about the condition of oil-paper insulation is of extreme importance for evaluating ageing status and maintenance requirement of transformers. Dielectric testing techniques in both time domain [Gubanski *et al.* 2003] and frequency domain [Zaengl *et al.* 2003, Fofana *et al.* 2010, Setayeshmehr *et al.* 2008] are being assessed in recent years as supplements to traditional techniques by researchers and utilities. The main impediment on dielectric testing techniques being widely adopted by practicing engineers has been the level of expertise necessary to interpret the test results. Interpretation of dielectric test data both in time and frequency domain have involved uses of expert systems, and modern classification tools [Saha *et al.* 2004, Hui *et al.* 2012]. Several attempts have thus been made to understand the physical processes of dielectric relaxation taking place during such tests by correlating conventional tests, time domain and frequency domain dielectric measurements, and chemical techniques [Saha *et al.* 1999, Saha *et al.* 2004, Dervos *et al.* 2006].Various models of the oil-paper insulation system, starting from the very basic Debye model [Debye 1945], have henceforth been proposed by various researchers. These simulation models are often useful for characterizing different components of the complex insulation

system from experimentally obtained data. The characteristic response function most commonly used in time domain to interpret the experimental data is the dielectric response function f(t) [Jonscher 1983]. Most of the dielectric response functions are however, based on the assumption of a single relaxation process. In practical insulation systems,it is difficult to separate processes that have inherently close relaxation times (frequencies), or those with non-exponential relaxation properties, or those with relaxation times (or frequencies) that are spread out in time (or frequency) domain. Such models thus cause practical processes to deviate from that predicted by the classical Debye theory. In such processes, it can be assumed that the dipoles relax according to a distribution of elementary Debye relaxation frequencies. Researchers [Tuncer *et al.* 2002, Macutkevic *et al.* 2001, Macdonald *et al.* 2004, Macdonald *et al.* 1962, Bello *et al.* 2008, Dias *et al.* 1997, Farag *et al.* 2003, Kliem *et al.* 1988, Liedermann *et al.* 1994] have proposed models that can simulate such many-body interaction processes with the help of suitable distribution density functions. Frequency Domain Spectroscopy (FDS) data can further be computed from the distribution density function thus formulated. Several researchers have reported their findings about the influence on operating temperature, moisture content, ageing etc. on the FDS characteristics [Fofana *et al.*

2010, Setayeshmehr *et al.* 2008, Hao *et al.* 2011, Liu *et al.* 2012, Martin *et al.* 2011].

The present contribution discusses result of dielectric tests performed on scaled down model of transformer oil-paper insulation system with varying ageing status, moisture content, geometry, excitation voltage, temperature etc. Results of tests on field transformers are also presented. Mathematical formulations used for transforming time domain data to distribution domain and further to frequency domain are summarized. Nature of FDS characteristics displaying the nature of distribution of relaxation frequencies of oil-paper insulation under different operating conditions is described.

2 EXPERIMENTAL DETAILS

Series of experiments were performed on a scaled down transformer insulation model under controlled laboratory conditions. A computer controlled instrument consisting of a Keithley 6517B electrometer, Siemens high voltage relays, timer and switching control blocks was developed to perform polarization and depolarization current measurements. Photograph of the transformer model, and temperature and humidity controlled furnace along with the dielectric test equipment, and Karl Fischer Titration (KFT) setup for oil and paper moisture measurement are shown in Fig. 1.

Test parameters such as voltage, temperature, Moisture Content (MC) of the insulation, ageing status etc. were varied during the experiments. Tests were also conducted on several real transformers in field. Details of all the test samples are provided in Table 1.

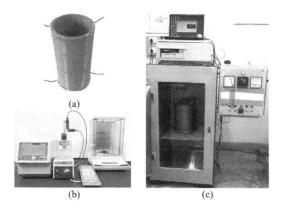

Figure 1. Experimental setup (a) Scaled down transformer insulation model, (b) KFT setup for moisture measurement in oil and paper, (c) control chamber for performing dielectric tests and the measuring instrument.

Table 1. Description of test samples.

Sample identifier	Description
A11	Charging voltage 100 V
A21	Charging voltage 200 V
A31	Charging voltage 300 V
B1	Test temperature 400 C
B3	Test temperature 700 C
G122	Paper MC 12.98%, Oil MC 10.5 ppm
G124	Paper MC 12.88%, Oil MC 28.2 ppm
D42	Dry paper (<2% MC), Aged Oil (30 years old, 16 ppm MC)
D43	Dry paper (<2% MC), New Oil (<5 ppm MC)
H8	70 MVA, 220/132 kV, Transformer before oil filtering and drying
H9	Same 70 MVA, 220/132 kV, Transformer after oil filtering and drying

3 TRANSFORMATION OF TIME DOMAIN DATA

In real dielectric materials, the range of frequency dispersion of the molecules during relaxation extends over much larger frequency region than that predicted by the Debye. In such cases, the assumption of a *distribution* of relaxation times or relaxation frequencies, rather than a single relaxation frequency, is one possible way of interpretation of the experimental observations. In this work, time domain dielectric test measurement data have been used to identify the nature of distribution of relaxation frequencies for more authentic interpretation of insulation behavior.

3.1 *Formulation of relaxation frequency distribution function*

The dielectric response function $f(t)$ according to the classical Debye model can be written as:

$$f(t) = \frac{A}{\tau_D} e^{-t/\tau_D} \qquad (1)$$

The time constant τ_D in (1) is called the 'Debye dielectric relaxation time' or simply the 'relaxation time'.

With a single relaxation time τ_D, or single relaxation frequency $\nu = \frac{1}{\tau_D}$, the polarization process can be expressed as:

$$\frac{1}{\nu}\frac{d}{dt}P(t) + P(t) = \lim_{t \to \infty} P(t) = \varepsilon_0 E_0 \chi_0 \qquad (2)$$

As discussed earlier, all practical dielectric systems consist of more than one relaxation processes going on at the same time, each having different relaxation frequencies, depending on its condition and operating state. The response function of such dielectric system cannot be described by a single differential equation as (1) but rather by a distribution of relaxation times or relaxation frequencies.

If the total polarization at equilibrium condition in such a practical dielectric is given by P_{ot} which consists of different polarization processes, then a distribution function f(v) can be defined as:

$$f(v) = \frac{1}{P_t} \frac{dP_t}{dv} \tag{3}$$

Where P_t is the total polarization corresponding to all the processes involved at the instant t. The frequency distribution function $f(v)$ is normally distributed with the property [Farag et al. 2003, Kliem et al. 1988]:

$$\int_0^\infty f(v)\,dv = \tag{4}$$

From (3) we have

$$dP_t(t) = P_t(t)f(v)dv \tag{5}$$

The current corresponding to the polarization $dP_t(t)$ belonging to the relaxation frequency v is thus:

$$
\begin{aligned}
di_t(t) &= \frac{d}{dt}dP_t \\
&= \frac{d}{dt}\Big[P_{ot}f(v)dv\big(1-e^{-tv}\big)\Big] \\
&= P_{ot}v.f(v)e^{-tv}dv
\end{aligned}
\tag{6}
$$

When all the individual relaxation processes present inside a composite dielectric material are considered to be acting simultaneously, the total current can be found out by simple superposition of the individual currents due to different relaxation processes:

$$i(t) = \int di_t(t) = \int_0^\infty P_{ot}v.f(v)e^{-tv}dv \tag{7}$$

Defining $vf(v) \equiv F(v)$, we have

$$i(t) = P_{ot}\int_0^\infty F(v)e^{-tv}dv \tag{8}$$

Also this total current can be related to the total polarization as:

$$i(t) = \frac{dP_t(t)}{dt} \tag{9}$$

Thus, combining (8) and (9):

$$\frac{dP_t(t)}{dt} = i(t) = P_{ot}\int_0^\infty F(v)e^{-tv}dv \tag{10}$$

$$\text{or,} \qquad i(t) = P_{ot}L\big[F(v)\big]$$

Where $L\big[F(v)\big]$ is the Laplace Transform of $F(v)$.

3.2 *Identifying frequency distribution function from time domain data*

In this case, a set of time domain experimental data points of depolarization current (t_{di}, i_{di}) is given and a distribution density function (v_k, F_k) need to be calculated.

Thus (8) can be re-written as:

$$i_{di}(t_{di}) = P_{ot}\int_0^\infty F_k(v_k)e^{-t_{di}v_k}dv_k \tag{11}$$

Setting $P_{ot} = 1$ [19], (11) can be expressed in discrete form by a linear set of discrete equation as

$$i_{di}(t_{di}) = \sum_{\substack{k=1 \\ i=1,2,\dots N}}^{N} \Delta v_k F_k(v_k)e^{-t_{di}v_k} \tag{12}$$

Equation (12) represent a set of equation with N measured (known) parameters $i_{di}(t_{di})$ and N unknown parameters $F_k(v_k)$, to be estimated. In practice, direct numerical solution of (12) is often unstable due to the fact that the corresponding matrix may be ill-conditioned [Tuncer et al. 2002, Farag et al. 2003, Kliem et al. 1988]. Certain numerical and analytical techniques for solution of such ill-conditioned equations have been presented in [Tuncer et al. 2002, Dias et al. 1997, Farag et al. 2003, Kliem et al. 1988]. In the present work, the iterative procedure explained by the authors of [Farag et al. 2003, Kliem et al. 1988] has been adopted to compute the relaxation frequency distribution spectra from time domain depolarization current data following (12). Figure 2 shows the nature of distribution density functions obtained when the test temperature was varied.

247

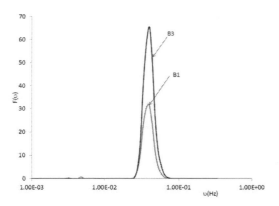

Figure 2. Relaxation frequency distribution functions of B1, and B3 at different test temperatures.

3.3 *Transformation from distribution domain to frequency domain*

The frequency domain parameter often used to characterize insulation condition is the complex susceptibility $\chi^* = \chi' + j\chi''$.

For obtaining the FDS data from the frequency distribution function, it is necessary to calculate the set of data points (ω_i, χ_i^*) from the given set (v_k, F_k). Set the an angular frequency $\omega_i = v_i = \dfrac{1}{\tau_k}$, with $k = N - i + 1$.

The imaginary part of susceptibility can be computed following [19] as a linear set of N equations:

$$\chi''(\omega_i) = \sum_{k=1}^{N} \Delta v_k \frac{F_k(v_k)(\omega_i / v_k)}{v_k \left[1 + (\omega_i + v_k)^2 \right]} \quad (13)$$

$i = 1 \ldots \ldots \ldots, N$

The real part of susceptibility can be computed following [19] as:

$$\chi'(\omega_i) = \sum_{k=1}^{N} \Delta v_k \frac{F(v_k)}{v_k \left[1 + (\omega_i / v_k)^2 \right]} \quad (14)$$

$i = 1 \ldots \ldots \ldots, N$

Plots of $\chi'(\omega_i)$ and $\chi''(\omega_i)$ for all test samples are provided in the next section.

4 ANALYSIS OF FDS RESULTS

FDS plots for different test samples are shown in Figure 3–7.

As expected, FDS plots are having higher values for higher operating voltages as can be seen from Figure 3. Loss peaks are also found to get more pronounced at higher operating voltages.

As seen in Figure 4, with the increase of temperature the values of χ' and χ'' increase at lower frequencies. Similar findings were reported by [Hao *et al.* 2011]. This is because, with the increased temperature, the ions mobility in oil gets enhanced. The dipole-dipole interactions due to the increased mobility of flexible portions in the cellulose chains decrease with the increase of testing temperature, which enhances the ease of rotation and polarizability of the side groups and the other flexible portion of cellulose. All the χ' and χ'' values decrease with the increasing frequency at each testing temperature. Presence of a loss peak is also visible in the χ'' plots (Figure 4b) at higher frequencies.

FDS plots of χ' and χ'' are lower for insulation with lower oil moisture content as compared to the one with higher oil moisture content. Authors in [Liu *et al.* 2012] have reported similar findings where the appearance of a loss peak at higher frequency has been demonstrated as has also been found in Figure 5(b).

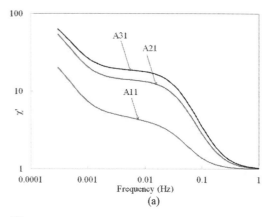

(a)

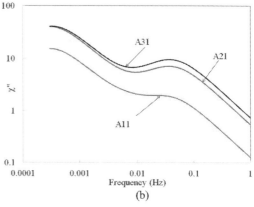

(b)

Figure 3. FDS plots for test samples at different operating voltages (a) χ', (b) χ''.

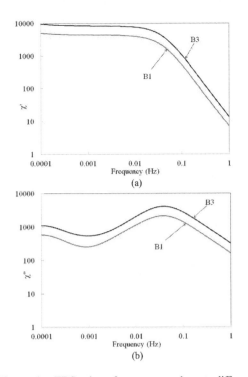

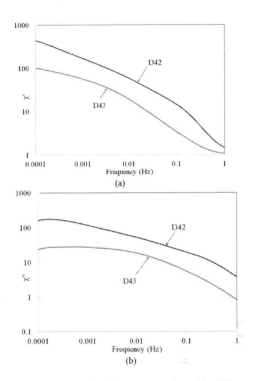

Figure 4. FDS plots for test samples at different operating temperatures (a) χ', (b) χ''.

Figure 6. FDS plots for test samples with different aging status of insulation (a) χ', (b) χ''.

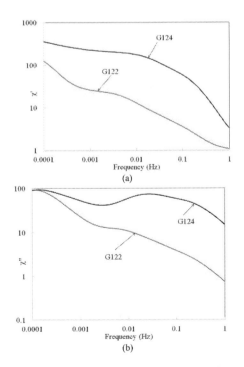

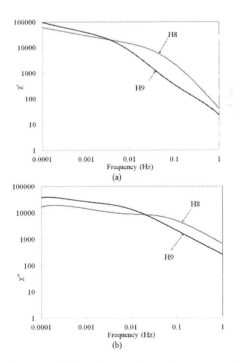

Figure 5. FDS plots for test samples with different oil moisture contents (a) χ', (b) χ''.

Figure 7. FDS plots for a transformer before and after oil filtering (a) χ', (b) χ''.

FDS plots of χ' and χ'' are lower for newer insulation as compared to aged insulation. [Martin *et al.* 2011] also reported similar findings where the plots for new oil were found to be lower than those for aged oil over the entire frequency range.

The intersection of plots at medium frequency range indicates that due to oil filtration and drying, the moisture could be driven out from the oil, but the paper condition could not be improved much. Also the loss peak is found to occur at higher frequencies for the case when the transformer oil was not filtered and dried.

5 CONCLUSION

This paper presents brief mathematical background of transforming time domain dielectric measurement data of transformer insulation first to distribution domain and then to frequency domain. Experimental results of transformer oil-paper insulation under various conditions of insulation and also field transformer are presented. Frequency Domain Spectroscopy (FDS) plots of different test samples are presented and are related to the operating state and condition of the oil-paper insulation.

REFERENCES

Bello. A., Laredo E., Grimau M., Nogales A., and Ezquerra. T.A., "Relaxation Time Distribution from Time And Frequency Domain Dielectric Spectroscopy In Poly (Aryl Ether Ether Ketone)", Journal of Chemical Physics, Vol. 113, No. 2, pp. 863–868, 2008.

Debye. P., *Polar Molecules*, Dover publication, New York, 1945.

Dervos. C.T., Paraskevas C.D., Skafidas P.D., and Stefanou N., "Dielectric Spectroscopy and Gas Chromatography Methods Applied on High-Voltage Transformer Oils", IEEE Trans. Dielectr. Electr. Insul., Vol. 13, No. 3, pp. 586–592, 2006.

Dias. C.J., "Determination of a Distribution of Relaxation Frequencies Using a Combination of Time and Frequency Dielectric Spectroscopies", in Proc. IEEE 1997 Annual Report Conf. Electr. Insul. Dielectr. Phenomena (CEIDP 97), Minnepolis, October, pp. 875–478, 1997.

Farag. N., Holten. S., Wagner. A., and Kliem. H., "Numerical Transformations of Wide-Range Time and Frequency Domain Relaxation Spectra", IEE Proc. Science, Measurement and Technology, Vol. 150, No. 2, pp. 65–74, 2003.

Fofana. I., Hemmatjou H., Meghnefi F., Farzaneh M., Setayeshmehr A., Borsi H., and Gockenbach E., "On the Frequency Domain Dielectric Response of Oil-Paper Insulation at Low Temperatures", IEEE Trans. Dielectr. Electr. Insul., Vol. 17, No. 3, pp. 799–807, 2010.

Gubanski. S.M., Boss. P, Csepes. G., Houhanessia V. der n, Filippini. J., Guuinic. P., Gafvert. U., Karius. V,. Lapworth. J, Urbani. G., Werelius. P., and Zaengl. W., "Dielectric Response Methods for Diagnostics of Power Transformers", IEEE Electr. Insul. Mag., pp. 12–18, 2003.

Hao, J, Ma, Z, Liao, R, Chen, G and Yang, L., " A comparative study of moisture and temperature effect on the frequency dielectric response behaviour of pressboard immersed in natural ester and mineral oil", International Symposium on high voltage engineering,Germany, 2011.

Hui Ma, Saha T.K., and Ekanayake C., "Statistical Learning Techniques and their Applications for Condition Assessment of Power Transformer", IEEE Trans. Dielectr. Electr. Insul., Vol. 19, No. 2, pp. 481–489, 2012.

Jonscher. A.K., *Dielectric Relaxation in Solids*, Chelsea Dielectric, Press, 1983.

Kliem. H., Fuhrmann. P. and Arlt. G., "A Numerical Method for the Determination of First-Order Kinetics Relaxation Time Spectra", IEEE Trans. Electr. Insul., Vol. 23, No. 6, pp. 919–927, 1988.

Liedermann. K., "The Calculation of a Distribution of Relaxation Times from the Frequency Dependence of the Real Permittivity with the Inverse Fourier Transformation", Journal on Non-crystalline Solids, Vol. 175, No 1., pp. 21–30, 1994.

Liu Jun, Zhou Lijun, Guangning Wu, Yingfeng Zhao, Ping liu, Qian Peng, "Dielectric frequency response of oil-paper composite insulation modified by nanoparticles", IEEE, transaction on dielectric and electrical insulation,Vol-19, pp. 510–520, 2012.

Macdonald. J.R., "Some Statistical Aspects of Relaxation Time Distributions", Physica, Elsevier, Vol. 28, No. 5, pp. 485–492, 1962.

Macutkevic. J., Banys J., and Matulis A., "Determination of Distribution of Relaxation Times from Dielectric Spectra", Juournal of Nonlinear Analysis, Modelling and Control, Vol. 9. No. 1, pp. 75–88, 2004.

Martin Sirueek, Vaclav Mentlik, Trukapavel, Bocek, Jiri, Pihera Josef, MrazPetr, "cole-cole diagram as diagnostic tool for dielectric liquids", IEEE, dielectric liquids (ICDL), pp. 1–4, 2011.

Saha. T.K., Darveniza M., Yao Z.T., Hill D.J.T., and Yeung G., "Investigating the Effects of Oxidation and Thermal Degradation on Electrical and Chemical Properties of Power Transformers Insulation", IEEE Trans. Power Delivery, Vol. 14, No. 4, pp. 1359–1367, 1999.

Saha. T.K. and Purkait. P., "Investigation of Polarization and Depolarization CurrentMeasurements for the Assessment of Oil-paper Insulationof Aged Transformers", IEEE Trans. Dielectr.Electr. Insul., Vol. 11, No. 1, pp. 144–154, 2004.

Saha. T.K., Purkait P., "Investigation of an Expert System for the Condition Assessment of Transformer Insulation Based on Dielectric Response Measurements", IEEE Trans. Power Delivery, Vol. 19, No. 3, pp. 1127–1134, 2004.

Saha. T.K., Purkait P., and Muller F., "An attempt to correlate time & Frequency Domain Polarisation Measurements for the Insulation Diagnosis of Power Transformer", in Proc. IEEE Power Engineering Society General Meeting, Vol. 2, pp. 1793–1798, 2004.

Setayeshmehr. A., Fofana I., Eichler C., Akbari A., Borsi H., Gockenbach E., "Dielectric Spectroscopic Measurements on Transformer Oil-Paper Insulation Under Controlled Laboratory Conditions", IEEE Trans. Dielectr. Electr. Insul., Vol. 15, No. 4, pp. 1100–1111, 2008.

Tuncer. E. and Gubanski S.M., "On Dielectric Data Analysis Using the Monte Carlo Method to Obtain Relaxation time Distribution and Comparing Non-linear Spectral Function Fits", IEEE Trans. Dielectr. Electr. Insul, Vol. 8, No. 3, pp. 310–320, 2001.

Tuncer. E., Serdyuk Y.V., and Gubanski S.M., "Dielectric Mixtures: Electrical Properties and Modeling", IEEE Trans. Electr. Insul. Vol. 9, No. 5, pp. 809–828, 2002.

Zaengl. W. S., "Applications of Dielectric Spectroscopy in Time and Frequency Domain for HV Power Equipment", IEEE Electr. Insul. Mag., Vol. 19, No. 6, pp. 9–22, 2003.

Frontiers in Computer, Communication and Electrical Engineering – Acharyya (Ed.)
© 2016 Taylor & Francis Group, London, ISBN: 978-1-138-02877-7

Generation of air gap rotating magnetic field using switched dc supply for sensor-less Brushless DC motor drive

Ujjal Dey, Ayandeep Ganguly & Sourav Tola
Haldia Institute of Technology, Haldia, East Medinipur, West Bengal, India

ABSTRACT: A Brushless DC (BLDC) motor is a permanent magnet synchronous motor that uses position detectors and inverter to control the armature current. The motor drive system requires a rotor position sensor to provide the proper commutation sequence. The position sensors such as resolvers, absolute position encoders and Hall sensors increase the cost and size of the motor. **A proposed scheme for rotor position sensor-less speed control of BLDC motors with different possible modes of stator pole generations is explained in this paper.** Proper stator voltage sequence has to be generated to match with the rotor magnetic polarity to eliminate the requirement of rotor position sensors. Since the detected position information on the sensor-less rotor has some uncertainty, the brushless DC motor cannot be driven with a maximum torque.

1 INTRODUCTION

The BLDC motor is a rotating electric machine where the stator is a classical three-phase stator similar to an induction motor. If the permanent magnets are mounted on the surface of the rotor, the construction is known as surface mounted permanent magnet rotor. If the magnets are placed on the slots on the rotor such that their surfaces are flush with the rotor surface, then those are known as interior PM rotors. These motors are generally controlled using a three-phase inverter (Bose Bimal K, 1988), requiring a rotor position sensor for starting. In brushless DC machines commutation is performed by power electronic devices forming part of Inverter Bridge. However, switching of the power electronic devices has to be synchronised with rotor position (Jang et al. 2002).

Unlike permanent magnet synchronous motors, the stator windings of BLDC motors are not energized by a 3 phase balanced ac supply to produce smooth rotating magnetic field; instead stator windings sequentially switched with a dc supply through the inverter. Each switching pattern remains constant for 60° electrical rotation of the rotor.

2 PROPOSED SCHEMES FOR STATOR FIELD GENERATION

Orientation of stator fields in different angular positions can be explained by basic electromag-

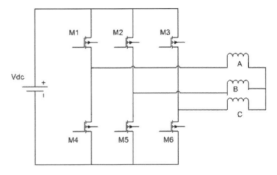

Figure 1. Switched DC generation using three phase inverter.

netic theory. The rotating magnetic field can be generated by controlling the direction of excitation current of the stator phases by using DC power source along with power electronics switches (Park et al. 2006).

It has also been shown here that the two or three winding configuration can also be used for the above purpose and different degree of rotation in each stapes is also possible. In the following paragraphs it is discussed in details:

2.1 Three winding configurations

2.1.1 2 pole machine:

1. 60° rotation
2. 120° rotation

2.1.2 *4 pole machine:*

1. 30° rotation
2. 60° rotation
3. 90° rotation

2.2 *Two winding configurations:*

2.2.1 *2 pole machine:*

1. 60° rotation
2. 120° rotation

2.2.2 *4 pole machine:*

1. 30° rotation
2. 60° rotation
3. 90° rotation

CONFIGURATION FOR RMF OF 3 WINDING 2 POLE 60° ROTATION

Let us consider a, b and c are the 3 stator phases and A–A', B–B', C–C', are the three coil which are connected as shown in the Fig. 2 to create the 2 pole stator winding. All possible switching sequences for 2 poles are described in Table 1.

For three winding configuration all three stator phases are excited. Let us start from the instant that the stator phases are excited such that positive terminal of dc source (+DC) is connected to the phase "A" & "B" and the negative terminal (–DC) with phase "C" by firing the appropriate semiconductor switches which result the excitation current to enter the stator through phase "A" & "B" and leaving through the phase "C".

Thus connectivity of supply to the terminal at any instant will decide the direction of flow of current through the stator winding which will determine the formation of polarity of the rotating magnetic field. Terminals of same polarity can be grouped together to create flux in clockwise or anticlockwise direction following cork-screw rule and it is observed from Sequence-1 in Figure 3, that two poles have been generated.

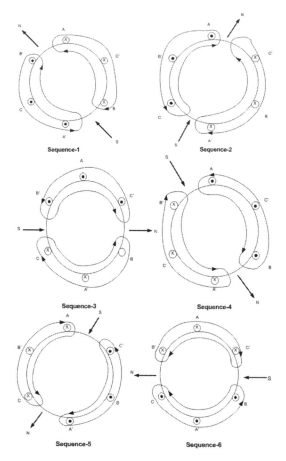

Figure 3. Switching sequences for three winding two pole 60° clockwise rotation of rotating magnetic field.

Now if the phase "B" is made positive and phase "A" & "C" are made negative by firing the appropriate semiconductor switches and terminals are again grouped together, as discussed before to create 2-poles, but compared to sequence-1 it can be seen that there is 60° rotation of poles in clockwise direction (Sequence-2).

Thus by exciting the coil combination by a semiconductor switching circuit from the 1st instant (Fig–3.1) to 2nd instant (Fig–3.2) will cause the stator pole to rotate by 60° clockwise, which is known as 60° switching algorithm.

Complete switching sequence to complete the rotation of the stator poles through the excitation sequence 3 to 6 as shown in Fig–3.3 to 3.6.

In case of two winding configuration one of the windings is always at No Connection (NC) state.

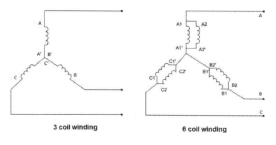

Figure 2. Different arrangement of three phase stator winding.

CONFIGURATION FOR RMF OF 3 WINDING 4 POLE 30° ROTATION

Let a, b and c are the 3 stator phases and A1–A1', A2–A2', B1–B1', B2–B2', C1–C1', C2–C2' are the six coil which are connected as shown in the Figure 2 to create the 4 pole stator winding.

Let us start from the instant that the stator phases are excited such that positive terminal of dc source is connected to the phase "A" & "C" and the negative terminal with phase "B" by firing the appropriate semiconductor switches which result the excitation current to enter the stator through phase "A" & "C" and leaving through the phase "B". The situation is shown in Figure 4. Similarly as in case of two pole generations, if similar polarities are grouped together, a 4 pole stator winding will be obtained.

All possible switching sequences for 4 poles are described in Table 2.

3 DESCRIPTION OF THE PROPOSED WORK

In the proposed scheme, 4 pole 30° rotation configurations are adopted. The block diagram of the scheme is shown in Figure 5 (Kim *et al.* 2003). Single phase variable ac supply is obtained from a 230 volt supply with an auto transformer. MOSFET bridge inverter circuit is employed to get the switched dc input for the BLDC motor. The switching instances are controlled by a micro controller based control circuit.

According to the orientation of four stator poles, in each step the stator field rotates by 30° mechanical angle, so 6 steps are required for 180°

rotation and for one complete rotation of 360° the same six steps are to be repeated.

The switching sequence of the MOSFET inverter for such orientation is:

(M1, M6), (M2, M6), (M2, M4), (M3, M4), (M3, M5) and (M1, M5) from Figure 1 and Table 2.

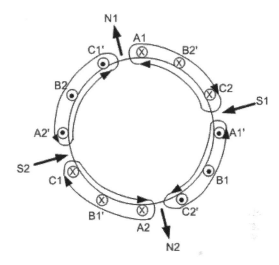

Figure 4. Starting sequence of 4 pole generation.

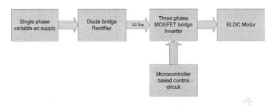

Figure 5. Block diagram of the proposed work.

Table 1. Different modes for two pole generation.

No. of windings employed	Angle of rotation of the RMF in degree	Polarity of stator terminals			
		Sequence of switching	A	B	C
3	120	Seq-1	+DC	+DC	−DC
		Seq-2	−DC	+DC	+DC
		Seq-3	+DC	−DC	+DC
2	60	Seq-1	+DC	−DC	NC
		Seq-2	+DC	NC	−DC
		Seq-3	NC	+DC	−DC
		Seq-4	−DC	+DC	NC
		Seq-5	−DC	NC	+DC
		Seq-6	NC	−DC	+DC
	120	Seq-1	+DC	−DC	NC
		Seq-2	NC	+DC	−DC
		Seq-3	−DC	NC	+DC

Table 2. Different modes for four pole generation.

No. of windings employed	Angle of rotation of the RMF in degree	Polarity of stator terminals			
		Sequence of switching	A	B	C
3	30	Seq-1	+DC	−DC	+DC
		Seq-2	−DC	−DC	+DC
		Seq-3	−DC	+DC	+DC
		Seq-4	−DC	+DC	−DC
		Seq-5	+DC	−DC	−DC
	60	Seq-1	+DC	−DC	+DC
		Seq-2	−DC	+DC	+DC
		Seq-3	+DC	+DC	−DC
	90	Seq-1	+DC	−DC	+DC
		Seq-2	−DC	+DC	−DC
2	30	Seq-1	+DC	−DC	NC
		Seq-2	NC	−DC	+DC
		Seq-3	−DC	NC	+DC
		Seq-4	−DC	+DC	NC
		Seq-5	NC	+DC	−DC
		Seq-6	+DC	NC	−DC
	60	Seq-1	+DC	−DC	NC
		Seq-2	−DC	NC	+DC
		Seq-3	NC	+DC	−DC
	90	Seq-1	+DC	−DC	NC
		Seq-2	−DC	+DC	NC
		Seq-3	+DC	−DC	NC

3.1 Description of the firmware

The speed of the rotating magnetic field depends on the time delay between two consecutive sequences. As such 12 numbers of steps required as discussed before, for a time delay of 3.3 ms between two consecutive steps, the speed of the rotating field will be 1500 rpm.

Initially a program was written for 3.3 ms delay but no rotation observed of the motor. After number of programs developed for different time delay it is observed that for a time delay of 0.078 sec and dc supply voltage of 30V at no load smooth starting is observed and the speed of the motor was recorded as 68 rpm. But for any further decrease in time delay the motor lost its self-starting property.

To overcome this problem a new program with dynamically decreased time delay is being developed. At starting the time delay is set to 0.078 sec and it is reduced by 0.00039 sec after one complete rotation of stator magnetic field i.e. the speed gradually increased to its final value which is set for the scheme as double the starting. For that the final time delay is set as 0.039 sec.

The set-up of control circuit of upper half MOSFET switches (only M1) and lower half switches are shown in Figure 6 and Figure 8 respectively.

Experimental set up for the proposed work is shown in Figure 7, Figure 9 and Figure 10.

3.2 Algorithm of variable time delay program

Main Program:

1. Define Port D, Port B, Counter 2 and Counter 3.
2. Configure all the pins as digital inputs and Port B is set as output and Port D as input port.
3. Set six output pins of port B.
4. Load counter3 by 200.
5. Delay subroutine (Xmsdelay) is called.
6. Reset output 1 and 6 and repeat step 5.
7. Set output 6 and reset output 5 and repeat step 5.
8. Set output 1 and reset output 3 and repeat step 5.
9. Set output 5 and reset output 4 and repeat step 5.
10. Set output 3 and reset output 2 and repeat step 5.
11. Set output 4 and reset output 6 and repeat step 5.
12. Set output 2 and reset output 1 and repeat step 5.
13. Repeat step 7 to step 12.

256

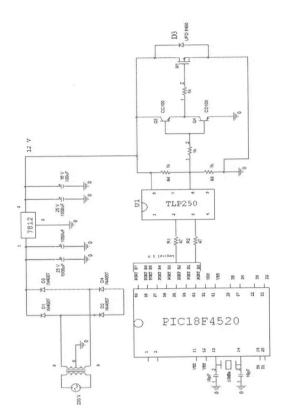

Figure 6. Power and gate triggering circuit of MOSFET M1.

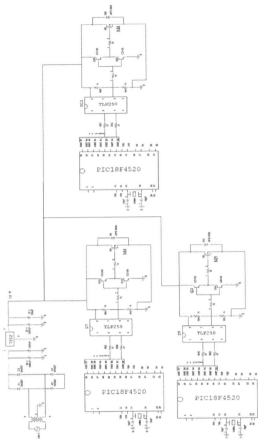

Figure 8. Power and gate triggering circuit of MOS-FET M4, M5 and M6.

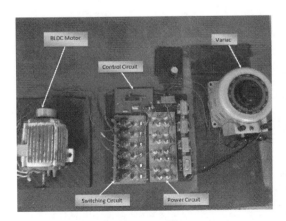

Figure 7. Overall set up of the proposed work.

Figure 9. MOSFET based inverter bridge circuit.

14. Compare counter 3 content with desired minimum count value and if condition is satisfied then go to step 16.
15. Decrease Counter 3 by one and repeat step 4 to step 14.
16. Repeat step 4 to step 14.

Xmsdelay Subroutine:

1. Counter 2 is loaded by content of counter 3.
2. 'Delay' subroutine is called.

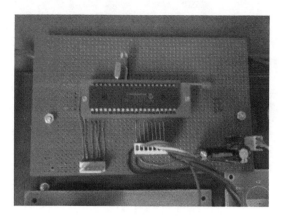

Figure 10.   Control circuit using PIC18F4520.

3. Decrement Counter2 and go to step 2.
4. Repeat step 3 until Counter 2 becomes zero.

Delay Subroutine:

1. Timer 0 register is loaded by hex equivalent of the result of calculation of the delay.
2. Set the status of Timer Control Register.
3. Monitor Interrupt controller second bit used as timer overflow flag.
4. If condition in step 3 is not satisfied repeat step 3.
5. Reset timer overflow flag.

4 CONCLUSION

The speed control scheme is developed using a modern microcomputer (PIC 18F4520) to minimize the measurement and calculation error results to the computation delay. Instead of varying the voltage manually, a chopper circuit can be used for automatic voltage control purpose which is effective to reduce torque pulsation and better transient response.

REFERENCES

Bose Bimal K, (November/December, 1988) A High-Performance Inverter-Fed Drive System of an Interior Permanent Magnet Synchronous Machine," *IEEE transactions on industry applications*, vol. 24, no. 6.
Jang G.H., Park J.H. and Chang J.H. (March 2002), Position detection and start-up algorithm of a rotor in a sensorless BLDC motor utilising inductance variation, *IEE Proc -Eleclr Power Appl,* Vol 149, No 2.
Kim Tae Heoung, Choi Jae-Hak, Ko Kwang Cheol, and Lee Ju (September 2003), Finite-Element Analysis of Brushless DC Motor Considering Freewheeling Diodes and DC Link Voltage Ripple, *IEEE TRANSACTIONS ON MAGNETICS*, VOL. 39, NO. 5.
Park Byoung-Gun, Kim Tae-Sung, Ryu Ji-Su, and Hyun Dong-Seok (2006), Fuzzy Back-EMF Observer for Improving Performance of Sensorless Brushless DC Motor Drive, *IEEE.*

Frontiers in Computer, Communication and Electrical Engineering – Acharyya (Ed.)
© 2016 Taylor & Francis Group, London, ISBN: 978-1-138-02877-7

Dependence of photonic bandgap on material composition for two-dimensional photonic crystal with triangular geometry

Shubhankar Mukherjee & Arpan Roy
Department of Electronics Science, A.P.C College, West Bengal, India

Arpan Deyasi
Department of Electronics and Communication Engineering, RCC Institute of Information Technology, West Bengal, India

Subhro Ghosal
Department of Electronics Science, A.P.C College, West Bengal, India

ABSTRACT: Complete photonic bandgap is formed inside the first Brillouin zone in two-dimensional triangular photonic crystal for TE mode propagation along either direction of refractive index variation. Maxwell's equations are solved using plane wave propagation method, from which TM and TE mode eigen-equations are obtained with appropriate boundary conditions. Different material compositions are considered to compute the width of photonic band between 'X' and 'Γ' points, and result is compared with that obtained for rectangular lattice (for TM mode propagation) with identical configuration. Result shows that width of bandgap increases with higher refractive index difference between the constituent materials whereas the variation is less for triangular geometry, i.e., triangular PC will provide better tuning of photonic bandgap. Analytical findings will become important in designing optical insulator in selected spectrum, and geometry of the structure will play crucial role in this regard.

1 INTRODUCTION

Photonic crystal is the subject of research for the last two decades (Loudon 1970, Yablonovitch et al., 1987, Andreani et al., 2003) due to its possible novel applications in device (Szczepański et al., 1988) and communication (D'Orazio et al., 2003, Russell 2006, Robinson et al., 2011) area which may replace the existing electronic and optoelectronic counterparts. It is the periodic array of dielectric/metal-dielectric composites in one, two or three dimensions; which exhibit the restriction of electromagnetic wave propagation in certain spectra (Fogel et al., 1998), and allowed in other ranges. This becomes possible due to the existence of photonic bandgap, in analogous to the electronic bandgap. Though one-dimensional photonic crystal has been investigated a lot during the last decade both theoretically (Prasad et al., 2012, Sanga et al., 2006) and experimentally (Xu et al., 2007, Stephanie et al., 2008) due to the ease of mathematical formalism and fabrication advantages respectively; but 2D structures are now gaining the interest of researchers due to novel properties (Limpert et al., 2004, Limpert et al., 2003) with superior performance. This is achieved precisely due to the higher confinement of propagating electromagnetic waves (Jiang et al., 1999). For analyzing property of 2D photonic crystal, knowledge of band structure is extremely important; and its accurate calculation inside the first Brillouin zone along with relative position and magnitude plays a major role for specific applications. Moreover, geometry of the structure also critically important as it controls the propagation mode inside the structure.

Different numerical techniques are already available for calculating the band structure of photonic crystal (Badaoui et al., 2011, He et al., 2000). In this paper, author choose the plane wave propagation method for computing the band structure as it provides near accurate result inside first Brillouin zone subject to the appropriate boundary conditions, and it also reduces the mathematical complexity. It does not require any grid structure for computation purpose. Authors choose different material composition to study the tuning of bandgap for triangular lattice, and result is compared with that obtained for rectangular lattice with identical structural parameters. It has been observed that very good optical insulator with precise control can be made by triangular geometry, whereas rectangular structure provides wider variation. Result is critically important for designing photonic isolator for optical integrated circuit.

2 THEORETICAL FOUNDATION

For computation purpose, we assume sinusoidal nature of electric field and magnetic fields propagating inside the structure. Under this condition, modified Maxwell equations can be written as:

$$\nabla . \varepsilon(r).E(r) = 0 \tag{1}$$

$$\nabla . B(r,t) = 0 \tag{2}$$

$$\nabla \times H(r) = i\omega\varepsilon_0\varepsilon(r)E(r) \tag{3}$$

$$\nabla \times E(r) = -i\omega\mu_0 H(r) \tag{4}$$

where ω is the oscillation frequency of the electromagnetic field, $\varepsilon(r)$ is the corresponding dielectric constant.

Assuming medium as non-dispersive with harmonic time dependence, Helmhotz equation may be put into the following form

$$\nabla \times \frac{1}{\varepsilon(r)}\nabla \times \vec{H}(r) = \frac{\omega^2}{C^2}\vec{H}(r) \tag{5}$$

If we consider that the dielectric function has full translational variance in either of the confinement direction, magnetic intensity can be expressed as a sum of infinite number of plane waves

$$H(r) = \sum_{\vec{G}_i,\lambda} h_{\vec{G}_i,\lambda} e^{i(\vec{k}+\vec{G}_i).\vec{r}} \hat{e}_\lambda \tag{6}$$

where $\lambda = 1, 2$, k denotes wave vector, $\vec{G}$ is the reciprocal lattice vector, $\hat{e}_\lambda$ is used to denote the two unit axes perpendicular to the propagation direction $\vec{k}+\vec{G}$, $h_{\vec{G}_i,\lambda}$ represents the coefficient of the H component along the axes $\hat{e}_\lambda$. It may be noted in this context that $(\hat{e}_1, \hat{e}_2, \vec{k}+\vec{G})$ are perpendicular to each other.

Dielectric function can also be written as

$$\varepsilon(\vec{G}) = \frac{1}{V}\oiiint_\Omega \varepsilon(r)\exp(-i\vec{G}.r) \tag{7}$$

where Ω is the unit cell and V is the volume of the unit cell. Using Eq. (6) and Eq. (7), one can write

$$\nabla \times \sum_{\vec{G}_i} \varepsilon^{-1}(\vec{G}_i)e^{i\vec{G}.\vec{r}} \nabla \times \sum_{\vec{G}_i,\lambda} h(\vec{G}_i,\lambda)e^{i(\vec{k}+\vec{G}_i).\vec{r}} \hat{e}_{\lambda,\vec{k}+\vec{G}_i}$$

$$= \frac{\omega^2}{C^2} \sum_{\vec{G}_i,\lambda} h(\vec{G}_i,\lambda)e^{i(\vec{k}+\vec{G}_i).\vec{r}} \hat{e}_{\lambda,\vec{k}+\vec{G}_i} \tag{8}$$

Decoupling the x-y plane along both of the basis vectors, one can obtain

$$\sum_{G_i,\lambda G_i'-G_i} \sum_{\lambda_i} \begin{bmatrix} h(G_i,\lambda)\varepsilon^{-1}(\vec{G}_i-\vec{G}_i)e^{i(\vec{k}+\vec{G}_i').r} \\ \times \left[(\vec{k}+\vec{G}_i)\times \hat{e}_{\lambda,\vec{k}+\vec{G}_i} \right] \end{bmatrix}$$

$$= \frac{\omega^2}{C^2} \sum h(G_i,\lambda)e^{i(\vec{k}+\vec{G}_i).\vec{r}} \hat{e}_{\lambda,\vec{k}+\vec{G}_i} \tag{9}$$

The final equation may be modified and written in the following form

$$\sum_{G',\lambda} \begin{bmatrix} (\vec{k}+\vec{G})\hat{e}_\lambda \\ \varepsilon^{-1}(\vec{G}-\vec{G}')h(\vec{G}',\lambda) \end{bmatrix}\begin{bmatrix} (\vec{k}+\vec{G}')\hat{e}'_\lambda \end{bmatrix} = \frac{\omega^2}{C^2}h(\vec{G},\lambda) \tag{10}$$

where $\lambda' = 1,2$. [for the two decoupled axes]. Eq. (10) may be put into the matrix form

$$\sum_{\vec{G}'} \begin{bmatrix} |k+G||k+G'|\varepsilon^{-1}(G-G') \\ \begin{bmatrix} \hat{e}_2\hat{e}'_2 & -\hat{e}_2\hat{e}'_1 \\ -\hat{e}_1\hat{e}'_1 & \hat{e}_1\hat{e}'_1 \end{bmatrix}\begin{bmatrix} h_1 \\ h_2 \end{bmatrix} \end{bmatrix} = \frac{\omega^2}{C^2}\begin{bmatrix} h_1 \\ h_2 \end{bmatrix} \tag{11}$$

Here, the 2×2 matrix provides the direction of wave propagation at the time of scattering with a lattice, 'h' matrix is an $N \times 1$ matrix which satisfies the number of plane waves propagating inside the crystal. Thus the 'e' matrix basically causes the diffraction inside the first Brillouin zone.

For oblique incidence, both K and G are in the same x-y plane, and one of the unit vectors can be constantly set to: $\hat{e}_{2,k+G}= \hat{z}$ the other unit $\hat{e}_{1,k+G}$ vector is also in the x-y plane. Therefore, $2N \times 2N$ equation group is decoupled into two equation groups: TE and TM. One group only contains H components along $\hat{e}_1$, no components along $\hat{e}_2 = \hat{z}$, and so is called TM wave. The other group only contains H component along $\hat{e}_2 = \hat{z}$, and E has no component along z which is why it is called the TE wave.

Now the TM mode can be formulated as:

$$\sum_{G'} |k+G||k+G'|\varepsilon^{-1}(\vec{G}-\vec{G}')h_1(\vec{G}') = \frac{\omega^2}{C^2}h_1(\vec{G}) \tag{12}$$

The TE mode can be formulated as

$$\sum_{G'} |k+G||k+G'|\varepsilon^{-1}(\vec{G}-\vec{G}')\times (\hat{e}_1\bullet\hat{e}'_1)h_2(\vec{G}') = \frac{\omega^2}{C^2}h_2(\vec{G}) \tag{13}$$

3 RESULTS AND DISCUSSION

Using Eq. (12) and Eq. (13), TM and TE mode propagations inside rectangular and triangular lattice can be obtained respectively. In this respect, it may be mentioned that TM mode propagation inside triangular lattice and TE mode propagation in rectangular lattice cause incomplete bandgap, and hence their discussions are omitted from the text.

Fig. 1 shows the behavior of photonic band gap in triangular lattice inside the first Brillouin zone for SiO_2/air composition. The dark region in the plot is called the photonic bandgap as it gives the region of forbidden frequency. From the figure, magnitude of MBF (mid-band frequency) is 0.35856 arb. unit, and magnitude of bandgap is 0.0221 arb. unit. where we consider lattice filling factor as 0.25. The figure suggests the photonic nature of the structure is exactly symmetric with the electrical nature.

Fig. 2 shows the bandgap profile for 2D triangular lattice with Air ($\varepsilon_1 = 1$)/Alumina ($\varepsilon_2 = 8.9$) interface. In this case, PBG becomes 0.0359 arb. unit., and magnitude of mid band frequency is 0.2494 arb. unit. Thus it may be concluded that with increase of index differences, magnitude of bandgap increases, whereas mid-band frequency decreases. This claim is justified once alumina is replaced by GaN. This is plotted in Fig. 3. In this case, PBG becomes 0.036 arb. unit., whereas MBF can be obtained as 0.2397 arb. unit.

But replacing GaN by Si shows opposite behavior. In this case, though refractive index contrast increases, but both PBG and MBF decreases. The magnitude of PBF is obtained as 0.0356 arb. unit., whereas MBF becomes 0.219 arb. unit. This is plotted in Fig. 4.

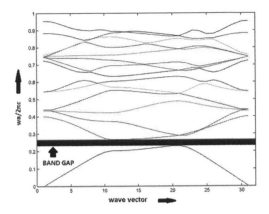

Figure 2. Band structure of 2D triangular lattice of Air ($\varepsilon_1 = 1$)/–Alumina ($\varepsilon_2 = 8.9$) system in TE mode with filing r/a = 0.25.

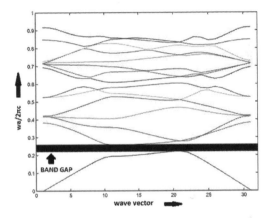

Figure 3. Band structure of 2D triangular lattice of Air ($\varepsilon_1 = 1$)/–GaN ($\varepsilon_2 = 9.7$) system in TE mode with filing r/a = 0.25.

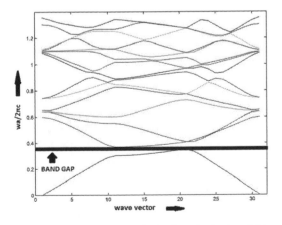

Figure 1. Band structure of 2D triangular lattice of Air ($\varepsilon_1 = 1$) / SiO_2 ($\varepsilon_2 = 3.9$) system in TE mode with filing r/a = 0.25.

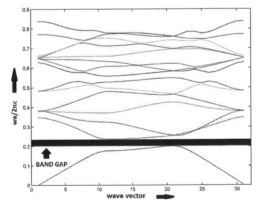

Figure 4. Band structure of 2D triangular lattice of Air ($\varepsilon_1 = 1$)/Si ($\varepsilon_2 = 11.8$) system in TE mode with filing r/a = 0.25.

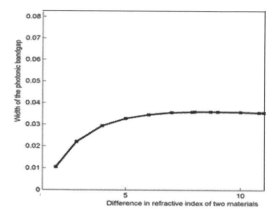

Figure 5. Width of photonic bandgap with gradient of refractive index of constituent materials.

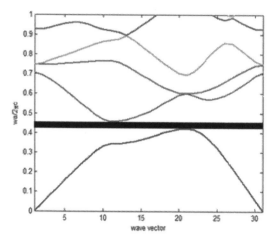

Figure 6. Band structure of 2D rectangular lattice of Air ($\varepsilon_1 = 1$)/SiO$_2$ ($\varepsilon_2 = 3.9$) system in TM mode with filing r/a = 0.25.

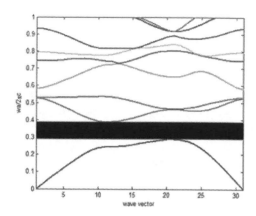

Figure 7. Band structure of 2D rectangular lattice of Air ($\varepsilon_1 = 1$)/Alumina ($\varepsilon_2 = 8.9$) system in TM mode with filing r/a = 0.25.

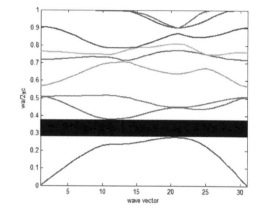

Figure 8. Band structure of 2D rectangular lattice of Air ($\varepsilon_1 = 1$)/GaN ($\varepsilon_2 = 9.7$) system in TM mode with filing r/a = 0.25.

The variation of the width of the photonic bandgap with respect to the difference in refractive indices of the two materials is plotted in Fig. 5. The refractive index of the inner layer (air) is kept fixed whether that of the outer material is increased. Here the filling factor is kept constant at 0.25. We have taken the difference of refractive index of two materials from 2.9 to 12.7 to get a smoother graph. The plot indicates that the width of photonic band gap initially increased and then becomes almost constant.

Next we made the comparative study with rectangular structure. Figure 6 to Figure 9 show the profiles for rectangular geometry for TM mode propagation.

Comparative study reveals that changing geometry increases the magnitude of PBG. For SiO$_2$,

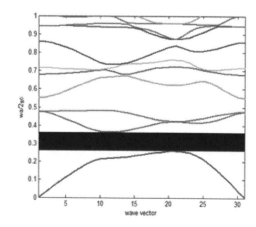

Figure 9. Band structure of 2D rectangular lattice of Air ($\varepsilon_1 = 1$)/Si ($\varepsilon_2 = 11.8$) system in TM mode with filing r/a = 0.25.

Alumina, GaN, Si; it is obtained as 0.0355, 0.1, 0.1021, 0.104 (all are in arb. unit.); whereas MBF decreases continuously. The magnitude of MBF are 0.44005, 0.3356, 0.32805, 0.3106 (all are in arb. unit.). Hence better tuning can be obtained for triangular geometry.

4 CONCLUSION

Photonic bandgap is computed for triangular 2D photonic crystal for TE mode propagation with different constituent materials, and result is compared with rectangular geometry with identical lattice filling factor. Simulated findings shows that controlled tailoring of bandgap can be obtained for triangular lattice, whereas for rectangular structure, bandgap varies in a wider range. Result is important for designing optical isolator in required frequency spectrum for different applications, more precisely for optical communication and information processing domain.

REFERENCES

Andreani. L.C., Agio. M., Bajoni. D., Belotti. M., Galli. M., Guizzetti. G., Malvezzi. A. M., Marabelli. F., Patrini. M., Vecchi. G., 2003, Optical Properties and Photonic Mode Dispersion in Two-Dimensional and Waveguide-Embedded Photonic Crystals, Synthetic Metals, 139, 695–700.

Badaoui. H., Feham. M., Abri. M, 2011, Photonic-Crystal Band-pass Resonant Filters Design using the Two-dimensional FDTD Method, International Journal of Computer Science Issues, 8, 127–132.

D'Orazio. A., De Palo. V., De Sario. M., Petruzzelli. V., Prudenzano. F., 2003, Finite Difference Time Domain Modeling of Light Amplification in Active Photonic Bandgap Structures, Progress In Electromagnetics Research, 39, 299–339.

Fogel. I.S., Bendickson. J.M., Tocci. M.D., Bloemer. M.J., Scalora. M., Bowden. C.M., Dowling. J.P., 1998, Spontaneous Emission and Nonlinear Effects in Photonic Bandgap Materials, Pure and Applied Optics, 7, 393–408.

He. S., Qiu. M., Simovski. C., 2000, An Averaged Field Approach for Obtaining the Band Structure of a Dielectric Photonic Crystal, Journal of Physics: Condensed Matter, 12, 99.

Jiang. Z., Niu. C.N., Lin. D.L., 1999, Resonance Tunneling through Photonic Quantum Wells, Physical Review B, 59, 9981–9986.

Limpert. J., Liem. A., Reich. M., Schreiber. T., Nolte. S., Zellmer. H., Tünnermann. T., Broeng. J., Petersson. A., Jakobsen. C., 2004, Low-Nonlinearity Single-Transverse-Mode Ytterbium-Doped Photonic Crystal Fiber Amplifier, Optic Express, 12, 1313.

Limpert. J., Schreiber. T., Nolte. S., Zellmer. H., Tunnermann. T., Iliew. R., Lederer. F., Broeng. J., Vienne. G., Petersson. A., Jakobsen. C., 2003, High Power Air-Clad Large-Mode-Area Photonic Crystal Fiber Laser, Optic Express, 11, 818.

Loudon. R., 1970, The Propagation of Electromagnetic Energy through an Absorbing Dielectric, Journal of Physics A, 3, 233–245.

Prasad. S., Singh. V., Singh. A.K., 2012, To Control the Propagation Characteristic of One-Dimensional Plasma Photonic Crystal using Exponentially Graded Dielectric Material, Progress in Electromagnetics Research M, 22, 123–136.

Robinson. S., Nakkeeran. R., 2011, Photonic Crystal Ring Resonator based Add-Drop Filter using Hexagonal Rods for CWDM Systems, Optoelectronics Letters, 7, 164–166.

Russell. P.S.J., 2006, Photonic-Crystal Fibers, Journal of Lightwave Technology, 24, 472–474.

Sanga. Z.F., Li. Z.Y., 2006, Optical Properties of One-Dimensional Photonic Crystals containing Graded Materials, Optics Communications, 259, 174–178.

Stephanie. A.R., Florencio. G.S., Paul. V.B., 2008, Embedded Cavities and Waveguides in Three-dimensional Silicon Photonic Crystals, Nature - Photonics, 2, 52–56.

Szczepański. P., 1988, Semiclassical Theory of Multimode Operation of a Distributed Feedback Laser, IEEE Journal of Quantum Electronics, 24, 1248–1257.

Xu. X., Chen. H., Xiong. Z., Jin. A., Gu. C., Cheng. B., Zhang. D., 2007, Fabrication of Photonic Crystals on Several Kinds of Semiconductor Materials by using Focused-Ion Beam Method, Thin Solid Films, 515, 8297–8300.

Yablonovitch. E., Gmitter. T.J., 1987, Photonic band structure: The face-centered-cubic case, Physical Review Letters, 63, 1950–1953.

Design and simulation of a power factor corrected boost converter

Sourav Tola, Ayandeep Ganguly, Ujjal Dey, Megha, Soumitra Mukherjee, Amit Kumar Das,
Tanushree Dey, Ishita Mondal & Sourav Mondal
Haldia Institute of Technology, Haldia, East Medinipur, West Bengal, India

ABSTRACT: The interest towards improving the quality of power, which are supplied to the electronic equipments has been increasing to ensure the optimum utilization of power generated. A low power factor is responsible for reducing the active power that's available from the grid. Thus, it causes a high harmonic distortion of the line current eventually, which results in electromagnetic interference problems and cross-interferences, through the line impedance, between different systems connected to the same supply. The standard rectifier employing a diode bridge followed by a filter capacitor for DC utilities gives unacceptable performances in terms of power quality according to the rising benchmark set by recent developments in the relevant field. The most popular topology for power factor correction is certainly the boost topology. In this work, a power factor corrected boost converter has been designed and simulated in MATLAB and the proposed control strategy has been implemented.

1 INTRODUCTION

The conventional AC/DC power converters that are connected to the line through diode rectifier draw a non-sinusoidal input current. Because of the large harmonic content, a typical diode rectifier is used for interfacing the power electronic equipment with the utility system and exceeds the limits on individual current harmonics and THD specified (IEEE guide 1992). These harmonic currents flowing through the impedances in the electrical utility distribution system can cause several problems such as voltage distortion, heating, and noises. These harmonics distort the local voltage waveform; potentially interfering with other electrical equipment connected to the same electrical service and reduces the capabilities of the line to provide energy. In addition to, the effect on the power line quality and the poor waveform of the input current also affects the power electronic component itself in different ways (Mohan *et al.* 2003). Due to this fact, the presence of standards or recommendations is forced to use power factor correction in power supplies.

Inductors and capacitors can be used in conjunction with the diode bridge rectifier to improve the waveform of the current, drawn from the utility grid. The obvious disadvantages of such arrangements are cost, size, losses, and the significant dependence of the average DC voltage on the power drawn by the load.

By using a power electronic converter for current shaping, it is possible to shape the input line current drawn by bridge rectifier is to be sinusoidal and in phase with the input voltage. The choice of the power electronic converter is based on different considerations. In most applications, it is acceptable, and in many cases desirable, to stabilize the DC voltage slightly in excess of the peak of the maximum of the AC input voltage. The input current drawn should ideally be at a unity power factor so that the power electronic interface emulates a resistor supplied by the utility source (Erickson *et al.* 2004). The cost, power losses, and size of the current shaping circuit should be as small as possible. Based on these considerations, the obvious choice for current shaping circuit is a step-up DC-DC converter.

2 PFC BOOST CONVERTER

2.1 *Basic principle of boost converter*

Figure 1 shows a boost converter. Its main application is in regulated DC power supplies and the regenerative braking of DC motors. As the name implies, the output voltage is always greater than the input voltage. When the switch is on, the diode is reverse biased, thus isolating the output stage. The input supplies energy to the inductor. When the switch is off, the output stage receives energy from the inductor as well as from the input. The output filter capacitor is assumed to be very large to ensure a constant output voltage at steady state.

The switch used in the boost converter configuration is typically a power BJT, power MOSFET,

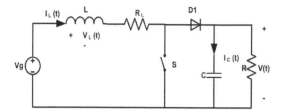

Figure 1. Boost converter operational circuit.

or IGBT. DC power is obtained by rectifying the power from an AC source using a diode bridge rectifier. The boost converter converts the rectified DC voltage to a higher output voltage.

The key principle that drives the boost converter is the tendency of an inductor to resist changes in the current. The operation of the boost converter has two distinct states.

- The switch is closed. The current $I_L(t)$ flows through L, R_L and back to the voltage source. In this mode, the inductor stores energy.
- The switch is opened. The path for the inductor current to flow is through the freewheeling diode $D1$, capacitor C, and load R. The stored energy in the inductor collapses and the polarity across the inductor is reversed. The energy is transferred from the inductor to the capacitor and the capacitor gets charged to a voltage higher than the source voltage.

At the end of On-state, the increase of I_L is

$$\Delta I_{LON} = \frac{1}{L} \int_0^{DT} V_i dt = \frac{DT}{L} V_i \qquad (1)$$

D is the duty cycle which represents the fraction of the total time period T that the switch is kept on. So, D varies from 0 to 1. During, the off period the change in I_L is:

$$V_i - V_o = L \frac{dI_L}{dt} \qquad (2)$$

So, the variation of I_L during the off period is

$$\Delta I_{LOFF} = \int_{DT}^{T} \frac{(V_i - V_0)dt}{L} = \frac{(V_i - V_o)(1-D)T}{L} \qquad (3)$$

The inductor current has to be the same at the start and end of the cycle at steady state. So,

$$\Delta I_{LON} + \Delta I_{LOFF} = 0 \qquad (4)$$

Substituting the values of ΔI_{LON} and ΔI_{LOFF} we get $\dfrac{V_o}{V_i} = \dfrac{1}{1-D}$ which can be rearranged to find

$$D = 1 - \frac{V_i}{V_o} \qquad (5)$$

So, from this expression we can see that the output voltage is always higher than the input voltage.

2.2 Proposed control strategy

The control strategy proposed in this work is hysteresis control. In hysteresis current control the current is made to vary within a specific band, as shown in Figure 2. Current–mode control can be achieved by sensing the current of the switching power devices or energy storage element and integrating it into main voltage control loop (Fagerstrom et al. 2012). Whenever the actual current reaches the upper limit of the band the switch of the boost converter must be off and whenever the current falls to the lower limit of the band, the switch must be ON. In this control scheme the switching frequency is not constant.

The block diagram of the implemented control strategy is shown in Figure 3.

2.3 Implementation of PI controller

Output voltage from the boost converter is compared with reference voltage level before feeding in the controller circuit. Proper choice of the PI controller parameters are decided by the transfer function obtained from the DC-to-DCconverter model.

From Figure1, when the switch is at position 1,

$$V_L(t) = V_g - i_L(t)R_L \qquad (6)$$

Using small ripple approximation, $i_L(t)$ replaced by its DC component I_L, hence

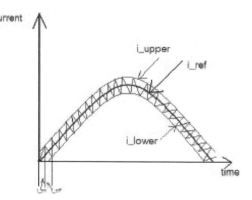

Figure 2. Input current waveform of PFC boost converter with hysteresis control.

266

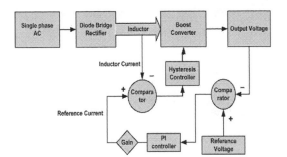

Figure 3. Block diagram of the proposed control strategy.

$$V_L(t) = V_g - I_L R_L \qquad (7)$$

Similarly,

$$i_c(t) = -\frac{V}{R} \qquad (8)$$

During the first subinterval, the switches are in position 1 for time dTs and during the second subinterval, of length of switching duration is $(1 - d)Ts$. The duty cycle $d(t)$ may now be a time-varying quantity. Using small signal approximation, the state equations of the circuit can be developed considering that natural frequency of the converter network is much smaller than the switching frequency.

Small signal control $(d(s))$ to output transfer function is described by (Mahmood et al. 2014),

$$G_d(s) = G_{d0} \frac{1 - \dfrac{s}{\omega_z}}{1 + \dfrac{s}{Q\omega_0} + \left(\dfrac{s}{\omega_0}\right)^2} \qquad (9)$$

With the condition $Vg(s) = 0$
Line to output transfer function can be obtained with $d(s) = 0$ as,

$$G_g(s) = G_{g0} \frac{1}{1 + \dfrac{s}{Q\omega_0} + \left(\dfrac{s}{\omega_0}\right)^2} \qquad (10)$$

where,

$$G_{g0} = \frac{1}{1 - D}, \omega_0 = \frac{1 - D}{\sqrt{LC}}$$

$$G_{d0} = \frac{V}{1 - D}, \omega_z = \frac{(1 - D)^2 R}{L}, Q = (1 - D)R\sqrt{LC}$$

Design of PI controller requires transient analysis and without detail analysis of each mode of switching cycle. Combined state equation model is formed for both ON and OFF state in the form of a set of two linear time-variant state equations ignoring ripple components. Linearization technique to optimize proportional and integral constant values using Ziegler-Nichols tuning method is applied.

3 RESULTS AND ANALYSIS

The simulation of both the normal boost converter and the power factor corrected boost converter has been done in MATLAB and the results have been compared.

Figure 4 shows the simulation diagram of the boost converter without power factor correction.

The rectified DC voltage, output DC voltage, and the inductor current waveforms for a boost converter without power factor correction are shown in Figure 5. Input AC voltage and current waveforms are shown in Figure 6. It can clearly be seen that the current waveform is not following the voltage waveform and the power factor obtained from the simulation is 0.822 and the THD is 0.5176. The current waveform is observed to be far from sinusoidal.

Circuit diagram for Power Factor Corrected boost converter operation has been shown in Figure 7. The rectified DC voltage, output DC voltage, and the inductor current waveforms for a power factor corrected boost converter are shown in Figure 8. Input AC voltage and current waveforms are shown in Figure 9. The inductor current is following the rectified voltage obtained from the diode bridge rectifier. Hence, current drawn from supply is in same phase with the voltage waveform. Consequently, the power factor obtained is 0.999

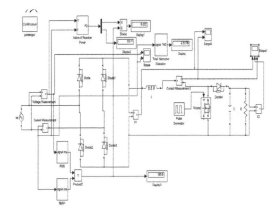

Figure 4. Circuit diagram for the rectifier fed boost converter without power factor correction.

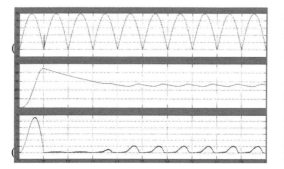

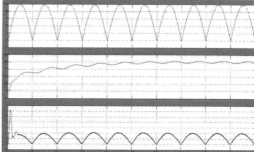

Figure 5. Rectified DC input voltage, output voltage across the capacitor and inductor current waveforms of the boost converter without power factor correction.

Figure 8. Rectified DC input voltage, output voltage across the capacitor and inductor current waveforms of the Power Factor Corrected boost converter.

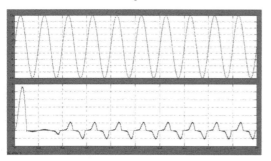

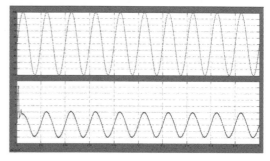

Figure 6. Input voltage and input current waveforms of the rectifier fed boost converter without power factor correction.

Figure 9. Input voltage and input current waveforms of the rectifier fed Power Factor Corrected boost converter.

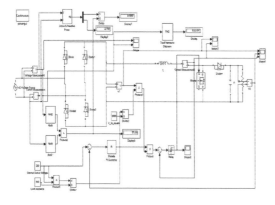

Figure 7. Circuit diagram for power factor corrected boost converter operation.

and the THD is 0.0319. Both of these values show improved operation compared to a boost converter without power factor correction.

4 CONCLUSION

The speed controlled scheme is developed using a modern microcomputer (PIC 18F4520) to minimize the measurement and calculation error results to the computation delay. Instead of varying the voltage manually, a chopper circuit can be used for automatic voltage control purpose which is effective to reduce torque pulsation and better transient response.

REFERENCES

Erickson, R.W., Maksimovic, D. (2004). Fundamentals of Power Electronics. Kluwer Academic Publishers.

Fagerstrom, S., Bengiamin, N. (2012). Modelling and Characterization of Power Electronics Converters Using Matlab Tools. InTech, http://www.intechopen.com.

Mohan, N., Undeland, T.M., Robbins, W.P. (2003). Power Electronics Converters, Applications and Design. John Wiley & Sons.

Mahmood, O.Y., Mashhadany Y.I., Khadim, J.S. (2014). High Performance DC-DC Boost Converter Based on Tuning PI controller", The Second Engineering Conference of Control, Computers and Mechatronics Engineering (ECCCM2).

(1992). IEEE Guide for Harmonic control and Reactive Compensation of Static Power converter. ANSI/IEEE Std. 519–1981, revised in 1992 to 519–1992.

Frontiers in Computer, Communication and Electrical Engineering – Acharyya (Ed.)
© 2016 Taylor & Francis Group, London, ISBN: 978-1-138-02877-7

Smart shoes with cueing system and remote monitoring of Parkinson's patients: SCARP

Piyali Das
The Ohio State University, Ohio, USA

Rupendra N. Mitra & Nuthapol Suppakitjarak
University of Cincinnati, Ohio, USA

ABSTRACT: In this paper, we present an innovated design and evaluation of SCARP (Smart shoes with Cueing system And Remote monitoring of Parkinson's patients), a system comprising of sensor based wearable pair of smart shoes for online-offline monitoring and detecting gait impairment of Parkinson's Disease (PD) patients. This novel system, hidden inside a shoe sole, is capable of providing visual cue on automatically recognizing Freezing of Gait (FoG), which then continues till the freezing is ceased. The shoe is equipped with a WiFi module to send the data from wearable sensors to a centralized server for remote monitoring. Health providers and doctors can access patients' gait data in real-time and could also use it later for offline analysis. The data in the server may be used to create an online gait repository. We also developed a User Interface (UI) to aid online monitoring of gait data. This UI is provisioned with user's selection of date-time as input for graphical representation of the data stored in database. This helps to graphically observe the variation of data over a period of days, week, and month. Thus, the work herein presented shows that wearable sensors combined with an embedded system makes real-time decision of initiating visual cue, provides reliable detection of FoG and sends the information that can be further used for clinical inferences.

1 INTRODUCTION

Parkinson's Disease (PD) is a neurodegenerative disorder that affects motor and non-motor functions of human body. The symptom of PD continues life long and it keeps on deteriorating with time. Personalized monitoring of motor status between in-clinic evaluations is expected to facilitate tailoring of therapy and improvement of health-related outcomes. Gait impairment, including shuffling, festination (progressive reduction in stride length with inefficient increase in cadence), and Freezing of Gait (FoG) are important variables in the assessment of the risk for falls and other ambulation-related injuries. FoG is defined as the paroxysmal inability to lift the feet off the ground and carry out effective stepping. Out of all the symptoms of PD, FoG has always been the major area of interest as it results in the analysis of several key factors in PD patients. How frequent the FoG is, how long it persists, how critical the tremor is during FoG event together tells the degree of severity of illness of a PD patient. The most commonly used rating scale for the clinical study of PD is the Unified Parkinson's Disease Rating Scale (UPDRS) (Unified Parkinson's Disease Rating Scale, 2015). The Hoehn and Yahr (H&Y) scale is used for

severity analysis of PD (Hoehn *et al.* 2015) and reports the advancement in patient's condition. Changes in footsteps, stride length, gait cycle, and walking pattern during FoG collectively provides a quantifiable platform to measure the gait impairment among patients. Gait cycle is the time elapsed between one of the foot making contact with the ground for two successions. The gait cycle starts when one foot makes contact with ground and ends when that same foot contacts the ground again (Definition and Description of Gait Cycle, 2015). We mapped this cycle to our results in section 4.

In our bid to address and classify one or more of the above mentioned stages, we propose a novel non-invasive hidden system in a shoe sole. It is a FoG monitoring system with the capability of providing visual cue where FoG will be is detected. A preliminary version of this work has been reported (Mitra *et al.* 2014). This system has been extended for storing data for future analysis as well as real time online data monitoring.

2 RELATED WORK

In last couple of years, we have seen a number of attempts to address the issues of Parkinson's patient well-being and better management. Many literatures report a wide spectrum of applications based on the different techniques that measure or monitor PD out of which sensor system is proven to be the most useful. Most of the commonly known systems that are currently available are based on wearable sensors like motion sensors, accelerometers, gyroscope, and pressure and force sensors. Patel *et al.* 2010, proposed a monitoring system for PD patients using accelerometer sensors and predicted the UPDRS score. In another work, Caldara *et al.* 2014, used five sensors across a body to quantify Gait quality, gait direction, bradykinesia, tremor etc., and thereby transmit the data using Bluetooth technology. Mariani *et al.* 2013, developed a wearable shoe strap that was enabled with accelerometers and gyroscope. Jamthe *et al.* 2013, proposed a home monitoring system using Received Signal Strength Indicator (RSSI) from sensor motes placed on walls and floors of a room and the base station located in the center of the room. A real time remote monitoring system has also been proposed for PD patients by Chen *et al.* 2014. They aimed at remote tuning of post operation deep brain stimulators though client-server architecture.

Another well-known project for improving self-management of PD patients is REMPARK 2015. The principle aim of this project is a state-of-art device for accurate and non-invasive monitoring of Parkinson's motor symptoms. This device is capable of detecting the rhythm of patient's gait and providing auditory cue as soon as the rhythm is falling than its usual tempo. Moreover, not many devices have been designed to continuously transmit the wireless gait data to any central processing unit for further analysis. SD-card has been used in some cases but it lacks much data storage and in this regard, real-time monitoring also seems to be challenging. Bluetooth has a limited range of transmission that normally restricts a patient's free movement. Even in REMPARK, the system itself is not hidden and needs to be carried by the patient.

3 PROPOSED SYSTEM

In this system we propose a wearable smart hidden shoe, which is a stand-alone device with the capability of detecting FoG, which then wirelessly transmits that information to a web server and projects LED light on the floor that acts as a visual cue. When a patient is aware that he is being monitored by someone, his motor-neural behavior may differ and his symptom does not seem to show properly, leading to an improper prediction. A patient's symptom varies considerably when

measured at doctor's clinic as compared with home environment. That is why a 'hidden' non-invasive monitoring system is one of the major needs for PD patients. Moreover, a wearable system capable of collecting a patient's gait data for a longer period say over weeks or month is needed. Also a method of assisting PD patients with automated visual cue at the time of their FoG and turning off the cue on a successful recovery from freezing is highly desired. It has been clinically proven that visual cue stimulates the patient in overcoming the freezing state more than any other cue mechanism, like auditory cue or impulsive cue.

Salient features of SCARP:

- Hidden monitoring and data collection platform in shoe,
- Light weight wearable sensor system,
- Low cost non-invasive system
- Automatic real-time detection of FoG,
- Remote monitoring of gait data,
- Online and Offline monitoring, and
- Upon detection of FoG, automatically turns on visual cue and keeps it on until freezing ceases.

We developed the prototype of SCARP by using two shoes that are having 4 sensors fitted in each. Each shoe has an Arduino UNO micro-controller. Micro-controller of one shoe equipped with LED, detects FoG episode and turns on the visual cue. On the other hand micro-controller in the other shoe drives the WiFi module to continuously transmit the sensed gait data.

The block diagram of SCARP architecture is shown in Figure 1. Data transmitted over the internet is stored in database and is retrieved by the server. A web application with a simple user interface (UI) is designed to retrieve this data from server. This application can be used to monitor real-time gait pattern of patient's gait. There is a provision to select the preferred start and end dates and times to obtain gait history of a

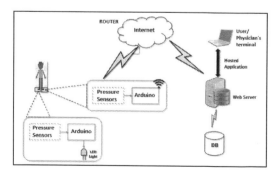

Figure 1. Block diagram representing architecture of SCARP.

patient. The patient data can be viewed remotely by any authorized person by logging into the web page thus enabling online and offline monitoring. The working principle of SCARP is shown in Figure 2 in a systematic flowchart. We gathered the gait data from six volunteers who simulated PD patient's gait pattern.

3.1 *Wearable hardware component*

The prototype of the shoe has been made up using the components mentioned in Table 1. The sensors are light in weight and flexible with high sensitivity and having a response time that is less than 5 μs. Arduino has a Flash Memory of 32 KB hence, a little onboard computation was performed and the pressure sensor data is sent to the central storage for further processing. Figure 4 represents the circuit diagram of the Arduino attached with four sensors and the WiFi breakout module in one shoe. An LED is used in the front of the other shoe that projects light beam in front of the other foot on detection of freezing acting as a visual cue.

3.2 *Sensing and monitoring*

We have used four sensors hidden inside the shoe sole. Based on the theories of R.W. Soames, 1985 the location of the sensors are chosen considering the points (heel, metatarsal heads, and mid foot) where in the selected zones the magnitude of pressure exerted is maximum. Scott *et al.* 1973 mentioned that heavier people tend to put more pressure on the lateral side of foot, hence we placed 1 sensor in the lateral area. Different other survey and literatures have proven this theory as well (Kapandji 2010, Soames 1985, Mazumder *et al.* 2012). We have used the first sensor to record the thumb pressure, second one just below the thumb area, third sensor in the lateral foot region and fourth one in the heel. The schematic diagram of the strategic positions of the sensors under foot is shown in Figure 3. Though the pressures in these points vary drastically in different patients as shown in Figure 9 but in general, these regions cumulatively gather the information of pressure variation to detect FoG. We have been interested only in taking into account the change in the

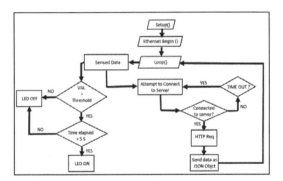

Figure 2.   A flow chart of SCARP.

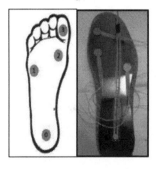

Figure 3.   Pressure points for sensor placement in shoe sole.

Table 1.   Components used for prototyping.

Components	Count	Specification
Flexiforce & FSR 402 Sensors	8	diameter of sensing circle: 14 mm; weight: 0.26 g
Arduino UNO Atmega 328 micro-controller	2	operating voltage: 5V; clock: 16 Mz; weight: 25g; flash memory: 32 KB
IL081 LED	1	5 mm
Onboard Power Supply	2	9V battery & power circuit
Pair of shoes	1	US Size 8
Adafruit CC3000 Wifi Breakout	1	IEEE 802.11 b/g; security mode: WPA2; weight: 3.4 g

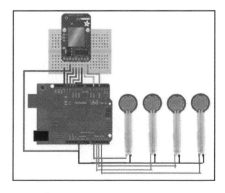

Figure 4.   Schematic circuit diagram of Arduino with WiFi and sensors.

pressure value instead of the absolute pressure value. Arduino scales the pressure measurements onto a scale of 0 to 1023 where 0 denotes absence of pressure and 1023 denotes maximum pressure. The pressure variation from different sensors cumulatively allows to determine whether freezing has occurred or not. Exceeding the threshold pressure value for a certain period of time (in our experiment 5 seconds) it enables the embedded system to make a decision whether freezing has occurred or not and thereby decision of firing LED is made. This threshold value of 950 is determined from the histogram of the FoG samples collected from the simulated patients as shown in Figure 5.

3.3 *Online data collection*

We used WiFi connectivity to send the data using TCP. We created windows compatible web development environment using WAMP server for this purpose. We collected the real time data from the Arduino and transferred it via HTTP POST request. MySQL database is used to store the data and an Apache server is used to process user request. Figure 6 shows the screen shot of real-time data getting recorded in the server. The data is recorded along with the original timestamp, which thereby helps in future analysis. This data can easily be accessed from anywhere across the world by

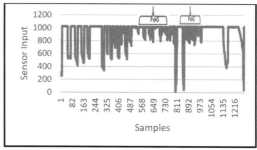

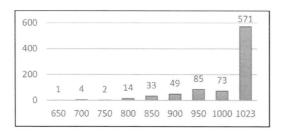

Figure 5. Histogram of FoG sample distribution.

Arduino Data Log

ID	Date and Time	Arduino Sensor Data 1	Arduino Sensor Data 2	Arduino Sensor Data 3	Arduino Sensor Data 4
1	[illegible]	1000	1020	1023	1023
2	2015-07-23 03:25:54	1000	1020	1023	1023
3	[illegible]	288	280	[illegible]	272
4	2015-07-23 12:54:23	403	350	315	298
5	[illegible]	294	[illegible]	278	272
6	2015-07-23 12:54:26	272	269	267	261
7	[illegible]	[illegible]	342	[illegible]	290
8	2015-07-23 12:54:30	284	281	273	271
9	[illegible]	273	272	270	264
10	2015-07-23 12:54:31	263	261	261	260
11	[illegible]	249	244	241	240
12	2015-07-23 12:54:32	254	252	253	232
13	[illegible]	246	242	241	240
14	2015-07-23 12:54:34	260	258	257	250
15	[illegible]	254	253	[illegible]	253
16	2015-07-23 12:54:35	254	253	254	253
17	[illegible]	245	241	239	239
18	2015-07-23 12:54:36	261	259	259	252
19	[illegible]	260	252	257	[illegible]
20	2015-07-23 12:54:38	255	254	255	253
21	[illegible]	261	260	259	254

Figure 6. UI displaying sensor values from server.

someone having a valid access rights. It gives the caregivers and health care providers to monitor patient's data on real time basis or even later.

The Graphical User Interface (GUI) that was designed also has the option of displaying the data based on a user's choice. The GUI provides choice of data filtering based on one to many days, weeks or even month. Figures 10 and 11 displays the screenshot of the system designed for displaying gait data for a date selection. Based on user's input, stored gait repository is searched for the gait data and is displayed in the form of graph. The pattern presented in the graph helps doctors and physicians in making decision on data and to graphically view a large dataset.

4 RESULTS AND DISCUSSIONS

In this section we briefly showcase the gait data collected from the sensors and analyze them. As mentioned earlier, we created a gait data repository from volunteers. These initial data helped us to determine the threshold pressure value to differentiate a normal gait pattern from an impaired one. Figure 8 shows different episodes of human gait

Figure 7. Pressor sensor data plotted against time (10 samples per second) showing FoG.

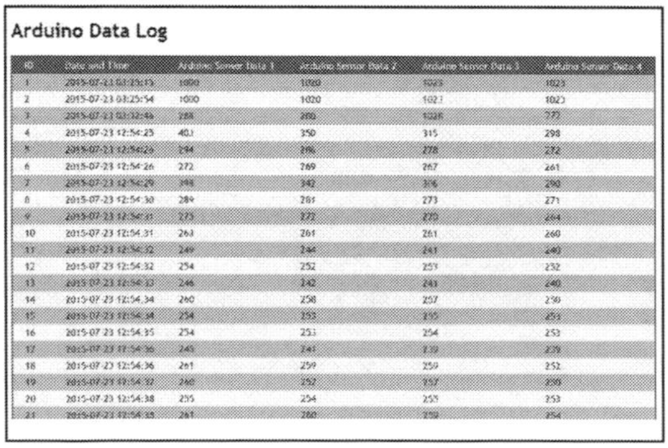

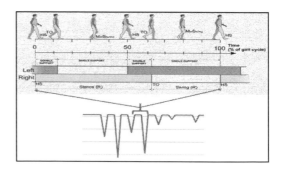

Figure 8. Gait cycle identified from pressure sensor data [Photo courtesy: ref.3].

and corresponding sensor data pattern collected from volunteers. In order to map the gait episodes to the recorded sensor data, we illustrate a complete gait cycle in the diagram. During the experiment we observed the subject's gait pause and noted the pattern of the graph that was being generated by the sensor data in real-time. The graph shown in Figure 7 confirmed that the rapid pressure variations in the data values is due to the gait pause. During this time he tries to continue his motion but is not being able to complete the gait properly and a rapid shaking and trembling is noticeable. This results in a quick change of a pressure value giving the thick lines in the shown graph. Apart from this the crest and troughs in the graphs is generated because of the pressure exerted during walking and having normal footsteps.

We used samples from gait data repository for analyzing gait behavior of the simulated patient volunteers. From these derived values we plotted a histogram in Figure 5 to study the frequency and range of the pressure values that is prevalent in these data.

These statistics were used to determine a good choice of threshold. Figure 5 shows the histogram

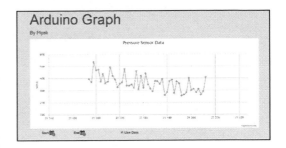

Figure 11. UI for graphical display of data and Live data display functionality.

that confirms the threshold value being 950; considering the FoG instances from the datasets we generated. The reason behind choosing the threshold value of 950 is that the occurrence of sensor value samples hence the information contained in the data beyond 950 is more frequent during FoG. That implies this is a critical value to for the dataset considered.

5 CONCLUSION

The work herein presented is a smart PD patient monitoring system that leverages wearable sensor technology, WiFi connectivity, and a client-server application to collect and analyze gait movement data. Our endeavor results in a smart shoe that demonstrates the capability of analyzing a wearable sensor data, detecting FoG and providing visual cue automatically. SCARP, could be considered for further upgrading by adding artificial intelligence into it. We can implement learning mechanism for personalized FoG detection with minimal latency in providing visual cue. The system can be even tested with a large number of PD patients and the huge gait movement data thus generated could be gathered in an online repository as a large variety of database for further gait data analysis and pattern mining.

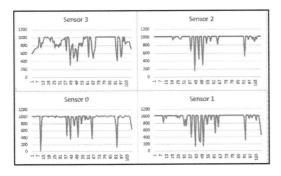

Figure 9. Variation in gait data collected simultaneously.

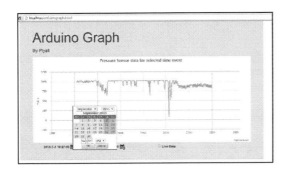

Figure 10. UI for graphical display of data based on date selection.

REFERENCES

Caldara, M. et al., "A novel Body Sensor Network for Parkinson's disease patients rehabilitation assessment", Wearable and Implantable Body Sensor Networks (BSN), 11th International Conference Zurich, published by IEEE, 2014, pp. 81–86.

Chen, Y. et al., "The Study on A Real-time Remote Monitoring System for Parkinson's Disease Patients with Deep Brain Stimulators", Engineering in Medicine and Biology Society (EMBC), 36th Annual International Conference of the IEEE, 2014, pp. 1358–1361.

Hoehn, M.M., Yahr, M.D. "Parkinsonism: onset, progression, and mortality", Neurology, 1967, Vol. 17, Number 5 Internet: http://www.neurology.org/content/17/5/427.full.pdf, [Feb 25, 2015].

Jamthe, A. S. Chakraborty, S.K. Ghosh, D.P. Agrawal, "An implementation of Wireless Sensor Networks in monitoring of Parkinson's Patients using Received Signal Strength Indicator", Distributed Computing in Sensor Systems (DCOSS), IEEE International Conference, 2013, ISBN: 978-1-4799-0206-4 (Print Version), pp. 442–447.

Kapandji, I.A. "The Physiology of the Joints",6th edition, Churchill Livingstone, Vol. 2, p.199.

Mariani, B., Jimenez, M., Vingerhoets, F., Aminian, K. "On-shoe wearable sensors for gait and turning assessment of patients with Parkinson's disease", Biomedical Engineering, IEEE Transactions, 2013, vol. 60, Issue 1, pp. 155–158.

Mazumder, O., Kundu, A.S., Bhaumik, S. "Development of Wireless Insole Foot Pressure Data Acquisition Device",International Conference onCommunications, Devices and Intelligent Systems (CODIS), IEEE, Kolkata, 2012, ISBN: 978-1-4673-4699-3 (Print Version), pp. 302–305.

Mitra, R.N., Das, P., Espay, A., Agrawal, D.P. "Recognition of Gait Impairment and Data Analysis for Parkinson's Patients Using Low-cost Sensor Based Embedded System", 36th Annual International IEEE EMBS Conference, 26 Aug'14 to 30 Aug'14, Chicago, USA.

Patel, S. et al., "Home Monitoring of Patients with Parkinson's Disease via Wearable Technology and a Web-based Application", EMBC, Annual International Conference of the IEEE, 2010, pp. 4411–4414.

Rempark Project Website, Internet: http://rempark.cetpd.upc.edu/project [Jul 05, 2015].

Soames, R.W. "Foot Pressure Patterns during Gait", Journal of Biomedical Engineering, 1985, Vol. 7, Issue 2, pp. 120–126.

Stott, J.R.R., Hutton, W.C., and Stokes, I.A.F. "Forces under the foot," The Journal of Bone and Joint Surgery, London, England, May 1973, vol. 55-B, no. 2, pp. 335–344.

Unified Parkinson's Disease Rating Scale, Internet:http://www.etas.ee/wp-content/uploads/2013/10/updrs.pdf, [Feb 20, 2015].

University of Glasgow, "Definition and Description of Gait Cycle", Internet: http://www.gla.ac.uk/t4/~fbls/files/fab/tutorial/anatomy/hfgait.html.

A survey on cloud computing and networking in the next generation

Piyali Das
The Ohio State University, Ohio, USA

Rupendra N. Mitra
University of Cincinnati, Ohio, USA

ABSTRACT: In this paper, we consolidate on the numerous research and development, whichendeavors towards seamless convergence of cloud computing, telecommunication, and networking. The advanced telecommunications and wireless networking technologies could further be accelerated successfully with the overlapping from cloud computing. Classically, networking, especially wireless and mobile networking, and cloud computing, have been taken under consideration by scientific literatures but through separate approaches. Recently, the need of virtual resource sharing is felt important through cloud computing in mobile networking due to the enhanced requirements of capacity and quality of services. As a result, we can see a large number of efforts are made to converge with wireless technology and cloud computing. It gives rise to the Service Oriented Architecture (SOA) for the purpose of network virtualization thereby making Network as a Service (NaaS) a prospective dimension of cloud based networking. This survey comprehensively summarizes the endeavors from the perspective of next generation cellular networking and also takes into account the open research areas to get the utmost benefit from the amalgamation of telecommunication and cloud computing.

1 INTRODUCTION

In recent years, cloud computing has become one of the biggest area of research and technological advancement. On the other hand, telecommunication and wireless networking industry have become an highly advanced industry catering to many people worldwide. Merging of wireless networks and cloud has helped in understanding the Mobile Cloud Computing (MCC) (Zin *et al.* 2015). As the demand of connected world is increasing, telecommunication and networking industry have faced several challenges that are addressed by the flexibility offered by cloud computing. Cloud computing is an architectural platform to get access to and share resources like software, applications, storage, or services by remote users. This special computing technique makes these resources available to users on basis of demand. The basic characteristic of the cloud enables virtualization of computing resources, thereby creating a partition among the use of the actual physical resources to map the logical functions. This abstract layer is responsible for this mapping of logical entities with that of the essential physical components.

As a result of persistent issues of connectivity control, trust, security, privacy, and local presence the cloud service faces numerous challenges. We are fortunate enough to have different network service providers to address these concerns. Different constraints of wireless networking like spectrum unavailability, resource allocation, capacity enhancement, energy efficiency, security, scalability, flexibility etc., have tried to overcome with proper integration of cloud technologies in the recent literature. We are on the verge of the 5th generation of cellular networking, which will cater to more than 50 billion connected devices and ten thousand times of the data requirements (Ericsson white paper, 2015). So in this article, we will discuss briefly how cloud computing is going to converge with the next generation of wireless networking.

After the introduction, this paper is organized in the following five sections. Section two and three gives the overview of the cloud architecture and Software Defined Networking (SDN) respectively. Section four describes the Service Oriented Architecture (SOA). The subsequent subsections further describe the advancements in different aspects of SOA. Since, it is an important part of modern wireless networks, as a part of service we have provided a detailed insight of Radio Access Network (RAN). Section five covers the security aspects of the futuristic converged system. Finally, we conclude this letter with section six.

2 OVERVIEW OF CLOUD COMPUTING

Following the footsteps of cloud computing, sharing of files, storing data along with processing power distribution, we have experienced a new dimension altogether. The three models of services that are inherent in cloud computing are Software as a Service (SaaS), Platform as a Service (PaaS), and Infrastructure as a Service (IaaS). They act as the business models of cloud computing. In essence, SaaS enables consumers to use applications that run on a cloud infrastructure. SaaS applications uses the web to provide applications that are managed by a third-party vendor. Google Apps for Work could be an example of SaaS. With PaaS, consumers can create and deploy applications onto the cloud infrastructure. PaaS provides the computing platform like Operating System, Database, Web Server, environment for running an application etc. It acts as a middleware system to help create applications using software components like Apprenda is a PaaS provider for Java. IaaS, on the other hand, provides the user with computing infrastructure, physical machines, or virtual machines, and resources. It acts as the remote delivery center or model of computing resources to let the users' access, monitor, and use resource center facilities like CPU, networking service, storage, and other essential resources. It enables the user to run software's like Operating System or deploy other applications. Amazon Web Services (AWS), Microsoft Azure are among the IaaS provider.

3 SOFTWARE DEFINED NETWORKING

The next generation network is truly going to be 'open'. That means no proprietary software will be used to ensure the open programmability of the networks. This will also help in getting rid of vendor specific hardware incompatibility issue. And this 'openness' is offered by Software Defined Networking (SDN).

3.1 What is new in SDN?

Software Defined Networking is a concept that cumulatively considers all software that is responsible in running a network and supports other applications. There are some network protocols and policies that together form SDN that enables the network to be open and programmable. SDN acts as the whole ecosystem. This ecosystem comprises of controller, network application, data plane or network forwarding devices, network services that manages the network as shown in Figure 1. The forwarding devices embedded with cached memory

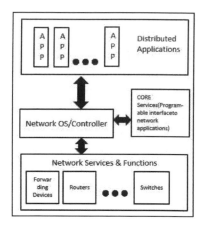

Figure 1. SDN architecture.

makes packet forwarding way faster through SDN. In SDN the network has been logically separated into controller and the data plane and yet it has a global view of the entire network (Banikazemi *et al.* 2013). The Network Operating System (NOS) or controller is the middleware of the SDN network. We will discuss the NOS in more details.

3.2 Need of agile networking

The type of networks we have today is sufficient but it needs to evolve. The major demand of such technology comes from the need of having custom security policy, constant monitoring and testing of network, use of architecture and features from multiple domains (Essens *et al.* 2007). An agile network also requires the system to be fault tolerant so that they can recover from both hard and soft failures, which can be due to fiber out or link failure, or other reasons. The future networking model needs to adapt network virtualization, become scalable with more traffic, store data in data repository with transparency and is enabled to use the distributed applications concurrently for ease, robustness, and faster execution of processes. This can help the network to evolve as fast as the software.

3.3 Network operating system

Network Operating System (NOS) is a special type of operating system dedicated to computer networking. NOS can be fed onto a router or other network component that conducts the functions of network layer. As shown in Figure 1 the NOS is responsible for coupling network services and distributed applications. It controls and manages the packet forwarding devices, routers, switches, and other hardware components to actually make a decision about forwarding a packet from source

276

to destination (Gude *et al.* 2008). Major feature of the NOS is the abstraction it provides to the other network components and the applications that interact with this SDN. It also maintains a global view of the entire network. The SDN controller can be used to serve multiple purposes like Topology service, Host tracking, Inventory services etc. It also manages the interaction of the north bound APIs above it. These API could be the Java API for making the network programmable. This makes the network open and programmable interface and easy to customize the network.

4 SERVICE-ORIENTED VIRTUAL NETWORKING

Service Oriented Architecture (SOA) is a powerful tool, which is used in network virtualization and cloud computing. SOA is a software designed approach in which the apps provide services independently to physical or other components connected through wide network via robust communication protocols. Web Service is a true example of SOA architecture. In this architecture we have a service provider, a registry, and a service consumer. For example, a web service is following the Simple Object Access Protocol (SOAP), where we have the service provider having its service available and published in the UDDI registry, which is then accessed by any service consumer.

The telecom industry has segregated the service based functionality and traffic transportation mechanism. Next Generation Network (NGN) and IP based network are the pivotal landmark in the future of telecom. NGN follows the above mentioned dissociation of the two layers, which actually creates the platform of virtual infrastructure. This effort has been made to make the system open rather than the traditional network, which was cohesive and was closely interdependent among the individual components that used to make the entire network difficult to manage. Use of some API's like Java Remote Method Invocation (RMI), Open Service Architecture (OSA), and Parlay are some of the intelligent ways to make a programmable network and efficient network resource abstraction.

4.1 *Cloud networking*

The next generation of networking is architecture, which aims at separating and abstracting the network controller and packet forwarding scheme and making the network easily scalable and programmable. For this, the SDN cloud based architecture plays a significant role. This model (Azodolmolky *et al.* 2013) includes:

- OpenFlow based peripheral or edge nodes that connects to the data center via a cloud backbone infrastructure
- Some core nodes that is responsible for traffic between these edge nodes,
- SDN controller to manage the forwarding of packets via the cloud backbone and achieve network virtualization,
- A software for resource management, data storage, retrieval process, and efficient utilization of network bandwidth.

4.2 *Networking-as-a-Service (NaaS)*

To virtualize the network the service providers need to bind the network services together into network infrastructure services and abstract it together. Till date network resources has been underutilized. Hence, sharing of resources was necessary and is done via Infrastructure-as-a-Service (IaaS). Network-as-a-Service (NaaS) is a platform that allows the network parameters to be controlled at real-time. In the NaaS domain, the end users of this virtual network access these platform and services without knowing the physical existence of the network devices and other infrastructure. The service providers are indeed responsible for this mapping of various on-demand network services with actual network resources. This is similar to the virtual mapping of cloud services to computing resources.

4.3 *Mobile Network-as-a-Service (MNaaS)*

Another significant role in cloud computing is provided by the network operators. They are responsible of providing reliable and low cost service to the end users. As mentioned in [1] Mobile Network-as-a-Service can be used to benefit low cost services with the help of mobile cloud computing and utility payment model. The Telecom Operator Cloud (TOC) acts as cloud connectivity management provider. At their end they apply both SaaS and PaaS model to become the cloud broker. TOC performs cloud connectivity, deliver cloud services, and efficiently utilize cloud assets which thereby makes user cost dynamic. A Schematic diagram of network cloudification has been shown in Figure 2.

4.4 *SoftRAN: RAN-as-a-Service*

This section summarizes the latest advancements on Cloudification of Radio Access Networks (C-RAN) of cellular communication system. This technology is expected to be prevalently available with the release of 5G telecommunication standards. All the cellular service providers are experiencing

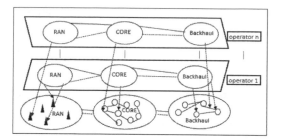

Figure 2. Schematic diagram of network cloudification.

an exponential increase in their capex (capital expenditure) and opex (operating expense).

Radio Access Network (RAN) is an important segment of cellular network. It provides wireless connectivity to the user equipped by radio coverage over a wide geographical region. Managing interference, balancing load among neighboring cells, and ensuring seem-less handovers becomes more difficult for the conventional RAN with distributed protocols and node architecture with the growth of network. This is where the concept of Cloud RAN or SoftRAN comes to play. C-RAN, as an integrated part of next generation access network, can minimize the operators' costs by eliminating most of the challenges faced by legacy RANs mentioned above. Moreover, the subscribers will get an uninterrupted services on demand.

China Mobile Research Institute introduced a cloud computing based architecture of RAN, unlike the conventional decentralized RAN (China Mobile 2010). It envisaged a centralized control plane protocol to manage the RAN as a single entity instead of the traditional decentralized protocols to control each node independently. Gudipati *et al.* argued for the convention architecture as suboptimal for Key Performance Indicators' (KPI's) of RAN and proposed their own framework SoftRan (Gudipati *et al.* 2008).

Moreover, the RAN part is gradually becoming more dense, chaotic, and heterogeneous in nature after the rapid implementations of small, pico, and femto cells. Since the cell radii are decreasing in dense network, there are more chances of inter-cell interference due to overshooting. The smaller the cells the faster the requirement of load balancing. Hence, to reduce the cost cellular service providers considered cloudification to form a much larger processing resource pool shared in a large geographical area.

4.4.1 *Key concept of C-Ran*
They key concept behind this disruptive C-RAN technology is to virtually unify all the RAN nodes (which includes the base stations, their base

band units, and inter node links) of an area in a centralized control plane. Then, logically allocate the radio resources as per the demand of a cell and improve the other KPIs viz. Handover, signal to interference ratio, throughput etc. The next generation access technology thus, will be able to serve the following needs.

- Virtual pool of Base-band unit: A pool of all available radio resources and nodes is logically formed at a centralized server. Now the processing resources and capabilities are dynamically allocated and reconfigured based on the real-time demands [3].

- SOARAN: He and Song presented SOARAN (Jun *et al.* 2015), a service oriented framework for RAN. SOARAN provides a set of abstract applications, which are configurable with respect to differentiated Quality of Experience (QoE) requirements. This may help operators to map their applications to the abstract pool of applications and allocate required bandwidth. So, it takes care of lower-layer resource allocation and makes radio allocation responsibilities by enabling resource virtualization.

- Support to multi-air interface Standards: Next generation access technology must support different air interfaces standards. Provision for easy software upgradation will be another feature.

4.4.2 *Architecture*
Many well-known companies and research bodies are engaged in C-RAN project to establish RAN-as-a-service. From operators' point of view spectrum utilization, unified RAN management and cost reductions are major issues to be address by the C-RAN architecture. Senjie *et al.* 2012, suggested a C-RAN architecture with three logical planes (service plane, control plane and PHY plane) that supports multi-mode (WiMax, LTE, WCDMA etc.) air interface. Similarly, Lin *et al.* 2009, reported their QoS report which is pretty impressive in terms of radio KPIs. Hong *et al.* 2012, implemented and reported PICASSO, a novel spectrum slicing that allows simultaneous trans-receiving using single RF front end circuitry and antenna system.

4.4.3 *Convergence of RAN and Cloud*
Virtualization technologies are well practiced in modern data centers. Network Function Virtualization (NFV) mainly consolidates the network nodes, links, and other resources. This cloud computation inspired the C-RAN for radio resource and computational resource cloudification so that dynamic allocation of radio capacity (base-band resource) can be done in more real-time and in an optimized way.

4.4.4 *Challenges & Advantages*

There are few challenges C-RAN may face and one of those is higher compression ratio for control and data plane. Because the resources are shared over the network connected with limited bandwidth links. Hence, higher rate of compression is required to provide more real-time services. Again, concept of virtualization is simple but actual implementation in a live network is more difficult. For an example, cloudification of multi-protocol, multi-vendor, multi-standard IT services requires a novel design of additional system accelerator. This accelerator in conjunction with web based cache, will make the computation-intensive Real-Time Operating System (RTOS) function properly (Primet *et al.* 2007). Moreover, optimization of the RTOS and the virtualization micro-management system are yet to be functional.

5 SECURITY AND THREAT ADMINISTRATION

Cloud computing offers a number of benefits and advantages as well as poses serious threats in different areas. Martucci *et al.* 2012, have discussed various trust and security problems that the communication industry had observed. Protection of data becomes a genuine issue in cloud architecture. Data retention, data availability, and legal complications are related to a flow of data beyond territory. Trust relationship among the cloud providers and their customers are also at stake in terms of security and privacy. Since, no more roaming and feebly international boundaries are considered, the cost policies that were to be charged to the customers is an issue as well. Most noteworthy of them is the trust worthy functionality of each network component.

To handle these security challenges a number of protocols like cryptographic techniques, encryption, and other regulatory acts have been enforced. (Modi *et al.* 2013) As transfer rate increases, IDS (intruder Detection System), which are software based, struggle to keep up and also requires more intelligence classifying flows. According to custom security policies, along with constant network testing, monitoring is also required.

6 CONCLUSION

This paper underlines that cloud computing solutions are obvious for the wireless networking and especially for telecom industry in near future. We brief different aspects of cloud computing, next generation network virtualization approaches, and research directions. However, we have also discussed the challenges like device privacy in cloud, security of the network etc. So, cloud computing solutions are to be converged with and deployed by telecommunication providers in the next generation network deployment. Telecommunication technology is awaiting its next generation to appear by 2020 and that is going to be a zero latency gigabit service oriented network (Nokia White Papers 2015). This article also surveys the cloud-migration and possibly the delivery model in multi-operator or multi-cloud environment.

Thus, overlapping among different areas like telecom, networking, cloud computing etc., provides a diction towards telecom-cloud convergence that will revolutionize cloud-based future networks.

REFERENCES

Azodolmolky, S., Wieder, P., and Yahyapour, R. "SDN-Based Cloud Computing Networking," Transparent Optical Networks (ICTON), Cartagena, 2013.

Banikazemi, M., Olshefski, D., Shaikh, A. *et al.*" Meridian: an SDN platform for cloud network services," IEEE Communications Magazine, vol. 51 no. 2, pp. 120–127, Feb. 2013.

Ericsson White Paper, Website: http://www.akosrs.si/files/Telekomunikacije/Digitalna_agenda/Internetni_protokol_Ipv6/More-than-50-billion-connected-devices.pdf, Sep. 2015

Essens, P., Spaans, M., and Treurniet, W., "Agile networking in command and control," The International C2 Journal, 2007.

Gude, N., Koponen, T., Pettit, J. *et al.*, "Nox: Towards an operating system for networks," ACM SIGCOMM Computer Comm. Review, vol. 38 no.3, pp.105–110, 2008.

Gudipati, A., Perry, D., Li, L.E., & Katti, S. SoftRAN: Software defined radio access network. In Proceedings of the second ACM SIGCOMM workshop on Hot topics in software defined networking (pp. 25–30). ACM, 2013.

He, Jun, and Wei Song. "SOARAN: A Service-oriented Architecture for Radio Access Network Sharing in Evolving Mobile Networks." arXiv preprint arXiv:1508.00306 (2015).

Hong, Steven S., Jeffrey Mehlman, and Sachin Katti. "Picasso: flexible RF and spectrum slicing." ACM SIGCOMM Computer Communication Review 42, no. 4 (2012): 37–48.

Martucci, L.A., Zuccato, A., Smeets, B. *et al.* "Privacy, Security and Trust in Cloud Computing: The Perspective of the Telecommunication Industry," Ubiquitous Intelligence & Computing UIC/ATC, Fukuoka, Sep. 2012.

Modi, C., Patel, D., Patel, H. *et al.*, "A survey of intrusion detection techniques in Cloud," Journal of Network and Computer Applications, vol. 36 no. 1, pp. 42–57, 2013.

Mobile, China. "C-RAN Road towards Green Radio Access Network." C-RAN Int'l. Wksp (2010).

Online resources, 2015 Nokia 5G white papers: http://networks.nokia.com/file/28771/5 g-white-paper

Primet, Pascale, Jean-Patrick Gelas Vicat-Blanc, Olivier Mornard, Dinil Mon Divakaran, Pierre Bozonnet, Mathieu Jan, Vincent Roca, and Lionel Giraud. "HIPCAL: State of the Art of OS and Network virtualization solutions for Grids." Sep 14 (2007): 38.

Siyu Lin; Zhangdui Zhong; Bo Ai, "Quality of Service Evaluation Scheme for Heterogeneous Wireless Network", Wireless Communications, Networking and Mobile Computing, 2009. WiCom '09. 5th International Conference on, Beijing, 2009, pp. 1–4.

Yin, Z., Yu, F.R., Shengrong, B., and Zhu, H. "Joint Cloud and Wireless Networks Operations in Mobile Cloud Computing Environments with Telecom Operator Cloud," IEEE Trans. on Wireless Communications, vol. 14, no. 7, July 2015

Zhang Senjie; Yang Xuebin; Liao Fanglan; Ngai Tinfook; Zhang, S.; Kuilin Chen, "Architecture of GPP based, scalable, large-scale C-RAN BBU pool", Globecom Workshops (GC Wkshps), 2012 IEEE, Anaheim, CA, pp. 267–272, 2012.

280

Frontiers in Computer, Communication and Electrical Engineering – Acharyya (Ed.)
© 2016 Taylor & Francis Group, London, ISBN: 978-1-138-02877-7

Tilaiya reservoir catchment segmentation using hybrid soft cellular approach

Kalyan Mahata
Government College of Engineering and Leather Technology, Kolkata, West Bengal, India

Rajib Das
School of Water Resources Engineering, Jadavpur University, Kolkata, West Bengal, India

Anasua Sarkar
SMIEEE, Government College of Engineering and Leather Technology, Kolkata, West Bengal, India

ABSTRACT: Image segmentation among overlapping land cover areas in satellite images is a very crucial task. Detection of belongingness is the important problem for classifying mixed pixels. This paper proposes an approach for pixel classification using a hybrid approach of Fuzzy C-Means and Cellular automata methods. This new unsupervised method is able to detect clusters using 2-Dimensional Cellular Automata model based on fuzzy segmentations. This approach detects the overlapping regions in remote sensing images by uncertainties using fuzzy set membership values. As a discrete and dynamical system, cellular automaton explores the uniformly interconnected cells with states. In the second phase of our method, we utilize a 2-dimensional cellular automaton to prioritize allocations of mixed pixels among overlapping land cover areas. We experiment our method on Tilaiya Reservoir Catchment on Barakar river for the first time. The clustered regions are compared with well-known FCM and K-Means methods and also with the ground truth knowledge. The results show the superiority of our new method.

1 INTRODUCTION

Cogalton and Green in 1999, defined remote sensing as "the art and science of obtaining information about an object without being in direct physical contact with the object" (Cogalton & Green, 1999). Canopy of methods exist for classifying pixels into known classes (for example, an urban area or turbid water) in satellite images. Theoretically, a remote sensing image can be defined as a set,

$$\mathcal{P} = \{p_{ijk} \mid 1 \le i \le r, 1 \le j \le s, 1 \le k \le n\} \quad (1)$$

of $r \times s \times n$ information units for pixels, where $p_{ij} \in \{p_{ij1}, p_{ij2}, \dots p_{ijk}\}$ is the set of spectral band values for n bands related with the pixel of coordinate *(i, j)*. In order to find similar regions, we segment this image by fuzzy sets, that considers both the spatial image objects and the imprecision attached to them.

Let $\mathcal{P}$ (usually $\mathbb{R}^n$ *or* $\mathbb{Z}^n$) denotes the space of the remote sensing image. Consequently, the points of $\mathcal{P}$ (pixels or voxels) are the spatial variables x, y. Let $d_{\mathcal{P}}(x, y)$ denotes the spatial distance between two pixels $\{x, y\} \in \mathcal{P}$. In existing works, $d_{\mathcal{P}}$ is taken as the Euclidean distance on $\mathcal{P}$ (Maulik & Sarkar, 2012), (Bandyopadhyay, 2005).

In the remote sensing image, a crisp object $\mathcal{C}$ is a subset of $\mathcal{P}, \mathcal{C} \subseteq \mathcal{P}$. Hence, a fuzzy object is defined as a fuzzy subset $\mathcal{F}$ of $\mathcal{P}, \mathcal{F} \subseteq \mathcal{P}$. This fuzzy object $\mathcal{F}$ is defined bi-univoquely by its membership function, μ. $\mu_{\mathcal{F}}(x) \in (0,1]$ is known as the membership function, which represents the membership degree of the point x to the fuzzy set $\mathcal{F}$. When the value of $\mu_{\mathcal{F}}(x)$ is closer to *1*, the degree of membership of x in $\mathcal{F}$ will be higher. Such a representation allows for a direct mapping of mixed pixels in overlapping land cover areas in remote sensing images. Let $\mathcal{F}$ denotes the set of all fuzzy sets defined on $\mathcal{P}$. For any two pixels x, y, we denote $d_{\mathcal{F}}(x, y)$ as their distance in fuzzy perspective. The definition of a new method utilizing the cellular automata over fuzzy segmentation solutions is the scope of this chapter.

In the unsupervised classification method, clustering is based on maximum intra-class similarity and minimum inter-class similarity. State-of-the-art clustering methods are–K-Means clustering (Hoon et al., 2004), simulated annealing (Lukashin & Futchs, 1999), graph theoretic approach (Xu, 1999), fuzzy c-means clustering (Dembele, 2003) and scattered object clustering (de Souto, 2008). Different perspectives methods like the supervised multi-objective learning approach (Maulik & Sarkar, 2012), also efficiently detect arbitrary shaped land cover areas in satellite images.

The membership functions of soft computing approaches like fuzzy sets, also efficiently detects overlapping partitions. Fuzzy set theory is a methodology to illustrate how to handle uncertainty and imprecise information in the dataset. The fuzzy models have been experimented for land cover detection of remote sensing images and pattern recognitions (Bandyopadhyay, 2005), (Dave, 1989). Applying the concepts of fuzzy membership function (Wang, 1997), (Pappis, 1993), fuzzy clustering (Huang, 2008), and fuzzy-rule based systems (Bardossy, 2002) in algorithms, the remote sensing image identification becomes more feasible.

A cellular automaton is a discrete and dynamical system composed of a very simple, uniformly and interconnected cells. A cellular automaton is a well-known method to detect states in cellular spaces. Therefore, to predict pixel classification of remote sensing imagery, we propose a 2-dimensional cellular automata to fuzzy set based initial clustering.

The present study focuses on the integration of fuzzy-set theoretic optimal classification with cellular automata based neighbourhood priority correction for pixel classification in remote sensing imagery. We demonstrate the performance of the new distance metric in pixel classification of a chosen LANDSAT remote sensing image of the catchment area of one Indian River. The quantitative evaluation over three existing validity indices indicates the satisfactory performance of our new hybrid Fuzzy set based and Cellular Automata (CA) Corrected Algorithm (FCA) to detect imprecise clusters. We compare our obtained solutions with those of K-Means and FCM algorithms to verify with the ground truth knowledge. The statistical tests also demonstrate the significance of our new FCA algorithm over K-Means and FCM algorithms.

2 FUZZY C-MEANS ALGORITHM

Clustering is an unsupervised pattern classification method based on maximum intra-class similarity and minimum inter-class similarity. In a well-known partitional clustering approach, named fuzzy clustering, points may belong to more than one cluster. Therefore, for each point in a cluster, one set of membership levels is associated. This set of levels indicates the amount of association between the point and each of the clusters. One of the most widely used fuzzy clustering algorithms is the Fuzzy C Means algorithm. Fuzzy set theory was introduced in 1965 by Zadeh (Zadeh, 1965)

as a mean to model the vagueness and ambiguity in complex systems. Fuzzy set theory handles the concept of partial membership to a set, with real valued membership degrees ranging from 0 to 1.

Introduced by Ruspini (Ruspini, 1970) and improved by Dune and Bezdek (Dunn, 1974) (Bezdek, 1981), the Fuzzy C-Means (FCM) algorithm partitions a finite dataset $X = \{x_1, x_2, ..., x_N\}$ into a collection of K fuzzy clusters, satisfying criterions (Reddi, 1984). Let m be the exponential weight of membership degree, $m \in (1, \infty]$. The objective function $W_{m\mu}$ of FCM is defined as:

$$W_m(U,C) = \sum_{i=1}^{N} \sum_{j=1}^{K} (\mu_{ij})^m (d_{ij})^2 \qquad (2)$$

Where μ_{ij} is the membership degree of point x_i to centroid C_j and d_{ij} is the distance between x_i and c_j. Let $U_j = (\mu_{1j}, \mu_{2j}, ..., \mu_{Kj})^T$.

Then $U = (U_1, U_2, ..., U_N)$ is the membership degree matrix and $C = (c_1, c_2, ..., c_K)$ is the set of cluster centroids. W_m indicates the compactness and uniformity degree of clusters. Generally, a smaller W_m reflects a more compact cluster set.

The algorithm of FCM is an iteration process mathematically described as follows:

1. Initialize m, M and initial cluster centroid set $C^{(0)}$. Set the iteration terminating threshold ε to a small positive value and iteration time q to zero. Calculate $U^{(0)}$ according to $C^{(0)}$ with the following equation:

$$\mu_{ij} = \frac{1}{\sum_{k=1}^{K} \left(\dfrac{d_{ij}}{d_{ik}} \right)^{\frac{2}{(m-1)}}} \qquad (3)$$

where $\sum_{j \in Ck, k=1,...,K} \mu_{ij} = 1$. If $d_{ij} = 0$, then $\mu_{ij} = 1$ and sets $\mu_{ik, k \neq j} = 0$ for membership of this pixel to other clusters.

2. Update $c^{(q+1)}$ according to $U^{(q)}$ with the following equation:

$$c_j = \frac{\sum_{i=1}^{N} (\mu_{ij})^m x_i}{\sum_{i=1}^{N} (\mu_{ij})^m} \qquad (4)$$

3. Calculate $U^{(q+1)}$ according to $c^{(q+1)}$.
4. Compare $U^{(q+1)}$ with $U^{(q)}$. If $\| U^{(q+1)} - U^{(q)} \| \leq \varepsilon$, stop iteration. Otherwise, go to (2).

3 CELLULAR AUTOMATA METHOD

Cellular Automata (CA) are defined to be the discrete spatially-extended dynamical systems to study models of physical systems (Smith, 1971). It evolves the computational devices in discrete space and time. A CA, which is seeded with any state from the set of states with all 0 and single 1 at different position, generates a fixed number of unique patterns.

Stephen Wolfram (Wolfram, 1986) proposes its simplest CA in a form of a spatial lattice of cells. Each cell stores a discrete variable at time t that refers to the present state of the cell. The next state of the cell at $(t + 1)$ is affected by its state and the states of its neighbors at time t. In the current work, we consider 3-neighborhood (self, left and right neighbors) CA, where a CA cell is having two states, either 0 or 1. The next state of each cell of such a CA is

$$S_i^{t+1} = f(S_{i-1}^t, S_i^t, S_{i+1}^t) \qquad (5)$$

where f is the next state function. S_{i-1}^t, S_i^t and S_{i+1}^t are the present states of the left neighbor, self and right neighbor of the i th CA cell at a time t. The f can be expressed as a look-up table as shown in Table 1. The decimal equivalent of the 8 outputs in Table 1, is called 'Rule' R_i (Wolfram, 1983). In a two-state 3-neighborhood CA, there can be a total of 2^8 (256) rules.

One such rule is 30 in Table 1. Rule 30 CA can generate a sequence of random patterns. The scientists observed in this experiment is that an n-cell rule 30 CA, seeded with a state all 0 and single 1, generates a state with fair distribution of 0 and 1 after n iterations. However, it is not guaranteed that after n iterations not all 0 pattern will come. It can be proved that for every, $n > 1$, an n-cell rule 30 CA seeded with any state with all 0 and single 1 at different positions, generates a non-zero states after $n/2$ iterations. Using Wolfram's classification scheme, rule 30 is a class III rule (Wolfram, 1986), displaying a periodic and chaotic behavior. Rule 30 of the elementary Cellular Automata (CA) was among the first rules, in which Stephen Wolfram noticed the appearance of intrinsic randomness in a deterministic system. When initialized with a single black pixel there is a pattern behavior down both sides of the unfolding CA which gives way to the randomly patterned center, is shown in Figure 1. This rule is of particular interest because it produces complex, seemingly-random patterns from simple, well-defined rules. In fact, Mathematica

uses the center column of pixel values as one of its random number generator.

If the leftmost and right most cells are neighbors of each other, the CA is defined to be with periodic boundary; otherwise it is a null boundary CA. Cellular automata can further be divided as deterministic and probabilistic (or, stochastic). Elementary CA as stated above is deterministic in nature. On the other hand, in case of probabilistic CA, the next state of each cell is updated based on not only the present states of its neighbors but also on a predetermined probability.

4 HYBRID FUZZY-CA CLUSTERING ALGORITHM

The new hybrid FCA algorithm consist of two phases–initial FCM clustering of remote sensing image to generate Fuzzy membership matrix U and fine-tuning using the cellular automata based on neighborhood priority correction method, is shown in Figure 2.

Initial random assignment put N pixels in K clusters for initializing FCM algorithm, as described in previous subsection. Then, we obtain the initial cluster centroids $C^{(0)}$. The iteration terminating threshold value ϵ is set to 1E–05. We initialize the membership degree matrix U from the initial random allocations. Then, we repeat the centroid updation method iteratively and compute the membership degree matrix $U^{(q)}$ for each of the q iterations. The iterations converge and stop, when the difference between the membership degree matrix in previous and current iterations becomes less than the iteration-terminating threshold.

After the first phase of FCM algorithm, we compute the new cellular automata based on neighborhood priority correction method over all pixels as described in next subsection. Depending on the final CA-corrected solutions, the validity indices values are also calculated.

4.1 Cellular automata based on neighborhood priority correction method

In our 2-dimensional CA model for neighborhood based priority correction, we have considered the

Table 1. Look-up table for rule 30.

Present state	111	110	101	100	011	010	001	000
Rule	(7)	(6)	(5)	(4)	(3)	(2)	(1)	(0)
Next State	0	0	0	1	1	1	1	0

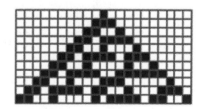

Figure 1. Example of Wolfram rule 30.

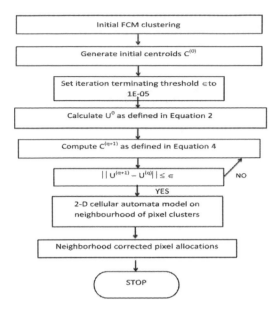

Figure 2. The flowchart of FCA algorithm for remote sensing classification.

states of the cells in each CA according to the clustering allocations from the fuzzy set based initial clustering phase. The cells in CA denote the pixels in their positions in the chosen remote sensing image. The state numbers of the cells in the model initially denote the assigned clusters from the first phase of our algorithm. State 0 denotes that the pixel has been assigned to cluster 0 in the first phase. In our proposed model, we adopt the CA with null boundary condition. We have used deterministic CA for our experiments. Our 2-dimensional CA model has been depicted in the flowchart in Figure 2.

The first stage CA is developed based on the cluster allocations outputs from the initial fuzzy set based clustering method FCM on the chosen remote sensing image. Now if among 4 neighbors (left, right, top, bottom), at least two neighbors show lower cluster values, the priority of current cell decreases to be of lower cluster value. Therefore, in this case, the present state of current cell decreases by 1 towards 0 priority. Similarly, if more than 2 neighbors have similar cluster values other than the current cell, then the next state of current cell is assigned to that value, and makes it belong to the similar clusters depending on the neighborhood pixels and reducing number of cluster outliers. We iterate through the CA matrices to obtain proper neighborhood corrections for pixels in the chosen remote sensing image.

5 APPLICATION TO PIXEL CLASSIFICATION

The new FCA algorithm is implemented using MATLAB 7.0 on HP 2 quad processor with 2.40 GHz. Comparing with the well-known K-Means and FCM methods are also being executed. Dunn (Dunn, 1973) and Davies-Bouldin (DB) (Davies & Bouldin, 1979) validity indices evaluate the effectiveness of FCA over K-Means and FCM quantitatively. The efficiency of FCA is also verified visually from the clustered images considering the ground truth information of land cover areas. The chosen LANDSAT image of the catchment region of Barakar River, which has been extracted for further research works, is available in 3 bands viz. green, red, and blue bands with original image as shown in Figure 3. Figure 3 shows the original LANDAST image of Barakar River catchment with histogram equalization with 7 classes: Turbid Water (TW), Pond Water (PW), Concrete (Concr.), Vegetarian (Veg), Habitation (Hab), Open space (OS), and roads (including bridges) (B/R). The River Barakar cuts through the image, in the middle of the catchment area. From the upper left corner of the catchment area the river is flowing through the middle part of the selected area. The river is shown as a thin line in the middle of the catchment area of blue and river colors. The Barakar Dam is shown as deep gray color in up-left corner in Figure 3. The segmented catchment area of Barakar River and the images obtained by K-Means and FCM algorithms respectively are shown in Figures 4 and 5 for (K = 7). In Figure 3, K-Means algorithm fails to classify the catchment area from the background. FCM clustering solutions in Figure 5 also fails to detect the catchment area properly from the background in the middle part. Some water bodies part and the background are mixed in both K-Means and FCM clustering solutions in Figures 4 and 5 respectively. However, our new FCA algorithm in Figure 6 is able to

Figure 3. Original image of the catchment area of Barakar river.

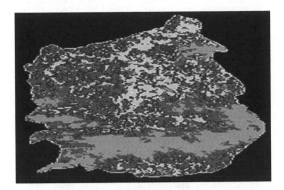

Figure 4. Pixel classification of Barakar river catchment area obtained by K-Means algorithm (with K = 7).

Figure 5. Pixel classification of Barakar river catchment area obtained FCM algorithm (with K = 7).

separate all catchment areas from the background. These indicate that FCA algorithm detects the overlapping arbitrary shaped regions significantly with better efficiency than K-Means and FCM algorithms.

5.1 Quantitative analysis

The clustering results have been evaluated objectively by measuring validity measures Davies-Bouldin (DB) and Dunn index, as defined in (Dunn, 1973) and (Davies & Bouldin, 1979) respectively, for K-Means, FCM and FCA algorithms on the Barakar River catchment remote sensing image in Table 1. It can be noticed that, FCA produces best final value for minimized DB index as 0.5135, while K-Means obtains a DB value of 0.7342 and FCM obtains 0.5780. Similarly the Dunn index produced by FCA algorithm (maximizing Dunn) is 1.3845, but K-Means algorithm provides smaller Dunn value.

These results imply that FCA optimizes DB and Dunn indices more than both K-Means and FCM. Hence, it is evident that FCA is comparable with good solutions to K-Means and FCM algorithms and even FCA sometimes outperforms to obtain superior fuzzy clustering results.

5.2 Statistical analysis

A non-parametric statistical significance test called Wilcoxon's rank sum for independent samples has been conducted at 5% significance level (Hollander & Wolfe, 1999). Two groups have been created with the performance scores, DB index values produced by 10 consecutive runs of K-Means, FCM and FCA algorithms on the chosen remote sensing Image. From the medians of each group on the dataset in Table 2, it is observed that FCA provides better median values than K-Means and FCM algorithms. Table 3 Median values of performance parameter DB index over 10 consecutive runs on different algorithms

Table 3 shows the P-values and H-values produced by Wilcoxon's rank sum test for comparison of two groups, FCA-K-Means and FCA-FCM. All the P-values reported in the table are less than 0.005 (5% significance level). The chosen remote

Figure 6. Pixel classification of Barakar river catchment area obtained by FCA algorithm.

Table 2. Validity indices values of the classified remote sensing image provided by K-means, FCM and FCA algorithms.

Index	Shanghai image		
	K-Means	FCM	FCA
Davies-Bouldin index	0.7342	0.5780	0.5135
Dunn index	0.6716	0.8406	1.3845

285

Table 3. Median values of performance parameter DB index over 10 consecutive runs on different algorithms.

Data	Algorithms		
	K-Means	FCM	FCA
Catchment Area of Barakar River Image	0.9754	0.5390	0.5135

Table 4. P-values produced by rank sum while comparing with FCA with k-means and FCA with FCM respectively.

Algorithm	Comparison with FCA	
	H	P-value
K-Means	1	2.00E-003
FCM	1	2.00E-003

sensing image on the catchment area of Barakar, is very small 2.00E-003 compared to P-value of rank sum test between FCA and K-Means. Thus, indicating the performance metrics produced by FCA to be statistically significant and not occurring by chance. Similar results are obtained for other group with FCM algorithm also. Hence, all results establish the significant superiority of FCA over K-Means and FCM algorithms.

6 CONCLUSIONS

In the realm of the remotely sensed imagery, the mixed pixel problems are common. This problem denotes the presence of multiple and partial class memberships for them. Therefore, the conventional crisp methodology fails to map land covers, properly to different regions similar to the ground truth information. The soft computing theory may overcome this problem. Therefore fuzzy sets may be applied to map overlapping regions in the image.

The contribution of this article lies in better detection of overlapping land cover areas in the remote sensing image than other crisp partitioning methodology by utilizing new fuzzy set based pixel segmentation in our FCA clustering algorithm. The primary contributions are–to utilize one new fuzzy set based initial clustering in remote sensing images with CA based neighborhood correction. The neighborhood correction phase helps to correct the wrong allocation of a single pixel to a cluster. It verifies the overall allocations with respect to the neighborhood, to obtain improved land cover areas.

The efficiency of the new FCA algorithm is demonstrated over one chosen remote sensing image of the catchment region of Barakar River.

Superiority of new FCA clustering algorithm over the widely used K-Means and FCM algorithms is established quantitatively over two validity indices. The verification with ground truth information also shows significant efficiency of new FCA algorithm over other two existing methods. Statistical tests also establish the statistical significance of FCA over K-Means and FCM algorithms.

REFERENCES

(Bandyopadhyay, 2005) Bandyopadhyay, S. 2005. Satellite image classification using genetically guided fuzzy clustering with spatial information", *Int J Remote Sens,* 26(3), 579–593.

(Bardossy, 2002) Bardossy, A. & Samaniego, L. 2002. Fuzzy Rule-Based Classification of Remotely Sensed Imagery, *IEEE TRANSACTIONS ON GEOSCIENCE AND REMOTE SENSING,* 40, 2.

(Bezdek, 1981) Bezdek, J.C. 1981. Pattern Recognition with Fuzzy Objective Function Algorithms, *Plenum Press,* New York.

(Cogalton & Green, 1999) Cogalton, R.G. & Green, K. 1999. Assessing the accuracy of remote sensed data: principles and practices. *Lewis publishers,* London, New York.

(Davies & Bouldin, 1979) Davies, D.L. & Bouldin, D.W. 1979 A Cluster Separation Measure, *IEEE Transactions on Pattern Analysis and Machine Intelligence. PAMI,* 1(2), 224–227.

(de Souto et al., 2008) de Souto, M.C.P., R RBCP, Soares, R.G.F, de Araujo, D.S.A, Costa, I.G., Ludermir, T.B. & Schliep, A. 2008. Ranking and Selecting Clustering Algorithms Using a Meta-Learning Approach. In *Proc. of IEEE International Joint Conference on Neural Networks.* IEEE Computer Society. 3728–3734.

(Dave, 1989) Dave, R.N. 1989. Use of the adaptive fuzzy clustering algorithm to detect lines in digital images. *Intell. Robots Comput. Vision VIII,* 1192, 600–611.

(Dembele & Kastner, 2003) Dembele, D. & Kastner, P. 2003. Fuzzy c-means method for clustering microarray data. *Bioinformatics.* 19, 973–980.

(Dunn, 1973) Dunn, J.C. 1973. A Fuzzy Relative of the ISODATA Process and Its Use in Detecting Compact Well-Separated Clusters, *Journal of Cybernetics,* 3(3), 32–57.

(Dunn, 1974) Dunn, J.C. 1974. A fuzzy relative of the ISODATA process and its use in detecting compact, well separated clusters, *Cybernetics,* 3, 95–104.

(Hoon et al., 2004) Hoon, M.J.L. de, Imoto, S., Nolan, J., & Miyano, S. 2004. Open source clustering software. *Bioinformatics.* 20(9), 1453–1454.

(Huang, 2008) Hung, C.C. Liu, W. & Kuo, B.C. 2008. A New Adaptive Fuzzy Clustering Algorithm for Remotely Sensed Images, in *Geoscience and Remote Sensing Symposium, 2008,.* 2, II-863-II-866.

(Lukashin & Futchs, 1999) Lukashin, A. & Futchs, R. 1999. Analysis of temporal gene expression profiles, clustering by simulated annealing and determining optimal number of clusters. *Nat. Genet.* 22, 281–285.

(Maulik & Sarkar, 2012) Maulik, U. & Sarkar, A. 2012. Efficient parallel algorithm for pixel classification in remote sensing imagery, *GeoInformatica*, 16(2), 391–407.

(Pappis, 1993), Pappis, C.P. & Karacapilidis, N.I. 1993. A comparative assessment of measures of similarity of fuzzy values, *Fuzzy Sets and Systems*, 56, 171–174.

(Ruspini, 1970) Ruspini, E. 1970. Numerical methods for fuzzy clustering, *Information Science*, 2., 319–350.

(Reddi, 1984) Reddi, S.S., Rudin, S.F. & Keshavan, H.R. 1984. An optimal multiple threshold Scheme for image segmentation, *IEEE–System, Man and Cybernetics*, 14, 611–665.

(Smith, 1971) Smith III, A.R. 1971. Two-dimensional Formal Languages and Pattern Recognition by Cellular Automata. In *IEEE Conference Record of 12th Annual Symposium on Switchinh and Automata Theory*.

(Wang, 1997) Wang, W.J. 1997. New similarity measures on fuzzy sets and on elements, *Fuzzy Sets and Systems*, 85, 305–309.

(Wolfram, 1986) Wolfram, S. 1986. Cryptography with cellular automata. *Lecture Notes on computer Science*, 218, 429–432.

(Wolfram, 1983) Wolfram, S. 1983. Statistical mechanics of cellular automata. *Rev. Mod. Phys.*, 55(3), 601–644.

(Xu et al., 1999) Xu, Y., Olman, V., & Xu, D. 1999. Clustering gene expression data using a graph theoretic approach, an application of minimum spanning trees. *Bioinformatics.* 17 309–318.

(Zadeh, 1965) Zadeh, L.A. 1965. Fuzzy sets, *Information and Control.*, 8, 338–353, 1965.

Frontiers in Computer, Communication and Electrical Engineering – Acharyya (Ed.)
© 2016 Taylor & Francis Group, London, ISBN: 978-1-138-02877-7

MIMO channel capacity in non-uniform phase distributed Nakagami channel with ZF Receiver

S.N. Sur, S. Bera & R. Bera
Department of Electronics and Communication Engineering, Sikkim Manipal Institute of Technology, Sikkim Manipal University, Sikkim, India

B. Maji
Department of Electronics and Communication Engineering, National Institute of Technology, Durgapur, West Bengal, India

ABSTRACT: Multiple-Input–Multiple-Output (MIMO) systems have appeared as a promising solution to provide high speed wireless services in extreme hostile environment. This paper deals with the performance analysis of Multiple Input Multiple Output (MIMO) system in non-uniform phase distributed Nakagami-m channel with zero forcing (ZF) receiver. The effect of Doppler shift and Nakagami-m fading parameter m on the channel ergodic and outage capacity have been analyzed here.

1 INTRODUCTION

To fulfil the demand of future generation wireless communication systems, MIMO systems have been recognized as one of the key and foremost technology. Starting from the research work by Foschini et al. [Gu *et al* 2006], lots of work have been done and also going on throughout the world to explore the capability of MIMO system in a better and sophisticated way.

With the growing demand of data seeking application under the spectral limitation and rich scattering environment, MIMO systems offer the possibility of spatial diversity which enables very high spectral efficiencies [Foschini 1996, Winters 1987 and Vucetic *et al.* 2003]. This features of MIMO encourage the researchers to the development of advanced signal processing architecture to support new age user demand. Vertical Bell laboratories Layered space-Time (V-BLAST) [Foschini 1996, Jiang *et al.* 2005, and Foschini 1998] is a MIMO architecture known to have high spectral efficiencies at average SNR without coding [Wolniansky *et al.* 1998].

Linear receivers like Zero-Forcing (ZF) which provide sub-optimal performance but offer significant computational complexity reduction with tolerable performance degradation. But these receiver required accurate Channel State Information (CSI) for its proper operation. In paper authors investigate the performance of MIMO ZF receivers in non-uniform phased distributed Nakagami-m channel.

A great deal of research has been carried out to the study of the ergodic capacity of MIMO channels under various environments, e.g., [Shin *et al.* 2003, Shin *et al.* 2008, Chiani *et al.* 2003, Shin *et al.* 2006, Oyman *et al.* 2003 and Kang *et al.* 2006]. In [Shin *et al.* 2003], the asymptotic performance of MIMO systems in Rayleigh-fading channels was investigated. Though Rayleigh and Rician are popular and commonly used statistical models for fading. But a more versatile model that correspond to variety of physical environments is the Nakagami fading model [Nakagami 1960, Zhong 2009 and Sur *et al.* 2013]. Most of the papers having the assumption that the phase of Nakagami-m channel coefficients follows a uniform distribution. As in [Yacoub *et al.* 2005] the phase distribution of a fading channel is generally non-uniform. Therefore it is required to explore the performance of MIMO system in non-uniform phase distributed Nakagami channel.

The objective of this paper is to study the ergodic capacity and outage capacity of MIMO non-uniform phased distributed Nakagami-fading channels with arbitrary finite number of antennas at both ends. This paper also include the effect of Doppler shift and Nakagami-m fading parameter m on the capacity of a MIMO system.

2 MATHEMATICAL MODEL

Let us consider a MIMO with t antennas at the transmitter and r antennas at the receiver. Also we assume that Channel State Information (CSI) is known perfectly at the receiver and equal power P is distributed throughout the transmitter antennas.

$$y = Hx + n \qquad (1)$$

H is the $r \times t$ channel matrix contains the channel coefficients. The column-vector y and x contains r received and t transmitted complex symbols respectively. And n is the complex AWGN vector with zero mean and. $E[nn\dagger] = N_0 I$

The ergodic capacity of a MIMO system is given by,

$$C = E\left[\log_2 \det\left(I + \frac{P}{N_0 t} HH^\dagger \right) \right] \qquad (2)$$

For convenience, we define $s = \min\{r,t\}$ and $k = \max\{r, t\}$ and

$$W = \begin{cases} HH^\dagger & r < t \\ H^\dagger H & r \geq t \end{cases} \qquad (3)$$

The capacity can be written in terms of the eigenvalues $\lambda_1 .... \lambda_n$ of W as

$$C = E\left[\sum_{i=1}^{n} \log_2 \left(I + \frac{P}{N_0 t} \lambda_i \right) \right] \qquad (4)$$

The channel is assumed to be independent and identically distributed (i.i.d) and which follows a Nakagami–m fading probability distribution function (pdf). Let γ represent the instantaneous SNR and it can be defined as

$$\gamma = \beta^2 \frac{E_s}{N_0} \qquad (5)$$

Where β is the fading amplitude, E_s is the energy per symbol, and N_0 is the noise spectral density. The probability distribution function of β for the Nakagami-m fading channel is given by

$$P_\beta(\beta) = \frac{2}{\Gamma(m)} \left(\frac{m}{\Omega} \right)^m \beta^{2m-1} \exp\left(\frac{-m\beta^2}{\Omega} \right),$$
$$\beta \geq 0, m \geq \frac{1}{2} \qquad (6)$$

Where $\Gamma(.)$ represents the gamma function, $\Omega = E[\beta^2$ and m is the parameter of fading depth that ranges from 0.5 to infinity and this parameter is responsible for the variation in fading condition. The fading parameter m is defined by the equation as given below

$$m = \frac{E^2[\beta^2]}{var[\beta^2]}. \qquad (7)$$

As in [Zhong 2009], the ergodic capacity of a MIMO system in Nakagami channel can be represented in terms of Meijer G-function [Prudnikov et al. 1990] as

$$C = \frac{s}{\Gamma(km) \, ln2} G_{3,2}^{1,3} \left(\frac{P}{N_0 \, t} \frac{\Omega}{m} \middle| {}^{1-km,1,1} \right) \qquad (8)$$

And also as in [Zhong 2009], at high SNR condition, we have

$$C_{hsnr} = s \log_2\left(\frac{P}{tN_0} \right) + \frac{s}{ln2}\left[\psi(km) - \ln\left(\frac{m}{\Omega} \right) \right] \quad (9)$$

Where $\psi(.)$ is the digamma function.

Now in zero forcing receiver, it estimate the transmitted symbols based on the pseudo inverse operation of the complex channel gain matrix H i.e, $H^\dagger = \left(H^H H \right)^{-1} H^H$. Therefore to have high post processing SNR, accurate channel state information (CSI) is essential but in practical scenario it is hardly available.

In this paper, we considered the effective post processing SINR of a ZF receiver in order to effectively analyze the impact of the time varying multipath channel. This analysis will help us to understand the impact of the Doppler shift on the channel capacity.

In presence of the time varying channel condition, the received signal can be represented as

$$y = Hx + E(t)x + n \qquad (10)$$

As in [Pohl et al. 2005], the pseudo inverse of the estimated channel can be represented

$$\hat{G} \cong H^\dagger \left(I_r - E(t)H^\dagger \right) \qquad (11)$$

Now, the zero-forcing estimation of the transmitted symbol vector can be given by,

$$\hat{r} \cong s + H^\dagger n - H^\dagger E(t)s - H^\dagger E(t)H^\dagger n \qquad (12)$$

As in above equation, last three terms represents additional noises and interferences generated by the post processing channel estimation error and noise and it can be expressed as

$$\hat{n} \cong H^\dagger n - H^\dagger E(t)s - H^\dagger E(t)H^\dagger n \qquad (13)$$

The noise covariance is

$$E\left[\hat{n}\,\hat{n}^{H} \right] = \left[N_0 + \frac{4E_s\pi^2\left(f_d t\right)^2}{3}I_t \right.$$

$$\left. + \frac{4\pi^2\left(f_d t\right)^2 N_0}{3}tr\left(H^H H\right)^{-1} \right]\left(H^H H\right)^{-1}$$

$$(14)$$

Therefore the post processing SINR per symbol of any i^{th} stream can be written as

$$\sigma_{i=} \frac{E_s/N_0}{\left[1+\frac{4E_s\pi^2\left(f_d t\right)^2}{3N_0}I_t + \frac{4\pi^2\left(f_d t\right)^2}{3}tr\left(H^H H\right)^{-1}\right]\left(H^H H\right)^{-1}\Big|_{ii}}$$

$$(15)$$

As in above equation, for a time varying channel the SINR per symbol for any stream received signal of a MIMO system is highly influenced by the Doppler shift. Therefore at high SINR condition the sub channel capacity of a MIMO system can be represented as

$$C_{h\sin r} = s\log_2\left(\frac{\sigma_i}{t}\right) + \frac{s}{ln2}\left[\psi(km) - \ln\left(\frac{m}{\Omega}\right)\right] \quad (16)$$

The above equation leads to the fact that with the increase in the Doppler shift value the channel capacity decreases.

3 RESULTS

In this paper, all the results are evaluated through the MATLAB simulation. Authors have consider MIMO system in non-uniformly phase distributed Nakagami-m channel condition.

Figure 1 shows the PDF and CDF of Uniformly Phase (UP) distributed and Non-Uniformly Phase (NUP) distributed Nakagami-m channel.

Figure 2 shows phase distribution of Non-Uniformly Phase (NUP) distributed Nakagami-m channel with the variation in m. As it is there in figure apart from m = 1, the phase distribution of the Nakagami-m channel is not uniform.

Figure 3 shows the ergodic capacity variation of MIMO system with the variation in transmitter and receiver antenna variation. Here, authors have consider Nakagami-m channel fade parameter $m = 0.5$. During the variation in antennas (receiver and transmitter side), authors have kept the number antennas at other side as 2. As in figure, at SNR = 1dB, for ($t = 10, r = 2$) the capacity of

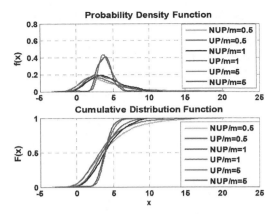

Figure 1. PDF and CDF of Uniform and Non-uniform phase distributed Nakagami-m channel.

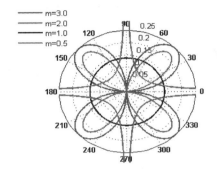

Figure 2. Phase distribution of Non-uniform phase distributed Nakagami-m channel.

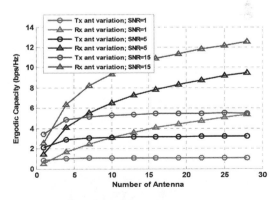

Figure 3. Ergodic capacity variation with the variation in number of antennas (Transmitter (t)/Receiver (r)).

the system is 1.084 bps/Hz and for ($t = 2, r = 10$) it is 3.112 bps/Hz. At SNR = 15dB, for ($t = 10, r = 2$) the capacity of the system is 5.311 bps/Hz and for ($t = 2, r = 10$) it is 9.364 bps/Hz. Therefore one can conclude that increment of number of antenna at the receiver side have significant effect on channel capacity in comparison to increment in transmitter side antenna. But the total capacity of the system saturate as the number of antenna increases to a large number.

As represented in figure 4, phase distribution of Nakagami-m channel have very negligible effect on ergodic capacity.

Figure 5 and 6 shows the change in outage capacity with the variation in Doppler shift. In both the cases authors have consider 4×4 MIMO system in Nakagami-m channel with non-uniform phased distribution. A Doppler frequency of 100 Hz corresponds to about 35 miles/hour for carrier frequency of 2 GHz. For figure 5, m = 1 is considered and for figure 6, Pout = 0.1 is considered. It is clear from the figure that Nakagami-m fading parameter m and Doppler shift having significant impact on the capacity of the system. The throughput decreases as Doppler frequency (Fd) value increases.

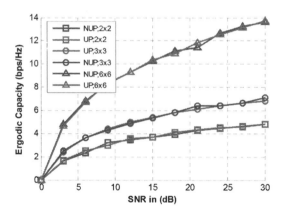

Figure 4. Ergodic capacity comparison for Uniform Phased (UP) and Non Uniform Phased (NUP) distributed Nakagami-m channel.

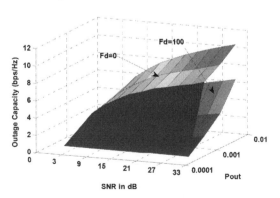

Figure 5. Outage capacity variation with Doppler shift.

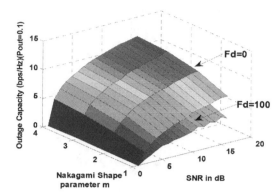

Figure 6. Outage capacity variation with Doppler shift.

4 CONCLUSION

The ergodic capacity and outage capacity of MIMO Nakagami-m fading channels is analyzed here. Results also show that the effect of non-uniform phase distribution on the PDF and CDF of the channel is negligible. Therefore its effect on the ergodic capacity is also negligible. And also in this paper, the mathematical formulation of Doppler effect and corresponding effect on the ergodic and outage capacity is presented. The simulated result shows the influence of Doppler shift and Nakagami channel fading parameter m over channel capacity.

REFERENCES

Chiani, M. Win, M.Z. & Zanella, A. 2003. On the capacity of spatially correlated MIMO Rayleigh-fading channels, IEEE Trans. Inf. Theory, vol. 49, no. 10, Oct. 2003, pp. 2363–2371.

Foschini, G.J. 1998. Layered space-time architecture for wireless communication, Bell Labs. Technology. Journal, Vol. 6, No.3, 1998, pp. 311–335.

Foschini, G.J. 1996. Layered Space-Time Architecture For Wireless Communication In A Fading Environment When Using Multi-Element Antenna, Bell Laboratories Technical Journal, Oct. 1996, pp. 41–59.

Gu, X. Peng, X-H and Zhang, G.C. 2006. MIMO systems for broadband wireless communications, BT Technology Journal, Vol. 24 No. 2, April 2006, pp. 90–96.

Jiang, Y. Zheng, X. & Li, J. 2005. Asymptotic Performance Analysis of V-BLAST, GLOBECOM, 2005, pp. 3382–3886.

Kang, M. and Alouini, M.S. 2006. Capacity of correlated MIMO Rayleigh channels, IEEE Trans. Wireless Commun., vol. 5, no. 1, Jan. 2006, pp. 143–155.

Nakagami, M. 1960. The distribution-a general formula for intensity distribution of rapid fading, in Statistical Methods in Radio Wave Prop, W.G. Hoffman, Ed. Oxford, U.K.: Pergamon, 1960.

Oyman, O. Nabar, R.U. Bolcskei, H. & Paulraj, A.J. 2003. Characterizing the statistical properties of mutual information in MIMO channels, IEEE Trans. Signal Process., vol. 51, no. 11, Nov. 2003, pp. 2784–2795.

Pohl, V. Nguyen, P.H. & Jungnickel,V. 2005. Continuous flat-fading MIMO channels: achievable rate and optimal length of the Training and data phases, IEEE Transactions on Wireless Communication, vol. 4, no. 4, July 2005, pp. 1889–1900.

Prudnikov, A.P. Brychkov, Y.A. & Marichev, O.I. 1990. Integrals and Series, Volume 3: More Special Functions. New York: Gordon and Breach, 1990.

Shin, H. and Lee, J.H. 2003. Capacity of multiple-antenna fading channels: Spatial fading correlation, double scattering, and keyhole, IEEETrans. Inf. Thoery, vol. 49, no. 10, Oct. 2003, pp. 2636–2647.

Shin, H. Win, M.Z. & Chiani, M. 2008. Asymptotic statistics of mutual information for doubly correlated MIMO channels, IEEETrans.Wireless Commun., vol. 7, no. 2, Feb. 2008, pp. 562–573.

Shin, H. Win, M.Z. Lee, J.H. and Chiani, M.2006. On the capacity of doubly correlated MIMO channels, IEEE Trans. Wireless Commun., vol. 5, no. 8, Aug. 2006, pp. 2253–2265.

Sur,S.N. & Ghosh, D. 2013. Channel Capacity and BER Performance Analysis of MIMO System with Linear Receiver in Nakagami Channel, I.J. Wireless and Microwave Technologies, vol. 3, no. 1, 2013, pp. 26–36.

Vucetic, B. & Yuan, J. 2003. Space-Time Coding, John Wiley & Sons Ltd, 2003.

Winters, J.H. 1987. On the capacity of radio communications systemswith diversity in rayleigh fading environments, IEEE J. Select. Areas Commun., vol. JSAC-5, June 1987, pp. 871–878.

Wolniansky, P.W. Foschini, G.J. Golden, G.D. and Valenzue, R.A. 1998. V-BLAST: An Architecture For Realizing Very High Data Rate Over The Rich-Scattering Wireless Channel, Proc. of URS International Symposium on Signals, Systems and Electronics (ISSS '98), Sep/Oct 1998, pp. 295–300. (doi: 10.1109/ISSSE.1998.738086).

Yacoub, M.D. Fraidenraich, G. and Santos Filho, J.C.S. 2005. Nakagami-m phase-envelope joint distribution, Elect. Letters, vol. 41, no. 5, Mar. 2005, pp. 259–261.

Zhong, C. Wong, K-K, & Jin, S. 2009. Capacity Bounds for MIMO Nakagami-Fading Channels, IEEE Transactions on Signal Processing, Volume 57, Issue 9, 2009, pp- 3613–3623.

Decision feedback equalization for large scale MIMO system

S.N. Sur & R. Bera
Department of Electronics and Communication Engineering, Sikkim Manipal Institute of Technology, Sikkim Manipal University, Sikkim, India

B. Maji
Department of Electronics and Communication Engineering, National Institute of Technology, Durgapur, West Bengal, India

ABSTRACT: Large Scale MIMO (LS-MIMO) is emerging as one of the basic requirements for the evolution of future generation wireless communication system. This paper represents the implementation of Decision Feedback Equalization (DFE)-based receivers for large scale MIMO systems in Nakagami-m channel. Moreover, sub-optimal, pre-equalization combining of DFE receivers, architectures are examined, and the performance is compared with its linear counter-parts.

1 INTRODUCTION

The need for a secured and high throughput communication system has forced the researchers towards the development of several signal processing and communications techniques for employing the resources efficiently [Foschini *et al.* 1998, and Alamouti. 1998]. And to deliver such a highly spectral efficient secure communication system, MIMO plays an important role [Verd´u, 1986]. To support high data rate under severe channel condition, the receiver architecture for the MIMO system should be designed in a proper way to compensate for both intersymbol and multi-access interstream interference.

Massive MIMO (also known as "Large-Scale Antenna Systems", "Very Large MIMO", "Hyper MIMO", "Full-Dimension MIMO" and "ARGOS") [Duel-Hallen 1992] emerge as a key technology for 5G communication. From the latest research, it has already been proven that massive MIMO is a fundamental requirement for the development of future broadband or 5G communication networks which will be energy-efficient, secure and robust, and spectrum efficient [Duel-Hallen 1992].

Now, to provide reliable and secure communications in such severe multipath channel condition, Maximum-Likelihood (ML) detection would be the optimum receiver. However, implementation of this receiver is practically impossible as the complexity of the receiver increases substantially with the increase in the number of transmit antennas [Larsson *et al.* 2014]. Therefore, it is required to design low complexity receiver such as linear receiver

[Lamare *et al.* 2008 and Kominakis *et al.* 2002] and Decision Feedback Equalization (DFE) receiver [Sur *et al.* 2013]. Generally in MIMO, Zero Forcing (ZF) and Minimum Mean Square Error (MMSE) based receivers are used as linear receiver. In this paper, the performance comparison between ZF and MMSE criterion based on linear (LE) and nonlinear (DFE) receiver has been carried out. Also in this paper, QR factorization is utilized to design DFE receiver.

This paper presents the performance analysis of different MIMO receiver (linear and non-linear) over the doubly correlated Nakagami-m fading channel. Here Channel State Information (CSI) is assumed to be present at the receiver side only. And it is also considered that the transmit power is uniformly distributed across the transmitting antennas.

2 SYSTEM MODEL

2.1 *Channel model*

We consider the MIMO system [Shin *et al.* 2004], as shown in Fig. 1, with N_t transmitting antenna and N_r receiving antennas. The received signal y can be described by

$$\psi = H\xi + v \qquad (1)$$

Where x is transmit symbols vector and it satisfies $E\left\{\|x\|^2\right\} \leq P$ (P is the total power), and n is the $N_r \times 1$ Additive White Gaussian Noise (AWGN) vector. The correlated channels are represented by

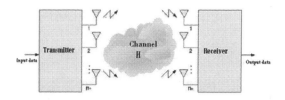

Figure 1. MIMO system architecture.

vector H. The envelope of the channel is distributed according to the Nakagami –m statistic. Let γ represent the instantaneous SNR and it can be defined as

$$\gamma = \beta^2 \frac{E_s}{N_0} \qquad (2)$$

Where β represents the fading amplitude, E_s is the energy per symbol, and N_0 is the noise spectral density. The Probability Distribution Function (PDF) of β for the Nakagami-m fading channel is given by

$$P_\beta(\beta) = \frac{2}{\Gamma(m)} \left(\frac{m}{\Omega} \right)^m \beta^{2m-1} \exp\left(\frac{-m\beta^2}{\Omega} \right), \quad \beta \geq 0, m \geq \frac{1}{2} \qquad (3)$$

Where $\Gamma(.)$ represents the gamma function, $\Omega = E\left[\beta^2 \right]$ and m is the parameter of fading depth that ranges from 0.5 to infinity. Different kinds of fading condition can be simulated by varying the m values. The fading parameter m is defined by the equation as given below

$$m = \frac{E^2[\beta^2]}{var[\beta^2]} \qquad (3)$$

Then, the pdf of the instantaneous SNR γ is given by [9]

$$P_\gamma(\gamma) = \frac{1}{\Gamma(m)} \left(\frac{m}{\overline{\gamma}} \right)^m \gamma^{m-1} \exp\left(\frac{-m\gamma}{\overline{\gamma}} \right), \quad \gamma \geq 0 \qquad (4)$$

2.2 Decision feedback equalizer

A simplified architectural overview of DFE [Sur et al. 2015, Dhahir et al. 1997], with QR decomposing algorithm has been investigated in this paper. There are different kind of MIMO channel matrix factorization techniques present, out of wich QR factorization is regorously used for DFE implementations [Dhahir et al. 1997].

The QR decomposed MIMO channel matrix can be represented as given below,

$$H = QR \qquad (6)$$

Where Q is unitary and R is upper triangular. Now, the received vector can be manupulated by the relation as mentioned below,

$$w = Q^* y = Rx + Q^* n = Rx + \overline{n} \qquad (7)$$

With $E\left[\overline{n}\, \overline{n}^* \right] = \sigma_x^2 I_{Nt}$

Then w is passed through the feedforward filter as represented by the equation given below,

$$K = \left[diag(R) \right]^{-1} \qquad (8)$$

where diag (R) denotes a diagonal matrix whose elements are equal to the diagonal elements of R.

The coefficients of the feedback filter can be calculated by

$$B = I - KR \qquad (9)$$

Here B represents an upper triangular matrix with zero diagonal elements.

The block level diagram for one-shot MIMO DFE systems in combination with QR factorization is shown in Fig. 2.

Similarly, the MMSE-DFE can be inplemented by using the QR decomposition of the augmented matrix,

$$\underline{H} = \begin{bmatrix} H \\ \sigma I \end{bmatrix} \qquad (10)$$

where $1/\sigma$ is the SNR.

2.3 Results

Fig. 3 shows the performance assessment of a 10×10 MIMO system in Nakagami-m channel with QPSK modulation. As represented in the figure the optimal pre-equalization combining DFE that forms soft estimates based on all received symbols performs better than its linear equalizer. Therefore, one can conclude that the performance improvement due to DFE receiver over LE receiver is clearly visible.

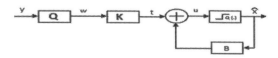

Figure 2. DFE architecture.

Fig. 4 represents the performance comparison of MMSE-DFE with different antenna configuration. In this simulation, authors have considered QPSK modulation, correlation coefficient = 0, and m = 0.5. As represented in the simulated results, fading parameter m have noteworthy impact over the MIMO system performance. And m = 0.5 can be considered as severe channel condition as it represents one sided Gaussian distribution. From the simulated results, it is clearly visible that with the increase in the number of antenna, the fading margin improve significantly. Therefore large MIMO with MMSE-DFE have the capability to improve the system performance in severe channel condition.

As presented in Table 1, the required SNR level decreases as the number of antennas increases. Figs. 5 and 6 represents performance comparison large scale MIMO systems with linear (ZF) and non-linear (DFE) receiver with respect to single antenna system. This performance analysis has

been carried out to point out the fading margin gain due to large scale MIMO system.

Fig. 7 represents the fading margin gap between the linear and non-linear receivers. Fading margin

Table 1. Required SNR comparison for different MIMO system.

BER = 10^{-3}	5×5	10×10	15×15	20×20	25×25
SNR (dB)	20	12.2	9.2	5	3.6

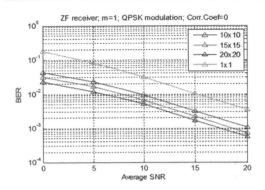

Figure 5. Performance comparison for ZF receiver.

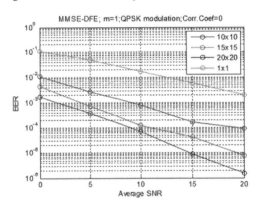

Figure 6. Performance comparison for MMSE-DFE receiver.

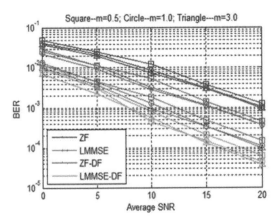

Figure 3. Comparison between LF and DFE.

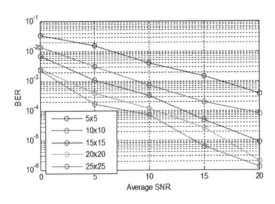

Figure 4. Performance comparison for MMSE DFE.

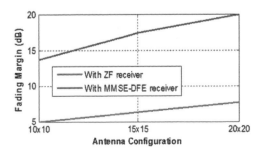

Figure 7. Fading margin comparison for linear and DFE receiver.

has been calculated for achieving BER of 10^{-3} and it is calculated with respect to a single antenna system. As depicted in the figure, it is clear that with Large MIMO system and low complex single shoot MMSE-DFE receiver, there is a large improvement in fading margin.

3 CONCLUSION

This paper focused on the design of Decision Feedback Equalization-based receivers for MIMO systems in correlated Nakagami-m channel. The equalizer coefficient matrices were derived using the QR decomposition along MMSE/ZF optimization. From the simulated results, it can be concluded that pre-equalization combining single shoot DFE achieve optimal performance and can be implemented to overcome severe channel condition. Therefore, low complex single shoot receiver is best suitable for large MIMO system, in order to reduce the interference effect.

REFERENCES

Alamouti, S.1998. A simple transmit diversity technique for wireless communications," IEEE J. Select. Areas Communication, vol. 16, no. 8, Oct. 1998, pp. 1451–1458.

Dhahir, A and Sayed, A.H. 1997. A parallel low-complexity coefficient computation processor for the MMSE-DFE, inProc. Thirty-First Asilomar Conference on Signals, Systems & Computers, Pacific Grove, CA, Nov 2–5, 1997, pp. 1586–1590.

Duel-Hallen, A. 1992. Equalizers for Multiple Input Multiple Output Channels and PAM Systems with Cyclostationary Input Sequences," IEEE Journal on Selected Areas in Communications, vol. 10, April, 1992, pp. 630–639.

Foschini, G.J. and Gans, M.J. 1998. On limits of wireless communications in a fading environment when using multiple antennas, Wireless Person. Communication, vol. 6, Mar. 1998, pp. 311–335.

Kominakis, C. Fragouli, C. Sayed, A.H. and Wesel, R. 2002. Multi-input multi-output fading channel tracking and equalization using Kalman estimation," IEEE Transactions on Signal Processing, vol. 50, no. 5, May, 2002, pp. 1065–1076.

Lamare, R.C. De. Hjørungnes A. and Sampaio-Neto, R. 2008. Adaptive Decision Feedback Reduced-Rank Equalization Based on Joint Iterative Optimization of Adaptive Estimation Algorithms for Multi-Antenna Systems",IEEE WCNC, 2008, pp. 413–418.

Larsson, E.G. Edfors, O. Tufvesson, F and Marzetta, T L. 2014. Massive MIMO For Next Generation Wireless S Ystems", IEEE Communications Magazine, Vol. 52, No. 2, Feb. 2014, pp. 186–195.

Shin, H and Lee, J. 2004. On the Error Probability of Binary and M-ary Signals in Nakagami-m Fading Channels, IEEE Transactions On Communications, Vol.52, No.4, April 2004, pp. 536–539.

Sur, S.N. Bera, R. and Maji, B. 2015.Decision Feedback Equalization for MIMO Systems", Intelligent Computing and Applications, Advances in Intelligent Systems and Computing Volume 343, 2015, pp. 205–212.

Sur, S.N. & Ghosh, D. 2013. Channel Capacity and BER Performance Analysis of MIMO System with Linear Receiver in Nakagami Channel, I.J. Wireless and.

Verd´u, S. 1986. Minimum Probability of Error for Asynchronous Gaussian Multiple-Access Channels, IEEE Transactions on Information Theory, vol.IT-32, no. 1, Jan., 1986, pp. 85–96.

Frontiers in Computer, Communication and Electrical Engineering – Acharyya (Ed.)
© 2016 Taylor & Francis Group, London, ISBN: 978-1-138-02877-7

Comparative study of DR image De-noising method based on quality parameters

Monisha Chakraborty
School of Bio-Science and Engineering, Jadavpur University, Kolkata, West Bengal, India

Manishankar Mondal
Student of Master of Bio-Medical Engineering, School of Bio-Science and Engineering, Jadavpur University, Kolkata, West Bengal, India

ABSTRACT: Automated Diabetic Retinopathy (DR) detection in an early stage is very necessary for the patients. Medical images taken from various sources are sometimes noisy. For this reason, it is very difficult for the doctors to identify the specific problem. The noise removal is very much essential for getting better quality of images. Noise removal of images is mainly done by using suitable filters. In this work, noise removal from images is done using Average, Gaussian, and Wiener filters. Noise removal from DR images which are collected from Standard Diabetic Retinopathy Database using Gaussian filtering method gives better image quality. The quality analysis is done by using image quality matrices PSNR, SSIM, and UIQI. The result of statistical analysis of the quality parameters signifies that the results of the parameters are statistically significant.

1 INTRODUCTION

The population of diabetic patients has been increasing day by day. Uncontrolled and prolonged diabetes can damage the microvasculature of the vital organs of the body such as eyes and kidneys. The damage caused to the tiny blood vessels in the retina of the human eye, is known as Diabetic Retinopathy (DR). Due to elevated amounts of glucose circulating through the body, the walls of blood vessels get damaged and several anomalies such as Microaneurysms, hemorrhages, Hard Exudates, Cotton wool spots start developing at various phases of retinopathy. The patient affected by DR may not experience visual impairment until the disease has progressed to a severe stage, when the treatment is less effective. Therefore the early detection and the regular follow-ups are necessary to treat DR (Paranjpe et al. 2014).

The process of Automatic Diabetic Retinopathy detection involves detection and segmentation of the abnormal features from the input images. Therefore, noise removal of the retinal images is necessary. Previously, there are several research works carried out on DR images like localization of retinal landmarks from DR images (Shivram et al. 2010), segmentation of red lesions in DR images (Kalaivaani et al. 2014). In this work, noise removal of DR images is done by using suitable filters like Average, Wiener, and Gaussian filters.

Image quality parameters e.g. PSNR, SSIM, and UIQI are obtained from noise free images and these results are subjected to statistical analysis. Here, 24 cases of DR images have been taken from Standard Diabetic Retinopathy Database.

2 METHOD

The de-noising and statistical analysis of the quality parameters of DR images has been done in this paper using the following steps as shown in Fig. 1.

3 FILTERING TECHNIQUES

There are different ways to remove or reduce noise in an image. Different methods are better for different kinds of noise. The methods considered in this study are:

- Average Filtering
- Wiener Filtering
- Gaussian Filtering

3.1 *Average filtering*

Average filter is a linear filter. We can use average filter for removing certain types of noise. For example, an averaging filter is useful for removing

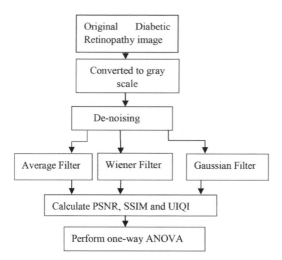

```
┌─────────────────────────┐
│ Original      Diabetic  │
│ Retinopathy image       │
└─────────────────────────┘
            ↓
┌─────────────────────────┐
│ Converted to gray       │
│ scale                   │
└─────────────────────────┘
            ↓
┌─────────────────────────┐
│ De-noising              │
└─────────────────────────┘
```

| Average Filter | Wiener Filter | Gaussian Filter |

Calculate PSNR, SSIM and UIQI

Perform one-way ANOVA

Figure 1. The proposed work sequence.

grain noise from a photograph. Because each pixel set to the average of the pixels in its neighborhood, local variations caused by grain are reduced but with blurring of edges (Kanagalakshi et al. 2011).

The mathematical form of average filter in an m x n region can be expressed as in Equation (1)

$$h(x,y) = \frac{1}{mn} \sum_{k=1}^{m} \sum_{l=1}^{n} f(k,l) \tag{1}$$

3.2 Wiener filtering

Wiener filter is a low pass filter that filters images that has been degraded by constant power additive noise (Advances in Artificial Intelligence 2011). This filter uses a pixel wise adaptive method based on statistical estimation from a local neighborhood of each pixel. Equations (2), (3), and (4) are used in the calculation of wiener filter in an m x n neighborhood

$$\mu = \frac{1}{mn} \sum_{n_1,n_2 \in \eta} a(n_1, n_2) \tag{2}$$

$$\sigma^2 = \frac{1}{mn} \sum_{n_1,n_2 \in \eta} a^2(n_1, n_2) - \mu^2 \tag{3}$$

$$b(n_1,n_2) = \mu + \frac{\sigma^2 - \nu^2}{\sigma^2}(a(n_1,n_2) - \mu) \tag{4}$$

where μ and σ^2 respectively are the local mean and variance around each pixel, and η and ν^2 are the n-by-m local neighborhood of each pixel in an image and noise variance.

3.3 Gaussian filtering

Gaussian filter is used for blurred images and for removing noise in detail. This filter works by using 2D distribution as a point–spread function. The Gaussian filter is a non-uniform low pass filter. The kernel coefficients diminish with increasing distance from the kernel's center (Hsiao et al. 2007).

2-D Gaussian filter can be expressed as given in Equation (5)

$$g(x,y) = \frac{1}{2\pi\sigma^2} e^{-\frac{x^2+y^2}{2\sigma^2}} \tag{5}$$

4 IMAGE QUALITY EVELUATION METRICES

Any image scanned or captured can never be fully noise free. These are affected by various types of noise. Depending on the noise density present in the image, pixel intensity values carrying significant information are altered from that of the original image thereby affecting the overall quality of the image (Bhattacharjee et al. 2012). There are various mathematically defined parameters in this context are Peak Signal-to-Noise Ratio (PSNR), Structural Similarity Index Measurement (SSIM), and Universal Image Quality Index (UIQI).

4.1 Peak Signal-to-Noise Ratio (PSNR)

Peak Signal-to-Noise Ratio (PSNR), is the mathematical measure of image quality based on the pixel difference between two images (Al-Najjar et al. 2012, Jean-Bernard Martens et al. 1998). PSNR can be defined as in Equation (6)

$$\text{PSNR} = 10 \log \frac{s^2}{MSE} \tag{6}$$

$$\text{Here, MSE} = \frac{1}{MN} \sum_{m=0}^{M-1} \sum_{n=0}^{N-1} e(m,n)^2$$

where s = 255 for an 8-bit image, $e(m, n)$ is the error difference between the original and the distorted images.

4.2 Structural Similarity Index Measurement (SSIM)

The mean Structural Similarity Index (Al-Najjar et al. 2012) is computed as follows:

$$\mu_x = \frac{1}{T} \sum_{i=1}^{T} x_i \,, \, \mu_y = \frac{1}{T} \sum_{i=1}^{T} y_i \tag{7}$$

$$\sigma_x^2 = \frac{1}{T-1}\sum_{i=1}^{T}\left(x_i - \bar{x}\right)^2, \ \sigma_y^2 = \frac{1}{T-1}\sum_{i=1}^{T}\left(y_i - \bar{y}\right)^2 \quad (8)$$

$$\sigma_{xy}^2 = \frac{1}{T-1}\sum_{i=1}^{T}\left(x_i - \bar{x}\right)\left(y_i - \bar{y}\right) \quad (9)$$

First, the original and distorted images are divided into blocks of size 8×8 and then the blocks are converted into vectors. Second, two means and two standard deviations, and one covariance value are computed from the images as in (7), (8), and (9). Third, luminance, contrast, and structure comparisons based on statistical values are computed like in UIQI, the structural similarity index measure between images x and y is given in Equation (10)

$$\text{SSIM}(x,y) = \frac{\left(2\mu_x\mu_y + c_1\right)\left(2\sigma_{xy} + c_2\right)}{\left(\mu_x^2 + \mu_y^2 + c_1\right)\left(\sigma_x^2 + \sigma_y^2 + c_2\right)} \quad (10)$$

Where c_1 and c_2 is constant.

4.3 Universal Image Quality Index (UIQI)

The Universal Image Quality Index (UIQI) (Wang et al. 2002) can be mathematically expressed as:

$$\text{UIQI}(x,y) = \frac{4\sigma_{xy}\overline{xy}}{\left(\sigma_x^2 + \sigma_y^2\right)\left(\left(\bar{x}\right)^2 + \left(\bar{y}\right)\right)} \quad (11)$$

Where $\bar{x} = \frac{1}{N}\sum_{i=1}^{N}x_i$, $\bar{y} = \frac{1}{N}\sum_{i=1}^{N}y_i$,

$$\sigma_x^2 = \frac{1}{N-1}\sum_{i=1}^{N}\left(x_i - \bar{x}\right)^2, \ \sigma_y^2 = \frac{1}{N-1}\sum_{i=1}^{N}\left(y_i - \bar{y}\right)^2,$$

$$\sigma_{xy} = \frac{1}{N-1}\sum_{i=1}^{N}\left(x_i - \bar{x}\right)\left(y_i - \bar{y}\right)$$

UIQI and SSIM are more accurate and consistent than PSNR.

5 RESULTS

In this work, 24 DR images are collected from the standard database. Out of these 24 cases, an example image, its gray scale image and subsequent filtered images using the various filters as considered in this work are shown in Figures 2, 3, 4, and 5 respectively.

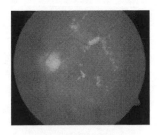

Figure 2. Original image in gray.

Figure 3. Average filtered.

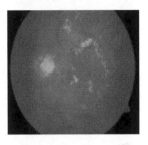

Figure 4. Gaussian filtered.

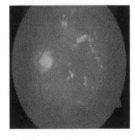

Figure 5. Wiener filtered.

5.1 Diseased image and their de-noised form

5.1.1 Values of PSNR, SSIM and UIQI after image de-noising

PSNR, SSIM, and UIQI values of 24 diseased cases are shown respectively in Table 1, Table 2, and Table 3.

Table 1. PSNR values.

Average filter	Gaussian filter	Wiener filter
24.2953	24.6697	24.3460
24.2996	24.6697	24.3471
24.2929	24.6697	24.3457
24.2823	24.6697	24.3453
24.2895	24.6697	24.3460
24.3747	24.7767	24.4462
24.2581	24.6697	24.3460
24.3010	24.6697	24.6697
24.3027	24.6697	24.3481
24.3027	24.6697	24.3481
24.2952	24.6697	24.3460
24.3013	24.6697	24.3474
24.3016	24.6697	24.3479
24.2870	24.6697	24.3445
24.2944	24.6697	24.3459
24.2907	24.6697	24.3451
24.2879	24.6697	24.3464
24.2988	24.6697	24.3473
24.2939	24.6697	24.3471
24.2639	24.6697	24.3382
24.2900	24.6697	24.3457
24.2938	24.6697	24.3454
24.2949	24.6697	24.3456
24.3002	24.6697	24.3475

Table 3. UIQI values.

Average filter	Gaussian filter	Wiener filter
0.9459	0.9773	0.9553
0.9474	0.9773	0.9536
0.9529	0.9772	0.9580
0.9564	0.9770	0.9616
0.9589	0.9772	0.9635
0.9416	0.9750	0.9515
0.9387	0.9763	0.9446
0.9446	0.9773	0.9487
0.9479	0.9774	0.9521
0.9479	0.9774	0.9521
0.9591	0.9772	0.9633
0.9511	0.9773	0.9549
0.9506	0.9774	0.9550
0.9554	0.9772	0.9596
0.9585	0.9772	0.9632
0.9550	0.9772	0.9606
0.9520	0.9771	0.9565
0.9540	0.9772	0.9582
0.9540	0.9772	0.9595
0.9474	0.9771	0.9544
0.9451	0.9771	0.9524
0.9407	0.9772	0.9502
0.9554	0.9773	0.9604
0.9405	0.9773	0.9460

Table 2. SSIM values.

Average filter	Gaussian filter	Wiener filter
0.9719	0.9785	0.9744
0.9729	0.9785	0.9748
0.9725	0.9784	0.9741
0.9726	0.9783	0.9746
0.9724	0.9784	0.9746
0.9715	0.9784	0.9733
0.9722	0.9779	0.9740
0.9743	0.9785	0.9754
0.9740	0.9786	0.9754
0.9740	0.9786	0.9754
0.9728	0.9785	0.9745
0.9736	0.9785	0.9749
0.9737	0.9786	0.9753
0.9723	0.9784	0.9739
0.9728	0.9784	0.9744
0.9719	0.9784	0.9739
0.9736	0.9784	0.9748
0.9736	0.9784	0.9748
0.9734	0.9785	0.9750
0.9706	0.9783	0.9733
0.9728	0.9784	0.9742
0.9719	0.9785	0.9740
0.9727	0.9785	0.9745
0.9734	0.9785	0.9748

5.2 Results of ANOVA calculation of PSNR, SSIM and UIQI

ANOVA is the statistical tool for the separation of variation due to a group of causes from the variation due to other groups (Das 2005). The results of ANOVA calculation of PSNR, SSIM, and UIQI are shown in the following tables respectively.

6 DISCUSSION

Results of this work reveal that the noise removal of DR images, are done by using Average, Gaussian, and Wiener Filters, which statistically gave significant results as seen in Tables 4, 5, and 6.

The image quality parameters have important role; higher values of PSNR, SSIM, and UIQI indicate better quality of image. The image analysis, as done in this study, show that, de-noising of DR images using Gaussian filter gives better quality of image.

7 CONCLUSION

This study concludes that Gaussian filter is the best amongst all other filters considered in this work

Table 4. Result of ANOVA calculation for PSNR.

Source	SS	df	MS	F	P value
Columns	1.95532	2	0.97766	528.2	1.47979e-42
Error	0.12771	69	0.00185		
Total	2.08303	71			

Table 5. Result of ANOVA calculation for SSIM.

Source	SS	df	MS	F	P value
Columns	0.0004	2	0.0002	507.68	5.332e-42
Error	0.00003	69	0		
Total	0.00043	71			

Table 6. Result of ANOVA calculation for UIQI.

Source	SS	df	MS	F	P value
Columns	0.00978	2	0.00489	220.24	1.10837e-30
Error	0.00153	69	0.00002		
Total	0.01131	71			

with respect to PSNR, SSIM, and UIQI values. Moreover, the results are statistically significant.

ACKNOWLEDGEMENT

The authors are thankful to all the members of the Standard Diabetic Retinopathy Database. All the images used in this paper have been taken from this database.

REFERENCES

Advances in Artificial Intelligence, Springer, 10th Mexican International Conference on Artificial Intelligence, Proceedings, Part 1, 2011.

Al-Najjar, Y.A.Y & Soong, D.C. (2012). Comparison of image quality assessment: PSNR, HVS, SSIM, UIQI. International Journal of Scientific & Engineering Research, Volume 3, pp.1–5.

Bhattacharjee, R. & Chakraborty, M. (2012). Brain tumor detection from MR images: image processing, slicing and PCA reconstruction.IEEE, Third International Conference on Emerging Applications of Information Technology, pp.97–101.

Das, N.G. (2005). Statistical Methods, M Das & Co.

Hsiao, P.Y. Chou, S.S. & Huang, F.C. (2007). Generic 2-D gaussian smoothing filter for noisy image processing. TENCON-IEEE Region 10 Conference, pp.1–4.

Jean-Bernard Martens, A.L.M. (1998). Image quality dissimilarity. Signal Processing, vol. 7, pp.155–176.

Kalaivaani, J. & Santhi, D. (2014).Segmentation of red lesions in Diabetic Retinopathy images. International Conference on Electronics and Communication System. http://www.it.lut.fi/project/imageret/diaretdb0/

Kanagalakshi, K. & Chandra, E. (2011). Performance evalution of filters in noise removal of fingerprint image. IEEE, pp.117–121.

Paranjpe, M.J. & Kakatkar, M.N. (2014). Review of methods for Diabetic Retinopathy detection and severity classification. International Journal of Research in Engineering and Technology, vol. 3, pp. 619–624.

Shivram, J.M. & Patil, R. (2010). Knowledge based framework for localization of retinal landmarks from Diabetic Retinopathy images.IEEE, vol. 2, pp.220–224.

Wang, Z. & Bovik, A.C. (2002). A Universal Image Quality Index. IEEE Signal Processing Letters, Vol. 9, pp.271–350.

Frontiers in Computer, Communication and Electrical Engineering – Acharyya (Ed.)
© *2016 Taylor & Francis Group, London, ISBN: 978-1-138-02877-7*

Energy efficient adaptive power control in indoor wireless sensor networks

Debraj Basu, Gourab Sen Gupta, Giovanni Moretti & Xiang Gui
School of Engineering and Advanced Technology, Massey University, New Zealand

ABSTRACT: In an energy-constrained wireless sensor node, the use of ARQ (Automatic Repeat reQuest) protocol will lead to retransmissions when an attempt to send a packet fails. In an indoor wireless environment, the channel conditions change very slowly. This block-fading effect may lead to subsequent retries also failing, which adds to the total energy expenditure of transmission. The received signal power and therefore the ratio of the average bit energy and noise power spectral density (E_b/N_0) values may also change over time, mainly due to obstructions, partitions and human movement between the transmitter and the receiver. The proposed adaptive power control algorithm does away with channel estimation methods before transmission, and was studied using just four of the output power levels available in CC2420 transceiver module. It is designed to meet the challenge of responding to an unknown and variable radio channel in an energy-efficient manner. The adaptive protocol uses past transmission experience or memory to decide the power level at which the new packet transmission will start. Signal strength measurements from real world indoor radio environment were combined with the adaptive algorithm and used as the basis for experiments described in this paper. The results obtained show that such a state-based approach can outperform the fixed-power transmissions in terms of cost and delay for a given packet success rate.

1 INTRODUCTION

Wireless sensor networks are increasingly used in a wide variety of indoor communications applications including smart home networks, office, industrial sensing and security surveillance. These sensors communicate with a central hub or base station, using either single or multi-hop links. Most of the sensors are battery powered with limited or no energy harvesting options. Battery life impacts the operational cost so it is important to minimize energy wastage as battery replacement is disruptive and expensive, as is the recycling and proper disposal of batteries.

Wireless channels have higher error rates and more bursty error patterns than do wire-line links. Due to limited lifetime of the sensor battery, minimizing power consumption in the sensors is crucial to reliable network operation. The retransmission of lost packets should be minimized as it increases the total cost of packet delivery.

There have been several energy models that have detailed the energy drainage pattern of wireless sensors (Karl & Willig 2005). Broadly speaking, a wireless node has three broad modes of operation: sensing, computational mode (involving the microcontroller) and the transmission/reception of data. Measurements indicate that communication is significantly more expensive than computation (Karl & Willig 2005) so the energy consumption calculations in this paper are based solely on current ratings during the transmission and reception of packets.

Indoor radio channels are considered to be very dynamic in time domain because of frequent movement of objects between (or near) a transmitter and a receiver. These can cause fluctuations of the received signal and therefore the received signal to noise ratio (Hashemi 1993, Lin *et al.* 2006). Static obstructions also have an effect. The focus of the paper is on single-hop communication between sensor nodes and a base station. It rules out any possibility of packet exchange between neighboring nodes for link/channel quality estimations in a multi-hop network. The performance of the proposed adaptive algorithm is evaluated based on its response to an unknown radio environment and compared with fixed-power transmission performance.

2 RELATED WORK

2.1 *Power saving solutions*

Power saving approaches can be broadly classified into Media Access Control (MAC) layer solutions and network layer solutions (Sheu *et al.* 2009).

In MAC layer solutions, energy efficiency is achieved by scheduling the sleep/wake-up times of the sensors. One major issue with energy-saving MAC protocols is the way they handle the data. If there is a fixed periodic sleep/wakeup time as in Sensor-MAC (S-MAC) (Ye et al. 2002), or dynamic SMAC which controls the periods of sleep/wake-up, then there can be latency issues, unlike in TDMA (Time Division Multiple Access), where the delay is constant. In wireless sensor networks, data transmission can be either periodic or event-driven. Under such circumstances, SMAC will fail completelybecause packets can get lost in between sleep/wake up time. In (Polastre et al. 2006), Polastre et al. propose aBerkeley-MAC (B-MAC) protocol that provides a flexible interface to obtain ultra-low power operation, effective collision avoidance, and high channel utilization. However, the inherent weakness of low-duty-cycled MAC protocols are that, in order for neighboring nodes to synchronize for receiving data, preamble packets are required before receiving message data. Even in low power wireless transceivers, such as theCC2420, Si4464/63/62/61/60 and nRF24 L01, the energy cost for reception is comparable to that fortransmission [Texas Instruments, Silicon Laboratories, Nordic Semiconductor]. Therefore thecost of a preamble exchange cannot be ignored, even if the preamble packet receive time is small compared to that of actual data packets.

The framework for adaptive transmission has been explained by Andrea Goldsmith in chapters 4 and 9 of (Goldsmith 2005). Link quality estimation is an integral part of adaptive transmission. This involves link monitoring, link measurement and metric evaluation (Baccour et al. 2012). The transceiver sends probe packets to the destination nodes, either in unicast or broadcast modes, and senses the channel usingfeedback packets before the actual transmission commences, which unfortunately consumes power.

2.2 Adaptive power control in wireless sensor networks

There are several transmission parameters that can be variedto suite the requirement of the application:

- transmission power
- modulation technique
- datarate
- error correction coding
- retransmission number control if using ARQ protocol

However, as mentioned in (Karl & Willig), transceivers use the mostenergy during transmission;so this paper focuses on adjusting transmission power to reduce the packet error rates, thereby minimizing the number of retries and extending the battery lifetime.

(Marabissi et al. 2008), present a power and retransmission control policy for use over a fading channel. The approach is not suitable for radios that can only switch between discrete transmit power levels. (Farrokh et al. 2008), utilized retransmission in a short range wireless network to reduce power consumption, butt heuse of AWGN (additive white Gaussian noise) channel is very simplistic. The use of Hybrid ARQ (HARQ) protocol in LTE (Long Term Evolution) has gained momentum in the past few years, with the general approach of dynamically adapting the coding rates for failed packet transmission to reduce packet losses. In (Soltani et al. 2011), an energy-efficient link layer protocol has been proposed that primarily targets delay sensitive applications such as real-time video communication. Unfortunately, there is a significant overhead in both data transmission and computational costs due to decoding (Karl & Willig).

In paper (Lin et al. 2006), Shan Lin and etal have introduced Adaptive Transmission Power Control (ATPC) that maintains a neighbor table at each node and a feedback loop for transmission power control between each pair of nodes. ATPC provided the first dynamic transmission power algorithm for WSN that uses all the available power output levels of CC2420 [Texas Instruments]. The CC2420 radio transceiver module operates in the ISM 2.4 GHz band and has 32 output power levels, ranging from 0 dBm to −25 dBm.

Practical-TPC (Lin et al. 2006), is a receiver oriented protocol that is considered robust against dynamic wireless environments and uses Packet Reception Rate (PRR) values to compute the transmission power that should be used by the sender in the next attempt. However, these power control protocols use initial probe packets for link estimation. In most of the adaptive power control algorithms for wireless sensors, the nodes exchange probe packets to build the model that relates packet reception ratio with LQI (Link Quality Indicator) and RSSI. While ATPC uses all 32 power levels, there are some algorithms that divide these 32 power levels into 8 levels, as in (Sheu et al. 2009). The work described is this paper aims to avoid the need for such probe packets.

2.3 Dynamic indoor radio channel

References from (Lin et al. 2006), and (Lai et al. 2003), suggest that the received E_b/N_0 value can vary cyclically over a period of 24 hours. Empirical values measured locally from RSSI readings of the beacon signals from a WiFi Access Point (AP) shows such a temporal variation. The Net Survey or tool [http://nutsaboutnets.com/netsurveyor-wifi-scanner/] was used to collect RSSI of beacon signals from a wireless access point. These RSSI values give a measurement of how thereceived E_b/N_0 over time. They are mapped with the received E_b/N_0 values with the maximum value of RSSI equal to 20 dB. The tool used samples the signals every 5 seconds. Therefore in our testing scenario, the model used the wireless sensors that send dataevery 5 seconds. Two sets of data werecollected from WiFi access points in public places: a University dining hall during busy hour (weekday), and a shopping mall during the busy hour (weekend).

Figure 1 showsa snapshot of the received E_b/N_0 values in these two radio environments. The duration of the plots is from 12:30 till 12:45 pm. The traditional power control protocols like ATPC and P-TPC use closed-loop feedback to adapt the link quality over time. While it is fine in a natural environment to exchange packets every hour to update the model, it can be extremely difficult to track the channel variation in a dynamic indoor radio channel environment.

Variations, even in the range of 5 to 10 dB, can cause a significant change in the packet success rate and the overall energy efficiency performance. The proposed algorithm is able to track these changes and find anenergy efficient way of switching between different states. In this paper 'states' refers to a*set of output power levels* that can be used are used to transmit packets. The next section describes the algorithm in detail.

3 ADAPTIVE POWER CONTROL ALGORITHM

The objectives of the energy-efficient adaptive power control algorithm are to:

- minimize retries inbad channel conditions by transmitting at the highest or maximum available output power level, and

- stay operating at low output power level when the channel condition is good because there is ahigh probability of successful packet transmission in that level.

3.1 *The adaptive power control model*

The adaptive power control has four states. In each state there is a predefined number of discrete output power levels. Four levels were used: −15 dBm, −10 dBm, −5 dBm and 0 dBm. These power levels are from the hardware specification of CC2420 transceiver module. Table 1 shows the states and the power levels used in each of the states.

3.2 *Description of the algorithm*

The output power levels are Low (L), −15 dBm: Medium (M), −10 dBm; High (H), −5 dBm, and Maximum (X), 0 dBm. Packet transmission always starts by using the lowest power level of astate. Each state represents one complete cycle of packet transmission. The state transition diagram is shown in Figure 2 and the corresponding state transition matrix is presented in Table 2. The state transition probabilities depend on the packet error probability in that state. It can be seen from the state transition table that the system uses transmission experience to determine the power level for the next fresh packet transmission. In each state, it will use a fixed number of power levels to send

Table 1. States and power levels.

State	1	2	3	4
Available power levels	L			
	M	M		
	H	H	H	
	X	X	X	X

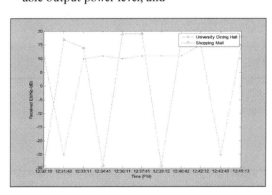

Figure 1. Snapshots of the variation of Eb/N0 in two locations.

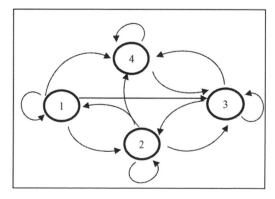

Figure 2. State transition diagram of the adaptive model.

Table 2. State transition matrix.

		Next state			
		1 (LMHX)	2 (MHX)	3 (HX)	4 (X)
Current State	1 (LMHX)	Succeed at level L	Succeed at level M	Succeed at level H	Failed or Succeed at level X
	2 (MHX)	Succeed at level M	Succeed at level H	Succeed at level X	Failed
	3 (HX)	No transition	Succeed at level H	Succeed at level X	Failed
	4 (X)	No transition	No transition	Succeed at level X	Failed

the packets but a failure will always retry at the next higher power level. Due to the limitations of the hardware, the output power step size is fixed at 5 dB.

3.3 Evaluation parameters

One of the parameters for the optimization is the energy consumed per useful bit transmitted over a wireless link (Lettieri *et al* 2002). Similarly, this paper uses the average cost of successful transmission as the optimization parameter. This is defined as:

$$Cost = \frac{Total\ Cost}{Total\ number\ of\ fresh\ packets\ sent - total\ number\ of\ failed\ packets}$$

(1)

All cost values are measured in *mA-sec*. The efficiency of the adaptive protocol is also defined by the average number of retries (Keshavarzian *et al.* 2007). This paper introduces *packet success rate with retries* as a new parameter for evaluation that takes into account the retries needed to achieve a particular packet success rate (PSR):

$$Packet\ Success\ Rate\ with\ retries = \frac{PSR}{1+ Average\ retry}$$

(2)

where

$$PSR = \frac{Total\ number\ of\ successfully\ transmitted\ packets}{Total\ number\ of\ fresh\ packets\ sent}$$

and

$$Average\ retry = \frac{total\ number\ of\ retries}{Total\ number\ of\ fresh\ packets\ sent}$$

Some applications of wireless sensors are sensitive to delays in data delivery. To evaluate the algorithm from this perspective, the cost and packet success rate with retries are compared to fixed-power transmissions in single-hop network. A retransmission penalty is added to the cost of transmission to account for the time needed to buffer the packet to be resent multiplied by the current consumption of that state of the transceiver. The buffered time is calculated based on the autocorrelation function in a fading channel that expresses how significantly the channel has changed over a time delay (T_d) for a given maximum Doppler spread (F_d). A high correlation value indicates that channel state has not changed enough to make packet error rate independent in each time slot of transmission. Due to this bursty behavior of the channel, errors tend to occur in blocks and the fading effect can affectthrough several packet transmissions (Sajadieh *et al* 1996). To avoid block fading effects, retransmitted packets are buffered and are sent in different time slots. The autocorrelation function for an indoor fading radio channel is a $sin(x)/x$ function where x is the product of F_d and delay (T_d) (Clarke & Khoo 1996).

This paper assumes an autocorrelation value of 0.75 or less for packet retransmission which accounts for the normalized delay value of 0.2 as shown from the autocorrelation plot of Figure 3. To find the buffer time for penalty calculations, it is necessary

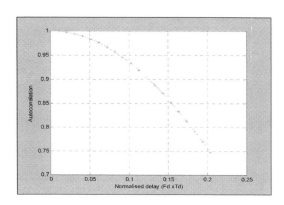

Figure 3. Autocorrelation plot in indoor radio fading channel.

to determine the maximum Doppler spread in the indoor radio environments. Results in (Howard & Pahlavan 1990) shows that the maximum Doppler spread value is around 20 Hz. Since we are considering all types of indoor radio environment, including factory setup as well as sensors with limited mobility options, the F_d in an indoor radio channel can be maximum around 40 Hz in 2.4 GHz ISM band. Therefore the delay or buffer time is approximately 5 milliseconds. The cost of buffering before retransmission in fixed-power transmission is the product of buffer time and the current used by the CC2420 while in theidle state (0.426 mA).

3.4 Results and discussions

The simulations are used to study the behavior of the adaptive algorithm when the channel condition is not changing with time and the average received E_b/N_0 is fixed. The simulation parameters are presented in Table 3. The evaluation parameters are compared with those withtransmissions at fixed power levels of –15 dBm, –10 dBm, –5 dBm and 0 dBm. Based on the path loss model for indoor radio communication [Universal Mobile], these power levels result inmean received E_b/N_0 values of 10 dB, 15 dB, 20 dB and 25 dB, respectively. In order to have a fair comparison, the transmission limit in fixed power mode is set at 4 as the adaptive system can use up to 4 power levels.

The algorithm wastested with these fixed E_b/N_0 values. The response of the system isshown in Figures 4 to 7, where it can be seenthat when channel E_b/N_0 is in the order of 25–20 dB, the system rarely switches to state 4, which will start withthe maximum output power level. However, when the received E_b/N_0 value has dropped to 10–15 dB, the system frequently switches to state 4.

The comparison of the packet success rate, cost and success rate with retries are presented in

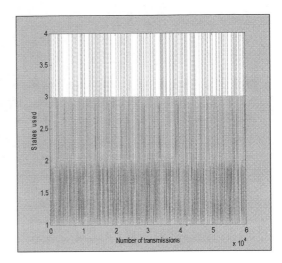

Figure 4. Adaptive 4 power levels used for fixed Eb/N0 = 10 dB.

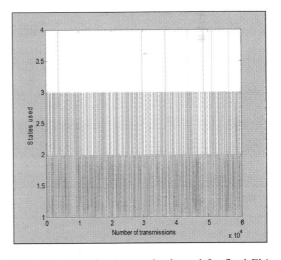

Figure 5. Adaptive 4 power levels used for fixed Eb/N0 = 15 dB.

Figures 8–10. These bar plots show that the adaptive protocol improves energy efficiency between 5% and 42%, while achieving a PSR of more than 99.97%. The average retry performance for a given PSR can also improve by up to 30%, depending on the output power level of transmission.

The next set of simulations used the empirical received E_b/N_0 values. Figures 11–16 compares the evaluation parameters in two different radio channel scenarios.

Figures 11 to 13 show that the adaptive system achieves its maximum energy efficiency when system has a PSR of around 94% and packet success

Table 3. Simulation parameters.

Modulation technique	DPSK
Channel data Rate	250 kbps
Maximum Doppler spread	20 Hz
Cyclic redundancy check	CRC-8
Multi-path Fading channel model	UMTS Indoor Office Test Environment [Universal Mobile]
Distance between sensor node and base station	20 meters (~65 ft)
Maximum number transmissions allowed when transmitting in fixed power	4

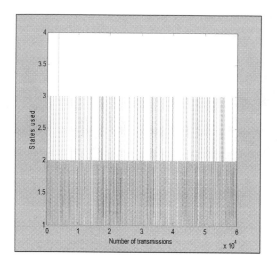

Figure 6. Adaptive 4 power levels used for fixed Eb/N0 = 20 dB.

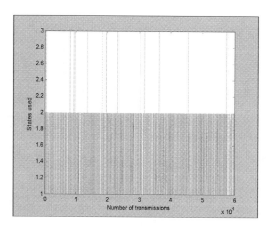

Figure 7. Adaptive 4 power levels used for fixed Eb/N0 = 25 dB.

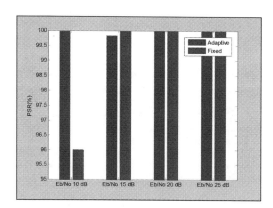

Figure 8. Comparison of the packet success rate (PSR) for the adaptive and fixed-power transmission protocols.

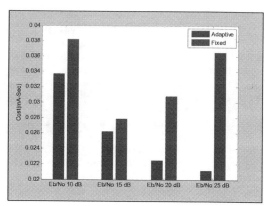

Figure 9. Comparing the power consumption of the adaptive and fixed-power transmission protocols.

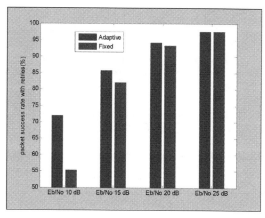

Figure 10. Comparing the packet success rate (with retries) of the adaptive and fixed-power transmission protocols.

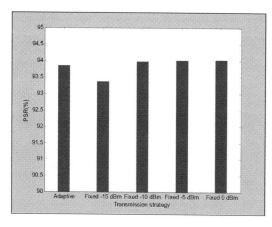

Figure 11. University dining hall: PSR comparison between the adaptive and fixed-power transmission protocols.

310

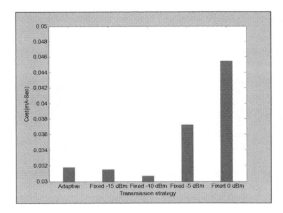

Figure 12.   University dining hall: Cost comparison of adaptive and fixed-power transmission protocols.

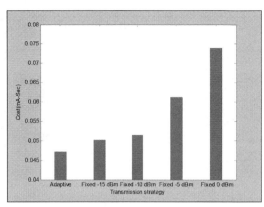

Figure 15.   Shopping mall: Cost comparison of adaptive and fixed-power transmission protocols.

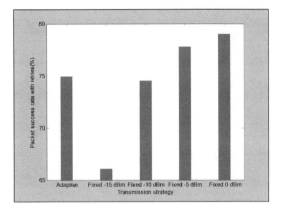

Figure 13.   University dining hall: Packet success rate with retries comparison of adaptive and fixed-power transmission protocols.

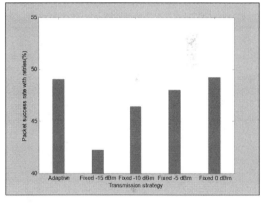

Figure 16.   Shopping mall: Packet success rate with retries comparison of adaptive and fixed-power transmission protocols.

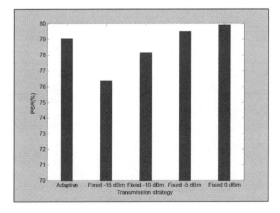

Figure 14.   Shopping mall: Comparing the PSR of the adaptive and fixed-power transmission protocols.

rate with retries of 75%. In fact, the energy efficiency has improved by a margin of up to 30%.

Figures 14 to 16 show that the adaptive system employed in the shopping mall scenario improves the energy efficiency by up to 36% for a given PSR of around 80% and packet success rate with retries of 49%.

4   CONCLUSION AND FUTURE WORK

This paper has proposed a state-based adaptive power control protocol that uses discrete power levels. This protocol was evaluated using real world channel characteristics derived from measurements of a populated indoor environment, and the current consumption of a widely used low-power transceiver. The experiments calculated the energy cost of transmission and the associated delay in delivery. Using the described protocol, there is marked improvement in the energy efficiency, especially when the radio channel conditions are varying widely and the transceiver is required to track the channel dynamics. The adaptive model can be represented using a state transition diagram and is therefore mathematically tractable and simple to implement. The practical implementation of the adaptive algorithm is planned to validate the results.

REFERENCES

Baccour Nouha, Koubaa Anis, Mottola Luca, Zuniga Marco Antonio, Youssef Habib, Boana Carlo Alberto, Alves Mario (2012), "Radio Link Quality Estimation in Wireless Sensor Networks: A Survey", *ACM Transactions on Sensor Networks*, Vol. 8, No. 4, Article 34, pp. 34:1–34:33.

Clarke Richard H. & Khoo Wee Lin (1997), "3-D Mobile Radio Channel Statistics", *IEEE Transactions on Vehicular Technology*, vol. 46, no. 3, pp. 798–799.

Farrokh Arsalan, Krishnamurthy Vikram, Schober Robert (2008), "Optimal Power and Retransmission Control Policiesover Fading Channels with Packet Drop Penalty Costs",IEEE *International Conference on Communications*, pp. 4021–4027.

Fu Yong, Sha Mo, Hackmann Gregory, Lu Chenyang (2012)," Practical Control of Transmission Power for Wireless Sensor Networks", *20th IEEE International Conference on Network Protocols (ICNP)*, pp. 1–10.

Goldsmith Andrea (2005), "Wireless Communication" *Cambridge University Press*, chapters 4 and 9.

[6] Hashemi H. (1993), "The Indoor Radio Propagation Channel", *Proceedings of the IEEE, vol. 81, No. 7* pp. 943–968.

Howard S.J., Pahlavan K. (1990), "Doppler Spread Measurements Of Indoor Radio Channel", *Electronics Letters* Vol. 26 No. 2, pp. 107–108.

Karl H. & Willig A. (2005),"Protocols and Architectures for Wireless Sensor Networks", *Wiley publishers*, chapter 2, pp. 36–44.

Keshavarzian A., Uysal-Biyikoglu E., Lal D., and Chintalapudi K. (2007), "From Experience with Indoor Wireless Networks: A Link Quality Metric that Captures Channel Memory", *IEEE Communications Letters*, Vol. 11, No. 9, pp. 72922–731.

Lai Dhananjay, Manjeshwar Arati, Herrmann Falk, Uysal-Biyikoglu Elif, Keshavarzian Abtin (2003), "Measurement and characterization of link quality metrics in energy constrained wireless sensor networks", *Global Telecommunications Conference*, pp. 446–452.

Lettieri Paul, Schurgers Curt, Srivastava Mani (1999), "Adaptive link layer strategies for energy efficient wireless networking", *Wireless Networks*, Volume 5, Issue 5, pp 339–355.

Lin S., Zhang J., Zhou G., Gu L., He T. & Stankovic J. A. (2006), "ATPC: Adaptive Transmission Power Control for Wireless Sensor Networks," SenSys'06, pp. 223–236.

Lin Shan, Zhang Jingbin, Zhou Gang, Gu Lin, He Tian, Stankovic John A. (2006), "ATPC: Adaptive Transmission Power Control for Wireless Sensor Networks," *SenSys'06*, November 1–3, pp. 223–236.

Marabissi Dania, Tarchi Daniele, Fantacci Romano & Balleri Francesco (2008), "Efficient Adaptive Modulation and Coding techniques for WiMAX systems", *IEEE communication society*, pp. 3383–3387.

Nordic Semiconductor, nRF24 L01 Single Chip 2.4 GHz Transceiver, https://www.sparkfun.com/datasheets/Components/SMD/nRF24 L01Pluss_Preliminary_Product_Specification_v1_0.pdf

Polastre Joseph, Hill Jason, Culler David (2004), "Versatile Low Power Media Access for Wireless Sensor Networks", *SenSys'04*, November 3–5, pp. 95–107.

Sajadieh M., Kschischang F.R., and Leon-Garcia A. (1996), "A Block Memory Model for Correlated Rayleigh Fading Channels", *IEEE International Conference on Communications, ICC '96*, Vol 1, pp. 282–286.

Sheu J., Hsieh K. & Cheng Y. (2009), "Distributed Transmission Power Control Algorithm for Wireless Sensor Networks", *Journal Of Information Science And Engineering 25*, pp. 1447–1463.

Silicon Laboratories, Si4464/63/62/61/60 datasheet, http://www.silabs.com/Support%20Documents/TechnicalDocs/Si4464–63–61–60.pdf.

Soltani Sohraab, Ilyas Muhammad U., Radha Hayder (2011)," An Energy Effcient Link Layer Protocol for Power-Constrained Wireless Networks", *Proceedings of 20th International Conference on Computer Communications and Networks (ICCCN)*, pp. 1–6.

Texas Instruments, CC2420 datasheet, http://www.mtl.mit.edu/Courses/6.111/labkit/datasheets/CC2420.pdf.

Universal Mobile Telecommunications System (UMTS); Selection procedures for the choice of radio transmission technologies of the UMTS (UMTS 30.03 version 3.1.0) http://www.etsi.org/deliver/etsi_tr/101100_101199/101112/03.01.00_60/tr_101112v030100p.pdf.

Ye Wei, Heidemann John, Estrin Deborah (2002), "An Energy-Effcient MAC Protocol for Wireless Sensor Networks", *IEEE INFOCOM*, pp. 1567–1576.

http://nutsaboutnets.com/netsurveyor-wifi-scanner/

Frontiers in Computer, Communication and Electrical Engineering – Acharyya (Ed.)
© 2016 Taylor & Francis Group, London, ISBN: 978-1-138-02877-7

The present energy scenario and need of microgrid in India

Arshdeep Singh & Prasenjit Basak
Department of Electrical and Instrumentation Engineering, Thapar University, Patiala, India

ABSTRACT: The rapid development of economy has led to an increase in energy crisis and environmental pollution. So, maximum utilization of renewable energy resources is in hour of need. There is a huge amount of various renewable energy resources that are available in India. In the present scenario, microgrid is one of the alternatives for electricity generation. It can satisfy the customers load demand at optimum cost by reducing the grid congestion. At present the paper deals with the current power generation scenario, availability, and utilization of renewable resources in India. Aspects of microgrid in India have been explored. Salient features of microgrid operation, such as renewable energy resources interconnection, meeting of load demand through grid connected operation, also including the scope of supplying surplus power to the grid have been verified through modeling and simulation using MATLAB software. Dispatch scheme has been studied in simulation and the results are analyzed.

1 INTRODUCTION

At present the demand of electricity is increasing with the need of improved energy utilization efficiency and reliability of the power system network. To cope up with this changing scenario, the new renewable energy based on power generation techniques are to be implemented by those having less carbon footprints. Microgrid is one of the expected local power supply system that consists of distributed generators, loads, power storage devices, and heat recovery equipment's (Li *et al.* 2007). Microgrid provides higher flexibility and reliability as it is able to run in both grid-connected and islanded mode of operation. It is connected to both the local generating units and the utility grid thus preventing power outages and the excess power can be sold to the utility grid. The microgrid has to be designed in such a manner so that there is ease in installation, commissioning, operation, and maintenance. The microgrid helps in reducing the congestion cost by reducing network congestion, line losses, and line costs and there by rising energy efficiency [Singh *et al.* 2014]. It has many other important aspects such as reduction in transmission losses, utilization of large land by installing PV generators and reduction in environmental impacts of existing fossil fuels are based on the generating systems.

The choice of Distributed Energy Resources (DER) for microgrid depends on the geographical conditions of the site, the climate and topology of the region, and fuel availability. DER technologies like Combined Heat and Power (CHP) systems, wind energy conversion system, solar PhotoVoltaic (PV) systems, small-scale hydroelectric generation, storage devices are all suitable for India [Sule *et al.* 2011]

According to the geographical location of India a good amount of solar energy potential is available. Moreover, most of the population is dependent upon agriculture and a huge amount of biomass is available. India has a large number of sugar industries. The biomass available with the sugar industry can be utilized in combined heat and power generation in the industry itself. In the southern region good amount of wind power is available along the coastal area. So the combination of DER's like wind energy conversion system, photovoltaic systems, biomass, CHP, and small-scale hydroelectric generation with storage devices are highly acceptable for India [Kaushal *et al.* 2014].

Different types of biomass can be burned or digested to produce energy. Biomass is a kind of versatile material and can be used to produce electricity, Additionally, combination of heat and power in a combined heat and power plant, generation of heat for space and hot water, a liquid fuel like biodiesel and vegetable oil all can be used to power vehicles or produce heat energy [Chouhan *et al.* 2014].

2 ASPECTS OF MICROGRID IN INDIA

The standard of living of low-income households in rural areas suffered from energy, poverty, and lack of human and economic development. Indian government is committed to implement reforms

Table 1. Installed capacity of energy resources in various regions of India.

Region	Coal (MW)	Gas (MW)	Diesel (MW)	Nuclear (MW)	Hydro (MW)	Rene-wable (MW)
North	39431	5331	12	1620	16666	5935
West	62384	10915	17	1840	7447	11271
South	28892	4962	939	2320	11398	13784
East	27727	190	17	0	4113	432
North East	60	1571	142	0	1242	256
Islands	0	0	70	0	0	11

Table 2. Energy resources available in various sectors of India.

Region	Coal (MW)	Gas (MW)	Diesel (MW)	Nuclear (MW)	Hydro (MW)	Renewable (MW)
Central	46775	7428	0	5780	10661	0
State	55890	6974	602	0	27482	3803
Private	55830	8568	597	0	2694	27888
All India	158495	22971	1199	5780	40867	31692

and encourage foreign investment to improve the living conditions of 1.1 billion citizens. This commitment increases the GDP growth rate to 8.2% (As per the data given by Trading Economics). However, because of its technically inefficient electrical grid, India losses money for every unit of electricity sold. So, it is mandatory for India to build a smart grid power infrastructure concept, as it has one of the largest grid networks in the world. This smart grid concept should be technically perfect and financially secured.

Instead of single source, the hybrid renewable energy has greater potential to provide higher quality to customers. A very large percentage of Indian population is dependent on agriculture. So, agriculture waste in the form of biomass can be utilized in producing electricity. These, renewable energy sources can complement each other in a healthy way. Hybrid energy sources are commercial, and can be used in remote areas for power applications where the grid supply is either not possible, or expensive. Despite of all the sources of renewable energy sources, the solar energy has got greater attention because of its availability in all regions of the world.

2.1 All India installed capacity (MW) as on 28-02-2015 region—wise

(Executive Summary report from Ministry of Power Central Electricity Authority New Delhi-February 2015)

Installed Capacity of Energy resources in various regions of India are listed in Table 2. In southern region, the production of electricity is maximum from renewable energy based plants as compared to other plants due to the utilization of

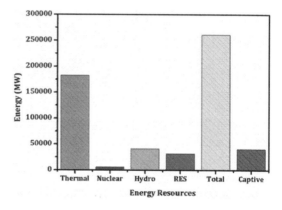

Figure 1. Graphical representation of installed capacity of energy resources in various regions of India.

wind energy. Also in western region, the renewable energy resources are playing a vital role in energy production.

Figure 1 shows the graphical representation of energy resources available in various regions in India. It depicts, that the total energy production is 261006.46 MW in India as on Feb 2015. Maximum energy of 182666.89 MW is produced using coal. Hydro, RES, and nuclear produces 40867.43 MW, 31692.14 MW, and 5780.00 MW respectively.

Figure 1 also reveals that approximately 12% of the total installed capacity of power generation in India is contributed by the renewable energy resources. Whereas, nuclear energy technology is contributes only 2.21% of the total present capacity. This national status of energy resources shows the great aspect of installation of renewable energy based microgrid system. Graphical representation of sector-wise pro-

314

duction of energy from various resources have been represented as a pi-chart in Figure 2.

2.2 *All India installed capacity (MW)*

Table 2 shows the production of energy from the different sectors such as Central, State, and Private using various sources.

At central level the major part of the energy 46775.01 MW is produced using coal, also it produces energy of 5780.00 MW using the nuclear resources which is zero in case of state and private sectors. Hydro energy is also produced in large amount of 10661.43 MW. State sector is utilizing maximum of the Hydro energy sources for producing 27482.00 MW and simultaneously it also produces 55890.50 MW using coal. Private sector is producing 27888.47 MW using RES which is more than the other sectors.

2.3 *Availability of renewable energy sources in India*

2.3.1 *Solar*

India is endowed with vast solar energy potential. Every year about 5,000 trillion kWh energy incidents over India's land area with most of the parts receiving 4–7 kWh per sq. m per day. Hence, both technology routes for conversion of solar radiation into heat and electricity, namely, solar thermal and solar photovoltaic's, can effectively be harnessed providing huge scalability for solar in India.

2.3.2 *Wind*

The potential for wind power generation for grid interaction has been estimated about 1,02,788 MW taking sites having wind power density greater than 200 W/m^2 at 80 m hub-height with 2% land availability in potential areas for setting up wind farms i.e. 9 MW/km^2.

2.3.3 *Biomass*

The current availability of biomass in India is estimated at about 500 million metric tons per year. Studies sponsored by the Ministry has estimated surplus biomass availability at about 120–150 million metric tons per annum covering agricultural and forestry residues corresponding to a potential of about 18,000 MW (As per the information given by Ministry of New and Renewable Energy, India).

From the above data it has been clear that renewable sources energy are abundantly available in India in various regions, so there is a dire need to utilize all these for the welfare of mankind. Table. 3 given below shows the installed capacity of various renewable energy sources.

From the above discussion it has been found that renewable sources of energy are available in greater

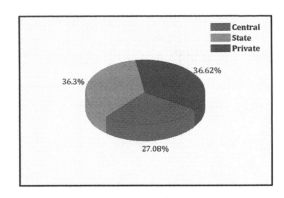

Figure 2. Graphical representation of sector-wise production of energy from various resources.

Table 3. Installed capacity of various renewable energy sources as on March, 2015 (according to the physical progress report by ministry of new and renewable energy, India, March 2015).

Installed capacity of various renewable energy sources as on March, 2015	
Type of resource	Capacity(MW)
Wind power	23444.00
Biomass power & gasification	1410
Solar power	3743.97

amount but we are utilizing only 7.8% of the available biomass and 22.8% of the available wind for the generation of power. Also, in case of solar as compared to availability the usage is very less.

3 SIMULATION OF MICROGRID

This work aims to develop a dispatching scheme for 24 hours of different distributed resources such as solar and wind. It depends upon the weather conditions. The microgrid model consisting of PV and wind turbine using MATLAB software.

3.1 *Modeling of a microgrid*

Figure 3 shows the simple model of a microgrid, which consists of two renewable energy resources i.e. PV and wind. The microgrid is connected to a particular load and utility grid.

Its operation depends upon the weather conditions affecting power output of PV array and wind turbine which are delivered to a microgrid. The operating cost of PV and wind turbine is very less and the power generated from the solar and wind energy is free from pollution. So maximum amount of power generated from these two

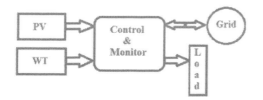

Figure 3.   Schematic of simple microgrid.

sources is delivered to the load. If at any time the total power generated by the PV and wind generator is less than the load demand then the rest of the power is taken from utility grid connected to the microgrid, so that at every hour the load demand is fulfilled. In case, at any hour the total power generated from these two sources is greater then the load demand then the microgrid will supply the surplus power to utility grid.

3.2   Problem formulation

3.2.1   Wind turbine

Wind turbine is used to convert the kinetic energy of wind into mechanical energy and then the mechanical energy is used to produce electrical energy. The power generated by the wind generator depends upon wind speed, size of turbine, air density, and the efficiency factor Cp [10].

$$P_{WT} = \frac{1}{2}\rho * \pi R_{WT}^2 * V^3 * C_P \qquad (1)$$

Where, P_{WT} is the quantity of power generated by wind turbine. ρ, V, R_{WT} are the air density, wind speed, and the blade radius of the turbine respectively. C_P represents the wind energy utilization efficiency.

For a fixed wind turbine and installation site, we can assume the air density and blade radius as constant. The relationship between P_{WT} and V can be written into a MATLAB function as:

$$P_{WT} = 0, V < V_{ci}$$

$$P_{WT} = aV^3 + bV^2 + cV + d, V_{ci} < V < V_r$$

$$P_{WT} = P_r, V_r < V < V_{co}$$

$$P_{WT} = 0, V < V_{co}$$

Where P_r, V_r, V_{ci}, V_{co} are the rated power, rated wind speed, cut-in, and cut-out wind speed respectively. Moreover, the constants a, b, c, and d depend on the type of WT and the installation site. The type of wind turbine in this article is selected as AOC 15/50 made by Seaforth Energy, Inc. and the

parameters used to model the power output-wind speed curve are as follows [Deng et al. 2010]:

$$V_{ci} = 5.26m/s, V_r = 14.68m/s, V_r = 22.40m/s$$

$$P_r = 50\ kw, a = -0.0609, b = 1.7882,$$

$$c = -10.8347, d = 16.3773$$

3.2.2   PV array

From a single module we cannot generate enough power to meet our load demand. A linked collection of solar panels used to convert the sunlight into electricity is called PV array. PV array has no moving parts and generate electricity without emission.

The output power of the module can be calculated as [10]:

$$P_{pv} = P_{STC}\ G_{ING}/G_{STC}\left((1 + K(T_c - T_r)\right) \qquad (2)$$

where: P_{pv} -Output power of the module at Irradiance (G_{ING}), P_{STC} -Module maximum power at Standard Test Condition (STC), G_{ING} -Incident Irradiance, G_{STC} Irradiance at STC 1000 (W/m²), k -Temperature coefficient of power, T_c -Cell temperature, T_r -Reference temperature. In the micro grid studied in this article the maximum power of pv array measured in STC is

$$P_{STC} = 25kw, K = 0.1067$$

3.2.3   Weather conditions

For mathematical modeling of PV and wind turbine in MATLAB, we need weather conditions including temperature, wind speed, light intensity along with its variation with time that is taken from [Mohamed et al. 2010]. Figures 4, 5, and 6 shows the variation of temperature, solar radiation and wind speed with time.

4   RESULTS AND DISCUSSION

Power output and load demand has been shown in Figure 6. It depicts that during first four hours there is less power consumption and the output power from the wind turbine is more than the demand. After 4th hour solar power also is also available and simultaneously the load is increasing. In first six hours power from PV and wind turbine is more than the load demand. During this interval of time the surplus power has been sent to the grid.

After 6th hour the power demand is increasing continuously and reach its maximum value at 15th hour, therafter load demand decreases as depicted in Figure 7. The power output of PV and wind

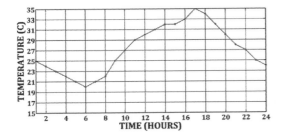

Figure 4. Temperature variation with time.

Figure 5. Light intensity variation with time.

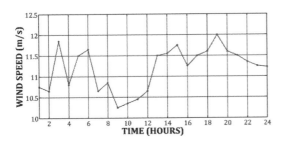

Figure 6. Wind speed variation with time.

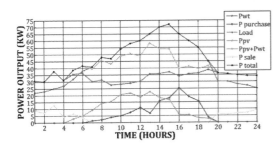

Figure 7. Power output and load demand.

turbine is not sufficient to meet the load demand from 6th to 20th hour. During this time interval microgrid provides rest of the power from the utility grid. From 20th to 24th hour solar power is not available, but power output of the wind turbine is more than the requirement. Also, at this time interval microgrid supply the surplus power to the grid.

5 CONCLUSION

Nowadays at central level the major part of the energy is produced using coal. As good potential of several renewable energy resources is available in India, but utilization is very less. Microgrid seems to be the reliable solution for the problems such as pollution and energy crisis. From the simulation result, it has been found that during 24 hour of load demand PV and wind turbine can contribute a good amount of power which is environment friendly. It reduces the overall load of utility grid. The surplus power can be sold to the utility grid which is economically favorable. Various distributed energy resources can be used as per availability. Using optimisation techniques the most suitable combination of DER's can be assembled for a particular load demand for successful implementation of this emerging technology.

REFERENCES

Chouhan, K., Ladhe, Y., Upadhayay. V. (2014). Biomass a Versatile Fuel for Energy and Power Generation. IOSR Journal of Mechanical and Civil Engineering (IOSR-JMCE). 8–11.

Deng, Q., Gao X., Zhou, H., Hu, W. (2011). System modeling and optimization of microgrid using Genetic Algorithm. The 2nd International Conference on Intelligent control and Information Processing.

Kaushal, J., Basak, P. (2014). The Deployment of Microgrid as an Emerging Power System in India and its Simulation using Matlab-Simulik. International Journal of Advanced Research in Electrical, Electronics and Instrumentation Engineering. 3(4), 9163–9170.

Li, Xi., Song, Y.J., Han, S.B. (2007). Study on power quality control in multiple renewable energy hybrid microgrid system. In: Proceedings of the Power Tech.

Mohamed, F.A., Koivo, H.N. (2010). System modelling and online optimal management of MicroGrid using Mesh Adaptive Direct Search, International Journal of Electrical Power & Energy Systems.32(5),398–407.

Singh, A., Surjan, B.S. (2014). MICROGRID: A REVIEW, International Journal of Research in Engineering and Technology 3(2), 185–198.

Sule, V., Kwasinski, A. (2011). Active Anti-Islanding Method Based on Harmonic Content Detection from Overmodulating Inverters Presented at Applied Power Electronics Conference and Exposition (APEC), 2011 Twenty-Sixth Annual IEEE\

http://www.tradingeconomics.com/india/gdp-growth-annual

http://www.cea.nic.in/reports/monthly/executive_rep/feb15.pdf

http://mnre.gov.in/schemes/grid-connected/solar-thermal-2/

http://mnre.gov.in/mission-and-vision-2/achievements/

Frontiers in Computer, Communication and Electrical Engineering – Acharyya (Ed.)
© *2016 Taylor & Francis Group, London, ISBN: 978-1-138-02877-7*

Influence of temperature on field emission from finite barrier quantum structures in presence of image force

Shubhasree Biswas Sett
The Institution of Engineers (India), Kolkata, India

Chayanika Bose
Department of Electronics and Telecommunication Engineering, Jadavpur University, Kolkata, India

ABSTRACT: The influence of temperature on the field emission from a finite quantum well is investigated in this communication. Here a constant carrier concentration is maintained for a particular quantum well by adjusting the Fermi Energy with variation in temperature. The emission is found to depend both on the position of the modified Fermi Energy and the number of electric subbands contributing to the emission.

1 INTRODUCTION

Field emission due to Fowler Nordheim Tunnelling from infinite quantum wells is well studied since late 80s (Majumdar *et al.* 1987, Bose *et al.* 1987, Sett *et al.* 2012 a,b). In all the cases, WKB approximation (Davis 1998, Harrison 2005) was used to estimate the probability of electron-tunnelling through the barrier. The current density is observed to vary exponentially with the applied field and oscillate with increasing well thickness. In (Sett *et al.* 2014), field emission from finite quantum structures were analysed in presence of image charge originating due to discontinuity in permittivity at the well boundary (Delerue *et al.* 2004). The emission is found to increase for finite structures with higher barriers and also due to incorporation of effect of image charge. In the present communication, we study the influence of temperature on the field emission from a finite quantum well in presence of image force.

2 THEORY

Let us consider a quantum well (QW) of width a ($\sim$ nm) in the z direction, where the width in the nanometre scale results in quantization of energy levels. When an electric field of strength F ($\sim 10^9$ V/m) is applied to the well in $-z$ direction, field emission is observed and the general expression of the current density is given by (Majumdar *et al.* 1987)

$$J = \frac{1}{2}\sum_n ev_n \Delta n_n t \qquad (1)$$

where e is electronic charge; the factor ½ includes electrons moving only along the field direction; Δn_n is the carrier concentration of the nth energy level and v_n is their velocity in the direction of the applied field given by

$$v_n = \frac{\hbar}{m^*}\left(\frac{n\pi}{a}\right) \qquad (2)$$

The transmission coefficient t, under the influence of image force, takes the form (Bose *et al.* 1987)

$$t = \exp\left[-\frac{4}{3}\left(\frac{2m^*}{\hbar^2}\right)^{1/2}(V_0 - E_n)^{3/2}\right]\theta(\alpha) \qquad (3)$$

where, V_0 is the height of the potential energy barrier, E_n is the energy of the nth quantum level and

$$\theta(\alpha) = \frac{1}{\sqrt{2}}\left(\frac{1-\alpha^2}{\sqrt{1+\alpha}}K(\lambda) + \sqrt{1+\alpha}E(\lambda)\right) \text{ with } K(\lambda)$$

and $E(\lambda)$ as complete elliptical normal integrals of first and second kind respectively.

Here the following substitutions are used

$$\lambda = \sqrt{\frac{2\alpha}{1+\alpha}}$$

$$1 - y^2 = \alpha^2,$$

$$y = \frac{e\sqrt{eF}}{\sqrt{4\pi\epsilon_s W_n}}, \text{ where } \epsilon_s \text{ is permittivity of the}$$

semiconductor barrier material and $W_n = V_0 - E_n$

At temperature $T = 0$ K, Δn_n can be written as

$$\Delta n_n = \frac{m^*}{\pi \hbar^2 a}\left(E_F - E_n\right) \tag{4}$$

whereas, for higher temperatures, the carrier concentration for nth level should be taken as (Bandyopadhyay *et al.* 2008)

$$\Delta n_n = \frac{m^* k_B T}{\pi \hbar^2 a}\ln\left(1 + \exp\left(1 + \frac{\left(E_F - E_n\right)}{k_B T}\right)\right) \tag{5}$$

The total carrier concentration considering all the subbands can therefore be expressed as

$$n_0 = \frac{m^*}{\pi \hbar^2 a}\sum_n \left(E_F - E_n\right), \quad \text{for } T = 0 \text{ K} \tag{6}$$

and

$$n_0 = \frac{m^* k_B T}{\pi \hbar^2 a}\sum_n \ln\left(1 + \exp\left(1 + \frac{\left(E_F - E_n\right)}{k_B T}\right)\right), \tag{7}$$

for higher temperatures

Substituting equation (2), equation (3) and equation (7) in equation (1), the current density can finally be formulated as

$$J = \frac{e k_B T}{2 \hbar a^2}\sum_n n \ln\left(1 + \exp\left(1 + \frac{\left(E_F - E_n\right)}{k_B T}\right)\right)$$
$$\times \exp\left(-\frac{4}{3}\left(\frac{2m^*}{\hbar^2}\right)^{1/2}\frac{1}{eF}\left(V_0 - E_n\right)^{3/2}\theta(\alpha)\right) \tag{8}$$

The effect of temperature T on the field emission from a finite quantum well is studied using equation (8).

3 RESULTS AND DISCUSSION

We consider an n-type GaAs film sandwiched between two $Al_x Ga_{1-x} As$ barrier layers to form the QW, where x is the Al-composition. The electronic energy levels in such finite barrier structures are computed by graphical solution of transcendental equations. The parameters used for simulation are taken from (Adachi 2008).

At $T = 4.2$ K, the carrier concentration for nth energy level Δn_n, as given by equation (4) is initially considered for a 10 nm QW. The Fermi energy is

computed using equation (6). As the temperature is increased the electrons are redistributed among the energy levels following Fermi Dirac distribution function $f(E)$. The highest contributing energy level is now limited to the energy (E_F') at which the Fermi Dirac distribution function attains $(1/e)$ of its value at $E = E_F$, i.e., E_F' should be determined from $f\left(E_F'\right) = \frac{1}{e}f\left(E_F\right)$. The resulting current density due to the applied field is computed using equation (8). The variations of current density for different temperatures are shown in Figure 1. The current density is found to increase with increasing temperature. For $T = 77$ K and $T = 4.2$ K the field emission currents are more or less same. At $T = 300$ K, the emission increases considerably due to contribution from an additional energy level.

However, while using the above method, the total carrier concentration increases at higher temperatures when the spread of Fermi Dirac distribution function increases. But this is not acceptable for a system having a specific doping level. To keep the carrier concentration unchanged, the Fermi energy itself needs to be modified with changing temperature (Davis 1998). Equation (7) is therefore employed to compute the modified Fermi energy for each temperature. Table 1 presents these Fermi energies for the cases considered in Figure 1. It is observed that the Fermi energies are lowered for higher temperatures to compensate the increase in carrier concentration.

These modified Fermi energies for the 10 nm QW are used for computation of current densities with increasing field strength and are plotted in Figure 2. The current density in this case decreases

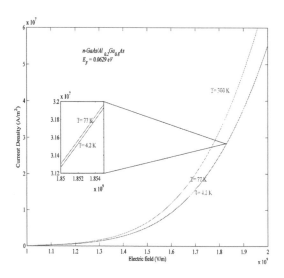

Figure 1. Field emission current density vs. electric field for a 10 nm QW at different temperatures.

Table 1. Fermi Energies at different temperatures for a n-GaAs/Al$_{0.2}$Ga$_{0.8}$ As QW with $n_0 = 10^{24}$ m^{-3}.

Sl No.	Temperature	$a = 10$ nm		$a = 20$ nm	
		Fermi Energy	Number of Contributing Energy Levels	Fermi Energy	Number of Contributing Energy Levels
	K	eV		eV	
1	4.2	0.062	1	0.059	2
2	77	0.056	2	0.056	3
3	300	0.037	2	0.046	4

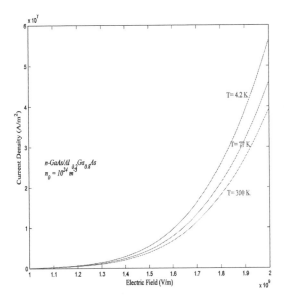

Figure 2. Field emission current density vs. electric field for a 10 nm QW at different temperatures.

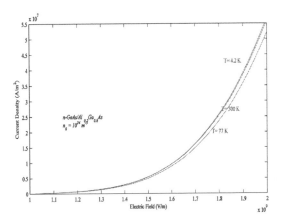

Figure 3. Variation of current density with electric field for a 20 nm n-GaAs/Al$_{0.2}$Ga$_{0.8}$ As QW at different temperatures.

with temperature. This can be explained with the help of equation (8), which clearly indicates that as Fermi energy decreases, the factor $(E_F - E_n)$ decreases, which in turn reduces the overall carrier concentration. This results in lowering of current density.

Now, we consider another QW of length 20 nm and the computed results are plotted in Figure 3. However, in this case, the emission for $T = 300K$ is higher than that for $T = 77$ K. Here the current density differs due to the change in number of contributing levels, as shown in Table 1. At $T = 300$ K, an additional energy level takes part in the emission, in comparison to that at $T = 77$ K. The number of energy levels taking part in the emission for both the temperatures remains same for the 10 nm well, as evident from Table 1.

4 CONCLUSION

In this paper, the influence of temperature on the field emission from a finite quantum well is formulated and the results are computed for a n-GaAs/Al$_{0.2}$Ga$_{0.8}$ As QW. It is observed that for maintaining a constant carrier concentration, as temperature is varied, the Fermi energy needs to be modified, which is found to decrease with increasing temperature. This results in reduction in the magnitude of current density for higher temperatures for the 10 nm QW. For a 20 nm QW, due to its larger width, inter subband separation decreases and thereby number of contributing level increases, resulting in increase in field emission current density with temperature.

REFERENCES

Adachi, S. 1985. GaAs, AlAs, and Al$_x$Ga$_{1-x}$ As: material parameters for use in research and device applications. *J. Appl. Phys.* 58(3): R1–R29.

Bandyopadhyay, S. & Cahay, M. 2008 *Introduction To Spintronics*. Boca Raton: Taylor & Francis.

Bose, M.K. Majumdar, C. Maity, A.B. & Chakravarti, A.N. 1987. Field emission from ultrathin films of semiconductors with image forces. *Phys. Stat. Sol. (B)* 143: 113–120.

Davies, J.H. 1998. *The physics of low-dimensional semiconductors an introduction*. Cambridge: Cambridge University Press.

Delerue, C. & Lannoo, M. 2004. *Nanostructures: theory and modelling*. New York: Springer-Verlag Berlin Heidelberg.

Harrison, P. 2005. *Quantum wells, wires and dots: theoretical and computational physics of semiconductor nanostructures*. England: John Wiley & Sons.

Majumdar, C. Bose, M.K. Maity, A.B. & Chakravarti, A.N. 1987. Effect of size quantization on field emission from ultrathin films of degenerate wide-gap semiconductors. *Phys. Stat. Sol. (B)* 141: 435–439.

Sett, S.B. & Bose, C. 2012a. Fowler-Nordheim tunneling from quantum wires of different cross-sections. *International Conference on Communications, Devices and Intelligent Systems* 172–175.

Sett, S.B. & Bose, C. 2012b. Influence of geometrical shape of quantum dot on field emission. *5th International Conference on Computers and Devices for Communication*. 1–4.

Sett, S.B. & Bose, C. 2014. Field emission from finite barrier quantum structures. *Physica B: Condensed Matter* 450: 162–166.

Frontiers in Computer, Communication and Electrical Engineering – Acharyya (Ed.)
© 2016 Taylor & Francis Group, London, ISBN: 978-1-138-02877-7

Tuning of fractional-order PID controller—a review

D. Shah, S. Chatterjee & K. Bharati
Department of Electrical and Electronics Engineering, NIT, Yupia Deemed University, Arunachal Pradesh, India

S. Chatterjee
Department of Electrical Engineering NERIST, Nirjuli, Arunachal Pradesh, India

ABSTRACT: Fractional-order PID controller is a generalization of classical integer order PID controller using the concept of fractional calculus. FOPID controller is far more superior to its integer order counterpart, besides setting proportional, integral, and derivative gain; it has two more parameters that is the order of integrator λ and differentiator μ, which widens the scope of control system design. This paper gives an overview of the various tuning method of FOPID controller available in literature. The tuning method include: Ziegler Nichol type tuning rule, Padula and Visioli method, frequency domain design, time domain design, and IMC tuning. The difficulties in the implementation of FOPID controller has also been indicated.

1 INTRODUCTION

Fractional order PID controller, proposed by Podlubny is a generalization (Podlubny, 1999) of the classical integer order PID controller using the concept of fractional calculus (Miller & Ross 1993). A FOPID controller is characterized by five parameters, i.e., the proportional gain, the integral gain, the derivative gain, the integral order, and the derivative order. Many researcher's results show that FOPID controller has better performance and robustness than classical PID controller (Chen, Hu & Moore 2003). Though it is so, the parameter tuning of FOPID controller is an important and critical issue. In literature, many approaches have been documented to design FOPID controller which is based on both analytic and heuristic search method.

2 FRACTIONAL ORDER PID CONTROLLER

Two more parameter of FOPID widens the scope of control system design. The fractional order controller generalizes the integer order PID controller design and expands it from point to plane (Monje, Chen, Vinagre, Xue & Feliu-Batlle 2010). This expansion adds more flexibility to control system design and we can control our real world process more accurately (Valerio & da Costa 2010). This is shown in Figure 1.

With the help of fractional order PID controller it is possible to design a controller to ensure that the

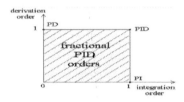

Figure 1. Expanding the order of controller from point to plane.

closed loop system is robust to gain variation and step response exhibits an iso-damping property (phase-bode plot of the system is flat at the gain crossover frequency) as a result system gets more robustness to gain variation and aging effects (Chen, Moore, Vinagre & Podlubny). The width of the flat phase region can be adjusted with the help of fractional order phase shaper (Monje, Vinagre, Feliu & Chen 2008) resulting into a more robust system. Most of the real world process can be efficiently modeled with the help of fractional order differential equation rather than integer ones. Therefore, fractional order controller is naturally suitable for these fractional order models to imply a better control.

3 REVIEW OF TUNING METHOD OF FOPID CONTROLLER

Recently lots of tuning techniques have been proposed by many researchers among which some are reviewed here. The various tuning method are organized in this section as:

1. Ziegler Nichol type tuning rule
2. Padula and Visioli method
3. Frequency domain design method
4. Time domain design
   - Dominant pole placement method
   - Optimal tuning based on time domain integral performance index
5. IMC Tuning

3.1 Ziegler-nichol tuning rule

This tuning rule (Valerió & da Costa 2005) assumes the dynamics of system to have s-shaped unit step response and have transfer function of the form given by:

$$G(s) = \frac{K}{1+sT} e^{-Ls} \tag{1}$$

When the minimization tuning method, proposed by Monje (Monje, Calderon, Vinagre, Chen & Feliu 2004) was applied to plants given by (1) for several values of L and T, with $K = 1$. Thus, the parameters of fractional PID obtained vary in a regular manner with L and T. Using the least-squares method; it is possible to translate that regularity into polynomial formulas to find acceptable values of the parameters from the values of L and T.

FIRST SET OF TUNING RULES: A first set of tuning rule is given in Table 1 of (Valerió & da Costa 2006), which is to be read as

$P = -0.0048 + 0.266L + 0.49820T + 0.0232L^2 - 0.0720T^2 - 0.0348LT$

And so on. They may be used if

$0.1 \leq T \leq 50$, and $L \leq 2$

And were designed for following specification:
$\omega_{cg} = 0.5 \ rad/s$
$\varphi = 2/3 \ rad = 38 \ degree$
$\omega_h = 10 \ rad/s$
$\omega_l = 0.1 \ rad/s$
$H = -10 \ dB$
$N = -20 \ dB$

SECOND SET OF TUNING RULES: A second set of rules is given in Table 2 of (Valerió & da Costa 2006). This second set of rules is applied for-

$0.1 \leq T \leq 50$ and

$L \leq 0.5$

Table 1. Value of parameter K_p, K_i and K_d for $M_s = 1.4$.

Value	a	b	c
K_p	0.2776	−1.10970	−0.1426
K_d	0.6241	0.5573	0.0442
K_i	0.4793	0.7469	−0.0239

Table 2. Value of parameter K_p, K_i and K_d for $M_s = 2$.

Value	a	b	c
K_p	0.1640	−1.449	−0.2108
K_d	0.6426	0.8069	0.0563
K_i	0.5970	0.5568	−0.0954

3.2 The padula and visioli method

Padula and Visioli proposed a set of tuning rules (Padula & Visioli 2011) based on first order plus time delay models. The tuning rules have been devised in order to minimize the integral absolute error with a constraint on the maximum sensitivity.

3.2.1 Tuning rules

First, the process is considered to have dynamics described by first order plus time delay model. The process dynamics can be conveniently characterized by the normalized dead time and defined as

$$\tau = \frac{L}{L+T} \tag{2}$$

Which represents a measure of difficulty in controlling the process. The tuning rules are devised for values of the normalized dead time in the range 0.05 to 0.8. The FOPID controller is modeled by the following transfer function-

$$G(s) = K_p \frac{K_i s^\lambda + 1}{K_i s^\lambda} \frac{K_d s^\mu + 1}{\frac{K_d s^\mu}{N} + 1} \tag{3}$$

The major difference of FOPID defined by this equation with the standard form of FOPID is that an additional first-order filter has been employed in this case, in order to, make the controller proper. The parameter N is chosen as $N = 10T^{\mu-1}$. The performance index is integrated absolute error which is defined as follows

$$IAE = \int_0^\infty |e(t)| dt \tag{4}$$

324

Using this equation as performance index yields a low overshoot and a low settling time at the same time. The maximum sensitivity is defined as

$$M_s = \max\left[\frac{1}{1+G_c(s)G(s)}\right] \qquad (5)$$

Which represent the inverse of the maximum distance of the Nyquist plot (Walter, 1986) from the critical point $-1 + j0$. Obviously, the higher value of M_s yields less robustness against the uncertainties. Tuning rules are devised such that the typical values of $M_s = 1.4$ and $M_s = 2.0$ are achieved. If only the load disturbance rejection task is addressed, we have

$$K_p = \frac{1}{K}\left(a\tau^b + c\right) \qquad (6)$$

$$K_i = T^\lambda\left(a(L/T)^b + c\right) \qquad (7)$$

$$K_d = T^\lambda\left(a(L/T)^b + c\right) \qquad (8)$$

Where the values of the parameters are shown in Tables 1, 2, 3, and 4.

3.3 Frequency domain design based on constrained optimization

It is an optimization based frequency domain tuning method proposed by Monje and Dorcak (Monje, Vinagre, Feliu & Chen 2008). This takes two extra specifications on the maximum value of magnitude of sensitivity and complementary sensitivity function along with the specifications presented in frequency domain design of FOPID controller.

3.3.1 Design specification and compensation problem

1. **No steady-state error**: Properly implemented a fractional integrator of order $k + \lambda$, $0 < \lambda < 1$ is,

Table 3. Value of parameter K_p, K_i and K_d for $M_s = 1.4$.

λ	μ
1	$1.0, if\ \tau \leq 0.1$
1	$1.1 if\ 0.1 \leq \tau \leq 0.4\ and\ 1.2, if\ 0.4 \leq \tau$

Table 4. Value of parameter K_p, K_i and K_d for $M_s = 2$.

λ	μ
1	$1.0, if\ \tau \leq 0.2$
1	$1.1 if\ 0.2 \leq \tau \leq 0.6\ and\ 1.2, if\ 0.6 \leq \tau$

for steady error cancellation as efficient as integer order integrator of order $k + 1$.

2. **Phase margin (φ_m) and gain crossover frequency (ω_{cg}) specifications:** The conditions must be fulfilled:

$$\left|G_c\left(j\omega_{cg}\right)G_p\left(j\omega_{cg}\right)\right| = 0\ dB \qquad (9)$$

$$Arg(G_c(j\omega_{cg})G_p(j\omega_{cg})) = -\pi + \varphi_m \qquad (10)$$

Where $G_c(s)$ is controller and $G_p(s)$ is the plant transfer function and ω_{cg} is the gain crossover frequency.

3. **Robustness to variations in the gain of the plant:** To this respect, the next constraint must be fulfilled is-

$$\left(\frac{d\left(Arg(G_c(j\omega)G_p(j\omega))\right)}{d\omega}\right)_{\omega=\omega_{cg}} = 0 \qquad (11)$$

With this condition the phase is forced to be flat at ω_{cg} and so, to be almost constant within an interval around ω_{cg}. It means that the system is more robust to gain changes and the overshoot of the response is almost constant within the interval.

4. **Robustness to high frequency noise:** To ensure a good measurement of noise rejection, it must fulfill the below condition:

$$\left|T(j\omega) = \frac{G_c(j\omega)G_p(j\omega)}{1+G_c(j\omega)G_p(j\omega)}\right|_{dB} \leq AdB \qquad (12)$$

$\forall \omega \geq \omega_t$ rad/sec

Where A is the desired noise attenuation for frequencies $\omega \geq \omega_t\ rad/s$.

5. **To ensure a good output disturbance rejection: The next constraint must be reached:**

$$\left|S(j\omega) = \frac{1}{1+G_c(j\omega)G_p(j\omega)}\right|_{dB} \leq BdB \qquad (13)$$

$\forall \omega \leq \omega_s$ rad/s

With B the desired value of the sensitivity function for frequencies $\omega \geq \omega_s rad/s$

3.3.2 The problem of nonlinear minimization

From the specifications above, a set of five nonlinear equations with five unknown parameters (K_p, K_i, K_d, λ, μ) is obtained. The complexity of this set of nonlinear equations is very significant. The function used for this purpose is called fmincon (.) (Rao & Rao 2009), which finds the constrained minimum of a function of several variables. In this case, the specification in (10) is taken as the main function to minimize, and the rest of specifications are taken as constrains for the minimization, all of them subjected to the optimization parameters defined within the function fmincon (.).

3.4 Time domain design

3.4.1 Dominant pole placement method

The approach is based on the root locus method of designing PID controllers (Petras, 2000). As, in the traditional root locus method, the peak overshoot M_p and rise time t_{rise} are specified, then from these specifications, damping ratio ζ and the undamped natural frequency ω_n are calculated according to (14) and (15).

$$\zeta = -\frac{\ln(M_p)}{\sqrt{\{\ln(M_p)\}^2 + \pi^2}} \qquad (14)$$

$$\omega_n = \frac{\pi - \tan^{-1}\left(\frac{\sqrt{1-\zeta^2}}{\zeta}\right)}{t_{rise}\sqrt{1-\zeta^2}} \qquad (15)$$

Using these computed values of ζ and ω_n, desired positions of the dominant poles $P_{1,2}$ of the closed loop pole of the system are determined as:

$$p_{1,2} = -\zeta\omega_n \pm j\omega_n\sqrt{1-\zeta^2} = -a \pm jb$$

Where $a = \zeta\omega_n$

$$b = \omega_n\sqrt{1-\zeta^2}$$

The closed loop transfer function of the controlled system is given as-

$$T(s) = \frac{G(s)}{1 + G(s)H(s)} \qquad (16)$$

$G(s) = G_c(s)G_p(s)$ Being the forward path transfer function and $G_c(s)$, the controller transfer function given by-

$$G_c(s) = K_p + T_i s^{-\lambda} + T_d s^{\mu} \qquad (17)$$

Assuming unity feedback, then $H(s) = 1$. In this case, the characteristic equation becomes $1 + G(s) = 0$,

$$1 + G_c(s)G_p(s) = 0 \qquad (18)$$

$$1 + G(s)\frac{P(s)}{Q(s)} = 0 \qquad (19)$$

$$Q(s) + G_c(s)P(s) = 0 \qquad (20)$$

Where $P(s)$ and $Q(s)$ are respectively the numerator and denominator polynomials of $G(s)$, $P(s)$ and $Q(s)$ have no common factor. As $P_{1,2}$ is poles of the closed loop system, each of them must be a root of the characteristic equation and hence must satisfy (20). Thus, putting $s = p_1 = -a + jb$ in (20), we get

$$Q(p_1) + G_c(p_1)P(p_1) = 0 \qquad (21)$$

$$Q(-a+jb) + (K_p + K_i(-a+jb)^{-\lambda} + K_d(-a+jb)^{\mu})P(-a+jb) = 0$$

Equation (21) is a complex equation in five unknowns, namely K_p, K_i, K_d, λ and μ. The problem of designing a controller that makes the closed loop dominant poles of the system coincide with $p_{1,2}$ now reduces the determination of the set values of $[K_p\ K_i\ K_d\ \lambda\ \mu]$ for which (21) holds good. But, as the number of unknowns exceeds the number of equations, there exists an infinite number of solution sets and the equation cannot be unambiguously solved by traditional methods. This necessitates the application of stochastic global search techniques, which, in turn, requires the formulation of a suitable objective function or cost function.

Let,

$R = real$ part of the L.H.S. of (21)

$I = imaginary$ part of the L.H.S. of (21) and

$$P = \tan^{-1}\frac{I}{R} \qquad (22)$$

We define $F[K_p, K_i, K_d, \lambda, \mu] = |R| + |I| + |P|$ as our objective function. Clearly, $f \geq 0$, in general and $f = 0$ if and only if $R = 0$, $I = 0$ & $P = 0$.

Now at this point, different search technique can be applied to scour the five-dimensional search space and find the optimal solution set of $[Kp\ Ki\ Kd\ \lambda\ \mu]$ for which $f = fmin = 0$. This method of finding the parameter of PID and FOPID is widely used in control engineering.

Deepyaman Maiti, et al. (Singhal, Padhee & Kaur 2012) uses particle swarm optimization for searching optimal value of parameter of FOPID. Subhransu Padhee, et al. (Das, Saha, Das & Gupta 2011) uses genetic algorithm for searching optimal value of K_p K_i K_d, λ and μ. Apart from this optimization technique many optimization technique is used by many researchers including differential evolution, chaotic ant swarm, simulated annealing, artificial bee colony etc.

The dominant pole placement tuning is only valid for strictly second order type systems and it does not give satisfactory result for higher order systems having several dominant poles and/or zeros (Maiti, Acharya, Chakraborty, Konar & Janarthanan 2008). Also, the dominant pole placement tuning gives inferior closed loop performance and often unstable response for time delay systems, since the Pade approximation of delay term effectively raises the order of the overall system.

3.4.2 Optmal tuning based on time domain integral performance

A typical unity feedback closed loop system is shown in Figure 2

- IAE index

$$J = \int_0^\infty |e(t)| dt \qquad (23)$$

- ISE index

$$J = \int_0^\infty |e(t)|^2 dt \qquad (24)$$

- ITAE index

$$J = \int_0^\infty t|e(t)| dt \qquad (25)$$

Let $e(kT)$ be the sampled value of the error $e(t)$ at an instant (kT), where T is the sampling interval. $k = 0,1,...,N$. For the given T, N is an integer which depends on the time span considered for computing $e(t)$. The following cost functions are defined corresponding to the performance indices

- IAE cost function

$$J_c = \sum_{k=0}^N |e(kT)| \qquad (26)$$

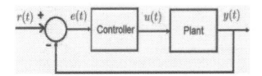

Figure 2. Block diagram of unity feedback closed loop system.

- ISE cost function

$$J_c = \sum_{k=0}^N |e(kT)|^2 \qquad (27)$$

- ITAE cost function

$$J_c = \sum_{k=0}^N kT|e(kT)| \qquad (28)$$

Each performance index emphasizes different aspects of the system response. Large errors contribute more to ISE than IAE. Consequently, the controller tuned for minimizing ISE ensures lower overshoot in the transient response than the IAE minimizing controller. The ISE, however, tends to give larger settling time. This is because, smaller errors ($e(t) <1$) are less quantified in ISE than IAE. The ITAE is the most sensitive of the three criteria. Due to the presence of the t (time) product term, ITAE weighs more heavily errors that occur later in the time (i.e. near to steady state). Therefore, the settling time is the shortest. Also, as compared to ISE, large errors occurring in the transient part (i.e. for smaller t values) are less quantified. This leads to larger overshoots than the ISE case.

The controller parameters $[K_p \, K_i \, K_d]$ are tuned by minimizing the selected performance index (Maiti, Acharya, Chakraborty, Konar & Janarthanan 2008).

3.5 Internal model control tuning

Valerio and Sa Da Costa show that the IMC methodology may be used in some cases to obtain PID or FOPID (Abadi & Jalali, 2012). IMC refer to the internal model control that corresponds to the control scheme which is shown in the figure below, where G^1 is an inverse of G. G^2 is a model of G and F is some judiciously chosen filter. If G^2 were exact, the error e would be equal to disturbance d. If, additionally, G^1 were the exact inverse of G and F were unity, control would be perfect. Since no models are perfect, e exactly will not be the disturbance. This is exactly why F exists and is usually a low-pass filter, to reduce the influence of high-frequency modeling errors. It also helps ensuring that product FG^1 is realizable.

This is equivalent to the second block diagram of Figure 3 if controller C is given by

$$C = \frac{FG^1}{1 - FG^1 G^2} \qquad (29)$$

C is not, in the general case, a PID or a fractional PID, but in some cases it will if

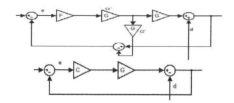

Figure 3. Block diagram of IMC (top) and block diagram equivalent to that above (bottom).

$$G = \frac{K}{1+S^{\alpha}T}e^{-Ls} \qquad (30)$$

Firstly, let $F = \dfrac{1}{1+sTF}$

$$G^1 = \frac{1+s^{\alpha}T}{K} \qquad (31)$$

$$G^2 = \frac{K}{1+s^{\alpha}T}(1-sL) \qquad (32)$$

Notice that the delay of G was neglected in G^1 but not in G^2, where an approximation consisting of a truncated McLaurin series has been used.

$$C = \frac{\dfrac{1}{K(T_F+L)}}{s} + \frac{\dfrac{T}{K(T_F+l)}}{s^{1-\alpha}} \qquad (33)$$

This is a fractional PID controller with the proportional part equal to zero.

Secondly, let $F = 1$

$$G^1 = \frac{1+S^{\alpha}T}{K} \qquad (34)$$

$$G^2 = \frac{K}{1+s^{\alpha}T}\frac{1}{1+sL} \qquad (35)$$

Equation (29) becomes

$$C = \frac{1}{K} + \frac{\dfrac{1}{KL}}{s} + \frac{\dfrac{T}{KL}}{s^{1-\alpha}} + \frac{T}{K}s^{\alpha} \qquad (36)$$

If one of the two integral parts can be neglected, will be a fractional PID controller. Finally, if a Pades approximation with one pole and one zero is used in G^2

$$F = 1$$

$$G^1 = \frac{1+s^{\alpha}T}{K}$$

$$G^2 = \frac{K}{1+s^{\alpha}T}\frac{1-sL/2}{1+sL/2} \qquad (37)$$

$$C = \frac{1}{2K} + \frac{\dfrac{1}{KL}}{s} + \frac{\dfrac{T}{KL}}{s^{1-\alpha}} + \frac{T}{2K}s^{\alpha} \qquad (38)$$

Again, (38) will be a fractional PID if one of the two integral parts is neglectable.

4 LIMITATION OF FRACTIONAL ORDER PID CONTROLLER

However, in spite of this research effort, the use of FOPID controllers in industry is still quite limited. This can be due to many reasons such as:

1. The performance improvement that can be obtained by using a FOPID controller has not been fully characterized yet, especially if all the control specifications such as set point following, load disturbance rejection, noise rejection, control effort are considered.
2. Simple, effective and robust tuning rules are still not available.
3. Additional functionality that are well established for standard integer order PID controllers have not been fully developed yet for FOPID controllers.
4. It is a matter of fact, in any case, that the design of FOPID controllers is a more complex task with respect to the setting of standard integerorder PID controllers. Indeed, there are five parameters to tune instead of three, if the basic controller expressions are considered, and the physical meaning of the parameters of a FOPID controller is not very intuitive.

This is a significant obstacle for the use of this kind of controllers by the process operators.

5 CONCLUSION

In this paper, we have presented a brief review of the FOPID controller and its important tuning methods available in literature. Some of the limitation of fractional order PID controller is also indicated in this paper.

REFERENCES

Abadi, M.R.R.M. & A.A. Jalali (2012). Fractional order pid controller tuning based on imc. *Int. J. Inf. Technol. Control Autom. 2*, 4.
Chen, Y., C. Hu & K.L. Moore (2003). Relay feedback tuning of robust pid controllers with iso-damping

property. In *Decision and Control, 2003. Proceedings. 42nd IEEE Conference on*, Volume 3, pp. 2180–2185. IEEE.

Chen, Y., K.L. Moore, B.M. Vinagre & I. Podlubny. Robust pid controller autotuning with an iso-damping property through a phase shaper.

Das, S., S. Saha, S. Das & A. Gupta (2011). On the selection of tuning methodology of fopid controllers for the control of higher order processes. *ISA transactions* 50(3), 376–388.

Maiti, D., A. Acharya, M. Chakraborty, A. Konar & R. Janarthanan (2008). Tuning pid and pi/λ d δ controllers using the integral time absolute error criterion. In *Information and Automation for Sustainability, 2008. ICIAFS 2008. 4th International Conference on*, pp. 457–462. IEEE.

Miller, K.S. & B. Ross (1993). An introduction to the fractional calculus and fractional differential equations.

Monje, C.A., A.J. Calderon, B.M. Vinagre, Y. Chen & V. Feliu (2004). On fractional pi λ controllers: some tuning rules for robustness to plant uncertainties. *Nonlinear Dynamics* 38(14), 369–381.

Monje, C.A., Y. Chen, B.M. Vinagre, D. Xue & V. Feliu-Batlle (2010). *Fractional-order systems and controls: fundamentals and applications*. Springer Science & Business Media.

Monje, C.A., B.M. Vinagre, V. Feliu & Y. Chen (2008). Tuning and auto-tuning of fractional order controllers for industry applications. *Control Engineering Practice* 16(7), 798–812.

Padula, F. & A. Visioli (2011). Tuning rules for optimal pid and fractional-order pid controllers. *Journal of Process Control* 21(1), 69–81.

Petras, I. (2000). The fractional-order controllers: methods for their synthesis and application. *arXiv preprint math/0004064*.

Podlubny, I. (1999). Fractional-order systems and pi/sup/spl lambda//d/sup/spl mu//-controllers. *Automatic Control, IEEE Transactions on* 44(1), 208–214.

Rao, S.S. & S. Rao (2009). *Engineering optimization: theory and practice*. John Wiley & Sons.

Singhal, R., S. Padhee & G. Kaur (2012). Design of fractional order pid controller for speed control of dc motor. *International Journal of Scientific and Research Publications* 2(6), 1–8.

Valerió, D. & J.S. da Costa (2005). Ziegler-nichols type tuning rules for fractional pid controllers. In *ASME 2005 International Design Engineering Technical Conferences and Computers and Information in Engineering Conference*, pp. 1431–1440. American Society of Mechanical Engineers.

Valerió, D. & J.S. da Costa (2006). Tuning of fractional pid controllers with ziegler–nichols-type rules. *Signal Processing* 86(10), 2771–2784.

Valerio, D. & J.S. da Costa (2010). A review of tuning methods for fractional pids. In *4th IFAC Workshop on Fractional Differentiation and its Applications, FDA*, Volume 10.

Walter, G. (1986). A review of impedance plot methods used for corrosion performance analysis of painted metals. *Corrosion Science* 26(9), 681–703.

Frontiers in Computer, Communication and Electrical Engineering – Acharyya (Ed.)
© 2016 Taylor & Francis Group, London, ISBN: 978-1-138-02877-7

Effect of void geometry on noise rejection in 1D photonic crystal with metamaterial/air interface

Bhaswati Das & Arpan Deyasi
Department of Electronics and Communication Engineering, RCC Institute of Information Technology, Kolkata, India

ABSTRACT: In one-dimensional photonic crystal, transmittiivity is analytically computed to solve coupled mode equation for different structural parameters and internal conditions. Two different nano-fishnet structures are considered for simulation with rectangular and elliptical void as photonic crystal, and computation is being carried out at 1550 nm for the sole purpose of optical communication, respectively. The structure is considered as Bragg grating where increase in grating length enhances the reflection of electromagnetic wave, and strong coupling provides larger bandgap spectral width. Simulated findings show that elliptical void can provide better noise rejection due to smaller spectral width, whereas rectangular void can be considered as efficient candidate for higher transmittivity at the central wavelength, when computed with identical structural parameters and coupling condition. Compared to the results obtained from conventional photonic crystal, it can be stated that DNG materials provide better spectral response for communication application.

1 INTRODUCTION

Photonic crystal is the periodic arrangement of dielectric/metal-dielectric materials (Yablonovitch 1987) where propagating electromagnetic waves can be localized in the desired spectrum, and can be allowed to transmit in the other region (Fogel *et al.*, 1998). This novel exhibition, is possible due to the formation of electromagnetic bandgap (D'Orazio *et al.*, 2003), analogous to electronic bandgap. The property is already effectively utilized in the design of optical transmitter (Szczepanski, 1988), optical receiver (Kalchmair *et al.*, 2011), photonic crystal fiber (Russel 2006), resonant tunneling device (Jiang *et al.*, 1999); and also in the domain of quantum information processing (Azuma 2008), photonic integrated circuit (Bayat *et al.*, 2010). This novel microstructure is one of the pioneering developments in the field of photonics, and has already become the promising candidate in the domain of optical communication by replacing its conventional counterpart optical fibre (Limpert *et al.*, 2004, Limpert *et al.*, 2003).

Research has already been carried out in the last decade regarding the various possible confinements in photonic crystal, but it is suggested that only 1D and 2D structures are efficient enough to fabricate and also implementation purpose for different optoelectronic integrated circuits. Among them, 1D structure is very convenient to study because of the ease of mathematical modeling and also from fabrication stand-point. Materials used so far by theoretical and experimental researchers in design and construction of photonic crystal and crystal-based devices are SiO_2/air (Andreani *et al.*, 2003, Robinson *et al.*, 2011) and semiconductor heterostructures (Maity *et al.*, 2013, Xu *et al.*, 2007), where refractive indices of them are positive. But these structures are not very suitable when design of the integrated circuit involves antenna, and noise rejection from the incoming electromagnetic wave at desired spectrum becomes very critical from application stand-point. As per the knowledge of the authors, research has not been conducted in this area so far. In the present paper, noise rejection property at the desired spectrum is analyzed using DNG material, which is already used in antenna design to improve SNR. Two different metamaterials, namely nano-fishnet with rectangular void and elliptical void are considered for simulation purpose when Bragg wavelength is set at 1550 nm. Coupled mode theory is used to solve the problem, whereas the simulated findings show that with the change of void geometry, structural parameters and coupling conditions; transmittiivity can be tuned at the desired central frequency. Spectral width of the structure, i.e., photonic bandgap can be measured from the reflection coefficient analysis.

2 THEORETICAL FOUNDATION

Inside a photonic crystal, the forward and backward propagating waves may be represented as

$$B(z,t) = b\exp\left[-j(\beta_b z - \omega t)\right] \quad (1.1)$$

$$A(z,t) = a\exp\left[j(\beta_a z + \omega t)\right] \quad (1.2)$$

where β_b and β_a are the propagation constants. Considering the propagation part along with the d.c magnitude, equations may be rewritten as

$$A(z,t) = a(z)\exp\left[j\omega t\right] \quad (2.1)$$

$$B(z,t) = b(z)\exp\left[j\omega t\right] \quad (2.2)$$

Hence, coupling of total power is dependent on $a(z)$ and $b(z)$ only. In absence of any coupling,

$$\frac{da(z)}{dz} = j\beta_a a(z) \quad (3.1)$$

$$\frac{db(z)}{dz} = -j\beta_b b(z) \quad (3.2)$$

When the waves are coupled, the Equation (3.1) and Equation (3.2) will be modified as

$$\frac{da(z)}{dz} = j\beta_a a(z) + \kappa_{ab} b(z) \quad (4.1)$$

$$\frac{db(z)}{dz} = -j\beta_b b(z) + \kappa_{ba} a(z) \quad (4.2)$$

Due to the fact that A(z,t) and B(z,t) propagate in opposite directions, net power can be given by

$$P = \left[\left|b(z)\right|^2 - \left|a(z)\right|^2\right] \quad (5)$$

From power conservation principle,

$$\frac{dP}{dz} = 0 \quad (6)$$

From Equation (4.1), we can write

$$b(z) = \frac{(D_a - j\beta_a)a(z)}{\kappa_{ab}} \quad (7)$$

From Equation (4.2), we can write

$$D_b\left[\frac{(D_a - j\beta_a)a(z)}{\kappa_{ab}}\right] = \kappa_{ba}a(z) - \frac{j\beta_b(D_a - j\beta_a)a(z)}{\kappa_{ab}} \quad (8)$$

This can give a second order differential equation

$$D_a D_b a(z) + j(D_a\beta_b - D_b\beta_a)(z) + (\beta_b\beta_a - \kappa_{ab}\kappa_{ba})a(z) = 0 \quad (9)$$

Solution of Equation (9) gives

$$a(z) = D\exp\left[\frac{jz(\beta_a - \beta_b)}{2}\right]$$
$$\times \exp\left[\frac{\sqrt{4\kappa_{ab}\kappa_{ba} - (\beta_a + \beta_b)^2}}{2}z\right]$$
$$+ E\exp\left[\frac{jz(\beta_a - \beta_b)}{2}\right]$$
$$\times \exp\left[\frac{-\sqrt{4\kappa_{ab}\kappa_{ba} - (\beta_a + \beta_b)^2}}{2}z\right] \quad (10)$$

Since, the wave is propagating along—ve Z direction, it will consider that component only. For such physical assumption, Equation (10) can be rewritten as

$$a(z) = D\exp\left[\frac{jz(\beta_a - \beta_b)}{2}\right]$$
$$\times \exp\left[\frac{\sqrt{4\kappa_{ab}\kappa_{ba} - (\beta_a + \beta_b)^2}}{2}z\right] \quad (11)$$

Subject to the appropriate boundary condition, Equation (11) may be modified as

$$a(z) = a\exp\left[\frac{jz(\beta_a - \beta_b)}{2}\right]$$
$$\times \exp\left[\frac{\sqrt{4\kappa_{ab}\kappa_{ba} - (\beta_a + \beta_b)^2}}{2}z\right] \quad (12)$$

Similarly, we can write

$$b(z) = b\exp\left[\frac{jz(\beta_a - \beta_b)}{2}\right]$$
$$\times \exp\left[\frac{-\sqrt{4\kappa_{ab}\kappa_{ba} - (\beta_a + \beta_b)^2}}{2}z\right] \quad (13)$$

Using Equation (6), we may write

$$\exp\left[\frac{4\kappa_{ab}\kappa_{ba}-(\beta_a+\beta_b)^2}{2}z\right]=\pm j\frac{b}{a} \quad (14)$$

As the right-hand side of Equation (14) is imaginary, so the term in the square-root should be less than zero, i.e., we can write this in the mathematical form

$$\kappa_{ab}\kappa_{ba}<\frac{1}{4}(\beta_a+\beta_b)^2 \quad (15)$$

This can be written as

$$\kappa_{ab}=\kappa_{ba}^*=\kappa \quad (16)$$

With the boundary conditions b(0) = b₀ and a(L) = 0 these equations may be solved analytically. Expressions of forward and backward waves thus can be given as

$$a(z)=b_0\frac{\kappa\exp[-j\Delta\beta z]\sinh[\alpha(z-L)]}{\Delta\beta\sinh\alpha L-j\alpha\cosh\alpha L} \quad (17)$$

$$b(z)=b_0\kappa\exp[j\Delta\beta z]$$
$$\times\frac{[\Delta\beta\sinh\{\alpha(z-L)\}+j\alpha\cosh\{\alpha(z-L)\}]}{-\Delta\beta\sinh\alpha L+i\alpha\cosh\alpha L}$$

$$(18)$$

where $\alpha=\sqrt{\kappa^2-\Delta\beta^2}$, L is the grating length.

Transmittiivity can be computed as a function of operating wavelength from the knowledge of forward and backward waves, given by

$$T=\left|\frac{a(0)}{b(0)}\right|^2=\frac{\kappa\sinh\alpha L}{\Delta\beta\sinh\alpha L-j\alpha\cosh\alpha L} \quad (19)$$

3 RESULTS AND DISCUSSION

Using Equation (19), transmittiivity of the proposed structure is calculated and plotted as a function of operating wavelength for different coupling conditions and different length of crystal structure. Fig 1 shows the profile for rectangular void geometry inside nano-fishnet, whereas Fig 2 exhibits for elliptical void geometry.

From the plot, it may be stated that transmittivity becomes higher for larger value of grating length. At resonance condition, due to the formation of photonic bandgap, most part of the light is transmitted, and smaller portion is reflected. Outside of this bandgap, transmittivity closes to zero,

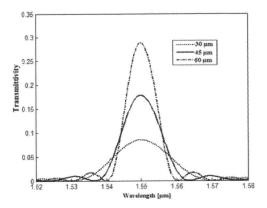

Figure 1a. Transmittivity profile with wavelength for different length of the structure for weak coupling condition with nano-fishnet (rectangular void)/air composition.

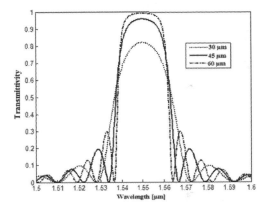

Figure 1b. Transmittivity profile with wavelength for different length of the structure for medium coupling condition with nano-fishnet (rectangular void)/air composition.

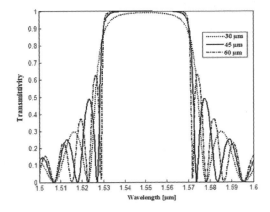

Figure 1c. Transmittivity profile with wavelength for different length of the structure for strong coupling condition with nano-fishnet (rectangular void)/air composition.

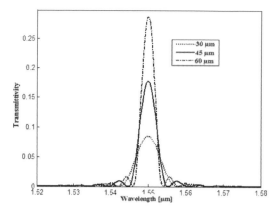

Figure 2a. Transmittivity profile with wavelength for different length of the structure for weak coupling condition with nano-fishnet (elliptical void)/air composition.

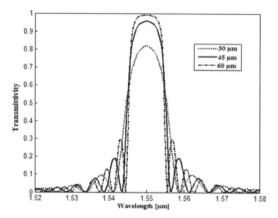

Figure 2b. Transmittivity profile with wavelength for different length of the structure for medium coupling condition with nano-fishnet(elliptical void)/air composition.

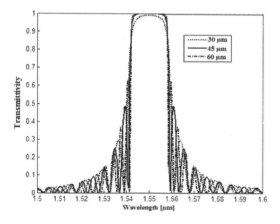

Figure 2c. Transmittivity profile with wavelength for different length of the structure for strong coupling condition with nano-fishnet (elliptical void)/air composition.

as we move away from Bragg wavelength. This phenomenon is also observed for medium and strong coupling conditions, and transmittivity reaches to unity for large grating length. This is plotted in Fig 1b and Fig 1c respec tively. A comparative study between these results indicates the fact that magnitude of transmittivity increases with increase of coupling coefficient.

Comparative analysis is carried out with nano-fishnet structure with elliptical void. Results are plotted in Fig 2. The change of void geometry increases the refractive index in negative scale, and hence the contrast ratio also increases. This reduces the magnitude of peak transmittivity, compared with that obtained for rectangular void geometry, with identical grating length and coupling condition. But the spectral width obtained for this case is lower compared to the previous case, which favors for its use in terms of noise rejection. The results are plotted for three different grating lengths.

4 CONCLUSION

Comparative analysis shows that void geometry in nano-fishnet structure is the critical parameter in rejecting noise for the optical communication application. Elliptical void provides lower spectral width, which is the key for rejecting noise, whereas rectangular void provides higher transmittivity in the desired domain. Hence, appropriate choice of DNG material plays the key role in design of photonic integrated circuit embedded with antenna.

REFERENCES

Andreani. L.C., Agio. M., Bajoni. D., Belotti. M., Galli. M., Guizzetti. G., Malvezzi. A.M., Marabelli. F., Patrini. M. and Vecchi. G. 2003, Optical Properties and Photonic Mode Dispersion in Two-Dimensional and Waveguide-Embedded Photonic Crystals, Synthetic Metals, 139, 695–700.

Azuma. H. 2008, Quantum Computation with Kerr-Nonlinear Photonic Crystals, Journal of Physics D: Applied Physics, 41, 025102.

Bayat K., Rafi. G.Z., Shaker. G.S.A., Ranjkesh. N., Chaudhuri. S.K. and Safavi-Naeini. S. 2010, Photonic-Crystal based Polarization Converter for Terahertz Integrated Circuit, IEEE Transactions on Microwave Theory and Techniques, 58, 1976–1984.

D'Orazio. A., De Palo. V., De Sario. M., Petruzzelli. V. and Prudenzano. F. 2003, Finite Difference Time Domain Modeling of Light Amplification in Active Photonic Bandgap Structures, Progress In Electromagnetics Research, 39, 299–339.

Fogel. I.S., Bendickson. J.M., Tocci. M.D., Bloemer. M.J., Scalora. M., Bowden. C.M. and Dowling. J.P. 1998, Spontaneous Emission and Nonlinear Effects in

Photonic Bandgap Materials, Pure and Applied Optics, 7, 393–408.

Jiang. Z., Niu. C.N. and Lin. D.L. 1999, Resonance Tunneling through Photonic Quantum Wells, Physical Review B, 59, 9981–9986.

Kalchmair. S., Detz. H., Cole. G.D., Andrews. A.M., Klang. P., Nobile. M., Gansch. R., Ostermair. C., Schrenk. W. and Strasser. G. 2011, Photonic Crystal Slab Quantum Well Infrared Photodetector, Appied Physics Letters, 98, 011105.

Limpert. J., Liem. A., Reich. M., Schreiber. T., Nolte. S., Zellmer. H., Tünnermann. T., Broeng. J., Petersson. A. and Jakobsen. C. 2004, Low-Nonlinearity Single-Transverse-Mode Ytterbium-Doped Photonic Crystal Fiber Amplifier, Optic Express, 12, 1313.

Limpert. J., Schreiber. T., Nolte. S., Zellmer. H., Tunnermann. T., Iliew. R., Lederer. F., Broeng. J., Vienne. G., Petersson. A. and Jakobsen. C. 2003, High Power Air-Clad Large-Mode-Area Photonic Crystal Fiber Laser, Optic Express, 11, 818.

Maity. A., Chottopadhyay. B., Banerjee. U. and Deyasi. A. 2013, Novel Band-Pass Filter Design using Photonic Multiple Quantum Well Structure with p-polarized Incident Wave at 1550 μm, Journal of Electron Devices, 17, 1400–1405.

Robinson. S. and Nakkeeran. R. 2011, Photonic Crystal Ring Resonator based Add-Drop Filter using Hexagonal Rods for CWDM Systems, Optoelectronics Letters, 7, 164–166.

Russell. P.S.J. 2006, Photonic-Crystal Fibers, Journal of Lightwave Technology, 24, 4729–4749.

Szczepanski. P. 1988, Semiclassical Theory of Multimode Operation of a Distributed Feedback Laser, IEEE Journal of Quantum Electronics, 24, 1248–1257.

Xu. X., Chen. H., Xiong. Z., Jin. A., Gu. C., Cheng. B. and Zhang. D. 2007, Fabrication of Photonic Crystals on Several Kinds of Semiconductor Materials by using Focused-Ion Beam Method, Thin Solid Films, 515, 8297–8300.

Yablonovitch. E. 1987, Inhibited Spontaneous Emission in Solid-State Physics and Electronics, Physical Review Letters, 58, 2059–2061.

A second-order bandpass response with a wideband frequency selective surface

Ayan Chatterjee & Susanta Kumar Parui
Indian Institute of Engineering Science and Technology, Shibpur, Howrah, West Bengal, India

ABSTRACT: In this paper, the design of a low-profile, bandpass Frequency Selective Surface (FSS) operating over a wide bandwidth is being presented. The FSS is obtained by cascading two hexagon shaped patch type metal layers with a cross grid metal layer in between for coupling and dielectrics for separation. The hexagon shaped patches are closely spaced to achieve compactness. The proposed FSS exhibits a second-order bandpass response with two closely spaced transmission poles near 3.5 GHz and 4.7 GHz leading to a wide transmission bandwidth of 48% with respect to 3 dB level. The FSS also gives a linear variation of transmission phase with a frequency that makes it useful for enhancing antenna radiation as a superstrate.

1 INTRODUCTION

Since 1960, Frequency Selective Surface (FSS) has always been an important field of research for its applications as radomes for reducing Radar Cross Section (RCS) of objects, absorber, polarizer, and subreflector. (Wu 1995, Bayatpur 2009). FSS is the wireless counterpart of traditional filters that are used in radio frequency circuits and it is a periodic array of metallic patches or apertures on metallic screen over a dielectric slab that gives bandstop (patch) or bandpass (aperture) response (Munk 2000, Bayatpur 2009). Bandpass FSSs are mainly used for radome structures but recently they are being used as superstrate above antennas to enhance their radiation (gain, bandwidth) at their resonating frequency or over a wide bandwidth. An Electromagnetic BandGap (EBG) antenna was proposed where a FSS with square loop cells was used as superstrate above the radiating patch, leading to enhanced directivity and broadband response (Pirhadi *et al.* 2012). The application of FSS superstrates for directivity enhancement over two separate frequency bands was presented (Lee *et al.* 2007). Broadband FSS is designed by cascading multiple layers of FSS with dielectric (Behdad et al. 2009, Chatterjee et al. 2013) or by cascading FSS cells of different shapes horizontally. FSS increases the order of the filter by cascading multiple layers of a patch and aperture type.

In this paper, two patch type FSS screens with closely spaced hexagon shaped unit cells are coupled by an aperture type FSS made of metallic grids to exhibit a bandpass response with wide bandwidth. The three layers are separated by two

dielectric layers of thickness much less than wavelength corresponding to resonating frequency. Capacitive patch type layers and inductive grids provide a second-order response of the filter. An equivalent circuit model for the cascaded FSS is also shown and explained.

2 DESIGN OF THE FSS

The proposed FSS structure is composed of three layers of capacitive patch type metallic layers and inductive grid layer as shown in the Fig. 1(a). The three metallic layers are separated by two thin dielectric layers of Arlon AD270 whose relative permittivity is 2.7 and loss tangent is 0.002. The upper and lower layers are arrays of hexagon shaped unit cells and they are closely spaced to reduce the

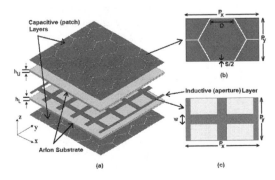

Figure 1. (a) Three dimensional view of the cascaded FSS. Unit cell dimensions of the metal (b) patch layers and (c) grid layer.

overall dimension whereas the middle layer is an array of cross shaped grids.

Both the unit cells of FSS are shown in Fig. 1(b) and (c) along with dimensions. The unit cell dimension used for simulation is P_X = 31mm and P_Y = 17.82 mm for both the layers. Side length of the hexagonal patch is D = 10 mm and spacing between two consecutive patches is 0.5mm whereas the width of the grids = 6mm. A thickness ($h_U = h_L$) of 1mm is used for dielectric layers.

3 EQUIVALENT CIRCUIT MODEL

An equivalent circuit of the multilayered FSS is given in Fig. 2(a), valid for normal incidence of the electromagnetic wave on it.

The first and third patch type layers are modeled as capacitors C_1 and C_3 respectively and are connected in shunt path, whereas the aperture type cross shaped grid layer is modeled as a parallel inductor L_2. The two dielectric substrates of height h_U and h_L ($h = h_U = h_L$) separating the metallic FSS layers can be modeled as two pieces of transmission line sections of characteristic impedances $Z_0/\sqrt{\varepsilon_{r1}}$ and $Z_0/\sqrt{\varepsilon_{r2}}$ respectively, where ε_{r1} and ε_{r2} are dielectric constants of the two substrates ($\varepsilon_{r1} = \varepsilon_{r2} = 2.7$) and $Z_0 = 377\Omega$ is the free space impedance. The transmission line sections are modeled by a series inductor (L_{D1}, L_{D2}) and a shunt capacitor (C_{D1}, C_{D2}). Their values can be found by the following Equations (1) and (2) (Chatterjee et al. 2015).

$$L_{D1} = L_{D2} = \mu_0 \mu_r h \tag{1}$$

$$C_{D1} = C_{D2} = \varepsilon_0 \varepsilon_r h \tag{2}$$

Here μ_0 and ε_0 are permeability and permittivity of free space. The spaces on both ends of the FSS are modeled as semi-infinite transmission lines with characteristic impedances of $Z_0 r_1$ and $Z_0 r_2$ respectively where r_1 and r_2 are the normalized source and load impedances and for free space $r_1 = r_2 = 1$. The T network (consisting L_{D1}, L_2 and L_{D2}) of the circuit in Fig. 2(a) is converted into a π network (consisting L_1, L_m and L_3). The values of L_1, L_m and L_3 can then be easily found from the values of L_{D1}, L_2 and L_{D2} using the following Equations (3), (4), and (5).

$$L_1 = L_{D1} + L_2\left(1 + \frac{L_{D1}}{L_{D2}}\right) \tag{3}$$

$$L_3 = L_{D2} + L_2\left(1 + \frac{L_{D2}}{L_{D1}}\right) \tag{4}$$

$$L_m = L_{D1} + L_{D2} + \frac{L_{D1}L_{D2}}{L_2} \tag{5}$$

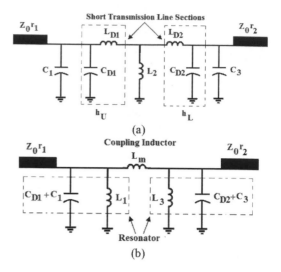

(a)

(b)

Figure 2. (a) Simple equivalent circuit model of the cascaded FSS (b) Second order schematic of the filter with π network (L_1, L_m and L_3) representation of T network (L_{D1}, L_2 and L_{D2}).

Second, order nature of the filter is clearly visible in the circuit as shown in Fig. 2(b) where two parallel LC resonators (L_1, $C_1 + C_{D1}$ and L_3, $C_3 + C_{D2}$) are coupled by a mutual inductance L_m and so the proposed FSS will act like a second-order bandpass filter with two transmission poles at two frequencies.

4 RESULTS AND DISCUSSION

Simulations have been performed for the proposed FSS using ANSYS HFSS using Finite Element Method (FEM). Simulation results for normalized transmission coefficient (dB) of the FSS are given below in the Fig. 3 for normal incidence of plane wave and different values of the inductive grid width w mm. It can be seen that the FSS exhibits two transmission poles at 3.5 GHz and 4.7 GHz that ensures the second order bandpass response. The 3 dB transmission band ranges from 3.2 GHz to 4.9 GHz for w = 5.4mm leading to a bandwidth of 48% with an insertion loss of about 2.8 dB. The parametric study shows reducing w lowers the first transmission pole. Reducing width of the inductive strips lowers its inductance and this accounts for the lowering of frequency.

Simulation results for reflection coefficient (dB) are also shown in Fig. 4 for different values of w mm and it is seen that the FSS exhibits upto −38 dB of reflectivity. Transmission phase (degree) variation with frequency for the FSS is given in Fig. 5 which shows an almost linear variation of phase with

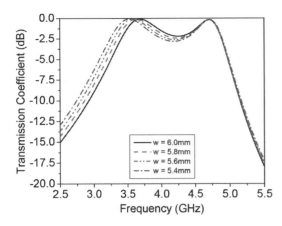

Figure 3. Simulated transmission coefficient of the FSS.

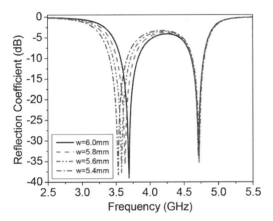

Figure 4. Simulated reflection coefficients of the FSS.

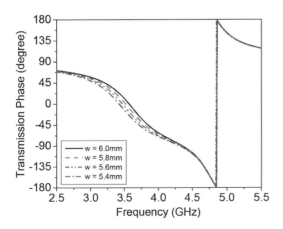

Figure 5. Simulated transmission phase response of the FSS.

frequency in the frequency band of operation. This makes the FSS useful for antennas with broadside radiation to enhance their radiation such as gain, directivity in the direction of radiation when used as a superstrate above the antenna. This is because when the FSS is placed above the antenna, a part of the radiated waves from antenna gets reflected by the lower patch type FSS thus facing multiple reflections between the antenna plane and lower layer of FSS and phase of these waves increases linearly with the frequency which gets cancelled by the linearly reduced transmission phase of FSS leading to constructive interference of waves in broadside direction of radiation.

5 CONCLUSION

A second-order, bandpass FSS with the broadband response has been studied in this paper. The FSS exhibits a wide transmission bandwidth (3 dB) of about 48% but with an insertion loss of about 2.8 dB for the separation between transmission poles by 1.2 GHz, lowering of which will reduce the insertion loss. Tuning of the first resonant frequency is done by changing the width of inductive wire grids which again changes the insertion loss and transmission bandwidth. The proposed design can be used to increase the gain, directivity of any antenna with broadside radiation over a wide bandwidth.

ACKNOWLEDGMENT

The authors would like to acknowledge Council of Scientific and Industrial Research, India (award no. 08/003(0101)/2014-EMR-I) for funding.

REFERENCES

Al-Joumayly, M. & Behdad, N. 2009. A New Technique for Design of Low-Profile, Second-Order, Bandpass Frequency Selective Surfaces, *IEEE Transactions on Antennas & Propagation*. 57(2). pp. 452–459.
Bayatpur, F. 2009. *Metamaterial-Inspired Frequency-Selective Surfaces*. Ph.D. Dissertation, University of Michigan.
Chatterjee, Ayan & Parui, S.K. 2013. A multi-layered bandpass frequency selective surface designed for Ku band applications, *In Proc. of IEEE Applied Electromagnetics Conference (AEMC)*. India. pp. 1–2.
Chatterjee, Ayan & Parui, S.K. 2013. A Multi-layered Broadband Frequency Selective Surface for X and Ku band Applications, *In Proc. of International Conference on Technical and Managerial Innovation in Computing and Communications in Industry and Academia (IEMCON)*. India. pp. 284–287.

Chatterjee, Ayan & Parui, S.K. 2015. Gain Enhancement of a Wide-Slot Antenna using Dual-Layer, Bandstop Frequency Selective Surface as a Substrate, *Microwave & Optical Technology Letters*. 57(9). pp. 2016–2020.

Chatterjee, Ayan & Parui, S.K. 2015. Gain Enhancement of a Wide-Slot Antenna using Second-Order Band-pass Frequency Selective Surface, *Radioengineering Journal*. 24(2). pp. 455–461.

Foroozesh, A. & Shafai, L. 2010. Investigation Into the Effects of the Patch-Type FSS Superstrate on the High-Gain Cavity Resonance Antenna Design, *IEEE Transactions on Antennas & Propagation*. 58(2). pp. 258–270.

Lee, D.H., Lee, Y.J., Yeo, J., Mttra, R., & Park, W.S. 2007. Design of novel thin frequency selective surface superstrates for dual-band directivity enhancement, *IET Microwaves Antennas and Propagation*. 1(1). pp. 248–254.

Munk, B.A. 2000. *Frequency Selective Surfaces—Theory and Design*. New York: Wiley Interscience.

Pirhadi, A., Bahrami, H., & Nasri, J. 2012. Wideband High Directive Aperture Coupled Microstrip Antenna Design by Using a FSS Superstrate Layer, *IEEE Transactions on Antennas & Propagation*. 60(4). pp. 2101–2106.

Wu, T.K. 1995. *Frequency Selective Surfaces and Grid Arrays*. New York: Wiley.

Zvere, A.I. 1967. *Handbook of Filter Synthesis*. New York: Wiley Interscience.

An acrylic sheet based frequency selective surface for GSM 1800 MHz band shielding

Pradipta Sasmal, Ayan Chatterjee & Susanta Kumar Parui
Indian Institute of Engineering Science and Technology, Shibpur, Howrah, West Bengal, India

ABSTRACT: This paper presents a Frequency Selective Surface (FSS) that functions as a reflector at the GSM 1800 MHz downlink band for mobile communication and allows the uplink frequency signals to pass through it without significant attenuation. The FSS is designed to exhibit a stopband in the 1805–1880 MHz band with a transmission level below −10 dB and a passband in the 1710–1785 MHz band. The structure is based on a transparent acrylic substrate and can be used in the wall, window of the top floors of high rise buildings close to the mobile phone towers to reduce the radiation hazards caused by cell tower radiation.

1 INTRODUCTION

With the enormous use of wireless communication, electromagnetic interference due to undesired radiation from different sources has become a serious problem. To overcome this problem, shielding has become an important technology (Celozzi 2008) that generally uses frequency selective filters such as Frequency Selective Surfaces (FSS) of both patch and aperture types to provide shielding against undesired radiation from external nearby sources. FSS, the wireless counterpart of the filters of traditional Radio Frequency (RF) circuit, is a periodic array consisting of conducting patch or apertures on metallic screen having band-stop or band-pass spectral behavior (Bayatpur 2009). In order to protect a flat array antenna system operating in the Ku-band from damage caused by the external environment such as rain, moisture, aerodynamic effects, etc., FSS coated on the glass (Gatti *et al.* 2008, Chen *et al.* 2012) has been used to effectively pass the Ku-band signals and stop interference of external noises. A band-pass shielding enclosure of dimensions $16 \times 10 \times 2.1$ cm for the portable digital wireless device was designed (Chiu *et al.* 2008) using single layer of multi-pole slot array FSS, with high transmittance in the specified wireless-signal band and high shielding effectiveness outside this band. This FSS wall is made by printing metal aperture array on the FR4 substrate.

Now-a-days with the increasing number of mobile phone users, density of cell phone towers has been increased very much especially in the urban areas with huge population. As a result, radiation from base station antennas are affecting people more who live in the high rise buildings near to these towers at a line of sight. In this paper, a FSS design has been proposed in the GSM 1800 frequency band that can be used in such building walls, curtains, window glass to reduce the radiation hazards at the same time allowing successful communication. A lot of research activities have been carried out on shielding mobile tower radiation using different substrate and FR4 is the leading one to be used for FSS design. The dielectric constant of FR4 is 4.4 and it is opaque which makes it non-popular to be used in house or in hospital, office and other areas. To use a FSS structure one would look whether it is transparent (for light to pass) or perforated (for air to pass through) or may be fabric based to be used as a window screen. The other important factor is that it should not disturb other band and shielding efficiency should be very high.

2 DESIGN OF THE FSS

The proposed FSS structure is an array of square loops on a dielectric substrate made of acrylic sheet and is shown in Figure 1. As can be seen the FSS cells are connected to form the array and thus unit cell dimension is same as the periodicity of the structure. All the dimensions of FSS unit cell are given in the figure. The acrylic sheet used here as substrate has a dielectric constant of 2.8 with a loss tangent of 0.02 and thickness of 1 mm. The periodicity of unit cell is 120×120 mm^2. The fabricated prototype is shown in figure 2.

3 RESULTS AND DISCUSSION

Full wave EM simulations have been performed for the proposed structure using Ansoft HFSS

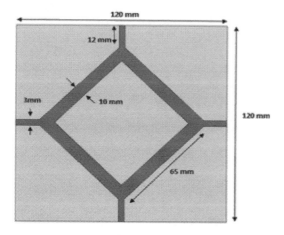

Figure 1.   Unit cell dimensions of the proposed FSS.

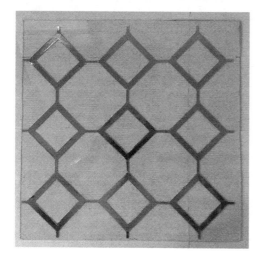

Figure 2.   A 3 × 3 fabricated prototype of the proposed FSS.

that uses Finite Element Method (FEM) for analysis. Analysis has been done for the composite structure and the simulation results for normalized transmission coefficient (dB) are given below in the Fig. 3 for two different angles of incidence of the plane wave and they are theta 0° and 60°.

From the plot of Figure 3 it is clear that for theta 0° i.e. for normal incidence the maximum attenuation is at around 1.85 GHz and the –10 dB band width is achieved in frequency range 1843–1885 GHz. Similarly for theta 60° there is not much variation in resonance frequency and band width but peak attenuation achieved is around –13dB. The important thing from this simulation is that

there is hardly any loss in the uplink frequency band which is from 1710 MHz–1780 MHz range.

A 3 × 3 array of the proposed FSS was fabricated in the lab by incorporating thin aluminum sheet over the acrylic sheet. Two monopole antennas were designed to operate in the GSM 1800 band for measurement of FSS transmission coefficient. The measurement was performed using vector network analyzer as a source and keeping the FSS screen in between two antennas. The transmission coefficients of the receiving monopole antenna before introducing FSS and after introduction of FSS were analyzed and it is shown in the Figure 4 that there is a shielding effectiveness of around 15 dB has been achieved with slightly shifting in resonance. The shift in resonating frequency may be attributed due to non-idle environment and finite FSS structure.

Surface current distribution (A/m) for unit cell of the FSS structure is shown in the Figure 5 at 1.85 GHz and the flow of current through the square loop confirms the peak attenuation at 1.85 GHz.

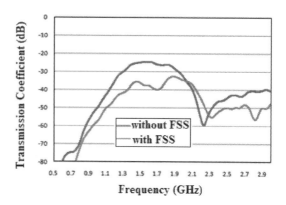

Figure 3.   Simulated transmission coefficient of the FSS.

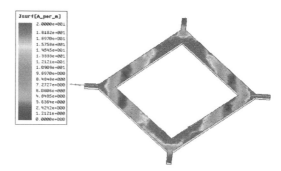

Figure 4.   Measured transmission levels without and with FSS.

342

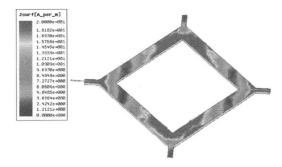

Figure 5. Surface current (A/m) for the FSS at 1.85 GHz.

4 CONCLUSION

A frequency selective surface for achieving shielding against the GSM 1800 MHz downlink band is proposed with the simulation and measurement results. The proposed design provides a reduction in power level by about 15 dB for the downlink frequency band only, allowing the uplink frequency signals to pass through it without significant attenuation. The proposed structure uses a transparent material as substrate for visibility making it useful for windows in the high rise buildings situated near cell phone towers. The future scope of this work will be implementing the design into fabrics for curtain as well as air as the substrate for the structure.

ACKNOWLEDGEMENT

The authors would like to acknowledge Council of Scientific and Industrial Research, India (award no. 08/003(0101)/2014-EMR-I) for funding. The authors are also grateful to DETCE, IIEST Shibpur for measurement.

REFERENCES

Al-Joumayly, M. & Behdad, N. 2009. A New Technique for Design of Low-Profile, Second-Order, Bandpass Frequency Selective Surfaces, *IEEE Transactions on Antennas & Propagation*. 57(2). pp. 452–459.

Bayatpur, F. 2009. *Metamaterial-Inspired Frequency-Selective Surfaces*. Ph.D. Dissertation, University of Michigan.

Celozzi, S. 2008. *Electromagnetic Shielding*. New York: Wiley Interscience.

Chatterjee, Ayan & Parui, S.K. 2013. A multi-layered bandpass frequency selective surface designed for Ku band applications, *In Proc. of IEEE Applied Electromagnetics Conference (AEMC)*. India. pp. 1–2.

Chatterjee, Ayan & Parui, S.K. 2013. A Multi-layered Broadband Frequency Selective Surface for X and Ku band Applications, *In Proc. of International Conference on Technical and Managerial Innovation in Computing and Communications in Industry and Academia (IEMCON)*. India. pp. 284–287.

Chatterjee, Ayan & Parui, S.K. 2015. Gain Enhancement of a Wide-Slot Antenna using Dual-Layer, Bandstop Frequency Selective Surface as a Substrate, *Microwave & Optical Technology Letters*. 57(9). pp. 2016–2020.

Chatterjee, Ayan & Parui, S.K. 2015. Gain Enhancement of a Wide-Slot Antenna using Second-Order Bandpass Frequency Selective Surface, *Radioengineering Journal*. 24(2). pp. 455–461.

Chen, H.Y., & Chou, Y.K. 2012. An EMI Shielding FSS for Ku-Band Applications, *In Proc. of IEEE Antennas and Propagation Society International Symposium (APSURSI)*.

Chiu, C.N. 2008. Bandpass Shielding Enclosure Design Using Multipole-Slot Arrays for Modern Portable Digital Device, *IEEE Transactions on Electromagnetic Compatibility*. 50(4). pp. 895–904.

Gatti, R.V., Marcaccioli, L., Sbarra, E. & Sorrentino, R. 2008. Flat array antennas for Ku-band mobile satellite terminals, *In Proc. of the 30th ESA Antenna Workshop on Antennas for Earth Observation, Science, Telecommunications and Navigation Space Missions*. PP. 534–537. Noordwijk.

Munk, B.A. 2000. *Frequency Selective Surfaces—Theory and Design*. New York: Wiley Interscience.

Sung, G.H.-h, Sowerby, K.W. Neve, M.J. & Williamson A.G. 2006. A Frequency selective Wall for Interface Reduction in Wireless Indoor Environments, *IEEE Antennas & Propagation Magazine*. 48(5). pp. 29–37.

Zvere, A.I. 1967. *Handbook of Filter Synthesis*. New York: Wiley Interscience.

Frontiers in Computer, Communication and Electrical Engineering – Acharyya (Ed.)
© 2016 Taylor & Francis Group, London, ISBN: 978-1-138-02877-7

Design of efficient second harmonics injection based solar inverter for standalone application

Sudip Mondal & Sumana Chowdhuri
Department of Applied Physics, University College of Science Technology and Agriculture, Kolkata, India

ABSTRACT: This proposed work is based on the design and development of pic-microcontroller and Second Harmonics Injection Based full bridge inverter with rated frequency output. The proposed hardware model shows a low THD with minimum filter requirement and reduction of sub-harmonics in the system. The developed inverter is tested with resistive, inductive, and motor load.

I INTRODUCTION

Inverters used in general applications are generally of quasi-square wave output, which requires a huge filter for converting this square wave to sinusoidal for different load applications. In this regard, a sinewave inverter which generates sinusoidal current is more efficient since there is no requirement of filter, and consequently, the power loss in the inverter decreases, and hence, the inverter efficiency increases. To generate sine waves at inverter output, in the control circuit, a high frequency carrier wave is compared with a modulating sine wave, and this output is fed to the inverter devices. This is called Sinusoidal Pulse Width Modulation (SPWM) technique. Similarly second harmonics are also generated in this technique. SPWM techniques have been the subject of intensive research during the last few decades toward the betterment of electrical power flow control to various applications (Rashid, M.H. 2004, Mohan, Ned, *et al.* 2003). In ref to (Pankaj H Zope1, *et al.* 2012, G. Eason, *et al.* 1955), the concept of Pulse Width Modulation (PWM) for inverters is described with analyses extending to different kinds of PWM strategies. Finally, the simulation results for a single-phase inverter (unipolar) with second harmonics injection using the PWM strategies described are presented. In general, we get 60.17% output from the inverter, but using the second harmonics injection technology we can increase the output by about 15.5%. So we can increase the efficiency.

In digital controller designs, this has been done by keeping the sampling values of sine + second harmonics and the triangles in the look-up table in RAM. Sinusoidal pulse +2nd harmonics modulation is the most used method in motor control and inverter application. This technique generally uses regular sampling of these three values (Hirak Patangia, *et al.* 2012). This makes output frequency constant. For controlling the speed of a single-phase induction machine with variable voltage, variable frequency is required as speed is dependent on supply frequency. Nowadays there is a growing interest in development of Pic-microcontroller based SPWM + 2 nd harmonics (Lalit Patnaik, *et al.* 2010) module system compared to other conventional ones like dedicated analog and digital control (IEEE Std 519–1992). The standalone model embedded the working features of a pic-microcontroller to simplify the hardware with reduced components, improved performance, enhanced reliability of the system, less aging than analog devices etc. More precisely the standalone pic-microcontroller based system offers modularity which can readily be integrated with power electronics with proper isolation in between. This single chip module can control the power electronics devices along with other features, like protection of the devices, and also all kind of annunciations. Thus, the flexibility in control with cost effectiveness is one of the major advantages of any micro controller based dedicated module.

This work is based on Microcontroller generated Sinusoidal Pulse Width Modulation (SPWM) + second harmonics injection technique to produce a pulsed waveform that can be filtered in a relatively easy way to achieve a good approximation to a sine wave for variable frequency application. The significant advantage of this SPWM + second harmonics (Lalit Patnaik, *et al.* 2010) injection approach is that easily variable frequency output can be obtained by single input control. In this proposed scheme, the inverter operates with an integral ratio of carrier to modulating signal frequency, where the modulating wave remains synchronized with the carrier wave for the entire region of frequency of operation (this causes low carrier frequency as the fundamental frequency goes down). This proposed inverter is a relatively highly efficient inverter. Without adding any equipment we can easily increase the output voltage level of the inverter which gives sinusoidal output with lower sub-harmonics for all kind of loads. The inherent advantage of this method is that lower order harmonics are almost eliminated. Instead small magnitude higher order harmonics are introduced in the system, which can easily be eliminated or minimized with the help of a smaller size filter. By increasing the number of pulses per half cycle, the process is also enhanced. The SPWM +2nd harmonics injection technique, however, inhibits poor performance with regard to maximum attainable voltage and power.

In this work, we used a 36-volt battery for the input source and from the output we get approximately 27 volts (rms) of output. Now for the grid tie-up application, we used a step-up transformer (27/230), and for the grid we used a replica of our laboratory busbar which is 230 volt and 50 Hz. Phase Lock Loop (PLL) technology is also used in this inverter for connecting the synchronizing grid.

2 DESIGN OF SPWM INVERTER

A. *Working principle*

The inverter taken for study is the H-bridge inverter shown in Figure 1 with unipolar control strategy and is preferred over other topologies of inverter in higher power ratings. With the DC input voltage, the maximum output voltage of the inverter increases in this topology.

In unipolar SPWM control strategy, a sinusoidal modulating signal is compared with a repetitive switching frequency, triangular or saw-tooth waveform, to generate the switching signals for the inverter devices (Fig 2). By changing the control signal magnitude, the width of the gate drive for the devices can be changed and so can output voltage magnitude. By changing the frequency of the modulating wave, the fundamental frequency at the inverter output changes. The inverter output voltage will not be of a perfect sinewave and will con-

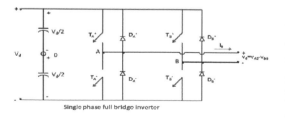

Figure 1.   Single phase full bridge inverter.

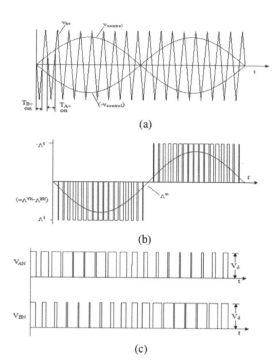

Figure 2.   a) Unipolar SPWM control strategy b) Gate-pulses c) Output voltage waveform.

tain voltage components at harmonics frequency of $f1$. f_1 is the desired fundamental frequency of the inverter voltage output (f_1 is also called the modulating frequency). The amplitude modulation ratio is defined as

$$m_a = \frac{v_{control}}{v_{Tri}} \qquad (1)$$

where $v_{control}$ is the peak amplitude of the control modulating signal and v_{Tri} the peak amplitude of triangular signal, which is generally kept constant. Generally, $m_a \leq 1$ to reduce overmodulation. Frequency modulation index, m_f, is defined as

$$m_f = \frac{f_{Tri}}{f_{mod}} \qquad (2)$$

where f_{tri} is the frequency of Carrier signal and f_{mod} is the frequency of modulating signal. Generally, $m_f \geq 1$ to reduce the harmonics at the output. Theoretically, the frequencies at which voltage harmonic occur will be for unipolar full bridge inverters at $f_h = (jm_f \pm k)f_{mod}$, where h is the order of harmonic, j is the even multiple of the frequency modulation index and k is the sideband number, and f_{mod} is the fundamental frequency.

B. Second harmonics injection scheme

In the case of second harmonics injection, we inject a common signal which also presents itself in the inverter. As a result, we can increase the output voltage upto a certain level. This waveform is also symmetrical about the axis. Lets assume that a common mode signal is injected in the modulating wave (m_A) for pole-A. In order to preserve the symmetry of waveforms at the two poles, the modulating wave (m_B) for pole-B is obtained by phase shifting m_A. To get the maximum peak value of the peak line-line voltage, this phase shift must equal 180° so the equation is

$$Sin\theta + k*Sin(2\theta + \varphi) \tag{3}$$

where from the Fourier analysis, we get the value of k is nearly {1/5}. So, with the addition of second harmonics component, we can increase the pole voltage about 16% and therefore the output voltage increases.

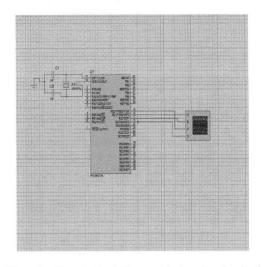

Figure 3.   Proteus block diagram of triggering circuit of inverter.

C. Schematic diagram of the control scheme and

The block diagram in Figure 3 shows that the pic-microcontroller 16F877a is designed to get the output of SPWM waveform shown in Figure 4, and it is fed to the four Gate pulses of MOSFET H-bridge. The inverter DC input is the battery itself. The inverter output is stepped up using a transformer for load connection. In the output of the transformer, a small low-pass filter circuit is provided to smoothen out the load current waveform. Now for the second harmonics injection shown in Figure 5, the pic-microcontroller program is modified, and it is fed to the MOSFET H-bridge driver circuit. Now from the graphical analysis we see that if we create phase difference between the fundamental and the second harmonics we get a better waveform. There are some waveforms obtained from the simulation results followed by the equation (3).

D. Designing the firmware

Here the algorithm of the pic-microcontroller is developed to perform the key features of the whole circuit, i.e., to generate SPWM + second harmonic (Lalit Patnaik, et al. 2010) signals from four pins of the controller. The sine wave for 0–60° is generated from stored look-up tables of sine functions of fixed number samples. The desired frequency can be obtained simply by changing the time delay in between the samples of the sine function. According to equation (3), the carrier triangular function of high frequency is generated from the triangular look-up table of the same number of samples. According to equation (4), in the software both

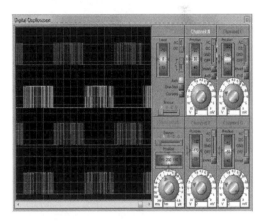

Figure 4.   Proteus output of SPWM for triggering circuit of inverter.

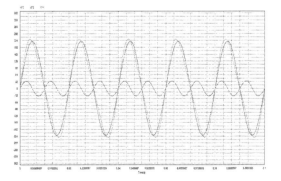

Figure 5.   Second harmonics injection (k = 0.20, $\varphi = 0°$).

347

these waves are compared for each cycle of the sine wave at each "Δt" interval.

$$f_{mod} = \frac{1}{(s \times \Delta t)} \quad (4)$$

The design of the controller for frequency control is such that it maintains accuracy at the inverter output. As already considered that the modulating signal is generated from a look-up table of "S" number of samples of sine function, to generate particular frequency (f_{mod}) of SPWM, the delay (Δt) between the samples is to be estimated from the controller ADC input.

The flowchart of the firmware has been demonstrated below. The technique is that it has to compare the sampled values of two modulating signals (sinwave) with sampled values of triangular wave.

$$f_{tri} = \frac{m_f}{(s \times \Delta t)} \quad (5)$$

where m_f is the frequency modulation index.

For unipolar inverters, one port pin is the complement to the other, which are used to generate

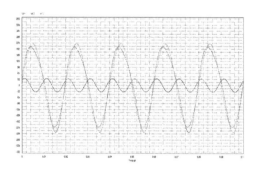

Figure 6. Second harmonics injection (k = 0.20, φ = 45°).

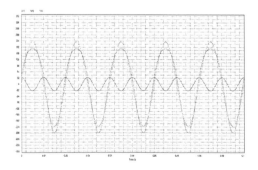

Figure 7. Second harmonics injection (k = 0.20, φ = 90°).

gate pulses to trigger the upper and lower MOSFETS of the same leg of full bridge inverter. To avoid "shoot through" faults, a certain amount of delay has been provided in the software.

Now for the second harmonics injection, we only make relative minor alterations to the already existing software needed to cater to this introduction.

The flow chart for the second harmonics injection method is given below

3 HARDWARE DESCRIPTION

A. *Circuit elements:*

MOSFET (4 NO.S), IR2110 Driver, 12V DC and 5V DC supply, Snubber (4 no.s), microcontroller (PIC16f877 A).

1. The power circuit for load has chosen H-Bridge inverter so that the voltage rating of the devices gets reduced to half of the maximum DC bus voltage. So the device rating and as well as cost will be reduced. Again for H-Bridge inverters, sinusoidal output waveform is better compared to Push Pull Inverter. Here, MOSFET IRF P250 N has been used (datasheet 1).

2. To protect the MOSFETs from transients here, the RC turn off Snubber (R = 10 ohm, 2 W C = 0.1μF) circuit has been used.

3. A freewheeling diode of fast recovery is used in anti-parallel with each MOSFET to provide freewheeling paths for inductive currents.

4. Here the firing pulses for SPWM have been generated by using programming in assembly language for microcontroller PIC16f877A. Two output ports are used to generate SPWM waveforms. Here unipolar SPWM topology has been taken.

IRFP250 NN Channel power MOSFET 200V, 30 A, 0.075Ω

The data sheet of the Mosfet used is provided in Table 1.

B. *Power MOSFET gate driver*

IR2110

This unit is 14-lead DIP package.

VDD: Logic supply

HIN: Logic input for high side gate driver output (HO), in phase

SD: Logic input for shutdown

LIN: Logic input for low side gate driver output (LO), in phase

VSS: Logic ground

VB: High side floating supply

HO: High side gate drive output

VS: High side floating supply return

VCC: Low side supply

LO: Low side gate drive output

COM: Low side return

FLOW CHART OF THE SPWM PROGRAMME

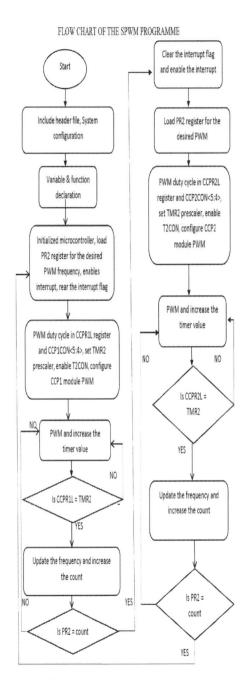

Figure 8. Flow chart of the program logic.

4 OUTPUT FROM HARDWARE

Tektronix TDS 2024B, a two-channel digital storage oscilloscope is used to measure the experimental results. SPWM1 and SPWM2 (in Fig. 13, applied to upper and lower switches) are the pulses generated from the controller for 50 Hz.

Fig. 14 shows that two MOSFET in the same leg cannot be fired at the same instant to avoid the shoot through fault. To prevent this fault, a certain amount of delay in microseconds has been incorporated in the logic of the firmware. Here the delay of 2 µs is sufficient enough to protect the devices.

The output is passed through a first-order filter to eliminate the harmonics and the output obtained is given below

SPWM1 is leads SPWM2 by a half cycle of the switching pulses in frequency command signals. Here the number of pulses per half cycle has taken an odd number. Hence Fig 6 shows the inverter voltage and current waveform for resistive loads of 50 Hz frequency.

For resistive loads, the THD in current is very high. The output current waveforms from a single-phase inverter with inductive load have been shown in Fig. 7. For inductive loads, the THD in current is 21.01% (in conventional SPWM technique, it is very high) as shown in Fig. 8.

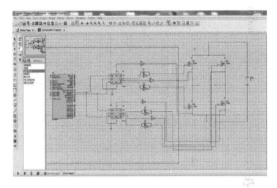

Figure 9. Circuit Diagram.

Figure 10. Mosfet.

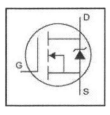

Figure 11. Mosfet symbol.

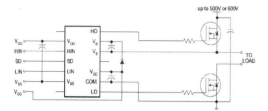

Figure 12. Pin diagram of IR2110.

Figure 13. Four gate pulses.

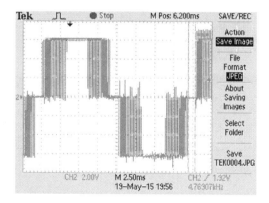

Figure 14. Delay to avoid "Shoot Through Fault".

Table 1. Mosfet data sheet.

Symbol	Parameter	Ratings	Unit
V_{DSS}	Drain to source voltage	200	V
V_{GS}	Gate to source voltage	+ −20	V
I_D	Drain Current continuous (Tc = 25°c, VGS = 10V)	30	A
	Continuous (Tc = 100°c, VGS = 10V)	21	A
E_{AS}	Single pulse avalanche energy	315	mJ
P_D	Power dissipation	214	W
	Derate above 25°C	1.4	W/°C
T_J, T_{STG}	Operating and Storage Temperature	−55 to 175	°C

Fig. 9 and Fig. 10 show 40-Hz and 75-Hz frequency outputs, respectively, for inductive loads. For inductive load with 50 Hz, the THD in voltage is 30.6% and for 75 Hz, THD is 24.01%. Fig. 11 and Fig. 12 show the frequency sweep from 17.48 to 75.76 Hz.

The inverter voltages with different PWM techniques have been implemented and the results are analyzed in Fast Fourier Transform (FFT) analysis to find the fundamental component and the harmonic component presented. The different voltages with different PWM techniques are shown in Figures 15 to 22 and tabulated in Table 2; it also increase the fundamental output to near 15% which is very important than others.

The snaps of the hardware setup are shown in Figure 23 and 24.

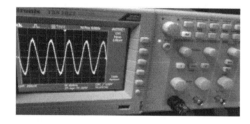

Figure 15. Output after passing through first-order filter.

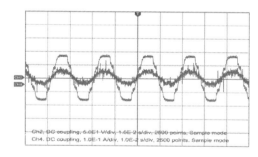

Figure 16. Voltage and current waveform for resistive load at 50 Hz.

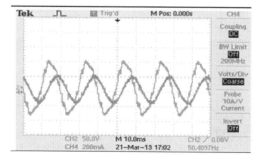

Figure 17. Voltage and current waveform for inductive load at 50 Hz.

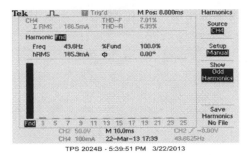

Figure 18.   THD for current in resistive load.

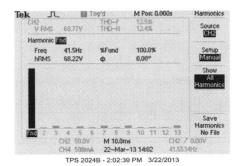

Figure 19.   THD of Voltage at 41.5 Hz in Inductive Load.

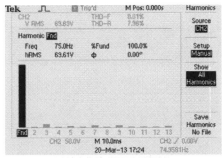

Figure 20.   THD of voltage at 41.5 Hz in inductive load.

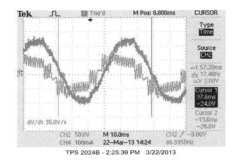

Figure 21.   Voltage and current waveform at 17.48 Hz.

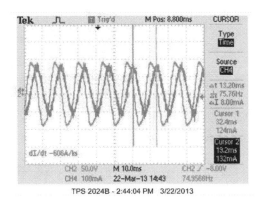

Figure 22.   Voltage and current waveform at 17.48 Hz.

Table 2.   FFT analysis of various techniques.

Types of inverters	THD (%)
Two level single bridge inverter (using SPWM)	85.49
Two level double bridge inverter (using SPWM)	70.43
Second harmonics + SPWM inverter	30.6

5 PICTURES OF HARDWARE SETUP

Figure 23.   Picture of experimental setup.

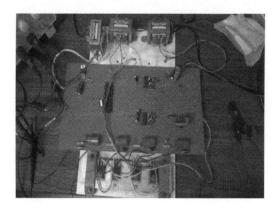

Figure 24.   Picture of main circuit board.

6   DISCUSSION

The observed voltage waveform shows that there is a distinct dead time effect for current transition from positive to negative and negative to positive value. This dead time effect can be compensated in future work.

For resistive load, THD of the current is much higher than the THD of current in inductive load, since the current in inductive load is more sinusoidal than the resistive load.

REFERENCES

Eason, G., B. Noble, and I.N. Sneddon, "On certain integrals of Lipschitz-Hankel type involving products of Bessel functions," Phil. Trans. Roy. Soc. London, vol. A247, pp. 529–551, April 1955.

Hirak Patangia, and Sri Nikhil Gupta Gourisetti, "A Novel Strategy for Selective Harmonic Elimination Based on a Sine-Sine PWM Model", MWSCAS, U.S.A, Aug 2012.

IEEE Std 519–1992, "IEEE Recommended Practices and Requirements for Harmonic Control In Electric Power Systems," Institute of Electrical and Electronics Engineers, Inc. 1993.

Ismail, B.S.T. "Development of a Single Phase SPWM Microcontroller-Based Inverter" First International Power and Energy Conference PEC, Putrajaya, Malaysia: IEEE, Nov, 28–29, 2006.

Lalit Patnaik, G. Narayanan, and L. Umanand "An Investigation into Even Harmonics Injection In Pole Voltages of A Single Phase Inverter", IIT ROORKEE, Nov 2010.

Mohan, Ned. M. Undiland, T.P. Robbins, William, and 2003. Power Electronics: Converter, Application and Design. United States of America: John Willey & Son.

Pankaj H Zope1, Pravin G.Bhangale2, Prashant Sonare3, S.R. Suralkar, "Design and Implementation of carrier based Sinusoidal PWM Inverter", International Journal of Advanced Research in Electrical, Electronics and Instrumentation Engineering, Vol. 1, Issue 4, October 2012, pp. 230–236.

Rashid, M.H. 2004. Power Electronics; Circuits, Devices and Applications. New Jersey Prentice Hall.

Proxima-Talk: A proposed framework for network assisted device-to-device communication

Rupendra N. Mitra & D.P. Agrawal
University of Cincinnati, Ohio, USA

ABSTRACT: Device-to-Device (D2D) wireless communication has been significantly researched in recent years due to its immense potential of multi-gigabit data transfer rate while using a 60-GHz unlicensed spectrum along with mmWave beamforming. Still, 60-GHz communications are yet to be available commercially due to its high loss and limited coverage. In this letter, we propose proxima-talk, a new D2D communication framework. The salient feature of proxima-talk is unlike many other D2D communication protocols. It can switch the data transfer link between D2D direct channel and conventional cellular network. It uses mmWave beamforming for D2D communication, reports Channel Quality Index (CQI) to the base station, and depending on the CQI of the D2D channel, it hands over the traffic to the cellular network for seamless data transfer. We conclude by showing possibilities and requirements of such futuristic D2D protocol.

1 INTRODUCTION

Device-to-Device (D2D) communication was first introduced to enhance the performance of conventional cellular networks. New data intensive apps for cell-phones are coming every day and requirement of higher data transfer rates becomes obvious in wireless communication. Apps like content distribution, proximity aware services, multi-player gaming, and cellular offloading of multimedia data are making D2D communications an integrated feature to modern digital devices. D2D communication provides improved spectral efficiency and very low latency since there is no networking involved with it. Earlier the infrared band was used in cell-phones to share files among devices (Dack *et al.* 1994). Then Bluetooth (BT) made an important contribution to change the way D2D communications was used on ISM band. It certainly added more agility in connecting devices, allowed increased distances between two communicating devices, and made possible wireless file transfer, wireless printing, and personal D2D communication easier and cheaper (Bray *et at.* 2002). But D2D was first proposed in cellular networks for multi-hop relaying by Lin and Hsu, 2000. Keeping in mind the features it enables in cellular communication, the next generation's 5G cellular standard may include D2D as an integrated feature to it.

In this letter, we discuss the various advancements of D2D communication since 2000 when it was first proposed. We identified one major open area of futuristic D2D communication with the agility of switching between D2D channel and conventional cellular channels for seamless data connectivity and thereafter propose Proxima-Talk, a novel network assisted D2D framework.

The rest of this letter is organized in following four sections. Section two highlights the theoretical foundations of D2D communication. Section three presents a brief survey of literatures addressing several issues and proposes advancements in D2D communication. Section four identifies an open possibility in D2D communication and proposes Proxima-Talk as a potential futuristic D2D communication framework to ensure uninterrupted data connectivity. Finally, section five concludes this letter with future research directions and possible applications of this technique in daily life.

2 MORE ON D2D COMMUNICATION

By definition, D2D communication is a point-to-point direct link between two end users without traversing the base station or the core network of cellular networks (Asadi *et al.* 2014). Generally this D2D channel can use the unlicensed spectrum or the cellular spectrum. Thus, D2D communication can be broadly classified into two categories, inband and outband [Figure 1]. Many researchers used cellular spectrum for D2D communication in their literatures because the licensed spectrum is well researched and highly controllable (Doppler *et al.* 2009). The supporting hardware is already advanced, and Quality of service (QoS) could be ensured therefore.

On the other hand, unlicensed spectrum is proposed by some researchers because it eliminates the chances of interference between cellular channels and direct D2D channels. In this case, the D2D channels use Bluetooth, Wifi-direct (*Wi-Fi Alliance*, 2010), etc.

This broad classification of D2D communication can further be categorized as follows: underlay and overlay communication come under inband D2D. Outband D2D are categorized into controlled and uncontrolled. In inband *underlay*, D2D communication framework, both cellular and D2D channels, use the same licensed radio resources. On the other hand, *overlay* communication D2D channels are dedicated cellular resources non-overlapping with cellular spectrum. Thus, inband D2D significantly improves spectrum efficiency of cellular networks but to avoid the key disadvantage of inband D2D, i.e., the interference between D2D channels and cellular channels, complex resource scheduling algorithms are implemented leading to significant computational overhead base stations and end user equipment. Many researchers who proposed outband D2D communication frameworks so that cellular spectra remain untouched by D2D channels have mainly two types of approaches. Some of them proposed the *controlled* outband framework where the link establishment is controlled by the cellular base station and other types that do not involve base station are called *uncontrolled*.

Tehrani *et al.* 2014, classified the D2D classification into four subsequent classes depending on the involvement of the cellular base station in assigning control channels in D2D link establishments and peer discovery. The second hierarchy of Figure 1 shows the four classes. In DR-OC, a user device in a poor cellular coverage area can communicate with the BS through relaying its information via other neighboring devices. In DC-OC, the peering devices communicate with each other independently but BS helps them to establish the link and initial discovery. In DR-DC, just like Ad-hoc network, several peering devices can communicate relaying their information through the neighboring peers without any involvement of BS. In DC-DC, two or more peering devices communicate directly without relay or without any involvement of BS.

In this letter, we propose Proxima-talk, a network assisted (DC-OC) D2D communication framework for futuristic cellular networks. The term "network assisted" is used because it is a supervised D2D communication and the cellular base station will monitor the quality and survivability of the D2D channel. The D2D link is being monitored by the cellular base station in terms of Channel Quality Index (CQI) and with other

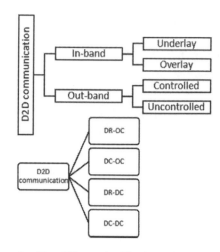

Figure 1. Two different classifications of D2D communication. First one is based on the spectrum usage and second one is based on the involvement of cellular base stations in D2D communication. OC stands for Operator Controlled and DC stands for Device controlled. DR stands for Device Relaying. OC means the channel assignment is taken care by the base station.

factors. If the base station observes a severe drop in the quality of the D2D communication channel, the data transfer gets switched through the cellular network. Base station is programmed to take the decision on detail supervision of the D2D channel quality logs. Thus, Proxima-Talk ensures uninterrupted data connectivity using both D2D link and cellular channels, and moreover, the connectivity between the two end users are constantly monitored by the base station during the entire period of the data transfer. Therefore, we suggest proxima-talk as a network assisted D2D communication. Although we ideate it as an outband D2D communication protocol, it may be extended as an inband underlay technique in future when the next generation of cellular networks will use higher frequency unlicensed spectrum, e.g., 60 Ghz, in a more controlled manner (Ohyun *et al.* 2015).

3 BACKGROUND WORKS

Since its introduction, many literatures discussed important aspects of D2D communication. *Spectrum efficiency, interference, resource allocation method, throughput improvement, power efficiency, and QoS performance* are major areas of D2D literatures to be discussed in a brief survey.

Kaufman *et al.* (2008) proposed to use the uplink channels of cellular licensed spectra for D2D link establishment to improve spectrum efficiency. Interference between cellular channels and

D2D channel is a key factor. In (Peng *et al.* 2009), authors proposed to read radio resource block assignment information from the cellular control channel and then cognitively use the cellular uplink channel for D2D communication to minimize the interference. Zhang *et al.* 2013, mathematically formulate resource allocation in an optimal way in D2D communication. Yu *et al.* 2012 suggested rate splitting techniques to improve the throughput of D2D communications. The key essence in rate splitting is that it divides the message in to two parts, public and private. The private message part can be decoded only by the intendent peer, and the public part is decodable by any peer. This scheme helps D2D channels to minimize the effect of interference. Doppler *et al.* 2009, studied session interference management between D2D and LTE networks. Moreover, several techniques, such as cloudification of resource (Fodor *et al.* 2012), non-cooperative game or bargaining game (Hossain *et al.* 2009), optimal power allocation algorithms (Feng *et al.* 2013), clustering based on SNR and relaying peer selection algorithms (Zhou *et al.* 2013), are proposed for interference management and power controlling. Xiao *et al.* 2011, proposed a heuristic algorithm for power control and subcarrier bit allocation in OFDMA-based cellular networks. Feng *et al.* 2013, mathematically formulated and optimized resource allocation problem, a non-linear constraint optimization problem, for D2D underlaying cellular network. This algorithm takes care of QoS requirements for both cellular and D2D users and cellular users.

4 PROPOSED FRAMEWORK

Most of the D2D techniques cannot guarantee uninterrupted data connectivity because of limited coverage. Specially, when two users are moving in comparatively high speed with respect to each other and the distance between them is varying randomly, D2D channel experiences poor channel quality and heavy data loss. This may lead to low data transfer rate or even complete loss of connectivity.

To address this issue, we propose Proxima-talk which ensures connectivity during the entire period of data transfer by handing over the connection to cellular network in case of poor D2D channel quality. So it is evident that Proxima-Talk is a network controlled (OC) protocol and BS plays the key role in peer discovery, link establishment, supervising the D2D channel quality, and decides when to take handover from D2D channel to cellular data network.

Neighbor discovery is an important step to initiate D2D communication between two end users.

At first, UE needs to get the knowledge of the location of the intended peer and decide if it is in an allowable range of D2D communication. Yang *et al.* 2013, mathematically modeled a random access based distributed UE discovery protocol for LTE-advanced cellular networks. The proposed algorithm also supports mobility while discovering peers with a discovery probability close to 0.99%. Hence, in proxima-talk we can use this peer discovery technology for initial simulation.

Millimeter wave (mmWave) communication is the next big technology to explore in cellular spectra because high frequency carriers give faster bit rate provisioning along with less interference with existing in-use spectra. Many researches are currently focused on 34-GHz and 60-GHz radio spectrum attenuations and hardware realization for the next generation of wireless communication. A recent survey forecasts that a 1500-million-dollar market will be created by the 60-GHz Wi-Fi chipset until 2018 (Solis and Cooney, 2013). The 60-GHz Wi-Fi radio signals suffer more signal attenuation than the legacy radio signal due to loss of air propagation and high absorption rate by oxygen. A mmWave array beamforming technology is thus adopted by the 60 GHz system to overcome the issues. Proxima-Talk can use this 60-GHz radio and beamforming for D2D channel establishment. This is an outband approach which will provide a high data transfer rate in D2D communication.

Beamforming is another signal processing technique which is used nowadays in directional communication. By means of beamforming, we can a direct majority of the transmitting signal energy from an array of antenna elements to a particular angular direction where the receiver is. Proxima-talk uses such directional antenna elements for constructive and destructive superpositions for beam steering. There are several beam searching algorithms to find out the best TX/RX beam pairs, viz., full search algorithm and sequential search with beam shaping or without beam shaping [8]. Hence, Proxima-talk can use this beamforming for D2D link establishment. Figure 2 sequentially shows a scenario of Proxima-talk how beamforming can be used in D2D transmission, how the obstacle makes the D2D channel quality poor to force data transfer to switch its path onto cellular network and then resumes after the obstacle is gone.

5 CONCLUSION

The first four generations of cellular standards did not consider the D2D communication functionality seriously. However, technologies like wifi-direct or BT are in use as unsupervised D2D communication

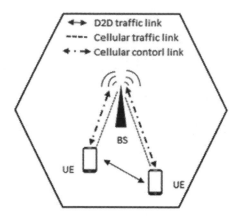

Figure 2. Schematic diagram of Proxima-Talk as a DC-OC D2D communication framework.

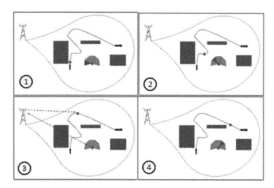

Figure 3. A comprehensive diagram to illustrate the working principle and beamforming in Proxima-Talk as a DC-OC D2D communication framework.

protocols. But these unlicensed band D2D interfaces cannot ensure QoS provisioning like cellular networks do. Operators, after closely following the current market trends, have now started insisting on D2D functionality in the next generation of cellular standards. Because highly congested data networks are envisaging exponential growth in data traffic, by numerous context aware services, location based cell phone applications, video dissemination etc., D2D communication can provide cellular data offloading provision to help reduce the cellular network congestion. D2D feature can play a vital role in mobile cloudification to facilitate effective sharing of base band resources. D2D can not only be used for commercial usage, but also during natural disasters since it may allow creating local relay based communication networks until the cellular networks get restored. A comprehensive diagram to illustrate the working principle and beamforming in Proxima-Talk as a DC-OC D2D communication framework is shown in Figure 3.

In this letter, we briefly discussed the progress of D2D communication and why it should be an integral part of upcoming cellular standards. We also proposed Proxima-Talk, a futuristic D2D communication framework which enables uninterrupted data connectivity between the peering devices, no matter how their inter-device distance or in-sight obstruction varies. This may be a very useful and required approach in future D2D functionality that is going to be an integrated feature of the next generation cellular networks.

REFERENCES

Asadi, Arash, Qing Wang, and Vincenzo Mancuso. "A survey on device-to-device communication in cellular networks." *Communications Surveys & Tutorials*, IEEE 16.4 (2014): 1801–1819.

Dack. D., Colin I'Anson and Graeme Proudler," Use of 115kb/s Infra-Red Interface for Mobile Multi-Media", *Personal, Indoor and Mobile Radio Communications, 1994. Wireless Networks—Catching the Mobile Future., 5th Ieee International Symposium on*, 1994, vol 3, pp. 980–985.

Doppler. K., Rinne, M.P., Janis. P., Ribeiro. C., and Hugl. K., "Deviceto-device communications; functional prospects for LTE-Advanced networks," *in Proc. IEEE ICC Workshops*, 2009, pp. 1–6.

Feng. D. *et al.*, "Device-to-Device Communications Underlaying Cellular Networks," *IEEE Trans. Commun.*, vol. 61, no. 8, 2013, pp. 3541–51.

Feng. D. *et al.*, "Device-to-device communications underlaying cellular networks," *IEEE Trans. Commun.*, vol. 61, no. 8, pp. 3541–3551, Aug. 2013.

Fodor. G. *et al.*, "Design Aspects of Network Assisted Device-to-Device Communications," *IEEE Commun. Mag.*, vol. 50, no. 3, Mar. 2012, pp. 170–77.

Hossain. E., Niyato. D., and Han. Z., "Dynamic Spectrum Access and Management in Cognitive Radio Networks", 1st ed., Cambridge Univ. Press, 2009.

Jennifer Bray and Charles F Sturman, "Bluetooth: connect without cables", 2nd ed, ISBN 0132442396, 2002.

Jo Ohyun. *et al,*" 60 GHz Wireless Communication for Future Wi-Fi", *ICT Express*, Volume 1, Issue 1, June 2015, Pages 30–33.

Kaufman. B. and Aazhang. B., "Cellular networks with an overlaid device to device network," in *Proc. Asilomar Conf. Signals, Syst. Comput.*, 2008, pp. 1537–1541.

Lin. Y.-D and Hsu. Y.-C., "Multihop cellular: A new architecture for wireless communications," in *Proc. IEEE INFOCOM*, 2000, vol. 3, pp. 1273–1282.

Peng. T., Lu. Q., Wang. H., Xu. S. and Wang. W., "Interference avoidance mechanisms in the hybrid cellular and device-to-device systems," in *Proc. IEEE P1MRC*, 2009, pp. 617–621.

Solis. P., Cooney. P., "Wi-Fi Semiconductors: 802.11 ac, MIMO, and 802.11ad", *ABI research,* 2013.

Tehrani, Mohsen Nader, Mustafa Uysal, and Halim Yanikomeroglu. "Device-to-device communication in 5G cellular networks: challenges, solutions, and future directions." *Communications Magazine, IEEE* 52.5 (2014): 86–92.

Wi-Fi Peer-to-Peer (P2P) Specification v1.1, *Wi-Fi Alliance*, 2010, vol. 1, pp.1–159.

Xiao. X., Tao. X., and Lu. J., "A QoS-aware power optimization scheme in OFDMA systems with integrated Device-to-Device (D2D) communications," in *Proc. IEEE VTC-Fall*, 2011, pp. 1–5.

Yang, Zhu-Jun, *et al.* "Peer discovery for device-to-device (D2D) communication in LTE-A networks." *Globecom Workshops (GC Wkshps), 2013 IEEE*, 2013.

Yu. C.-H. and Tirkkonen. O., "Device-to-device underlay cellular network based on rate splitting," in *Proc. IEEE WCNC*, 2012, pp. 262–266.

Zhang. R., Cheng. X., Yang. L., and Jiao. B., "Interference-aware graph based resource sharing for device-to-device communications underlaying cellular networks," in *Proc. IEEE WCNC*, 2013, pp. 140–145.

Zhou. B. *et al.*, "Intracluster Device-to-Device Relay Algorithm with Optimal Resource Utilization," *IEEE Trans. Vehic. Tech.*, vol. 62, no. 5, 2013, pp. 2315–2326.

Frontiers in Computer, Communication and Electrical Engineering – Acharyya (Ed.)
© 2016 Taylor & Francis Group, London, ISBN: 978-1-138-02877-7

Effect of slow decaying trapped charges on PDC data and associated diagnosis of power transformer insulation

Nitanshi Verma & Arijit Baral
Indian School of Mines, Dhanbad, Jharkhand, India

ABSTRACT: The operational age of a significant number of power transformers have already exceeded their designed lives. Though prone to failures, such transformers are still in use to maximize asset utilization and defer major capital investment. Ensuring fault free operation of such crucial HV power equipment by strategies based on reliable insulation diagnosis can mean substantial saving for the utilities. Modern non-invasive techniques based on measurement and analysis of Time Domain Spectroscopy data provides information regarding the behavior of solid insulation in power transformer insulation. Commercially available equipment (capable of analyzing dielectric response function) being extremely costly, researchers are actively engaged in devising methods to assess insulation conditions. However, very few of such reported techniques consider the effect of trapped charge and its influence on the results of insulation diagnosis. The focus of this paper is aimed at addressing this issue. The data used for the present work is obtained from a real-life power transformer making the findings extremely relevant to the field of insulation diagnosis.

1 INTRODUCTION

The demand for a reliable electricity supply has significantly increased during the last few decades. The reliability of service provided by power utilities is heavily dependent on satisfactory operation of high-voltage equipment. Due to high cost of power system equipment, like transformers, they are not replaced to increase reliability of the system by considering their age only. On the contrary, a large number of such equipments that are operating today have already exceeded their designed life. Extending the life of such equipments by policies based on proper condition assessments can mean substantial saving for the utilities besides providing fault-free operation. Therefore, correct condition assessments based on reliable non-invasive diagnosis of equipments, like power transformers, is required before taking any decisions about replacement of such capital intensive equipment.

It is reported that the value of paper moisture in cellulosic parts of power transformer insulation can be predicted using the transfer function of Modified Debye Model (Baral *et al.* 2014). However, the data used for predicting paper moisture is recorded from equipment which have negligible trapped charge content. Such a condition was ensured at the time of data measurement by allowing sufficient relaxation time for the equipment after shutdown. PDC measurement is an offline measurement technique. Equipment is generally provided for a relaxation time t_r after shutdown with its terminals short circuited. This allows the equipment to reach thermo-hydrodynamic equilibrium. This phase also helps to neutralize trapped charge, i.e., generated mechanisms like charge separation and flow-electrification. It is reported that such charge decays at an extremely slow rate especially if the walls of the tank are insulated (Roach *et al.* 1988). Furthermore, due to unavailability of standards, the value of t_r is decided by the operator. Involvement of the human factor combined with slow decay rate may result in a finite amount of trapped charge within the insulation prior to PDC measurement. It is understood that it is practically difficult to ensure that the insulation system will contain negligible trapped charge every time PDC measurement is performed.

The characteristics of trapped and space charge along with the location of the same have been investigated by several researchers. The popular acoustic pulsed method is used to determine the space charge profile in dielectrics (Tang *et al.* 2010). It is reported that the applied DC voltage affects the amount of space charge whereas the temperature affects the mobility and distribution of space charge (Lewiner *et al.* 1986). Application of DC voltage causes homocharge injection (Tang *et al.* 1986). It is also reported that magnitude of the applied voltage maintains a direct correlation with the amount of positive charge that gets accumulated at the interface of multilayer insulation.

Effect of temperature on the characteristics of space charge has also been reported (Wang *et al.* 2011). Researchers have observed that at elevate temperatures (during charging the test sample), charge injection occurs at a much faster rate and causes deeper space charge distribution (Wang *et al.* 2011). Oil property is also known to be an important factor in determining the overall distribution of trapped charge within the insulation (Hao *et al.* 2011). In the case of deteriorated oil, charge density injected into the sample is observed to increase.

2 THEORY

2.1 *Effect of radial temperature gradient on insulation model parameters*

In the case of a real-life power transformer, while viewing from core to tank (Houhanessian *et al.* 1988), it can be observed that the temperature of the insulation decreases as the distance from the winding (closest to the core) increases. After attaining a minimum value, the temperature value again increases due to the presence of the second winding. Though the value of minimum temperature and distance of the insulation region exposed to it is influenced by the physical parameter of the insulation concerned, values of maximum and minimum temperatures rarely undergo any change during the operating life of the equipment.

A long time of exposure to this radial temperature gradient causes maximum aging to occur in regions close to either winding while minimum aging occurs in the region exposed to the minimum temperature. It is a known fact that this non-uniform aging of oil-paper insulation is an irreversible process and affects all dielectric measurement data including PDC measurement. This further implies that the characteristic of a particular dipole group present close to the HV and LV winding will be different from that of the same dipole group exposed to minimum temperature. It is shown by the second author of this paper that such a non-uniform aging in insulation can be modeled using the Modified Debye Model (Baral *et al.* 2013). The parameters of such model can be easily derived from the polarization current of the insulation.

2.2 *Modeling PDC data affected by slow decaying trapped charge*

It can be observed that the presence of trapped charge will influence the characteristics of dipole characteristics. If the decay rate of the trapped charge is sufficiently slow, as in the case of transformers in an insulated tank, the effect of the trapped charge will create a non-zero initial condition for the capacitors present in a particular branch of MDM. Such non-zero initial conditions can be simulated by placing a constant magnitude DC voltage source in series with the branch sub-elements of MDM.

It is reported that trapped charge mainly resides at the interface of solid and liquid insulation. Furthermore, the polarity of the slow decaying trapped charge is primarily positive in nature (Tang *et al.* 2010). This implies that the effect of the trapped charge will be to reduce the effect of the applied DC field (during PDC measurement) for dipoles present in the interfacial region and solid dielectric while keeping the effect on dipoles present in oil relatively unaffected. Hence, in order to simulate the effect of such trapped charge, a DC voltage source, V_i, having a suitable magnitude, is placed in branches of MDM whose time constant is greater than 100 s. The pictorial representation of such a model is illustrated in Figure 1.

In Figure 1, the element $Z_0(s)$ represents the effect of insulation geometry. On the other hand, the values of resistance and capacitance, models $Z_i(k,s)$ in Figure 1, are related to the time constant of the *i*th branch τ_i by equation (1)

$$\left.\begin{array}{l} R_i(k) = \text{Re}[Z_i(k,s)] \\ C_i(k) = \text{Im}[Z_i(k,s)] = \tau_i \times \dfrac{1}{R_i(k)} \end{array}\right\} \quad (1)$$

In the present work, the author simulated the presence of trapped charge by considering the poling field (Tang *et al.* 2010) to be generated by V_{poling} = 2000 V DC source. In the present work, the poling field is assumed to be applied for t_{poling} s. It can be understood that the trapped charge gets introduced into the insulation by a process, which is similar to the polarization phase in PDC measurement. Hence, the magnitude of voltage source V_i is given by equations (2a) and (2b)

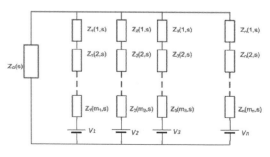

Figure 1. MDM structure with effect of trapped charge.

$$V_i = V_{poling} \times \left(1 - \exp\left(\frac{-t_{poling}}{\tau_i}\right)\right); \ \tau_i \geq 100s \qquad (2a)$$

$$V_i = 0; \ \tau_i < 100s \qquad (2b)$$

In equations (2a) and (2b), the time constant τ_i of the ith branch is related to MDM branch parameters by equation (3).

$$\tau_i = \Lambda_R^i / \Lambda_C^i$$
$$\Lambda_R^i = \sum_{k=1}^{no\_el} R_i(k) \qquad (3)$$
$$\Lambda_R^i = \sum_{k=1}^{no\_el} 1/C_i(k)$$

The parameter of the insulation model considered for the present case is already reported in (Baral *et al.* 2013). In the present work, the value of t_{poling} varies within the limits of 0–400 s in steps of 100 s. It is worth mentioning here that the range of t_{poling} chosen is exemplary in nature. The proposed analysis is valid for any range of t_{poling} chosen.

3 RESULTS AND DISCUSSION

It is reported by the second author of this paper that the value of paper moisture can be estimated reliably by investigating the performance parameter evaluated from the transfer function of MDM. The performance parameter mentioned above refers to the magnitude of system zero (Z_1) located farthest away from the origin in the *LHS* of the *s-plane*. It can be observed from Figure 1 that dipole groups present in interfacial and solid insulation will be subjected to a charging voltage which is less than V_{dc}. In fact, equations (2a) and (2b) show that different dipole groups will be subjected to different charging voltage ($V_{dc} - V_i$). It can be understood that the polarization current recorded under such a condition will not provide proper information related to dipole characteristics. As a result, MDM parameters (evaluated from recorded PDC data) and consequently the value of Z_1 is also likely to get affected. This implies that the value of dielectric dissipation factor and paper moisture evaluated using Z_1 will become unreliable and inaccurate. Figure 2 shows the values of polarization current for different quantities of trapped charge.

Polarization current at a larger value of time is influenced by the condition of dipoles present in the interfacial region and solid insulation (Saha *et al.*. 2005). It can be observed from Figure 2 that the effect of the trapped charge on polariza-

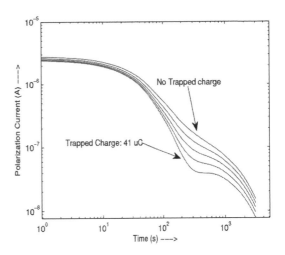

Figure 2. Influence of trapped charge on polarization current.

tion current is more prominent in a larger value of time. In order to demonstrate the effect of trapped charge on MDM parameters, the value of equivalent resistance and capacitance in the largest time constant branch of the insulation model evaluated for the transformer under test is given in Table 1.

Table 2 shows the value of Z_1, trapped charge quantity, and paper moisture value (evaluated using the reported relation (Baral, A. et al 2014)). It should be mentioned here that these parameters are evaluated using the information of excitation voltage V_{dc} (used for measurement of PDC data) and the trapped charge affected polarization current. The data presented in Table 2 is expected as majority of trapped charge (having positive polarity) is assumed to reside in the interfacial region and solid part.

It can be understood that evaluation of MDM parameters using affected PDC data will invariably lead to inaccurate results. Figure 3 shows the pictorial representation of the data presented in Table 2. Data presented in Table 2 along with Figures 1 and 2 show that the effect of trapped charge may significantly affect the measured Time Domain Spectroscopy data especially at larger values of time. This in turn will invariably lead to inaccurate parameterization of MDM and incorrect diagnosis of the insulation concerned. It can be further observed that depending on the magnitude of the trapped charge, the error in insulation diagnosis can reach a substantial value. Hence, it can be reasoned that insulation diagnosis using PDC data is always associated with a level of uncertainty if the influence of trapped charge (present in the insulation) at the time of data measurement is ignored.

Table 1. Influence of trapped charge on the largest time constant branch of MDM.

Charge (μC)	$\Lambda_R^i(G\Omega)$	$1/\Lambda_C^i(\mu F)$
0	17.06	8.51
11.33	19.67	7.38
21.90	22.96	6.32
31.77	27.20	5.34
40.99	32.88	4.42

Table 2. Influence of trapped charge on performance parameter.

Zero Z_1	Charge (μC)	Paper moisture (%)
0.988	0	1.71
0.943	11.33	1.69
0.913	21.90	1.67
0.893	31.77	1.67
0.879	40.99	1.66

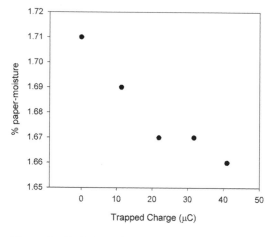

Figure 3. Variation of paper moisture and Z_1 with trapped charge.

4 CONCLUSIONS

The analysis presented in this paper suggests that the effect of slow decaying trapped charge cannot be neglected while modeling the dielectric response function of transformer insulation. A methodology is proposed in this paper using which the effect of positive trapped charge (located at the solid-liquid interface) on insulation model parameters can be studied. The presented technique can also be used for studying the influence of trapped charge on recorded PDC data and hence on the dielectric response of the insulation. Related analysis presented in this paper shows that the use of PDC data affected by trapped charge for estimation of paper moisture will invariably lead to inaccurate diagnosis.

REFERENCES

Baral, A. & Chakravorti, S. (2013). Modified Maxwell Model for characterization of relaxation processes within insulation system having Non-uniform Aging due to temperature gradient, *IEEE Transactions on Dielectrics and Electrical Insulation*, 20(2), 524–534.

Baral, A. & Chakrovarti, S. (2014). Condition Assessment of Cellulosic Part in Power Transformer Insulation using Transfer Function Zero of Modified Debye Model *IEEE Transactions on Dielectrics and Electrical Insulation*, 21(5), 2028–2036.

Houhanessian, V. & Zaengl, W.S. (1988). On-Site diagnosis of power transformers by means of relaxation current measurements, Int'l. Sympos. Electr. Insul., 28–34.

Lewiner, J. (1986). Evolution of experimental techniques for the study of the electrical properties of insulating materials. *IEEE Transactions on Electrical Insulation*, EI-21(3), 351–360.

Roach, J.F. & Templeton, J.B. (1988). An Engineering Model for Streaming Electrification in Power Transformers, *Electrical Insulating Oils, STP 998, H.G Erdman Ed., American Society for Testing and Materials. Philadelphia*, 119–135.

Saha, T.K., Purkait, P. & Müller, F. (2005). Deriving an Equivalent Circuit of Transformers Insulation for Understanding the Dielectric Response Measurements, *IEEE Transactions on Power Delivery*, 20(1), 149–157.

Tang, C., Chen, G., Fu, M. & Liao, R. (2010). Space charge behavior in multilayer oil paper insulation under different DC voltages and temperatures'. *IEEE transactions on Dielectric and electrical Insulation*, 17(3), 775–784.

Wang, D., Wang, S., Lei, M., Mu, H. & Zhang, G. (2011). Temperature Effect on Space Charge Behavior in oil impregnated paper insulation. *In Proc. of Conf. Electrical Insulating Material (ISEIM)*, 378–382.

Frontiers in Computer, Communication and Electrical Engineering – Acharyya (Ed.)
© *2016 Taylor & Francis Group, London, ISBN: 978-1-138-02877-7*

Priority based service scheduling in Enterprise Cloud Bus architecture

G. Khan & S. Sengupta
B. P. Poddar Institute of Management and Technology, Kolkata, India

A. Sarkar
National Institute of Technology, Durgapur, India

ABSTRACT: Cloud computing has emerged as a high performance dynamic computing environment that leverages on-demand enterprise service applications. Since, there has been a rapid increase in number of clouds and its services so the demand for delivering services to a large number of users has also increased. This leads to efficient and dynamic scheduling of services as a major and challenging issue in cloud computing domain. Under this situation the cloud service providers need to use resource efficiently and gain maximum profit by delivering the service to the users as per their quality requirements. To accomplish this, in this paper we have proposed an approach for scheduling service in Enterprise Cloud Bus (ECB) based on Analytical Hierarchy Process (AHP) theory. Based on this approach an algorithm called Priority Based Service Scheduling (PSS) is also discussed where each of the services are assigned a fixed priority based on some set of criteria before the services are processed in a specific order.. The proposed algorithm also helps to reduce waiting time of the services in the scheduler and maximizes the quality of service before delivering it to the client. We illustrate the approach with the help of a case study on airline reservation system.

1 INTRODUCTION

Cloud computing is an emerging distributed computing paradigm that uses Service Oriented Architecture (SOA) for delivering on demand service over the network as per user's requirement. With the advancement of this technology, the enterprise software application has become a predominant domain. Nowadays, the growing interest of enterprise towards cloud computing technology leads to the problem of delivering services to the users as per their quality requirements. An SOA application is composed of multiple services that communicate with each other via messages over a distributed Enterprise Service Bus (ESB) (Bhattacharya et al. 2011) where various cloud publish their services in the ESB platform independently and the consumers can discover their services according to their demand made at the initial stage of request publish in the Service Registry.

In the domain of SOA, Enterprise Service Bus (ESB) technology provides an abstraction layer on top of an implementation of an enterprise messaging system, but still it is expected to have limitations like virtualization, resource pooling, scalability and it also increases overhead and slower the communication speed of many compatible cloud services. Therefore, with the increase in no of clouds and no of services in a cloud registration, discovery,

scheduling, negotiation and composition of services are facing complexity and performances issues. To address these issues, earlier we have proposed an approach by integrating agent technology in an Enterprise Cloud Bus, an abstraction layer framework ([1]Khan et al. 2014, [2]Khan et al. 2014, [3]Khan et al. 2015, [4]Khan et al. 2015, [5]Khan et al. 2015) that automatically allocates service resources suitable for various cloud consumers bolstering some challenges of cloud services like, service registration, service discovery, service scheduling and service evaluation and performance.

To increase the efficiency of delivering services in inter cloud architecture, scheduling is one of the most prominent activities performed to get maximum profit. This paper proposes a service scheduling approach in Enterprise Cloud Bus (ECB) that helps to reduce the service overhead and minimize the waiting time of the service. This paper focuses on the Priority based Service Scheduling algorithm called (PSS) based on set of criteria which required for assigning the priority to each service.

2 REVIEW OF RELATED WORK

Several job scheduling algorithms have been proposed in distributed computing area. Most of them can be applied in the cloud environment with suitable verifications (Patel et al. 2013, Assuncao et al. 2012, Khan et al. 2014 Ghanbari et al. 2012). The main goal of service scheduling is to achieve a high performance computing and the best system throughput. In (Patel et al. 2013), the author proposes a systematic review of various priority based job scheduling that focus on priority of jobs and reduces service response time and improving performance etc. In (Assuncao et al. 2012), the author discusses about the rationalizing the resource utilization in cloud computing environment that leads to significant improvement of quality of service. In (Khan et al. 2014), the author proposes a conceptual framework to address these challenges for citizens and public administrations of smart cities, identify the artifacts and stakeholders involved at both ends of the spectrum.

In (Ghanbari et al. 2012), the authors have proposed a new priority based job scheduling algorithm (PJSC) in cloud computing based on multiple criteria decision making model. (Shimpy et al. 2014) study different scheduling algorithms in different environments with their respective parameters. (Sharma et al. 2013) proposed an algorithm which is based on 3-tier cloud architecture which benefits both the user (QoS) and the service provider (Cost) through effective schedule reallocation based on utilization ratio leading to better resource utilization. In (Li et al. 2014) the authors have proposed Greedy-Based Algorithm in cloud computing. (Paxar et al. 2012) proposes an algorithm which considered Preempt able task execution and multiple SLA parameters such as memory, network bandwidth, and required CPU time. In (Agawam et al. 2014) the author presented a Generalized Priority algorithm for efficient execution of task and comparison with FCFS and Round Robin Scheduling. For the scheduling model, a solving method based on multi-objective genetic algorithm (MO-GA) is designed in (Liu et al. 2013).

Recently several researches (Bhoi et al. 2013, Dubey et al. 2013, Sun et al. 2013, Jayadivya et al. 2012) are based on scheduling of services in cloud environment. A unique modification of Improved Max-min task scheduling algorithm is proposed in (Bhoi et al. 2013). (Dubey et al. 2013) explores various methods of task scheduling done in cloud computing. (Sun et al. 2013) focuses on the task scheduling algorithms based on comprehensive QoS and constraint of expectation. In (Jayadivya et al. 2012) the authors introduce a strategy, QoS based Workflow Scheduling (QWS) to schedule many workflows based on user given QoS parameters like Deadline, Reliability, Cost etc.

Traditional service scheduling algorithms does not provide support in multi cloud architecture. Therefore, this paper proposes a new priority based service scheduling algorithm called PSS which schedule the services in ECB platform.

3 A BRIEF DESCRIPTION OF ECB FRAMEWORK

Cloud Enterprise Service Bus (CESB) and Enterprise Cloud Bus (ECB) in ([1]Khan et al. 2015) are the abstraction layers of Software as Service (SaaS) architecture in cloud computing environment. CESB is the extension of ESB that enhance the ESB's to register their services for single cloud environment. ECB is a hierarchical layer of SaaS architecture that extends the CESB's to register their services from various locations through cloud agent for inter cloud environment and is shown in Figure 1.

The following subsections portray briefly the building blocks of the ECB system:

Service Consumer: Service Consumer is the end-users in cloud computing environment. Here, the consumer placed the request for service in Service Level Agreement (SLA) that is made between the consumer and the service provider. In the other hand, a provider agent is deployed in ECB to invoke the request from the customer and publish it in Cloud Universal Description Discovery and Integration (CUDDI), a meta-service registry. During this process a timestamp is maintained to track time for several consumer registration requests made by the provider agent in CUDDI.

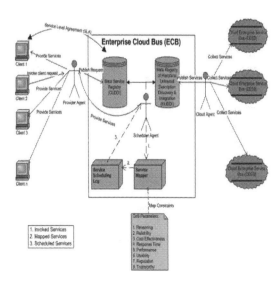

Figure 1. Enterprise Cloud Bus framework.

Service Provider: A Cloud Service Provider is the entity responsible for providing web services available to customer. A cloud agent is deployed for collecting various services from different cloud service providers based on various locations, context, etc. in SaaS platform and publishes the services in Hierarchical Universal Description Discovery and Integration (HUDDI), an extended meta-Service Registry of CESB's in ECB.

Service Scheduler: A Scheduling Agent is deployed in ECB to configure, discover and schedule the cloud services as per Quality of Service (QoS) parameters.

4 PROPOSED SERVICE SCHEDULING APPROACH IN ECB

4.1 Service scheduling framework in ECB

Figure 2 represents a service scheduling framework where Scheduler Agent schedules the discovered services [4Khan et al. 2015] before delivered it to the client. The Cloud Mapper of Figure 2 maps the set of discovered cloud services so that the client receives the optimal cloud service. The optimal cloud service is found based on some service criteria like availability, budget, facility, features. And subsequently the list of discovered services is scheduled.

4.2 Priority based Service Scheduling Algorithm in ECB: PSS

Scheduling an appropriate Web Service published in the HUDDI registry of ECB framework is a

critical issue. Therefore, in this section an algorithm for service scheduling is proposed based on AHP theory so that users can find the appropriate service as they offered. In this algorithm we consider some service criteria like Budget, Facility and Features with respect to a set of discovered services for a particular client request. Initially, the matrix Δ for each nxn discovered services is calculated based on Eigen values. Secondly, using the Eigen values the priority matrix λ of all dxd criteria is evaluated. Finally, the Priority Value Scheduling is calculated by multiplying the matrix Δ and λ. It is observed that the higher the Eigen value of PVS equation (5) greater will be the service priority value.

```
1.   Input d, n [where d is the criteria and n is
the services]
2.   Enter all d criteria & n services
3.   Create a priority comparison matrix (d x d)
for all criteria where the value will be inserted by
the user based on the priority from best to worst.
All value will be inserted using the formula:
     Pij=1/pji wherei  i =j
                              =1where i=j
4.   Create the priority comparison matrices (n x
n) separately for all criteria (d)
5.   Calculate the Priority Vector value for all
(d+1) matrices.
6.   Form a new Matrix  Δ using the Eigen Values
of all d matrices. Where the order of this matrix is
n x d
7.   Form a new matrix  λ using the Eigen value of
d x d matrix. Where the order of this matrix is d x
1
8.   Calculate the PVS (Priority Value
Scheduling) using the formula:
                              PVS = Δ * λ
9.   Matrix-Multiply( Δ, λ)
10.  if columns [ Δ] ≠ rows [λ] then
11.  error "incompatible dimensions"
12.  else
13.     for i =1 to rows [ Δ]
14.       for j = 1 to columns [ λ]
15.          PVS[i, j] =0
16.          for k = 1 to columns [ Δ]
17.             PVS[i, j]=C[i, j]+  Δ [i, k]*
λ[k, j]
18.  return PVS
19.  Arrange the value of PVS in descending order
using Bubble Sort
20.  For i = 1:m,
21.  swapped = false
22.  for j = m:i+1,
23.  if a[j] < a[j-1],
24.  swap a[j,j-1]
25.  swapped = true
26.  ⊠ invariant: a[1..i] in final position
27.  break if not swapped
28.  Show the output of all services of according
to their
     Corresponding PVS values
29.     End
```

4.3 Mathematical representation of PSS

Let n be the number of criteria, and m be the number of services.

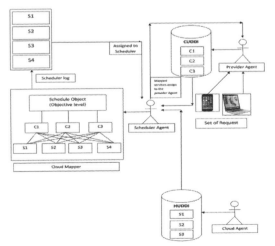

Figure 2. Service scheduling framework.

Let the Criteria PCM = $\begin{bmatrix} c_{1,1} & c_{1,2} & .. & c_{1,n} \\ c_{2,1} & c_{2,2} & .. & c_{2,n} \\ .. & .. & .. & .. \\ c_{n,1} & c_{n,2} & .. & c_{n,n} \end{bmatrix}$ (1)

Corresponding Criteria EV $(\lambda) = \begin{bmatrix} \lambda_1 \\ \lambda_2 \\ .. \\ \lambda_n \end{bmatrix}$ (2)

Let the Service PCMs and their EVs

$= \left\{ \begin{bmatrix} s_{1,1}^1 & s_{1,2}^1 & .. & s_{1,m}^1 \\ s_{2,1}^1 & s_{2,2}^1 & .. & s_{2,m}^1 \\ .. & .. & .. & .. \\ s_{m,1}^1 & s_{m,2}^1 & .. & s_{m,m}^1 \end{bmatrix} \begin{bmatrix} \Delta_1^1 \\ \Delta_2^1 \\ .. \\ \Delta_m^1 \end{bmatrix} \right.$

$\begin{bmatrix} s_{1,1}^2 & s_{1,2}^2 & .. & s_{1,m}^2 \\ s_{2,1}^2 & s_{2,2}^2 & .. & s_{2,m}^2 \\ .. & .. & .. & .. \\ s_{m,1}^2 & s_{m,2}^2 & .. & s_{m,m}^2 \end{bmatrix} \begin{bmatrix} \Delta_1^2 \\ \Delta_2^2 \\ .. \\ \Delta_m^2 \end{bmatrix}$ (3)

$..$

$\left. \begin{bmatrix} s_{1,1}^n & s_{1,2}^n & .. & s_{1,m}^n \\ s_{2,1}^n & s_{2,2}^n & .. & s_{2,m}^n \\ .. & .. & .. & .. \\ s_{m,1}^n & s_{m,2}^n & .. & s_{m,m}^n \end{bmatrix} \begin{bmatrix} \Delta_1^n \\ \Delta_2^n \\ .. \\ \Delta_m^n \end{bmatrix} \right\}$

where cij is the input of set of criteria in equation (1), PCM is the Priority Comparison in equation (2) Matrix for all Criteria, and λ is the Eigen Vector in equation (3).

Then $\Delta = \begin{bmatrix} \Delta_1^1 & \Delta_1^2 & .. & \Delta_1^n \\ \Delta_2^1 & \Delta_2^2 & .. & \Delta_2^n \\ .. & .. & .. & .. \\ \Delta_m^1 & \Delta_m^2 & .. & \Delta_m^n \end{bmatrix}$ (4)

$PVS = \Delta \cdot \lambda = \begin{bmatrix} p_1 \\ p_2 \\ .. \\ p_m \end{bmatrix}$ (5)

Eigenvector (EV) calculation:

Let A = $\begin{bmatrix} a_{1,1} & a_{1,2} & .. & a_{1,n} \\ a_{2,1} & a_{2,2} & .. & a_{2,n} \\ .. & .. & .. & .. \\ a_{n,1} & a_{n,2} & .. & a_{n,n} \end{bmatrix}$ (6)

Firstly, we compute B to be the square of A, then B

$= A2 = \begin{bmatrix} b_{1,1} & b_{1,2} & .. & b_{1,n} \\ b_{2,1} & b_{2,2} & .. & b_{2,n} \\ .. & .. & .. & .. \\ b_{n,1} & b_{n,2} & .. & b_{n,n} \end{bmatrix}$ (7)

Then we compute the row sums (S) = $\begin{bmatrix} s_1 \\ s_2 \\ .. \\ s_n \end{bmatrix}$ (8)

where, $s_i = \sum_{j=1}^{n} b_{i,j}$ Therefore, sum of the row-sums (T) = $\sum_{i=1}^{n} s_i$

Normalized eigenvectors E = $\begin{bmatrix} e_1 \\ e_2 \\ .. \\ e_n \end{bmatrix}$, (9)

where $e_i = \frac{s_i}{T}$ (ei and si is taken as general parameters for e1, e2... en, and s1, s2 .... sn). We now set B equation (7) to be the new A equation (6) for the next round and using the value of the equation (8) we calculate E equation (9). This process is repeated till there is negligible change between the E values from two successive rounds. When E becomes stable, it is assumed to be the eigenvector for the original A matrix.

5 CASE STUDY

In order to validate the proposed service scheduling methodology in our proposed architecture, we establish our approach with the help of a case study of an airline reservation system.

Steps for service scheduling in airline reservation system:

- Input number of Criteria (Budget, Feature, and Facility and available service (Indigo, Spice Jet, Kingfisher, Emirates) for a request->3 (d), 4(n). Input criteria using pair wise priority matrix (dxd) and compute Eigen value for the entire d+1 matrix.
- Form a matrix Δ equation (4) and λ using the Eigen values of priority matrix of all nxn service and dxd criteria and calculate the PVS value which helps to identify the service priority.

The following Figure 3 shows the snapshot for airline reservation system:

Criteria:

```
1.Budget
2. facility
3. Features

Airlines:
1. Indigo
2. Spicejet
3. Kingfisher
4. Emirates

Criteria Matrix          Priority-Vector
1      3      5           0.64
0.33   1      3           0.26
0.2    0.33   1           0.1

Criteria:Budget
                                  Priority-Vector
1      0.5    4      0.17         0.14
2      1      4      0.25         0.2
0.25   0.25   1      0.2          0.06
6      4      5      1            0.6

Criteria:facility
                                  Priority-Vector
1      0.5    0.67   5            0.21
2      1      0.5    7            0.31
1.5    2      1      9            0.43
0.2    0.14   0.11   1            0.04

Criteria:Features
                                  Priority-Vector
1      0.2    1      0.5          0.14
5      1      0.5    2            0.34
1      2      1      3            0.36
2      0.5    0.33   1            0.16

PVS
0.16
0.25
0.19
0.41

Airlines based on your criteria:
1. Emirates
2. Spicejet
3. Kingfisher
4. Indigo
```

Figure 3. Service Scheduling in Airline Reservation System.

6 CONCLUSION

In this paper we have proposed a priority based service scheduling algorithm in ECB that helps in decreasing overhead and minimize service waiting time. Future work includes simulation of Service Scheduling process for ECB architecture for verification and validation of the proposed system. In addition, improving the proposed algorithm in order to gain less response time will be a challenging domain to work in.

REFERENCES

Agarwal, Dr, & Saloni Jain. (2014). Efficient optimal algorithm of task scheduling in cloud computing environment." *arXiv preprint*, 1404–2076.

Assuncao, Marcos D., Marco Netto, Friedrich Koch, & Silvio Bianchi. (2012). Context-aware job scheduling for cloud computing environments. *In Utility and Cloud Computing (UCC), IEEE Fifth International Conference on*, 255–262.

Bhattacharya Swapan, Chanda Jayeeta, Sengupta, Sabnam & Kanjilal Ananya, (2011). Dynamic Service Choreography using Context Aware Enterprise Service Bus. *In SEKE*, 319–324.

Bhoi Upendra, & Purvi N. Ramanuj. (2013). Enhanced Max-min task scheduling algorithm in cloud computing. *International Journal of Application or Innovation in Engineering and Management*, 259–264.

Dubey Sonal, & Sanjay Agrawal. (2013). QoS Driven Task Scheduling in Cloud Computing. International Journal of Computer Applications Technology and Research 2, no. 5 595-meta.

Ghanbari Shamsollah, & Mohamed Othman. (2012). A priority based job scheduling algorithm in cloud computing. *Procedia Engineering 50*, 778–785.

Jayadivya, S.K. & S. Mary Saira Bhanu. (2012). Qos based scheduling of workflows in cloud computing. *International Journal of Computer Science and Electrical Engineering*, 2315–4209.

Khan Zaheer, Saad Liaquat Kiani, & Kamran Soomro. (2014). A framework for cloud-based context-aware information services for citizens in smart cities. *Journal of Cloud Computing 3, no. 1*, 1–17.

[1]Khan Gitosree, Sengupta Sabnam, Sarkar Anirban, & Debnath C. Narayan. (2014). Modeling of Inter-Cloud Architecture using UML 2.0: Multi-Agent Abstraction based Approach. *In Proc. of 23rd International Conference on Software Engineering and Data Engineering (SEDE), New Orleans, Louisiana, USA*, 149–154.

[2]Khan Gitosree, Sengupta Sabnam, Sarkar Anirban, & Debnath C. Narayan. (2014). WSRM: A Relational Model for Web Service Discovery in Enterprise Cloud Bus (ECB). *In proc of Int. Conference at NITK Surathkal, Mangalore*, 117–122.

[3]Khan Gitosree, Sengupta Sabnam, Sarkar Anirban, & Debnath C. Narayan. (2015). Modeling of Services and their Collaboration in Enterprise Cloud Bus (ECB) using UML 2.0. *In proc. Of 2015 International Conference on Advances in Computer Engineering and Applications (ICACEA)*, IMS *Engineering College, Ghaziabad, India*, 207–213.

[4]Khan Gitosree, Sengupta Sabnam, Sarkar Anirban, & Debnath C. Narayan. (2015). Web Service Discovery in Enterprise Cloud Bus Framework: T Vector Based Model. *In proc.of 13th IEEE International Conference on Industrial Informatics, Cambridge, UK*, 1672–1677.

[5]Khan Gitosree, Sengupta Sabnam, Sarkar Anirban, & Debnath C. Narayan. (2015). XML based Service Registration System for Enterprise Cloud Bus. *In proc. Of 3rd IEEE International Conference on Computing, Management and Communications DaNang, Vietnam*.

Li, Ji, Longhua Feng, & Shenglong Fang. (2014).An greedy-based job scheduling algorithm in cloud computing. *Journal of Software 9, no. 4*, 921–925.

Liu, Jing, Xing-Guo Luo, Xing-Ming Zhang, Fan Zhang, & Bai-Nan Li. (2013). Job scheduling model for cloud computing based on multi-objective genetic algorithm. *IJCSI International Journal of Computer Science Issues 10, no. 1*, 134–139.

Patel, Swachil, S. Upendra Bhoi. (2013). Priority Based Job Scheduling Techniques In Cloud Computing:

A Systematic Review. *International journal of scientific & technology research*, 147–152.

Pawar, Chandrashekhar S. & Rajnikant B. Wagh. (2012). Priority based dynamic resource allocation in cloud computing. *In Cloud and Services Computing (ISCOS), 2012 International Symposium* on, 1–6.

Sharma, Ram Kumar, & Nagesh Sharma. (2013). A Dynamic Optimization Algorithm for Task Scheduling in Cloud Computing With Resource Utilization.

International Journal of Scientific Engineering and Technology, Volume 2, 1062–1068.

Shimpy, Er, & Mr Jagandeep Sidhu. (2014). Different Scheduling Algorithms In Different Cloud Environment.

Sun, Hong, Shi-ping Chen, Chen Jin, & Kai Guo. 2013). Research and simulation of task scheduling algorithm in cloud computing. *TELKOMNIKA Indonesian Journal of Electrical Engineering 11, no. 11*, 6664–6672.

Frontiers in Computer, Communication and Electrical Engineering – Acharyya (Ed.)
© 2016 Taylor & Francis Group, London, ISBN: 978-1-138-02877-7

Real-time vehicle safety monitoring and recording system using microcontroller Atmega16 and ultrasonic sensor

S.S. Thakur
Department of Computer Science and Engineering, MCKV Institute of Engineering, Liluah, Howrah, Kolkata, India

J.K. Sing
Department of Computer Science and Engineering, Jadavpur University, Kolkata, India

ABSTRACT: The anti-collision system is proposed here in order to avoid vehicular head to head/back collision that estimates the distance between the two vehicles running under extreme traffic conditions. It incorporates distance finding between two vehicles using an ultrasonic range finder. The vehicle collision and its impact emerged as a major problem in the last two decades when the use of automobiles increased significantly. In order to avoid vehicle collision/road accidents, this system will work in two stages: a range finder will continuously track the distance between two vehicles moving and sends it to the display using these inputs; if it finds the vehicle in the vicinity of the other, it will automatically display the LED/lamp and put the sound buzzer on. This system is reliable, cost-efficient, and fault tolerable. These characteristics enable the vehicle anti-collision system to be effectively used in local traffic environment.

1 INTRODUCTION

Among all the greatest achievements of the history, the invention of automobiles is most probably the one that significantly changed human life. The periodic improvement in technology gives the human race a new height. In the later years after independence, the number of vehicles subsequently increased. But in the last two decades it spread drastically in every level of the society; hence, safety becomes the main concern. Road accidents pose a severe threat to lives in both ways physical as well as financial, even after digital control of vehicles. However, due to human avoidance, circumstantial error, and negligible accidents occur. A large number of vehicle accidents occur each year. Road accidents account pose a severe threat to human lives from both an injurious as well as a financial perspective. Given that vehicles are designed to facilitate a smooth means of transportation, manufacturers have long been in the process of designing vehicles based on principles of reliability and safety. Today, special attention is focused on the technologies that can reduce traffic accidents [1]. Many people lose their lives every year in vehicle collisions due to the driver's inability to keenly observe the vehicles' vicinity while driving [2].

Safety is a necessary part of man's life. According to the statistics provided by National Highway Traffic Safety Administration (NHTSA), there were 5,811,000 vehicle crashes reported in 2008 by police across US; 37,261 people were killed and 2,346,000 people were injured [1]. Road safety has been considered important the world over the past few years. If vehicle drivers were provided with early warnings, a large number of crashes could have been avoided. Only the drivers' observation and reaction may not be sufficient to avoid accidents. Thus, if a device for observation is designed and incorporated into the cars it will reduce the incidence of accidents on our roads and various premises. A lot of research has been conducted to develop collision warning systems to aid driving. Therefore, several initiatives, such as Cooperative Intersection Collision Avoidance systems and integrated vehicle based safety systems, have been proposed in USDOT's intelligent transportation system program. As a collision avoidance system [2, 3, 4], an automatic braking system that operates under critical conditions would be ideal. Practically, it is impossible to develop such a braking system which operates only in extreme emergencies [3, 4]. In the 1970s, a number of systems used microwave radars to avoid collisions [5, 6]. But such radar systems are not practical due to large antenna size, high cost, and difficulties in getting approval with regard to the Radio Law in Japan. Ultrasonic sensors are used for distance measurement between the cars. Communicating systems use well-defined formats for exchanging messages. Such a system should have the capability of supporting real-time systems that can warn drivers. Thus, ultrasonic sensors are used with a transmitter and transmit messages to LCD outputs on the drivers' side.

Services provided by the Intelligent Transportation System (ITS), including collision warning; collision avoidance; and automatic control, are eventually expected to result in a reduction of critical traffic accidents. The data are provided by sensors, information systems, and analyzer devices located inside the vehicles. Low-cost vehicular enhancements are an impediment for large scale deployment. What is desired is a simple in-service upgradeable method for avoiding collisions amongst moving vehicles. Vehicular communication (V2V) resulting from ad hoc and peer-to-peer networking has recently gathered significant attention [7, 8, 9] as both a communication technology as well as for providing possible collision avoidance. V2V technologies are also expected to augment the Intelligent Transportation System (ITS) services. V2V technologies are simple to implement primarily because of their reliance on wireless communication. A wireless location aware ad hoc network of mobile nodes (vehicles) facilitates a framework for collision avoidance. Creating a wireless ad hoc location aware communicating [10, 11, 13] infrastructure involves several components—location awareness, real-time communication, mapping of mobile entities, and taking appropriate action upon detection of collision courses.

2 HISTORICAL BACKGROUND

The earliest research into inter-vehicular communications was conducted by JSK (Association of Electronic Technology for Automobile Traffic and Driving) of Japan in the early 1980s (Tsugawa, 2005). This work treated inter-vehicular communications primarily as traffic and driver information systems incorporated in ATMS (Asynchronous Transfer Mode). From the 1990s through 2000, American PATH (Hedrick et al., 1994) and European "Chaffeur" (Gehring et al., 1997) projects investigated and deployed automated platooning systems through the transmission of data among vehicles. Recently, the promises of wireless communications to support vehicular safety applications have led to several national/international projects around the world. Since 2000, many European projects (CarTALK2000, FleetNet, etc.), supported by automobile manufacturers, private companies, and research institutes, have been proposed a common goal to create a communication platform for inter-vehicle communication [5, 10, 11,

12, 14]. The IST European Project CarTALK2000 focused on new driver assistance systems which are based upon inter-vehicle communication. The main objectives were the development of co-operative driver assistance systems and the development of a self-organizing ad-hoc radio network as a communication basis with the aim of preparing a future standard. The FleetNet project in Germany (FleetNet project—Internet on the road, supported by six manufacturers and three universities from the 2000 through 2003) produced important results in several research areas, including the experimental characterization of VANETs, the proposal of novel network protocols (MAC, routing), and the exploration of different wireless technologies.

Recently some research was carried out on the anti-collision device using an ad hoc wireless network, V2V communication [15, 16] GPS, and Radar implementation. However, all these efforts were informatory in nature which only give signals to the driver or produce some buzzing sound, but finally the action will be taken by the driver in which there are chances of the collision [17, 18]. This work is originally motivated from the local traffic condition of Howrah specially where slow moving traffic on roads often leads to minor or major accidents. We have developed a system which will provide the drivers a safe warning. The paper is organized as follows. The block diagram of the complete system is given in Section 3. Section 4 deals with the details of hardware developed. Software development is explained in detail in Section 5. Section 6 explains the results obtained from the developed system. Conclusion and future work is mentioned in Section 7.

3 BLOCK DIAGRAM

The block diagram of the developed system is given in Figure 1. A micro controller (ATMEGA 16) receives echo signals from the ultrasonic range finding sensor. Ultrasonic sensors continuously read distances between two vehicles and the output is displayed on

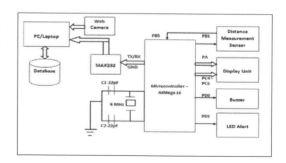

Figure 1. Block Diagram of the complete system.

the dashboard of the vehicle. If the distance reduces to a certain level, the data is recorded and displayed. The developed Circuit is interfaced with a PC/Laptop with the help of USB to serial cable. The power measuring unit measures the power from the main line. This data is fed to the microcontroller. The program is written on an embedded C using a coder and debugger AVR studio 4 and compiling is done.

4 HARDWARE DEVELOPMENT

The following is the list of the components used in the proposed model:

□ ATmega16
□ MAX232
□ Power supply units
□ Ultrasonic Sensor
□ LCD Display

4.1 *Atmega16*

It is a microcontroller from Atmel which is powered by the AVR core. It is an 8-bit, low powered microcontroller with 16-kilobyte built in self-programmable flash. This core is capable of running 16 MIPS with a 16-MHz crystal. It has an advanced RISC architecture with 32 X 8 general purpose working registers. The microcontroller features programmable serial USART and master/slave SPI serial interface. It has 32 programmable I/O lines and a 40-pin PDIP. It is capable of executing one instruction per cycle.

The AVR core combines a rich instruction set with 32 general purpose working registers. All the 32 registers are directly connected to the Arithmetic Logic Unit (ALU), allowing two independent registers to be accessed in one single instruction executed in one clock cycle. The resulting architecture is more code efficient while achieving throughputs upto ten times faster than the conventional CISC microcontrollers.

The ATmega16 provides the following features: 16-K bytes of In-System Programmable Flash Program memory with Read-While-Write capabilities, 512 bytes EEPROM, 1 K byte SRAM, 32 general purpose I/O lines, 32 general purpose working registers, JTAG interface for boundary scan, on-chip programming and debugging support, three flexible timer/counters with compare modes, internal and external interrupts, a serial programmable USART, a byte oriented a two-wire serial interface, an 8-channel 10-bit ADC with optional differential input stage with programmable gain (TQFP package only), a programmable Watchdog timer with internal oscillator, an SPI serial port, and six software selectable power savings modes.

The idle mode stops the CPU while allowing the USART, a two-wire interface, A/D converter,

SRAM, timer/counters, SPI port, and interrupt system to continue functioning. The power-down mode saves the register contents but freezes the Oscillator, disabling all other chip functions until the next external Interrupt or hardware resets. In the power-save mode, the asynchronous timer continues to run, allowing the user to a time base while the rest of the device is sleeping. The ADC noise reduction mode stops the CPU and all I/O modules except an asynchronous timer and ADC to minimize the switching noise during ADC conversions. In standby mode, the crystal/resonator oscillator is running while the rest of the device is sleeping. This allows a rapid start-up combined with less power consumption. In the extended standby mode, both the main oscillator and the asynchronous timer continue to run.

4.2 *MAX232*

This is the level converter IC from MAXIM which is used to create logic compatibility between TTL and RS232 logic. The IC converts the 5-V logic into an 8-V negative logic. This converter is located between the almega16 microcontroller and the zigbee module. The microcontroller uses TTL logic whereas the zigbee module uses RS logic. The main purpose of this converter is to convert the TTL logic to RS logic.

4.3 *Power supply unit*

This unit is basically designed to power up node 1 and node 2. This provides 5 V a 500-mA output to drive the nodes. Here, the AC voltage at 220 V is stepped down to 20 V using a 220/20-V step-down transformer. This AC voltage at 20 V is fed to the rectifier that converts it to DC voltage and is then filtered using a 40-Farad shunt capacitor. The filtered DC voltage is then regulated using a 7805-regulator, and is then supplied to the microcontroller at 5 V, 500 mA.

4.4 *Ultrasonic sensor*

The HC-SR04 ultrasonic sensor uses sonar to determine the distance from an object like bats or dolphins do. It offers excellent noncontact range detection with high accuracy and stable readings in an easy-to-use package. From 2 cm to 400 cm or 1" to 13 feet. Its operation is not affected by sunlight or black material like Sharp rangefinders are (although acoustically soft materials like cloth can be difficult to detect). It comes complete with ultrasonic transmitter and receiver module as shown in Figure 2.

To start the measurement, the Trig of SR04 must receive a high pulse (5 V) for at least 10 us. This will

Figure 2. Ultrasonic Sensor—Transmitter and Receiver Module.

initiate the sensor and transmit 8 cycles of ultrasonic bursts at 40 kHz and wait for the reflected ultrasonic burst. When the sensor detects an ultrasonic burst from the receiver, it will set the echo pin to high (5 V) and delay for a period (width) proportionate to the distance. To obtain the distance, measure the width (Ton) of the Echo pin.

4.5 LCD display

This is the most widely used display device for embedded systems. The LCD unit receives character codes (8 bits per character) from a microprocessor or microcomputer, latches the codes to its display data RAM (80-byte DD RAM for storing 80 characters), transforms each character code into a 5′ 7 dot-matrix character pattern, and displays the characters on its LCD screen.

5 SOFTWARE DEVELOPMENT

Embedded C is asset of language extensions for the C programming language by the C standards committee to address the commonality issues that exist between C extensions for different embedded systems. Historically, embedded C programming requires nonstandard extensions to the C language in order to support exotic features such as fixed-point arithmetic, multiple distinct memory banks, and basic I/O operations. In 2008, the C standards committee extended the C language to address these issues by providing a common standard for all implementations to adhere to. It includes a number of features not available to normal C, such as fixed-point arithmetic, named address space, and basic I/O hardware addressing. Embedded C uses most of the syntax and semantics of standard C, e.g., main () function, variable definition, datatype declaration, conditional statement (if, switch, case), loops (while, for), functions, arrays and strings, structures and union, bit operations, macros, unions etc. A technical report was published

in 2004 and a second revision in 2006. Embedded systems programming is different from developing applications on a desktop computer. Key characteristics of an embedded system as compared to PC are as follows:

- Embedded devices have resource constraints (limited ROM, RAM, stack space, and less processing power).
- Components used in embedded systems and PCs are different: embedded systems typically use smaller, less power-consuming components. Embedded systems are more tied to the hardware.
- Two salient features of embedded programming are core speed and code size. Code speed is governed by the processing power and timing constraints, whereas code size is governed by the available program memory and use of programming language. The goal of embedded system programming is to get maximum features, using minimum space and time.

The front end of the vehicle safety and monitoring system is developed using Microsoft Visual C++ (often abbreviated as MSVC or VC++), a commercial (free version available), Integrated Development Environment (IDE) product for Microsoft for the C, C++, and C++/CLI programming languages. It features tools for developing and debugging the C++ code, especially the code written for Microsoft Windows API, the DirectX API, and the Microsft.NET framework.

Many applications require redistributable Visual C++ packages to function correctly. These packages are often installed independent of applications, allowing multiple applications to make use of the package while only having to install it once. The predecessor to Visual C++ was called Microsoft C/C++. Visual C++ ships with different versions of C runtime libraries. This means users can compile their code with any of the available libraries. However, this can cause some problems when using different components (DLLs, EXEs) in the same program. Microsoft recommends using the multithreaded, dynamic link library to avoid possible problems.

Although the product originated as an IDE for the C programming language, the compilers' support for that language conforms only to the original edition of the C standard from 1989. The later revisions of the standard c99 and C11 are not supported. According to Herb Sutter, the C compiler is only included for "historical reasons" and is not planned to be further developed. Users are advised to either use only the subset of C language that is also valid C++, and then use the C++ compiler to compile their code, or to just use a different compiler, such as Intel C++ Compiler or the GNU Compiler collection, instead.

5.1 Coding/debugging

Coding/debugging in a high-level language (such as C or Java) or assembler. A compiler for a high level language helps to reduce production time. To program the microcontrollers, the WinAVR was used. Although inline assembly was possible, the programming was done strictly in the C language. The source code has been commented to facilitate any occasional future improvement and maintenance. Test Source Code has been written in the C Language to test the microcontroller.

5.2 Compiling

The compilation of the C program converts it into machine language file (.hex). This is the only language the microcontroller will understand because it contains the original program code converted into a hexadecimal format. During this step there were some warnings about eventual errors in the program.

5.3 Burning

Machine language (hex) files of compile programs burned into the microcontroller's program memory is achieved with a dedicated programmer, which attaches to a PC's peripheral. A PC's serial port has been used for the purpose. Its purpose is reading and writing every serial device. It supports I²C Bus, Micro wire, SPI eeprom, and the Atmel AVR and Microchip PIC microcontroller. The microcontrollers were programmed in approximately two seconds with a high speed-programming mode. The program memory, which is of Flash type, has, just like the EEPROM, a limited lifespan. The Atmega16 Programmer (ISP) is used to burn the program into AVR microcontrollers.

5.4 Central processor

Every vehicle is equipped with a central processor. Its function is to extract information from the data received (on the information frequency) and then execute the protocol described in the next section. Based on the protocol, the central processor computes a collision course and then undertakes a recourse action that results in collision avoidance. Recourse is done by sending signals to either the subsystems of a vehicle (automatic recourse) or to the driver enabling collision avoidance.

6 RESULT ANALYSIS

The module was successfully developed and tested in the roads with actual traffic conditions. The ultrasonic was able to measure the data up to 1300 cms distance. The results are accurate with a minor tolerance value.

As shown in Figure 3, it can be noted that the data was recorded from Bally to Salkia at 9.15 am, and the speed of the vehicle at that moment of time was 20–40 km/hour and the recording was done for 21 minutes, in which 63 data points were recorded. The result shows that 27 times the other vehicles following the vehicle (in which developed system is installed) entered the danger zone, i.e., an unacceptable limit; 21 times it was in the tolerable zone; and 15 times it was in the allowable zone.

As shown in Figure 4, it can be noted that the data was recorded from Salkia to Bally at 1.30 pm, and the speed of the vehicle at that moment of time was 20–25 km/hour. The recording was done for 35 minutes, in which 63 data points were recorded. The result shows that 34 times the other vehicles following the vehicle (in which developed system is installed) entered the danger zone, i.e., unacceptable limit, 18 times it was in the tolerable zone, and 11 times it was in the allowable zone.

As shown in Figure 5, it can be noted that the data was recorded from Bally to Salkia at 6.30 pm and the speed of the vehicle at that moment of time was 20–30 km/hour. The recording was done for 27 minutes, in which 63 data points were recorded. The result shows that 31 times the other vehicles following the vehicle (in which developed

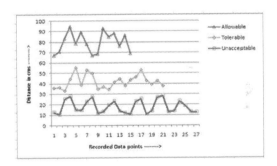

Figure 3.   Plot of recorded data at 9.15 am.

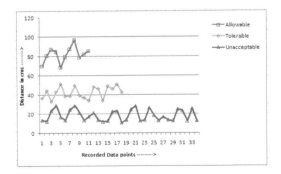

Figure 4.   Plot of recorded data at 1.30 pm.

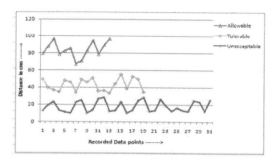

Figure 5. Plot of recorded data at 6.30 pm.

system is installed) entered the danger zone, i.e., unacceptable limit, 19 times it was in the tolerable zone, and 13 times it was in the allowable zone.

6 CONCLUSION AND FUTURE SCOPE

In this paper we have proposed and analyzed the effectiveness of an active vehicle anti-collision system, in which the ultrasonic range finder helps to find the gap between two vehicles. Based on the distance, the lamp in the system glows either red, yellow, or green which alerts the driver following the said vehicle as it is working as a signaling system. From the above results, it can be concluded that during rush hours, i.e., between afternoon 1.30 pm and 6.30 pm mostly the other vehicles following the vehicle (in which the developed system is installed) are entering the danger zone and hence gives an indication or high possibility that accidents may takes place. In case of accidents, it is possible to see recorded images captured by the camera as stored in the developed database which helps to find which vehicle is at fault. In future, we are developing this system with GPS to log the position of collided vehicles to help transmit the location of the accident to the emergency helpline.

REFERENCES

[1] U.S. Dept. of Transportation "Advanced Vehicle Collision Safety Systems," Intelligent Transportation Systems, 2005.
[2] A. Gamester, R. Singhai and A. Sahoo, IntelliCarTS: Intelligent Car Transportation System," Proc. IEEE LANMAN, June 2007.
[3] Kiyoshi Minami, Tohru Yasuma, Shigeru Okabayashi, Masao Sakata and Itsuro Muramoto, Tadao Kohzu, "A collision—avoidance warning system using Laser Radar", SAE international paper,19.
[4] Huang Zhu, Gurdip Singh, "A Communication Protocol for a Vehicle Collision Warning System", 2010 IEEE International Conference on Green Computing and Communications & 2010 IEEE International Conference on Cyber, Physical and Social Computing, 2010.

[5] Nobuyoshi Mutoh, Yusuke Sasaki, "A Driver Assisting System for Eco-Vehicles with Motor Drive Systems Which Avoids Collision with Running Vehicles by Using Inter-Vehicle Communications", proceedings of the 2007 IEEE Intelligent Transportation Systems Conference Seattle, WA, USA, Sept. 30–Oct. 3, 2007.
[6] A. Gumaste and A. Sahoo, "VehACol: Vehicular Anti-Collision Mechanism," Technical Report, ITB/KReSIT/2006/April/2, April 2006.
[7] Dr. A.G. Keskar, S.S. Dorle, et.al, "Design of Protocol for Intersection Collision Detection & Warning In Intelligent Transportation System (ITS)", Proc. IEEE ICETET, July 2008.
[8] P. Bahl and V.N. Padmanabhan, "RADAR: An in-building RF-based user location and tracking system," Proc. IEEE INFOCOM, March 2000.
[9] Y. Morioka,T Sota, M.Nakagawa," Anti-car collision system using DGSP and 5.8 GHz Inter-Vehicle Communication at an off-sight Intersection " Technical report of IEICE ITS 2000-4, 2000, pp. 19–24.
[10] X. Yang et al., "A Vehicle-to-Vehicle Communication Protocol for Cooperative Collision Warning," Proc. 1st Annual Int'l. Conf. Mobile and Ubiquitous Syst: Networking and Services, 2004.
[11] S. Biswas et. al."Vehicle-to-Vehicle Wireless Communication Protocols for Enhancing Highway Traffic Safety," IEEE Commun Mag Vol. 22 No. 1 Jan 2006.
[12] Yusuke Takatori, Hiroyuki Yashima," A study of driving assistance system based on a fusion network of inter-vehicle communication and in-vehicle external sensors", 14th International IEEE Conference on Intelligent Transportation Systems Washington, DC, USA. October 5–7, 2011.
[13] Benliang Li, Houjun Wang Bin Yan and Chijun Zhang, "The Research of Applying Wireless Sensor Network to intelligent Transportation System Based on IEEE 802.15.4", 6th international Conference on ITS Telecommunication proceeding, 2006, pp. 939–942.
[14] Mimoza Durresi, Arjan Durresi and Leonard Barolli,"Sensor Inter-vehicle Communication for Safer Highways", in proceedings of the 19th International Conference on Advanced Information Networking and Applications, AINA 2005.
[15] Samer Ammoun, Fawzi Nashashibi, Claude Laurgeau "Real-time crash avoidance system on crossroads based on 802.11 devices and GPS receivers", in proceedings of the IEEE ITSC 2006, IEEE Intelligent Transportation Systems Conference Toronto, Canada, September 2006.
[16] X. Yang, J. Liu, F. Zhao and N.H. Vaidya, "A Vehicle-to-Vehicle Communication Protocol for Cooperative Collision Warning," Mobile and Ubiquitous Systems: Networking and Services, pages 114–123, August 2004.
[17] Qiang Ji, Zhiwei Zhu, and Peilin Lan, Real-Time Nonintrusive Monitoring and Prediction of Driver Fatigue. IEEE Transactions on Vehicular Technology, Vol. 53, No. 4, July 2004, pp. 1052–1068.
[18] Zutao Zhang, Jiashu Zhang, "A Novel Vehicle Safety Model: Vehicle speed Controller under Driver Fatigue", "IJCSNS International Journal of Computer Science and Network Security", Vol. 9 No. 1, January 2009.

Frontiers in Computer, Communication and Electrical Engineering – Acharyya (Ed.)
© 2016 Taylor & Francis Group, London, ISBN: 978-1-138-02877-7

Innovative structures for Stepped Impedance Resonator filters for wireless applications

Amit K. Varshney, Sumit K. Varshney, R.K. Saw, S.K. Shaw & T. Biswas
Supreme Knowledge Foundation Group of Institutions, Mankundu, West Bengal, India

ABSTRACT: Two innovative techniques for achieving miniaturization and wide band in the design of Inward Folded Stepped Impedance Resonator band-pass filter are proposed. The filter is designed at a center frequency of 2.4 GHz. A bandwidth enhancement of 570 MHz around the center frequency 2.4 GHz is achieved by introducing interdigital capacitors between the two resonators and etching slots in the ground plane compared to that obtained without using IDCs and etched slots for the given inter-resonator spacing. A new method of achieving miniaturization is proposed where two resonators are placed one below the other on either side of the ground plane and the coupling between the two resonators is achieved by etching slots in the ground plane. A size reduction by almost half in the transverse plane and the bandwidth enhancement of 280 MHz around the center frequency 2.4 GHz is achieved with this new technique compared to that obtained when the two resonators are placed side by side without using Inter-Digital Capacitors (IDCs) and etched slots.

1 INTRODUCTION

Designing broadband band-pass filters using microstrip filters is a challenging task because any filter designed using microstrip lines tends to be narrow band. Band-pass filters are generally used in the receiver's front end. When filters are used for wireless applications, achieving miniaturization becomes a challenging task because achieving miniaturization and broadband simultaneously is a difficult task. If conventional structures of designing filters are used, such as hairpin-like filters (Frankel 1971), end coupled filters, edge coupled filters etc., then for enhancing the bandwidth we have to increase the number of resonators, and miniaturization is jeopardized. So it is necessary to go for non-conventional techniques where the structures are optimized in full-wave simulation software and various miniaturization techniques are used. However, initial design formulation is required to start with.

Techniques for achieving miniaturization and broadband characteristics have been explored in the past. A band-pass filter based on internally coupled $\lambda g/2$ stepped impedance resonator has been proposed (Varshney, *et al.* 2009). New methods of extracting the coupling coefficient has been introduced that is found suitable for these resonators. The impedance bandwidth obtained is 280 MHz with an overall dimension of 17 mm by 10.2 mm. A modified SIR-based band-pass filter has been proposed (Varshney *et al.* 2010) with slots in the ground plane. The total electrical length is kept

less than 180 degrees with an optimum impedance ratio of 0.54665. The filter is realized on a dielectric substrate of height 0.795 mm and permittivity 2.2, the etching of the slot in the ground plane just below the BPF improves the bandwidth as magnetic coupling is increased. The use of slots in the ground plane of a conventional edge coupled filter in enhancing the bandwidth has also been demonstrated (Medina *et al.* 2004). The cutting slot in the ground plane enhances magnetic coupling thereby enhancing the bandwidth. Square open loop resonators have been proposed that help in making the band-pass filter compact in size (Jiang et al. 2007). Several others have proposed various methods to miniaturize the filter and/or improve bandwidth (Cohn *et al.* 1958, Makimoto *et al.* 1980, Castillo *et al.* 2004, Ming et al. 2008).

The authors in this paper have presented two innovative structures for achieving miniaturization and broadband to some extent.

2 SIR FORMULATION

The basic structure of a Stepped Impedance Resonator (SIR) filter is shown in Fig. 1(a). The most important parameter of SIR is R_z, the impedance ratio, expressed as (Lancaster *et al.* 1958)

$$R_Z = \frac{Z_2}{Z_1} \tag{1}$$

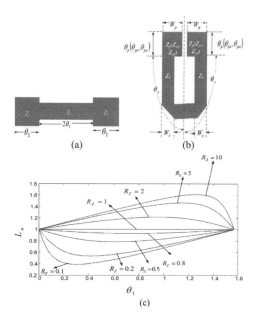

(a)   (b)

(c)

Figure 1. (a) Basic SIR and its parameters (b) An (internally coupled) inward folded SIR with all its parameters (c) Diagram illustrating the change of normalized length L_n of SIR with θ_1 for different values of R_z

Resonance conditions demand

$$R_Z = \tan\theta_1 \tan\theta_2 \tag{2}$$

The normalized length L_n can be written as

$$L_n = \frac{2}{\pi}\left(\theta_1 + \theta_2\right) \tag{3}$$

$$L_n = \frac{2}{\pi}\left(\theta_1 + \tan^{-1}\left(\frac{R_Z}{\tan\theta_1}\right)\right) \tag{4}$$

It can be shown that (Fig. 1(c)) L_n attains minimum values for different values of R_z only when $R_z<1$ and maximum value when $R_z>1$ and the condition for yielding the minimum resonator length for a constant R_z is given as

$$\theta_1 = \theta_2 = \theta_o \tag{5}$$

$$R_Z = \tan\theta_1 \tan\theta_2 = \tan^2\theta_o \tag{6}$$

$$\theta_o = \tan^{-1}\sqrt{R_Z} \tag{7}$$

To further miniaturize the structure, capacitive coupling can be enhanced to reduce inductance introduced by the length of the microstrip lines (Fig. 1(b)).

3 RESULTS AND DISCUSSION

Considering the optimum value of Impedance Ratio $R_Z = 0.5$, we obtain

$$\theta_o = \tan^{-1}\sqrt{R_Z} \text{ or, } \theta_o = 35.26°.$$

As per the band-pass filter design specifications, the electrical parameters of the microstrip realization obtained using the above relations are given below using the following specifications of the substrate:

Type: Dielectric; Material: FR4_epoxy
Thickness: 0.795 mm

$$R_z = Z_p/Z_s = 0.5, Z_s = 100\ \Omega, Z_p = 50\ \Omega$$

Using the synthesis expression for the microstrip realization [Pozar 2007], we obtain the physical parameters as
l_P = 6.74152 mm, W_P = 1.49621 mm, $l_S = 7.11816$ mm
$W_S = 0.32563$ mm, where l_p is the physical length of the strip of electrical length θ_p and physical width W_p, and l_s is the physical length of the strip of electrical length θ_s and physical width W_s.

The dimensions are further optimized (Fig. 2) in full wave electromagnetic software and suitable results are obtained (Fig. 3). It is found that a wider bandwidth is obtained if two resonators are antiparallel.

However due to restriction in the fabrication, the distance between the two resonators cannot be reduced much. In order to relax the spacing, the coupling length between the two resonators is increased.

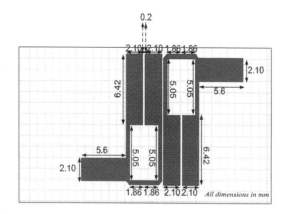

Figure 2. The layout of the designed SIR filter consists of two internally coupled SIR resonators anti-parallel to each other. All dimensions shown are in mm. Input and output ports are designed to provide 50 Ω impedance.

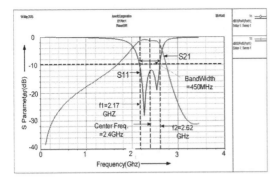

Figure 3. S11 and S21 parameters of the proposed structure whose layout is shown in Fig. 4(a) and 4(b).

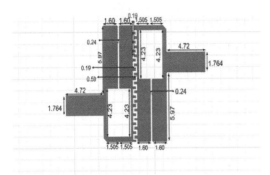

(a)

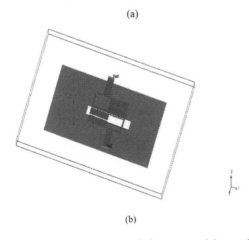

(b)

Figure 4. (a) The layout of the proposed innovative structure showing the introduction of IDC between resonators with etched slots in the ground plane. Etched slots in the ground plane is at the center of the entire microstrip structure with dimensions of 12.84 mm by 2.35 mm. All dimensions are in mm, (b) 3D view of the proposed structure shown in (a).

There are several techniques to do so. One such technique is to use Interdigital Capacitors (IDC); by doing so, the effective length available for coupling increases which increases the bandwidth. Etching

slots in the ground plane further enhance the bandwidth. The dimension of the IDC between the two resonators and the slot in the ground plane beneath the IDC are optimized for the center frequency and bandwidth (Fig. 4). The results have clearly shown the improvement of bandwidth by 570 MHz (Fig. 5)

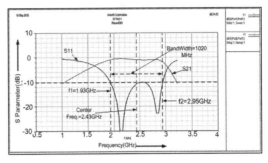

Figure 5. S11 and S21 parameters of the proposed innovative structure shown in Fig. 6(a) showing enhancement of the bandwidth by about 570 MHz.

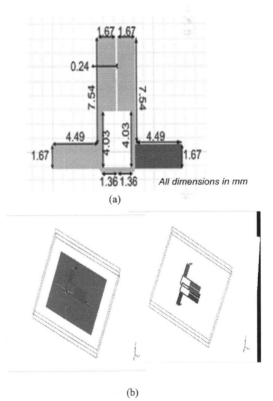

(a)

(b)

Figure 6. (a) Layout of another proposed innovative structure in which the two resonators are placed one below the other with the ground plane in between and an etched slot in the ground plane, (b) 3D views of (a). The dimension of the etched slot is 19.0 mm by 0.26 mm. All dimensions are in mm.

377

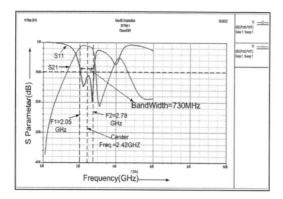

Figure 7. S11 and S22 parameters of the proposed structure shown in Fig. 6(a) and 8(b).

compared to the structure where neither IDC is included nor slots are etched in the ground plane (Fig. 3).

In order to further miniaturize the dimension, the two resonators are placed one below the other instead of placing side by side (Fig. 6). The coupling is achieved through slots in the ground plane. A size reduction by almost half in the transverse plane and the bandwidth enhancement of 280 MHz around the center frequency 2.4 GHz is achieved (Fig. 7) with this new technique compared to that obtained when the two resonators are placed side by side without using IDCs and etched slots (Fig. 3).

4 CONCLUSION

Inward folded Stepped Impedance Resonator Filter is an obvious advantage over the conventional uniform impedance resonator filter as far as miniaturization of the overall structure is concerned. It has been found that the impedance ratio should be less than one in order to achieve miniaturization. This is a must in wireless communication especially when the communication is in the microwave region. Moreover, it is the non-conventional technique that governs a successful design. The closer the gap between two resonators, the higher is the coupling and consequently the wider is the bandwidth. IDC enhances the bandwidth by increasing the coupling length without reducing the coupling gap. Etched slots in the ground plane increases coupling. Placing one resonator below the other reduces the size.

ACKNOWLEDGMENT

The authors are thankful to B.N. Basu and B.N. Biswas for encouraging us to carry out the work.

REFERENCES

Castillo Maria del, *et al.*, "Parallel Coupled Microstrip Filters With Ground-Plane Aperture for Spurious Band Suppression and Enhanced Coupling", IEEE transactions on microwave theory and techniques, vol. 52, no. 3, march 2004, pp. 1082–1086.

Cohn S.B., "Parallel-coupled Transmission Line Resonator Filters", IRE Trans. MTTS, vol. MTT-6, April 1958, pp. 223–231

Cristal E.G. and Frankel S., "Design of hairpin-line and hybrid hairpin parallel coupled line filters", IEEE-MTTS Digest, pp. 12–13, 1971.

Hong Jia-Sheng and Lancaster M.J., "*Microstrip filters for RF/Microwave Applications*", John Wiley and Sons, Inc.

Jiang Xin-Hua, *et al.*, "Compact designs of a Band Pass Filter With Novel Open-Loop Resonators", Microwave and Optical Technology Letters, vol. 49, No. 11, November 2007, pp. 2755–2757.

Makimoto Mitsuo, *et al.*, "Bandpass Filters Using Parallel Coupled Stripline Stepped Impedance Resonators", IEEE transactions on microwave theory and techniques, vol. Mit-28, no. 12, December 1980, pp. 1413–1417

Medina Francisco, *et al.*, "Parallel Coupled Microstrip Filters With Ground-Plane Aperture for Spurious Band Suppression and Enhanced Coupling", IEEE transactions on microwave theory and techniques, vol. 52, no. 3, march 2004, pp. 1082–1086.

Pozar David M., Microwave Engineering, John Wiley and Sons, 2007, pp. 144–145.

Varshney Amit Kumar, Das Chirantan, Mukherjee Paramita, Chakrabarty Joydeep, and Ghatak Rowdra, "miniaturization wide band stepped impedance resonator filter for wireless application national conference on computing and communication system, CoCoSys-09, Jan 2009 organized by the University Institute of Technology, The University of Burdwan, pp. 38–40.

Varshney Amit Kumar, Manimala Pal, Chakravorty J and Ghatak Rowdra, "A Wideband Stepped Impedance Resonator Band Pass Filter With Ground Slots For Dcs And Ieee 802.11b Application", ICRPA 2010 organized by Department of Physics, Burdwan-713104, India, 16th and 17th January, 2010, pp. 63–65.

Yu Ming, *et al.*, "Adjustable Bandwidth Filter Design Based on Interdigital Capacitors", IEEE microwave and wireless components letters, vol. 18, no. 1, January 2008, pp. 16–18.

Frontiers in Computer, Communication and Electrical Engineering – Acharyya (Ed.)
© 2016 Taylor & Francis Group, London, ISBN: 978-1-138-02877-7

Effect of variation of load demand on bus voltage magnitude in radial distribution system

Vishwanath Gupta & Manas Mukherjee

Department of Electrical Engineering, Supreme Knowledge Foundation Group of Institutions, West Bengal, India

ABSTRACT: In the proposed work, a radial distributed system (IEEE 33 Bus radial system) is considered to study the effects on the bus voltage magnitude of the weakest bus due to variation of load demand on individual buses. The weakest bus is first recognized and then the variation of its voltage magnitude is studied due to the variation of load demand on individual buses. This study will help to determine the range of load demand for satisfactory distribution of electric power.

1 INTRODUCTION

To study radial distribution system, load flow solution is the most efficient tool. The steady operating point of the system is obtained by load flow analysis at fixed load demand. Due to radial structure and high R/X ratio, popularly used Newton-Raphson and Fast Decoupled solution technique can be used for this system (Goswami & Basu 1991). Many solution techniques are available to solve this problem (Das 2006, Das 2002, Das *et al.* 1995, Goswami & Basu 1991, Baran & Wu 1989).

But in most of the literature, load flow solutions are done at a fixed load demand which differs from the reality because load demand at each and every bus may vary from the forecasted load data. Due to this variation the voltage magnitude (and phase angle) for each and every bus should also vary and it may also violate the permissible bus voltage limit. This work reports the variation of bus voltage magnitude for variation of load demand. First, the weakest bus i.e. the bus having lowest voltage is identified and the effect of change in voltage magnitude is observed in this bus by the variation of load demand on other bus.

2 PROBLEM FORMULATION

The IEEE 33 bus radial system shown in Figure 1 is considered in the proposed work to study the effect of variation of load demand on bus voltage magnitude on radial distribution system.

The designated values of the various parameters of the system are given in Table 1. Bus 1 is the substation node, the voltage at which is fixed at 12.66 kV.

The load flow solution is done by using forward and backward sweep algorithm (Das 2002) at fixed

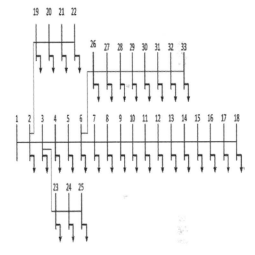

Figure 1. IEEE 33 bus radial system.

load demand. In this work, the weakest bus which is the bus having lowest voltage magnitude is first recognized. Then the load demands on the other buses both at the same layer (2, 3, 6, 13, and 17) as well as on different layers (22, 25 and 33) are varied in steps of 10% one at a time to observe the variation of voltage on the weakest bus with change in load demand at different buses.

3 RESULTS AND DISCUSSIONS

Under normal operating conditions Bus 18 is obtained as the weakest bus. The magnitude of voltages at different buses under normal condition is given in Table 2.

Table 1. IEEE 33 Bus Radial System.

From bus	To bus	Impedance between 'From Bus' and 'To Bus' (Ω) R	X	Load connected at 'To Bus' kW	kVAR
1	2	0.0922	0.0477	100	60
2	3	0.4930	0.2511	90	40
3	4	0.3660	0.1864	120	80
4	5	0.3811	0.1941	60	30
5	6	0.8190	0.7070	60	20
6	7	0.1872	0.6188	200	100
7	8	1.7114	1.2351	200	100
8	9	1.0300	0.7400	60	20
9	10	1.0400	0.7400	60	20
10	11	0.1966	0.0650	45	30
11	12	0.3744	0.1238	60	35
12	13	1.4680	1.1550	60	35
13	14	0.5416	0.7129	120	80
14	15	0.5910	0.5260	60	10
15	16	0.7463	0.5450	60	20
16	17	1.2890	1.7210	60	20
17	18	0.7320	0.5740	90	40
2	19	0.1640	0.1565	90	40
19	20	1.5042	1.3554	90	40
20	21	0.4095	0.4784	90	40
21	22	0.7089	0.9373	90	40
3	23	0.4512	0.3083	90	50
23	24	0.8980	0.7091	420	200
24	25	0.8960	0.7011	420	200
6	26	0.2030	0.1034	60	25
26	27	0.2842	0.1447	60	25
27	28	1.0590	0.9337	60	20
28	29	0.8042	0.7006	120	70
29	30	0.5075	0.2585	200	600
30	31	0.9744	0.9630	150	70
31	32	0.3105	0.3619	210	100
32	33	0.3410	0.5302	60	40

Table 2. Magnitude of voltages at different buses under normal condition.

Bus no.	Voltage magnitude in pu
2	0.997032325318559
3	0.982938397954372
4	0.975457067395126
5	0.968060132460940
6	0.949659700098616
7	0.946174252256271
8	0.941330239067279
9	0.935061404870701
10	0.929246673535228
11	0.928386516078442
12	0.926886992671924
13	0.920774146242436
14	0.918507484523395
15	0.917095230661316
16	0.915727367504944
17	0.913700241247472
18	0.913093200044191
19	0.996503961710157
20	0.992926368806490
21	0.992221865726037
22	0.991584447333775
23	0.979352695363598
24	0.972681583033030
25	0.969356617055071
26	0.947730502852262
27	0.945166851943039
28	0.933727694050550
29	0.925509900711659
30	0.921952618631052
31	0.917791607995075
32	0.916876221967295
33	0.916592589340710

Table 3. Voltage magnitude of bus 18 for variation of load on different buses

Percentage change	Voltage at bus 18 for load variation on Bus 2	Voltage at bus 18 for load variation on Bus 3	Voltage at bus 18 for load variation on Bus 6	Voltage at bus 18 for load variation on Bus 13	Voltage at bus 18 for load variation on Bus 17	Voltage at bus 18 for load variation on Bus 22	Voltage at bus 18 for load variation on Bus 25	Voltage at bus 18 for load variation on Bus 33
10	0.913099556	0.913129991	0.913186036	0.913404473	0.913541365	0.913099	0.91327	0.913192
20	0.913105913	0.913166779	0.913278849	0.913715479	0.913988936	0.913105	0.913446	0.91329
30	0.913112269	0.913203563	0.91337164	0.914026218	0.914435913	0.913111	0.913621	0.913389
40	0.913118625	0.913240345	0.913464409	0.914336691	0.914882301	0.913116	0.913797	0.913487
50	0.913124981	0.913277123	0.913557155	0.914646899	0.915328099	0.913122	0.913972	0.913586
60	0.913131337	0.913313898	0.913649879	0.914956842	0.915773311	0.913128	0.914146	0.913684
70	0.913137692	0.91335067	0.913742581	0.91526652	0.916217938	0.913134	0.91432	0.913782
80	0.913144048	0.913387439	0.91383526	0.915575935	0.916661982	0.913139	0.914494	0.91388
90	0.913150404	0.913424205	0.913927917	0.915885087	0.917105445	0.913145	0.914668	0.913978

Table 4. Percent change in Voltage magnitude of bus 18 for variation of load on different buses.

Percentage change	Percent change in Voltage at bus 18 for load variation on Bus 2	Percent change in Voltage at bus 18 for load variation on Bus 3	Percent change in Voltage at bus 18 for load variation on Bus 6	Percent change in Voltage at bus 18 for load variation on Bus 13	Percent change in Voltage at bus 18 for load variation on Bus 17	Percent change in Voltage at bus 18 for load variation on Bus 22	Percent change in Voltage at bus 18 for load variation on Bus 25	Percent change in Voltage at bus 18 for load variation on Bus 33
10	−0.00069613	−0.00403	−0.01017	−0.03409	−0.04908	−0.00063	−0.01933	−0.01081
20	−0.00139225	−0.00806	−0.02033	−0.06815	−0.0981	−0.00127	−0.03861	−0.0216
30	−0.00208836	−0.01209	−0.03049	−0.10218	−0.14705	−0.0019	−0.05785	−0.03239
40	−0.002784459	−0.01611	−0.04065	−0.13618	−0.19594	−0.00253	−0.07705	−0.04316
50	−0.003480549	−0.02014	−0.05081	−0.17016	−0.24476	−0.00316	−0.09621	−0.05393
60	−0.004176629	−0.02417	−0.06097	−0.2041	−0.29352	−0.0038	−0.11532	−0.06469
70	−0.004872698	−0.0282	−0.07112	−0.23802	−0.34221	−0.00443	−0.1344	−0.07543
80	−0.005568758	−0.03222	−0.08127	−0.2719	−0.39085	−0.00506	−0.15343	−0.08617
90	−0.006264807	−0.03625	−0.09142	−0.30576	−0.43941	−0.00569	−0.17243	−0.0969

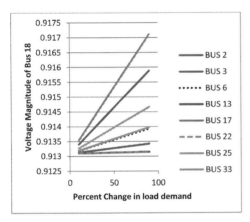

Figure 2. Voltage magnitude of Bus 18 for variation in load demand.

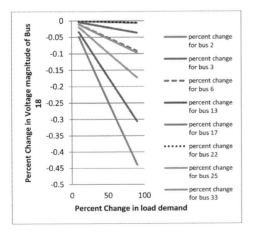

Figure 3. Percent Change in Voltage magnitude of Bus 18 for variation in load demand.

After the weakest bus has been identified as was obvious from practical findings, the effect of varying the load on different buses was studied. The load on buses 2, 3, 6, 13, 17, 22, 25 and 33 were varied in steps of 10% one at a time. The values of voltage magnitude and the percentage change of voltage magnitude of bus 18 for variation of load on each of the above mentioned bus is given in Table 3 and Table 4 respectively. The variation of voltage magnitude and the percentage change of voltage magnitude of bus 18 for variation of load are shown in Figure 2 and Figure 3 respectively.

From the obtained results it is observed that the effect of change in load demand at buses nearer (Bus 17 and Bus 13) to the weakest bus (Bus 18) has a more pronounced effect on the voltage magnitude of the weakest bus than the change in load demand at buses far away (Bus 2, Bus 3, Bus 5) from it on the same layer and the buses on other layers (Bus 22 and Bus 33). It is also observed that the change in voltage magnitude of the weakest bus varies linearly with variation in load demand.

4 CONCLUSIONS

From the proposed work, it can be concluded that the change of load demand on buses nearer to the weakest bus affect the voltage magnitude of the weakest bus. So in order to maintain minimum operating voltage on weakest bus, the load demand on buses nearer to it should be kept within appropriate range.

REFERENCES

Baran M. E & Wu F.F. (1989), "Optimal Sizing of Capacitors Placed on a Radial Distribution System", *IEEE Trans. Power Delivery*, vol.no.1, pp. 1105–1117.

Das B. (2002), "Radial distribution system power flow using interval arithmetic", *Electric Power Energy System*, 24:827–36.

Das B. (2006), "Consideration of input parameter uncertainties in load flow solution of three-Phase unbalanced radial distribution system", *IEEE Trans Power System*, 21(3):1088–95.

Das D., Kothari D P., Kalam A. (1995), "Simple and efficient method for load flow solution of radial distribution networks", *Electrical Power & Energy Systems*, Vol. 17, No. 5, pp. 335–346

Goswami S. K., Basu S.K. (1991), "Direct solution of distribution system", *IEE Proc C*, 138(1):78–88.

A hybrid distribution state estimation algorithm with regard to distributed generations

Ujjal Sur & Gautam Sarkar
Department of Applied Physics, University College of Science and Technology,
University of Calcutta, Kolkata, India

ABSTRACT: Nowadays, power flow is no longer the same as in the conventional systems, since the dispersed generating plants provide power at the distribution grid level too. Connecting generating plants to distribution grids is impossible unless some special control and monitoring tools are available and utilized. The state estimation in these kinds of networks is often called mixtribution. Actually, state estimation is an optimization problem including discrete and continuous variables, whose objective function is to minimize the difference between calculated and measured values of variables, i.e., node voltage and active and reactive powers, in the branches. In this paper, a new hybrid approach based on Genetic Algorithm (GA) and Particle Swarm Optimization (PSO) is proposed to solve this optimization problem using Distributed Generation (DG). The feasibility of the proposed approach is demonstrated and compared separately for the IEEE 34 bus radial distribution test systems.

1 INTRODUCTION

Deregulation creates the opportunity for private sectors to increase their share in power system investments and gives the opportunity to the customers to be responsible for their own power consumption with new types of small-sized generations and user-friendly performance. Due to this, the flow of power is no longer the same as in the conventional systems, as distribution systems have some kinds of generations by themselves. Therefore, the development of a new system and algorithms to enhance the performance of whole system will be required.

Here in this paper, the distribution system's state estimation including DGs has been analyzed. It is clear that while a huge number of small-sized generations are connected to distribution systems, appropriate monitoring and controlling algorithms are required to control the state variables of the system and minimize the operational cost of the whole network. For this, an estimation algorithm is needed as in the case of a distribution network. All the nodes cannot be measured and therefore has to be estimated based on some measured data. This paper presents an algorithm for a distribution state estimation where there are limited measuring devices used.

The typical state estimation method based on the Weighted Least Square (WLS) approach is used which solves the system states by minimizing the mean square error of an overdetermined system of equations (Abur *et al.* 2004). In the conventional estimation methods, it is assumed that the objective functions and constraints should be continuous and differentiable, respectively. Existence of distributed generations, Static VAR Compensators (SVCs), and transformer tap changers having discrete performances showed that compared to ordinary mathematical methods, evolutionary methods would be good alternatives. Many stochastic algorithms are available which can be used to implement DSE problems, but here we use the most reliable optimization technique; Particle Swarm Optimization (PSO) (Kennedy *et al.* 1995) and a hybrid operation of Genetic Algorithm (GA) (Bies *et al.* 2006).

In this paper, a practical distribution state estimation, including DGs, Static VAR Compensators (SVCs), and Voltage Regulators (VRs), is presented and loads of variable outputs are considered as state variables where the difference between measured and calculated values is assumed as the objective function.

2 OBJECTIVE FUNCTION CREATION

The above said estimation problem can be mathematically expressed as an optimization problem with equality and inequality constraints. The objective function is the summation of difference between the calculated and measured values. Therefore, Distribution State Estimation (DSE) can be expressed as follows:

$$\min f(\bar{x}) = \sum_{i=1}^{N} w_i (z_i - h_i(\bar{x}))^2$$
$$\bar{x} = [\overline{P_D}, \overline{P_L}]$$
$$\overline{P_D} = [P_D^1, P_D^2, \ldots\ldots, P_D^{N_d}]$$
$$\overline{P_L} = [P_L^1, P_L^2, \ldots\ldots, P_L^{N_l}] \tag{1}$$

The above equations must satisfy some constraints as follows:

$$\min\left(P_D^i\right) \le P_D^i \le \max\left(P_D^i\right), i = 1, 2, \ldots, N_d$$

$$\min\left(P_L^i\right) \le P_L^i \le \max\left(P_L^i\right), i = 1, 2, \ldots, N_l$$

$$\| P_L^{ij} | < \max\left(P_L^{ij}\right)$$

$$\min(Vr^i) \le Vr^i \le \max(Vr^i), i = 1, 2, \ldots, N_v$$

$$0 \le Q_c^i \le \max\left(Q_c^i\right), i = 1, 2, \ldots, N_c$$

where $\bar{x}$ is the state variables' vector including load and DG outputs; z_i is measured values; W_i is weighting factor of the ith measured variable; h_i is the state equation of the ith measured variable; N is number of measurements; N_d and N_l are number of DGs and loads with variable outputs, respectively; P_D^i is active power of the ith DG; P_L^i is active power of the ith load; N_v and N_c are number of VRs and capacitors installed along the feeder; P_L^{ij} is transmission line power flow; Vr^i is tap position of ith VR, and Q_c^i is reactive power of the ith capacitor.

3 DISTRIBUTED GENERATION MODELING

Depending on the control status of a generator, the DGs can be operated in any one of the following modes:

- Constant output power at a specified power factor.
- Constant output power at a specified terminal voltage.
- In "parallel operation" with the feeder, i.e., the generator is located near the loads and designed to supply a large load with a fixed real and reactive power output.

An approach based on the compensation method to model the generator as a PV node is given here (Niknam *et al.* 2002). In this paper, DGs are considered as PQ nodes.

4 HYBRID GA-PSO ALGORITHM

At first some basic idea about GA and PSO is described here to have a better understanding of the

hybrid one. Genetic Algorithm (GA) is a computational method designed to simulate the evolution processes and natural selection in organisms, which follows the sequence as generating the initial population, evaluation, selection, crossover, mutation, and regeneration (Banga *et al.* 2007). The initial population becomes important as this represents the solution and it is normally randomly generated. Using a problem-specific function, the population is then evaluated. GA will select some of them based on certain probability that they will mate in the next process. Crossover and Mutation will be performed on them to get new and better ones.

Particle Swarm Optimization (PSO) algorithm is represented as a simulation of social activities of insects, birds, and animals which simulates the natural procedure of group communication to share individual knowledge when such swarms flock or migrate. In fact PSO simulates a case that a member of a swarm recognizes as a favorite path to go, and the other members follow it soon. For each particle i, the velocity and position of particles can be updated by the following equations (Kennedy *et al.* 1995):

$$V_i(t+1) = \omega V_i(t) + C_1 r \, and(.)(P_{best} - X_i(t)) + C_2 rand(.)(G_{best} - X_i(t))$$
$$X_i(t+1) = X_i(t) + V_i(t+1)$$
$$\omega(t+1) = \omega_{\max} - \frac{\omega_{\max} - \omega_{\min}}{t_{\max}} * t \tag{2}$$

where,

$i = 1, 2, \ldots, n$ = number of particle; P_{best} is the previous best value of the ith particle; G_{best} is the best value of the entire population; $rand(.)$ is the random number; ω is the inertia weight; and C_1 and C_2 are the weighting factors for the algorithm. Below, the steps of hybrid algorithm are given based on the theoretical approach of both the stochastic algorithms used here.

Step 1: Define the input data like line parameters, value of load and DG, initialization of different GA and PSO parameters.

Step 2: Calculate the objective function based on Equation 1 and calculate the G_{best} value using the initial input data. According to the input state variables, a randomly generated initial population and velocity is written as follows:

$$Velocity = \begin{bmatrix} v_1 \\ \vdots \\ v_n \end{bmatrix}$$
$$v_i = \left[v_j \right]_{1*n}$$
$$v_j = rand(.) * (P_{D_{j\max}} - P_{D_{j\min}}) + P_{D_{j\min}}$$
$$v_j = rand(.) * (P_{L_{j\max}} - P_{L_{j\min}}) + P_{L_{j\min}} \tag{3}$$

$$Population = \begin{bmatrix} X_1 \\ \vdots \\ X_n \end{bmatrix}$$

$$X_i = \begin{bmatrix} X_j \end{bmatrix}_{1*n}$$

$$X_j = rand(.) * (P_{D_{jmax}} - P_{D_{jmin}}) + P_{D_{jmin}}$$

$$X_j = rand(.) * (P_{L_{jmax}} - P_{L_{jmin}}) + P_{L_{jmin}} \qquad (4)$$

where n is the number of measurements and P_D, P_L are the state variables.

Step 3: Now Genetic Algorithm (GA) is applied to find a suitable best value of G_{best} using selection, crossover, mutation, and reproduction processes.

Step 4: Then compare the new G_{best} value with the old one and replace the old one if the new value is superior.

Step 5: Now select the ith particle and its corresponding best solution P_{best}, locally then evaluate new velocity and position using these G_{best} and P_{best} values. Check the given limit for the new value of position.

Step 6: Check whether all particles have been selected or not, if not then repeat step 5, else go to step 7.

Step 7: Now to check the convergence criteria for the algorithm and if it is not converging then repeat steps 2–6, else collect the global optimum value G_{best}.

Here, two flowcharts of Genetic algorithm and hybrid GA-PSO algorithm are given which have been used in this estimation problem.

5 SIMULATION RESULTS

The flowcharts of the Generic Algorithm as well as hybrid GA-PSO algorithm are shown in Figures 1 and 2. The given hybrid algorithm is simulated using MATLAB software and tested over the IEEE 34-bus radial distribution system network (Kersting 1991) shown in Figure 3 with the following parameters.

For a GA population of 50, crossover 0.8, mutation rate 0.001, and for a PSO swarm of 30, C_1 and C_2 value is 2, ω_{min} is 0.5, ω_{max} is 1, and the number of iterations for both methods is 500. Three DGs have been installed at buses 9, 20, and 31 of values 60, 70, and 90 KW, respectively, and power factor

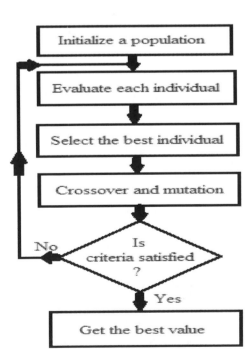

Figure 1.   Flowchart of Genetic Algorithm.

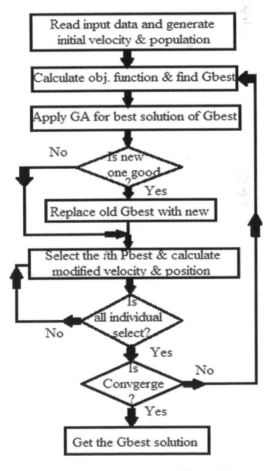

Figure 2.   Flowchart of hybrid GA-PSO algorithm.

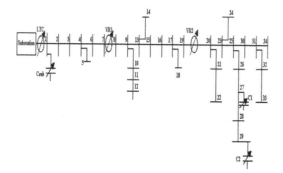

Figure 3.   IEEE 34-bus radial distribution network.

Table 1.   Comparison of estimated values.

	Method	True value (KW)	Calculated values (KW) Worst	Best
G1	GA	65	64.73	66.03
	PSO	65	64.32	65.54
	Hybrid	65	64.12	65.31
G2	GA	76	76.14	77.09
	PSO	76	75.85	76.47
	Hybrid	76	75.64	76.28
G3	GA	85	85.23	85.97
	PSO	85	84.73	85.50
	Hybrid	85	84.35	85.22
L1	GA	81	80.23	81.87
	PSO	81	81.39	81.79
	Hybrid	81	81.11	81.40
L2	GA	90	89.66	91.11
	PSO	90	90.48	90.69
	Hybrid	90	90.25	90.09

Table 2.   Comparison of different parameters.

Method	Execution time (s)	Average (10^{-3})	Standard deviation (%)
GA	100	0.03	12
PSO	40	0.24	10
Hybrid	20	0.18	8

of 0.8 and with a standard deviation of 10%. Also, two variable loads of value 85 and 95 KW are connected at buses 17 and 25, respectively, of power factor 0.75 and with a standard deviation of 10%. Three measurement devices are installed at buses 6, 17, and 30; therefore, n is 3 in this case.

Also, it is assumed that the following information is available:

- Value of output for each of the constant loads and DGs.
- Average value and standard deviation for each of the variable DGs and loads.
- Values of the measured points
- Power factors of loads and DGs
- Set points of VRs and local capacitors

Comparison of estimated values have been listed in Table 1. As shown in Table 2, the average evaluation value of the proposed algorithm for the different runs is 0.18×10^{-3}; this is approximately 60% and 15% of that obtained by GA and PSO, respectively.

6   CONCLUSION

With the increase in the number of DGs, their impacts on power system have to be studied more. The most important issue in distribution systems is Distribution Management System (DMS), which is severely affected by installing DGs. State estimation in DMS plays a major role in estimating the system in a real-time state. In this paper, a reliable approach to estimate distribution state variables in the presence of DGs is presented. The simulation results indicate that the method can estimate the targeted system conditions accurately. Not only that, the proposed algorithm could be applied to similar problems associated with state estimation. The execution time of the proposed method is remarkably short and pointed that the method can be implemented without any restriction in practical networks.

REFERENCES

Abur, A., & Exposito, A.G. (2004). *Power System State Estimation: Theory and implementation*, Marcel Dekkered., New York.
Banga, V.K., Singh, Y. and Kumar, R. (2007). Simulation of Robotic Arm using Genetic Algorithm & AHP, *Journal of Engineering and Technology*, 25, 95–101.
Bies, R.R., Muldoon, M.F., Pollock, B.G., Manuck, S., Smith, G. and Sale, M.E. (2006). A Genetic Algorithm Based Hybrid Machine Learning Approach to Model Selection, *Journal of Pharmacokinetics and Pharmacodynamics*, 196–221.
Kennedy, J. and Eberhart, R. (1995). Particle swarm optimization, *IEEE International Conference on Neural Networks*, Piscataway, NJ, 1942–1948.
Kersting, W.H. (1991)Radial distribution test feeders, *IEEE Trans. On Power Systems*, 6(3), 975–985.
Niknam, T. and Ranjbar, A.M. (2002). Impact of Distributed Generation on Distribution Load Flow, *International Power System Conference*, Tehran, Iran.

Frontiers in Computer, Communication and Electrical Engineering – Acharyya (Ed.)
© 2016 Taylor & Francis Group, London, ISBN: 978-1-138-02877-7

Electric stress control on post-type porcelain insulators using a coating of RTV Silicone Rubber with BaTiO$_3$ nanofillers

Argha Kamal Pal
Department of Electrical Engineering, Birbhum Institute of Engineering and Technology, Birbhum, West Bengal, India

Arijit Baral
Department of Electrical Engineering, Indian School of Mines, Dhanbad, Jharkhand, India

Abhijit Lahiri
Department of Electrical Engineering, Supreme Knowledge Foundation Group of Institutions, Mankundu, West Bengal, India

ABSTRACT: In order to ensure continuous supply of power, the reliability of power transmission line must be high. One of the requirements is to minimize the flashover of the insulators when stressed under extreme harsh environmental conditions. To ensure the reliability of the ceramic insulators, coating of Room Temperature Vulcanized Silicone Rubber (RTV SIR) on the surface of the insulator has gained a lot of attention due to its several mechanical and electrical properties. In the recent years, use of nano-composites as dielectric materials have gained edge because of their high permittivities and low dielectric losses. The present work examines the effect of using nanofillers of Barium Titanium Oxide (BaTiO$_3$) with RTV SIR coating on the resultant electric stress distribution over the surface of a post type insulator made of porcelain.

1 INTRODUCTION

Insulators that are used in overhead transmission lines are not only exposed to environmental stresses but are also electrically stressed throughout their service period. Because of these, the insulators should be designed in such a way that the electrical stress distribution along the insulator surface must be uniform to ensure uninterrupted power transmission.

In the case of ceramic insulator, various methods have been adopted for minimizing the electric stress over its surface. In the recent past, RTV SIR has gained importance worldwide because of its electrical and mechanical properties.

E.A Cherney and R.S Gorur (Cherney & Gorur 1999) observed the application of RTV Silicone rubber coatings for outdoor ceramic insulators. R. Hackam (Cherney & Hackam 1990) showed that the formation of leakage current strongly depends on the intensity of water droplets on the insulator surface.

The effect of wetting conditions on the flashover voltage of non-ceramic insulating materials was studied by R.S Gorur et al. (de la O 1993). The effect of water salinity and temperature on Ethylene Propelene Diene Monomer (EPDM) was studied by R. Hackam (Tokoro & Hackam 1996).

Insulators possess different characteristics under different environmental conditions. Taking this into account, calculation of electric stress on a HV insulator under wet condition was performed by B. Sarang et al. (Sarang & Basappa 2009). Considering RTV SIR, the effect of natural tropical climate to the surface properties of silicone rubber was observed by H.C Kerner et al. (Sirait & Kerner 1998).

At 100 Hz frequency, the effect of filler concentration on the dielectric properties of nanocomposites was evaluated by R. Balakumar et al. (Babu 2012). F. Madidi et al. (Madidi & Farzaneh 2013) reported the effect of filler concentration of Titanium Dioxide (TiO$_2$) on dielectric properties of RTV silicon rubber.

In the present case, a study has been carried out to find out the effect of $BaTiO_3$ nanofillers of different concentrations and having different Aspect Ratios (AR) on electric stress distribution along the surface of a post type insulator made of porcelain with a RTV SIR coating over it. Three different combinations of concentrations and AR of $BaTiO_3$ are considered in the present study those are, 10% vol $BaTiO_3$ with AR = 6, 20% vol $BaTiO_3$ with AR = 6 and 20% vol $BaTiO_3$ with AR = 15.

2 MODEL DESCRIPTION

The electrode-insulator arrangement that has been considered in the present work is shown in Figure 1. It is a post-type insulator which is made of porcelain whose relative permittivity 6 and is stressed between the two electrodes. The surrounding medium is taken as air. A normalized voltage of 1 V is applied to the live electrode for ready reference. The other electrode is grounded. The total arrangement is rotationally symmetric.

The major dimensions of the arrangement are shown in Figure 1. The entire system is simulated by using COMSOL Multiphysics 4.3 b, software.

3 ELECTRIC FIELD DISTRIBUTION ALONG THE SURFACE OF THE POST INSULATOR

Studies on electric field distribution on the insulator surface have been carried out for the following three cases:

1. When the insulator is made of porcelain
2. When the insulator is made of porcelain and having an outer coating of RTV SIR
3. When the insulator is made of porcelain and having an outer coating of RTV SIR loaded with $BaTiO_3$ nanofillers as mentioned earlier

The dielectric properties of each of the above materials are presented in Table 1.

4 RESULTS AND DISCUSSIONS

The variation of the potential along the surface of the insulator has been studied. It has been observed that for each of the cases under study, potential along the insulator surface has varied uniformly from 1 V to zero when measured from live electrode to ground electrode. Since the nature

Table 1. Permittivities of different materials used.

Serial no.	Material	Relative permittivity (ε_r)
1.	Porcelain	6.0
2.	RTV SIR	3.1
3.	RTV SIR with 10 vol. % $BaTiO_3$; AR = 6	5.0
4.	RTV SIR with 20 vol. % $BaTiO_3$; AR = 6	8.9
5.	RTV SIR with 20 vol. % $BaTiO_3$; AR = 15	11.7

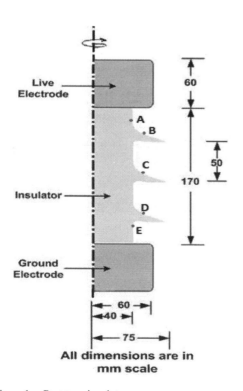

All dimensions are in mm scale

Figure 1. Post type insulator.

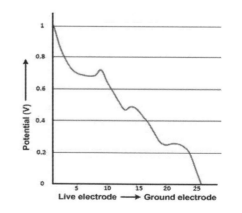

Figure 2. Potential distribution along the insulator surface.

of the variation of the potential along the insulator surface is same in each of the cases hence this variation has been plotted only when porcelain is used as the insulating material and the variation has been presented in Figure 2.

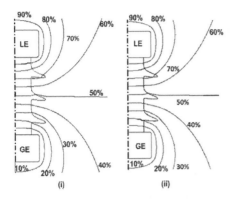

Figure 3. Equipotential lines when the insulator is made of (i) Porcelain and (ii) Porcelain with RTV SIR coating loaded with 20% BaTiO$_3$ of AR = 15.

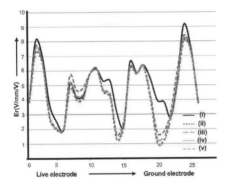

Figure 4. Resultant along the Insulator Surface (i) Porcelain (ii) RTV SIR (iii) 10% BaTiO$_3$, AR = 6 (iv) 20% BaTiO$_3$, AR = 6 and 20% BaTiO$_3$, AR = 15.

The equipotential lines are plotted for each of the three cases under study. From the plots of the equipotential lines no significant deviation has been observed either with RTV SIR coating or when coating of RTV SIR with nanofiller loading is used as compared with the case when the insulator is simply made of porcelain. Since all the plots are same hence equipotential lines have been shown in Figure 3 only for two cases (i) when the insulator is made of porcelain and (ii) when 20% BaTiO$_3$ with AR = 15 is present in RTV SIR coating. From the plot of the potential distribution along the insulator surface and the plots of the equipotential lines, it may be said that the coating of RTV SIR or the coating of RTV SIR with nanofillers can be used with porcelain insulators provided they serve as better electrical stress reducing materials over the insulator surface.

Electric stress distribution over the insulator surface has been studied to observe the effects of RTV SIR coating and the effects of the presence of nanofillers in RTV SIR coating. The distribution of electric stress along the surface of the insulator has been presented in Figure 4. From Figure 4 it may be observed that resultant stress along the insulator surface is more near the electrodes compared to the other regions. It may also be observed from Figure 4 that the resultant stress is maximum when the insulator is made only of porcelain. A decrease in the resultant stress near the two electrode ends is observed when the insulator has a outer coating of RTV SIR which further decreases when nanofillers are used with RTV SIR.

Table 2 presents the resultant stress at five different points on the insulator surface including those the points A and E where the value of the resultant stress is maximum. The value of the resultant stress at these points are tabulated for each of the cases under consideration. From Table 2 it may be observed that as compared to porcelain, the % reduction in stress is maximum at each of the

Table 2. E_r (V/mm/V) along the insulator surface.

Insulating material	E_r at A	% reduction	E_r at B	% reduction	E_r at C	% reduction	E_r at D	% reduction	E_r at E	% reduction
Porcelain	8.109	–	3.65	–	3.13	–	3.96	–	9.15	–
RTV SIR	7.7	5.04	2.84	22.19	1.61	48.5	1.69	57.3	8.40	8.19
10% vol. BaTiO$_3$ AR = 6	7.55	6.89	2.75	24.65	1.38	55.9	1.28	67.6	8.25	9.83
20% vol. BaTiO$_3$ AR = 6	7.32	9.72	2.69	26.30	1.24	60.38	0.99	75.0	8.14	11.03
20% vol. BaTiO$_3$ AR = 15	7.29	10.09	2.67	26.84	1.20	61.66	0.906	77.27	8.05	12.02

points under consideration when 20% vol. BaTiO$_3$ with AR = 15 is used with RTV SIR coating over the porcelain.

5 CONCLUSIONS

The electric field computations performed here is based on the capacitive field using COMSOL Multiphysics 4.3 b. It is to be noted that the entire observation is performed at power frequency. From the obtained results the following conclusions may be made.

i. On the insulator surface, maximum electric stress occurs in the regions near the live and the ground electrodes.
ii. Use of RTV SIR coating over the insulator surface causes a reduction in electric stress distribution over the insulator surface particularly at the electrode ends.
iii. Use of BaTiO$_3$ with RTV SIR does not distort electric field on and around the insulator surface.
iv. Use of BaTiO$_3$ with RTV SIR proves to be a more effective means for stress control over the insulator surface than using RTV SIR coating only.

ACKNOWLEDGEMENT

The author also acknowledges the financial support provided by The Institution of Engineers (India) through awarding the project entitled "Studies on Designing of Durable Superhydrophobic Surfaces for Outdoor Insulators" by Project ID No. PG2015012 in the academic year 2015–16.

REFERENCES

Babu, B.G., S.E. e. a. (2012). Analysis of relative permittivity and tan delta characteristics of Silicone Rubber based nanocomposites. *International Journal of Scientific Engineering and Technology 1*, 201–206.

Cherney, E. & R. Gorur (1999). RTV Silicone Rubber coatings for outdoor insulators. *IEEE Transactions on Dielectrics and Electrical Insulation 6*, 605–611.

Cherney, E.A., K.S. & R. Hackam (1990). Evaluation of RTV Silicone Rubber insulator coatings in a salt-fog chamber. Annual Report 10.1109/CEIDP.1990.201365, Dept. of Electr. Eng., Windsor Univ., Ontario.

de la O, A., G.R.S.C.J. (1993). Effect of wetting conditions on the flashover voltage of non-ceramic insulating materials. Conference report, Electrical Insulation and Dielectric Phenomena.

Madidi, F., M.G. & M. Farzaneh (2013). Effect of filler concentration on dielectric properties of RTV Silicone Rubber / TiO2 nanocomposite. Conference report, 2013 Electrical Insulation Conference, Ottawa, Ontario, Canada.

Sarang, B., L.V. & P. Basappa (2009). Electric field calculations on a high voltage insulator under wet conditions. Conference report, 2009 IEEE Electrical Insulation Conference, Montreal, QC, Canada.

Sirait, K.T., S.S. & H. Kerner (1998). The effect of natural tropical climate to the surface properties of Silicone Rubber. In *Proc. of 1998 International Symposium on Electrical Insulating Materials*, Toyohashi, Japan.

Tokoro, T. & R. Hackam (1996). Effect of water salinity and temperature on the hydrophobicity of Ethylene Propelene Diene monomer insulator. Annual report, Conference on Electrical Insulation and Dielectric Phenomena, San Fransisco.

Mapping forest cover of Gautala Autramghat ecosystems using geospatial technology

Yogesh Rajendra
Srinivasa Ramanujan Geospatial Chair, Geospatial Technology Research Laboratory,
Dr. Babasaheb Ambedkar Marathwada University, Aurangabad, Maharashtra, India
Department of Computer Science and Information Technology, Geospatial Technology Research Laboratory,
Dr. Babasaheb Ambedkar Marathwada University, Aurangabad, Maharashtra, India

Sandip Thorat & Ajay Nagne
Department of Computer Science and Information Technology, Geospatial Technology Research Laboratory,
Dr. Babasaheb Ambedkar Marathwada University, Aurangabad, Maharashtra, India

Rajesh Dhumal
Srinivasa Ramanujan Geospatial Chair, Geospatial Technology Research Laboratory,
Dr. Babasaheb Ambedkar Marathwada University, Aurangabad, Maharashtra, India
Department of Computer Science and Information Technology, Geospatial Technology Research Laboratory,
Dr. Babasaheb Ambedkar Marathwada University, Aurangabad, Maharashtra, India

Amol Vibhute & Amarsinh Varpe
Department of Computer Science and Information Technology, Geospatial Technology Research Laboratory,
Dr. Babasaheb Ambedkar Marathwada University, Aurangabad, Maharashtra, India

S.C. Mehrotra
Srinivasa Ramanujan Geospatial Chair, Geospatial Technology Research Laboratory,
Dr. Babasaheb Ambedkar Marathwada University, Aurangabad, Maharashtra, India
Department of Computer Science and Information Technology, Geospatial Technology Research Laboratory,
Dr. Babasaheb Ambedkar Marathwada University, Aurangabad, Maharashtra, India

K.V. Kale
Department of Computer Science and Information Technology, Geospatial Technology Research Laboratory,
Dr. Babasaheb Ambedkar Marathwada University, Aurangabad, Maharashtra, India

ABSTRACT: Monitoring of the forest area has gained recognition in the worldwide scenario due to the understanding of its role in carbon appropriation and global warming. Gautala Autramghat is a protected area of the Marathwada region of Maharashtra state, India. It is situated in the Satmala and Ajantha hill ranges of the Western Ghats; it is administrated by Aurangabad and Jalgaon District. The wildlife sanctuary was established in an existing reserved forest area in 1986. The area is a southern tropical dry deciduous forest with scattered bush and grasslands. India, being a mega-biodiversity country, supports diverse forest vegetation of tropical, sub-tropical and temperate types. The remote-sensing techniques are employed to map the forest cover. IRS RS2 LISS III sensor data are used to generate a medium-scale vegetation cover map. One scene of 01-Feb-2015 with minimum cloud cover was acquired, preprocessed, and georeferenced to Survey of India toposheets. The satellite image was then subjected to the Maximum Likelihood Classification, assuming six types of objects, namely Open Forest, Moderate Dense Forest, Shrubs Forest, Barren Land, Crop, and Water. A standard forest vegetation cover classification legend was used for this purpose. The vegetation classes were visited on ground to collect information on their structure and composition, which was utilized in the classification exercise.

1 INTRODUCTION

Satellite images are an important basis for vegetation mapping and monitoring of ecosystem functions, mainly through relationships between reflectance from vegetation structure and its composition. Mapping is a method of depicting nature, and the classification permits the mapper to estimate true environments as clearly as possible. Although maps show objects with respect to characteristics, their purpose is to represent objects in terms of their relative location (Thakker, P. S 1999). A good and useful mapping exercise requires a large amount of information that comes from various sources such as satellite images and ground truths (Behera, M. D 2000).

Forest ecosystem plays a very important role in the global carbon cycle. Increased human population demanding development in diverse sectors has resulted in unexpected disturbances in forest ecosystems. The drastic changes in forest ecosystems erode the capability to provide the goods and services (Costanza, R 1997, Lodhi, M. S (2014). In India, the most widely adopted classification system for forest cover is derived from the classification (Champion, H. G 1968).

The present study highlights the significance of remote sensing in the vegetation mapping of Gautala Autramghat of the Marathwada region using the satellite image from the LISS III sensor. A supervised Maximum Likelihood Classification was implemented in our approach (Das S. 2013). The final classification product provided the identification and mapping of dominant forest cover types, including forest types and non-forest vegetation. Remote-sensing datasets were calibrated using a variety of field verification measurements. Field methods included the identification of dominant forest species, forest type, and relative health of selected tree species. Information from an extensive ground truth survey was used to assess the accuracy of the classification. The vegetation type map was prepared from the classified satellite image. The dry deciduous forests constitute a major portion of the total forest area.

The Maximum Likelihood Classifier calculates the probability that a pixel belongs to a class. Data from the training sets are assumed to be usually distributed, which allows the mean vector and the covariance matrix of the spectral cluster of each category of brightness values to be calculated (Lillesand T. M. 2000).

2 STUDY AREA

Gautala Autramghat is mega-biodiversity regions of Marathwada. It is covered in three districts in Maharashtra. These districts are Aurangabad, Nashik, and Jalgaon. The present study was

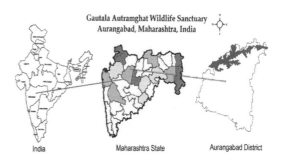

Figure 1. Study area location of Gautala Autramghat Wildlife Sanctuary.

conducted in the area shown in Figure 1. Gautala Autramghat Wildlife Sanctuary is a protected area of Maharashtra state. It is situated in the Satmala and Ajantha hill ranges of the Western Ghats. The wildlife sanctuary was established in 1986 in an existing reserved forest area (Bhatt Shankarlal C. 2006).

It covers a total area of 26,061.19 hectares (64,399 acres) with reserved forest areas of 19706 ha. in Aurangabad and 6355.19 ha. in Jalgaon Coordinator Conservator of Forests, Aurangabad Forest Division (2010).

It is located between the north latitude 20°36'2.93"N and 74°48'16.50"E and the east longitude 20°13'41.09"N and 75°59'11.26"E. It has a semi-tropical climate and remains humid throughout the year. Its average annual precipitation varies from 700 mm to about 800 mm, while the annual mean temperature varies from 16°C to 32°C. The altitude of the study area varies from 300 to 380 m. The summer is extremely hot, lasting from March to June, with May being the hottest month, while the winter lasts from November to January.

3 DATA

The spatial data for the year 2015 were selected for the study. The vector database for the forest was generated from the forest vegetation maps prepared by the Forest Survey of India at the 1:50,000 scale. The map for the year 2015 was generated from the cloud-free IRS LISS III data of 2015 (path/row: 96/58) obtained from National Remote Sensing Centre, Hyderabad. The radiometric and geometric corrections were already taken into account in the data by the agency. The LISS III camera provides multispectral data in four bands. The spatial resolution for visible (two bands) and near infrared (one band) is 23.5 meters with a ground swath of 141 km. The fourth band (short-wave infrared band) has a spatial resolution of 70.5 meters with a ground swath of 148 km. The temporal resolution

Table 1. LISS III sensor characteristics.

Type	Linear Imaging Self Scanner (LISS) III
Spectrum	VIS (~0.40μm to ~0.75μm)
	NIR (~0.75μm to ~1.3μm)
	SWIR (~1.3μm to ~3.0μm)
Resolution class	Medium (20m–200m)
MS resolution	23.5m
Swath	141km
Revisit time	24 days
On-board	IRS 1C, 1D, P6, P7

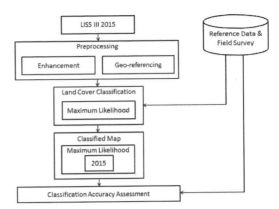

Figure 2. Workflow of methodology.

of LISS III is 24 days. The LISS III sensor characteristics are presented in Table 1.

The vector map was overlaid with the 2015 image database to study the forest vegetation change. An interpretation key, prepared from the ground survey and image analysis for different classes, was used for discriminating different vegetation types. All the data generation and analysis were performed using the ERDAS Imagine 2014 and ArcGIS 10.2v software environment.

4 METHODOLOGY

The entire methodology included pre-processing and classification of satellite images followed by accuracy assessment; the final map preparation, quantification, and analysis were also made. The workflow for the digital image classification included image segmentation and sample selection of target land use/cover classes; The Maximum Likelihood and Mahalanobis Classification were performed. The LISS III image was obtained in GeoTIFF format for processing. The study area was extracted by subsetting from the whole image.

The flowchart of the research methodology can be divided into five stages, as shown in Figure 2.

(i) Procurement of satellite data of IRS P7 LISS III sensor data of the year 2015, (ii) preprocessing and georeferencing of the satellite data, (iii) applying the Maximum Likelihood Classifier based on area of interest, (iv) creation of classified maps, and (v) accuracy assessment of classified maps based on reference data collected from an extensive field survey. The spectral variation of each vegetation type was extracted from digital data by interpreting satellite images based on interpretation elements such as image color, texture, tone, pattern, and association information. In the standard "false color" composite, vegetation looks in shades of red. Trees will appear darker red than hardwoods. This is a very popular band combination and useful for vegetation studies. Usually, deep red hues indicate broad leaf and/or healthier vegetation,

while lighter red hues signify grasslands or sparsely vegetated areas (M. Jakubauskas 1998).

Diverse methods were developed for this study. Those methods can be broadly grouped into unsupervised classification or supervised classification depending on whether or not true ground data are inputted as references. The general steps involved in vegetation mapping include image preprocessing and image classification (S. Saura 2002). The image preprocessing comprises certain necessary preparatory steps such as radiometric correction, geometric correction, and image enhancement in order to improve the quality of original images, which then results in the assignment of each pixel of the scene to one of the vegetation groups defined in a vegetation classification system. The supervised classification was done with the Maximum Likelihood Classifier algorithm in Erdas Imagine software 2014 (Thorat, S.S 2015, Rajendra, Y.D. 2014). Supervised classification techniques are based on external knowledge of the area displayed in the image. The supervised classification can be specified generally as the method of samples of known identity to classify pixels of unknown identity. Samples of known identity are those pixels placed within the training areas (K.V.S. Badarinath 2009).

Pixels located within these areas are included in the training samples used to guide the classification algorithm to assign particular spectral values to suitable information classes. Supervised classification techniques require prior knowledge of the number, and in the case of statistical classifiers, certain aspects of the statistical nature of the information classes with which the pixels establishing an image are to be discovered (Pant, D.N 1992). Training sites are areas representing each known land cover category that appear fairly homogeneous on the image classification, and subjected to statistical processing in which every pixel is compared with the various signatures and assigned to the class whose signature matches closely. A few pixels in a scene do not match and remain unclassified, because these may belong to a class not recognized or defined (A. Kumar 2007).

5 RESULTS AND DISCUSSION

The results clearly demonstrated the forest cover of the state of Gautala Autramghat forest regions, Aurangabad, Maharashtra, India. Satellite monitoring of forests has detected the events of forest cover change in Gautala Biosphere Reserve. Geospatial analysis has addressed the distribution of forest cover. Among the six class types mapped (Open Forest, Moderate Dense Forest, Shrubs Forest, Barren Land, Crop, and Water), as shown in Figure 3, the Moderate Dense Forest is the most predominant forest type of the biosphere reserve.

Figure 4 shows the classified map of Gautala Autramghat Wildlife Sanctuary of the year 2015. The result of the accuracy assessment is summarized in Table 2, and the area covered by each class

Table 2. Accuracy assessment.

Class name	Ref total	Class total	Number correct	Producer accuracy	User accuracy
Shrubs forest	19	19	17	89.47%	89.47%
Moderate dense forest	13	12	12	92.31%	100.00%
Barren land	20	21	18	90.00%	85.71%
Open forest	24	24	20	83.33%	83.33%
Water	9	10	9	100.00%	90.00%
Crop	15	14	12	80.00%	85.71%
Total	100	100	88		

Classification Accuracy Assessment Report.
Maximum Likelihood Classification 2015.

Table 3. Land cover area statistics.

Class name	Area (in hectare)
Shrubs forest	26942.4
Moderate dense forest	5754.59
Barren land	33663.6
Open forest	42757.8
Water	2388.84
Crop	14676

Overall classification accuracy = 88.00%.
Overall kappa statistics = 0.8534.

type is presented in Table 3. To validate the classification accuracy, the stratified random sampling method was used to collect the sample data on the field. The equal number of sample points was selected to each stratum. Using these field data, the confusion and error matrices were determined. The present study was carried out at the 1:50,000 scale, which is suitable for regional level interpretation. Therefore, it is recommended to carry out detailed targeted studies at finer resolutions to fine-tune the present findings and prioritize the conservation of forest cover at the local level. The change in forest cover exhibits a great deal of variation in both spatial and temporal contexts, maybe a result of different strategies and efforts made by the forest department and due to the change in climatic conditions and other socio-economic factors.

Figure 3. Original FCC image of Gautala Autramghat of the year 2015.

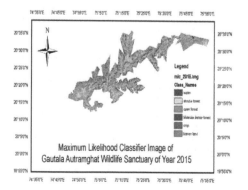

Figure 4. Classified image of Gautala Autramghat of the year 2015.

6 CONCLUSION

The results show the forest cover in the Gautala Autramghat region for the year 2015. The study provides useful direction for future monitoring efforts to support organization policies and initiatives taken by the Government of India such as Green India Programme. In this study, we estimated the actual area covered by the forest from the LISS III image, and we found that the results are very satisfactory, after comparing the results achieved by the maximum likelihood with the ground reference data. The overall accuracy of the system determined using the Maximum Likelihood approach were 88%. It is concluded that the Maximum Likelihood Classifier is more reliable and satisfactory for the classification of the forest area using the remote-sensing images. The support from local public will be important for applying conservation actions and long-term supervision. However, there is a need to consider other threats such as forest fires and illegal tree cutting for the complete management of forest biodiversity.

ACKNOWLEDGMENTS

The authors would like to thank Dr Babasaheb Ambedkar Marathwada University, Aurangabad, Maharashtra, INDIA for financial assistance through DST/NRDMS Srinivasa Ramanujan Geospatial Chair, the State Forest Department, Maharashtra, India for reference information and to UGC-DRS SAP Phase II and DST-FIST programmes for infrastructure facilities.

REFERENCES

Badarinath, K. V. S., T. R. Kiran Chand & V. Krishna Prasad 2009, Emissions from grassland burning in Kaziranga National Park, India—analysis from IRS-P6 AWiFS satellite remote sensing datasets, Geocarto International, 24:2, 89–97.

Behera, M. D., Srivastava, S., Kushwaha, S. P. S. and Roy, P. S., 2000, Stratification and mapping of Taxus baccata L. bearing forests in Tale valley using RS and GIS. Current Science 78(8): 1008–1013.

Bhatt, Shankarlal C. and Bhargava, Gopal K. 2006, "Wildlife". Land and people of Indian states and union territories in 36 volumes, volume 16 Maharashtra. Delhi, India: Gyan Publishing House.

Champion, H. G., & Seth, S. K. 1968, Revised survey of forest types of India. Nasik: Govt. of India publisher. pp. 404.

Co-ordinator Conservator of Forests, Aurangabad Forest Division 2010, "Vision 2020: Aurangabad Forest, Wildlife & Social Forestry" (pdf). Aurangabad (Maharashtra). p. 13.

Costanza, R., et al. 1997, The value of the world's ecosystem services and natural capital, Nature, 387, 253–260.

Das, S., & Singh, T. P. 2013, Mapping Vegetation and Forest Types using Landsat TM in the Western Ghat Region of Maharashtra, India. IJCA, 76(1), 33–37.

Jakubauskas, M., K. Kindscher & Diane Debinski 1998, Multitemporal characterization and mapping of montane sagebrush communities using Indian IRS LISS-II imagery, Geocarto International, 13:4, 65–74.

Kumar, A., S. K. Uniyal & B. Lal 2007, Stratification of forest density and its validation by NDVI analysis in a part of western Himalaya, India using RS and GIS techniques, IJRS, 28:11, 2485–2495.

Lillesand T. M. & Kiefer R. W., 2000, RS and Image Interpretation, 4th ed. Wiley & Sons.

Lodhi, M. S., Samal, P. K., Chaudhry, S., Palni, L. M. S., & Dhyani, P. P. 2014, Land Cover Mapping for Namdapha National Park (Arunachal Pradesh), India Using Harmonized Land Cover Legends. Journal of the ISRS, 42(2), 461–467.

Pant, D. N., Das, K. K., & Roy, P. S. 1992, Mapping of tropical dry deciduous forest and landuse in part of Vindhyan range using satellite remote sensing. JISRS, 20(1), 9–20.

Rajendra, Y. D., Mehrotra, S. C., Kale, K. V., Manza, R. R., Dhumal, R. K., Nagne, A. D., & Vibhute, A. D. 2014. Evaluation Of Partially Overlapping 3d Point Cloud's Registration By Using Icp Variant And Cloudcompare. ISPRS-International Archives of the Photogrammetry, Remote Sensing and Spatial Information Sciences, 1, 891–897.

Saura, S. & J. San Miguel-Ayanz 2002, Forest cover mapping in Central Spain with IRS-WIFS images and multi-extent textual-contextual measures, IJRS, 23:3, 603–608.

Thakker, P. S., Sastry, K. L. N., Kandya, A. K., Kimothi, M. M. and Jadav, r. N., 1999, Forest vegetation mapping using remote sensing and GIS—a case study in Gir (Sasan) forest. Proceedings of ISRS. Hyderabad, January 18–20.

Thorat, S. S., Rajendra, Y. D., Kale, K. V., & Mehrotra, S. C. 2015, Estimation of Crop and Forest Areas using Expert System based Knowledge Classifier Approach for Aurangabad District. International Journal of Computer Applications, 121(23).

Wang L, Sousa W, Gong P 2004, Integration of object-based and pixel based classification for mapping mangroves with IKONOS imagery. IJRS 25(24): 5655–68.

Frontiers in Computer, Communication and Electrical Engineering – Acharyya (Ed.)
© 2016 Taylor & Francis Group, London, ISBN: 978-1-138-02877-7

Synthesis of linear array antenna of a large number of elements using Restricted Search Particle Swarm Optimization

Archit Ghosh, Tamal Das & Soumyo Chatterjee
Department of ECE, Heritage Institute of Technology, Kolkata, India

Sayan Chatterjee
Department of ETCE, Jadavpur University, Kolkata, India

ABSTRACT: In this paper, synthesis of a large number of linear arrays with Restricted Search Particle Swarm Optimization (RSPSO) is presented. The multi-objective synthesis problem of a reduced side-lobe level and a narrow beamwidth is considered. Accordingly, excitation amplitude of each array element and uniform inter-element spacing are optimized using the RSPSO. Restriction in search space is defined using the nested product algorithm. The effectiveness of the proposed method is highlighted through two design examples of 51- and 60-element linear arrays. The zero convergence value of the RSPSO algorithm in the design of the 60-element array indicates superiority over other contestant algorithms.

1 INTRODUCTION

In present-day communication systems, ultra-low side-lobe level and narrow beamwidth are one of the most important design criteria. In the conventional synthesis, simultaneous reductions in peak side-lobe level and beamwidth are somewhat conflicting in nature, as improvement in one parameter introduces degradation in the other. Thus, analytical array design techniques distinctly offer a side-lobe suppression or beamwidth reduction but are ineffective in the multi-objective situation (Dolph *et al.*, 1946, Taylor *et al.*, 1955, Bresler *et al.*, 1980). Hence, evolutionary approaches are applied to solve the aforementioned objectives (Bhargav *et al.*, 2013, Basu *et al.,* 2011). According to the open literature, the most popular and efficient approaches in solving the aforementioned multi-objective problem are Particle Swarm Optimization (PSO) and Differential Evolution (DE) (Das *et al.*, 2012, Pal *et al.*, 2010).

Generally, the aforementioned approaches suffer from the randomness in the initial search space. Recently, restriction in search space in PSO has been introduced with the improvement in optimization efficiency (Chatterjee *et al.*, 2014 & 2012; Archit *et al.*, 2015). Chatterjee et al. (2012) reported the synthesis of a 16-element linear array with the reduced side-lobe level of −20 dB, with restriction defined from the Dolph–Chebyshev distribution. However, such distribution introduces distortion for a large number (>45) of arrays. Bresler's Nested Product (NP) algorithm (Bresler *et al.*, 1980)

overcomes such limitation but is still ineffective in the multi-objective design situation. Consequently, it is necessary to apply algorithms with restriction in search space to achieve multiple objectives of SLL suppression and beamwidth reduction.

In this work, the synthesis problem of reduced SLL and narrow beamwidth is addressed for a large number of element arrays with excitation amplitude and inter-element spacing as optimization parameters for PSO algorithms with and without restriction. Restriction is developed using Bresler's method (Bresler *et al.*, 1980) of Chebyshev amplitude distribution and newly developed design expression for the FNBW of the Dolph–Chebyshev array. In order to prove the effectiveness of restriction in search space, a comparative study involving random and restricted search DE was carried out.

2 PROBLEM FORMULATION

In order to realize the maximum SLL reduction along with narrow beamwidth for higher number of linear arrays, two independent error functions (Archit *et al.*, 2015) were defined in (1) and (2). The first error function (EF1) was introduced as the difference between the desired SLL and the peak SLL obtained at each iteration:

$$EF1 = \left| SLL_d - \max\left\{ 20\log_{10}\frac{AF_{SLL}(\theta)}{\max(AF_{SLL}(\theta))} \right\} \right| \quad (1)$$

SLL$_d$ is the desired SLL and AF$_{SLL}(\theta)$ is the array factor in the side-lobe region. The second error function (EF2) is the difference between the desired beamwidth (FNBW$_d$) and the calculated FNBW (FNBW$_c$) at each iteration:

$$EF2 = |FNBW_d - FNBW_c| \qquad (2)$$

The overall Cost Function (CF) can be formulated by combining the two error functions by the weighted sum method as follows:

$$CF = \alpha EF_1 + \beta EF_2 \qquad (3)$$

Here, both α and β are chosen to be 1 as two different error functions are perturbed by two different sets of optimization parameters. According to the open literature (S. Chatterjee *et al.*, 2014; Bhargav *et al.*, 2013, Archit *et al.*, 2015), the SLL reduction and narrow beamwidth problem can be handled separately by varying excitation amplitude and inter-element spacing, respectively. Hence, in this particular case, EF1 is primarily affected by excitation amplitude variation, whereas EF2 by inter-element spacing. Hence, keeping 100% influence on both the error functions does not introduce any subjective nature and, consequently, no trade-off is needed.

3 EVOLUTIONARY ALGORITHMS AND RESTRICTION IN SEARCH SPACE

Evolutionary algorithms use mechanisms inspired by biological evolution, such as reproduction, recombination, mutation, and selection of the population and then processed through repeated iterations, and fitness function determines the quality of solutions (Kennedy *et al.*, 1995, Storn *et al.*, 1997). Mathematical background of PSO and DE algorithms is explained in the following sections.

3.1 *Particle Swarm Optimization algorithm*

The PSO is a population-based algorithm proposed by Kennedy and Eberhart, inspired mainly by a flock of birds. In the standard PSO (SPSO), a swarm consists of a set of particles and each particle represents a potential solution of an optimization problem. The position and velocity at iteration t of the ith particle are denoted by X$_i$(t) and V$_i$(t), respectively. The velocity of each particle can be updated using (4):

$$V_i^d(t+1) = wV_i^d(t) + C_1 rand()(Pbest_i^d(t) - x_i^d(t))$$
$$+ C_2 rand()(Gbest^d(t) - x_d^i(t))$$
$$(4)$$

where rand () is a uniformly distributed random numbers in the interval [0,1]; acceleration coefficients C$_1$ and C$_2$ are non-negative acceleration coefficients; Pbest$_i$(t) is the personal best solution representing the best solution found by the ith particle itself until iteration t; Gbest^D(t) is called the global best solution, representing the best solution found by all particles globally until iteration t; w is the inertia weight to balance the global and local search abilities of particles in search space, which is given by:

$$w = \frac{w_{max} - w_{min}}{T} \times t \qquad (5)$$

where w$_{max}$ is the initial weight; w$_{min}$ is the final weight; t is the current iteration number; and T is the maximum iteration number. The updated particle position is represented as follows:

$$x_i^d(t+1) = x_i^d(t) + V_i^d(t+1) \qquad (6)$$

3.2 *Restriction in search space*

The concept of restriction in search space is to first report with the PSO algorithm (S. Chatterjee *et al.*, 2012, 2014). Later, in Archit *et al.* (2015), it is also extended in the DE algorithm. With the restriction, the initial randomness of search space is eliminated, and in the first generation, the set of the parameter values is much closer to the optimal solution. Once again, with the restriction, it is not necessary that the designer should have an idea about the optimal solution.

In this work, all the restrictions are derived from Chebyshev array parameters. Excitation amplitudes are restricted from the Chebyshev amplitude distribution of the NP algorithm, whereas the upper bound of the inter-element spacing (d) is calculated from the newly defined closed-form expression of FNBW of the Chebyshev array (Tamal *et al.*, 2015). The lower bound of d is set to 0.5λ. Restriction is applied to both PSO and DE algorithms, and restricted search PSO (RSPSO) and RSDE are used as acronyms for their restricted search versions.

4 EXPERIMENTAL RESULTS

In order to prove the effectiveness of restriction in search space on both PSO and DE, 51- and 60-element linear arrays with the SLL objective of −20 and −30dB and the FNBW of 3.2 and 3.8 degrees, respectively, were considered.

4.1 *Experimental setup*

In this table, r$_i$ is the difference between the maximum and minimum possible values of optimization parameters at the ith iteration. The dimension D

Table 1. Experimental setup for SPSO/RSPSO and DE/RSDE algorithms.

SPSO/RSPSO		DE/RSDE	
Parameters	Value	Parameters	Value
Iteration	100	Iteration	100
Swarm size	100	No. of agents	100
C_1	Linearly increased from 0.5 to 2.5	CR (cross-over)	Linearly increased from 0.2 to 0.95
C_2	Linearly decreased from 2.5 to 0.5	F (mutation)	Exponentially decreased from 0.6 to 0.1
W (inertia weight)	Linearly decreased from 0.9 to 0.4		
$V_{d, max}$	$0.9 \times r_i$		

Table 2. Comparison of p-values for RSPSO.

No. of elements	Method comparison	p-value
51	RSPSO/SPSO	0.0884
	RSPSO/DE	1.49×10^{-5}
	RSPSO/RSDE	0.0073
60	RSPSO/SPSO	5.12×10^{-27}
	RSPSO/DE	1.95×10^{-26}
	RSPSO/RSDE	1.11×10^{-24}

Table 3. Comparison of statistical and antenna parameters for each contestant algorithm.

Design instance	Statistical/ antenna parameters	RSPSO	SPSO	DE	RSDE
51	SD	0.0001	0.032	0.002	0.0018
	MCF	0	0	0	0
	Exec. Time	80.1s	119.8s	249.8s	194.4s
$SLL_d = -20dB$	Excitation Amplitude Ratio	2.462	3.792	9.011	6.5605
$FNBW_d = 3.2^0$	Optimum Inter-element Spacing	0.8078	0.8516	0.9309	0.9262
	HPBW	1.4	1.4	1.4	1.4
60	SD	0.00002	0.6512	1.1215	0.8373
	MCF	0	4.8071	3.263	1.600
	Exec. Time	74.51s	121.3s	248.5s	183.03s
$SLL_d = -30dB$	Excitation Amplitude Ratio	4.3270	2.9600	3.467	2.275
$FNBW_d = 3.8^0$	Optimum Inter-element Spacing	0.7802	0.8345	0.9563	0.9206
	HPBW	1.4	1.0	1.0	1.0

depends on the number of elements (N) considered. Consequently, for even number of elements $D = (N + 1)$ and for odd number of elements, it is $D = ((N + 1)/2 + 1)$. The number of independent runs considered in each case is 20.

4.2 Parametric analysis

In this article, datasets were randomly varied and required a statistical hypothesis test. Wilcoxon's rank-sum test was done with a null hypothesis of no difference between the set of minimum cost function values for different optimization objectives, at each iteration of the best independent run obtained using conventional DE and PSO and their restricted counterparts (RSDE and RSPSO). The p-values are given in Table 2, reflecting the significant difference in the best set of minimum cost function data obtained by the contestant algorithms over 20 independent trials.

A further comparative study of Standard Deviation (SD), Mean Cost Function (MCF), and other antenna parameters revealed the effectiveness of restriction in search space.

From Table 3, it can be concluded that among all the design instances, RSPSO had achieved the least MCF and SD. Consequently, the quality of optimal solution was observed with a minimum CPU execution time. Also, the excitation amplitude ratio and optimal spacing was minimum (in the case of the 60-element array, though the excitation amplitude ratio was less for the subsequent algorithms, it had not reached the given objective) for RSPSO, whose effectiveness can be anticipated

subject to design and fabrication of array. It can also be observed that restriction was effective for both the algorithms, but RSPSO was more robust compared with RSDE.

4.3 Graphical comparison

For graphical comparison, the convergence graph and the array-factor pattern are shown in Figures 2 and 3. From the convergence graph, it can be concluded that only RSPSO had completely achieved the objective at the 47th iteration. It is also evident that RSPSO was the fastest to

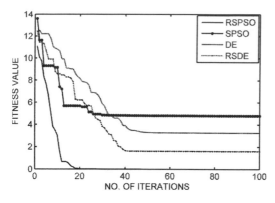

Figure 1. Convergence graph for all the four algorithms for the 60-element array, with desired $SLL_d = -30$ dB and $FNBW_d = 3.8$ degrees.

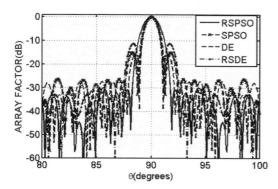

Figure 2. Array factor for N = 60 synthesized with $SLL_d = -30$ dB and $FNBW_d = 3.8$ degrees by all the four algorithms with the main beam region.

converge with an average CPU execution time of 1.5358s/iteration. From Figure 3, it can be observed that only RSPSO provided all the side lobes below the desired SLL of −30 dB. Also, FNBW reduction was achieved successfully.

5 CONCLUSION

The present work focused on the synthesis of high number of linear arrays with and without restriction in search space on the PSO algorithm. The design objective of SLL suppression and FNBW reduction was given with array excitation amplitudes and uniform inter-element spacing that were considered as optimization parameters. Previously, restriction using the Chebyshev distribution was not defined for high number of arrays as Dolph's solution deviated for number of elements exceeding 45. In this work, restriction was extended for high number of arrays with the help of the NP algorithm. The results were obtained using two design instances of the 51- and 60-element arrays. A comparative study with DE and RSDE justified the need and advantage of restriction in terms of speed, quality of solution (i.e., MCF and SD), and elimination of initial randomness of search space. Further, it was also concluded that the concept of restriction was more effective on PSO than on DE, highlighting the superiority of the PSO algorithm. Also, RSPSO had greater flexibility from the design point of view as it achieved the best design parameters compared with the other contestant algorithms.

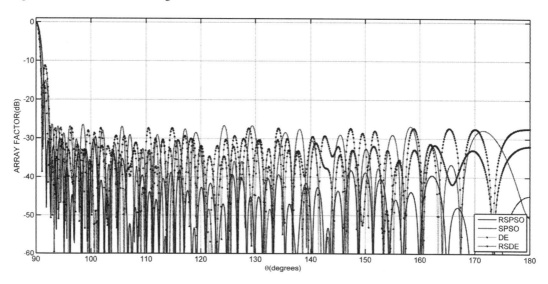

Figure 3. Side-lobe region for the aforementioned design instance of the 60-element array, $SLL_d = -30$ dB and $FNBW_d = 3.8$ degrees.

REFERENCES

Basu B & Mahanti G K, "Fire Fly and Artificial Bees Colony Algorithm for Synthesis of Scanned and Broadside Linear Array Antenna", Progress in Electromagnetics Research B, Vol. 32, 2011, pp. 169–190.

Bhargav A & Gupta N (2013) "Multiobjective Genetic Optimization of Nonuniform Linear Array With Low Sidelobes and Beamwidth," *IEEE Antennas and Wireless Propagation Letters, Vol. 12, 2013, pp. 1547–1549.*

Bresler A.D. (1980) "A new algorithm for calculating current distribution of Dolph-Chebyshev Arrays", *IEEE Transactions on Antennas & Propagation, November, 1980.*

Chatterjee S & Chatterjee Sayan (2014) Pattern synthesis of centre fed linear array using Taylor one parameter distribution and restricted search Particle Swarm Optimization, *Journal of Communications Technology and Electronics, Vol. 59, No. 11, 2014, 1112–1127.*

Chatterjee S, Chatterjee S & Poddar D R (2012) "Side lobe level reduction of a linear array using Chebyshev polynomial and particle swarm optimization," *IJCA Proc. Of Int. Conf. on Communication, Circuits and Systems, Bhubaneshwar, India, 2012, Oct. 6–7, Bhubneswar, India.*

Das T, Ghosh Archit, Chatterjee Soumyo & Chatterjee Sayan (2015) "Design Expression for First Null Beamwidth of Broadside Dolph Chebyshev Antenna Array," *Proc. of C3IT, Adisaptagram, India, Feb. 7–8, 2015.*

Das S, Bhattacharya M, Sen A & Mandal D, "Linear Antenna Array Synthesis with Decreasing Sidelobe and Narrow Beamwidth," *ACEEE International Journal on Communications, Vol. 3, No. 1, 2012, pp. 10–14.*

Dolph C. (1946) "A Current Distribution for Broadside Arrays which optimizes the Relationship between Beamwidth and Side Lobe Level", *Proceeding IRE, Vol.34, No. 5, 1946.*

Ghosh A, Das Tamal, Chatterjee Soumyo & Chatterjee Sayan (2015) Linear Array Pattern Synthesis Using restriction Search Space for evolutionary Algorithms: A Comparative Study, *Proc. of ReTIS-2015, pp 92–97, Jadavpur, Kolkata, India, July 9–11, 2015.*

Kennedy J & Eberhart R, "Particle swarm optimization," *Proc. Conf. IEEE Int. Conf. Neural Networks, 1995.*

Pal S, Qu B Y, Das S & Suganthan P N, "Optimal Synthesis of Linear Antenna Arrays with Multi-objective Differential Evolution," *Progress In Electromagnetics Research B, Vol. 21, 2010, pp. 87–111.*

Storn R & Price K, "Differential Evolution—A Simple and Efficient Heuristic for Global Optimization Over Continuous Spaces", *Journal of Global Optimization, Vol. 11, No. 4, 1997, pp. 341–359.*

Frontiers in Computer, Communication and Electrical Engineering – Acharyya (Ed.)
© 2016 Taylor & Francis Group, London, ISBN: 978-1-138-02877-7

Effect of a high-k dielectric material on the surface potential and the induced lateral field in short-channel MOSFET

Akash Ganguly, Chandrima Ghosh & Arpan Deyasi
Department of Electronics and Communication Engineering, RCC Institute of Information Technology, West Bengal, India

ABSTRACT: In this paper, the surface potential and the induced lateral electric field in the short-channel MOSFET with a high-k dielectric material were analytically computed by solving the two-dimensional Poisson's equation. TiO_2 was considered as the dielectric material for the simulation purpose, and the result was compared with that obtained for the Si–SiO_2 system. Doping concentration and dielectric thickness were varied within the practical limit to observe the deviation of electrical parameters as a function of normalized channel length. The result indicated that better tuning of the threshold voltage was possible for the Si–TiO_2-based system than the conventional Si–SiO_2 system, which was predicted from the variation of the surface potential. Also, higher induced field was obtained for the high-k dielectric material in the presence of the virtual cathode even for low doping and higher dielectric thickness. This result is critically important for designing the short-channel MOSFET with the required threshold voltage and induced field with smaller dielectric thickness and moderate doping.

1 INTRODUCTION

Research on short-channel MOSFET has always been the choice of VLSI engineers in order to find the solution of the problems that arise due to the short-channel effect (Kawaura *et al.*, 2000, Doris *et al.*, 2002, Hokazono *et al.*, 2002), although recent work has suggested that scaling beyond 20 nm gate length is possible with a remarkable result on electrical performance (Frank *et al.*, 2001, Solomon *et al.*, 2002, Likharev *et al.*, 2003). Achievement of the ITRS roadmap projections 2001 (International Technology Roadmap for Semiconductors, 2001 Edition) is the main objective of the decade, where gate length can be scaled beyond 10 nm. One of the major modifications already considered to achieve the target is the replacement of a conventional dielectric material by a high-k material (Frank *et al.*, 2009, Mohapatra *et al.*, 2014). With the introduction of novel dielectric materials, better performance can be achieved (Zaunert *et al.*, 2007). Among other electrical parameters, determination of threshold voltage is extremely crucial as it controls the magnitude of sub-threshold current (Kumar *et al.*, 2012). This is essentially determined by the surface potential, and its accurate estimation plays a pivotal role in controlling the short-channel effect in the VLSI circuit.

Limit on gate dielectric thickness has already been studied (Hirose *et al.*, 2000) to analyze the magnitude of tunneling current flow and possible dielectric breakdown (Cellere *et al.*, 2005). Scaling to the sub-nanometer dielectric thickness is possible, thanks to the microelectronics fabrication technology; however, this puts restriction on the doping of source and drain regions. High doping of these regions will lead to dielectric breakdown, which essentially makes burn-out of the MOSFET. Hence, low-to-moderate doping is required for lower dielectric thickness. Again, lower channel length increases the lateral induced electric field. This along with moderate doping and lower dielectric thickness are the constraints in designing the short-channel MOSFET.

In this paper, the surface potential and the lateral electric field in the short-channel MOSFET were analytically computed by solving the two-dimensional Poisson's equation with appropriate boundary conditions. Low-to-moderate doping concentrations and lower dielectric thickness were considered for the simulation purpose. Calculation was made for the Si–TiO_2 system, and the result was compared with that obtained for the Si–SiO_2 system. The results were very important in analyzing the performance of the short-channel MOSFET and the corresponding VLSI circuit.

2 THEORETICAL FOUNDATION

For a short-channel MOSFET, charge density in the channel is characterized by the two-dimensional Poisson's equation:

$$\frac{\partial E_x}{\partial x} + \frac{\partial E_y}{\partial y} = -\frac{qN_A}{\varepsilon} \tag{1}$$

where E_x and E_y are the field distributions inside the depletion region. For the solution of the surface potential, boundary conditions can be written as

$$\phi_{ss}(0) = \phi_{bi} \tag{2.1}$$

$$\phi_{ss}(L) = V_{DS} + \phi_{bi} \tag{2.2}$$

where we assume that the device is free of the body effect.

If we consider that $E_d(0,y)$ is the electric field in the dielectric region, and $E_x(0,y)$ is the electric field at the dielectric–semiconductor interface, then continuity of the displacement vector is given by

$$\varepsilon_d E_d(0,y) = \varepsilon_x(0,y) \tag{3}$$

where the fields are represented in the following form:

$$E_d(0,y) = \frac{1}{t_d}[V_{GS} - V_{fb} - \phi_{ss}(y)] \tag{4.1}$$

$$E_x(0,y) = \frac{\varepsilon_d}{\varepsilon_s t_d}[V_{GS} - V_{fb} - \phi_{ss}(y)] \tag{4.2}$$

For the computation purpose, we can safely assume that the normal electric field component in the substrate region just below the channel is zero, i.e., $E_x(W_d, y) = 0$. Then, from Poisson's equation

$$\frac{\partial E_x}{\partial x} = \frac{E_x(0,y) - E_d(W,y)}{W} \tag{5}$$

This can be written in the following manner:

$$\frac{\partial E_x}{\partial x} = \frac{\varepsilon_d}{\varepsilon_s t_d W_d}[V_{GS} - V_{fb} - \phi_{ss}(y)] \tag{6}$$

Substituting the magnitude of E_x in Equation (1), Poisson's equation can be represented in the following form:

$$\frac{\partial E_y}{\partial y} + \frac{1}{l^2}[V_{GS} - V_{fb} - \phi_{ss}(y)] = -\frac{\eta qN_A}{\varepsilon} \tag{7}$$

where l is the characteristic length of the channel. Since

$$E_y = -\frac{d\phi_{ss}}{dy}$$

So, we can modify Equation (7) as

$$\frac{d^2\phi_{ss}}{dy^2} - \frac{\phi_{ss}(y)}{l^2} = \frac{\eta qN_A}{\varepsilon} - \frac{(V_{GS} - V_{fb})}{l^2} \tag{8}$$

Solution of Equation (8) is given by

$$\phi_{ss}(y) = \phi_{ss}(L) + [V_{DS} + \phi_{bi} - \phi_{ss}(L)]\frac{\sinh(y/l)}{\sinh(L/l)}$$
$$+ [\phi_{bi} - \phi_{ss}(L)]\frac{\sinh[(L-y)/l]}{\sinh(L/l)} \tag{9}$$

Thus, the induced lateral electric field is

$$E_y = E_s(L) - \frac{1}{l}[V_{DS} + \phi_{bi} - \phi_{ss}(L)]\frac{\cosh(y/l)}{\sinh(L/l)}$$
$$- E_s(L)\frac{\sinh(y/l)}{\sinh(L/l)} + E_s(L)\frac{\sinh[(L-y)/l]}{\sinh(L/l)}$$
$$\times \frac{1}{l}[\phi_{bi} - \phi_{ss}(L)]\frac{\cosh[(L-y)/l]}{\sinh(L/l)} \tag{10}$$

3 RESULTS AND DISCUSSION

Using Equations (9) and (10), the surface potential and the induced lateral electric field were calculated and plotted as a function of normalized length for different structural and internal parameters. TiO$_2$ was considered as the high-k dielectric material, and the result was compared with that obtained for the conventional SiO$_2$ dielectric system. Doping concentrations and dielectric layer thickness were varied to observe the effect on the electrical parameters, and a comparative study was also performed.

Figure 1 shows the variation of the surface potential for different doping concentrations for both the dielectric materials under consideration. From Figure 1a, it can be observed that for lower concentrations, the surface potential becomes a linear function of normalized length, whereas nonlinearity arises when the concentration increases. A comparative study with Figure 1b reveals the fact that even for very high concentrations, the surface potential remains constant for a wider range of channel length; and it also exhibits minima. Thus, the high-k dielectric system provides a better response than the conventional SiO$_2$ dielectric system for a wider range of channel length with a

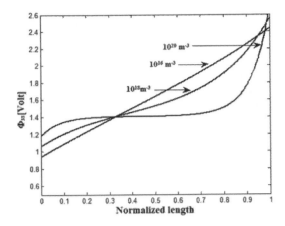

Figure 1a. Variation of the surface potential with the normalized channel length for different doping concentrations of TiO$_2$.

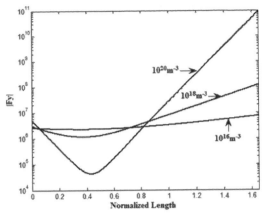

Figure 2a. Variation of the induced lateral electric field with normalized channel length for different doping concentrations of TiO$_2$.

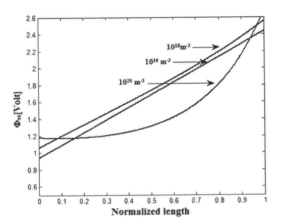

Figure 1b. Variation of the surface potential with the normalized channel length for different doping concentrations of SiO$_2$.

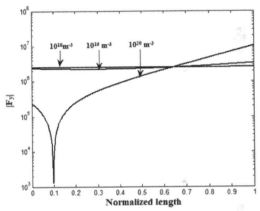

Figure 2b. Variation of the induced lateral electric field with normalized channel length for different doping concentrations of SiO$_2$.

constant surface potential even for a high doping condition. This variation helps us to conclude that for the high-k dielectric material, the threshold voltage increases much faster with channel length variation in a lower range, whereas for higher dimensions, it almost saturates.

Figure 2 shows the effect of doping concentration on the induced electric field. For the high-k dielectric material, induced field variation shows the existence of minima at a particular channel length, whereas it increases for either reducing or increasing the magnitude. This is plotted in Fig. 2a. Moreover, it shows a higher magnitude than that obtained for the conventional SiO$_2$ system. For lower doping, the induced field remains almost constant for SiO$_2$. Thus, the existence of the virtual cathode can be determined for the high-k

dielectric material at moderate and higher concentrations, but it is only traceable for high doping when a low-k dielectric material is used.

Figure 3 shows the variation of the surface potential with channel length for different oxide thicknesses. The result is plotted for both TiO$_2$ and SiO$_2$ in Figures 3a and 3b, respectively. It can be observed from the plot that the thickness of the dielectric material hardly affects the surface potential except the higher channel length for the TiO$_2$-based system, whereas it modifies the surface potential from a moderate to higher channel length range for the SiO$_2$-based system. This suggests that the threshold voltage tuning is possible at the desired range when the high-k dielectric material is used (as the constant surface potential indicates an increase in the threshold voltage) compared

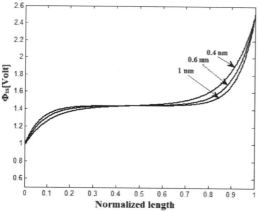

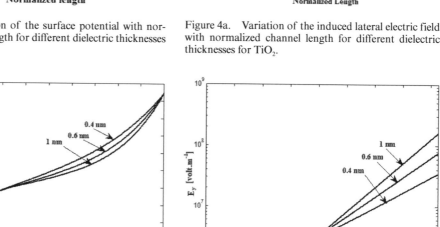

Figur 3a. Variation of the surface potential with normalized channel length for different dielectric thicknesses for TiO$_2$.

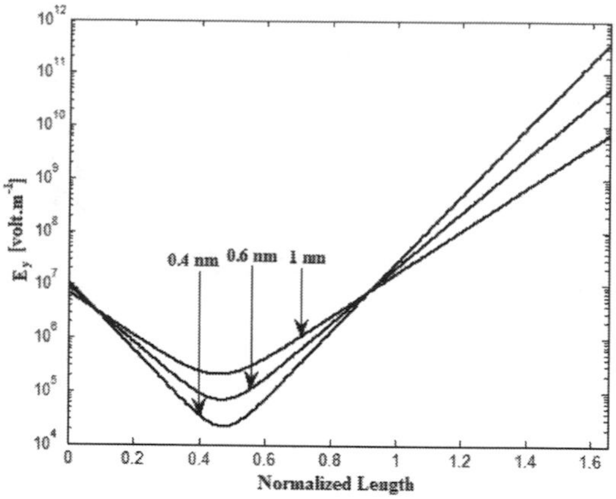

Figure 4a. Variation of the induced lateral electric field with normalized channel length for different dielectric thicknesses for TiO$_2$.

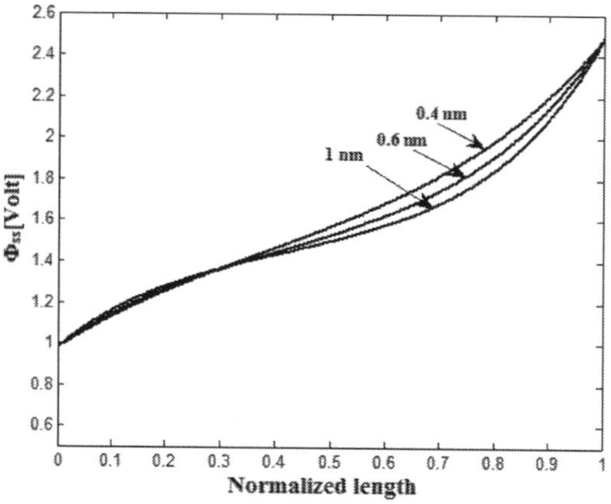

Figure 3b. Variation of the surface potential with normalized channel length for different dielectric thicknesses for SiO$_2$.

Figure 4b. Variation of the induced lateral electric field with normalized channel length for different dielectric thicknesses for SiO$_2$.

with that obtained for the conventional material system.

Figure 4 shows the variation of the lateral induced electric field for different dielectric thicknesses. Comparative analysis reveals that the magnitude of the induced field is higher for the high-k dielectric system. The position for the field minima shifts right as the thickness of the dielectric material decreases, whereas the magnitude of the shift (with respect to channel length) remains negligible for the low-k dielectric system. However, for both the material systems, it can be observed that the rate of increase in field is lower for higher thickness with the increase in channel length. Thus, higher induced field generation is possible for the high-k dielectric system, and its tuning in a wider

range can be made by selecting the dielectric thickness prior to fabrication.

4 CONCLUSION

The surface potential and the induced lateral electric field for the short-channel MOSFET were analytically computed as a function of normalized channel length. Doping concentration and dielectric thickness were varied to observe the effect on those parameters. The high-k dielectric material was used for the simulation purpose, which resulted in the increase of the induced field and also better tuning of the surface potential. The results indicated that the use of the high-k dielectric material enhanced the tuning ability of the threshold voltage compared with that obtained for the conventional low-k dielectric material.

REFERENCES

Cellere. G., Paccagnella. A., Valentini. M.G., 2005, Influence of dielectric breakdown on MOSFET drain current, IEEE Transactions on Electron Devices, 52, 211–217.

Doris. B., Meikei. I., Kanarsky. T., Ying. Z., Roy. R.A., Dokumaci. O., Ren. Z., Jamin. F.F., Shi. L., Natzle. W., Huang. H.J., Mezzapelle. J., Mocuta. A., Womack. S., Gribelyuk. M., Jones. E.C., Miller. R.J., Wong. H.S.P., Haensch. W., 2002, Extreme scaling with ultrathin Si channel MOSFETs, IEDM Technical Digest, 267–270.

Frank. D.J., Dennard. R.H., Nowak. E., Solomon. P.M., Taur. Y., Wong. H.S.P., 2001, Device scaling limits of Si MOSFETs and their application dependencies, Proceedings of IEEE, 89, 259–288.

Frank. M.M., Kim. S.B., Brown. S.L., Bruley. J., Copel. M., Hopstaken. M., Chudzik. M., Narayanan. V., 2009, Scaling the MOSFET gate dielectric: From high-k to higher-k? Microelectronic Engineering, 86, 1603–1608.

Hirose. M., Koh. M., Mizubayashi. W., Murakami. H., Shibahara. K., Miyazaki. S., 2000, Fundamental limit of gate oxide thickness scaling in advanced MOSFETs, Semiconductor Science and Technology, 15, 485–490.

Hokazono. A., Ohuchi. K., Takayanagi. M., Watanabe. Y., Magoshi. S., Kato. Y., Shimizu. T., Mori. S., Oguma. H., Sasaki. T., Yoshimura. H., Miyano. K., Yasutake. N., Suto. H., Adachi. K., Fukui. H., Watanabe. T., Tamaoki. N., Toyoshima. Y., Ishiuchi. H., 2002, 14 nm gate length CMOSFETs utilizing low thermal budget process with poly-SiGe and Ni salicide, IEDM Technical Digest, 639–642.

International Technology Roadmap for Semiconductors, 2001 Edition: http://public.itrs.net/.

Kawaura. H., Sakamoto. T., Baba. T., 2000, Observation of source-to-drain direct tunneling in 8 nm gate electrically variable shallow junction MOSFETs, Applied Physics Letters, 76, 3810–3812.

Kumar. K.K., Rao. N.B., 2012, Characterization of Variable Gate Oxide Thickness MOSFET with Non-Uniform Oxide Thicknesses for Sub-Threshold Leakage Current Reduction, International Conference on Solid-State and Integrated Circuit, 78–83.

Likharev. K., Greer. J., Korkin. A., Labanowski. J., 2003, Electronics below 10 nm, Nano and Giga Challenges in Microelectronics, Eds. Amsterdam, The Netherlands: Elsevier.

Mohapatra. S.K., Pradhan. K.P., Sahu. P.K., 2014, Influence of High-k Gate Dielectric on Nanoscale DG-MOSFET, International Journal of Advanced Science and Technology, 65, 19–26.

Solomon. P.M., Luryi. S., Xu. J., Zaslavsky. A., 2002, Strategies at the end of CMOS scaling, Future Trends in Microelectronics, Eds. New York: Wiley, 28–42.

Zaunert. F., Endres. R., Stefanov. Y., Schwalke. U., 2007, Evaluation of MOSFETs with crystalline high-k gate-dielectrics: device simulation and experimental data, Journal of Telecommunications and Information Technology, 2, 78–85.

Frontiers in Computer, Communication and Electrical Engineering – Acharyya (Ed.)
© *2016 Taylor & Francis Group, London, ISBN: 978-1-138-02877-7*

Effect of carrier-carrier collisions on RF performance of millimeter-wave IMPATT sources

Prasit Kumar Bandyopadhyay
Department of Electronics and Communication, Dumkal Institute of Engineering and Technology, Murshidabad, West Bengal, India

Subhendu Chakraborty
Department of Electronics and Communication, Supreme Knowledge Foundation Group of Institutions, Mankundu, Hooghly, West Bengal, India

Arindam Biswas
Department of Electronics and Communication, NSHM Knowledge Campus, Durgapur, West Bengal, India

Aritra Acharyya
Department of Electronics and Communication, Supreme Knowledge Foundation Group of Institutions, Mankundu, Hooghly, West Bengal, India

A.K. Bhattacharjee
Department of Electronics and Communication, NIT, Dugrapur, West Bengal, India

ABSTRACT: In this paper, the authors have studied the effect of carrier-carrier collisions on the static (DC) and large-signal characteristics of Double-Drift Region (DDR) IMPATTs based on Si designed to operate at mm-wave atmospheric window frequencies such as 94, 140 and 220 GHz. Simulations are carried out to study the RF performance of those diodes by taking into account both empirical relation of field dependence of ionization rates developed from experimental data as well as a appropriate analytical model of same parameter which takes into account the energy loss of charge carriers due to carrier-carrier collisions. Studies have been carried out to investigate how far extent the carrier-carrier interactions affects the RF performance of the device especially at higher mm-wave frequencies.

1 INTRODUCTION

Impact Avalanche Transit Time (IMPATT) diodes are well recognized two terminal solid-state sources to generate sufficiently high RF power with adequate DC to RF conversion efficiency at both microwave (1–30 GHz) and millimeter-wave (mm-wave) (30–300 GHz) frequency regimes (Luy *et al.* 1987, Wollitzer *et al.* 1996, Midford *et al.* 1979). The principle of operation of IMPATT diode and its negative resistance property is based on impact ionization followed by avalanche multiplication and transit time of charge carriers to cross the depletion layer of the device. Therefore the ionization rates of charge carrier in the semiconductor base material are the key parameters which govern the RF performance of IMPATT oscillators. In the year 1973, W.N. Grant (Grant 1973) reported ionization rates of electrons and holes by extracting those from photo multiplication measurements on Si p^+-n mega diodes as functions of both electric field and junction temperature. At present also the experimental data of W.N. Grant are considered as the most acceptable and authentic ionization rate values of Si. W.N. Grant reported empirical relations representing the ionization rates of electrons and holes fitted from his experimental data as functions of both field and temperature. Those are given by (Grant 1973)

$$\alpha_{e,h}(\xi,T_j) = A_{e,h}\exp\left(-\frac{B_{e,h}(T_j)}{\xi}\right), \tag{1}$$

where $\alpha_e(\xi,T_j)$, $\alpha_h(\xi,T_j)$ are the ionization rate of electrons and holes respectively at the electric field of ξ and for the junction temperature of T_j, A_e, $B_e(T_j)$ and A_h, $B_h(T_j)$ are the ionization coefficients associated with electrons and holes respectively. Both coefficients A_e and A_h are independent of T_j; however, both $B_e(T_j)$ and $B_h(T_j)$ are the functions of T_j.

Two years later in 1975, Ghosh et al. (Ghosh et al. 1975) shown that the higher carrier density in a semiconductor enhances the energy loss due to carrier-carrier collisions which causes significant deterioration in ionization rate of charge carriers. This influence of carrier-carrier interactions at higher doping levels of the semiconductor material is not reflected in equation (1) reported by Grant (Grant 1973) which was based the Shockley's model. The doping densities taken into account by Grant were 10^{23} m^{-3} (Grant 1973) which is more or less appropriate for the IMPATTs operating around 100 GHz. Thus the empirical relation given in equation (1) can be used for analyzing IMPATT diodes designed to operate at 94 GHz or slightly higher frequency than it. However, for analyzing higher mm-wave frequency diodes such as 140 or 220 GHz IMPATTs in which the doping densities of epitaxial layers are appreciably more than those taken into account by Grant (Grant 1973), i.e. the equation (1) is insufficient.

In the year 2014, Acharyya et al. (Acharyya et al. 2014) formulated a generalized analytical model to evaluate the ionization rate of charge carriers in semiconductors which takes into account the multistage scattering phenomena. All possible combinations of optical phonon scattering as well as carrier-carrier collision events prior to the ionizing collision have been taken into account in that model to obtain the probability of impact ionization. And finally from the energy-balance equation, the analytical expressions of ionization rates of charge carriers have been obtained (Acharyya et al. 2014). In the present paper, the authors have adopted the ionization rate model proposed by Acharyya et al. (Acharyya et al. 2014) to carry out both static (DC) and large-signal simulation of DDR IMPATTs based on Si designed to operate at mm-wave atmospheric window frequencies such as 94, 140 and 220 GHz. Simulations are carried out to study the RF performance of those diodes by taking into account both ionization rate models reported by Grant (Grant 1973) as well as Acharyya et al. (Acharyya et al. 2014) to investigate how far extent the carrier-carrier interactions affects the RF performance of the device especially at higher mm-wave frequencies.

2 IONIZATION RATES

The analytical expressions impact ionization rates of electrons and holes developed by Acharyya et al. (Acharyya et al. 2014) are given by

$$\alpha_{e,h}(\xi) = \frac{P_{T_{e,h}}(\xi)\left(1+\dfrac{l_r}{l_{ee,hh}}\right)\left(q\xi-\left\langle\dfrac{dE_{ee,hh}}{dx}\right\rangle\right)}{\left(1-P_{T_{e,h}}(\xi)\right)E_r + P_{T_{e,h}}(\xi)\left(1+\dfrac{l_r}{l_{ee,hh}}\right)E_{i(e,h)}},$$

(2)

where $P_{T_e}(\xi)$ and $P_{T_h}(\xi)$ are the probability of impact ionization of electrons and holes respectively obtained by considering the multistage scattering phenomena (Acharyya et al. 2014), E_r is the energy of optical phonons, E_{ee} and E_{hh} are the average energy loss due to electron-electron and hole-hole collision events respectively, $E_{i(e)}$ and $E_{i(h)}$ are the ionization threshold energies of electrons and holes respectively, l_r, l_{ee}, l_{hh} are the mean free paths associated with optical phonon scattering, electron-electron and hole-hole collisions, q is the charge of an electron ($q = 1.6 \times 10^{-19}$ C). The terms $\langle dE_{ee}/dx\rangle$ and $\langle dE_{hh}/dx\rangle$ in equation (2) represents average energy loss per unit length due to electron-electron and hole-hole collisions respectively (Acharyya et al. 2014).

The variations of α_e and α_h in Si obtained from equations (2) with inverse of the electric field for different electron and hole densities are shown in Figure 1. The other material parameters of Si such as band-gap $E_g = 1.0659$ eV, $E_{ie} = 1.5989$ eV, $E_{ih} = 1.3324$ eV, $E_r = 0.063$ eV, $l_r = 65–100$ Å, mean free path of ionizing collision for electrons $l_{ie} = 1300$ Å and for holes $l_{ih} = 1120$ Å at 500 K are taken from published reports (Electronic Archive 2015). The temperature

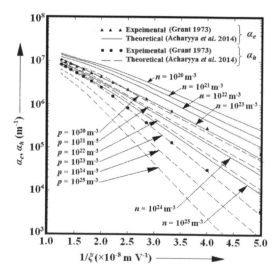

Figure 1. Ionization rate of electrons and holes in Si versus inverse of the applied electric field at 500 K. Points (filled triangle and filled circle) represent plot of the experimental data of Grant (Grant 1973) (electron and hole concentrations in the experiment were $n, p = 10^{23}$ m^{-3}).

is taken to be 500 K. Since the junction temperature of the IMPATT devices are kept nearly 500 K (just below the burnout temperature of Si, i.e. 575 K) by appropriate heat sinking arrangement in order to obtain maximum RF Power output (Acharyya *et al.* 2013). The $\alpha_{e,h}$ versus $1/\xi$ graphs plotted from the equations (1) and Grant's experimental data (Grant 1973) are also shown in Figure 1. The best fittings of the α_e and α_h described by equations (2) with the Grant's α_e and α_h values for electron (n) and holes (p) densities of 10^{23} m^{-3} have been obtained by adjusting four fitting parameters relating n and p with l_{ee} and l_{hh} respectively; those relations are $l_{ee} = (E)^c n^{-1/3}$ and $l_h = (H)^d p^{-1/3}$, where E, c, H and d are the above mentioned fitting parameters. Best fittings of equation (2) with equations (1) for n, $p = 10^{23}$ m^{-3} (which are the n and p values taken into account in Grant's experiment (Grant 1973)) are obtained for $E = 6.13$, $c = 0.81$, $H = 4.79$, $d = 1.02$.

It is interesting to observe from Figure 1 that the analytical model proposed by Acharyya *et al.* (Acharyya *et al.* 2014) provides close arrangement with the experimental data (Grant 1973) for n, $p = 10^{23}$ m^{-3} and $T_j = 500$ K within the electric field range of 2.0×10^7–8.0×10^7 V m^{-1}. However, it is noteworthy from Figure 1 that impact ionization rate of charge carriers decrease significantly when the carrier densities are increased. Degradation of impact ionization probabilities of charge carriers as a consequence of increase of energy loss per unit length due to increased amount of carrier-carrier collision is the cause of reduction in impact ionization rates at higher carrier densities. The carrier densities in epitaxial layers of IMPATT diodes operating at higher mm-wave frequencies are much larger than 10^{23} m^{-3}, thus the ionization rate of charge carriers are reduced significantly within the active region of those. Therefore a detail investigation is required to estimate how far extent this reduced impact ionization rates affects the RF performance of the device at different mm-wave frequency bands.

3 RESULTS AND DISCUSSION

The n- and p-epitaxial layer widths (W_n and W_p) and corresponding doping concentrations (N_D and N_A) of DDR IMPATT diodes are designed

for operation at 94, 140 and 220 GHz frequencies respectively by following a well established method reported earlier (Acharyya *et al.* 2013). The doping concentrations of the n^+- and p^+- contact layers (N_{n+} and N_{p+}) are taken to be in the order of 10^{25} m^{-3} (Luy *et al.* 1987). Bias current density parameters (J_0) are optimized subject to obtain maximum DC to RF conversion efficiency (η_L) at peak optimum frequency (f_p) (Acharyya *et al.* 2013). The design parameters such as W_n, W_p, N_D, N_A, N_{n+}, N_{p+} and J_0 are given in Table 1. The junction diameters of 94, 140, and 220 GHz diodes are chosen to be 35, 25 and 20 μm keeping in mind the rigorous thermal analysis reported earlier (Acharyya *et al.* 2013). The realistic field dependence of drift velocity of charge carriers (v_e and v_h) and other material parameters such as intrinsic carrier concentration (n_i), effective density of states in conduction and valance bands (N_c and N_v), diffusivities (D_e and D_h), motilities (μ_e and μ_h) and diffusion lengths (L_e and L_h) of charge carriers in Si for realistic junction temperature of 500 K (Acharyya *et al.* 2013) have been taken the published experimental reports (Electronic Archive 2015) to carry out both DC and large-signal simulations. Important breakdown voltage (V_B), avalanche zone voltage drop (V_A), ratio of drift zone voltage drop to breakdown voltage (V_D/V_B; $V_D = V_B - V_A$), avalanche zone width (x_A) and ratio of avalanche zone width to total depletion layer width (x_A/W; $W = W_n + W_p$) of DDR Si IMPATTs designed to operated at 94, 140 and 220 GHz at optimum bias current densities (J_0) have been obtained from the present DC simulation. It is observed that ξ_p, V_B, V_A and x_A obtained in present simulation are significantly larger especially in higher frequency diodes as compared to those given by the earlier simulation. Percentage of charges in the said parameters observed in the present simulation with respect to the earlier counterpart are shown as bar graphs in Figure 2. The percentage of charges ($\Delta\chi = \left[\left(\chi^{(PS)} - \chi^{(ES)} \right) \big/ \chi^{(ES)} \right] \times 100\%$; where $\chi \equiv \xi_p$ or V_B or V_A or x_A, the superscripts (ES) and (PS) stand for earlier and present simulation obtained values respectively) of the said parameters are observed to be increasing as the frequency of operation of the device is increasing.

Table 1. Structural, doping and bias current density parameters.

Serial number	f_d (GHz)	W_n (μm)	W_p (μm)	N_D ($\times 10^{23}$ m^{-3})	N_A ($\times 10^{23}$ m^{-3})	N_{n+} ($\times 10^{25}$ m^{-3})	N_{p+} ($\times 10^{25}$ m^{-3})	J_0 ($\times 10^8$ Am^{-2})
1	94	0.400	0.380	1.200	1.250	5.000	2.700	3.40
2	140	0.280	0.245	1.800	2.100	5.000	2.700	5.80
3	220	0.180	0.160	3.950	4.590	5.000	2.700	14.5

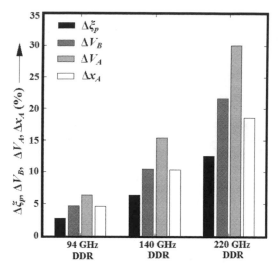

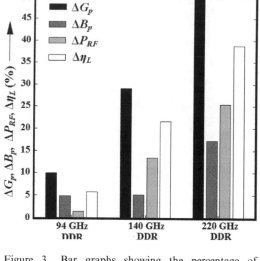

Figure 2. Bar graphs showing the percentage of changes in peak electric field, breakdown voltage, avalanche zone voltage drop and avalanche zone width of 94, 140 and 220 GHz DDR Si IMPATTs obtained in present simulation with respect to the earlier work (Acharyya *et al.* 2013).

Figure 3. Bar graphs showing the percentage of changes in peak negative conductance, corresponding susceptance, RF power output and DC to RF conversion efficiency of 94, 140 and 220 GHz DDR Si IMPATTs obtained in present simulation with respect to the earlier work (Acharyya *et al.* 2013).

The important large-signal parameters of the diodes such as peak optimum frequency (f_p), avalanche resonance frequency (f_a), peak negative conductance (G_p), corresponding susceptance (B_p), quality factor or Q-factor $\left(Q_p = -\left(B_p/G_p\right)\right)$, RF power output ($P_{RF}$) and DC to RF conversion efficiency (η_L) for 50% voltage modulation are obtained from the large-signal simulation. It is observed that due to the enhanced energy loss per unit length as a result of higher carrier-carrier collisions at higher doping densities, f_p, f_a, $\left|G_p\right|$ and B_p decrease significantly. Since the doping densities of the higher frequency diodes are more, thus the above mentioned effect is found to more pronounced in those diodes. Percentage of decrease in B_p of all diodes are found to be smaller as compared to those of $\left|G_p\right|$. As a result the Q-factors increase significantly causing degradation of oscillation growth rate and less stability of oscillation (Acharyya *et al.* 2013). Both P_{RF} and η_L are found to be decreased in the present simulation as compared its earlier counterpart.

Percentage of changes in $\left|G_p\right|$, B_p, P_{RF} and η_L with respect to the earlier reported values (Acharyya *et al.* 2013) are shown as bar graphs in Figure 3. It is observed from Figures 2 and 3 that percentage of decrease in $\left|G_p\right|$ is more pronounced than percentage of increase in V_B due to the effect under consideration. Consequently the values of P_{RF} decreases ($\because P_{RF} = (1/2)(V_{RF})^2 \left|G_p\right| A_j$) and the input DC power ($P_{DC} = J_0 V_B A_j$) increases for all

diodes which ultimately leads to deterioration in DC to RF conversion efficiency ($\eta_L = (P_{RF}/P_{DC})$).

Figure 4 shows the variations of RF power output of DDR Si IMPATTs, with optimum frequency obtained from present simulation as well as the earlier reported results (Acharyya *et al.* 2013) for different values of R_S. Figure 4 also show the experimentally measured values of P_{RF} of 94, 140 and 220 GHz DDR IMPATTs (Luy *et al.* 1987, Wollitzer *et al.* 1996, Midford *et al.* 1979). In the year 1987, Luy *et al.* (Luy *et al.* 1987) experimentally obtained peak RF power of 600 mW with 6.7% efficiency from 94 GHz DDR Si IMPATT oscillators. He used molecular bean epitaxy (MBE) growth technique to fabricate 94 GHz diodes he also measured the maximum series resistance of 0.2 Ω from the diodes under experimentation. Large-signal simulation of 94 GHz DDR Si IMPATT presented in this paper provides variations of P_{RF} and η_L from 625.30 to 593.04 mW and 7.49 to 7.10% respectively the variation of R_S from 0 to 0.3 Ω. The simulation results are very close to the experimental results of Luy *et al.* (Luy *et al.* 1987). Especially for $R_S = 0.2$ Ω (which is the measured value of R_S in 94 GHz DDR (Luy *et al.* 1987)), simulation provides $P_{RF} = 603.84$ mW and $\eta_L = 7.10\%$ which are very much close in agreement with the experimental results. Wollitzer *et al.* (Wollitzer *et al.* 1996) and Midford *et al.* (Midford *et al.* 1979) reported around 225 and 50 mW RF power outputs from 140 and 220 GHz DDR Si IMPATT

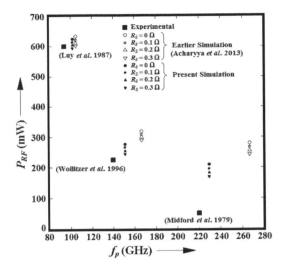

Figure 4. RF power output versus optimum frequency of DDR Si IMPATTs obtained from experimental measurement (Luy *et al.* 1987, Wollitzer *et al.* 1996, Midford *et al.* 1979), earlier simulation (Acharyya *et al.* 2013) as well as present simulation.

oscillators respectively. Present simulation shows that RF power outputs of 140 and 220 GHz DDR Si IMPATTs can vary from 276.65 to 243.28 mW and 209.07 to 169.36 mW respectively for R_S values ranging from 0.0 to 0.3 Ω. Simulation obtained power output in 140 GHz diode is fairly close in agreement with the experimental results of Wollitzer *et al.* (Wollitzer *et al.* 1996). However, the large-signal simulation of 220 GHz diode is predicting larger power can be drawn from the source as compared to the experiment carried out by Midford *et al.* (Midford *et al.* 1979). This discrepancy is may be due to the un-optimized device structure, different biasing conditions, inappropriate experimental arrangements, etc. adopted by the experimentalists at 220 GHz (Midford *et al.* 1979).

4 CONCLUSION

The authors have studied the effect of carrier-carrier collisions on the DC and large-signal characteristics of DDR Si IMPATTs Si designed to operate at mm-wave atmospheric window frequencies such as 94, 140 and 220 GHz. Simulations are carried out to study the RF performance of those diodes by taking into account both empirical relation of field dependence of ionization rates developed from experimental data as well as a appropriate analytical model of same parameter which takes into account the energy loss of charge carriers due to carrier-carrier collisions. Studies have been carried out to investigate how far extent the carrier-carrier interactions affects the RF performance of the device especially at higher mm-wave frequencies. It is observed that the simulation results obtained in the present simulation- (i.e. by taking into account the carrier-carrier interactions) are significantly closer in arrangement as compared its earlier counterpart (i.e. by taking into account the empirical relation of field dependence of ionization rates obtained from experiment). Therefore, the effect of reduced ionization rates at higher doping densities must be incorporated in the simulation in order to obtain better and more realistic results.

REFERENCES

Acharyya, A. and Banerjee, J.P. (2014) A Generalized Analytical Model Based on Multistage Scattering Phenomena for Estimating the Impact Ionization Rate of Charge Carriers in Semiconductors. *Journal of Computational Electronics* 13, 917–924.

Acharyya, A., J. Chakraborty, K. Das, S. Datta, P. De, S. Banerjee and J.P. Banerjee, "Large-Signal Characterization of DDR Silicon IMPATTs Operating in Millimeter-Wave and Terahertz Regime," *Journal of Semiconductors*, vol. 34, no. 10, 104003–8, 2013.

Electronic Archive: New Semiconductor Materials, Characteristics and Properties. Available from: http://www.ioffe.ru/SVA/NSM/Scmicond/index.html (Last accessed on: October 2015).

Grant, W.N. (1973) Electron and hole ionization rates in epitaxial Silicon at high electric fields. *Solid-State Electronics* 16, 1189–1203.

Luy, J.F., Casel, A., Behr, W., and Kasper, E. (1987) A 90-GHz double-drift IMPATT diode made with Si MBE. *IEEE Trans. Electron Devices* 34, 1084–1089.

Midford, T.A., and Bernick, R.L. (1979) Millimeter Wave CW IMPATT diodes and Oscillators. *IEEE Trans. Microwave Theory Tech.* 27, 483–492, 1979.

Scharfetter, D.L., and Gummel, H.K. (1969) Large-Signal Analysis of a Silicon Read Diode Oscillator. *IEEE Trans. on Electron Devices* 6, 64–77.

Wollitzer, M., Buchler, J., Schafflr, F., and Luy, J.F. (1996) D-band Si-IMPATT diodes with 300 mW CW output power at 140 GHz. *Electronic Letters* 32, 122–123.

Frontiers in Computer, Communication and Electrical Engineering – Acharyya (Ed.)
© 2016 Taylor & Francis Group, London, ISBN: 978-1-138-02877-7

Removal of the baseline wander and the power line interference from ECG signals using the Median–Kalman filter

Kiron Nandi & Abhijit Lahiri
Supreme Knowledge Foundation Group of Institutions, Mankundu, Hooghly, West Bengal, India

ABSTRACT: In order to improve the accuracy and reliability of the diagnosis of a patient's cardiac condition, the first important requirement is to denoise the Electrocardiograph (ECG) signal before recording it. The most common noises that are present in an ECG signal are due to the baseline wander and the power line interference. So far, different types of adaptive and non-adaptive digital filters have been proposed to remove these noises from an ECG signal. In this paper, the Kalman filter coupled with the median filter was applied to denoise ECG signals. To evaluate the performance of this method, a simulated ECG signal corrupted with 0.25 Hz and 50 Hz signals was considered. The method was then extended to the ECG signals available in the MIT-BIH arrhythmia database. The accuracy of the proposed Median–Kalman (M-K) filter was measured by the detection rate of the R-peak of the denoised signal and is discussed in this paper.

1 INTRODUCTION

Electrocardiograph (ECG) is the graphical representation of the electrical signal that is generated by the depolarization and repolarization of the atria and ventricles in the human heart, which appears as a nearly periodic signal. The direction and magnitude of temporal differences in potential electrical forces provides information about the heart rate, rhythm, and morphology. ECG varies from person to person due to the difference in position, size, anatomy of the heart, chest configuration, age, and relative body weight, altered by cardiovascular diseases and abnormalities.

Each portion of the ECG waveform carries valuable clinical information for proper diagnosis. The ECG signal taken from a patient can be corrupted by an external noise (Yeh *et al*. 2008, Al. Mahamdy *et al*. 2014). Two types of predominant sources of interference in ECG signals are the Baseline Wander (BLW) and the 50–60 Hz Power Line Interference (PLI). Breathing electrode impedance changes due to perspiration and increased movements of the body are the important sources of the BLW. Usually, the BLW is expected to have undesired frequency components that normally lie in the range of 0–0.5 Hz, which is the same as the frequency range of the ST segment and influences the visual interpretation of the ECG. Detection of ischemia is done by analyzing the ST segment of the ECG. So, before analyzing the ECG signal, the removal of the BLW is essential.

Removing the undesired frequency components due to the BLW using linear filtering is not generally preferred, since this operation also removes the information-bearing frequency components of the ECG signal that lie in the same frequency range. Kalman filter-based BLW removal is not a good idea because of the adaptability and convergence factor of the Kalman filter and the unpredictability of the BLW (Mneimneh *et al*. 2006).

The 50–60 Hz PLI also often corrupts ECG signals, which comes from the power supply lines to the measurement systems despite proper grounding, shielding, and amplifier design. This may be either due to the change in either coupling capacitance or coupling inductance, or both. The value of coupling capacitance decreases with the increase in the separation distance of the electrode and the patient's body. It is responsible for the high-frequency noise. On the other hand, inductive coupling is caused by mutual inductance between two conductors, which introduces low-frequency noise.

Although various precautions can be taken to reduce the effect of the PLI, it may still be necessary to perform signal processing to remove such noise characterized by the 50–60 Hz sinusoidal interference, possibly accompanied by a number of harmonics (Bond *et al*. 1993, Visinescu *et al*. 2006, Bushra *et al*. 2010).

Traditionally, one ECG cycle is labeled using the alphabets P, Q, R, S, and T for individual peaks. The P wave arises from the depolarization of the atrium. The QRS complex arises from the depolarization of the ventricles. The T wave arises from the repolarization of the ventricle muscle. The R-peak is always positive and is the most striking point among all the waves, as shown in Figure 1.

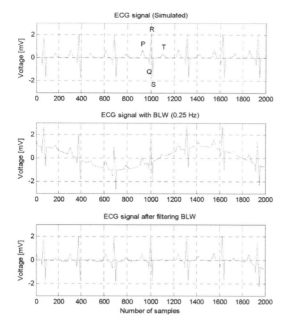

Figure 1. Median filtering to reduce the BLW.

The detection of the R-peak is the first step of feature extraction. In fact, R-peak detection is very difficult because of the noisy ECG signal with the BLW and the PLI. To get the correct information from an ECG signal, the above problems must be eliminated (Kumar *et al.* 2010).

2 DATASETS

Evaluation of performance is even more difficult because of the absence of a normal ECG file in the database for comparison. The proposed approach was applied to two datasets. The first dataset was a simulated ECG signal corrupted with two signals, one with a frequency of 0.25 Hz and the other with a frequency of 50 Hz. Since there is no quantitative measurement of noise, the simulated ECG can be used to measure the performance of the filters used in the present work.

The second dataset was the ECG signals taken from the MIT-BIH arrhythmia database, which is available at http://www.physionet.org/physiobank/database/mitdb/. Those ECGs were sampled at the rate of 360 samples/sec for 30 min duration.

3 METHODOLOGY

3.1 *Median filtering*

The BLW can be eliminated without changing or disturbing the characteristics of ECG signals by median filtering. To compute the output of the median filter, an odd number of samples of the input signal was taken, i.e., the length of the window should be an odd number so that the value of the middle most sample within the window corresponds to the median value of all the samples within the window. Thus, if the window length is $L = 2M + 1$, the filtering procedure is denoted as (Yin *et al.* 1996):

$$Y(n) = MED[X(n - M), ..., X(n), ..., X(n + M)] \quad (1)$$

where $X(n)$ and $Y(n)$ are the nth sample of the input and output sequences, respectively.

In the present case, the simulated ECG signal corrupted with the 0.25 Hz signal was processed with a window of length 45 to remove the R-peak. The output signal was again processed through a window of length 301 to remove the P and T waves. The resultant signal represents the noise responsible for the BLW. Subtracting this noise signal from the simulated ECG signal corrupted with the 0.25 Hz signal, the signal without BLW can be obtained, as shown in Figure 1.

3.2 *Kalman filtering*

The presence of the PLI greatly affects the feature extraction of an ECG signal. To remove this noise without affecting the underlying physiological information of an ECG signal, the Kalman filtering technique was proposed in the present case because of its ability to estimate the information accurately from inaccurate data. It essentially estimates the state of a system based on the system output measurements that contain random errors (Sameni *et al.* 2005).

An autoregressive model is assumed for the ECG signal, which is not contaminated by noise contaminated, such that

$$y_i = a_1 y_{i-1} + a_2 y_{i-2} + ... + a_{n-1} y_{i-n} \quad (2)$$

and the 50 Hz PLI is represented as follows:

$$e = a_{e1} k + a_{e2} \quad (3)$$

where, k is time sample of the signal. Therefore, the signal with the 50 Hz PLI is

$$y_n = a_1 y_{i-1} + ... + a_{n-1} y_{i-n+1} + a_{e1} k + a_{e2} \quad (4)$$

The general form of the updated equations can be written as follows:

$$X_{k+1} = A \cdot X_k \quad (5)$$
$$Z_k = H \cdot X_k \quad (6)$$

where X_k is an n-dimensional vector and Z_k is the measured value.

The solution can be represented in terms of Equations (5) and (6) as follows (Haykin 2010):

$$K_k = A \cdot P_k \cdot H^T / (H \cdot P_k \cdot H^T + \varepsilon)$$
$$\dot{X}_{k+1} = A \cdot \dot{X}_k + K_k \cdot (Z_k - H \cdot \dot{X}_k) \qquad (7)$$
$$P_{k+1} = (A - K_k \cdot H) \cdot P_k \cdot (A - K_k \cdot H)^T + \varepsilon K_k K_k^T$$

where K is the Kalman gain; k is the discrete time sample; P is the uncertainty covariance matrix; and ε is added to model the measurement noise. Equations (7) are iterated over the input signal. The performance of the method for removing the power line hum is shown in Figure 2.

Once the performances of both the filters were tested with the simulated ECG signal, the technique was applied to nine different ECG signals obtained from the MIT-BIH arrhythmia database.

Figure 3 shows the performance of the median filter in removing the BLW from the ECG signal of record number 103 in the MIT-BIH database. As shown in Figure 3, the BLW can also be eliminated from other EGC signals by using the median filtering technique.

The output of the median filter is then passed through the Kalman filter for suppressing the PLI effect that is still present in the signal. The pre-processed and post-processed signals are shown in Figure 4. As shown in Figure 4, the Kalman filtering technique can be adopted to eliminate the noise due to the PLI from the ECG signals.

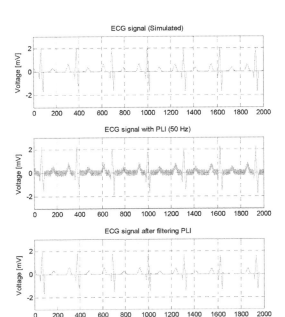

Figure 2. Kalman filtering to reduce the 50 Hz PLI.

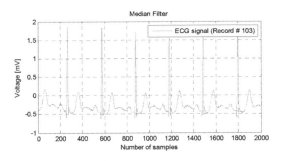

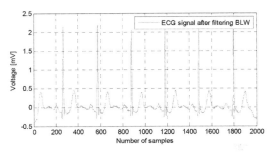

Figure 3. Comparison of an ECG signal with the BLW and the ECG signal after the BLW is removed.

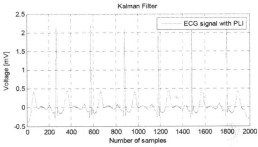

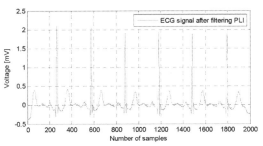

Figure 4. Comparison of an ECG signal with the PLI and the ECG signal after the PLI is removed.

3.3 Detection of the R-peak

The efficiency of the Median–Kalman (M-K) filter was tested in terms of its capability of detecting the R-peaks after eliminating the noises. Detection of the R-peaks was done in two ways. The first was to identify whether any R-peak was eliminated in the process of eliminating the noises and the second was to identify whether any R-peak was shifted along the time axis after the suppression of the noises. Normally, the R-peak is that point where the heartbeat has the highest amplitude. Hence, the accuracy of the proposed work was evaluated in terms of accurate detection of the R-peaks after noise filtering.

To detect the R-peaks of the signal finally obtained from the M-K filter, the output of the M-K filter was again passed through the median filter of window length equal to 45 to suppress the R-peaks present in the signal that was obtained from the M-K filter.

The output of the median filter was then subtracted from the output of the M-K filter to extract the number of R-peaks present in the output of the M-K filter, which was actually the denoised signal. Figure 5 shows the result of such a removal from the ECG signal of record number 103 in the MIT-BIH database. From Figure 5, it can be observed that the R-peaks are predominantly visible approximately over 311 cycles.

Table 1 presents the percentage of the R-peaks that were retained by each of the nine signals that were obtained from the MIT-BIH database after denoising by the M-K filtering technique. From Table 1, it can be seen that the detection rate of the R-peaks is as high as 99.7% with the ECG signal of record number 103, and the lowest detection rate is 97.4% with record number 219. The average successful rate of detection of the R-peaks is 99.3%.

Next, the locations of these R-peaks were compared with those of the R-peaks of the original signal to check whether there was any distortion in local time-scale characteristics. The study was carried with the ECG signal of record number 219, and the comparison is shown in Figure 6. From

Table 1. Results on the detection of the R-peaks for the signals in the MIT-BIH arrhythmia database.

Record number	Total beats	Detected beats	% of detection
103	2084	2078	99.7121
112	2539	2530	99.6455
115	1953	1946	99.6416
117	1535	1530	99.6743
122	2476	2468	99.6769
123	1518	1510	99.4730
219	2154	2097	97.3538
230	2256	2247	99.6011
234	2753	2721	98.8376
Average	**2141**	**2125**	**99.2907**

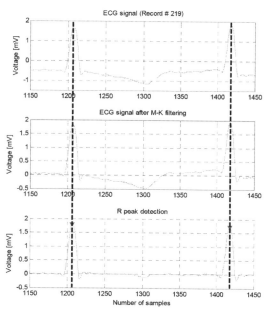

Figure 6. Comparison of R-peak locations in local time-scale characteristics.

Figure 6, it can be observed that denoising an ECG signal by the M-K filtering technique satisfactorily restores the local time-scale characteristics of the signal.

Among the other suggested methods observed, the wavelet-based zero-crossing detector misinterpreted when finding out the exact locations of the R-peaks (Sasikala *et al.* 2010). In the case of the wavelet-based zero-crossing R-peak detector, the analysis was performed on a frame-by-frame basis until the end of the data was reached. Therefore, it is not suitable for use in any real-time application.

However, the proposed method may produce discontinuities in the resulting signal at the initial

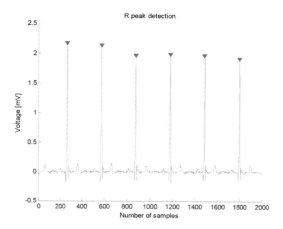

Figure 5. R-peak detection in the QRS complex.

and final points. Therefore, for a reliable detection of the signal, this issue may be fixed by implementing the weighted median filter.

4 CONCLUSION

A two-stage algorithm was applied in this work for the removal of noises from an ECG signal that originates due to the BLW and the PLI. Although the proposed method uses the M-K filtering technique, M-K filtering can also be done without affecting the efficiency of the algorithm. In that case, first, the noise due to the PLI will be eliminated followed by the elimination of the noise caused by the BLW.

The accuracy of the proposed method was measured in terms of the successful detection of the R-peaks. The validation was done using nine ECG records obtained from the MIT-BIH arrhythmia database. In the method adopted in this work, the average rate of successful detection of the R-peaks was as high as 99.29%. Also, the method did not cause any shift in the signal along the time axis.

REFERENCES

AlMahamdy, M. & Bryan R.H. (2014). Performance Study of Different Denoising Methods for ECG Signals. *ELSEVIER, 4th International Conference on Current and Future Trends of Information and Communication Technologies in Healthcare, Procedia Computer Science*, 37, 325–332.

Bond, A.B., Greco, E.C., Bowser, R., Kadri, N.N. & Sketch, M.H. (1993). Robust and computational efficient QRS detection. *IEEE Conference Publications, Computers in Cardiology, Proceedings*, 507–510.

Bushra, J., Beya, O., Fauvet, E. & Laligent, O. (2010). QRS Complex Detection by Non Linear Thresholding of Modulus Maxima. *IEEE Conference Publications, Pattern Recognition (ICPR), 20th International Conference*, 4500–4503.

Haykin, S. (2002). *Adaptive Filter Theory*. Fourth Edition, Prentice-Hall, Inc., Englewood Cliffs, NJ.

Kumar, Y. & Malik, G.K. (2010). Performance Analysis of different filters for power line interface reduction in ECG signal. *International Journal of Computer Applications*, 3(7), 1–6.

Mneimneh, MA., Yaz, EE., Johnson, MT. & Povinelli. (2006). An Adaptive Kalman Filter for Removing Baseline Wandering in ECG Signals. *IEEE Conference Publications, Computers in Cardiology*, 253–256.

Sameni, R., Shamsollahi, M.B., Jutten, C. & Babaie-Zadeh, M. (2005). Filtering noisy ECG signals using the extended Kalman filter based on a modified dynamic ECG model. *IEEE Conference Publications, Computers in Cardiology*, 1017–1020.

Sasikala, P. & Wahidabanu, R.S.D. (2010). Robust R. Peak and QRS detection in Electrocardiogram using Wavelet Transform. *International Journal of Advanced Computer Science and Applications*, 1(6), 48–53.

Visinescu, M., Bashour, C.A., Bakri, M. & Nair, B.G. (2006). Automatic detection of QRS complexes in ECG signals collected from patients after cardiac surgery. *Proceedings of the 28th IEEE, EMBS Annual International Conference*, 3724–3727.

Yeh, Y.C. & Wang, W.J. (2008). QRS complexes detection for ECG signal: The Difference Operation Method. *ELSEVIER, computer methods and programs in biomedicine*, 9I, 245–254.

Yin, L., Yang, R., Gabbouj, M. & Neuvo, Y. (1996). Weighted Median Filters: A Tutorial. *IEEE Transactions on Circuits and Systems-II: Analog and Digital Signal Processing*, 43(3), 157–192.

Frontiers in Computer, Communication and Electrical Engineering – Acharyya (Ed.)
© 2016 Taylor & Francis Group, London, ISBN: 978-1-138-02877-7

A practical approach for power system state estimation based on the hybrid Particle Swarm Optimization algorithm

Ujjal Sur & Gautam Sarkar
Department of Applied Physics, University College of Science and Technology, University of Calcutta, Kolkata, India

ABSTRACT: Power system dynamic state estimation is an important tool for the analysis, planning, and operation of a power system. In this paper, a new hybrid state estimation method based on the Nelder–Mead (NM) simplex search method and Particle Swarm Optimization (PSO) is proposed. State estimation is an optimization problem including continuous variables, whose objective is to minimize the difference between the calculated and measured values of variables. Here, the NM-PSO hybrid method tries to find the global optima solutions much efficiently in comparison with other artificial intelligence techniques such as genetic algorithm. The two main state estimation methods, namely weighted least square and weighted least absolute value, were used to construct the objective function over which this hybrid stochastic algorithm was applied, and a comparison was made over the data obtained from a case study on the IEEE 14 bus test system for a better understanding of this paper.

1 INTRODUCTION

To solve the Power System State Estimation (PSSE) problem, several authors have used traditional approaches based on derivative methods such as Newton, gradient descent, and linear programming methods (Kamireddy *et al.* 2008). However, it has been shown that these traditional methods present some inadequacies. The Newton and gradient methods both suffer from the difficulty in handling inequality constraints. On the other hand, the linear programming method suffers from oscillation and slow convergence problems. A common limitation of traditional optimization methods is the fact that they are local minima-based algorithms while the PSSE is a highly nonlinear problem. The advantage of Artificial Intelligence (AI) algorithms is due to the fact that they only require the fitness function (performance index or objective function) to guide the search, unlike the traditional algorithms that need gradient (derivative) information. One of the AI algorithms is the Particle Swarm Optimization (PSO), which has many advantages over other AI algorithms such as Genetic Algorithm (GA) (Bies *et al.* 2006). However, here, we used the Nelder–Mead (NM) simplex search method with the PSO to have the best global optimum value with less computational time. With this in mind, this paper applies the advantages of PSO in solving the PSSE problem. The corruption of telemetric raw data measurements is simulated by introducing statistically a random error in the measurements obtained after running the Newton–Raphson (NR) load flow algorithm. The solutions obtained from load flow (before introducing errors) are, therefore, considered as true solutions and used as benchmark solutions in evaluating the effectiveness of NM-PSO solutions. GA is used in order to compare the performance over other AI algorithms.

2 STATE ESTIMATION

The general state estimation equation involves the estimation of the state vector x in the presence of an error e based on a set of measurements z. A mathematically synthesized model describing the functional relations between the parameters z, x, and e is given by Equation (1). This model is expressed in a set of nonlinear equations in relation to the measurements z and the true state vector x:

$$z_i = f_i(x) + e_i \quad \text{or,} \quad e_i = z_i - f_i(x) \quad (1)$$

where, $i = 1, 2, 3, \ldots, n$; z_i is the ith measurement (measurement vector of dimension n); and $f_i(x)$ is the nonlinear function relating state variables with measurements. This is basically the power flow or power injection equations, where e_i is the ith measurement error vector; x is the state vector of dimension m; and n is the number of measurements. The state estimation problem is basically interpreted as a Weighted Least Square (WLS) problem, which is solved using different numerical techniques

(Abur et al. 2004). Another method that we use is the weighted least absolute value (WLAV) method.

2.1 WLS state estimation

The WLS estimation method can be formed by minimizing the objective function:

$$\text{Min } J(x) = \sum_{i=1}^{n} [W_i^2 (z_i - f_i(x))^2]$$
$$= \sum_{i=1}^{n} \left[\frac{1}{\sigma_i^2} (z_i - f_i(x))^2 \right] \quad (2)$$

where W_i is the weight of the measurement i and σ_i^2 is the variance of the ith measurement.

2.2 WLAV state estimation

The WLAV estimation method can be formed as that of the WLS as follows:

$$\text{Min } J(x) = \sum_{i=1}^{n} [W_i | z_i - f_i(x) |] \quad (3)$$

The nonlinear function $f_i(x)$ contains the active and reactive power flows and power injections whose equations can be obtained. The active and reactive power injections in the polar form are given as follows:

$$P_i = V_i \sum_{j=1}^{nb} [V_j | Y_{ij} | \cos(\delta_i - \delta_j - \theta_{ij})] \quad (4)$$

$$Q_i = V_i \sum_{j=1}^{nb} [V_j | Y_{ij} | \sin(\delta_i - \delta_j - \theta_{ij})] \quad (5)$$

where P_i is the active power injection at bus i; Q_i is the reactive power injection at bus i; V_i and V_j are voltage magnitudes at buses i and j; δ_i and δ_j are voltage angles at buses i and j; $| Y_{ij} |$ is the magnitude of the bus admittance elements i, j; θ_{ij} is the angle of the bus admittance elements i, j; and nb is the number of buses.

3 NM SEARCH METHOD

The NM method (Chelouah et al. 2005) is a non-linear optimization algorithm, where the numerical method is used to optimize an objective function in a multidimensional space using four different procedures, namely reflection, expansion, contraction, and reduction, which are described in detail below.

3.1 Initialize

Form an m-dimensional simplex by generating $(m + 1)$ individuals randomly. At each vertex point of the simplex, we calculate the objective function

value. Based on the objective function values, individual sorting proceeds as follows:

$$\begin{bmatrix} z_l & f_l \\ \vdots & \vdots \\ z_h & f_n \end{bmatrix}_{(m+1)*(n+1)} \quad (6)$$

where z_l and z_h are the vertices with the lowest and the highest function values, respectively; f_l and f_h are the corresponding objective function values; and n is the number of state variables.

3.2 Reflection

To find z_c, the center of the simplex in the minimization problem without using z_h and generates a new vertex z_r by reflecting the worst point by the following equation:

$$z_c = \frac{1}{m} \sum_{\substack{i=1 \\ z_i \neq zh}}^{m+1} z_i$$
$$z_r = (1 + \alpha) * z_c - \alpha * z_h \quad (7)$$

where α is the reflection coefficient ($\alpha > 0$). Now, according to the NM method, four different judgments are considered:

1. If $f_l < f_r < f_h$, then, by reflection, replace z_h with z_r and the reflection process will be repeated for the next iteration;
2. If $f_r < f_l$, then perform the expansion process stated below;
3. If $f_l > f_r > f_h$, then replace z_h with z_r and go to the contraction process below; and
4. If $f_h < f_r$, go to the contraction process below without replacing z_h with z_r.

3.3 Expansion

If $f_r < f_l$, then the reflection process is expanded to extend the search space in the same direction, and the expansion point is given as

$$z_e = \beta * z_r + (1 - \beta) * z_c \quad (8)$$

where β is the expansion coefficient ($\beta > 1$). Now, if $f_e < f_l$, then, by reflection, replace z_h with z_e and the reflection process will be repeated for the next iteration.

3.4 Contraction

The contraction vertex is now calculated using the following equation:

$$z_{co} = \beta * z_h + (1 - \gamma) * z_c \quad (9)$$

422

where γ is the contraction coefficient ($0 < \gamma < 1$). Now, if $f_{co} < f_l$, then, by reflection, replace z_h with z_{co} and the reflection process will be repeated for the next iteration, else $f_{co} > f_h$ and the reduction process will begin as stated below.

3.5 *Reduction*

Here, in this step, the entire simplex will be shrinking except z_l given by

$$z_i = \beta * z_l + (1 - \gamma) * z_l \qquad (10)$$

If the criterion given for convergence is met, then it is the end or otherwise go to the reflection step and start the algorithm from there again. Now, according to the NM method, the values of the different coefficients α, β and γ are $\alpha = 1$, $\beta = 2$, and $\gamma = 0.5$. Here, in this paper, we use these same values for the simulation process.

4 PARTICLE SWARM OPTIMIZATION

PSO is one of the best AI applications in a population-based random stochastic algorithm domain. This algorithm is actually based on the simulation of animals' social activities such as insects and birds, which shows the natural habit of the group or swarm communication to share an individual's idea when such swarms flock or migrate. For each particle i, the velocity and position of the particles can be updated by the following equations (Kennedy *et al.* 1995):

$$
\begin{aligned}
V_i(t+1) &= \omega V_i(t) + C_1 rand(.)(P_{best} - X_i(t)) \\
&\quad + C_2 rand(.)(G_{best} - X_i(t))
\end{aligned}
$$

$$X_i(t+1) = X_i(t) + V_i(t+1) \qquad (11)$$

$$\omega(t+1) = \omega_{max} - \frac{\omega_{max} - \omega_{min}}{} * t$$

where $i = 1, 2, ..., n$, with n being the number of particles; P_{best} is the previous best value of the ith particle; G_{best} is the best value of the entire population; $rand(.)$ is the random number; ω is the inertia weight; $\omega_{max} = 1$, $\omega_{min} = 0.5$, and C_1 and $C_2 = 2$ are the weighting factors of the algorithm.

5 HYBRID PSO ALGORITHM

By combining PSO with the NM algorithm, we obtain a new PSO-NM hybrid optimization algorithm. The application of the hybrid algorithm is detailed below.

5.1 *Input data statement*

Here, in this step, all the input data relating to the real and pseudo-measurement values, the distribution line network, the RES and load output, and the standard and average deviations of RES and loads are stated and incorporated in the search algorithm.

5.2 *Generation of initial population and velocity*

According to the input state variables, a randomly generated initial population and velocity can be written as follows:

$$
Velocity = \begin{bmatrix} v_1 \\ \cdots \\ v_n \end{bmatrix}, \; population = \begin{bmatrix} X_1 \\ \cdots \\ X_n \end{bmatrix}
$$

$$
\begin{aligned}
v_i &= [v_j]_{1*n}, X_i = [X_j]_{1*n} \\
v_j &= rand(.) * (V_{j\max} - V_{j\min}) + V_{j\min} \\
v_j &= rand(.) * (\delta_{j\max} - \delta_{j\min}) + \delta_{j\min} \\
X_j &= rand(.) * (V_{j\max} - V_{j\min}) + V_{j\min} \\
X_j &= rand(.) * (\delta_{j\max} - \delta_{j\min}) + \delta_{j\min}
\end{aligned}
\qquad (12)
$$

where n is the number of measurements, and V and δ are the state variables.

5.3 *Objective function calculation and selection of G_{best} and initial population*

In this step, the objective function based on the WLS method (Equation (2)) or the WLAV method (Equation (3)) is to calculate for each individual using the result of load flow analysis. Selection of the initial population value is in ascending order based on the value of the objective function. Also, the minimum objective function is to select the best global position (G_{best}).

5.4 *Application of the NM search method*

We apply the NM simplex search method to search the global solution. Here, in this step, the NM method searches the G_{best}, and if the obtained solution is better than the G_{best}, then it swaps it.

5.5 *Application of the PSO algorithm*

First, select the ith individual and then the best local position P_{best} for the ith individual. Next, calculate the modified velocity for each individual based on P_{best} and G_{best}. Then, calculate the modified position for each individual based on Equation (11) and then check its limit. Next, check whether all individuals are selected or not, and if not, then go back to the first step of the PSO algorithm, where the ith individual is selected. Finally, check the convergence criteria if it is ok and then end. G_{best} is the required solution and if not satisfied, then go to the initial state of the objective function calculation.

6 SIMULATION RESULTS

The application of both the traditional WLS and WLAV state estimation methods employing the NM-PSO hybrid algorithm approaches for the state estimation of the IEEE 14 bus test case is presented (shown in Figure 1) here. Here, the NR load flow results are taken as the benchmark values

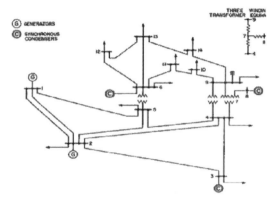

Figure 1.   IEEE 14 bus test case.

Table 1.   Comparison of the estimated values of the WLS objective function.

Bus no.	True value V (pu)	True value δ (degree)	GA V (pu)	GA δ (degree)	NM-PSO V (pu)	NM-PSO δ (degree)
1	1.060	0	1.041	0	1.045	0
2	1.045	−04.99	1.026	−05.16	1.029	−05.13
3	1.010	−12.75	0.988	−13.24	0.992	−13.16
4	1.013	−01.24	0.994	−10.65	0.997	−10.59
5	1.017	−08.76	0.998	−09.12	1.001	−09.06
6	1.070	−14.45	1.053	−15.01	1.055	−14.92
7	1.046	−13.24	1.027	−13.76	1.030	−13.68
8	1.080	−13.24	1.063	−13.76	1.065	−13.68
9	1.031	−14.82	1.012	−15.40	1.015	−15.31
10	1.030	−15.04	1.012	−15.62	1.014	−15.53
11	1.046	−14.86	1.028	−15.44	1.031	−15.34
12	1.053	−15.30	1.036	−15.89	1.038	−15.79
13	1.047	−15.33	1.029	−15.92	1.004	−15.83
14	1.019	−16.08	1.001	−16.69	1.004	−16.60
MAPE (%)			1.78	3.62	1.51	3.05
Computational time (s)			0.6490		0.4381	

Table 2.   Comparison of the estimated values of the WLAV objective function.

Bus no.	True value V (pu)	True value δ (degree)	GA V (pu)	GA δ (degree)	NM-PSO V (pu)	NM-PSO δ (degree)
1	1.060	0	1.043	0	1.046	0
2	1.045	−04.99	1.027	−05.15	1.030	−05.12
3	1.010	−12.75	0.989	−13.21	0.993	−13.13
4	1.013	−01.24	0.996	−10.63	0.998	−10.56
5	1.017	−08.76	0.999	−09.10	1.002	−09.04
6	1.070	−14.45	1.055	−15.07	1.058	−14.99
7	1.046	−13.24	1.038	−14.66	1.040	−14.58
8	1.080	−13.24	1.074	−14.62	1.076	−14.54
9	1.031	−14.82	1.023	−16.20	1.025	−16.09
10	1.030	−15.04	1.020	−16.24	1.023	−16.15
11	1.046	−14.86	1.033	−15.69	1.035	−15.64
12	1.053	−15.30	1.038	−15.95	1.041	−15.87
13	1.047	−15.33	1.031	−15.98	1.034	−15.90
14	1.019	−16.08	1.008	−17.16	1.010	−17.06
MAPE (%)			1.34	5.61	1.08	5.06
Computational time (s)			0.6137		0.4133	

of the state variables. The NR load flow method can be applied to obtain the load flow results of the system. To have a fair comparison of the state estimation accuracy for both the WLS and WLAV approaches using GA and NM-PSO, the Mean Absolute Percentage Error (MAPE) is introduced as follows:

$$MAPE = \frac{1}{n}\sum_{i=1}^{n}\left|\frac{A_i - F_i}{A_i}\right| *100\% \qquad (13)$$

where A_i is the actual value and F_i is the final calculated value.

The comparison of the estimated values of the WLS and WLAV objective functions are given in Tables 1 and 2.

7 CONCLUSION

In this paper, the application of the NM-PSO hybrid algorithm for the PSSE is discussed. Here, the NR load flow algorithm was used to generate the required data to be taken as the true values of the state variables to be estimated from the corrupted measurements present in the power system network. GA was used as another Artificial Intelligent (AI) algorithm to evaluate the performance of the NM-PSO hybrid algorithm, using two formulations of the objective functions, which are the WLS and the WLAV techniques. The simulation result showed a higher accuracy with lower computational time for the NM-PSO hybrid algorithm compared with the GA algorithm. In future work, it can be extended to the distribution network state estimation, where distributed generations and renewable energy connections play an important role, which can be handled smoothly by using this NM-PSO hybrid algorithm.

REFERENCES

Abur, A., & Exposito, A.G. (2004). *Power System State Estimation: Theory and implementation*, Marcel Dekkered., New York.

Bies, R.R., Muldoon, M.F., Pollock, B.G., Manuck, S., Smith, G. and Sale, M.E. (2006). A Genetic Algorithm Based Hybrid Machine Learning Approach to Model Selection, *Journal of Pharmacokinetics and Pharmacodynamics*,196–221.

Chelouah, R. and Siarry, P. (2005). A hybrid method combining continuous tabu search and Nelder-Mead simplex algorithms for the global optimization of multi minima functions. *European Journal of Operational Research*, 161(3), 636–654.

Kamireddy, S., Schulz, N.N. and Srivastava, A.K. (2008). Comparison of state estimation algorithms for extreme contingencies, *40th North American Power Symposium, NAPS'08*, 1–5.

Kennedy, J. and Eberhart, R. (1995). Particle swarm optimization, *IEEE International Conference on Neural Networks*, Piscataway, NJ, 1942–1948. http://www.ee.washington.edu/research/pstca/, Power System Test Case Archive.

Frontiers in Computer, Communication and Electrical Engineering – Acharyya (Ed.)
© 2016 Taylor & Francis Group, London, ISBN: 978-1-138-02877-7

Design of the Leaky-Wave Antenna using the Composite Right/Left-Handed Transmission Line metamaterial at the center frequency of 2.33GHz

Sumit K. Varshney, Amit K. Varshney, A.M. Chowdhury, A. Ganguly & A. Samanta
Supreme Knowledge Foundation Group of Institutions, Mankundu, Hooghly, West Bengal, India

ABSTRACT: In this paper, the author tries to simulate a transmission line-based Composite Right/Left-Handed (CRLH) metamaterial Leaky Wave Antenna (LWA) with the center frequency of 2.33 GHz, where it will radiate in the broadside direction only when it is operated in the fast wave region. Also, its frequency scanning capability is tested by changing the excitation frequency of the antenna and observing the far-field radiation pattern. A CRLH metamaterial LWA has the unique property of continuous frequency scanning from the backward end fire to the forward end fire through the broadside direction unlike in the conventional Right-Handed (RH) LWA, where the radiation angles are highly restricted and also there is no broadside radiation. It finds its application in scanning radar where the loss of target cannot be tolerated because of the absence of the broadside radiation lobe in the RH frequency scanning LWA.

1 INTRODUCTION

Metamaterials are the artificial or engineered materials that exhibit properties not readily found in nature. Its theory was first given by Veselago in 1967, which states that metamaterials exhibit a negative refractive index due to both ε and μ being negative (Veselago V. 1968). The Transmission Line (TL) approach (Caloz C. *et al*. 2002) towards the design of a metamaterial is necessary as it can be used to analyze and design novel wideband microwave components at high frequency with minimum loss as TLs are non-resonant structures unlike split-ring resonator–thin wire-based metamaterials, which are resonant structures with a narrow band and offer high losses at the microwave frequency. Conventional materials or Right-Handed (RH) materials are non-dispersive, exhibiting a straight line in the ω-β diagram, while the Composite Right/Right-Handed (CRLH) TL dispersion diagram exhibits a nonlinear nature. It has left-handed properties at low frequencies and right-handed properties at high frequencies, hence the name. The incremental model of a TL can be practically realized by a bandpass LC filter (Eleftheriades G.V. *et al*. 2002) whose physical dimension 'p' is infinitesimally small. We used the distributed component realization of the LC network, i.e., the microstrip form which when cascaded forms the CRLH TL. So, each LC network will be called a unit cell.

A right-handed TL operates in the slow wave region (guided mode) when $\beta > k_0$, where β is the propagation constant in the direction of the propagation of the wave and k_0 is the free space wave number. On the other hand, it operates in the fast wave region when $\beta < k_0$. In this fast wave region, it progressively leaks out power as it travels down the line, giving rise to radiation. β determines the angle θ_{MB} of radiation (measured from vertical) of the main beam following the simple relation:

$$\theta_{MB} = \sin^{-1}\left(\frac{\beta}{k_0}\right) \tag{1}$$

And the width of the mail lobe Δθ by

$$\Delta\theta = 2\sin^{-1}\left(\frac{1}{\frac{l}{\lambda_0}\cos\theta_{MB}}\right) \tag{2}$$

where *l* is the length of the leaky wave antenna (LWA) and λ_0 is the free space wavelength.

Leaky-wave antennas (Itoh Tatsuo *et al*. 2006) belong to the class of traveling wave antenna and exhibit superior directivity owing to their large radiating aperture when compared with any other resonant antennas. If they are designed using metamaterials, then they can be made to radiate both in the forward and backward directions as well as in the broadside direction, i.e., $-90° \le \theta_{MB} \le +90°–0°$. Negative values of angles are obtained as β is negative for the left-handed region and positive for the right-handed region. This seamless transition from the left- to right-handed region is possible only when the CRLH TL is a balanced one, i.e., there is no band gap region during the transition from left to right during which β becomes imaginary and no propagation takes place.

2 DESIGN AND SIMULATION OF THE BALANCED CRLH TL UNIT CELL

The interdigital capacitor shorted stub CRLH TL unit cell was designed in the microstrip form in Ansoft Designer. Design values for the Inter-digital capacitor shorted stub based one dimensional CRLH balanced unit cell are given in Table 1. The voltage gradient between the digits of the unit cell gives the effect of the left-handed capacitance C_L, the current flow through the via (for shorted stub) realizes the left-handed inductance L_L, the current flow through the digits of the interdigital capacitor and the associated flux linkage realizes the right-handed inductance L_R and the voltage gradient developed between the trace and the ground plane of the microstrip realizes the right-handed capacitance C_R. The schematic of the Inter-digital capacitor shorted stub based one dimensional CRLH balanced unit cell is shown in Figure 1. Simulation results are presented in Figures 2–6.

Table 1. Design values for the Inter-digital capacitor shorted stub based one dimensional CRLH balanced unit cell.

Parameters	Symbols	Design values mm
Unit cell period	p	11.4
Stub length	l_s	10.9
Stub width	w_s	1.00
Interdigital finger length	l_c	10.2
Interdigital finger width	w_c	0.30
Spacing between fingers	S	0.20
Via radius	r	0.12
Substrate* heigh	h	1.57
Substrate* permittivity	ε_r	2.2**

*Rogers Duroid 5880; **Unit less.

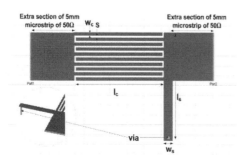

Figure 1. Inter-digital capacitor shorted stub based one dimensional CRLH balanced unit cell. (Inset) 3D view of via.

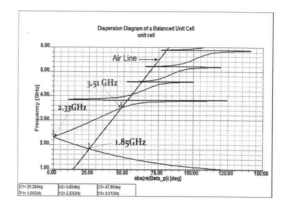

Figure 2. One-dimensional CRLH balanced unit cell dispersion diagram.

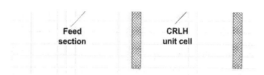

Figure 3. 20 unit cell CRLH balanced TL with feed sections at both the ends.

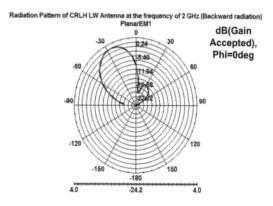

Figure 4. Radiation Pattern of CRLH Leaky Wave Antenna at the frequency of 2 GHz (Backward radiation).

β is negative in the left hand region, so $\theta_{MB} = \sin^{-1}\left(\frac{\beta}{k_0}\right)$ is negative indicating backward angle radiation.

The dispersion diagram is extracted from the unit cell, and it is plotted over the free space dispersion diagram with the following equation (air line equation):

$$airline = \beta p = \frac{\omega p}{c_0} \qquad (3)$$

where c_0 is the free space velocity of the wave. The airline is plotted in order to get the slow wave and fast wave regions.

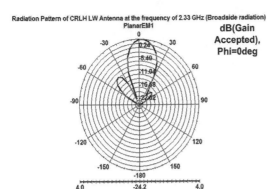

Radiation Pattern of CRLH LW Antenna at the frequency of 2.33 GHz (Broadside radiation)
PlanarEM1

dB(Gain Accepted), Phi=0deg

Figure 5.   Radiation pattern of the CRLH LW antenna at the frequency of 2.33 GHz (broadside radiation). β is 0 at the center frequency, so $\theta_{MB} = \sin^{-1}\left(\frac{\beta}{k_0}\right)$ is 0, indicating broadside radiation.

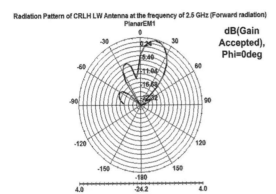

Radiation Pattern of CRLH LW Antenna at the frequency of 2.5 GHz (Forward radiation)
PlanarEM1

dB(Gain Accepted), Phi=0deg

Figure 6.   Radiation Pattern of CRLH LW Antenna at the frequency of 2.5 GHz (Forward radiation). β is positive in the right hand region, so $\theta_{MB} = \sin^{-1}\left(\frac{\beta}{k_0}\right)$ is positive indicating forward angle radiation.

The dispersion diagram can be directly obtained from the de-embedded S parameters using the relation

$$\beta p = \cos^{-1}\left(\frac{1 - S_{11}S_{22} + S_{12}S_{21}}{2S_{21}}\right) \quad (4)$$

From the dispersion diagram,
Forward radiation: 2.33 to 3.51GHz
Backward radiation: 1.85 to 2.33GHz
Broadside radiation: 2.33 GHz

These 20 unit cells are cascaded along with the feed sections (to avoid abrupt discontinuity) to form a CRLH TL, which starts radiating when operated in the fast wave (radiating) region, i.e.,

from 1.85 to 3.51 GHz, and its far-field radiation pattern is simulated. The simulation results clearly show the forward, backward, and broadside lobes.

3   DISCUSSION AND CONCLUSION

From the radiation patterns it is clear that when CRLH TL based Leaky Wave antenna is operated at a frequency which falls in the left hand fast wave region of the dispersion curve of the balanced two dimensional unit cell, i.e., from 1.85 to 2.33 GHz, it radiates in the backward direction, i.e., from θ = 0° to −90°. In this left-handed fast wave region (region above the airline), β ranges from −k_0 when θ becomes −90° to 0 when θ becomes 0 as per the equation (1). Similarly, when the CRLH TL-based LW antenna is operated with the frequency that falls in the RH fast wave region, i.e., from 2.33 to 3.51 GHz, it radiates in the forward angle direction, i.e., from θ = 0° to +90°. In this RH fast wave region (region above the airline), β ranges from 0 when θ becomes 0° to +k_0 when θ becomes +90°. Interestingly, for the case of a balanced unit cell, at β = 0, the group velocity of the wave in non-zero, i.e., there is radiation of energy and we get a broadside lobe. Hence, in case of a balanced design, there is continuous scanning of the beam from the backward end fire to the forward end fire through the broadside direction, a feature very unique to the metamaterial LWA unlike the uniform conventional or purely RH LWA that radiates in very small restricted angles in the forward direction, and periodic conventional LWA that can radiate in the backward direction (if excited by negative space harmonics) but not in the broadside direction. Such conventional LWA used in the frequency scanning radar may result in the loss of target due to the absence of broadside radiation. LWA exhibit superior directivity compared with conventional resonant antennas if the leakage factor can be made as small as possible.

REFERENCES

Caloz, C., Okabe, H., Iwai, T. & Itoh, T. 2002. "Transmission line approach of left-handed (LH) materials", in *Proc. USNC/URSI National Radio Science Meeting*, San Antonio, TX, vol. 1, p. 39.

Eleftheriades, G.V., Iyer A.K. & Kremer P.C. 2002, "Planar negative refractive index media using periodically L-C loaded transmission lines", *IEEE Trans. Microwave Theory Tech.*, vol. 50, pp. 2702–2712.

Itoh Tatsuo & Christophe Caloz. 2006, Electromagnetic Metamaterials: Transmission Line Theory and Microwave Applications, the Engineering Approach, A John Wiley & Sons, Inc.

Veselago, V. 1968. "The electrodynamics of substances with simultaneously negative values of ε and μ", *Soviet Physics Uspekhi*, vol. 10, no. 4, pp. 509–514.

Frontiers in Computer, Communication and Electrical Engineering – Acharyya (Ed.)
© 2016 Taylor & Francis Group, London, ISBN: 978-1-138-02877-7

An artificial neural network-based temperature measurement system using a thermistor in astable multivibrator circuit

Saket Kumar Sahu & Kiron Nandi
Supreme Knowledge Foundation Group of Institutions, Mankundu, Hooghly, West Bengal, India

ABSTRACT: This paper presents a novel, cheap, and unique adaptation technique of a PC-based temperature measurement system. This adaptation revolves around a 555 timer, made to operate in an astable multivibrator configuration. The negative temperature coefficient thermistor is used in this mode, which acts as a timing resistor of the astable multivibrator using the 555 timer. The 'HIGH' and 'LOW' level durations from the multivibrator output are taken for consideration, and these values are used to determine the temperature by the appropriate training of artificial neural network.

1 INTRODUCTION

Measuring temperature has been a scientific procedure conducted since aeons because it is one of the building blocks of science, technology, and industry. The SI unit of temperature measurement is degree centigrade. Temperature measurement has a plethora of applications including domestic, industrial, manufacturing, and medical applications. Temperature measurement dates back to the 17th Century. In 170 AD, Claudius Galenus made a yardstick for measuring the temperature by mixing an equal proportion of ice and boiling water to formulate a neutral mixture to measure the temperature. In 1600s, Galileo constructed a device to measure the relative change in the temperature called the thermoscope. The modern-day thermometer received its first shape in the 18th century by Gabriel Fahrenheit by calibrating a mercury thermometer against a scale. After the rapid development of electronic instrumentation systems and as the requirement of temperature sensing spread to newer domains, the necessity of having electronic temperature sensors was felt consistently. This triggered the development of temperature sensors, such as thermocouples, thermistors, Resistance Temperature Detectors (RTDs), fiber optic sensors, and IR sensors.

A 'thermistor' is a portmanteau of 'thermal' and 'resistor', i.e., the resistance of a thermistor depends on its temperature. In this respect, a thermistor is generally of two types: Negative Temperature Coefficient (NTC) whose resistance decreases with an increase in the temperature and positive temperature coefficient whose resistance increases with an increase in the temperature. A main advantage of a NTC thermistor is that they are highly temperature sensitive, which can detect a very minute temperature change in comparison with a RTD or thermocouple. NTC thermistors are prepared from metal oxides through ceramic or film technology and used in any medium such as gas, liquid, or solid; however, in all cases, the medium should not be allowed to chemically, electrically, or physically contaminate.

A NTC thermistor has gained popularity among the researchers who study the linearity schemes of transducers because of its non-linear transfer characteristics (Brady et al. 1951, Kay et al. 1970). Linearization can be achieved by two schemes: the analog hardware-based scheme and the digital software cum hardware-based scheme. Obviously, the latter has an advantage over the former with respect to accuracy and efficiency and is constantly replacing the former (Rathore 2003, Khan et al. 1984, Sengupta et al. 1988). One of the soft computing techniques is Artificial Neural Network (ANN), which employs the numerical linearization of sensor characteristics.

This paper is divided into four main parts. The first part contains a design approach for thermistor sensing systems based on a 555 timer circuit working in the astable multivibrator mode. The second part includes the development of a data acquisition system employed for capturing the output of the timer circuit using LabVIEW. The third part describes an ANN (linearizer) for obtaining the temperature value from the pulse train parameters. The last part of the paper provides results, conclusions, and future research directions.

2 METHODOLOGY

Characteristic linearization of a NTC thermistor is possible by converting temperature into frequency or time intervals (Stankovic *et al.* 1977). This is achieved by using the thermistor as one of the timing resistors for an astable multivibrator employing a 555 timer IC. It is the most common of the IC chip that is used in complex electronic circuit systems. In the present system as shown in Figure 1, the frequency or time intervals of the output pulses from the 555 timer operating in the astable multivibrator mode depend and vary with R_1, R_{th}, and C as follows (Coughlin 2002):

$$f = 1.44/[(R_1+R_{th})/C] \qquad (1)$$

The duration of the LOW level and that of the HIGH level of the pulse train are indicated as follows:

$$T_{OFF} = 0.69R_{th}C \qquad (2)$$

$$T_{ON} = 0.69(R_1+R_{th})C \qquad (3)$$

Therefore, the total time period is given by

$$T = T_{OFF}+T_{ON} = 0.693(R_1+2R_{th})C \qquad (4)$$

where R1 = 1000 Ω; R_{th} is the variable thermistor resistance, and C = 470 μF.

An astable multivibrator is simply used as an oscillator. It generates a continuous rectangular 'LOW-HIGH' pulse varying between two voltage

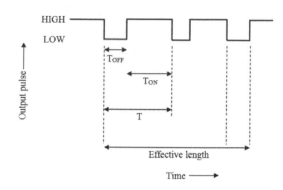

Figure 2. Output waveform of the 555 timer.

levels, as shown in Figure 2. Here, the resistance R_1 and the capacitance C are constant. So, the only variable parameter is the thermistor resistance R_{th}. As this resistance varies with change in the temperature, the timer will generate the output with different values of T_{OFF}, T_{ON}, and T.

3 DATA ACQUISITION SYSTEM USING LABVIEW

LabVIEW contains a comprehensive set of tools for acquiring, analyzing, displaying, and storing data. In the present scheme, pin no. 3 and no. 1 of the 555 timer are connected to pin no. 2 and ground pin no. 18 of the parallel port, respectively. So, the communication between the 555 timer and the parallel port is a single-bit communication. Data acquisition is done with the help of Lab-VIEW for the status record as a Text file (.txt) in the desired address. In the calibration phase, the values of T_{OFF}, T_{ON}, and T from the multivibrator output are recorded at known temperatures, which can be determined using a mercury-in-glass thermometer. A program code is written in MATLAB, which reads the stored file and finds the average T_{OFF}, T_{ON}, and T at the corresponding temperature.

As shown in Figure 2, it is evident that the starting of the pulse train may be either LOW or HIGH. It is also clear that the starting and ending pulses of the train cannot be considered for calculation, as it is not a complete pulse. The rest of the 'LOW-HIGH' sequences of the available pulses can be used for the calculation of average T_{OFF}, T_{ON}, and T. For this purpose, the MATLAB code calculates the LOW number of samples and the HIGH number of samples present in the effective length. The program also calculates the number of times the LOW and

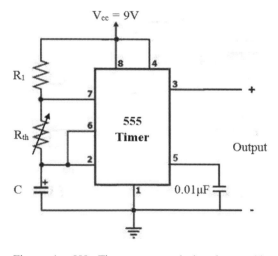

Figure 1. 555 Timer connected in the astable multivibrator mode.

HIGH states occurs. Therefore, the average T_{OFF} and the average T_{ON} can be calculated as

$$\text{Average } T_{OFF} = \frac{\text{Total number of LOW samples}}{\text{Total number of LOW states}} \times 1 \text{ ms} \quad (5)$$

$$\text{Average } T_{ON} = \frac{\text{Total number of HIGH samples}}{\text{Total number of HIGH states}} \times 1 \text{ ms} \quad (6)$$

While calculating the average T, the number of LOW states and the number of HIGH states in the effective length have to be the same. If not, then the extra one (LOW or HIGH) is eliminated for further calculations.

4 ANN-BASED LINEARIZATION SYSTEM

NTC thermistors have one of the most nonlinear transfer characteristics among all the commonly used sensors. The overwhelming use of the computer-based data acquisition system acts as a booster for software-based linearizers. However, the quest for achieving more and more accurate results similar to the actual values is still a benchmark problem in all research work (Aggoune *et al.* 1993, Pau *et al.* 1990).

With application of the ANN as a competent soft computing technique, it is quite natural that this would also be used for the numerical linearization of signal processing. Although some work has been reported in this field (Dey *et al.* 2010, Chatterjee *et al.* 2000, Mondal *et al.* 2009), there is still a wide scope for further advancement. This paper explores the feasibility of employing ANN-based linearization techniques for a single thermistor thermometer.

In the present work, after testing many combinations for the number of neurons per layer, a

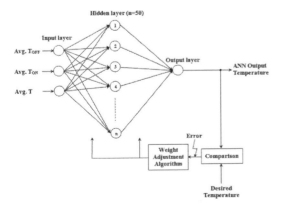

Figure 3. A 3-50-1 multilayer feedforward neural network.

multilayer neural network (3-50-1) was found to be the best combination, as shown in Figure 3. That is, it has three neurons at the input layer, 50 neurons at the hidden layer, and one neuron at the output layer. The three parameters, namely the average T_{OFF}, T_{ON}, and T, are used as network inputs, where the temperature is used as the network output.

A feedforward back propagation network performs the function of input–output nonlinear mapping. With a set of input data patterns, the network can be 'trained' to give the corresponding desired pattern at the output. It is a processing element in the brain's nervous system that receives and combines the input patterns (T_K) with the desired output (D) to find the weights (W_K) and the bias signal (b). The weights can be positive or negative. Mathematically, the output expression can be given as detailed below (Bose 2001).

This input–output supervised learning continues up to 500 iterations to measure the appropriate weights by a training algorithm called the Levenberg–Marquardt algorithm until the pattern matching occurs. At the end of the training, not only the network should be capable of recalling all the trained input–output patterns, but also for each new input pattern of the signal, there should be a corresponding ANN output pattern by altering these weights (Haykin 1994).

5 RESULTS AND DISCUSSION

An ANN requires supervised training by recorded data. This is similar to the training of a biological neural network from experience. At the beginning, the average T_{OFF}, T_{ON}, and T as the input data pattern and the corresponding temperature as the output data pattern are used for the training. Transfer functions (TFs) for hidden and output layers are TANSIG (tan-sigmoid TF) and PURELIN (linear TF), respectively. The derivative dF(S)/dS is maximum at S = 0, and gradually decreases with an increasing value of S in either direction (Charalambous *et al.* 1992, Masters *et al.* 1997).

With one input pattern, the output is calculated (defined as the forward pass) and compared with the desired output pattern. The weights are then altered in the backward direction (defined as the reverse pass) by a back propagation algorithm until the error between the calculated pattern and the desired pattern is very small and acceptable. A round-trip (forward and reverse passes) of the calculation is defined as an epoch (iteration). Similar training is repeated with all the input patterns, so that the mean square error (MSE) is minimum and acceptable, as shown in Figure 4.

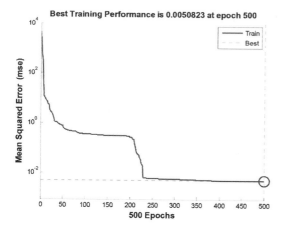

Figure 4.   MSE reduces with the increase in epochs.

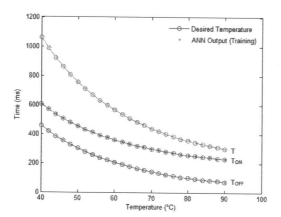

Figure 5.   Time vs. temperature curve (ANN training response).

Table 1.   Performance of an ANN-based linearizer at training.

Sl No.	ANN Output Temp (°C)	Desired Temp (°C)	Error	% Error
1	40.0321	40.0	0.0321	0.0803
2	41.9535	42.0	−0.0465	−0.1107
3	44.0254	44.0	0.0254	0.0577
4	45.9605	46.0	−0.0395	−0.0859
5	48.0025	48.0	0.0025	0.0052
6	50.0259	50.0	0.0259	0.0519
7	51.9978	52.0	−0.0022	−0.0043
8	54.0019	54.0	0.0019	0.0035
9	56.0003	56.0	0.0003	0.0005
10	58.0000	58.0	−0.0000	−0.0001
11	60.0031	60.0	0.0031	0.0052
12	62.0126	62.0	0.0126	0.0204
13	64.0476	64.0	0.0476	0.0744
14	66.0146	66.0	0.0146	0.0222
15	67.8977	68.0	−0.1023	−0.1505
16	70.0774	70.0	0.0774	0.1105
17	71.9466	72.0	−0.0534	−0.0742
18	73.8965	74.0	−0.1035	−0.1399
19	75.8879	76.0	−0.1121	−0.1475
20	78.1522	78.0	0.1522	0.1952
21	80.1518	80.0	0.1518	0.1897
22	82.0700	82.0	0.0700	0.0854
23	83.9089	84.0	−0.0911	−0.1084
24	85.9815	86.0	−0.0185	−0.0215
25	87.8757	88.0	−0.1243	−0.1412
26	90.0749	90.0	0.0749	0.0832
MSE	0.0051	0.0093		
Root MSE			0.0713	0.0962

In Figure 5 and Table 1, the network is trained and its output plotted against the desired targets.

$$D = F(S) = F\left[\sum_{k=1}^{N} T_K W_K + b\right] \qquad (7)$$

Once the neural network is trained properly, it is tested with intermediate data to verify that the training is efficient and capable of giving acceptable performance with unknown inputs, as shown in Figure 6 and Table 2.

From the results summarized in Tables 1 and 2, it can be found that the assessed measurement errors are substantially low, which is well within the acceptable range. Thus, the performance of the proposed thermistor-based scheme with respect to the temperature measurement is satisfactory.

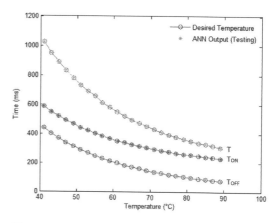

Figure 6.   Time vs. Temperature curve (ANN testing response).

Table 2. Performance of an ANN-based linearizer at testing.

Sl No.	ANN Output Temp (°C)	Desired Temp (°C)	Error	% Error
1	40.8949	41.0	−0.1051	−0.2564
2	43.0869	43.0	0.0869	0.2021
3	44.9669	45.0	−0.0331	−0.0735
4	47.0308	47.0	0.0308	0.0656
5	49.0994	49.0	0.0994	0.2029
6	51.0463	51.0	0.0463	0.0908
7	52.9621	53.0	−0.0379	−0.0716
8	55.1380	55.0	0.1380	0.2510
9	56.8566	57.0	−0.1434	−0.2516
10	58.9522	59.0	−0.0478	−0.0810
11	61.1563	61.0	0.1563	0.2562
12	63.1003	63.0	0.1003	0.1591
13	65.1051	65.0	0.1051	0.1617
14	66.8797	67.0	−0.1203	−0.1795
15	68.9656	69.0	−0.0344	−0.0498
16	71.0767	71.0	0.0767	0.1080
17	73.1417	73.0	0.1417	0.1942
18	74.8872	75.0	−0.1128	−0.1503
19	77.1298	77.0	0.1298	0.1686
20	79.0415	79.0	0.0415	0.0525
21	80.9663	81.0	−0.0337	−0.0416
22	83.2835	83.0	0.2835	0.3415
23	84.8581	85.0	−0.1419	−0.1670
24	86.7151	87.0	−0.2849	−0.3275
25	89.1383	89.0	0.1383	0.1554
MSE	0.0159	0.0335		
Root MSE			0.1259	0.1829

6 CONCLUSIONS

The aim of this study was to develop an ANN-based real-time temperature measurement technique using a thermistor as a temperature-sensing device in an astable multivibrator circuit. The output of the multivibrator was fed to a PC where the features of this signal were extracted. A pre-trained ANN was used to compute the unknown temperature from the above-mentioned features. The computed temperature was displayed on the user interface using LabVIEW. The results revealed that the developed scheme gives acceptable accuracy in determining the unknown temperature over the specified range. However, it is evident that the range of operation can be extended by increasing the training data set of the ANN.

ACKNOWLEDGMENTS

The authors would like to express their sincere acknowledgement to Prof. Sugata Munshi and Dr Debangshu Dey, Department of Electrical Engineering, Jadavpur University, Kolkata for their endless encouragement and valuable suggestions.

REFERENCES

Aggoune, M.E., Boudjema, F., Bensenouci, A., Hellal, A., Vadari, S.V., & El Mesai, M.R. (1993). Design of an adaptive-structure voltage regulator using artificial neural networks. *Proceedings of the 2nd IEEE Conference on Control Applications*, 337–343.

Bose, B.K. (2001). *Modern Power Electronics and AC Drives*. Prentice Hall PTR, Upper Saddle River, NJ.

Brady, A.P., Huff, H., & McBain, J.W. (1951). Measurement of Vapour Pressure By Means of Matched Thermistors. *Journal of Physical Colloid Chemistry*, 55, 304–311.

Charalambous, C. (1992). Conjugate gradient algorithm for efficient training of artificial neural networks. *IEE Proceedings (Circuits, Devices and Systems)*, 301–310.

Chatterjee, A., Munshi, S., & Rakshit, A. (2000). ANN based linearising technique for constant temperature anemometer system. *Journal of the Instrument Society of India*, 109–114.

Coughlin, R.F., & Driscoll, F.F. (2002). *Operational Amplifiers and Linear Integrated Circuits*. Pearson Education, Inc., NJ.

Dey, D., & Munshi, S. (2010). Simulation studies on a new intelligent scheme for relative humidity and temperature measurement using thermistors in 555 timer circuit. *International Journal on Smart Sensing and Intelligent Systems*, 3(2), 217–229.

Haykin, S. (1994). Neural Networks, A Comprehensive Foundation, Macmillan.

Kay, B.D., & Low, P.F. (1970). Measurement of the Total Suction of Soils by a Thermistor Psychrometer. *Soil Sciences Society of America, Proceedings*, 34, 373–376.

Khan, A.A., & Sengupta, R. (1984). A linear temperature/voltage convertor using thermistor in log-network. *IEEE Transactions on Instrumentation. and Measurement*. 2–4.

Masters, T., & Land, W. (1997). A new training algorithm for the general regression neural network. *IEEE International Conference on Systems, Man., and Cybernetics, Computational Cybernetics and Simulation*, 3, 1990–1994.

Mondal, N., Abudhahir, A., Jana, S.K., Munshi, S., & Bhattacharya, D.P. (2009). A Log Amplifier Based Linearization Scheme for Thermocouples. *Sensors & Transducers Journal*, 100(1), 1–10.

Pau, L.F., & Johansen, F.S. (1990). Neural Network Signal Understanding for Instrumentation. *IEEE Transactions on Instrumentation & Measurement*, 39(4), 558–564.

Rathore, T.S. (2003). *Digital Measurement Techniques*. Narosa Publishing House, New Delhi.

Sengupta, R.N. (1988). A widely linear temperature-to-frequency converter using a thermistor in a pulse generator. *IEEE Transactions on Instrumentation. and Measurement.*, 62–65.

Stankovic, D., & Elazar, J. (1977). Thermistor multivibrator as the temperature-to-frequency converter and as a bridge for temperature measurement. *IEEE Transaction on Instrumentation and Measurement*, 1M-26(1), 41–46.

Valence Band Anticrossing model for InAs$_{1-x}$Bi$_x$ by k·p method

D.P. Samajdar
Department of ECE, Heritage Institute of Technology, Anandapur, Kolkata, India

T.D. Das
Department of Electronic Science, National Institute of Technology, Yupia, Arunachal Pradesh, India

S. Dhar
Department of Electronic Science, University of Calcutta, Kolkata, India

ABSTRACT: The band gap reduction and the increase in spin-orbit splitting energy in InAs$_{1-x}$Bi$_x$ is explained by the Valence Band Anticrossing (VBAC) model. The valence band structure is modified due to the interaction of the Bi related impurity levels with the extended states of the valence band of the host semiconductor InAs. The VBAC band gap reduction in InAs$_{1-x}$Bi$_x$ is found out to be 42.8 meV/at%. The spin-orbit splitting energy in InAsBi increases by 18.2 meV/at%Bi which is responsible for the suppression of Auger recombination processes making it a potential candidate for optoelectronic applications in the 3–5 µm regime.

1 INTRODUCTION

III-V-Bi alloy systems have been a source of interest for the researchers worldwide for potential near- and mid-infrared optoelectronic applications. A recent Springer Series volume [Li *et al.* (Eds.) 2013] reviewed the surprising yet interesting properties of this novel class of materials. The most prominent property of III-V-bismides is the substantial reduction in band gap caused by the incorporation of a dilute concentration of Bi. This property makes the bismides similar to comparatively more studied alloy systems-dilute III-V-nitrides [Henini *et al.* (Ed.) 2005]. In dilute nitrides, the interaction of the N related resonant state with the conduction band of the host semiconductor causes the reduction in band gap. Unlike the dilute nitrides, the band gap reduction in bismides occurs due to the anticrossing interaction of the host valence band matrix with the Bi related impurity levels formed close to its valence band edge due to Bi incorporation into the III-V alloys [Alberi *et al.* 2007]. These impurity states interact with the valence band states, i.e. the Light Hole (LH) and Heavy Hole (HH) bands, resulting in an reduction of the material bandgap [Alberi *et al.* 2007] and the interaction with the spin-orbit Split-Off (SO) band causes the increase in spin-orbit splitting energy [Okamoto *et al.* 1998]. This results in the formation of $\Delta_{SO} > E_g$ regime which helps to reduce or even suppress Auger recombination and Inter Valence Band Absorption (IVBA) [Li *et al.* (Eds.) 2013] loss

processes that degrade the efficiency of LASERs and photodetectors. Degradation of electron mobility in bismides is also reported to be less than nitrides since the Bi incorporation results in least perturbation of the conduction band in the bismides [Kini *et al.* 2009, Cooke *et al.* 2006].

There are a number of reports on the growth and characterization of InAs$_{1-x}$Bi$_x$. InAs$_{1-x}$Bi$_x$ is a narrow band gap semiconductor and it is investigated for applications in the 3–5 µm spectral region [Okamoto *et al.* 1998]. In the recent years, InAs has become a source of attraction for the researchers for making possible the fabrication of optoelectronic devices with single carrier multiplication, low excess noise [Marshall *et al.* 2008, 2011] and low dark currents [Maddox *et al.* (2012), Ker *et al.* (2011) and Sandall *et al.* 2013]. However, it is not suitable for Mid Wave Infrared (MWIR) applications due to its cut-off wavelength of 3.5 µm. This necessitates the incorporation of Bi into InAs. Okamoto *et al.* (1999) reported a band gap reduction of 42 meV/%Bi in InAsBi layers, grown by MOVPE. A band gap reduction of 58 meV/%Bi is obtained from PL measurements for thick InAsBi layers (Svensson *et al.* (2012)). Ma *et al.* [1990] reported a band gap decrease of 55 meV/% Bi in InAsBi using PL measurements. In this paper, we have calculated the details of the valence band structure of InAsBi using a 12 × 12 Hamiltonian. The VBAC calculated values of bandgap energy are compared with the Virtual Crystal Approximation (VCA) calculated values. The spin-orbit coupling energy is calculated to find out the Bi at% required to obtain the $\Delta_{SO} > E_g$ regime.

2 MATHEMATICAL MODELLING

The incorporation of Bi atoms into InAs leads to the formation of localized defect states near the valence band edge of the host semiconductor. These Bi related impurity states interact with the extended states of the host resulting in the restructuring of the valence band by splitting the LH, HH and SO bands into a series of E_+ and E_- sub bands which are reduces the band gap in one hand and enhances spin-orbit splitting energy on the other. The valence band structure of InAsBi in the <100> direction can be best described by a 12×12 Hamiltonian given below [Alberi et al. 2007]:

$$H_v = \begin{pmatrix}
H & \alpha & \beta & 0 & 0 & -i\sqrt{2}\beta & V(x) & 0 & 0 & 0 & 0 & 0 \\
0 & L & 0 & \beta & \frac{iD}{\sqrt{2}} & 0 & 0 & V(x) & 0 & 0 & 0 & 0 \\
\beta & 0 & L & 0 & 0 & \frac{iD}{\sqrt{2}} & 0 & 0 & V(x) & 0 & 0 & 0 \\
0 & \beta & 0 & H & -i\sqrt{2}\beta & 0 & 0 & 0 & 0 & V(x) & 0 & 0 \\
0 & \frac{-iD}{\sqrt{2}} & 0 & i\sqrt{2}\beta & S & 0 & 0 & 0 & 0 & 0 & V(x) & 0 \\
i\sqrt{2}\beta & 0 & \frac{-iD}{\sqrt{2}} & 0 & 0 & S & 0 & 0 & 0 & 0 & 0 & V(x) \\
V(x) & 0 & 0 & 0 & 0 & 0 & E_{Bi} & 0 & 0 & 0 & 0 & 0 \\
0 & V(x) & 0 & 0 & 0 & 0 & 0 & E_{Bi} & 0 & 0 & 0 & 0 \\
0 & 0 & V(x) & 0 & 0 & 0 & 0 & 0 & E_{Bi} & 0 & 0 & 0 \\
0 & 0 & 0 & V(x) & 0 & 0 & 0 & 0 & 0 & E_{Bi} & 0 & 0 \\
0 & 0 & 0 & 0 & V(x) & 0 & 0 & 0 & 0 & 0 & E_{Bi-so} & 0 \\
0 & 0 & 0 & 0 & 0 & V(x) & 0 & 0 & 0 & 0 & 0 & E_{Bi-so}
\end{pmatrix}$$

At the Γ point, where $\mathbf{k} = 0$, the solution of the Hamiltonian yields four distinct eigenvalues corresponding to the E_+ and E_- sub bands of the LH/HH and SO energy levels. The solutions of the Hamiltonian for $\mathbf{k} = 0$ are given by the following set of equations [Samajdar et al. 2014],

$$E_{LH/HH\pm} = \frac{1}{2}\left(H + E_{Bi} \pm \sqrt{H^2 - 2HE_{Bi} + E_{Bi}^2 + 4V^2} \right) \quad (1)$$

$$E_{SO\pm} = \frac{1}{2}\left(S + E_{Bi-so} \pm \sqrt{S^2 - 2SE_{Bi-so} + E_{Bi-so}^2 + 4V^2} \right)$$

$$(2)$$

where $H = L = \Delta E_{VBM}\, x$, $S = \Delta E_{VBM}\, x - \Delta_0 \Delta E_{SO}$ x and $V = C_{Bi}\sqrt{x}$. ΔE_{VBM} and ΔE_{SO} respectively denote the valence band and spin-orbit split-off band offset between the end point compounds. Δ_0 is the spin orbit splitting energy InAs. V takes into account the coupling between the host valence band and the Bi-related impurity levels with C_{Bi} used as a fitting parameter [Jefferson et al. 2006]. E_{Bi} and E_{Bi-so} are the Bi related impurity levels corresponding to the LH/HH and SO bands.

3 RESULTS AND DISCUSSIONS

The value of the spin orbit splitting energy for InAs and InBi is reported as 0.39 eV [Vurgaftman et al. 2001] and 2.2 eV [Carrier et al. 2004] respectively. The direct E_0 energy gap for InBi is calculated as -1.62 eV [Barnett et al. 2014]. The values of the Valence Band Offset (VBO), Conduction Band Offset (CBO) and spin orbit split-off band offset between InAs and InBi are calculated as 0.94 eV, -1.03 eV and -0.87 eV respectively [Samajdar et al. (2016)]. Valence Band Offset (VBO) for InBi is obtained by extrapolating the graph between VBO and lattice constant for In containing binary compounds using Reference [Samajdar et al. 2014]. The location of the heavy/light hole levels for Bi E_{Bi} and corresponding spin-orbit split-off level E_{Bi-so} is found to be 0.4 eV and 1.9 eV respectively below the valence band maximum (VBM) of InAs. The value of the fitting parameter C_{Bi} is calculated to be 1.00 eV [Samajdar et al. 2016]. The coupling parameter C_{Bi} determines the magnitude of shift of the E_+ and E_- levels of the HH,LH and SO sub bands as is evident from equations (1) and (2) [Samajdar et al. 2015].

Figure 1 shows the restructuring of the valence band structure of InAs due to the incorporation of bismuth. The E_+ and E_- energy levels corresponding to the LH, HH and SO sub bands are obtained by solving the 12×12 Hamiltonian discussed above.

Figure 2 shows the band structure of $InAs_{1-x}Bi_x$ as a function of Bi mole fraction, calculated at Γ point. The positions of the E_+ and E_- related energy levels are calculated using equations (1) and (2). The E_+ and E_- levels corresponding to the HH/LH and SO energy bands repel each other. The HH/LH E_+ band moves upward by about 32.5 meV/at% Bi

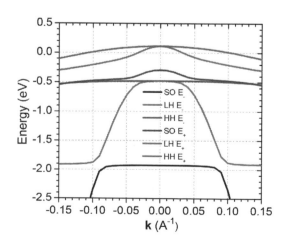

Figure 1. Valence Band Structure for $InAs_{0.96}Bi_{0.04}$ in the <100> direction.

which is primarily responsible for the most part of energy gap reduction in III-V bismides. The lowering of the conduction band minimum for $InAs_{-x}Bi_x$ is obtained from the VCA calculations, E_{CB-VCA} can be written as [Alberi *et al.* 2007],

$$E_{CB-VCA} = E_g - \Delta E_{CBM}x \qquad (3)$$

where E_g is the band gap energy of the InAs and ΔE_{CBM} is its conduction band offset. For InAs, $Eg = 0.35$ eV and $\Delta E_{CBM} = 1.03$ eV at T = 300 K. It has been calculated from the plot that for a Bi mole fraction of 0.5 at%, Δ_{SO} becomes greater than E_g which suppresses the Auger recombination loss processes in InAsBi based optoelectronic devices.

Figure 3 shows a comparison of the VCA and VBAC calculated values of bandgap for InAsBi.

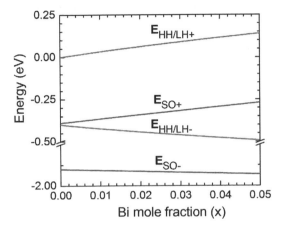

Figure 2. Relative positions of the E_+ and E_- energy levels corresponding to the HH/LH and SO sub bands for InAsBi as a function of Bi mole fraction.

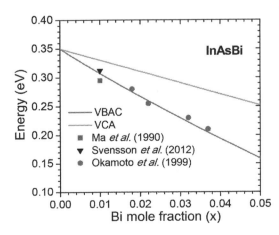

Figure 3. Comparison of band gap values for different Bi mole fractions using VCA and VBAC Models.

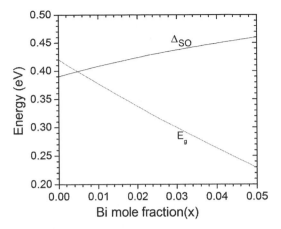

Figure 4. Plot of VBAC band gap and spin-orbit splitting energy versus Bi mole fraction for InAsBi.

Indeed, the plot shows a fair agreement of the VBAC calculated values and the experimental data. However, the VBAC values deviate slightly from the VCA calculated values indicating the existence of mismatch between InAsBi and InAs due to Bi incorporation.

Figure 4 shows the variation of VBAC calculated band gap and spin-orbit splitting energy as a function of Bi mole fraction. It can be seen that for a Bi concentration of about 0.05 at%, spin-orbit splitting energy becomes greater than band gap energy which results in the formation of $\Delta_{SO}>E_g$ regime.

4 CONCLUSIONS

VBAC model has been used to draw the band structure of $InAs_{1-x}Bi_x$ for x = 0.04 and the reduction in band gap occurring due to the upward movement of the HH/LH E_+ energy level. VBAC model predicts the upward movement of the HH/LH E_+ level by about 32.5 meV/at%Bi and downward movement of the conduction band minimum by 10.5 meV/at%Bi resulting in a band gap reduction of 42.8 meV. at%Bi. The upward shift in the SO E_+ energy level by about 18.2 meV/at%Bi results in $\Delta_{SO}>E_g$ regime at Bi mole fraction of 0.005 which suppresses the Auger and IVBA recombination processes.

REFERENCES

Alberi, K. *et al*, (2007). Valence-band anticrossing in mismatched III-V semiconductor alloys. *Phys. Rev. B* 75, 045203.

Alberi, K., Dubon, O.D., Walukiez, W., Yu, K.M., Bertulis, K. & Kroktus, A. (2007). Valence band anticrossing in GaBi$_x$As$_{1-x}$. *Appl. Phys. Lett.* 91, 051909.

Barnett, S.A., (1987). Direct E. 0 energy gaps of bismuth-containing III-V alloys predicted using quantum dielectric theory. *J. Vac. Sci. Technol. A*, 5 2845.

Carrier, P. & Wei, S.H., (2004). Calculated spin-orbit splitting of all diamondlike and zinc-blende semiconductors: Effects of $p_{1/2}$ local orbitals and chemical trends. Phys. Rev. B 70, 035212.

Cooke, D.G., Hegmann, F.A, Young, E.C, Tiedje, T., (2006). Electron mobility in dilute GaAs bismide and nitride alloys measured by time-resolved terahertz spectroscopy. *Appl. Phys. Lett.* 89, 122103.

Henini M. (Ed.), *Dilute Nitride Semiconductors*, Elsevier, Oxford, 2005.

Ker, P.J., Marshall, A., Krysa, A., David, J.P.R & Tan, C.H. (2011). Temperature Dependence of Leakage Current in InAs Avalanche Photodiodes. *IEEE J. Quantum Electron.* 47, 1123.

Kini, R.N., Bhusal, L., Ptak, A.J., France, R. & Mascarenhas, A., (2009). Electron Hall mobility in GaAsBi. *J. Appl. Phys.* 106, 043705.

Li H. & Wang Z.M. (Eds.), Bismuth-containing compounds, Springer series in Materials Science, vol. 186, Springer Science + Business Media, New York, 2013.

Ma, K.Y., Fang, Z.M., Cohen, R.M & Stringfellow, G.B., (1990). Organometallic vapor-phase epitaxy growth and characterization of Bi-containing III/V alloys. *J. Appl. Phys.* 68, 4586.

Maddox, S.J., *et al.* (2012). Enhanced low-noise gain from InAs avalanche photodiodes with reduced dark current and background doping. *Appl. Phys. Lett.* 101, 151124.

Marshall, A.R.J, Tan, C.H, Steer, M.J. & David, J.P.R., (2008). Electron dominated impact ionization and avalanche gain characteristics in InAs photodiodes. *Appl. Phys. Lett.* 93, 111107.

Marshall, A.R.J, Vines, P., Ker, P.J, David, J.P.R & Tan, C.H., (2011). Avalanche Multiplication and Excess Noise in InAs Electron Avalanche Photodiodes at 77 K. *IEEE J. Quantum Electron* 47, 858.

Okamoto, H. & Oe, K., (1998). Growth of Metastable Alloy InAsBi by Low-Pressure MOVPE. *Jpn. J. Appl. Phys.* 37, 1608.

Okamoto, H., & Oe, K., (1999). Structural and Energy-Gap Characterization of Metalorganic-Vapor-Phase-Epitaxy-Grown InAsBi. *Jpn. J. Appl. Phys.* 38, 1022.

Samajdar, D.P. & Dhar, S., (2014).Valence Band Structure of $InAs_{1-x}Bi_x$ and $InSb_{1-x}Bix$ Alloy Semiconductors Calculated Using Valence Band Anticrossing Model. *The Sci. World. J.* 2014, 704830.

Samajdar, D.P., Das, T.D. & Dhar, S., (2015). Valence Band Anticrossing model for $GaSb_{1-x}Bi_x$ and $GaP_{1-x}Bi_x$ using k.p Method. *Mater. Sci. Semicond. Proc.* 40, 539.

Samajdar, D.P., Das, T.D. & Dhar, S., (2016). Calculation of Valence Band Structure and Band Dispersion in Indium containing III-V Bismides by k·p method. *Comp. Mater. Sci.* 111, 497.

Sandall, I., Zhang, S., & Tan, C.H., (2013). Linear Array of InAs APDs operating at 2 μm. *Opt. Express* 21, 25780.

Svensson, S.P, Hier, H., Sarney, W.L, Donetsky, D., Wang, D. & Belenky, G., (2012). Molecular beam epitaxy control and photoluminescence properties of InAsBi. *J. Vac. Sci. Technol. B* 30, 02B109.

Vurgaftman, I., Meyer, J.R., & Ram-Mohan, L.R., (2001). Band parameters for III–V compound semiconductors and their alloys *J. Appl. Phys,* 89, 5815.

Frontiers in Computer, Communication and Electrical Engineering – Acharyya (Ed.)
© 2016 Taylor & Francis Group, London, ISBN: 978-1-138-02877-7

Assessment of different transient conditions in a radial feeder by THD- and DWT-based Skewness analysis

A. Maji & A. Chattopadhyaya
S.K.F.G.I., West Bengal, India

B. Das
Government College of Engineering and Textile Technology, West Bengal, India

S. Chattopadhyay
G.K.C.I.E.T., West Bengal, India

ABSTRACT: This study deals with the assessment of transient conditions in a radial feeder system. At first, different transient conditions, including capacitor switching transient condition in a radial feeder, were simulated in a MATLAB environment; then R-phase current signal at the load bus was captured and different analyses were performed to assess the transient conditions in that feeder. Total Harmonic Distortion (THD) was calculated on the captured current signals in different conditions and then Discrete Wavelet Transform (DWT) was performed on those captured signals where significant differences in results were observed which clearly suggest the transient conditions of radial feeder. After DWT decomposition, Skewness values for DWT detail and approximate level coefficients have been calculated to assess the transient conditions of the radial feeder. Using these techniques the results which have come out are very much optimistic for proper discrimination of transient conditions in a radial feeder.

1 INTRODUCTION

In recent years, capacitor placement in radial feeder has become very important to reduce line losses and improve the power factor. Energizing a capacitor bank into the feeder is a bit tricky as it introduces a large inrush current and transient voltage. As the inrush current is much higher than the normal current, discriminating them from faults is also necessary as capacitor switching can mal-operate the relays. So it has drawn separate attention and a lot of research works are being done.

Some studies done earlier showed the necessity of capacitor bank and its direct effects on bus voltages and power factor. A proposal on selecting different capacitor banks for different commercial or industrial loads and their effects on the harmonics level had been analyzed. The transient effects on voltages and inrush currents due to capacitor switching were proposed by ShehabAbdulwadood Ali et al. The same was shown in various studies and further researches showed how an inductor can limit the transient effect. Later on an analysis showed the effects on Total Harmonic Distortion (THD) due to capacitor switching using Fast Fourier Transform (FFT). Further investigation showed the effects of transients on radial feeder using MATLAB/Simulink

software. A new approach of Discrete Wavelet Transform (DWT) for analyzing the transient was proposed by M. A. Beg, Dr. M. K. Khedkar, and Dr. G. M. Dhole. Later they made some modifications over the previous one and solved the same problem using DWT-PAC method, showing how easily fault current and switching current can be discriminated with their distinguishing nature. A study had been done by Sayed A. Ward hesham Said et al. and Mahmoud N. Ali estimating the nature of transient overvoltage, and a model was developed using an Artificial Neural Network (ANN) tool to estimate the different types of over—voltage switching. Various investigations were done on discrimination and protection of feeder to enhance reliability, and different techniques were proposed, one such useful method that discriminates the fault current from switching current is by using the Phase Locked Loop (PLL) method. Some methods to limit this current have also been proposed like installing reactor, zero-voltage closing control, three DC reactors type etc. Some other techniques proposed by different researches are on Clarke plane, Park plane, symmetrical components, Skewness, and Kurtosis based Discrete Wavelet Transform to analyze signals for detection and classification of different faults in electrical systems.

This paper emphasizes on the analysis of transient currents due to capacitor switching and different types of faults using different techniques like THD-and DWT-based Skewness analysis. A simple radial feeder is chosen for this purpose and from the analysis proper discrimination of transient conditions has been achieved for proper protection of the radial feeder.

2 MATLAB MODELING

Total analysis has been done in MATLAB. The above circuit of Figure 1 was simulated in MATLAB simulink environment. Here generation voltage is considered as 33 kV and transmission line voltage is considered as 132 kV. A 100-km transmission line and five buses are considered for this network. Load is connected at bus 5, and capacitor bank is added at bus 4. Using the above circuit, four different conditions have been created: (i) Normal condition (ii) Capacitor switching transient condition (iii) Fault condition (three phase, L-L, L-G fault) (iv) Capacitor switching and L-G fault condition. All the faults are created in between bus 4 and bus 5 and in all the cases, current and voltage is considered at bus 4 for assessment of the transient conditions in that radial feeder.

3 PROPOSED METHODS FOR ANALYSIS

In this paper to assess the transient conditions three different methods have been used, which are (i) THD-based technique (ii) DWT-based technique (iii) DWT-based Skewness analysis.

3.1 *Total Harmonic Distortion (THD)*

In this paper, THD is used as a method to assess the transient conditions in the above mentioned radial feeder for proper protection purpose. Total harmonic distortion is the degree to which a waveform deviates from its pure sinusoidal values due to summation of harmonics present in a signal. THD can be used in voltage and as well as current signal to measure the harmonic distortion present in that particular signal and it can be defined as the ratio of the square root

of the sum of the currents of all harmonic components to the current of fundamental frequency. It is also used for linearizing the power quality in electric power systems. To calculate the THD of bus 4 current, we consider up to the seventh harmonics.

$$THD = \frac{\sqrt{(I_2^2 + I_3^2 + I_4^2 + \ldots\ldots I_n^2)}}{I_1} X \ 100 \qquad (1)$$

where, n = nth harmonics

3.2 *DWT based technique*

Using Wavelet Transform (WT), better time frequency representation can be achieved from non-stationery signals which were the shortcomings of other signal processing techniques like, FFT, STFT etc. Continuous Wavelet Transform (CWT) and Discrete Wavelet Transform (DWT) are the two classifications of WT. Due to some computational difficulties of CWT DWT is considered as a technique in this work to assess the transient conditions for protecting a radial feeder. In DWT, signals pass through high pass and low pass filter banks in each decomposition level to get different frequency bands present in the signal. In this paper, nine levels of DWT decomposition and 'db4' mother wavelet is considered for DWT analysis.

3.3 *DWT-based Skewness analysis*

Skewness is a statistical parameter. Skewness can be mathematically defined as the average cubed deviation from the mean divided by the standard deviation cubed. If the result of the computation is greater than zero, the distribution is positively skewed. If it's less than zero, it's negatively skewed, and equal to zero means it's symmetric. For univariate data Y_1, Y_2, ..., Y_N, the formula for Skewness is:

$$Skewness = \frac{\sum_{i=1}^{N}(Y_i - \bar{Y})^3 / N}{S^3} \qquad (2)$$

where $\bar{Y}$ is the mean, S is the standard deviation, and N is the number of data points. In this work, the signal is first decomposed by 'db4'-based DWT decomposition. Then the Skewness value was calculated for detail along with approximate (approximation) level coefficients up to nine levels of DWT decomposition.

4 RESULTS

The above circuit of Figure 1 was simulated in MATLAB, and three phase voltages and currents

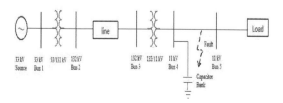

Figure 1. Simulation model of radial feeder system.

at bus 4 due to the addition of capacitor banks are depicted in Figures 2 and 3 respectively. Switching transient occurred due to the sudden addition of capacitor bank which is clearly observed in bus 4 voltages and currents. Transient inrush current is observed in three phases due to switching transient but for assessment of transient inrush current, R-phase current at bus 4 is considered for all the cases. Figure 4 is used to show the transient inrush current due to capacitor bank switching.

4.1 Case 1: Results of THD analysis

Using the MATLAB model, the abovementioned four different conditions simulated in MATLAB and all the cases of R-phase current at bus 4 was captured at the proper sampling frequency and then THD was calculated. THD results are mentioned in Table 1. There is a clear difference of THD values in the four different conditions; amongst them, THD value is maximum in capacitor switching condition.

4.2 Case 2: Results of DWT analysis

For assessment of the transient condition, DWT has been done on R-phase current in all the

conditions. In DWT, up to level three is considered in all the cases. Figures 5–9 are used to depict the DWT results of bus 4 current in all the above mentioned conditions.

4.3 Case 3 Results of DWT-based skewness analysis

R-phase signal is decomposed by 'db4' mother wavelet-based DWT decomposition (up to level nine of decomposition) and each level for detail and approximation coefficients Skewness values has been calculated which are depicted in Figures 11 and 12.

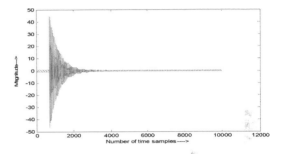

Figure 3. Three phase currents (at bus 4) due to capacitor bank switching.

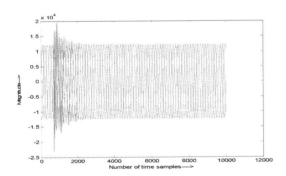

Figure 2. Three phase voltages (at bus 4) due to capacitor bank switching.

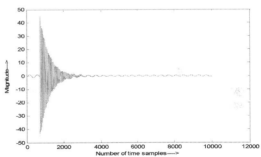

Figure 4. R phase current (at bus 4) due to capacitor bank switching.

Table 1. Calculation of THD in different conditions.

Sl no	Different Conditions	1st harmonic (%)	2nd harmonic (%)	3rd harmonic (%)	4th harmonic (%)	5th harmonic (%)	6th harmonic (%)	7th harmonic (%)	THD in percentage (%)
1	Normal	100	0	0	0	0	0	0	0
2	Capacitor switching	100	14.62	14.42	10.17	8.11	10.54	11.31	28.75
3a	3 Phase fault	100	6.06	4.83	2.60	1.62	1.71	1.60	9.05
3b	L-L fault	100	4.55	3.51	1.49	1.85	1.43	0.93	7.14
3c	L-G fault	100	6.15	4.89	2.58	1.62	1.72	1.60	9.05
4	Switching & fault	100	1.03	0.12	0.11	0.13	0.09	0.03	6

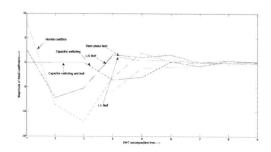

Figure 5. DWT of R-phase current in normal condition.

Figure 9. DWT of R-phase current in L–G condition.

Figure 6. DWT of R-phase current in capacitor switching transient condition.

Figure 10. DWT of R-phase current in capacitor switching transient and L–G fault condition.

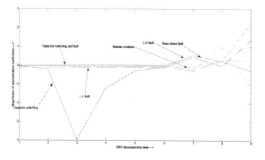

Figure 7. DWT of R-phase current in three-phase fault condition.

Figure 11. Skewness values of detail coefficients with respect to DWT decomposition level.

Figure 8. DWT of R-phase current in L-L fault condition.

Figure 12. Skewness values of approximation coefficients with respect to DWT decomposition level.

5 OBSERVATIONS

Figures 2 and 3 are used to depict the bus voltage and currents respectively after the capacitor bank switching, and Figure 4 is used to show R-phase current at bus 4 due to capacitor switching. The maximum value of R-phase inrush current due to capacitor bank switching is almost 50 amp which is shown in Figure 4.

THD of R-phase current at bus 4 in different conditions have been recorded in Table 1, where THD is maximum in capacitor switching condition and THD is minimum (0.0) in normal condition. In fault conditions, THD values are in between 7 and 9. THD value is 6 (in percentage) when capacitor switching and fault, both the transient conditions, were present in the system. Considering the THD values in Table 1, all the conditions can be classified properly.

To asses and classify the transient inrush current at bus 4 R-phase current, 'db4'-based DWT analysis has been done for R-phase current at different conditions which are depicted in Figures 5–10, where maximum changes have been observed in level coefficients. DWT decomposition is considered up to level nine (9), though DWT result is shown up to level three (3) in all these figures. Using all these DWT results, different conditions of the radial feeder can be assessed and classified which will be helpful for proper protection of radial feeder.

Figures 11 and 12 are used to depict the Skewness values for detail and approximate level coefficients of DWT decomposition at different conditions. In both the figures maximum difference is observed at DWT decomposition level 2 from where inrush current at bus 4 due to capacitor switching can be easily discriminated from other conditions properly.

6 CONCLUSION

In this paper, assessment of transient conditions in a radial feeder has been assessed by THD—and DWT-based Skewness analysis techniques. Assessment was done by capturing the R-phase current at bus 4 of the abovementioned radial feeder. THD has been calculated in normal condition and other different conditions; then DWT was done on the captured R-phase current at bus 4 on those conditions and comparing the results, transient conditions of radial feeder have been detected and discriminated. Further Skewness values of DWT detail and approximate level coefficients were calculated from where transient conditions of that radial feeder were detected and assessed which will be helpful for proper protection of radial feeder.

REFERENCES

Abdelsalam. H.A. Hamz, Eimanali al-jazzaf, Mohamed and a. H. A badr, 2013. "*Inrush current and Total Harmonic Distortion Transient of Power Transformer with Switching Capacitor Bank,*" Engineering Science and Technology: An International Journal (ESTIJ), ISSN: 2250-3498, Vol.3, No.3, June 2013.

Ananwattanaporn, S., Patcharoen, T. Yoomak, S. Ngaopitakkul, A. and Pothisarn, C. 2014. "*Characteristics and Behavior of Voltage Transient and Inrush Current due to Switching Capacitor Bank,*" Proceedings of the 2nd International Conference on Intelligent Systems and Image Processing 2014.

Beg, M.A., Dr. Khedkar M.K. and Dr. Dhole, G.M. 2012. "*Discrimination of Capacitor Switching Transients Using Wavelet,*" pratibha: international journal of science, spirituality, business and technology (ijssbt), vol. 1, no.1, march 2012 issn (Print) 2277–7261.

Beg, M.A., Khedkar, M.K. Paraskar S.R. and Dhole, G.M. 2011. "*Classification of fault originated transients in high voltage network using DWT–PCA approach,*" International Journal of Engineering, Science and Technology Vol. 3, No. 11, 2011, pp. 1–14.

Chattopadhyay, S., Chattopadhyaya, A. and Sengupta, S. 2013. "*Analysis of Stator Current of Induction Motor Used in Transport System at Single Phasing by Measuring Phase Angle, Symmetrical Components, Skewness, Kurtosis and Harmonic Distortion in Park plane*", IET Electrical System in Transportation, September, 2013, pp. 1–8.

Chattopadhyaya, A., Chattopadhyay S. and Sengupta, S. 2013. "*Stator Current Assessment of an Induction Motor at Crawling in Clarke Plane,*" International Journal of Electronics & Communication Technology, Volume 4, ISSUE-SPL1, ISSN: 2230-7109 (Online), 2230–9543 (Print), Jan.–Mar., 2013.

Chattopadhyaya, S., Chattopadhyaya, A. and Sengupta, S. 2014. "*Measurement of Harmonic Distortion and Skewness of Stator Current of Induction Motor at Crawling in Clarke plane*", IET Science, Measurement and Technology, March, 2014, pp. 1–9.

Chattopadhyaya, S., Chattopadhyaya, A. and Sengupta, S. 2014. "*Harmonic Power Distortion Measurement in Park Plane*", ELSEVIER, Measurement 51, February, 2014, pp. 197–205.

Chattopadhyaya, A., Chattopadhyay, S., Mitra, M. and Sengupta, S. 2012. "*Wavelet Analysis for Assessment of Crawling of an Induction Motor in Clarke Plane*" International Journal of Electrical, Electronics and Computer Engineering, (IJEECE ISSN: ONLINE 2277 2626), Special Edition for Best Papers of Michael Faraday IET India Summit-2012, MFIIS-12, November, 2012, pp. 71–76.

Chattopadhyaya, A., Ghosh, A., Chattopadhyay, S. and Sengupta, S. 2013. "*Stator Current Harmonic Assessment of Induction Motor for Fault Diagnosis*" International Journal of Electronics & Communication Technology, Volume 4, ISSUE-SPL1, ISSN: 2230-7109 (Online), 2230–9543 (Print), Jan–Mar, 2013.

NIST, 2012 "*Engineering Statistics Handbook*", NIST/SEMATECH e-Handbook of Statistical Methods, Retrieved 18th March 2012.

Omar, A.S., Badr, M.A.L. Abdel-Hamid, W.H. and Abdin, A.M. 2014. *"Selection of Industrial Capacitor Banks for Power Factor Correction in Industrial Load Application,"* International Journal on Power Engineering and Energy (IJPEE) Vol. (5) – No. (3) July 2014.

OodoOgidiStephen, Liu Yanli and Sun Hui,2011. *"Application of Switched Capacitor banks for Power Factor Improvement and Harmonics Reduction on the Nigerian Distribution Electric Network,"* International Journal of Electrical & Computer Sciences IJECS-IJENS Vol: 11 No: 06 Dec' 2011.

Rajkumar Nagar, Yadav, D.K. 2015. *"Analysis of switching transient and fault detection of strong power system using PLL,"* International Journal of Engineering Development and Research Volume 3, Issue 2, 2015, ISSN: 2321-9939.

Sayed A. Ward, Hesham Said and Mahmoud N. Ali, *"Estimation of capacitor bank switching overvoltages using artificial neural network,"* Journal of Electrical Engineering.

Shehab Abdulwadood Ali, 2011 *"Capacitor Banks Switching Transients in Power Systems,"* Energy Science and Technology Vol. 2, No. 2, 2011, pp. 62–73.

Surendar J., and Dr. Loganathan, N. 2014. *"Development and Analysis of Radial Feeder Protection for Capacitor Switching Transients,"* International Conference on Engineering Technology and Science-(ICETS'14) Volume 3, Special Issue 1, February 2014.

Teenu Jose, Divya K, Marymol Paul and Ann Sonia M, 2015. *"A Method for Reducing Three-Phase Power Capacitor Switching Transients,"* International Journal of Advanced Research in Electrical, Electronics and Instrumentation Engineering Vol 4, Issue 3, March 2015 ISSN (Print): 2320–3765.

Frontiers in Computer, Communication and Electrical Engineering – Acharyya (Ed.)
© 2016 Taylor & Francis Group, London, ISBN: 978-1-138-02877-7

Study of dynamic responses of an interconnected power system using a zero-order hold circuit

Subhankar Mukherjee, Soumya Kanti Bandyopadhyay & Aveek Chattopadhyaya
Supreme Knowledge Foundation Group of Institutions, West Bengal, India

Bikash Das
Government College of Engineering and Textile Technology, Berhampore, India

ABSTRACT: Power system engineering was and still has the maximum preoccupation and concern of controlling mega-watt power since it is the basic governing element of revenue. This paper is entirely focused on the study of dynamic responses of single-area and two-area interconnected power systems. Here two area systems with reheat turbine, non-reheat turbine, and hydro-turbine are considered for study. The state space model of an interconnected system is developed for the studies of dynamic responses. The main objective of the present work is to demonstrate the effect of change of speed regulation parameters and sampling time constant. Dynamic responses are analyzed in MATLAB-SIMULINK environment with 10% load (severe case) change in any control area.

1 INTRODUCTION

Power system engineering was and still has the maximum preoccupation and concern to controlling mega-watt power since it is the basic governing element of revenue. In the earlier days, power system network, transmissions, and utilities were not so complicated; therefore, there was no obvious great concern of controlling the system. Due to the advent of modern technology, power system networks become more and more extensive because of long transmission lines, more load demand, frequent load fluctuations etc. Therefore, we must pay more attention on control strategy, which minimizes disturbances occurring in the system while maintaining the other system's constraints like as security, reliability, and stability. Nowadays our power system networks are connected through transmission lines called Tie lines. All generating units contribute their generations in interconnected systems. Change of load in any unit results in changes in steady state frequency. In a large interconnected power system, the generation comprises a suitable mix of thermal, hydro and nuclear generations. Nuclear units has higher efficiency, usually operated at its base load, i.e., it is close to the maximum load. Therefore, it has no participation in system frequency control. Thus, the natural duty for load frequency control falls on thermal and hydel systems only.

In the early days, the distribution of electric energy was only meant for limited areas. Now as days are passing on, it can be seen that the need of electricity has increased. Today we have to serve electricity in rural areas also which has made the supply as well as production of electricity a challenge. As a result the concept of interconnection of two areas has come into the picture. We are also thinking of some alternative sources of energy with which we can generate power to meet the desired level. As thermal is a nonrenewable source of energy, we are also thinking of hydro (water) as a medium for the generation, using which the cost of production will be reduced.

Planning, operation, and control of an interconnected power system is a challenging and important problem where planning, scheduling, and operation of the system in unit wise, plan wise and interconnection wise must be optimal for proper operation of the power system in terms of absolute economy. Economic operation of power system along with the suitable analytical model results in meaningful savings.

2 SYSTEM INVESTIGATED

This paper modeled MW frequency control system assuming a small signal assumption. In this paper, a two-area interconnected power system with one thermal and one hydel unit has been considered. For simplicity, different nonlinear elements and other parametric uncertainties, such as dead zone of turbines, GRC of governor etc., are discarded. Literature survey shows that in practice measurement of frequency or tie line power errors are done

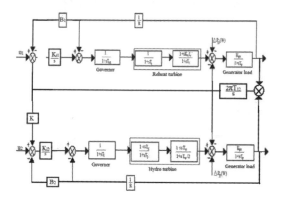

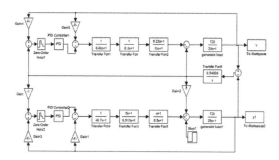

Figure 1. Block Diagram of interconnected hydro-thermal system.

Figure 2. Simulink model of a two-area hydrothermal system using a zero-order hold.

in a fixed interval of time using a zero-order hold circuit. Based on the above knowledge and assumption, this paper modeled the two-area hydrothermal system in discrete data form. Fig. 1 shows the proposed model of an interconnected hydrothermal system with a Zero-Order Hold circuit. Simulink model of a two-area hydrothermal system using a zero-order hold is shown in Fig. 2.

3 MODELING OF AN INTERCONNECTED SYSTEM

$$\dot{X} = Ax + Bu + Gw \qquad (1)$$

Equation (1) gives a differential equation which defines the dynamic behavior of a continuous time system. A, B and G are called as system matrix, input matrix, and disturbance matrix, respectively, which depend on different system parameters and operating conditions of the system. Whereas x, u, and w are known as state variables, controlled input, and disturbance input, respectively.

The state space matrix of a two-area system is given by:

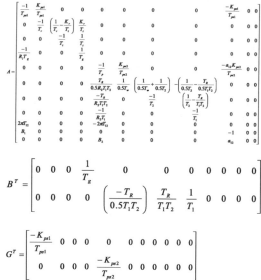

$$B^T = \begin{bmatrix} 0 & 0 & 0 & \dfrac{1}{T_g} & 0 & 0 & 0 & 0 & 0 & 0 & 0 \\ 0 & 0 & 0 & 0 & \left(\dfrac{-T_R}{0.5T_1T_2}\right) & \dfrac{T_R}{T_1T_2} & \dfrac{1}{T_1} & 0 & 0 & 0 & 0 \end{bmatrix}$$

$$G^T = \begin{bmatrix} \dfrac{-K_{ps1}}{T_{ps1}} & 0 & 0 & 0 & 0 & 0 & 0 & 0 & 0 & 0 \\ 0 & 0 & 0 & 0 & \dfrac{-K_{ps2}}{T_{ps2}} & 0 & 0 & 0 & 0 & 0 \end{bmatrix}$$

4 SIMULATION RESULTS

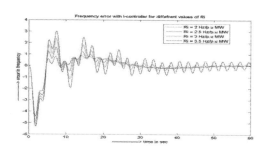

Figure 3. Frequency error of a two—area system with I controller.

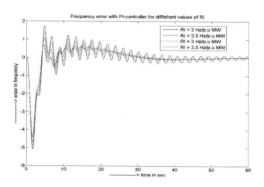

Figure 4. Frequency error of a two-area system with PI controller.

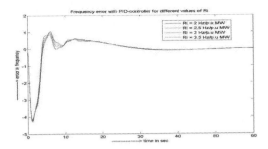

Figure 5. Frequency error of a two area system with PID controller.

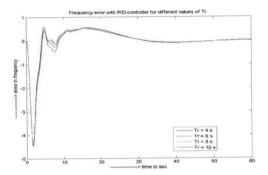

Figure 6. Frequency error of a two-area system with PID controller.

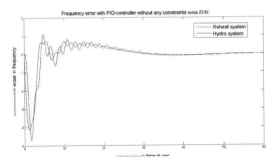

Figure 7. Frequency error of a two-area system for PID-controller using ZOH.

5 OBSERVATIONS

The interconnected two-area and multi-area power system theory and hydrothermal system is studied in MATLB–SIMULINK environment both in continuous time and discrete time form.

The following things have been observed:

1. In Fig (3), the settling time of hydro and Reheat thermal systems are 7 sec and 11 sec, respectively (approximately), which shows that reheat system has sluggish dynamic response compare to non-reheat system.

2. Interconnected power systems provides more stable operation compared to control areas operating individually. It takes 6–8 sec to settle down the oscillation which is depicted in Fig. (6).

3. In Fig. (5), the system has high peak overshoot which is highly minimized with proper choice of frequency bias constraint (B_i) as shown in Fig (5).

4. It is observed from Fig. (4) that tie-line power flow deviates from nominal setting, with sudden load changes in both the areas and it takes 10–12 sec to settle down.

5. From the investigation carried out in this work, it has been observed that PI-controller strategy offers an ameliorated dynamic performance of a system compared to P-controller.

6. Peaks of undershoots of hydrounits are less compares to thermal unit.

7. Frequency errors in hydel units get damped faster than that of thermal areas.

8. With the replacement of conventional integral controller by PI and PID controller, overshoots of a dynamic performance is highly minimized.

9. The results reveal that for a small gain of integral controller gain, (K_I) gives better performance so far as out-of-system oscillation is considered, Fig. (7).

10. Fig. (7) shows that a higher value of speed regulation parameter (R_i) causes deviation of frequency as well as tie-line power that leads to unstable operation of system dynamics.

6 CONCLUSION

In this paper, an attempt has been made to evaluate the state space model of a two-area interconnected power system. Further, the study is extended to obtain a discrete model of a power system. This method is very much realistic since both frequency and tie-line power flow is measured at discrete intervals of time.

A comparison has also been made for the analysis of the dynamic performances of a hydrothermal coordinated system. The reheat type system has sluggish performances compare to the nonreheat type system though the efficiency of the power plant is high when it is operated with a reheat turbine. In real situations, a large power system network comprises different power generating sources such as hydro, thermal, gas, and nuclear. From the study, it is observed that the dynamic performances of hydro-thermal system is quite better than the thermal-thermal system. The optimum sampling period of zero-order-hold circuit is to be set judiciously considering the large value of sampling time as long as the system dynamics maintains its stability. From the analysis it is also observed that the range of sampling time period is better in hydro-thermal system compared to thermal-thermal system, and the effectiveness of PI and PID controllers in lieu of conventional integral controller on dynamics of power system is better. Finally, the importance of change of K_r and T_r on dynamics of power system is understood which can be implemented in real-time power systems.

REFERENCES

Dipayan Guha et al., "*Dynamic Response Analysis of Automatic Generation Control in a 2–area (Reheat and Non–reheat) Interconnected Power System and a scheme for Improvement for the same*", International Journal of Modern Engineering Research, Vol. 3, Issue 1 Jan–Feb 2013.

Dipayan Guha, T.K. Sengupta, A. Das, "*Study of Sampled data analysis of Dynamic Response of an interconnected Hydro—Thermal System*", International Journal of Electronics and Communication Technology, Vol. 1, Issue 1, pp. 157–161. [Impact Factor 0.306]

Dipayan Guha, P.K. Prasad, "Problem analysis in MW frequency control of an Interconnected Power system using sampled data technique", IJLTET, Vol.2, Issue 2, March – 2013, pp: 54–62.

Hossein Shayeghi et. al., "Automatic generation control of Interconnected power system using ANN technique based on μ synthesis", Journal of Electrical Engg., vol. 55, No. 11–12, 2004, pp: 306–313.

Hemin Golpira et. al., "*Effect of physical constraints on AGC dynamic behavior in an interconnected power system*", Int. Journal of Advanced Mechartronics Syst., vol. 3, no. 2, 2011, pp: 79–87.

Ignacio Egido et al., "*Modeling of Thermal Generating units for Automatic Generation Control Purpose*", IEEE Trans. On Control System Technology, vol. 12, No. 1, Jan 2004, pp: 205–210.

Indulkar, C.S. "Analysis of MW frequency control problem using sampled data theory", IEEE trans., January 1, 1992.

Mohamed et al., "*Decentralized model predictive based load frequency control in an interconnected power system*", Energy conversion and Management, vol. 52, 2011, pp: 1208–1214

Nanda, J., Kothari, M.L. "Sample data AGC of Hydro-Thermal system considering GRC", IEEE-trans., September 25, 1989

Prabhat Kumar, Ibraheem, "*Dynamic performance evaluation of 2-area interconnected power system—a comparative study*", IEEE-trans, August 14, 1996.

Sengupta, T.K. "*Studies on assessment of power frequency in interconnected grid – its computer based control & protection*", 2008, thesis paper in JU.

Un-Chul Moon et al., "*A Boiler—Turbine System Control using Fuzzy Auto—Regressive Moving Average Model*", IEEE Trans. On Energy Conversion, Vol. 18, No. 1, March 2003, pp: 142–148

APPENDIX

Nominal values of all parameters for an Interconnected system:

f = 50 Hz	P_{ri} = 2000 MW
T_{12} = 0.086	B_i = 0.425
K_{PS} = 120	T_{PS} = 20 sec
T_t = 0.3 sec	T_{sg} = 0.08 sec
R_i = 2.4 Hz/p.u. MW	K_r = 0.5
T_r = 10 sec	T_R = 5 sec

T_1, T_2 = 48.75 sec and 0.513 sec, respectively.
T_w = 1 sec
K_P, K_D, K_I = 0.8036, 0.6356, and 0.1832, respectively.

Frontiers in Computer, Communication and Electrical Engineering – Acharyya (Ed.)
© *2016 Taylor & Francis Group, London, ISBN: 978-1-138-02877-7*

Nanorobot—the expected ever reliable future asset in diagnosis, treatment, and therapy

Orijit Biswas & Anindya Sen
Department of Electronics and Communication Engineering, Heritage Institute of Technology, Kolkata, West Bengal, India

ABSTRACT: Nanorobot or nanobot, an emerging technology of the current time, is a robotic machine of nanoscale dimensions that can be programmed and viewed under a microscope. Nanorobot, an application of nanotechnology, has versatile applications in areas such as material defect detecting and repairing, oil resource extraction, oil spill clean-up, solar power harnessing, and repairing the depletion of the ozone layer, thus protecting the environment. It also plays an essential role in critical care applications of medicine and health. Its superior role as a substitute to conventional medical care was investigated, as well as the absence of its side effects while effecting the cure was also investigated. This work evaluates the potential of inorganic and organic nanobots for their diagnostic and therapeutic roles under different critical care conditions with examples.

1 MOTIVATION

Nanorobot plays an essential role in critical care applications. It is an ultramicroscopic machine of suitable dimensions, capable of working upon living cells, and acting like a universal solvent for prevention of all kinds of diseases. It is programmable and, according to the set program, will act upon all the diseased cells by biophysical and biochemical actions to bring them back to a natural, healthy state. Nanotechnology and nanorobotics guarantee the cure while avoiding the choice of taking several medicines at different times[1]. This work demonstrates the current applications of nanorobots in therapies of cancer, myocardial infarction, genetic disorders, and tetanus, and highlights the advantages of methods to overcome the disadvantages.[1,2]

2 INTRODUCTION

Nanorobot or nanobot, an emerging technology of the current time, is literally a robot or machine of nanoscale dimensions that can be viewed under a microscope. Such devices are constructed of nanoscale or molecular components. The device size ranges from 0.1 µm to 10 µm.[1]

3 BACKGROUND

Nanorobot or nanobot, also known as nanoids, nanites, nanomachines, or nanomites, is the most innovative manifestation of one of the prime nonconventional technologies of the modern times in nanotechnology[1]. It can control, manipulate, and implement the properties of a matter at the molecular level[1] within the maximum dimension of 150 nm. For a particular substance, its properties vary widely from discrete atomic levels to the bulk macroscopic levels. However, in the domain of a certain definite number of atoms or molecules within the maximum dimension of 150 nm, the properties of the substance or material are of a unique kind matching neither with the properties of the discrete atom or molecule nor with the properties of the bulk material or substance at the macroscopic domain. The study of the unique properties of matter in this domain is nanoscience, and the domain of application of nanoscience is nanotechnology. Nanotechnology is the study, design, creation, synthesis, manipulation, and application of materials, devices, and systems at the nanometer scale. Currently, nanotechnology is becoming increasingly important in fields such as engineering, agriculture, construction, microelectronics, and healthcare. Nanorobot, one of the most revolutionary applications of nanotechnology, has versatile applications in areas such as material defect detecting and repairing, oil resource extraction, oil spill clean-up, solar power harnessing, and repairing the depletion of the ozone layer, thus protecting the environment.[4] This work focuses on its applications to medicine and health care. We studied the constructional and constitutional features, functional components and parts, scopes, and limitations of this device, and the ways to overcome the limitations by utilization of the basic laws of science and nature, and eventually applying this technology to fight targeted diseases, which are discussed below.

4 APPLICATION IN MEDICAL TREATMENT

Nanorobot, a machine of nanoscale dimensions, is comparable with the dimension of the cell of a living organism. The amazing aspect of this nanorobot is that it can be exploited for diagnosis, treatment, and therapy at the cellular level, which is in contrast to the conventional medication methods at the macroscopic level of organs and systems.

5 TYPICAL APPLICATION SCENARIOS

5.1 *Tetanus*

Tetanus is caused by puncture of the surface of the body by rusty nails lying unnoticed due to negligence or aging. Once the mishap occurs, mortality is imminent if there is a delay in the initiation of treatment. The factor responsible for this deadly condition is the pathogen (bacteria) called *Clostridium tetani* that is naturally present on the surfaces of rusting nails and metallic objects. This bacteria entering the body releases the neurotoxin pair TeTx within a short time, which causes paralysis or locking of the whole body from head to foot, leading to subsequent death. The conventional treatment involves injecting anti-tetanus vaccine to counteract the *Clostridium tetani* bacteria and the TeTx neurotoxin pair released within a limited time duration. An alternative treatment to avoid mortality is to use a programmable nanobot. When these nanorobots are injected into the body, they will annihilate the *Clostridium tetani* bacteria and the released deadly neurotoxin pair TeTx by biophysical and biochemical actions, which leads to total healing at the cellular level and also spares the side effects of conventional vaccination. Thus, we can shift to a nonconventional treatment method, which will mark the all-round revolution in the field of medical treatment in the new era.

5.2 *Myocardial infarction*

This is generally caused by blockage due to the cholesterol deposit in arteries, veins, and blood vessels. These vessel blocking factors or plaques can be detected and removed by nanorobots, molecule by molecule, thus healing the ailment. While the current plaque moving techniques, such as angioplasty, depend on surgical skills and sometimes have some side effects, the process using nanobots is free from these effects, as shown in Figure 1.[1]

5.3 *Cancer*

Nanorobots can distinguish between the malignant and benign cells. They sense surface antigens and their moves are directed by chemotactic sensors that are embedded in the machine. Nanorobots have excellent specificity, i.e., they target only cancerous cells and heal them by acting upon them both biochemically and biophysically. Thus, a cell's abnormal biochemical behavior of overgrowth could be checked, and the cell could be repaired to a natural healthy state. Nanobots do not harm the healthy and normal cells as conventional methods do, thus preventing the sickening and damaging side effects. Nanobots do not harm the immune system (being localized), thus making patients healthier throughout the treatment. Nanorobots are highly effective and can be reused and reprogrammed for the treatment of various cancer-affected cells.

5.4 *Tumors*

Tumors are usually caused by the excessive unwanted growth of the tissues. This can be treated by the first set of nanorobots inflaming the tumor-infested tissues and the second set of nanorobots detecting the inflamed tissues and releasing the payload, which is the chemotherapy drug to heal the tissue, as shown in Figure 2.[2]

Figure 1. Nanorobots preventing heart attacks.

Figure 2. Nanobot targeting the tumor site.

452

5.5 Genetic disease

This is due to do the presence of defective genes in the genetic sequence. Nanorobots approach the gene sequence and detect the defective genes. After that, they dissociate them and place the payload, which is the healthy gene in its place, thus curing the generations of long-genetic defect.

5.6 Damaged tissue repair

Nanorobots can easily repair and heal the damaged tissues by taking existing molecules, replicating them, and assembling new molecules into new layers of the tissues. Nanorobots can slowly regrow the portions of the damaged bone and might one day be able to reproduce the bone marrow.

- Other functional possibilities in the aspect of damaged tissue repair include:
  - Closing of a split vein
  - Reforming the damaged skin
  - Reducing dead flesh (from a wound), resulting in little scarring.

6 ADVANTAGES OF NANOROBOTS WITH RESPECT TO CONVENTIONAL MEDICINE, RADIATION THERAPY, AND SURGERY

Nanobots heal at the cellular level. Thus, if each and every cell gets healed, then the entire body gets healed. If we look at the advantages of this, then we cannot help without accepting this newly emerged technology over the conventional existing one. In conventional medication methods, the medicine applied obviously affects and heals the diseased cells; however, its application is not targeted but acts on all the cells of the whole body so that along with the diseased cells, the healthy and normal cells are also subjected to the biochemical reaction of the applied medicine with the cells having the serious occurrence of side effects. But the action of nanorobots is targeted. By virtue of being a programmable machine, the nanorobot is programmed to detect the diseased cells only and to act upon them leaving aside the healthy cells, thus rendering no medicinal side effects.[1,2] This method of healing by nanorobot can also be a safe substitute for radiation therapy. The current ongoing research on radiation therapy is conducted to selectively work upon the defective cells, leaving aside the healthy and normal cells. Conventional medicine cannot guarantee to act upon all the diseased cells of the body, since the action of the medicine is based on the probability of association with the cells. However, the nanorobot, which is a programmable machine, is programmed before being injected into the body and, upon entering into the body, acts according to the programming upon all the diseased cells and roams throughout the body by navigation, thus guarantying the complete elimination of the disease.[1,2]

Surgical methods in treatments are invasive and liable to injury. Depending on the ailment, surgical methods can be expensive and time consuming, with the success depending on the skill and efficiency of the operating surgeon and his team. It carries high risk: a minor error made by one surgical team member may cost the patient's life. But the action of nanorobots on the defective cells evades this risk. In the conventional surgical procedure, the treatment of the interior of the body is externally controlled by the surgical team using therapeutic tools, involving instruments, machines, and devices, to invade the body. However, when nanorobots treat the interior of the body, the controlling is preset to be done by the single machine itself, thus guarantying much safer and exact treatment.[1,2]

7 FEATURES (CONSTRUCTIONAL AND FUNCTIONAL) OF NANOROBOTS

Nanorobots are often less than 100 nm in length and, the pieces or components that make up the machine are generally as small as 1 nm. The component pieces are primarily created of carbon, most often in the form of diamondoid or fullerene nanocomposites, i.e., carbon nanotubes. A very smooth exterior passive diamond coating shields the device from being attacked by the host's immune system.[4]

Nanorobot is a programmable machine that can be programmed or set in order to perform the desired operation inside the body. After being injected into the body, it is steered or remote controlled from outside the human body, according to which the device roams or navigates throughout the whole body, accesses the remote sites where access is impossible without invasive surgical techniques, and performs its desired operation.

There are two basic kinds of nanorobots: assemblers and self-replicators. Assemblers are simple cell-shaped nanorobots that are able to interpret molecules or atoms of different types and are controlled by specific specialized programs. Self-replicators are fundamentally assemblers that are capable of duplicating themselves at a very large, fast rate, the sort of duplication that aids the deployment of nanobots for large-scale applications and tasks.[1]

In order to perform the desired functions, the nanorobots require the structural-cum-functional features as follows[1]:

I. *Size and shape* – The size and shape will depend on the intended function and operating environment of the nanorobot. A major design criterion is the minimum size.[1]

II. *Sensors* – For the nanorobot to fulfill its desired operation, it must be capable of sensing different parameters of the cells upon which it will act, such as the various physical, chemical, biophysical, biochemical, and physiological properties, based on which the nanorobot will distinguish between the normal healthy cells from the diseased cells.[1]

III. *Means of mobility/propulsion* – Propulsion is necessary for the nanorobot in order to travel throughout the whole body and to perform its desired operation, as shown in Figure 3.[1] One kind of propulsion method is to have a simple ball-shaped nanorobot, which has no propulsion mechanism of its own, but rather flows with the pressure of the fluid flow inside the blood vessel into which the nanorobot is injected. Another kind of propulsion method involves the nanorobot having a propulsion method of its own, imitating the principle of locomotion of microorganisms in nature where bacteria, notably, move with the help of flagella, where the waving motion of the flagella causes a net motion in the forward direction.[1,2]

IV. *Power generation*[1,3] – Power generation is the major requirement of the nanorobot device in order to perform its desired operation. Power generation method becomes a very sensitive issue for nanoscopic devices, as in nanorobots the power requirement is in the nanoscale. Taking this into consideration, various methods of power generation can be used for the nanorobot device as follows:

a. In-built battery inside the nanorobot device.

b. Induction of electrical power inside the nanorobot device by an externally induced electric, magnetic, or electromagnetic field, e.g., the electromagnetic field generated by the Magnetic Resonance Imaging (MRI) device. It induces considerable electrical power inside the nanorobot device compatible with nanoscale dimensions.

c. Another inexpensive method of generating power is to convert the body temperature (thermal energy) or fluid flow (blood flow) motion inside the body (mechanical energy) into the electrical power. A pair of electrode terminals, i.e., cathode and anode, which are placed inside the nanorobot, break down the glucose in the blood by the electrolysis method and become the positive and negative terminals of a battery.

V. *Data acquisition and storage*[1] – Upon monitoring the cells, various conditions of the

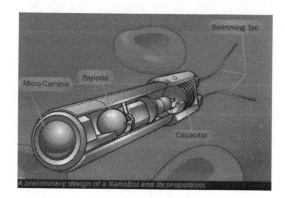

Figure 3. Propulsion in the nanorobot.

cells are collected or acquired as data by the nanorobot, which need to be stored in the device. Thus, data acquisition and storage are crucial for the nanorobot, and this aspect will be fulfilled by the quantum computing that is appropriate to the ultra-small dimensions, i.e., nanoscale dimensions.

VI. *Telemetry and data storage*[1] – Nanobots, which are injected into the body, must be steered all along the body to monitor the cells, to sense or detect the condition of the cells, and to collect the associated information, as the data should be sent to the controller outside the body, who is steering the nanobots, and the data should be stored there using the method of telemetry. The data storage for the nanorobot performance is a tough design function that needs to be incorporated into these ultra-small domains. Recent findings by the researchers in the University of California Irvine have unveiled a working radio built from carbon nanotubes of dimension of only few atoms that show some possibilities of transmitting data at the nanoscale.

VII. *Control and navigation* – Nanobots after being injected into the body are required to be controlled and steered throughout the body from outside, thus marking the necessity of the control and navigation system[1]. A remarkable method to achieve this is the "Swarm" method or the "Swarm Intelligence" method, in which the intelligent and programmable nanobots work together in a synchronized manner (the term "Swarm" implying collectiveness or togetherness). The same programming is set by the controller to all the nanorobots together and then the nanorobots are injected into the body and travel all over, synchronized among themselves, steered together by the external controller, and act upon the body cells at various sites simultaneously according to the set program.[2]

454

Navigation systems can be implemented externally by ultrasonic signals, MRI device, injecting a radioactive dye, X-rays, radio waves, microwaves, or heat and internally or on-board by chemical sensors and miniature television camera.[3]

8 LIMITATIONS OF NANOROBOTS AND THE SCOPES TO OVERCOME THEM

Notably, two usable approaches related to the limitations and scopes are the "Top–Down" approach and the "Bottom–Up" approach.[2] In the "Top–Down" approach, the currently used machines in the microscopic scale, known as MEMS (micro electro mechanical systems), the devices, and the components are scaled down to the nano level, thus starting from the bigger scale and progressing to the smaller scale ("Top–Down"). But this scaling down to the smaller scale is obviously not unbounded, ultimately meeting the quantum mechanical constraints at the molecular and atomic levels (in the domain of atom quantum mechanics acts). Also, the smallest constituents of matter are atoms and molecules, so the miniaturization cannot be done beyond the level of molecules and atoms. Thus, ultimately, there lies the margin that needs to be overcome by alternative methods. Because of this limitation, two vital components of nanorobots, i.e., sensors (which measure and store the measured parameters as data) and actuators (which enable the device to act according to the measured parameters), cannot be incorporated into the nanorobot device, but only one of them can be incorporated.[2]

Also, power-generation compatibility to such ultra-small dimension becomes a big problem. Although power at such a small scale can be generated by alternative methods, but the storage of power becomes another big problem because at that dimension, the sizes of the devices and the components are so small that the generated power is comparably of bigger magnitude, implying the increase of storage power density to such a level that the shrinking down of the size of storage batteries becomes unfeasible and practically impossible. This implies that in-built power supply in the nanorobot is not feasible because of the battery size of the nanorobot device. One suggested method against this is fabrication, engineering, and embedding such materials that facilitate the ultra-high power storage density.[2]

Though there are alternative methods such as Peltier effect (utilization of thermal energy or temperature of the body), utilization of glucose in the blood as fuel and natural flow of fluids such as blood in the body are used to generate power. However, they are also not of much use.[2]

Also to note that at the nanoscale, a kind of resistive force known as viscous drag force becomes predominant, and to overcome such high resistance, a lot more power needs to be generated.[2]

Therefore, to generate a lot more power to avoid the problem of power storage, the feasible solution is to generate power externally rather than internally. This would solve the problem of energy generation, as well as provide a better guidance system for the nanomachine. The inspiration for this technique comes from Mother Nature, which shows that a certain kind of bacteria, called as magnetotactic bacteria (discovered in 1975 by Richard Blakemore), have been found to be sensitive to magnetic fields,[2] most of them tending to naturally orient themselves along the magnetic north and indicating a similar guidance system for future nanobots containing small magnets as steering wheels under the influence of an external magnetic field generated by, for example, an MRI device. Another way of generating power is by electromagnetic induction where an external magnetic field could be used to charge the internal batteries of our nanorobots by induction.[2]

In the "Top–Down" approach involving scaling down of MEMS technology to nanoscale, using NEMS (Nano Electro Mechanical Systems) technology, we ought to know about the very prominent limitations. Though newer materials and technology could bypass this, the molecular approach seems to be the most promising one, which involves building up the nanomachine molecule by molecule or atom by atom, thus progressing from the minute basis to the bigger structure. Molecular chemistry has played a significant role in this field in the last decade. Nanoscale versions of macroscopic actuators like motors could be built from the bottom-up approach in future. Montemagno and Bachand were the pioneers in this field, who created the first artificial hybrid motor. Recent developments suggest that we could soon have nanogears and nanobearings at our disposal. Considering the assembly part, molecular self-assembly is possible using various methods already being employed, such as biological or chemical programming. One of the methods of self-assembly is called SAM (self-assembled monolayers). It involves a monolayer of molecules, fixed to a surface upon which a structure can be built vertically, like a building being built floor by floor. An innovative idea is to "program" biobots to use *in situ* molecules to create copies of themselves. This would help keep a healthy bot count, just like our body replaces worn-out blood cells.[2]

In 2005, a nanoswimmer was developed using an external magnetic field to beat a filament attached to the red blood cells. Also, a nanocar was developed using buckyballs (fullerenes) as wheels. However, these developments are a few among many in recent years, and could lead to the first nanorobots in the near future. The development of nanofabrication techniques such as electron beam lithography and scanning probe lithography has been a huge leap forward, allowing for fabrication of features as small as 3 nm in size. We also have a lot of aid from the nature in making artificial biobots with respect to the availability of biological resources. A lot of the materials that are needed to build a nanorobot are available naturally such as rhodopsin and bacteriorhodopsin, proteins that capture solar energy. After having an energy source, an actuator is needed, which can be implemented by ATP synthesized by rhodopsin or bacteriorhodopsin, to be used by certain molecules such as F1-ATPase to rotate nanoscopic shafts. Research is ongoing to develop more efficient motors, which can be actuated by a variety of signals, such as certain chemicals, which detect and provide data by reacting with the constituents inside the body and body cells, thus acting as sensors.[2] However, the limitations that creep in naturally are due to the physics of the nanodomain that is entirely different from that of the macroscopic domain. Fluid effects such as viscous forces about five orders of magnitude greater now and surface effects such as electrostatic forces dominate over conventional forces due to mass that is now negligible. Frictional force is dependent on load and velocity, which is not the case in the classical mechanics in the macroscopic domain. Also, nanorobots, being of about the size of molecules, act as molecules or pseudo-molecules rather, being subjected to collisional forces among the molecules such as 'Brownian motion' that is tremendously frequent and of titanic magnitude comparatively, causing deformation in the nanorobots that alter the mechanics of the motion considerably, which can be corrected by artificial intelligence that is, however, unlikely to be embedded in such a small scale. So, the possible way to overcome this is the biobot, which would be essentially rigid under all the circumstances by virtue of being built by molecular self-assembly and rendered practically rigid by strong intermolecular forces. As it can be observed, to overcome the predominant resistive forces in the nanodomain, the innovative solution comes from the nature, where the microorganisms such as bacteria move with the help of cilia or flagella that beat asymmetrically, causing net motion/displacement in one direction.[2]

Generation and storage of power remains a major challenge. Another challenge is developing engineering materials usable for the manufacture of nanobots using NEMS or NEMS-like technology and being biocompatible at the same time. Effective ways of power generation still have to be figured out. A much bigger challenge is on-board power storage (as in the case of NEMS bots) or continuous power generation (as in the case of biological bots).[2]

The nanorobots that first come into the scenario are the inorganic nanorobots that are constituted of those materials and components from which larger machines are made, with the difference being in the size. Inorganic nanorobots have the advantages of well-understood component behavior, ease of programming, and ease of external control. However, what vitally required is several nanobots rather than a single one to be injected into the body to perform the treatment in the whole body. But the inorganic nanorobot is difficult and expensive to make, leaving behind the making of several nanorobots. So, the convenient way is to replicate one nanorobot, which is not having the capability of the inorganic nanorobot. Also, inorganic nanorobots are not biocompatible; they have difficulty in communicating with the organic systems. Another point is that inorganic nanorobots can carry limited payload i.e., the drug to be delivered to the diseased cells.[3]

The disadvantages of inorganic nanorobots can be overcome by organic or bionanorobots, which are made of organic or biological components that make up the body. Organic nanorobots have the advantages of ease in making use of genetic engineering, and capability of self-reproducing, thus making it cheaper, easily communicable with other organic systems, and the payload being manufactured by protein factories. However, organic nanorobots have the difficulty of poorly understood component behavior (behavior of proteins), difficulty in programming, and limited external control mechanisms such as navigation, telemetry, and data transmission. These limitations need to be overcome by exploiting the natural biological properties of cells and tissues, for which much further research is utmost and indispensably necessary.[3]

REFERENCES

[1] Nagal, D., Mehta, Dr. S.S., Sharma, S., Singh Mehta, G., & Mehta, H. 2012. Nanobots and their application in Biomedical Engineering. In Dr. R.K. Singh (ed.), *Proc. of the Intl. Conf. on Advances in Electronics, Electrical and Computer Science Engineering—EEC 2012*: 215–219.

[2] Bhat, A.S. 2014. Nanobots: The Future of Medicine, In *International Journal of Engineering and Management Sciences* 5 (1): 44–49.

[3] Electrical Engineering Lab of Biomedical Engineering, Department of Electrical Engineering, National Taiwan University, 2012. *Medical Revolution – Nanobots*: 1–24. http://cc.ee.ntu.edu.tw/~ultrasound/belab/midterm_oral_files/2012_101_1/101-1-mid–5.pdf.

[4] Armstrong, L., Arnold, C., Banjara, K., & Aldouah, A. 2000. *Nanomachines and Robots – Developments and Applications*. http://research.che.tamu.edu/groups/Seminario/materials/G01_Nanomachines.pdf.

Frontiers in Computer, Communication and Electrical Engineering – Acharyya (Ed.)
© *2016 Taylor & Francis Group, London, ISBN: 978-1-138-02877-7*

Fabrication and resilience measurement of thin aluminium cantilevers using scanning probe microscopy

Aviru Kumar Basu
Design Programme, Indian Institute of Technology, Kanpur, India

Hansaraj Sarkar
Department of Electrical Engineering, Indian Institute of Technology, Kanpur, India

Shantanu Bhattacharya
Design Programme, Indian Institute of Technology, Kanpur, India
Department of Mechanical Engineering, Indian Institute of Technology, Kanpur, India

ABSTRACT: Metallic cantilevers have found major applications in MEMS devices. Of significant mention in the use of such cantilevers in bio-sensor devices for the diagnostic industry. In this work we report the fabrication ofsmall Aluminium cantilevers without the use of any hard sacrificial layer for the releasing step of the cantilever structures. Furthermore, the fabrication process just uses a coating of positive photoresist (acting as the sacrificial soft polymeric layer) (M/s Shipley) as a masking agent for releasing the aluminium patterns so that they can formulate hanging structures. We have also performed a characterization study on the aluminum cantilever filmsby performing Scanning Probe Microscopy in Nanoindentation and Tapping mode over the films.

1 INTRODUCTION

Metallic micro and nano structures have a wide range of applications in Sensing and diagnostic applications. Of high significance is the metallic cantilever where wide ranging research has been carried out using such structures in biosensing, gas sensing chemical/biochemical sensing etc. Easy fabrication with as less steps as possible forms a major challenge to pattern and structure such cantilever structures. The most complex step involved in the fabrication of such structures is the release of the cantilever structure from the remaining substrate surface where a lot of techniques like wet or dry etching using sacrificial hard films like silicon oxide or nitride are commonly used (Azad et al. 2013). Some earlier research that has been carried out of cantilever fabrication without the use of hard sacrificial layer always reports of stiction or damage of these structures during the release process (Yan et al. 2001).

Xiao et al (1999) has used polysilicon as a hard sacrificial layer for releasing the anchor and Phosphosilicate Glass (PSG) as the sacrificial layer for final release of cantilever beams using oxygen plasma etching and Hydrofluoric acid (40%) for complete etching of the PSG layer. DeVoe et al (2001) has reported releasing the accelerometers using SiO_2 as the hard sacrificial layer. However, use of hard sacrificial layer has an additional drawback

of longer release time because the sacrificial layer material is removed through wet or dry etching processes which involve additional steps. Moreover, it may also lead to stiction due to the presence of surface tension forces mainly capillary forces while release using wet etching (Bartek et al. 1997).

In this paper, a new method of etching and releasing aluminium microcantilevers on by using a photoresist sacrificial layer has been reported. The cantilever like hanging structures are further assessed for the resilience. The modulus of resilience is measured of the aluminium cantilevers by use of Scanning Probe Microscopy in Nanoindentation and Tapping mode.

2 MATERIALS AND METHODS

A p-type (100) silicon wafer is properly cleaned using TCE, Acetone, I.P.A. (Isopropyl Alcohol) and D.I (deionized) water. Several films of Aluminum of thicknesses ranging between 200–800 nm are deposited on cleaned Si-wafers through sputtering. On the top surface of this deposited Al film, cantilever structures are patterned through positive photoresist S-1813 (M/s Shipley). The photoresist is coated at 1000 rpm for 10 secs and at 3000 rpm for 25 secs. The thickness of the film obtained 1.4 µm. The thin photoresist is hard baked for 1 hr at

95°C. The positive photoresist structure acts as a mask or protective layer for the TMAH etching process that is used subsequently for releasing of the cantilever structures. If the positive photoresist is prebaked for a higher duration 1 to 1.5 hr it takes a higher time to get dissolved. The Al thin film is first etched off through the vias opened up for the remaining portion of the film. [Refer to Figure 1(a) for the top elevation of the mask drawing]. The etching of the aluminum from the opened positions are further etched using Transene solution with a composition of 80% H_3PO_4 + 5% HNO_3 + 5% Acetic Acid + 10% DI Water which is commonly used for Al etching. The temperature used while carrying out the wet etching is around 50°C. For releasing the Al Micro-cantilever a second step of masking and anisotropic wet etching is done by TMAH Etch process [Composition of the etching process (5 ml TMAH + 95 ml water) ++(0.4–0.7 ml Ammonium Peroxidisulphate $(NH_4)_2S_2O_8$ + 99.5 ml water)] (Yan et al. 2001). Figure 1(b) shows a fabrication flowchart of the process used for etching of silicon cantilevers. Mask layout for the microcantilever fabrication and the schematic representation of Al microcantilever fabrication have been shown in Figures 1(a) and (b) respectively.

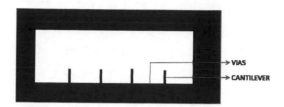

Figure 1(a). Mask layout for the microcantilever fabrication.

3 CANTILEVER CHARACTERIATION SET-UP

Scanning probe microscopy has been used for characterization of the hanging cantilever structures formulated in the last step. A Hystiron TI 750 Ubi L, D and H machine was used for the measurements. For our purpose we have used the standard load indentation with a maximum load of 10 mN. Standard nanoindentation probes Berkovich probes with total included angle from one edge to the opposite side 142.35° with half angle of 65.35° and radius of curvature of approximately 150 nm are deployed to carry out the force deflection testing of the cantilevers.

During the indentation test, the probe is indented into a sample and then slowly retracted or withdrawn by decreasing the applied force.

The applied load versus displacement plot is acquired and plotted on the cantilever structure shown in Figure 2 and the data plotted in Figure 3.

4 RESULTS AND DISCUSSION

Micro cantilever structures 70 μm × 20 μm were used for evaluating the resilience. We observed all the cantilever structures formulated to have a pre-strained induced by the fabrication process which helped in self releasing of these structures from the substrate surface. We hypothesize that this prestrain

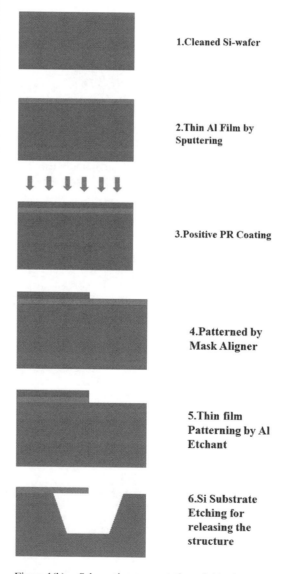

1.Cleaned Si-wafer

2.Thin Al Film by Sputtering

3.Positive PR Coating

4.Patterned by Mask Aligner

5.Thin film Patterning by Al Etchant

6.Si Substrate Etching for releasing the structure

Figure 1(b). Schematic representation of Al microcantilever fabrication.

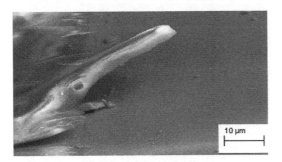

Figure 2.   FESEM image of Al microcantilever.

Table 1.   Mechanical parameters of Aluminium thin film obtained during Nanoindentation.

E_r (GPa)	7.79
Hardness (MPa)	262.61
Contact depth (nm)	338.6
Contact Stiffness (μN/nm)	17.1
Max Force (μN)	997.8
Max depth (nm)	382.3
Contact area (nm²)	3799603.6

can be possibly caused by the liquid surface tension during contact etching of both the top and bottom surface of the cantilever. The resolution of the hanging structures still needs improvement but this process of pre-straining definitely seems to have value and problems like stiction of the lip etc. may get resolved if we are able to pre-strain the films as in our case. Detailed studies are further on as to ascertain the real cause of this pre-strain and are not reported here for the sake of brevity.

Figure 3 reflects the acquired data on the force displacement as acquired by the nano-indentor using a Berkovich tip. The curve reflects a flat in-between the loading and unloading cylces which signifies hold duration between the loading and unloading cycle. This is indicative of the reversal of forces as provided by the berkovich tip on the concerned surface. After that the tip unloads or retracts we observe a hysteresis between the loading and unloading cycle which only points out a plastic bending of the cantilevers in our case due to the loading cycle. [Refer to Figure 3 indicating similar load values although the overall displacement shows a general shift].

Huson et al. (2006) has performed similar testing over a polymeric material and in our case we have focused to thin metal films and we find out the modulus of resilience by looking at the area under the loading or the unloading curve. Area under the loading curve means the work done considering upto maximum load = 2.4596e+05 μN-nm, Area under the unloading curve = 1.7474e+05 μN-nm. R value computed as percentile using the formula as mentioned by Huson et al. (2006) is calculated as 71.2. The maximum load applied in our case for freestanding thin Al films is 997.8 μN while Huson et al. (2006) has used a maximum load of 10e-05 N on polymeric materials. The resilience studies are performed on thin cantilevers of aluminum metal show similar resilience properties to a polymer like Natural Rubber whose resilience has been report by Huson et al. (2006) as 86%. The rate of deflection is 53.8 nm/sec of the cantilever structure in our case.

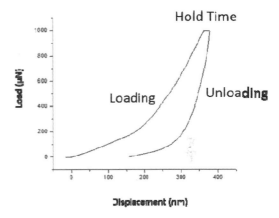

Figure 3.   Load-Displacement curves obtained from nanoindentation.

Obviously resilience is the ability of a material to store and return energy when subjected to rapid deformation. In polymers this property has been earlier explained to happen because of the ability of the polymer chains to rotate freely whereas in the said case as the cantilever film is thin to the meso-scopic case it may have a very large surface area in comparison to the bulk volume. We hypothesize that the surface of the Al cantilever film may demonstrate a reversible movement of dislocation leading to a polymer like behavior in the resilience of the cantilever films.

5   CONCLUSION

Fabrication methodology of thin film alumunium microcantilevers have been studied without using any sacrificial layer. So in our case the problem of stiction is completely removed. The cantilevers are pre-strained as can be seen in Figure 3 which gives an advantage of reduced stiction in our cantilevers.

The resilience measurements are carried out in accordance to SPM studies done earlier by Huson et al. (2006) in which a indentation through an SPM probe is performed on the polymer surface. The measurements indicate similar resilience values for a lower load and a similar rate of deformation on the Al Cantilever film. More studies are carried out to ascertain the exact reason of the similar resilience values in comparisons to polymers but it seems that the surface dislocations and their reversible movement within the grain boundaries may be possibly responsible for comparative values of resilience to polymers like Natural rubber.

REFERENCES

Azad, J.B., Rezadad, I, Nath, J, Smith, E, Peale, R.E. (2013) Release of MEMS devices with hard-baked polyimide sacrificial layer. *SPIE Proceedings*, 8682, 26.

Barteky, M, & Wolffenbuttel, R.F. (1998) Dry release of metal structures in oxygen plasma process characterization and optimization. *J. Micromech. Microeng*, 8(1998), 91–94.

DeVoe, D.L., Pisano, A.P. (2001) Surface micromachined piezoelectric accelerometers. *Journal of Microelectromechanical System*, 10, (2001), 180–186

Huson, M.G., & Maxwell J.M. (2006) The measurement of resilience with a scanning probe microscope. *Polymer Testing*, 25, 2–11.

Raccurt, O., Tardif, F., Arnaud, D., Vareine, T. (2004) Influence of liquid surface tension on stiction of SOI MEMS. *Journal of Micromechanics and Microengineering*, 14, 1083–1090.

Xiao, Z., Hao, Y., Li, T., Zhang, G., Liu, S., Wu, G. (1999) A new release process for polysilicon surface micromachining using sacrificial polysilicon anchor and on thin films. photolithography after sacrificial etching. *IOP, Journal of Micromechanics and Microengineering*, 9, 300–304.

Yan, G., Chan, P.C.H., Hsing, I.M., Sharma, R.K., Sin, J.K.O., Wan, Y. (2001) An improved TMAH Si-etching solution without attacking exposed aluminum. *Sensors and Actuator*, 89,135–141

Frontiers in Computer, Communication and Electrical Engineering – Acharyya (Ed.)
© *2016 Taylor & Francis Group, London, ISBN: 978-1-138-02877-7*

Comparative analysis of AODV routing protocols based on network performance parameters in Mobile Adhoc Networks

Mamata Rath
C.V. Raman Computer Academy, Bhubaneswar, Odisha, India

Binod Kumar Pattanayak
Department of Computer Science and Engineering, Siksha 'O' Anusandhan University, Bhubaneswar, Odisha, India

Bibudhendu Pati
Department of Computer Science and Engineering, C.V. Raman College of Engineering, Bhubaneswar, Odisha, India

ABSTRACT: Power deficiency of the nodes, multi-hop wireless connectivity, and high frequency of variation in topology of the network are major challenges that a Mobile Adhoc Network (MANET) has to face due to which it is very demanding to develop a dynamic and efficient routing protocol for this network. Adhoc On Demand Distance Vector (AODV) is an efficient reactive routing protocol in MANET in which the route is established only when it is required, so this protocol minimizes the overhead routing by not consuming enough energy in continuous route maintenance and repairing at time if there is no route request. In resource constraint environments, like an MANET, AODV protocol satisfies the necessary conditions of routing by considering stimulating matters such as highly mobile nodes, frequent link failure, limited battery power, variation in topology, satisfying quality of service. There are many improved AODV protocols, which are implemented for better network performance and intelligent routing. This paper presents a detailed study and comparative analysis of three leading protocols of MANET.

1 INTRODUCTION

Mobile Adhoc Network is a collection of autonomous mobile stations which can communicate with each other in wireless environments. Those mobile nodes which are not in the same range can communicate with other through multiple intermediate nodes. These nodes are flexible, self configurable, robust, and part of a distributed network. So, though routing in such mobile networks is efficient, it is still a challenging task to maintain the routing by applying efficient routing strategies. Though many improved routing protocols have already been implemented considering the above issues, high level research is still going on in this technical area of MANET routing. Officially, an MANET can be defined as a communication network that combines a group of mobile nodes with limited resources taking part in data communication. The neighboring stations behave as routers or relay nodes that carry and forward the data packets while transmitting from one source station to destination.

Due to change in network physical organization, high node mobility, and limited power level of nodes, very robust and highly reliable routing protocols are necessary for communication. There are basically two types of routing protocols in MANET, such as proactive and reactive routing protocols. Routing information regarding every possible destination node is stored in routing tables of proactive routing protocols. Any little change in the position of nodes due to their highly mobile nature is immediately updated in all the related nodes with broadcasting by the neighboring nodes. Whereas in reactive routing protocols, the network bandwidth is better utilized by avoiding unnecessary continuous broadcast of information and by creating routes only when desired, hence performing efficient bandwidth utilization.

2 LITERATURE SURVEY

Many valuable works have been carried out in this direction till date. Maintaining Quality of Support in the network is a challenging issue, which is to be considered during the design of AODV protocol. The objective of QoS support is satisfaction of the required flow characteristics during transmission. QAODV [6] fulfills network parameters such as available bandwidth estimation, Cumulative delay calculation occurs during transmission network load in particular routes and hop counts. The nodes in MANET suffer from energy restriction since a mobile station uses a battery with limited power [5]. For reliable data delivery in MANET, before possibility of a link failure, a warning message can be generated by every alternate node in a path to alert the neighbor before forwarding the packet to failed links, thus eliminating the chance of delay due to route repair or retransmission. In this regard, two eminent protocols are designed AODV-RD(AODV with Reliable Delivery) and AODV-BR(AODV for Broken Route) [7]. Simulation results when compared between AODV-RD, AODV-BR, and AODV signifies that there is significant improvement in Packet Delivery Ratio in AODV-RD with a short end-to-end delay than AODV-BR. There is enhancement in the performance of the network and improved throughput.

3 ANALYSIS AND OVERVIEW OF AODV PROTOCOL

It is the basic objective of AODV protocol to find the route from a specific source station to a destination station as and when required without continuously maintaining the existing route and repairing it when not in an active state. Link failures and change in network topology are handled properly by AODV protocol. In the route discovery process of AODV, it is checked if a direct path from source to destination node exists in the routing table or not. If there is no such information then a route request process gets started which includes broadcasting of RREQ packet by the source node; then all other nodes after increasing their sequence numbers, pass the RREQ packet which is a control packet such as a hello message for alert. Any node which has a direct route to the destination node in its routing table, forwards the RREQ towards the destination. An RREP packet in the form of a reply to the route, is sent back to the source node, hence conforming the route. During the data transmission in the reserved route, if there is any link failure, then it is taken care by the RERR and route maintenance mechanism of AODV.

4 BASIC ISSUES OF AODV DESIGN AND IMPLEMENTATION

4.1 *QoS Support in AODV*

Maintaining Quality of Support in the network is a challenging issue, which is to be considered during the design of AODV protocol. Objective of QoS support is satisfaction of required flow characteristics during transmission. QAODV [6] considers to fulfill network parameters such as available bandwidth estimation, cumulative delay calculation during transmission network load in particular route, and hop count. Depending on the number of hops, if there are less number of hops, and light congestion, then it has a smaller routing cost and if there are more number of hops, and more congestion, then routing cost is high.

4.2 *Identification of link interference*

Two types of interferences exit. One is called Link Interference which is a routing metric based on any interference that affects the links in the transmission of a source to a destination and another metric called node interference [8] which is based on global interference calculated by nodes used[8] to determine the path in an improved method of AODV. This proposal was validated by changing most of the concurrent connections keeping stable the number of nodes in the network.

4.3 *Multicast tree based on node mobility*

In a multicast tree-based Wireless Network, the leader of the group creates and does the maintenance of low cast multicast trees that contain all member nodes of the group[1]. Only the members of the group actively forward the data to the next nodes, so this scheme has a strong forwarding efficiency. The group information is maintained by the group leader. This DMT-Based AODV proposal selects the forwarding routes which further connect multicast receivers to satisfy the routing condition.

4.4 *MANET AODV in ANDROID OS*

Basic Standards of Wifi connection are formulated by Wifi and IEEE 802.11 alliance. Using these provisions, a wifi card may be joined with an Access Point which serves as the connection. Other services, such as communication and sharing of resources, can also be availed in this approach [9]. Every node in an infrastructure-based network can access the AP by co-ordinating the communication and resources can be shared among all the nodes.

4.5 *Balanced energy consumption among all participating nodes*

The nodes in MANET suffer from energy restriction since a mobile station uses a battery with limited power [5]. Therefore, the performance of a network mostly depends on its energy efficiency mechanism. So, developing an optimized routing protocol becomes the basic factor for network performance. A Local Energy Aware protocol uses balanced energy consumption at every node which all take part in routing. This approach is useful in applications like business partners sharing information during conference and communication in military applications, persons co-ordinating during disaster relief, and emergency scenarios.

4.6 AODV with reliable delivery

For reliable data delivery in MANET, before possibility of a link failure, a warning message can be generated by every alternate node in a path to alert the neighbor before forwarding the packet to failed links, thus eliminating the chance of delay due to route repair or retransmission. In this regard, two eminent protocols are designed AODV-RD(AODV with Reliable Delivery) and AODV-BR(AODV for Broken Route) [7]. Simulation results when compared between AODV-RD,AODV-BR and AODV signifies that there is significant improvement in Packet Delivery Ratio in AODV-RD with a short end-to-end delay than AODV-BR. There is enhancement in performance of the network and improved throughput.

4.7 Security based AODV

Security based routing protocol, a useful algorithm is presented [2] based on Wait time. A Request Reply table is constructed to check the black hole attacks and a method is used to prevent the attack which is based on AODV protocol.

Table 1 describes summary of protocols and issues considered in AODV implementation in this article.

Table 2 describes protocol names along with methods used for routing and their advantages.

Selecting three protocols for analysis

Out of the studied protocols, three protocols QAODV, PHAODV, and EAODV are selected for

Table 1. Issues implemented in protocols.

ISSUEs considered in AODV	Protocol
QoS Support in AODV	Q AODV [6]
Identification of Link Intereference	IA AODV [8]
Multicast tree based on Node Mobility	DMT AODV [1]
AODV in Android	FB AODV [9]
Balanced Enrgy consumption	PC AODV [10]
Reliable Delivery in MANET	AODV RD [7]
Security based AODV	M AODV [2]

Table 2. Summary of studied routing protocols.

Protocol	Mechanism	Advantages
Q-AODV	Considers Network parameter Delay and Bandwidth	Improved throughput
IA-AODV	Global interference and local interference	Fast idea about interference by all nodes
EAODV	Distributed minimum transmission technique used	Supports multi cast routing
PC AODV	Required power is checked before node selection	Minimum power level used
FB-AODV	WCIM* mechanism used	Improved performance in Smart phones
PH-AODV	Residual battery power considered	Even load distribution among nodes
MAODV	Checks security of network	Delay avoidance

*Weighted contention Interface Routing.

Table 3. Comparison of network performance parameters in three leading protocols.

Protocol	Power Consumption	End to end delay	Network life time	PDR	QoS Support
PH AODV	Low	Low	More	High	Medium
E AODV	Medium	Medium	Medium	Medium	Low
Q AODV	High	High	Low	low	High

comparison of performance as they strongly represent QoS aware, Power Efficient, and Delay Managed protocols.

Table 3 below describes network performance parameters considered in different protocols as per our study and analysis. We have selected three main protocols which suitably represent power and delay-based AODV protocols. The following parameters are considered for performance evaluation.

Power consumption

Nodes in MANET have limited battery power. They use this power during transmission, forwarding, and receiving of the data packets. When the nodes are in sleeping state or idle state in waiting mode to receive packets, they are also in active state and consume some power. The power gradually gets reduced as the mobile nodes increase; this is taken as a routing metric.

End to end delay

End to end delay refers to the time taken (average) by the packet to reach the destination, which includes the route discovery time and queue handling time during transmission. A low value for end to end delay shows better protocol performance. It can be calculated as $-\Sigma$(reaching time—sending time)/Σ(No. of hops).

Network lifetime

It is the time period starting from the implementation of the protocol till the end of protocol functionality, which is always application dependent. This depends on the simulation time taken. Many factors greatly affect the network life time, such as sudden death of mobile nodes, link failures etc., due to which the transmission time is more. The increase in battery power directly increases the network lifetime.

PDR (Packet Delivery Ratio)

It is the ratio of the total number of packets successfully delivered to the destination to the total number of packets sent. It can be calculated as $-\Sigma$ no. of packets received at destination/Σ no. of packets initially sent.

A high value of packet delivery ratio ensures better protocol performance.

QoS support

The objective of Quality of Service support is to guarantee the delivery according to specific requirement satisfaction such as bandwidth, timely delivery, Channel Quality, speed, throughput, error rate, prioritized service, etc.

To measure the parameters we have considered three possible values, high, medium, and low, as we are considering three leading AODV protocols for evaluation purpose. So out of the three candidate protocols, the one which exhibits excellent performance is considered as High/more and similar measurements are taken for Low and Medium.

5 SIMULATION AND RESULTS

Ns 2.33 tool was used for simulation purpose. In total, 120 of nodes were scattered randomly in an area of 2000×2000 m during the beginning of the simulation. In the mobility scenario, mobiles nodes were moving in six different speeds within a range of 0–10 m/s. Table 4 shows the simulation parameters used in the simulation.

Table 4. Simulation parameters.

Parameters	Value
No. of Nodes	120
Area Size	2000×2000
Mobility model	Random Way Point
Traffic Type	CBR
Channel capacity	2 Mbps
Communication System	MAC/IEEE 802.11 G
Routing Protocols	QAODV,PHAODV,EAODV

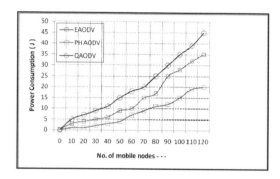

Figure 1. Avg. power consumption vs. node mobility.

From the graph in Figure 1 it can be observed that out of three AODV-based MANET routing protocols studied above, QAODV consumes more power in comparison to PH AODV and EAODV. The reason is that it handles the QoS requirement of the flow, so for processing, more residual energy is utilized by the node as per the protocol requirement. Next, in comparison to PH AODV and EAODV, PHAODV performs better by utilizing less energy and hence, preserving more energy as a result of which the lifetime of the network extends. EAODV consumes more power than PHAODV and less than QAODV and hence exhibits an average network performance showing an average end-to-end delay as compared to QAODV and PHAODV.

Fig. 2 shows end-to-end delay analysis among three leading protocols and the output graph shows that QAODV exhibits higher delay specifically when the speed of the moving nodes gradually increases, This is due to the relocation of nodes due to which path finding and route maintenance takes more time in addition to taking care of QoS processing parameters. PHAODV shows minimum delay out of three protocols which proves that it is more power efficient and delays managed protocol. But it does not take care of the Quality of Service support; hence, its performance can be further

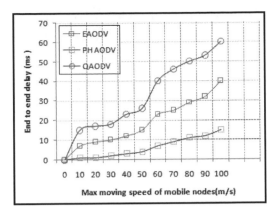

Figure 2. End to end delay analysis.

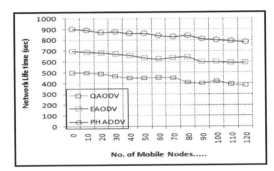

Figure 3. Network lifetime vs. node mobility.

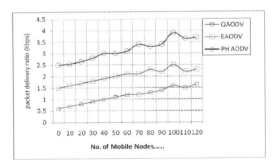

Figure 4. PDR vs. node mobility.

evaluated to check the Quality of Service Support. EAODV handles delay up to some extent so it also exhibits less delay when compared to QAODV.

Fig. 3 shows that PHAODV performs better by lengthening the network's lifetime due to use of better power management techniques in comparison to EAODV and QAODV.

Fig. 4 shows that with increase in node mobility, packet delivery ration steadily increases up to 60

nodes in three protocols, whereas after 60 mobile nodes, there is a slight varation in the increase of PDR in the three protocol cases.

6 CONCLUSION

This research paper first discusses about the need of an energy-efficient routing protocol for MANET. Then it discusses about the issues that are considered by many novel proposals which are implemented as extended AODV protocols and then, three leading protocols are selected on the basis of their power and delay mechanism to compare their performance on the basis of powr consumption and delay handling capability. We conclude that PHAODV, which is based on power management technique, performs better than the other two QAODV and EAODV, as it requires more processing time for Quality of Service satisfaction. This analysis will be helpful for developing a more robust AODV protocol for Mobile Adhoc Networks. In future work, we are planning to develop an innovative AODV-based protocol as per the analysis done in this research work.

REFERENCES

[1] De-gan Zhang, Xiao-dong Songa, b, Xiang Wanga, b, Yuan-ye. 2015, "Extended AODV routing method based on distributed minimum transmission (DMT) for WSN, *Int. J. Electron. Commun. (AEÜ)*, 69 Page No 371–381.

[2] D. Roy Choudhury, Dr. Leena Ragha, Prof. Nilesh Marathe, 2015, "Implementing and improving the performance of AODV by receive reply method and securing it from Black hole attack", *International Conference on Advanced Computing Technologies and Applications (ICACTA-2015), Procedia Computer Science*, Vol.45, Pages 564–570.

[3] H. Safa, MarcelKaram, Bassam Moussa, 2014, "PHAODV: Power aware heterogeneous routing protocol for MANETs" Journal of Network and Computer Applications, 46, Pages 60–71.

[4] Haldar T.K., 2014 "Power Aware AODV Routing Protocol for MANET", Advances in Computing and Communications (ICACC), Fourth International Conference on, 27–29 Aug., Pages 331–334.

[5] R. Madhan Mohan, K. Selvakumar, 2013 "Power controlled routing in wireless ad hoc networks using cross layer approach," *Egyptian Informatics Journal*, Vol. 13, Pages 95–101.

[6] Ling Liu1, Lei Zhu1, Long Lin2, Qihui Wu1, 2012, "Improvement of AODV Routing Protocol with QoS Support in Wireless Mesh Networks" 2012 International Conference on Solid State Devices and Materials Science, Physics Procedia 25 Pages 1133–1140.

[7] A Gabri Maleka, Chunlin LIb, Hiyong Yangc, Naji Hasan. A. Hd, Xiaoqing Zhang, 2012. "Improved the Energy of Ad Hoc On-Demand Distance Vector

Routing Protocol" 2012 International Conference on Future Computer Supported Education, IERI Procedia, 2, Pages 355–361.

[8] Floriano De Rango, Fiore Veltri, Peppino Fazio, 2011, "Interference Aware-based Ad-Hoc On Demand Distance Vector (IA-AODV) ultra wideband system routing protocol" Computer Communications 34, Pages 1475–1483.

[9] Nicola Corriero and Emanuele Covino and Angelo Mottola, 2011, "An approach to use FB-AODV with Android", The 2nd International Conference on Ambient Systems, Networks and Technologies (ANT), Procedia Computer Science 5, Pages 336–343.

[10] Zhu Qiankun, XuTingxue, Zhou Hongqing, Yang Chunying, Li Tingjun, 2011 "A Mobile Ad Hoc Networks Algorithm Improved AODV Protocol," *Procedia Engineering*, Vol. 23, Pages 229–234.

Frontiers in Computer, Communication and Electrical Engineering – Acharyya (Ed.)
© 2016 Taylor & Francis Group, London, ISBN: 978-1-138-02877-7

Pseudo mesh schema based data warehouse architecture employing encryption request algorithm and intelligent sensor algorithm for secured transmission and performance enhancement

Rajdeep Chowdhury
Department of Computer Application, JIS College of Engineering, Kalyani, India

Prasenjit Chatterjee
Wipro Limited, Kolkata, India

Soupayan Datta
Department of Computer Science and Engineering, JIS College of Engineering, Kalyani, India

Mallika De
Department of Computer Science and Engineering, Dr. Sudhir Chandra Sur Degree Engineering College, Dumdum, Kolkata, India

ABSTRACT: Data warehouse is a set of integrated databases purposeful to inflate decision-making and problem solving, espousing exceptionally condensed data. Data warehouse happens to be increasingly more established theme for contemporary researchers with respect to current proclivity towards business and decision-making purview.

A Data Warehouse is a repository which includes all facts and figures of the organization. With the latest inclination of hacking along with the advent of novel hacking expertise, the security for Data Warehouse has befallen to be a decisive fraction.

1 INTRODUCTION

In the proposed Data Warehouse Architecture, appliances of Pseudo Mesh Schema Architecture (Chowdhury, Pal, Ghosh, De, 2012) have been referenced and instantiated, as an alternative to conventional Star Schema Architecture or Snowflake Schema Architecture (Chowdhury, Pal, Ghosh, De, 2012. Chowdhury, Pal, 2010). Initially, a translucent discussion on the said schema is requisite.

As the name entails, the concept of Mesh Topology from Computer Networking is the basis of the Pseudo Mesh Schema Architecture and its functionality (Chowdhury, Pal, Ghosh, De, 2012).

A set of Dimension Tables interlinked with one another, ensuring improvement in the time complexity of a Data Warehouse is presented through the formulation. Fact Tables and Dimension Tables have been employed in their normalized form to diminish redundancy.

The notion of Pseudo Mesh Schema Architecture, conforming interlinking of all Dimension Tables and Fact Tables to one another, have been proposed at the very inception of the referenced work.

The number of links could be precisely calculated using n*(n−1)/2, wherein n embodies the number of tables present within the structure.

The structure is evidently a flexible one, as any increase or decrease of one or more databases within the structure, does not affect the entire schema structure of the Data Warehouse in concern. The structure employs the concept of Fact Views to connect with the existing Dimension Tables.

In a nutshell, the couple of stated figures conjure up the very essence of the originated concept and its impact, on the whole.

The Figures 1 (a) and (b) illustrate the View concept amid various Dimension Tables.

Improvement in time complexity is an allied component and could be adhered with utmost ease. The figure illustrates Pseudo Mesh Schema Architecture with View Table amid intermediary nodes.

The Fact Table consists of k_n number of keys, with those set of keys being copied in the View Table. If the schema consists of n number of Dimension Tables, then for setting up relation amid apiece Dimension Table with others, n*(n−1)/2 numbers of View Tables are requisite.

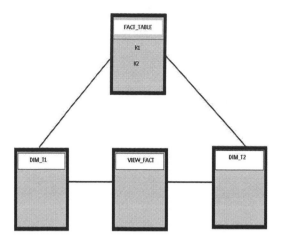

Figure 1a. Mesh Schema having Two Dimension Tables.

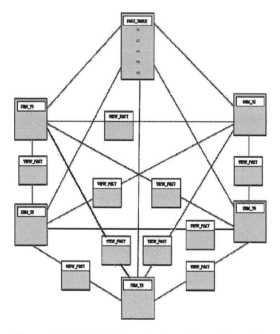

Figure 1b. Mesh Schema having Many Dimension Tables.

A View Table comprises of k_n set of keys, from which a specific k_i key has to be selected, for setting up relation amid two Dimension Tables.

Apiece Dimension Table is anticipated to be exceedingly normalized, in lieu with the architectural design formulated. A Dimension Table could be conked out into n number of sub-dimension tables. The architectural design comprises only of

one Fact Table and n number of Dimension Tables attached to the Fact Table. Nevertheless, the speed could be optimized as there are unswerving relationships amid the various Dimension Tables, which in the process diminishes gratuitous comparisons amid the various Dimension Tables in a Data Warehouse.

The concept not only diminishes the query processing time but also diminishes storage complexity.

Establishment of relationships amid various Dimension Tables requires some space and for massive Data Warehouses, the space gradually intensifies. The relation amid various tables is established through View, which is a logical concept and requires no memory space.

Conclusively, it could be inferred that maintaining a Data Warehouse using Mesh Schema could be relatively intricate, as with increase in Dimension Tables, the number of View Tables also increases. However, increase in View Tables does not engender the problem of space complexity rather it further enhances direct relation amid Dimension Tables.

2 PROPOSED WORK

The prime focus is on how the proposed work could be applied on the proposed schema. The proposed work depicts that whenever any user attempts to access the data, it sends a query to the Warehouse. After a successful execution of the query by the Warehouse Query Processor, the data in the encrypted form would be returned to the apt View Table of the Dimension Table (Chowdhury, Dey, Datta, Shaw, 2014. Chowdhury, Chatterjee, Mitra, Roy, 2014. Chowdhury, Pal, Ghosh, De, 2012).

In the compiled paper, it has been proposed that apiece Dimension Table has an allied Intelligent Sensor Algorithm, which senses and determines whether the data requested by the user is already encrypted or not.

If the data is encrypted, then it dispatches the data unswervingly to the apt View Table, otherwise it sends a request to the Encryptor for encrypting the data.

Here Encryptor is a hardware device which is built into the Data Warehouse Architecture. For the entire Data Warehouse, there is only one Encryptor and hence it might be busy for most of the time.

It is duty of the Intelligent Sensor Algorithm to check whether the Encryptor is busy or not before sending any request. Subsequently, the Intelligent Sensor Algorithm employs another Encryption Request Algorithm.

2.1 General work

Prior to stating any sort of theoretical elucidation, the proposed work necessitates to be ascertained pictorially, followed by indulgence in fragmented discussion.

In Figure 2, apiece Dimension Table contains an Intelligent Sensor Programming which is evident by the acronym IS in the rectangular box alongside apiece Dimension Table. The overview of the Entire Process is illustrated in Figure 3.

The focal rationale for Intelligent Sensor Programming is to determine whether the data that is presently being requested by a user query is encrypted or not. If the data is already encrypted by the Encryptor, then it needs to be dispatched to its apt View Fact, otherwise the data needs to be send to the Encryptor for encryption. For presiding over aptly, the Intelligent Sensor employs an algorithm called Decision Making Algorithm.

Encryptor is a hardware device which would be entrenched along with the Data Warehouse Architecture. Additionally, Encryptor contains the proposed algorithm for encrypting data. For complete Data Warehouse, apiece Encryptor is employed. Hence, when the Intelligent Sensor of the Dimension Table desires to send any data, it might so transpire that Encryptor is busy. To prevail over such an intricate circumstance, Intelligent Sensor employs another algorithm coined Request Algorithm.

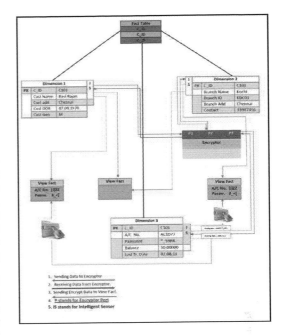

Figure 3.   Overview of the Entire Process.

Encryptor would contain certain keys which would be exclusive for apiece user and for apiece user those keys would be amassed with the Encryptor.

Presumably there are 'n' users; the Encryptor would comprise 'n' keys, that is, K_1, K_2, K_3...K_n (K_ith key for ith user).

The formulation of the immaculate Encryption Request Algorithm and Intelligent Sensor Algorithm ensures that there would be substantial diminution of access time, keeping in mind for the secured transmission, improved intrusion prevention and performance enhancement in due course.

2.2 Encryption request algorithm

Initially Encryptor Semaphore is 0.

After a user is verified, its query is taken into consideration.

Whenever any data enters the Encryptor, the Encryptor_Semaphore becomes 1.

Step–1:
Access the Encryptor_Semaphore variable state.
Step–2:
If the Encryptor_Semaphore is 0, then finally enter the Encryptor.
Now, Encryptor does the following:
i. Set the Encryptor_Semaphore to 1.
ii. Call the Encryption Function.

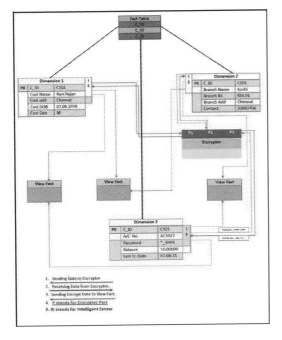

Figure 2.   Functioning of the Intelligent Sensor.

Else

Enter into the Encryptor waiting queue and set the Priority as per the entry position in the waiting queue and wait till the arrival of turn.

When a user is authorized and endorsed to use the data warehouse, an administrator would amass a unique set of keys in the warehouse's Encryptor.

2.3 *Intelligent sensor algorithm*

Initially flag is set to 0.

Step–1:

If flag is 0 then prepare to send the requested data from Dimension Table to Encryptor and do the following:

i. Increment the flag by 1.

ii. Call the Encryption Request Algorithm.

Else

Send the Dimension Table data to its appropriate View Table and Decrement the flag by 1, so that it is again 0.

3 CONCLUSION

The proposed architectural design endows with a proficient and much secure procedure for performing transaction in Data Warehouse.

With the emergent significance of Data Warehouse, it could be effortlessly envisaged that what sort of function the proposed design should engage in.

REFERENCES

Chowdhury, R., Dey, K., S., Datta, S., Shaw, S., 2014, "Design and Implementation of Proposed Drawer Model Based Data Warehouse Architecture Incorporating DNA Translation Cryptographic Algorithm for Security Enhancement", Proceedings of International Conference on Contemporary Computing and Informatics: 55–60.

Chowdhury, R., Chatterjee, P., Mitra, P., Roy, O., 2014, "Design and Implementation of Security Mechanism for Data Warehouse Performance Enhancement Using Two Tier User Authentication Techniques", International Journal of Innovative Research in Science, Engineering and Technology, 3(6): 441–449.

Chowdhury, R., Pal, B., Ghosh, A., De, M., 2012, "A Data Warehouse Architectural Design Using Proposed Pseudo Mesh Schema", Proceedings of First International Conference on Intelligent Infrastructure: 138–141.

Chowdhury, R., Pal, B., 2010, "Proposed Hybrid Data Warehouse Architecture Based on Data Model", International Journal of Computer Science and Communication, 1(2): 211–213.

Saurabh, A., K., Nagpal, B., 2011, "A Survey on Current Security Strategies in Data Warehouses", International Journal of Engineering Science and Technology, 3(4): 3484–3488.

Vieira, M., Vieira, J., Madeira, H., 2008, "Towards Data Security in Affordable Data Warehouse", 7th European Dependable Computing Conference.

Patel, A., Patel, J., M., 2012, "Data Modeling Techniques for Data Warehouse", International Journal of Multidisciplinary Research, 2(2): 240–246.

Chaudhuri, S., Dayal, U., 1997, "An Overview of Data Warehousing and OLAP Technology", ACM SIGMOD Record, 26(1).

Simulation study on variation of attenuation and power handling capacity of micro-coaxial line for different characteristic impedances

Manish Kumar Sharma, Adrish Bhattacharya & Raka Ramona Day
Supreme Knowledge Foundation Group of Institutions, Mankundu, West Bengal, India

ABSTRACT: Rectangular micro-coaxial lines having low loss, high power handling capacity and broad bandwidth have been demonstrated. Such lines of different characteristic impedances are simulated and studied. On comparing them, it is found they behave similar to circular coaxial lines, having minimum attenuation (0.12 dB/cm when simulated upto 20 GHz) for 70 Ω line. But maximum power handling limit is 314.5 W for 8.8 Ω line unlike circular coaxial lines.

1 INTRODUCTION

Recent years show minimization of millimeter wave circuits to micron order. The transmission lines used previously were microstrip line and coplanar wave-guide, which were dispersive and suffered losses. But the introduction of rectangular micro-coaxial line (Brown et al. 2004, Filipovi'c et al. 2006, Ehsan et al. 2005) fabricated through multi-layer photolithography process, evolved transmission lines which are non dispersive, low loss, high power handling capability and broad bandwidth. The various millimeter wave components implemented with rectangular micro-coaxial lines, designed by polystrata process are passive devices such as couplers (Vanhille et al. 2007), antennas (Luki'c et al. 2007), resonators (Vanhille et al. 2006), and filters (Chen et al. 2004).

In this paper, the design and simulation of micro-coaxial line at different characteristic impedances are demonstrated. From the simulation result, comparison of losses and calculation of power handling capacity of different impedance lines are done. In this paper, four impedance (8.8 Ω, 50 Ω, 30 Ω and 70 Ω) line design and simulation have been demonstrated.

2 DESIGN AND SIMULATION OF MICRO-COAXIAL LINE

2.1 *Structural design of micro-coaxial line*

The fabrication of micro-coaxial line involves sequential deposition of copper layers and photoresist on a silicon wafer. Copper layer thickness ranges from 10 μm to 100 μm. The inner conductor is supported by 100 μm long dielectric

straps with periodicity of 700 μm. After copper layer deposition, the photoresist filled in all spaces unoccupied by copper and dielectric straps, is rinsed away ("released") through 200 μm × 200 μm release holes. The structure of micro-coaxial line is shown in Figure 1.

The characteristic impedance of micro-coaxial line can be varied by changing their dimensions (w, b, g, h) as shown by Chen 1960, in equation 1.

$$Z = \frac{376.62}{2\left(\dfrac{b}{g} + \dfrac{w}{h}\right) + 4\left(\dfrac{C_{f1}}{\epsilon} + \dfrac{C_{f2}}{\epsilon}\right)} \tag{1}$$

where, 'Z' is characteristic impedance of the line in ohm, 'b' is height of inner conductor in μm, 'w' is width of inner conductor in μm, 'g' is lateral

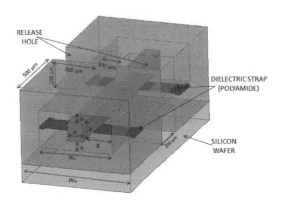

Figure 1. Isometric view of the rectangular micro-coaxial line.

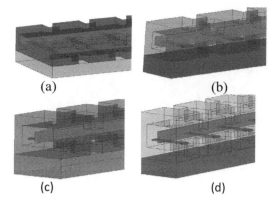

(a)   (b)

(c)   (d)

Figure 2. Isometric view of rectangular micro-coaxial lines of different characteristic impedances: a) 8.8 Ω, b) 30 Ω, c) 50 Ω and d) 70 Ω.

spacing in μm, 'h' is vertical spacing in μm, 'C_{f1}' is fringing capacitance produced by flux disturbance along half of the horizontal side in farads per meter, 'C_{f2}' is fringing capacitance produced by flux disturbance along half of the vertical side in farads per meter. The dimension parameters are labeled in Figure 1.

Thus the dimensions have been calculated for four different characteristic impedances (8.8 Ω, 30 Ω, 50 Ω, 70 Ω). The dimensions of the rectangular micro-coaxial lines of different characteristic impedance are as following: (i) 8.8 Ω line (w = 960 μm, b = 100 μm, g = 120 μm, h = 50 μm, W_a = 1200 μm, W_o = 1400 μm), (ii) 30 Ω line (w = 125 μm, b = 100 μm, g = 37.5 μm, h = 50 μm, W_a = 200 μm, W_o = 400 μm), (iii) 50 Ω line (w = 82 μm, b = 100 μm, g = 159 μm, h = 50 μm, W_a = 400 μm, W_o = 600 μm) and (iv) 70 Ω line (w = 40 μm, b = 75 μm, g = 180 μm, h = 150 μm, W_a = 400 μm, W_o = 600 μm). W_a is the inner width of the outer conductor and W_o is the outer width of the outer conductor of the micro-coaxial line, as shown in Figure 1.

These lines are analyzed with Ansoft High Frequency Structure Simulator (HFSS) with setup frequency ranges from 1 to 100 GHz. The simulated structure of the four micro-coaxial lines with different characteristic impedances (8.8 Ω, 30 Ω, 50 Ω and 70 Ω) are shown in Figure 2.

2.2 Simulation results of micro-coaxial line

A comparison study on the simulation results of micro-coaxial lines of different characteristic impedances (i.e. 8.8 Ω, 50 Ω, 30 Ω and 70 Ω) have been done to understand the difference in their behavior. From the comparison study on input return loss as shown in Figure 3(a), we see

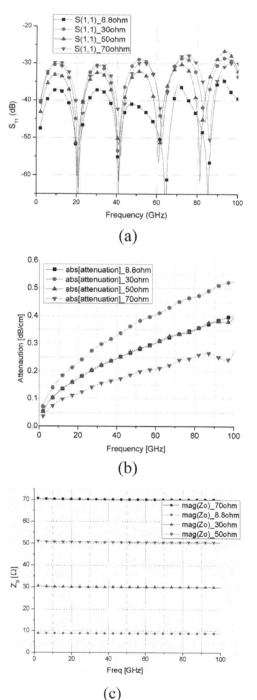

(a)

(b)

(c)

Figure 3. (a) Comparison of input return loss for different characteristic Impedances (8.8 Ω, 30 Ω, 50 Ω and 70 Ω) lines. (b) Comparison of attenuation for different characteristic Impedances (8.8 Ω, 30 Ω, 50 Ω and 70 Ω). (c) Plot of characteristic impedance of the simulated lines (8.8 Ω, 30 Ω, 50 Ω and 70 Ω).

the return loss for different characteristic impedances is below 29.5 dB upto 20 GHz. While the 8.8 Ω line has return loss 37.3 dB up to 20 GHz. When simulated upto 100 GHz, return loss of all lines is below 27.6 dB. Thus the micro-coaxial lines are very broadband circuits.

Since the line is reciprocal and symmetric, the result of S11 and S22 are same. Figure 3(b) shows the comparison study on the attenuation for different characteristic impedances. The attenuation of all lines is below 0.23 dB/cm upto 20 GHz, whereas the minimum attenuation is 0.12 dB/cm for 70 Ω line, when simulated upto 20 GHz. When all lines are simulated upto 100 GHz, maximum attenuation is 0.47 dB/cm (for 30 Ω impedance) and minimum attenuation is 0.25 dB/cm (for 70 Ω line). Thus 70 Ω line has minimum loss similar to circular coaxial line. Figure 4(a) and 4(b) shows the magnitude of electric field distribution and electric field vector distribution of a rectangular micro-coaxial transmission line of characteristic impedance 50 Ω, respectively. The field lines show TEM mode propagation with maximum intensity around the central conductor.

Table 1. Attenuation and maximum power handling limit for different characteristics impedance lines.

Impedances	Attenuations (dB/cm) at 20 GHz	Maximum power handling limit
8.8 Ω	0.1732	314 W
30 Ω	0.2306	88.45 W
50 Ω	0.1731	127.675 W
70 Ω	0.1164	128.75 W

2.3 Measurement of power handling capacity of simulated micro-coaxial line

Maximum Power handling limit is measured by giving maximum power to the line until the air break down voltage (3×10^6 V/m) or polyamide breakdown voltage (2.5×10^7 V/m) is reached. Electric field breakdown was estimated based on magnitude of electric field distribution in HFSS simulations. The transverse field distribution is strongest in the narrow air gaps around the inner conductor corners as shown in Figure 4(a). The maximum power handling limit and attenuation of each micro-coaxial line of different impedances, simulated at 20 GHz is shown in Table 1. It clearly depicts that the 8.8 Ω lines has maximum power handling limit (314 W) and 30 Ω line has minimum power handling limit (88.45 W) unlike circular coaxial line. So power handling capacity depends on structural dimensions.

3 CONCLUSION

The rectangular micro-coaxial lines designed and simulated over 1 to 100 GHz frequency range, have minimum loss 0.25 dB/cm for 70 Ω and maximum loss 0.47 dB/cm for 30 Ω lines. The 8.8 Ω line is found to handle maximum power at 20 GHz. The micro-coaxial lines are very wide bandwidth having maximum loss of 28.655 dB at 100 GHz for 30 ohm impedance. Thus these micro-coaxial lines are broad bandwidth, low loss, and high power handling transmission lines carrying TEM mode propagation.

ACKNOWLEDGEMENT

The authors would like to acknowledge Arijit Majumder of SAMEER Kolkata Centre, Sayan Chatterjee of Jadavpur University and authorities of Supreme Knowledge Foundation Group of Institutions for the help and encouragement provided to carry out the work.

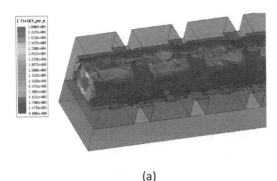

(a)

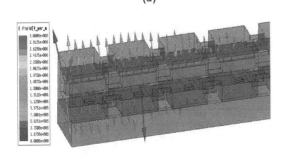

(b)

Figure 4. (a) Magnitude of electric field distribution of micro-coaxial lines of characteristic impedance 50 Ω (isometric view). (b) Electric field vector distribution of micro-coaxial lines of characteristic impedance 50 Ω (isometric view).

REFERENCES

Brown, E.R., et al. (2004). Characteristics of microfabricated rectangular coax in the Ka band. *Microwave and Optical Technology Letters. vol. 40, no. 5*, 365–368.

Chen, R.T., E.R. Brown, & C.A. Bang (2004). A compact low loss Kaband filter using 3-dimensional micromachined integrated coax. *17th IEEE International Conference on MEMS*, 801–804.

Chen, T. (1960). Determination of the capacitance, inductance, and characteristic impedance of rectangular lines. *Microwave Theory and Techniques, IRE Transactions*, vol.8, no. 5, 510–519.

Ehsan, N., et al. (2009). Micro-Coaxial Lines for Active Hybrid-Monolithic Circuits. *IEEE Conference*.

Filipović, D.S. et al. (2006). Modelling, design, fabrication, and performance of rectangular μ-coaxial lines and components. *IEEE MTT-S International Microwave Symposium Digest*.

Lukić, M.V., & D.S. Filipović (2007). Surface-micromachined dual Ka-band cavity backed patch antenna. *IEEE Trans. Antennas Propagat.*, vol. 55, no. 7, 2107–2110.

Vanhille, K.J., D.L. Fontaine, C. Nichols, D.S. Filipović, & Z. Popović (2006). Quasi-planar high-Q millimeter-wave resonators. *IEEE Trans. Microwave Theory Tech.* vol. 54, no. 6, 2439–2446.

Vanhille, K.J., et al. (2007). Balanced low-loss Ka-band μ-coaxial hybrids. *IEEE MTT-S International Microwave Symposium.* 1157–1160.

Frontiers in Computer, Communication and Electrical Engineering – Acharyya (Ed.)
© 2016 Taylor & Francis Group, London, ISBN: 978-1-138-02877-7

Design and development of a web-based teaching performance assessment tool with student feedback and fuzzy logic

Sonali Banerjee, Rajib Bag & Atanu Das
Supreme Knowledge Foundation Group of Institutions, Mankundu, West Bengal, India

ABSTRACT: Assessment of teaching performance is a significant part of any quality education system. Such performance is normally taken in to account by manual student feedback often found biased due to human interventions. The present work is toward designing an automated web-based tool for collection and processing of student feedback. The collected information is habitually imprecise and it is uncertain to arrive at a conclusion while processing that feedback. This work used a fuzzy logic based scheme for integrating the accumulated teaching attributes perceived and supplied by student users for improvement of the teaching quality. The design is evaluated through some case studies where the prototype of the tool is tested in-house and if found useful by the college level education system, it could be advocated for use by others.

1 INTRODUCTION

The education system has become entirely outcome based throughout the world. Outcome-based education is enforced by continuous evaluation and improvement which enhance the quality of education. So, continuous evaluation of the teaching-learning process is very important and could be applicable in the existing education system. There is a huge shortage of properly qualified and experienced teachers particularly in rural India. Teaching performance based evaluation of teachers is usually treated as a mandate to enhance the quality of the education setup. There are various performance evaluation techniques (Adim et al. 2009, Djam et al. 2013, Jamsandekar et al. 2013, Khan et al. 2012, Gupta et al. 2012, Mago et al. 2014, Wei et al. 2009) which have been developed in the recent past. None of these approaches focused on processing of students' imprecise and uncertain feedback processing. It is also observed that users' imprecise and uncertain information processing could be taken care of by using fuzzy logic (Zhou and Huang, 2004; Mitra and Das 2015) This paper proposes a design and advocates for the development of a web based automated student feedback processing tool (named henceforth Teaching Performance Assessment Tool or TPAT), where the teaching performance information is integrated using fuzzy logic to reach a conclusion on the same.

A computer based expert system can be designed based on a set of rules to determine what action to perform when a certain situation is triggered (Khan et al. 2012). To analyze the efficiency of the teaching ability and prevent occurrence of academic uncertainties, this paper also presents the prototype development issues of the said TPAT based on fuzzy approach. Academic auditors and administrators need such a mechanized tool to keep teachers on correct direction. The traditional method of teacher performance evaluation often suffers due to involvement of a lot of qualitative (subjective or formative) data which is required to be transformed in to quantity (objective) for transparent and interpretable conclusion. The considered fuzzy approach does not only automate the decision-making process but can effectively provide periodic confidential feedback mechanisms for decision-makers in any institutions. The TPAT takes a set of parameters as input and accordingly analyzes teaching performance by assigning weights to the corresponding parameters.

The rest of the paper is organized as follows: The following section presents the proposed system, the methodology of teachers' competency analysis, and the proposed method for integrating user feedbacks or inputs using fuzzy rule. Ultimately, the paper ends with conclusions indicating the limitations, contributions, and future scope of the present design and development.

2 FUZZY ASSESMENT METHOD

The teacher assessment processes are described in natural terms to avoid complexity. A fuzzy assessment method is presented here. Let G be a set of grades (For example, G = {Excellent, Very Good, Good, Average, Below Average, Poor}. The evaluator's evaluation can be represented as fuzzy relation (matrix) E:

$$\begin{array}{c} \textit{Grades} \\ \begin{array}{ccccc} & g_1 & g_2 & \cdots\cdots & g_d \end{array} \\ \begin{array}{c} s_1 \\ s_2 \\ \cdots\cdots \\ \cdots\cdots \\ s_n \end{array} \left[\begin{array}{cccc} \mu_{11} & \mu_{12} & \cdots\cdots & \mu_{1d} \\ \mu_{21} & \mu_{22} & \cdots\cdots & \mu_{2d} \\ \cdots\cdots & \cdots\cdots & \cdots\cdots & \cdots\cdots \\ \cdots\cdots & \cdots\cdots & \cdots\cdots & \cdots\cdots \\ \mu_{n1} & \mu_{n2} & \cdots\cdots & \mu_{nd} \end{array} \right] \end{array}$$

where $s_1, s_2,\ldots, s_n$ are the sub-features. $\mu_{ij} \in [0,1]$; $1 \le i \le n$, and $1 \le j \le d$. $\mu_{i,j}$ reflects the relationship between pairs of sub-features and grade. It may be impractical or not easy for evaluators to specify their subjective judgments using such a matrix. Therefore, in order to simplify this process for those evaluators that cannot or do not wish to assign specific numerical values μ_{ij}. A judgment term is an ordered d-tuple $<A_1, A_2, A_3, \ldots\ldots, A_d>$, where $A_j \in [0, 1]$, $1 \le j \le d$. The values of the judgment term, i.e., d-tuple, of a sub-feature, represent the degrees that a specific sub-feature belongs to the grades $g_1, g_2, g_3, \ldots\ldots, g_d$, respectively.

For example, let G = {Excellent, Very Good, Good, Average, Below Average, Poor}, then a statement "The grade of a sub-feature is less than Very Good" can be expressed as < 0.7,0.3,0,0,0,0>, and a statement "The grade of a sub-feature is Average" can be expressed as < 0,0,0.2,0.8,0,0>. Due to the different importance of the sub-features, the evaluators may want to assign different levels of importance to the sub-features. The weights are expressed as a "weight vector" $W = \{w_1, w_2, w_3, \ldots\ldots, w_n\}$.

To calculate the competence, we derived the fuzzy evaluation relation E and the fuzzy weight W. The composition relation is applied to calculate the final grade of the teachers by $W \circ E$. The result is a fuzzy vector (evaluation vector), denoted as Y, containing the membership values in each of the evaluation grades $g_1, g_2, g_3,\ldots, g_d$:

$$Y = W \circ E = (y_1, y_2, \ldots\ldots, y_d)$$

where

$$y_j = (w_1 \circ \mu_{1j}) \oplus (w_2 \circ \mu_{2j}) \oplus \ldots\ldots \oplus (w_n \circ \mu_{nj})$$

and "$\circ$", "$\oplus$" are defined as: algebraic product, $a \circ b$: $s = ab$ and bounded sum, $a \oplus b$: $s = a \oplus b = \min \{1, a+b\}$.

According to the principles of fuzzy classification, we have $y_i = \text{Max}(y_1, y_2, y_3, \ldots\ldots, y_d)$. Thus, the corresponding grade g_i is the grade of the teacher. In the practical evaluation process, we set a parameter δ to check if $|y_i - y_{i+1}| < \delta$, $i < d$. If it happens, then we can say this teachers evaluation is between grades g_i and g_{i+1}. If $|y_i - y_j| < \delta$, and $i < j$,

$j \ne i+1$, $i < d-1$, that means the teacher has strong contradicting sub-feature performance.

3 PROPOSED SYSTEM ARCHITECTURE PROPOSED METHODOLOGY

STEP 1: Gradation
1.1 Initially the set of abstract data values are considered, on which the "Fuzzyfication" will be applied.
1.2 We choose the input variables as the set of abstract data values converted to true values.
1.3 Respective weights are entitled to the input values based on the evaluation criteria.
1.4 Fuzzy logic (Yes/No) is applied in calculating the output true values of the concerned inputs.

STEP 2: Qualitative Evaluation Relation
2.1 The second set of input values is considered and assigned weightage as per the quality.
2.2 Based on the quality inferred on the Teachers, the Evaluation Relation is implied and the judgment value is computed

Case Study:
Assume the set of Grades of a student Feedback is G1 = {YES(1),NO(0)} and G2 = {Excellent (5), Very Good (4), Good (3), Average (2), Below Average(1), Poor (0)}. The Student is asked to evaluate the Teacher's performance in the College.

For Grade 1 (G1), the Criteria are (Table 1):

• Has the teacher covered entire syllabus as prescribed by University?
• Has the teacher covered relevant topic beyond the syllabus?

The weights of the two criteria W1 = {0.75, 0.25} and the 2-tuple is denoted as A1 and A2.

For Grade 2(G2), the Criteria are:

• Delivered lectures with emphasis on fundamental concepts and with illustrative examples?
• Whether any audiovisual aids were used?
• Technical Content
• Communication Skills
• Overall effectiveness
• How do you rate the contents of the lecture delivered?

The weights of the six criteria W2 = {0.3, 0.3, 0.1, 0.1, 0.1, 0.1}

Table 1. Judgment terms and corresponding 2-tuple.

Judgment term	2-tuple
YES	<1,0>
NO	<0,1>

476

Evaluation Result:

$A1 = \begin{pmatrix} 1 \\ 0 \end{pmatrix}$, $W1 = (0.75, 0.25)$

Result 1 = W1 * $A1$ = (0.75, 0.25) * $\begin{pmatrix} 1 \\ 0 \end{pmatrix}$

$A2 = \begin{pmatrix} 0 \\ 1 \end{pmatrix}$, $W1 = (0.75, 0.25)$

Result2 = W1 * $A2$ = (0.75, 0.25) * $\begin{pmatrix} 0 \\ 1 \end{pmatrix}$

From Table 2: Evaluation relation E:

$$\begin{pmatrix} 1.0 & 0.0 & 0.0 & 0.0 & 0.0 & 0.0 \\ 0.0 & 0.0 & 0.2 & 0.8 & 0.0 & 0.0 \\ 0.0 & 0.0 & 0.2 & 0.8 & 0.0 & 0.0 \\ 0.7 & 0.3 & 0.0 & 0.0 & 0.0 & 0.0 \\ 0.0 & 0.0 & 1.0 & 0.0 & 0.0 & 0.0 \\ 0.0 & 0.0 & 1.0 & 0.0 & 0.0 & 0.0 \end{pmatrix}$$

$W2 = (0.3, 0.3, 0.1, 0.1, 0.1, 0.1)$
$Y = W2 \degree E = (0.3, 0.0, 0.0, 0.07, 0.0, 0.0)$
MAX{Y} = 0.3 The position is first. Hence, the positional point is 1.0 (From Table 3)
RESULT = ((Result1+Result2+Positional point of MAX{Y})/m*MAX{W1}+1))*100
= ((.75+.25+1.0)/(2*.75+1))*100 = 80
(m = no of criteria for Grade 1, here m = 2)

From the above result it is found that the performance of the Teacher is VERY GOOD (From Table 4).

Now the calculations on the basis of examples 1 and 2 have been listed in Tables 5 and 6.

Table 2. Judgment terms and corresponding 6-tuple.

Judgment term	6-tuple
Excellent (5)	<1,0,0,0,0,0>
Very Good (4)	<0.7,0.3,0,0,0,0>
Good (3)	<0,0,1,0,0,0>
Average (2)	<0,0,0.2,0.8,0,0>
Below Average (1)	<0,0,0,0,1,0>
Poor(0)	<0,0,0,0,0.2,0.8>

Table 3. Judgment position and corresponding positional points.

Judgment position	Positional point
1st	1.0
2nd	0.9
3rd	0.8
4th	0.7
5th	0.6
6th	0.5

Table 4. Judgment terms and corresponding ranges.

Judgment term	Range
Excellent	90 <= result <= 100
Very Good	80 <= result <= 89
Good	70 <= result <= 79
Average	60 <= result <= 69
Below Average	50 <= result <= 59
Poor	Result 50

Evaluation Result:

$A1 = \begin{pmatrix} 1 \\ 0 \end{pmatrix}$, $W1 = (0.75, 0.25)$

Result 1 = W1 * $A1$ = (0.75, 0.25) * $\begin{pmatrix} 1 \\ 0 \end{pmatrix}$

$A2 = \begin{pmatrix} 0 \\ 1 \end{pmatrix}$, $W1 = (0.75, 0.25)$

Result2 = W1 * $A2$ = (0.75, 0.25) * $\begin{pmatrix} 0 \\ 1 \end{pmatrix}$

From Table 2: Evaluation relation E:

$$\begin{pmatrix} 0.0 & 0.0 & 0.2 & 0.8 & 0.0 & 0.0 \\ 0.0 & 0.0 & 0.2 & 0.8 & 0.0 & 0.0 \\ 0.0 & 0.0 & 0.2 & 0.8 & 0.0 & 0.0 \\ 0.0 & 0.0 & 1.0 & 0.0 & 0.0 & 0.0 \\ 0.0 & 0.0 & 0.2 & 0.8 & 0.0 & 0.0 \\ 0.0 & 0.0 & 1.0 & 0.0 & 0.0 & 0.0 \end{pmatrix}$$

$W2 = (0.3, 0.3, 0.1, 0.1, 0.1, 0.1)$
$Y = W2 \degree E = (0.0, 0.0, 0.36, 0.64, 0, 0)$
MAX{Y} = 0.64. The position is 4th whose positional point is 0.7 (From Table 3)
RESULT = ((Result1 + Result2 + Positional point of MAX{Y})/m * MAX{W1}+1)) * 100
= ((.75 + .25 + 0.7)/(2 * .75 + 1)) * 100 = 68

where, m = no of criteria for Grade 1(here m = 2)

From the above result, it is found that the performance of the Teacher is AVERAGE (From Table 4).

4 SYSTEM EVALUATION AND DISCUSSION

The proposed designed is evaluated by a simple prototype development and case studies are presented above. The flow diagram of the proposed algorithm (shown in Fig. 1) is applied for the assessment of teaching by computer simulation. The student feedback form as a processing tool is also designed with the help of computer programming (shown in Fig. 2). These show that simple functionalities are working even more efficiently than our expectations. It was also found useful in an undergraduate engineering institution functioning around 50 Km away from the main city. In-house testing

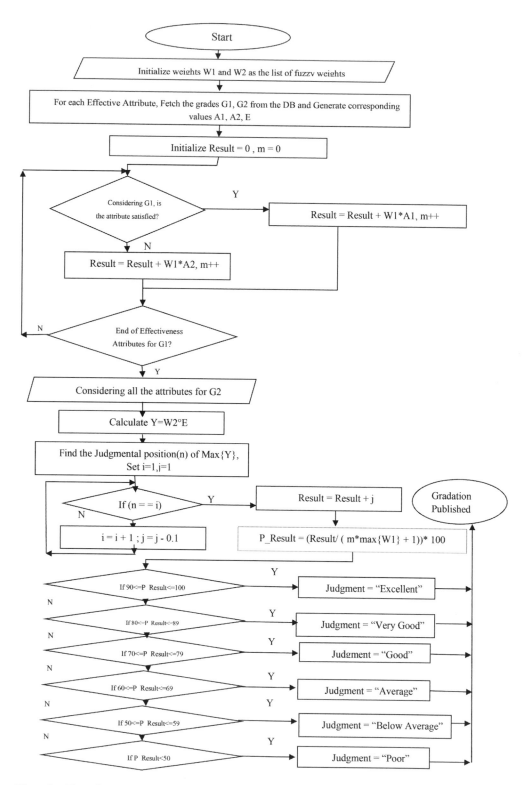

Figure 1.   Flow diagraom of the proposed algorithm.

Figure 2.   Design of the student feedback form.

Table 5.   Calculation on the basis of Example 1.

Sl. no	Effectiveness of teachers in terms of	Linguistic
1	Has the teacher covered entire syllabus as prescribed by University?	Yes
2	Has the teacher covered relevant topic beyond syllabus	No
3	Delivered lectures with emphasize on fundamental concepts and with illustrative examples?	Excellent
4	Whether any Audio visual Aids used?	Average
5	Technical Content	Average
6	Communication Skills	Very Good
7	Overall effectiveness	Good
8	How do you rate the contents of the lecture delivered?	Good

Table 6.   Calculation on the basis of Example 2.

Sl. no	Effectiveness of teachers in terms of	Linguistic
1	Has the teacher covered entire syllabus as prescribed by University?	Yes
2	Has the teacher covered relevant topic beyond syllabus	No
3	Delivered lectures with emphasize on fundamental concepts and with illustrative examples?	Average
4	Whether any Audio visual Aids used?	Average
5	Technical Content	Average
6	Communication Skills	Good
7	Overall effectiveness	Average
8	How do you rate the contents of the lecture delivered?	Good

of the prototype demonstrated that the institutional administration could identify a considerable number of limitations of the existing teaching practices. Even, the teachers themselves can identify some of their limitations along with the existing infrastructural shortage.

5   CONCLUSION

This paper identified a fuzzy logic based approach to assess teaching practices in an academic institution. An assessment tool, called TPAT, has been designed and developed to support the proposed approach. Students' imprecise formative feedback (input) has been transformed automatically to summative judgment terms using fuzzy logic applications in this present research. These judgments are expected to be exceptionally useful for academic audits and formulation of corresponding corrective measures. The proposed design may be enhanced to incorporate various levels and streams of teaching practice assessment. The proposed algorithm may also help as an assessment tool in modern Learning Management Systems (LMS) and Intelligent Tutoring Systems (ITS).

REFERENCES

Amin, H., Khan, A.R. (2009). Acquiring Know ledge for Evaluation of Teachers 'Performance in Higher Education—using a Questionnaire. Int. Journal of Computer Science and Information Security (IJCSIS), 2180–187.

Djam, X., Mishra, A. (2013). Fuzzy Cognitive Map Based Approach for Teachers Performance Evaluation. The Pacific Journal of Science and Technology.

Gupta, L.R. & Dhawan, A.K. (2012). Diagnosis, Modeling and Prognosis of Learning System using Fuzzy Logic and Intelligent Decision Vectors. International Journal of Computer Applications by IJCA Journal 37 (6).

Jamsandekar, S.S., Mudholkar, R.R., (2013). Performance Evaluation by Fuzzy Inference Technique. International Journal of Soft Computing and Engineering (IJSCE), ISSN: 2231–2307, 3(2).

Khan, A., Amin, H., & Rehman, Z. (2012). Application of Expert System with Fuzzy Logic in Teachers' Performance Evaluation. (IJACSA) Int. Journal of Advanced Computer Science and Applications, 2(2).

Mago, J., & Sandhu, P. (2014). Model to Evaluate Education System in India using Fuzzy Logic. Apeejay Journal of Computer Science and Application. ISSN: 0974–5742.

Mitra, M., Das, A. (2015), A Fuzzy Logic Approach to Assess Web Learner's Joint Skills, I.J. Modern Education and Computer Science, 2015, 9, 14–21.

Wei, W.G. (2009). Some geometric aggregation functions and their application to dynamic multiple attribute decision making in intuitionistic fuzzy setting. International Journal of Uncertainty, Fuzziness and Knowledge-Based Systems, 17(2), 179–196.

Zhou, D., Huang, W. (2004), Fuzzy Set Approach to Assessing E-commerce Websites, Proc. of the 10th Americas Conf. on Inf. Sys., NY, pp. 2376–2383, August.

Frontiers in Computer, Communication and Electrical Engineering – Acharyya (Ed.)
© *2016 Taylor & Francis Group, London, ISBN: 978-1-138-02877-7*

A comparison of the performance analysis between the PWM- and SVPWM-Fed induction motor drive

S.K. Bandyopadhyay, S. Mukherjee & Aveek Chattopadhyaya
Supreme Knowledge Foundation Group of Institutions, Mankundu, Hooghly, West Bengal, India

S. Naha
Regent Education and Research Foundation, Kolkata, India

R. Bhadra
Heritage Institute of Technology, Kolkata, India

ABSTRACT: This study presents a comparison between the Space Vector Pulse Width Modulation (SVPWM) and the sinusoidal Pulse Width Modulation (PWM) technique for the induction motor drive application. For this purpose, MATLAB/SIMULINK-based simulation was used. Also, a study of steady-state and transient performance characteristics was carried out for both the cases. Different methods for PWM generation and SVPWM were broadly classified to reduce the switching losses and harmonic content in the system. The simulation results indicate that the transient response is similar for both schemes, but the SVPWM technique has the advantage because it has less harmonic content, which is useful in real-time applications of SVPWM. In addition, due to the low harmonic content, the SVPWM system can avoid overheat and malfunction in the proper system operation. Total harmonic distortion—and continuous wavelet transform-based techniques were used for this purpose to analyze the performance of the PWM and SVPWM techniques for the induction motor drive application.

Keywords: Pulse Width Modulation; Space Vector Pulse Width Modulation; MATLAB/SIMULINK; three phase inverter; induction motor

1 INTRODUCTION

The use of power converters has been increasing day by day in many applications. The electric utility in many cases has used power electronics-based devices because finer and more intelligent controls are needed to improve the electrical grid performance at different voltage levels. Three-phase voltage source inverters are widely used in variable-speed AC motor drive applications since they provide a variable voltage and variable frequency output through the pulse width modulation control. Continuous improvement in terms of cost and high switching frequency of power semiconductor devices and development of machine control algorithms leads to the growing interest in more precise Pulse Width Modulation (PWM) techniques. The PWM technique has been studied extensively during the past decades by researchers. Many different PWM methods have been developed to achieve the following aims:wide linear modulation range; less switching loss; less Total Harmonic Distortion (THD) in the spectrum of switching waveform; and easy implementation and less computation time.

Space Vector Pulse Width Modulation (SVPWM) is one of the most important PWM methods for the applications of a three-phase inverter as it uses the space vector concept to compute the duty cycle of the switches. In other words, it is simply the digital implementation of PWM modulators. The relationship between space vectors and fundamental modulation signals is derived. In this paper, a comparison of the performance analysis between the PWM and SVPWM techniques for induction motor drive applications was made.

2 PWM INVERTER

The DC–AC converter, also known as the inverter, converts DC power to AC power at a desired output voltage and frequency. A complete three-phase PWM inverter consists of three single-phase inverters with sinusoidal control voltages shifted by 120° between phases. Frequency control in a PWM inverter is accomplished by changing the frequency of the input control voltage.

The following parameters needs to be defined for the design of the PWM inverter system:

i. model that defines different processes of pulses,
ii. implementation of the zero-order hold and the relational operator to the sine wave, and
iii. a periodic scalar signal having a waveform that is specified using the time values and output values.

The following parameters also need to be defined to implement a PWM Inverter:

i. sine wave operation in the time-based or sample-based mode,
ii. repeating sequence of a periodic scalar signal having a waveform that is specified using the time values and output value parameters,
iii. zero-order hold in the input for the sample period, and
iv. relational operator for comparing two inputs.

3 SVPWM SCHEME

Space Vector Pulse Width Modulation (SVPWM) was originally developed as a vector approach to the PWM for three-phase inverter applications. The SVPWM method is an advanced, computation-intensive PWM method and possibly the best technique for induction motor drive applications. In this paper, we used this method for less switching losses and thus less THD.

3.1 *Basic principle of SVPWM*

SVPWM is a special switching scheme of six power transistors of a three-phase power converter. A three-phase voltage source PWM inverter model is shown in Figure 1. S1 to S6 are the six power switches of the inverter that shape the output waveform. When an upper transistor is switched on, i.e., S or S3 or S5 is 1 (for the switch-on condition), the corresponding lower transistor is switched off, i.e.,

S2 or S4 or S6 is 0 (for the switch-off condition). Hence, the on and off states of the upper transistors S1, S3, and S5 can be used to control the output waveform.

The main idea behind the SVPWM is to divide the 2D plane into six equal areas, and each of them is called a sector. Each sector is determined by four vectors via V_{i+1}, where $I \in \{1..5\}$. These vectors are called active vectors because when these vectors are applied to the power module, the output voltage of the power module will be greater than zero, i.e., one of the switches will not be switched off. The other two vectors V0 and V7 are called the inactive vector because all switches will be switched off or on. These two vectors allocate the center of the circle C, as shown in Figure 1.

V_{ref} vector will scan all sectors with the time. For every sample time, the sector can be determined to contain V_{ref} for calculating the time period for each vector of the determined sector. Sector implementation is shown in Figure 2; whereas the eight switching states are shown in Figure 3.

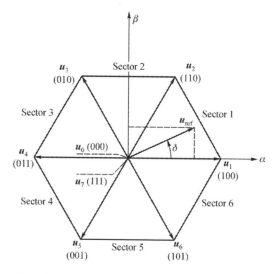

Figure 2. Sector implementation.

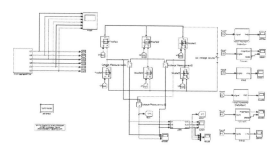

Figure 1. PWM inverter.

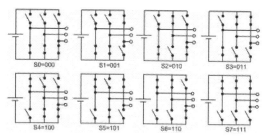

Figure 3. Eight switching states.

Table 1 presents the space vectors and switching states.

$$T_1 = T_S \, x(V_S)/Vdc \, x2/\sqrt{3} \, \sin(60 - \alpha) \qquad (1a)$$

$$T_2 = T_S \, x(V_S)/Vdc \, x2/\sqrt{3} \, \sin(\alpha) \qquad (1b)$$

$$T_0 = T_S - (T_1 + T_2) \qquad (1c)$$

$$V_A(avg) = [(V_{dc}/2)/T_S] \, (T_1 + T_2) \qquad (2a)$$

$$V_B(avg) = [(V_{dc}/2)/T_S] \, (-T1 + T2) \qquad (2b)$$

$$V_C(avg) = [(V_{dc}/2)/T_S] \, (-T1 - T2) \qquad (2c)$$

$$T_{AS} = V_A(T_S/V_{dc}) \qquad (3a)$$

$$T_{BS} = V_B(T_S/V_{dc}) \qquad (3b)$$

$$T_{CS} = V_C(T_S/V_{dc}) \qquad (3c)$$

The procedure of sector selection has been illustrated in Figure 4. Figure 5 shows the block diagram of SVPWM inverter.

Table 1. Space vectors and switching states.

Space vectors	Switching state	On-state switch	Vector definition
V0	[000]	S4,S6,S2	$V0 = 0$
V1	[100]	S1,S6,S2	$V1 = 2/3 \, V_{dc}e^{j0}$
V2	[110]	S1,S3,S2	$V2 = 2/3 \, V_{dc}e^{j\pi/3}$
V3	[010]	S4,S3,S2	$V3 = 2/3 \, V_{dc}e^{j2\pi/3}$
V4	[011]	S4,S3,S5	$V4 = 2/3 \, V_{dc}e^{j3\pi/3}$
V5	[001]	S4,S6,S5	$V5 = 2/3 \, V_{dc}e^{j4\pi/3}$
V6	[101]	S1,S6,S5	$V6 = 2/3 \, V_{dc}e^{j4\pi/3}$
V7	[111]	S1,S3,S5	$V7 = 0$

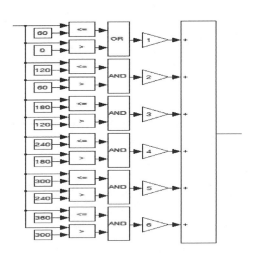

Figure 4. Sector selection.

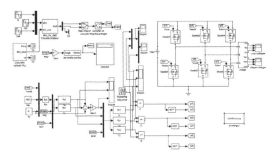

Figure 5. SVPWM inverter.

4 SIMULATION RESULTS FOR THE PWM INVERTER

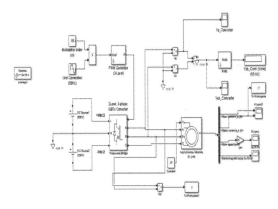

Figure 6. Circuit model.

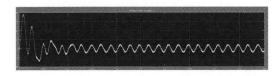

Figure 7. Line current waveform.

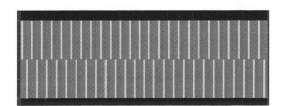

Figure 8. Line voltage waveform.

5 SIMULATION RESULTS FOR THE SVPWM INVERTER

The circuit model is shown in Figure 9.

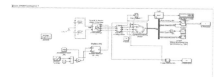

Figure 9. Circuit model.

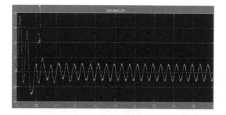

Figure 10. Line current waveform.

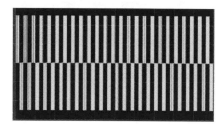

Figure 11. Line voltage waveform.

6 RESULT ANALYSIS OF THE PWM AND SVPWM INVERTERS

Calculated results are presented in Table 2.

Table 2. Percentage of harmonic distortion of the PWM and SVPWM inverters.

Harmonic	SVPWM	PWM
1st (50 Hz)	100	100
2nd (100 Hz)	11.9	16.2
3rd (150 Hz)	2.13	8.16
4th (200 Hz)	0.87	5.79
5th (250 Hz)	0.45	4.48
6th (300 Hz)	0.29	3.8
7th (350 Hz)	0.17	3.27
8th (400 Hz)	0.15	2.85
9th (450 Hz)	0.15	2.51
THD	12.15	40.89

7 RESULT ANALYSIS OF THE PWM AND SVPWM INVERTERS WITH MATHEMATICAL MODEL OF INDUCTION MOTOR

The THD of the PWM inverter is 40.89% and that of the SVPWM inverter is 12.14%. To observe the lower-order harmonic content 'db4', the mother wavelet-based Continuous Wavelet Transform (CWT) was conducted in both the cases. CWT, which is used to achieve the time–frequency representation from a signal x (t), can be defined as follows:

$$X_{WT(\tau,s)} = \frac{1}{\sqrt{|s|}} \int x(t) \cdot \psi * \left(\frac{t-\tau}{s} \right) dt \qquad (1)$$

The transformed signal $\mathbf{X_{WT\,(\tau,\,s)}}$ is a function of the translation parameter τ and the scale parameter s. The mother wavelet is denoted as $\mathbf{\psi\,(t)}$ and * (asterisk) indicates the complex conjugate, which is used in the case of a complex wavelet.

From the graph presented above, the following observations can be made:

i. In Figures 7 and 8, the line current and line voltage waveforms of the PWM inverter are shown, where the presence of harmonic content can be observed, whereas in Figures 10 and 11 showing the waveforms, it is clear that using the SVPWM inverter, the harmonic content for both the line current and voltage is reduced.

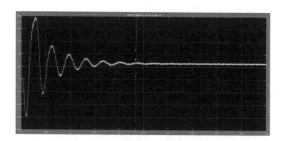

Figure 12. Electromagnetic torque of the induction motor for SVPWM.

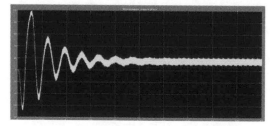

Figure 13. Electromagnetic torque of the induction motor for PWM.

ii. In Figures 12 and 13, the electromagnetic torque waveform is shown, where it can be easily observed that the SVPWM-fed induction motor drive has less harmonic content in comparison with the PWM-fed induction motor drive.

iii. FFT analysis of both SVPWM and PWM inverter fed induction motors is shown in Figures 14, 15, 16 and 17.

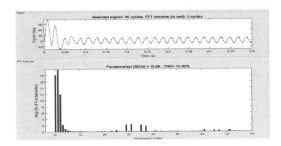

Figure 14. FFT analysis of the current of the SVPWM inverter with induction motor.

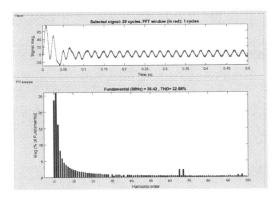

Figure 15. FFT analysis of the current of the PWM inverter with induction motor.

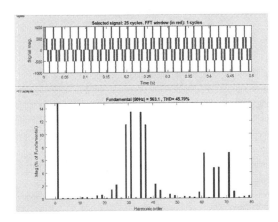

Figure 16. FFT analysis of the voltage of the SVPWM inverter with induction motor.

iv. In this paper, CWT analysis shows that more lower-order harmonic content is present in the PWM inverter than in the SVPWM inverter, as shown in Figures 18–21.

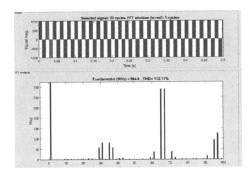

Figure 17. FFT analysis of the voltage of the PWM inverter with induction motor.

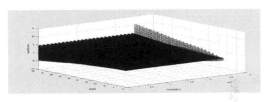

Figure 18. Wavelet analysis of the current of the SVPWM inverter with induction motor.

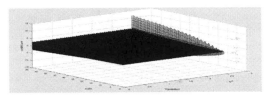

Figure 19. Wavelet analysis of the current of the PWM inverter with induction motor.

Figure 20. Wavelet analysis of the voltage of the SVPWM inverter with induction motor.

Figure 21. Wavelet analysis of the voltage of the PWM inverter with induction motor.

8 CONCLUSION

In this paper, simulation studies clearly show the comparison of the performance analysis between the PWM and SVPWM inverters for induction motor drive applications. To analyze the performance of the PWM and SVPWM techniques, the THD-and CWT-based techniques were used. The analysis revealed that in the PWM technique, THD was higher than that in the SVPWM technique due to the presence of more harmonic content in the PWM inverter to control the induction motor drive. CWT was also conducted in both the cases, which clearly depicted that more lower-order harmonic content are present in the PWM technique than in the SVPWM technique, and that this is the advantage of the SVPWM technique over the conventional PWM technique to control the induction motor drive for the proper system operation.

REFERENCES

Bose, B.K. 1986. Power electronics and ac drives. Prentice hall Inc., Englewood Cliffs, New Jersey.

Erfidan, T., S. Urugun, Y. Karabag and B. Cakir.2004. New Software implementation of the Space Vector Modulation. Proceedings of IEEE Conference. pp. 1113–1115.

Holmes D.G. and T.A. Lipo. 2003. Pulse Width Modulation for Power Converters: Principles and Practice. M.E. El-Hawary, Ed. New Jersey: IEEE Press, Wiley Interscience. pp. 215–313.

Shao-Liang An, Xiang-Dong Sun Member, IEEE, Qi Zhang, Yan-Ru Zhong, and Bi-Ying Ren, "Study on the Novel Generalized Discontinuous SVPWM Strategies for Three-Phase Voltage Source Inverters", IEEE Transactions on Ind. App. May-2011.

Sinusoidal and Space Vector Pulse Width Modulation for Inverter K. Mounika, B. Kiran Babu International Journal of Engineering Trends and Technology (IJETT)—Volume 4 Issue 4- April 2013.

Frontiers in Computer, Communication and Electrical Engineering – Acharyya (Ed.)
© *2016 Taylor & Francis Group, London, ISBN: 978-1-138-02877-7*

A study on the switching life test of low-power-factor Compact Fluorescent Lamps

S. Naha
Department of Electrical Engineering, Regent Education and Research Foundation, Hooghly, India

S.K. Bandyopadhyay, S. Mukherjee & S. Datta
Department of Electrical Engineering, Supreme Knowledge Foundation, Hooghly, India

ABSTRACT   In this paper, we reveal the real picture of a parametric study designed to test key hypotheses regarding the impact of ballast parameters on Compact Fluorescent Lamp (CFL) life. Here, samples of 5 W to 14 W CFL lamps from different lamp manufacturers were operated on duty cycles of 5 min on and 5 min off. The results indicate which parameters seem to have the biggest effect on lamp life and can be used in establishing new performance standards for CFL. The changes made during the life cycle test are remarkable.

1 INTRODUCTION

Fluorescent lamp systems became commercially available in the late 1930s. Since then, they have been used as an alternative to incandescent lamp systems in commercial and industrial applications because they have a higher output, are more efficient, and have longer life. Fluorescent lamps were first developed in the shape of a linear tube. Over the years, these lamps have gradually acquired different shapes, sizes, wattages, and color characteristics. The use of rare-earth activated phosphors allowed the development of Compact Fluorescent Lamps (CFLs) that have a tube diameter of ≤16 mm. They are available in different wattages, shapes, and sizes. CFL products are used to replace incandescent lamps in luminaires with medium screwbase sockets, such as ceiling and wall-mounted luminaires, exterior luminaires, recessed downlights, track lighting, and floor and table lamps. Extended lamp life is one of the main justifications for using CFL products, compared with incandescent lamps. However, a recent survey has revealed that early burnout of CFLs accounted for 22% of the complaints received about the product. If lamp life is one of the main reasons for the consideration of CFLs, a consumer will spend more for a CFL than for an incandescent lamp. Thus, factors that lead to early lamp failures should be better understood. The failure of fluorescent lamps is caused mainly by the loss of electron emissive coating of lamp electrodes, and the electrode temperature directly determines the rate of loss of this emissive coating. Thus, it is important for a ballast to provide an appropriate electrode heating during lamp starting and operation to reduce the damage to lamp electrodes and maintain a long lamp life. However, the electrode temperature is relatively difficult to measure.

The average life of a CFL is defined as the time during which 50% of the lamps reach the end of their individual life. A large number of previous studies have shown that when linear fluorescent lamps are started more frequently than the standard 3 h on/20 min off, they will have a statistically shorter life than their average life. However, none of the previous studies has reported on the effects of frequent switching of CFL products.

2 PROCEDURE

Several numbers of retrofit CFLs were used in this experiment. These lamps were purchased from the market. They were of different wattages (from 5 W to 15 W) and obtained from different companies. The companies were Philips, Havells, and Osram. First, eight lamps were selected for the experiment. More lamps were then used gradually. As in the life testing equipment, the number of lamp holders is even, thus an even number of lamps was selected. Moreover, this selection of even number of lamps is suitable for determining the average life of lamps. At the initial stage, illuminance of different lamps was measured. Then, lamp electrodes were detached from the ballast part of lamps. The resistance value of lamp electrodes was measured by a multimeter at the cold condition or room temperature (generally 25°C), and then these values were recorded. The values were generally in the range of 6 Ω–15 Ω. The electrode and the ballast part of lamps were considered to measure electrical characteristics such as input voltage, input current, output voltage, output current, input watt, lamp watt, input power factor, lamp power factor, total output, and efficiency. This measurement was done using the High-Frequency Data Acquisition System (HFDAS) with computer interface. These parameter values were recorded and saved at a particular location in the computer. V-I characteristics of each lamp were also determined from the HFDAS. After obtaining the electrical characteristics, the lamp electrode and the ballast part were reconnected to test whether the lamp glowed or not. These lamps were placed at the specific holder of the life testing equipment for switching operation, and programming was done for 100 cycles. The counter indicated how many cycles were completed during the experiment. Each cycle consisted of 5 min on and 5 min off duration. Thus, each cycle was of 10 min duration. The above-mentioned measurements were repeated after completing 100 cycles, and these processes were iterated after each 100 cycle of measurement.

In this operation, one question may arise: how many times will this process of measurement take place? The answer is, first, this is an experimental study and thus the time at which the lamps burn out is not predictable, so due to the limitation of the time, the whole process was decided to be run for 1000 cycles. However, it should be noted that while performing the experiment, some lamps failed to start even after about 300 or 400 cycles.

Therefore, the main objective of this study was to determine the effect of some lamp parameters on lamp life and V-I characteristics from time to time. The average life of lamps of different wattages can be predicted in some cases.

3 ANALYSIS

The analysis involved the measurement of lamp parameter values. The lamp parameters were R_c, R_h, R_h/R_c, T_h, and lamp illuminance. In addition to these parameters, V-I characteristics of lamps were also included in the analysis of the lamp life. The lamp parameters are defined as follows:

R_c = resistance of the electrode of the lamp measured at room temperature (25°C), without switching on the lamp.

R_h = hot cathode resistance at the time of ignition of the lamp and R_c is the cold cathode resistance measured without switching on the lamp. It is measured by the ratio of the time-integrated value of the lamp voltage at the starting period to the time-integrated value of the lamp current at the starting period.

$\int_0^{start} Vdt$ correlates well with high-frequency switching life of the lamps tested.

R_h/R_c = ratio which is used to determine the temperature of the lamp electrode at the time of the ignition of the lamp.

T_c = temperature of the lamp electrode at the cold condition (25°C or 298 K).

T_h = temperature of the lamp electrode at the time of the ignition of the lamp. In general, it is measured by the following mathematical expression:

T_h = room temperature $\times (R_h/R_c)^{0.814}$.

The equation is taken as a reference equation from a lamp life predictor.

Simulation results are presented in Figures 1–14 and Tables 1–3.

Comparative analysis was carried out with the following different wattages of lamps for the same parameter values: data 1–5W(1) CFL Philips; data 2–5W(2) CFL Philips; data 3–11W(1) CFL Philips; data 4–11W(2) CFL Philips; data 5–14W(1) CFL Philips; data 6–14W(2) CFL Philips.

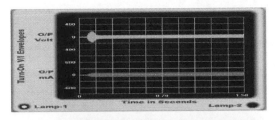

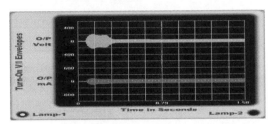

Figure 1. Initial voltage and current waveform for 5W(1) and 5W(2) Philips lamps.

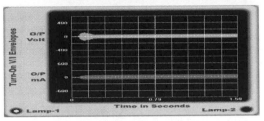

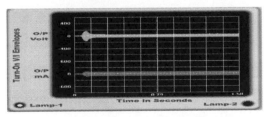

Figure 2. Output voltage and current waveform for 5W(1) and 5W(2) Philips lamps after 400 cycles.

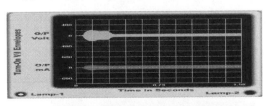

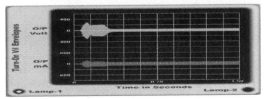

Figure 3. Output voltage and current waveform for 5W(1) and 5W(2) Philips lamps after 1000 cycles.

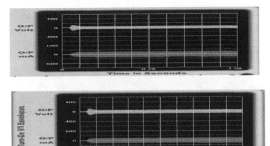

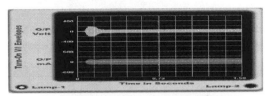

Figure 4. Initial voltage and current waveform for 11W(1) and 11W(2) Philips lamps.

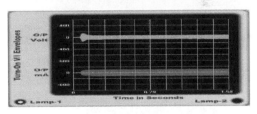

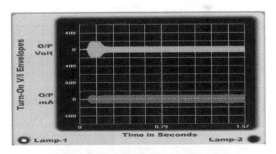

Figure 5. Output voltage and current waveform for 11W(1) and 11W(2) Philips lamps after 400 cycles.

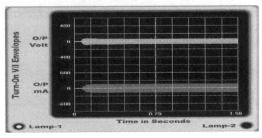

Figure 6. Output voltage and current waveform for 11W(1) and 11W(2) Philips lamps after 700 cycles.

489

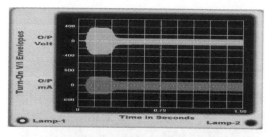

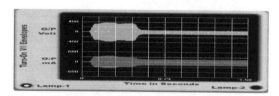

Figure 7. Initial voltage and current waveform for 14W(1) and 14W(2) Philips lamps.

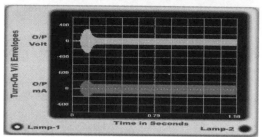

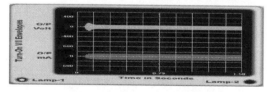

Figure 8. Output voltage and current waveform for 14W(1) and 14W(2) Philips lamps after 500 cycles.

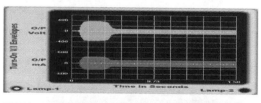

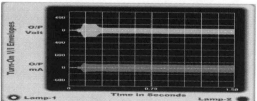

Figure 9. Output voltage and current waveform for 14W(1) and 14W(2) Philips lamps after 1000 cycles.

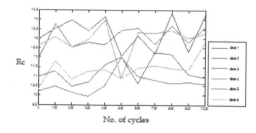

Figure 10. Rc vs. no. of cycles.

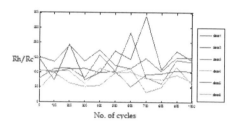

Figure 11. Rh vs. no. of cycles.

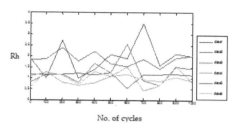

Figure 12. Rh/Rc vs. no. of cycles.

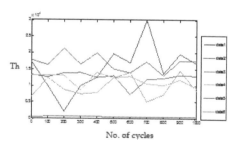

Figure 13. Th vs. no. of cycles.

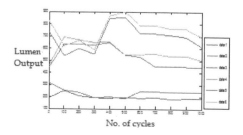

Figure 14. Lumen output vs. no. of cycles.

490

Table 1. Calculated parameters (1st set).

Serial no.	No. of cycles	Ignition time for voltage curve (ms)		Ignition time for current curve (ms)	
1	Initial	0.1896	0.2212	0.1580	0.1896
2	400	0.3160	0.4740	0.3160	0.4424
3	1000	0.3792	0.3634	0.3160	0.3002

Table 2. Calculated parameters (2nd set).

Serial no.	No. of cycles	Ignition time for voltage curve (ms)		Ignition time for current curve (ms)	
1	Initial	0.3160	0.0158	0.3160	0.1738
2	400	0.1580	0.3160	0.1580	0.3160
3	700	0.2528	0.1580	0.2212	0.1580

Table 3. Calculated parameters (3rd set).

Serial no.	No. of cycles	Ignition time for voltage curve (ms)		Ignition time for current curve (ms)	
1	Initial	0.2212	0.2212	0.2054	0.2054
2	500	0.5688	0.1896	0.5372	0.1580
3	1000	0.3950	0.3160	0.3792	0.3634

4 CONCLUSION

Although corroborating data from the standard 5 min on/ 5 min off cycle are not yet available, the switching-cycle testing results indicate that the current ANSI standards are not adequate for selecting ballasts to ensure good lamp life. The R_h/R_c ratio appears to be a very good predictor for lamp life based on lamp starting. A low R_h/R_c ratio indicates that the lamp electrodes have not been heated sufficiently during lamp starting, resulting in reduced lamp life. The standard cycle results also provide an insight into the impact of the lamp operating parameters (e.g., lamp operating current, ballast factor) on lamp life. For higher wattages of lamps (11–15 W), the rate of failure is higher than that for lower wattages of lamps, as revealed by the experiment. The probable factors that the quality of the ballast components is poor or that the heat dissipation is improper reduces the life of the components used in CFLs.

REFERENCES

[1] Hand Book on Lighting Engineering: IESNA
[2] *Frequently Switched Instant Start Fluorescent Systems* by N Narendran, T Yin, C O Rourke, A Bierman and A Malyagoda Lighting Research Center, Rensselaer Polytechnic Institute, Troy, NY 12180.

Frontiers in Computer, Communication and Electrical Engineering – Acharyya (Ed.)
© 2016 Taylor & Francis Group, London, ISBN: 978-1-138-02877-7

Demonstration of edge detection technology using the coupling of the IR sensor and robotics

Subhankar Mukherjee
Supreme Knowledge Foundation Group of Institutions, Mankundu, Hooghly, India

Suranjan Ghosh
Department of Electronics and Communication Engineering, West Bengal University of Technology, West Bengal, India

ABSTRACT: Edge detection technology is best demonstrated by the edge avoider robot, a mobile device, which "senses" and "avoids" the absence of the surface below it. Demonstration of the concept of edge detection technology is done with the help of a microcontroller that also helps the robot to prevent it from falling. The specific microcontroller used here is the P89V51RD2 microcontroller, also known as the "bit and byte processor", which is programmed accordingly for the specified action of edge avoiding. The goal of the robot is to drive forward as long as it does see the floor, turn away from the side when one IR sensor does not detect an obstacle, and withdraw when neither IR sensor detects an obstacle. This activity could be done on a tabletop, but the results might be disastrous to the robot if something goes wrong.

1 INTRODUCTION

Robotics is defined as "A reprogrammable, multifunctional manipulator designed to move material, parts, tools, or specialized devices through various programmed motions for the performance of a variety of tasks." Robots in movies are portrayed as fantastic, intelligent, and even dangerous forms of artificial life. However, today's robots are not exactly the walking and talking intelligent machines of movies, stories, and our dreams. Today, we find most robots working for people in factories, warehouses, and laboratories. In future, robots may show up in other places: our schools, our homes, and even our bodies. Robots have the potential to change our economy, health, standard of living, knowledge, and the world in which we live. As technology progresses, we find new ways to use robots. Each new use brings new hope and possibilities, but also potential dangers and risks. An Edge Avoider Robot (EAR) is a robot which avoids the edges of, for example, a table and does not fall down. The robot will be autonomous. This robot working depends on the sensor used to detect the edge and warn the brain of the robot to change the direction. The main idea behind edge detection using Infrared (IR) is to send IR light in a certain direction: if an edge is present not too far from the sensor, then IR will be reflected back and detected by the sensor. Generally, an IR-based obstacle detection sensor is used to perform the edge-avoiding operation. The mechanical assembly consists of chassis and two tires mounted on a two-geared DC motor. The DC motor is controlled by the main board. The main board consists of a microcontroller circuit and also a motor driver circuit with the remaining ports expansion for sensor interfacing. Sensor is an IR proximity sensor that generates a high signal pulse when something obstructs its path. This high signal pulse is passed to the microcontroller ports by which the microcontroller decides where to rotate the chassis or the robot as programmed in it.

2 A CONCEPT OF EDGE DETECTION

Edge detection technique is demonstrated using the IR-based line detecting module. The modules are connected in front of the iBOT so as to detect the edge early and take a proper action in time. The distance maintained between the sensors is greater than the width of the iBOT, considering the turning radius of the wheels. When the surface is detected, the IR module sends a high pulse signal to the microcontroller, and when the edge is detected, the IR module does not reflect the light, thus giving a low pulse to the microcontroller.

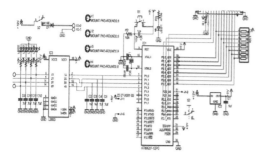

Figure 1.   Main circuit board.

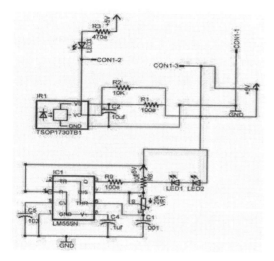

Figure 2.   Sensor board.

3   CIRCUIT DIAGRAM

Figures 1 and 2 illustrates the schematics of the main circuit board and sensor board.

4   PCB TESTING

Step 1: Testing the Power Supply Section

Switch on the supply and check whether power LED glows or not. If it does not glow, then check its polarity. Check the voltage at the terminal block whether it receives the power or not. Then, check the voltage at the voltage regulator IC. See the reading of the multimeter at the time of testing of its both pins, i.e., input and output. Its input pin has AC supply through the adaptor and its output is +5 V.

Step 2:   Testing the Microcontroller Section

Check the voltage at pin 40, i.e., Vcc whether it gets +5 V or not. If the multimeter shows the 5 V

reading, then it means that it is normal, otherwise check the tracks of PCB and connections. Check the ground connection at pin 20 of the microcontroller. Check the voltage at pin 31 of the microcontroller. It must show the +5 V reading on the multimeter, which indicates that the microcontroller uses the internal memory. Check the voltage at pin 9 of the microcontroller. First, it shows 0 V and then after pressing the Reset switch only once, it starts showing the +5 V reading on the multimeter.

Step 3: Testing the communication between R_x and T_x

Test and program the microcontroller supported over the robot microcontroller chassis using the software named 'Flash Magic' in a Window-installed PC or laptop. A serial cable for programming as well as a serial port are needed at the back of the PC. If there is no serial port in the PC or laptop, then use the serial to USB converter and program the microcontroller through the USB port.

The robot microcontroller's controller board is tested and programmed by connecting it to a PC running Microsoft Windows. For this purpose, the serial port (9-pin male connector) should be used. A 9-pin male–female cable is included to connect the PC's serial port to the robot microcontroller. It may be necessary to remove the modem or other devices (except the mouse) to release the serial port. Switch boxes are also available from computer stores that allow one serial port to serve two devices. Most computer stores also offer serial port cards that can be added to the computer.

4.1   Testing the communication between the RX and TX of the microcontroller

Plug in the DB9 serial cable male connector into the female serial connector over the board. Plug in the DB9 serial cable female connector into the male serial port at the back of the PC. If there is no serial port in the PC, use the serial to USB connector and plug in USB into the PC.

Short pin 10 and pin 11 of the microcontroller with the help of a screwdriver to check that the transmitted data should be received by the microcontroller.

4.2   Checking the COM port

Set the device to 89C51RD2 and then set COM port to COM3 (COM port of the PC or laptop) (See Figure 3).

Go to the "Tools" and then to the "Terminal." (See Figure 4)

A window is opened. If the input is obtained as it is at the output, then the pins of the microcontroller

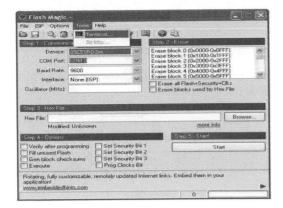

Figure 3.　Flash Magic software.

Figure 4.　Terminal.

Figure 5.　IR sensor testing.

are working properly. This means that the microcontroller receives the same data that are transmitted. If output is not obtained, then check the communication through MAX232 IC.

4.3　*Testing the communication between the RX and TX of the MAX232 IC section*

If we want to check the communication between the R_x and T_x of MAX232 IC, then we have to short its T_x and R_x pins.

Short its pin 11 and pin 12, i.e., T_x and R_x, respectively, with the help of a screwdriver.

Plug in the DB9 serial cable male connector into the female serial connector over the board.

Plug in the DB9 serial cable female connector into the male serial port at the back of the PC.

If there is no serial port in the PC, use the serial to USB connector and plug in USB into the PC.

Repeat the same process with Flash Magic software.

If no output is obtained, then check the communication in the DB9 connector.

4.4　*Testing the communication between the RX and TX of the DB9 connector section*

If we want to check the communication between the R_x and T_x of the DB9 connector cable, then we have to short its T_x and R_x pins. Short its pin 2 and pin 3, i.e., R_x and T_x, respectively, with the help of fly leads, and repeat the same process with Flash Magic software. If no output is obtained, then replace the connector. If no output is obtained while testing with MAX232 IC and the output is obtained while testing the DB9 connector, then replace the MAX232 IC after checking the PCB tracks. Also, check whether there is any unwanted shorting on the PCB and then unshort them. If no output is obtained from testing the communication with the microcontroller, then check MAX232 and get the output. Thereafter, replace the microcontroller after checking the unwanted shorting between the PCB tracks.

Step 4: Testing the L293D IC section

Checking the voltage through the multimeter at the following pins of L293D IC:

Pin 1, pin 2, pin 9, pin 15, and pin 16 to +5 V
Pin 4, pin 5, pin 7, pin 10, pin 12, and pin 13 to ground.

Check the continuity between pin 8 of L283D IC and pin 1 of the voltage regulator IC 7805.

5　IR SENSOR TESTING

First, we need a small screwdriver to adjust the potentiometer VR1. We then connect the ground and the supply with CON1. In the circuit board, pin no. 1 is for GND, pin no. 2 is for output, and pin no. 3 is for supply (+5 V DC (preferred), +9 V

DC, or +12 V DC). Next, we turn the potentiometer VR1 towards the left position and then place our hand at a curtain distance away from the IR sensor LED. We then start turning VR1 towards the right position until LED3 just glows, and then we remove our hand from the IR sensor LED and it will turn off LED3. Again, when we place our hand or any object near the IR LED3, it will just glow. If the IR LED does not glow, we have to adjust VR1. Figure 5 shows the arrangement for IR sensor testing.

6 IR SENSOR AND MAIN PCB CONNECTION

Figure 6 illustrates the sensor board and main board connection. First, we connect the CON1 terminal to the IR sensor PCB with a 3-pin terminal named SENSOR over the main board PCB using a 3-wire cable connector. Then, we place the microcontroller over the 40-pin IC base, which is pre-programmed with the edge avoider program. We then connect +12 V power supply to the POWER terminal block. Next, we press the POWER ON switch to start the EAR.

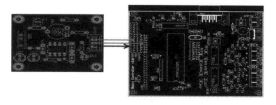

Figure 6.   Sensor board and main board connection.

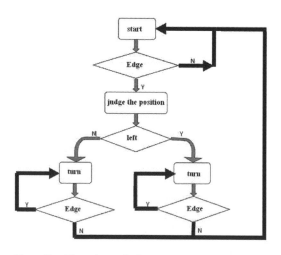

Figure 7.   Flow chart of edge avoidance.

7 FLOW CHART OF EDGE AVOIDANCE

The flowchart of the edge avoidance algorithm is shown in Figure 7.

8 OBSERVATION

The following points were observed from the hardware experiment:

1. The EAR was found to avoid edges.
2. The EAR was found to turn and withdraw away from edges.
3. Back and forth movements.
4. Right and left movements were found to be functional.
5. Speed of the motors attached to the wheels can be changed by changing the microcontroller programming.

9 CONCLUSION

The EAR was found to avoid edges and thus prevented it from falling. Therefore, it can be seen that it demonstrates the concept of edge detection along with the implementation of robotics to some extent. When the surface is detected, the IR module sends a high pulse signal to the microcontroller, and when the edge is detected, the IR module does not reflect the light, thus giving a low pulse signal to the microcontroller. Thus, this would greatly help in detecting edges when vehicles are driven in hilly areas and thus would reduce the chances of accidents.

REFERENCES

Basavaraj Prof. B. & Dr. H.N. Shivashankar., "BASIC ELECTRONICS" Second Edition VIKAS publication.
LMP90100/LMP90099/LMP90098/LMP90097 Sensor AFE System: Multi-Channel, Low Power 24-Bit Sensor AFE with True Continuous Background Calibration, SNAS510P–JANUARY 2011–REVISED MARCH 2013, Texas Instruments.
Resistor Array, Document Number: 31043248 Revisions: 13-Oct-08.
Kasap, S.O. Principles of Electrical Engineering Materials and Devices, McGraw-Hill. New York. 2002.
Marshall Leach, Jr., W. Professor, Georgia Institute of Technology, School of Electrical and Computer Engineering.

Frontiers in Computer, Communication and Electrical Engineering – Acharyya (Ed.)
© 2016 Taylor & Francis Group, London, ISBN: 978-1-138-02877-7

Measurement and analysis of harmonics contribution by HID lamp systems

S. Datta, S.K. Bandyopadhyay & S. Mukherjee
Supreme Knowledge Foundation Group of Institutions, Hooghly, West Bengal, India

ABSTRACT: Harmonic contributions of electrical devices are one of the major reasons behind the poor quality of power. High Intensity Discharge (HID) Lamps are known as non linear electrical loads. Hence, these types of lamps are a major contributor of harmonics in electrical distribution systems. The work presented in this paper, emphasizes on measurement of harmonic contribution by 250 W High Pressure Sodium Vapour (HPSV) lamp and 250 W Metal Halide (MH) lamps when driven by electromagnetic ballast at 240V, 50 Hz power supply. These types of lamp systems are widely used in outdoor lighting especially in road lighting & sports lighting applications and also for indoor lighting such as in shopping malls. The purpose is to know the harmonic contribution by such type of lamp systems and to draw a comparison.

Keywords: high intensity discharge lamp; lamp voltage harmonics; lamp current harmonics; electromagnetic ballast; Total Harmonic Distortion (THD); power quality

1 INTRODUCTION

Harmonics in electrical circuits mainly depend upon the nature of the load connected with the circuit. There are two types of electrical loads viz. linear and non-linear. Non-linear types of loads do not follow linear relation between voltage and current. V-I characteristics of such loads cause the current waveform to distort and it results in production of harmonics. All discharge lamps (both low and high pressure) are non-linear loads owing to its discharge phenomena [1]. When the supply voltage itself is distorted, considerable amount of rise in lamp current Total Harmonic Distortion (THD) has been observed and along with reduction in light output. Hence, it can be said that in presence of supply voltage harmonics both electrical and photometric performances of lamps get degraded[2]. In comparison of conventional incandescent lamps, discharge are harmful to power quality. Mass deployment of CFLs can cause poor power quality [3]. In case of High Intensity Discharge (HID) lamp systems, use of electronic ballasts reduce the harmonic contribution of the lamp systems [4]. In most of the earlier works, only one lamp system, viz. CFL, FTL or HPSV is considered separately but the comparison of harmonic contribution of different lamp systems had not been evaluated. This work emphasizes on the methods of measurement of harmonic contribution, both in voltage and current signals, by 250 W High Pressure Sodium Vapour and Metal Halide lamp systems. To compare the harmonic contributions of these types of lamp systems, an analysis has also been carried out with recorded voltage and current waveforms.

2 LAMPS AND BALLASTS USED FOR EXPERIMENTATION

For the purpose of above said measurements different lamps, electromagnetic ballasts and ignitors are used. In Tables 1, 2 and 3 lists of the lamps, electromagnetic ballast and ignitors used for the measurement are shown.

Table 1. List of lamps.

Lamp type	Lamp wattage (Watt)	Maker's name	Model/ trade name	Supply voltage rating (Volt)
High Pressure Sodium Vapour	250	Philips	SON-T Plus 250 W	200–250
Metal Halide	250	Philips	HPI-T Plus 250 W	240

Table 2. List of ballasts.

Ballast type	Maker's name	Sl. No.	Rated voltage & frequency	Lamp watt (Watt)	Rated current (Amp)	Rated pf
Electro-magnetic Ballast	Philips	C8325-X100	240V, 50 Hz	250 (SON +MH)	3	0.4

Table 3. List of ignitors.

Ignitor trade name	Maker's name	Specified lamp wattage	Lamp voltage (Volt)	Frequency (Hz)
SN 58/01	Philips	**SON(-T):** 100–150–250– 400 W	220–240	50/60
		MH: 35–70–150–250 W		
		CDM: 35–70–150 W		
SI 51/02	Philips	**HPI(-T):** 250–400 W	220–240	50/60

As pointed out in Table 2, only one ballast is used to drive both HPSV and MH lamps of same rating.

3 AQUISITION AND ANALYSIS OF VOLTAGE AND CURRENT WAVEFORMS OF LAMP SYSTEM

In a normal HID lamp circuit there are mainly two sources of harmonics; discharge lamp itself and the inductive electromagnetic ballast. Here the harmonic contribution of lamps in presence of electromagnetic ballast in the circuit is examined. Capacitor is not used for power factor correction.

3.1 Circuit diagram

The circuit diagram of an HID lamp system driven by electromagnetic ballast, is shown in Figure 1. A very low resistance (shunt) of value 0.194 Ω is put in series in the circuit and two terminals are provided for the measurement of voltage waveform across it with the help of an oscilloscope. This resistive element being a linear electrical load has a voltage waveform which is similar to the waveform of current following through it. Hence, by recording the waveform of voltage across it we can record the nature of the waveform of circuit current. For analysis of data values of current was found out from the recorded voltages as the value of resistance of that shunt was known.

3.2 Experimental steps

The steps given below are followed to obtain the wave forms and record data:

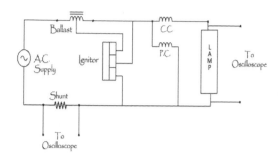

Figure 1. Circuit diagram of HID lamp system with conventional control gear.

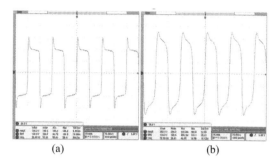

(a)   (b)

Figure 2. Voltage waveform of (a) 250W HPSV lamp and (b) 250W MH lamp.

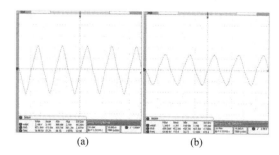

(a)   (b)

Figure 3. Current waveform of (a) 250W HPSV lamp and (b) 250W MH lamp.

- STEP 1: Supply voltage is set at 240V and frequency kept constant at 50 Hz. An AC programmable power source is used for feeding the lamp.
- STEP 2: At stable condition of lamp, the waveforms of lamp voltage and current are recorded with oscilloscope.
- STEP 3: The data obtained are then analyzed with the help of a developed MATLAB program to evaluate the different components of the harmonics.

- STEP 4: From the evaluated harmonic components, Total Harmonic Distortion is determined.

3.3 Acquired waveform

Lamp voltage and circuit current waveforms recorded with oscilloscope are shown below.

In Figure 2(a) and 2(b), the voltage waveforms of 250 W HPSV and MH lamps driven by electromagnetic ballast are shown respectively. In Figures 3(a) and 3(b), current waveforms of 250 W HPSV and MH lamps driven by electromagnetic ballast are given respectively.

3.4 Analysis of acquired waveforms

The analyses of above measured waveforms (Figs. 2 and 3) have been done with developed MATLAB programme to compute the harmonic contents of the voltage and current. The analyzed data are tabulated below for comparison between the harmonic contents in voltage and current waveforms of different types of lamps having same wattage ratings.

It is shown in Table 4, that 250 W HPSV lamp driven by electromagnetic ballast has much higher harmonic content in its voltage waveform compared to that present in voltage waveform of 250 W MH.

Table 4. Voltage Harmonics of 250 W HPSV & 250 W MH lamps when driven by electromagnetic ballast (at 240V, 50 Hz).

| Harmonic order | Voltage Harmonics | | | |
| | 250 W HPSV | | 250 W MH | |
	Volt	% of Fundamental	Volt	% of Fundamental
Fundamental Component		127.463		179.676
2nd Harmonic	0.089	0.069824	0.7982	0.444244
3rd Harmonic	47.4827	37.25214	52.8446	29.41105
4th Harmonic	0.0134	0.010513	0.205	0.114094
5th Harmonic	30.4385	23.88026	21.7715	12.11709
6th Harmonic	0.0153	0.012003	0.0836	0.046528
7th Harmonic	20.4167	16.01775	8.8961	4.95119
8th Harmonic	0.0094	0.007375	0.0713	0.039683
9th Harmonic	14.2638	11.19054	4.6793	2.604299
THD (%)		44.6		22.9
RMS (Volt)		90.19		131.5

Table 5. Current Harmonics of 250 W HPSV & 250 W MH lamps when driven by electromagnetic ballast (at 240V, 50 Hz).

| Harmonic order | Current Harmonics | | | |
| | 250 W HPSV | | 250 W MH | |
	Ampere	% of Fundamental	Ampere	% of Fundamental
Fundamental Component		3.771744		2.24155
2nd Harmonic	0.001061	0.028121	0.002828	0.126183
3rd Harmonic	0.421086	11.16423	0.201527	8.990536
4th Harmonic	0.000707	0.018748	0.000354	0.015773
5th Harmonic	0.06364	1.687289	0.028992	1.293375
6th Harmonic	0.000707	0.018748	0.001061	0.047319
7th Harmonic	0.076722	2.034121	0.018031	0.804416
8th Harmonic	0.000707	0.018748	0.000707	0.031546
9th Harmonic	0.003182	0.084364	0.002121	0.094637
THD (%)		11.49		9.1
RMS (Ampere)		3.8		2.25

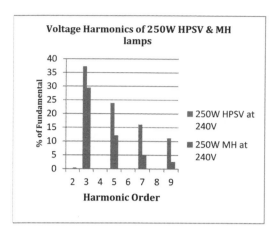

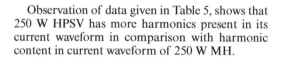

Figure 4a. Comparison between voltage harmonics of 250W HPSV & MH.

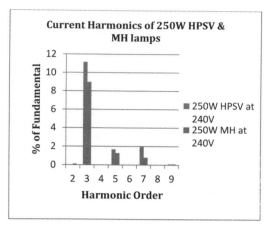

Figure 4b. Comparison between current harmonics of 250W HPSV & MH.

Observation of data given in Table 5, shows that 250 W HPSV has more harmonics present in its current waveform in comparison with harmonic content in current waveform of 250 W MH.

4 COMPARISON OF HARMONIC CONTRIBUTION OF DIFFERENT LAMPS USING ANALYZED DATA

With the help of data provided in Tables (4 and 5), charts have been prepared to compare harmonic contribution by different lamps driven by electro-magnetic ballast.

Comparison done in Figure 4(a) reveals that at 250 W rating MH lamp is having considerably lower harmonic content in their voltage waveforms in comparison to 250 W HPSV lamp. From Figure 4(b) it is seen that at 250 W rating MH lamp is having considerably lower harmonic content in their current waveforms in comparison to 250 W HPSV lamp.

5 CONCLUSION

From this work the following conclusions can be drawn:

- Metal Halide (MH) lamp working with electro-magnetic ballast generates much less harmonics in their voltage and current waveforms than that of High Pressure Sodium Vapour lamps.
- In both the cases of HPSV and MH; for higher power rated lamps (400 W), harmonic content in the current waveform is less than that of the lower power rated lamps (250 W).
- In all the cases voltage waveforms are more distorted than current waveforms.

REFERENCES

Arseneau R., Ouellette M, "The Effects of supply harmonics on the performance of CFL", IEEE Transactions on Power Delivery, Vol. – 8, 2002.

Chang, Y.N., Moo, C.S., Jeng, J.C., "Harmonic Analysis of FTL with Electromagnetic Ballast", Tencon '93. Proceedings. Computer, Communication, Control and Power Engineering.1993 IEEE.

Faranda R., Guzzetti S., Leva S., "Power Quality in Public Lighting Systems", IEEE 978–1-4244–7245—1/10, 2011.

Richard M.K., Sen P.K., "CFL and Their Effects on Power Quality and Application Guidelines", IEEE 978–1-4244–6395–4/10, 2010.

Frontiers in Computer, Communication and Electrical Engineering – Acharyya (Ed.)
© 2016 Taylor & Francis Group, London, ISBN: 978-1-138-02877-7

Optimization of electrode-spacer geometry of a gas insulated system for minimization of electric stress using SVM

Suryendu Dasgupta & Abhijit Lahiri
Department of Electrical Engineering, Supreme Knowledge Foundation Group of Institutions, Mankundu, West Bengal, India

Arijit Baral
Department of Electrical Engineering, Indian School of Mines, Dhanbad, Jharkhand, India

ABSTRACT: In order to design a High Voltage (HV) system, it is compulsory to have a thorough knowledge of electric field distribution on and around the system. The design should ensure reliability of the system to withstand the electric stress and simultaneously it should not be over dimensioned. The dielectric properties of the insulating material is also of great importance while designing the insulation of HV apparatus between phases and the earth and also between phases. The electric field intensity on the surface of the electrodes also determines the withstand voltage of the external insulation of HV apparatus. Thus, a comprehensive study of the electric field distribution on and around HV equipments is of great significance. For better economy and increased system reliability, it often becomes essential to optimize the dimensions of the HV equipments. In this paper live electrode contour of a gas insulated substation (GIS) is done using support vector machine (SVM). The GIS system considered in the present case is normally preferred for 12 kV, 36 kV, 72.5 kV, 145 kV, 245 kV, 420 kV and above.

1 INTRODUCTION

A considerable effort has been given by different researchers to optimize electrode-spacer contours in order to minimize electric field intensity on and around the HV equipments. P. Grafoner and H. Singer (Grafoner & Singer 1975) studied on optimization of electrode and insulator contours. H. Singer (Singer 1979) optimized clectrode geometries. Both of these methods were based on classical approaches.

A computer based design of electrode and insulator profile was reported by H. Gronewald (Gronewald 1983) where a functional relationship was established between the electric field and the geometry of the dielectric boundary considering the mathematical expression of tangential field intensity. Algorithm for optimization of insulator contour was developed based on the given tangential stress distribution over the surface of the insulator.

Stih (Stih 1986) did a study on design of HV insulators. The method was based on Integral Equation Method (IEM) where the surface charge density was approximated with bi-cubic spline function. For contour optimization circular arcs were considered for smooth approximation of the modified contours which thereby provided a relatively simple geometrical description of the modified contour. The design combined the classical approach of changing the distances between system elements and simultaneously optimizing their contours.

A. Lahiri and S. Chakravorti applied Genetic Algorithm (Lahiri & Chakravorti 2004) and Simulated Annealing Algorithm (Lahiri & Chakravorti 2005) both coupled with a trained Artificial Neural Network and optimized asymmetric electric field intensity of complex GIS geometries. S. Banerjee, A. Lahiri and K. Bhattacharya (Banerjee, Lahiri, & Bhattacharya 2007) optimized support insulators that are used in HV applications using SVM algorithm. Dot kernel was used to generalize the Support Vector Algorithm (SVA).

The objective of the present work is to optimize the electric field distribution on an axi-symmetric electrode-spacer configuration used in GIS. The contour of the electrode under consideration is a curved surface. Modeling the configuration and subsequently calculating the electric field is done by using COMSOL Multiphysics 4.3b Software.

2 MODEL DESCRIPTION

The electrode-insulator arrangement considered in the present work is shown in Figure 1. It is a GIS, a technology that has been developed to meet

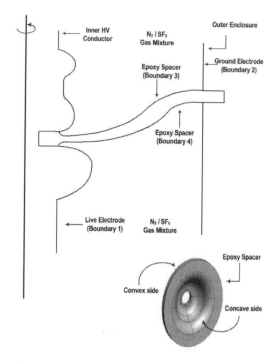

Figure 1. Electrode-Spacer Arrangement.

the requirement of this century. It is normally preferred for 12 kV, 36 kV, 72.5 kV, 145 kV, 245 kV, 420 kV and above. In this arrangement, various apparatus like current transformers, voltage transformers, load break switches, circuit breakers, bus bars, isolators, earthling switches etc. are housed in different metal enclosures and each of these modules is filled with a gas mixture of Nitrogen (N_2) and Sulphur Hexafluoride (SF_6). These enclosures are made of Aluminum or Stainless Steels which are non-magnetic materials and are earthed on both sides to provide protection to shock.

All the components within the enclosures are maintained at live potential, in this case it is a normalized voltage of 1V used for reference. The components within the enclosure are supported by insulators made of epoxy cast resin whose permittivity is 5. For centering theconductors inside the enclosure, the insulators are placed at regular intervals of which some are designed as barriers between adjacent modules so as to prevent the flow of gas through them.

Gaseous insulation in GIS requires a bulk amount of gas. Though SF_6 has good insulation capability, it is very expensive. Carbon Tetrafluoride or Tetrafluoromethane (CH_4) and Octafluorocyclobutane (C_4F_8) have higher insulation capabilities than SF_6 but owing to high toxicity

and low withstand capability under partial discharge, they are not generally not considered in GIS. Considering all aspects, a gas mixture containing 80% N_2 and 20% SF_6, having permittivity of 1.005 is considered to be the most suitable gaseous dielectric for GIS (Christophorou & Burnt 1995).

The electrode-spacer arrangement considered in this paper consists of four boundaries:

1. The first one is the boundary between the live electrode and the insulating gas mixture.
2. The second boundary is the boundary between the ground electrode and the insulating gas mixture.
3. The third one is between the epoxy resin spacer and the gas mixture above the spacer.
4. The fourth boundary is the one between the epoxy resin spacer and the gas mixture below the spacer.

At the first boundary, the curved portion of the live electrode on the convex side of the insulator consists of three elliptical segments having three different radius of curvature $r_1 = 0.1$ m, $r_2 = 0.2$ m and $r_3 = 0.05$ m, while the curved portion of the live electrode on the concave side of the insulator comprises of four elliptical segments whose radii of curvature are denoted as r_4, r_5, r_6 and r_7 respectively. In the present work, r_1, r_2 and r_3 are kept constant at the values mentioned above while r_4 to r_7 are varied over a wide range to study the effect of electric field distribution on the live electrode boundary.

3 BASIC THEORY OF SVM

SVM, introduced by Vapnik (Christophorou & Burnt 1995), has two aspects viz., classification and regression.

3.1 Support vector classification

For classification, the goal is to classify the classes by introducing a function. If the training data are linearly separable then this function will be the optimal separating hyperplane. If a two class data is considered then this plane is the one that will maximize the Euclidian distance between the nearest data point of each class and itself. This linear boundary is expected to offer the best possible generalization in contrary to the other possible classifiers that can be defined for the given set of data. For a given class of data this plane is an unique one. For a linearly separable class of data the generalized optimal separating hyperplane can be determined by the vector ω (Vapnik 1995) that minimizes the function

$$\Phi(\omega)=0.5\|\omega\|^2 \qquad (1)$$

subject to the constraints of the equation:

$$y^i[<\omega,x^i>+\rho]\geq1 \quad i=1,2,\dots k \qquad (2)$$

The parameters ω and ρ are constrained by:

$$min\,|<\omega,x^i>+\rho|=1 \qquad (3)$$

But in most of the real time problems, the data classes are not linearly separable. In that case equation (1) is modified as:

$$\Phi(\omega,\psi)=0.5\|\omega\|^2+\lambda\sum_i \psi_i \qquad (4)$$

and accordingly the constraint is modified as:

$$y^i[(\omega,x^i)+\rho]\geq1-\psi_i, i=1,2,\dots k \qquad (5)$$

where, λ is a given constant and $\psi_i\geq0$ are the misclassification errors.

The solution of the optimization problem given by equation (4) under the constraints given by equation (5) is given by the saddle point of the Lagrangian (Minoux 1986) as:

$$\phi=arg\,min\,0.5\sum_{i=1}^{k}\sum_{j=1}^{k}\phi_j y_i y_j <x_i,xj>-\sum_{l=1}^{k}\phi_l \qquad (6)$$

subject to the constraints:

$$0\leq\phi_i\leq\lambda\ i=1,2,\dots,k$$
$$\sum_{j=1}^{k}\phi_j y_j=0 \qquad (7)$$

The solution is same as that obtained for the case when the data are linearly separable. The only difference is in modification of the Lagrangian multipliers. The parameter λ strengthens the controlling ability of the classifier but it has to be defined first in such a way that λ should represent the noise present in the data.

But typically the data are linearly separable only in higher dimensional space. When the data are corrupted by noise then only with a finite number of training data it is not possible to classify the data without ambiguity. This leads to the adoption of the method of classifying the data considering the data set as linearly non-separable type putting an upper limit on the Lagrangian multipliers but that again leads to a problem of defining λ.

3.2 Kernel functions

As mentioned in the previous section, to ensure proper functioning of the machine, one of the methods to have linearly separable data is to map the data into higher dimensional feature space. But this mapping into higher dimensional space requires a large training set to make the vector machine an efficient classifier. This problem is obviated by using Kernel functions such that an inner product in feature space which is potentially higher dimensional has an equivalent kernel in the input space (Schölkopf, Burges, & Smola 1999).

A proper choice of kernel is now the only task to have a linearly separable data for proper functioning of SVM as a classifier. Among the several kernels that are commonly used, only dot kernel and radial kernel are found to be suitable for the present set of data.

A dot kernel is defined as:

$$\kappa(x,y)=x\bullet y \qquad (8)$$

i.e., it is the inner product of the input-output data set x and y respectively. A radial kernel is defined as:

$$\kappa(x,y)=e^{-\gamma\|x-y\|^2} \qquad (9)$$

where γ is a parameter of the kernel which has to be tuned properly for proper mapping of the data.

3.3 Support vector regression

By introducing an alternative loss function, SVM can also be applied to regression problems (Smola 1996). If a set of data given by

$$\mathbb{S}=\{(x_1,y_1),(x_2,y_2),..(x_k,y_k)\},\ x\in\mathbb{R}^n, y\in\mathbb{R} \qquad (10)$$

is approximated using a linear function:

$$\varphi=<\omega,x>+\rho \qquad (11)$$

then the optimal regression function is given by the minimum of the function:

$$\Phi(\omega,\psi^+,\psi^-)=0.5\|\omega\|^2+\lambda\sum_i(\psi_i^++\psi_i^-) \qquad (12)$$

subject to the conditions:

$$((\omega\bullet x_i)+\rho)-y_i\leq\varepsilon+\psi_i^+ \qquad (13)$$

$$y_i-((\omega\bullet x_i)+\rho)\leq\varepsilon+\psi_i^- \qquad (14)$$

$$\psi_i^+,\psi_i^-\geq0 \qquad (15)$$

where, $i = 1, 2, ..k$, the parameter λ should have a pre-specified value and ψ_i^+, ψ_i^- are the two slack variables that defines the upper and the lower constraints on the outputs of the systems. For any error smaller than ε then ψ_i^+, ψ_i^- need not to be introduced in the objective function represented by equation (12).

SV Regression (SVR) faces the same problem as that of a SV Classifier (SVC) for a non-separable data. Similar to the SVC approach, a non-linear mapping may be applied to map the data in a higher dimension feature space to perform linear regression but the kernel trick again is a solution to a non-linear SVR problem.

4 TRAINING AND TESTING OF THE SVM

The input-output data for training the SVM is obtained by varying the radii of the elliptical segments of the curved portion i.e. r_4 to r_7 and noting down the maximum resultant electric stress, Er_{max}, on the live electrode boundary. The radii are varied by adjusting the span of the segments while maintaining the smooth profile of the electrode boundary. The radii are varied by considering any one at a time, any two at a time, any three at a time and all four at a time.

The input data set i.e. the different combination of radii, are normalized according to the following equations:

$$x_{i,norm} = \frac{x_i - x_{min}}{x_{max} - x_{min}} \tag{16}$$

where,

$x_{i, norm}$ is the normalized value of the ith input variable.

x_{max} is the maximum value of the input variable.
x_{min} is the minimum value of the input variable.

Similarly, the output data set i.e. the corresponding maximum resultant stress on the live electrode boundary, is normalized according to the following equation:

$$y_{j,norm} = \frac{y_j - y_{min}}{y_{max} - y_{min}} \tag{17}$$

where,

$y_{j,norm}$ is the normalized value of the jth output variable.

y_{max} is the maximum value of the output variable.

y_{min} is the minimum value of the output variable.

During training, total number of input-output data considered is 100. The accuracy of the train-

ing is measured in terms of Root Mean Squared Error (RMSE). For the purpose of testing, only six data set have been considered and the accuracy of testing is measured in terms of % mean absolute error (% MAE). A representative training data set is shown in Table 1. The data set for testing is presented in Table 2.

Table 1. Representative training data set.

Variation of radius	r_4	r_5	r_6	r_7	Er_{max}
r_4 (+%2.5)	1.6	2.0	10.0	2.4	11.4
r_4 (−%2.5)	0.9	2.0	10.0	2.4	11.4
r_5 (+%5.0)	1.3	3.5	10.0	2.4	10.0
r_5 (−%5.0)	1.3	0.5	10.0	2.4	13.3
r_6 (+%7.5)	1.3	2.0	17.5	2.4	4.9
r_6 (−%7.5)	1.3	2.0	2.5	2.4	18.0
r_7 (+%4.0)	1.3	2.0	10.0	3.3	11.8
r_7 (−%4.0)	1.3	2.0	10.0	1.6	12.2
r_4 & r_5 (+%2.5)	1.6	2.7	10.0	2.4	10.1
r_4 & r_5 (−%2.5)	1.6	1.2	10.0	2.4	12.4
r_4 & r_6 (+%2.5)	1.6	2.0	12.5	2.4	8.9
r_4 & r_6 (−%2.5)	0.9	2.0	7.5	2.4	13.9
r_4 & r_7 (+%6.0)	2.1	2.0	10.0	3.3	11.5
r_4 & r_7 (−%7.5)	0.5	2.0	10.0	0.8	11.8
r_5 & r_6 (+%5.0)	1.3	3.5	15.0	2.4	6.8
r_5 & r_6 (−%5.0)	1.3	0.5	5.0	2.4	18.7
r_5 & r_7 (+%7.5)	1.3	4.2	10.0	4.0	9.5
r_5 & r_7 (−%2.5)	1.2	4.2	10.0	1.9	12.7
r_6 & r_7 (+%2.5)	1.3	2.0	12.5	3.0	9.0
r_6 & r_7 (−%2.5)	1.3	2.0	7.5	1.9	14.6
r_4, r_5 & r_6 (+%5.0)	1.9	3.5	15.0	2.4	5.8
r_4, r_5 & r_6 (−%5.0)	0.7	0.5	5.0	2.4	17.5
r_4, r_5 & r_7 (+%10.0)	2.5	5.0	10.0	4.0	8.6
r_4, r_5 & r_7 (−%5.0)	0.7	0.5	10.0	1.6	14.0
r_4, r_6 & r_7 (+%7.5)	2.2	2.0	17.5	3.9	5.5
r_4, r_6 & r_7 (−%7.5)	0.5	2.0	2.5	1.2	19.2
r_5, r_6 & r_7 (+%5.0)	1.3	3.5	15.0	3.5	4.8
r_5, r_6 & r_7 (−%5.0)	1.3	0.5	5.0	1.5	17.8
r_4, r_5, r_6 & r_7 (+%7.5)	2.2	4.2	17.5	3.9	2.0
r_4, r_5, r_6 & r_7 (−%5.0)	0.7	0.5	5.0	1.5	17.3

Table 2. Data set for testing.

r_4	r_5	r_6	r_7	Er_{max} (Calculated)
2.1	2.0	10.0	2.4	12.4
1.3	4.2	10.0	2.4	9.5
1.6	2.0	10.0	3.3	11.7
1.3	2.0	15.0	3.5	7.5
0.7	2.0	5.0	1.5	16.5
1.3	1.2	7.5	1.9	15.5

The parameter λ that appears in the objective function of the optimization problem of SVC or SVR is to be chosen by the user to assign penalties to the errors. A high value of the parameter implies that more penalty is assigned to the errors and thus the SVM, which is meant to minimize the error, is considered to have lower generalization capability. In contrary to this, if a low value is assigned to λ then less penalty is assigned to the errors so in turn SVM is trained to minimize the margin while allowing the error will and thus will have better generalization capability. Thus, it can be said that a judicious choice of the kernel and a proper selection of the parameter λ will finally define the sensitivity of the performance of the SVM.

5 RESULTS AND DISCUSSIONS

In the present work, both dot kernel and radial kernel are used. To optimize the performance of the SVM, the parameter λ is optimized in each case. With dot kernel, λ is varied over a wide range from 1 to 100 and the corresponding % MAE are noted. Table 3 presents the % MAE for each value of λ. From Table 3 it may be observed that for λ equal

Table 4. Sensitivity analysis using radial Kernel.

γ	λ	%MAE	γ	λ	%MAE
0.001	47	0.817983	0.040	14	1.354354
0.002	50	0.762827	0.050	17	1.349619
0.003	50	0.782900	0.060	15	1.319088
0.004	19	0.805337	0.070	17	1.226336
0.005	50	0.803303	0.080	14	1.311884
0.006	48	0.788243	0.090	11	1.401832
0.007	40	0.803662	0.1	16	1.477675
0.008	50	0.816138	0.2	5	1.989675
0.009	42	0.819160	0.3	4	2.366003
0.010	19	1.283168	0.4	4	2.707852
0.020	13	1.344594	0.5	3	3.035025
0.030	22	1.380147	0.6	3	3.303871

to 35, the value of % MAE is the least and is equal to 0.721275. Since it is observed that the value of % MAE remains constant at 0.726271 from λ equal to 69 and onwards hence it is not been reported in Table 3 after λ equal to 75.

With radial kernel, two parameters are tuned to obtain the least value of % MAE viz., λ and γ. In this study, γ is varied over a wide range and for each value of γ, the parameter λ is varied from 1 to 100. The optimum combination of these two parameters will be the one for which % MAE will be least. Table 4 presents that value of λ which gives the minimum value of % MAE for a given value of γ and also presents the corresponding minimum value of the % MAE. From Table 4 it may be observed that with radial kernel, the optimum combination is achieved with γ and λ equals to 0.002 and 50 respectively corresponding to which % MAE is equal to 0.762827. From Table 3 and Table 4, it may be concluded that for the given problem, though both the kernels, dot and radial, are almost equally efficient in training the SVM but dot kernel has an edge on the error as compared to radial kernel.

6 CONCLUSIONS

SVM is generalized with four critical dimensions of the GIS arrangement considered in the present work. This generalized architecture of SVM is achieved with dot kernel and radial kernel. The %MAE with which the accuracy of testing SVM is measured is 0.05% more when radial kernel is used instead of dot kernel. Both Artificial Neural Network (ANN) and SVM need problem specific generalization but with SVM the task is easy as SVM has got very less tunable parameters compared to ANN. %MAE with which the degree of accuracy is

Table 3. Sensitivity analysis using dot Kernel.

λ	%MAE	λ	%MAE	λ	%MAE
1	0.745534	26	0.722162	51	0.722328
2	0.734834	27	0.722325	52	0.721865
3	0.745994	28	0.721935	53	0.721917
4	0.744845	29	0.721999	54	0.722437
5	0.727155	30	0.722161	55	0.721966
6	0.746034	31	0.721959	56	0.721819
7	0.722024	32	0.721959	57	0.721278
8	0.722163	33	0.721942	58	0.725979
9	0.721967	34	0.722514	59	0.721955
10	0.722173	35	0.721275	60	0.727013
11	0.721960	36	0.725982	61	0.722441
12	0.721915	37	0.722241	62	0.722299
13	0.721973	38	0.722427	63	0.721960
14	0.727524	39	0.721904	64	0.722170
15	0.727643	40	0.721975	65	0.721779
16	0.727100	41	0.721817	66	0.721964
17	0.721729	42	0.722242	67	0.722160
18	0.721779	43	0.721837	68	0.721975
19	0.721975	44	0.722163	69	0.726271
20	0.721962	45	0.722436	70	0.726271
21	0.722058	46	0.722105	71	0.726271
22	0.721780	47	0.721961	72	0.726271
23	0.726478	48	0.721898	73	0.726271
24	0.722148	49	0.722437	74	0.726271
25	0.721865	50	0.722335	75	0.726271

measured reflects the potency of SVM in optimizing electric field intensity of HV arrangements.

REFERENCES

Banerjee, S., A. Lahiri, & K. Bhattacharya (2007). Optimization of support insulators used in hv systems using support vector machine. *IEEE Transactions on Dielectrics and Electrical Insulation 19*(2).

Christophorou, G. & R.J. Burnt (1995). SF_6/N_2 mixtures, basic and hv insulation properties. *IEEE Transactions on Dielectrics and Electrical Insulation 2*(5).

Grafoner, P. & H. Singer (1975). Optimization of electrode and insulator contours. Conference report, *2nd* International Symp. on HV Engg., Zurich.

Gronewald, H. (1983). Computer aided design of hv insulators. Conference Report 11.01, *4th* International Symp. on HV Engg., Athens.

Lahiri, A. & S. Chakravorti (2004). Electrode-spacer contour optimization by ann aided genetic algorithm. *IEEE Transactions on Dielectrics and Electrical Insulation 11*(6).

Lahiri, A. & S. Chakravorti (2005). A novel approach based on simulated annealing coupled to artificial neural network for 3d electric field optimization. *IEEE Transactions on Power Delivery 20*(3).

Minoux, M. (1986). *Mathematical Programming: Theory and Algorithms.* New York: John Wiley and Sons.

Schölkopf, B., C.J.C. Burges, & A.J. Smola (1999). *Advances in Kernel Methods—Support Vector Learning.* Cambridge, MA: MIT Press.

Singer, H. (1979). Computation of optimized electrode geometries. Conference Report 11.06, *3rd* International Symp. On HV Engg., Milan.

Smola, A.J. (1996). Regression estimation with support vector learning machines. Technical report, Technische Universität *München, München,* Germany.

Stih, Z. (1986). High voltage insulating system design by application of electrode and insulator contour optimization. *IEEE Transactions on Electrical Insulation 21*, 579–584.

Vapnik, V. (1995). *The Statistical Learning Theory.* New York: Springer.

Study of electrical and thermal stress distribution using boron nitride with Silicon Rubber in HV cable termination

Mayur Basu, Vishwanath Gupta & Abhijit Lahiri
Department of Electrical Engineering, Supreme Knowledge Foundation Group of Institutions, West Bengal, India

Arijit Baral
Department of Electrical Engineering, Indian School of Mines, Dhanbad, Jharkhand, India

ABSTRACT: Due to high electrical and thermal stresses in High-Voltage (HV) cable terminations, breakdown of insulation has become a common fault nowadays. This can be avoided by using high permittivity and good thermal conductive material at the stress cone. To ensure the reliability of the stress cone, the use of Silicon Rubber (SIR) and polymer nanocomposites as the stress control material has gained a lot of attention in recent time. In this proposed work, different volumes of nano-sized boron nitride-reinforced SIR composites were used as the stress control material. The use of these nanocomposite materials showed a substantial amount of reduction in electrical as well as thermal stress in the cable termination compared with pure SIR, which is practically significant for any HV equipment.

1 INTRODUCTION

The majority of the High-Voltage (HV) cable line failures in underground distribution systems are caused by the defects in cable joints or cable terminations as it is the weakest link in it. The breakdown of cable terminations mostly occurs due to the strong electric field in the cable insulation, close to the cable screen end. Excessive electrical stress on cable terminations can cause local discharge, flashover, or puncture, and may further lead to the system failure (Wei *et al.* 2005). Recent research has experimentally shown that the thermal failure of the stress grading layer due to the high-frequency components of the applied voltage waveform has become very common (Ming *et al.* 2004).

Several approaches such as application of geometrical potential shaping, non-linear resistive field grading coatings, refractive field grading coatings, and capacitive coatings have been used (Revenc & Lebey 1999, Nikolajevic *et al.* 1997). Commonly, it is controlled by the deflector's cones (conventional stress relief cones), which are the geometric solution to the problem.

An ideal stress cone must possess high thermal conductivity, low coefficient of thermal expansion, high dielectric constant, and low loss tangent (Tanaka *et al.* 2004). Thus, elastomeric polymers, such as Silicon Rubber (SIR), reinforced with highly thermally conductive but electrically insulating nanofillers, such as Aluminum nitride, Boron Nitride (BN), silicon carbide, and alumina (Al_2O_3), can be useful materials for HV insulation applications.

To improve the thermal conductivity of SIR, the study of electrical and thermal properties of Al_2O_3- and ZnO-reinforced SIR that can be used for HV applications has also been carried out (Sim *et al.* 2005).

A high aspect ratio of barium titanate fiber and graphene platelet mixed into a polymer matrix to improve the dielectric constant of the polymer composite has also been presented in another study (Wang *et al.* 2012).

The aim of the proposed work was to study the effect of reinforcing BN fillers with silicon elastomers (Kemaloglu *et al.* 2010) used as the stress cone material in the cable termination when compared with conventional SIR.

2 MODEL DESCRIPTION

The model of the cable termination considered in the proposed work is shown in Figure 1. It is an axis-symmetric model consisting of a copper conductor, XLPE insulation, a SIR stress cone, and a semiconductor screen. The whole arrangement is immersed in silicon oil. Normally, the material used in the stress cone is SIR. Relative permittivity and thermal conductivity of XLPE, silicon oil, and SIR are detailed in Table 1. The potential of the conductor is considered to be 1 V, which is the normalized potential used as the reference, and that of the semiconductor screen is zero.

The major dimensions of the arrangement are shown in Figure 1. The entire system is

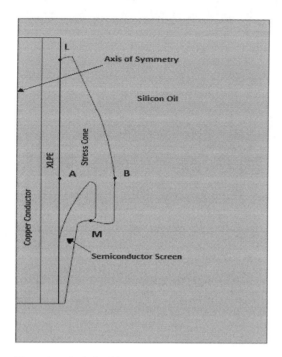

Figure 1.   Typical cable termination model.

Table 1.   Electrical and thermal properties of the materials used in the cable termination model.

Material	Permittivity (ε_r)	Thermal conductivity [W/(m.k)]
XLPE	2.3	0.38
Silicon oil	2.7	0.1
SIR	3.3	0.2

simulated by using COMSOL Multiphysics version 4.3b software.

3   DIFFERENT NANOCOMPOSITES USED

Silicon elastomers with considerable high aspect ratio nanofillers have gained attention as the stress cone material in the cable termination due to their high permittivity and low dielectric loss (Wang et al. 2012). Specific heat and thermal conductivity are also another criterion for improvement, which are determined using different nanofiller loadings of SIR to reduce the thermal stress.

In the present work, BN powder named MK-hBNN70 (Kemaloglu et al. 2010) reinforced with SIR was used as the stress control material to study the electrical and thermal stress distribution at the

Table 2.   Electrical and thermal properties of nano-sized BN-reinforced SIR composites (Kemaloglu et al 2010).

Material	Permittivity (ε_r)	Thermal conductivity [W/(m.k)]
10% MK-hBNN70+SIR	3.46	0.45
20% MK-hBNN70+SIR	3.55	0.63
40% MK-hBNN70+SIR	3.69	0.86

surface from L to M of the stress cone as well as along the line AB of the cable termination. MK-hBNN70 is irregularly shaped, nano-sized partially aggregated particles of BN with an average particle size of 70 nm (Kemaloglu et al. 2010). The permittivity and thermal conductivity at different loading levels by weight are given in Table 2.

4   EFFECT OF NANOCOMPOSITE MATERIALS ON ELECTRIC FIELD AND TEMPERATURE DISTRIBUTION ALONG THE SURFACE OF THE STRESS CONE AT POWER FREQUENCY

The variation in electrical and thermal stresses along the surface of the stress cone on the dielectric–dielectric boundary between the stress cone and silicon oil from L to M for different materials, listed in Table 2, is shown in Figures 2 and 3, along with SIR to find out the effect of reinforcing BN nanofillers in the stress cone.

In the proposed work, the temperature is considered as the measure of the thermal stress. The more the temperature is, the more the thermal stress generated.

From Figures 2 and 3, it can be observed that as the loading of the nanofillers is increased, the percentage reduction of electrical as well as thermal stress is decreased. The material 40% MK-hBNN70+SIR showed the maximum reduction in both electrical and thermal stresses.

5   EFFECT OF NANOCOMPOSITE MATERIALS ON ELECTRIC FIELD AND THERMAL STRESS DISTRIBUTION ALONG THE RADIAL DIRECTION FROM A TO B OF THE STRESS CONE AT POWER FREQUENCY

Comparison of electrical resultant stress and thermal stress at the maximum point along the surface from M to N of the stress cone with different stress cone materials has been given in Table 3.

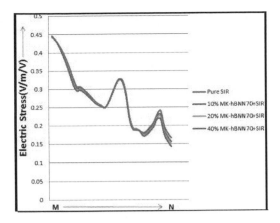

Figure 2. Resultant stress distribution along the stress cone surface for different materials.

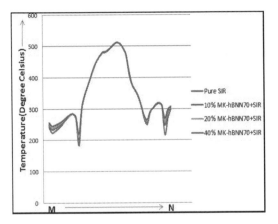

Figure 3. Thermal stress distribution along the stress cone surface for different materials.

Table 3. Comparison of electrical resultant stress and thermal stress at the maximum point along the surface from M to N of the stress cone with different stress cone materials.

Material	Resultant stress, E_r (V/m/V) at maximum point	Reduction (%)	Thermal stress, T (degree Celsius) at maximum point	Reduction (%)
SIR	0.421094	–	511.9639	–
10% MK-hBNN70+SIR	0.419235	0.45	511.9624	0.01
20% MK-hBNN70+SIR	0.41845	0.63	511.9446	0.01
40% MK-hBNN70+SIR	0.417563	0.84	511.9072	0.03

The variation in electrical and thermal stresses along the radial direction from A to B of the stress cone along with SIR is shown in Figures 4 and 5. The comparison of the resultant electrical stress and thermal stress at the maximum point along the radial direction from A to B of the stress cone with different stress cone materials is given in Table 4. From Table 4, it can be observed that a subsequent amount of change in the percentage reduction of electrical and thermal stresses occurs by increasing the loading level of BN fillers at the very end of the cable screen where both the electrical and thermal stresses are expected to be highest.

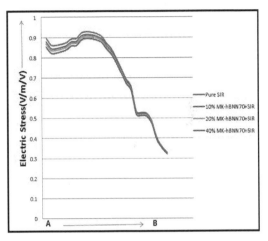

Figure 4. Resultant stress distribution in the radial direction between A to B within the stress cone material for different materials.

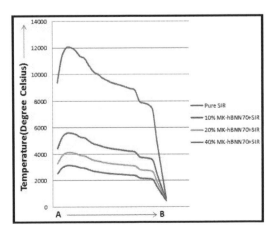

Figure 5. Thermal stress distribution in the radial direction between A to B within the stress cone material for different materials.

Table 4. Comparison of electrical resultant stress and thermal stress at the maximum point along the radial direction from A to B of the stress cone with different stress cone materials.

Material	Resultant stress, E_r (V/m/V) at maximum point	Reduction (%)	Thermal stress, T (degree Celsius) at maximum point	Reduction (%)
SIR	0.929174	–	12021.57	–
10% MK-hBNN70+SIR	0.915188	1.54	5594.404	53.47
20% MK-hBNN70+SIR	0.907567	2.33	4125.234	65.69
40% MK-hBNN70+SIR	0.896047	3.57	3142.871	73.86

Again, the material 40% MK-hBNN70+SIR with the highest loading level showed a maximum reduction of 3.57% of electrical stress and 73.86% of thermal stress distribution at their corresponding maximum point.

6 CONCLUSIONS

An electro-thermal model using nano-sized BN-reinforced SIR as the stress cone material was proposed in this paper. For obtaining the electrical and thermal stress distribution on and around the stress cone, corresponding differential equations were solved using proper boundary conditions. The convection and radiation heat transfer mechanisms were applied to the model. A substantial amount of reduction in electrical as well as thermal stress was found in the BN-reinforced SIR compared with pure SIR.

REFERENCES

Wei H.J., Jayaram S. & Cherney E.A. (2005), "A study of electrical stress grading of composite bushings by means of a resistive silicone rubber coating", *Journal of Electrostatics 63.*

Ming L., Sahlen F., Halen S., Brosig G., & Palmqvist L. (2004), "Impacts of High-frequency Voltage on Cable-termination with Resistive Stress grading", *International Conference on Solid Dielectric*, Toulouse, France.

Revenc J. P. & Lebey T. (1999), "An overview of electrical properties for stress grading optimization", *IEEE Trans. Dielec. & Elec. Insul.,* Vol. 6, pp 309–318.

Nikolajevic S.V., Pekaric-Nadj N.M. & Dimitrijevic R.M. (1997), "Optimization of cable terminations", *IEEE Trans. Power Deliv.* pp. 527–532.

Tanaka T., Montanari G.C. & Mulhaupt (2004), "Polymer nano-composites as dielectrics and electrical insulation-perspectives for processing technologies, material characterizations and future applications", *IEEE Electr. Insul. Mag., Vol. 11, No.5,* pp. 863–784.

Sim L.C., Ramanan S.L., Ismail H. & Seetharamu K.N. (2005), "Thermal characterization of Al_2O_3 and ZnO reinforced silicone rubber as thermal pads for heat dissipation purposes", ***Thermochimica Acta 430.***

Wang Z., Nelson J., Miao J., Linhardt R., Schadler L., Hillborg H. & Zhao S. (2012), "Effect of high aspect ratio filler on dielectric properties of polymer composites: A study on barium titanate fibers and grapheme platelets", *IEEE Trans. Dielec. & Elec. Insul., Vol. 19, No.3,* pp. 960–967.

Kemaloglu S., Ozkoc G., Aytac A. (2010), "Properties of thermally conductive micro and nano size boron nitride reinforced silicon rubber composites", *Thermochimica Acta 499,* pp. 40–47.

Frontiers in Computer, Communication and Electrical Engineering – Acharyya (Ed.)
© 2016 Taylor & Francis Group, London, ISBN: 978-1-138-02877-7

A hybrid compartmental epidemic model for predicting the Ebola outbreak

Suman Roy & Soham Basu Chaudhury
Department of Electrical Engineering, Jadavpur University, Kolkata, India

ABSTRACT: The deadliest epidemic of the Ebola virus disease is currently ongoing in some of the countries of Africa. The most affected countries are Guinea, Liberia, and Sierra Leone. Using mathematical models (e.g., SIR, SIRS, SEIR), it is possible to simulate this ongoing outbreak. All the data used in this study were from Liberia. The main purpose of this model was to detect the epidemic at an early stage in future and predict the possible outbreak in an area. In doing so, two main parameters were considered, namely β and μ. β is the daily contact rate, i.e., the average number of adequate contacts per infective per day. μ is the daily death removal rate. R is the ratio of β and μ, and expressed as the basic reproductive number.

1 INTRODUCTION

Over the last half of the century, Ebola has struck the human population and victimized the lives of countless people. Over the last 50 years, the advent of epidemic Ebola has shown a considerable increase in its virulence and thus claiming more lives with each strike. Thus, Ebola forecasting has become critically important. Equipped with the knowledge of its movement and growth parameters, prior planning and prior preparation to combat this lethal disease can be taken. With a successive increase in the virulence of Ebola, correspondingly, the incubation period of the disease has drastically decreased over the years and hence medical science needs to progress at an equal or greater velocity to combat this lethal disease.

Thus, for the manufacture of proper drugs and to combat Ebola, the knowledge of certain parameters, such as growth rate, disease-related death rate, and population class ratio, and other physical parameters becomes crucially important. The knowledge about whether Ebola will propagate in a certain region or will die out with time will ensure the best and proper use of human resources/medical/engineering resources.

In this paper, we compare various epidemic models and try to generate a hybrid model combining the various epidemic related compartmental models with the ulterior aim of generating a mathematical model, which will help us predict the growth and future advances of Ebola, and will give us a rough idea of the total lives it might claim in the future and hence help us to be prepared to fend off its virulence.

2 VARIOUS MODELS ON EBOLA OUTBREAK IN LIBERIA IN 2014

2.1 *SIR model*

The population of a specific area or country taken into consideration is classified into three disjoint classes as follows:

1. Class of Susceptible (S): this class consists of those individuals who are not yet infected by the disease, but can develop the disease and become infected at any time. Such an individual is known as susceptible.
2. Class of Infective (I): this class consists of those individuals who have been infected by the disease and can transmit the disease to others. Such an individual is known as infective.
3. Class of Removal (R): this class consists of those individuals who have had the disease and are removed from the population. The individual can be dead or might have recovered from the disease.

These three classes will form the total population N. It is assumed that the size of each class is a continuous variable depending on time (Astacio *et al.* 1996).

So,

$$N = S(t) + I(t) + R(t)$$

We assume that as N is very large, N is considered to be constant throughout the time, i.e., the assumption is that births and natural deaths occur at an equal rate.

Two parameters are used in this model, namely β and μ. Individuals are removed from the class "Infective" due to death at a rate proportional to the size of class having proportionality constant μ, which is called the daily death removal rate. $1/\mu$ is the average time for an individual from the infection to death. For the parameter β, it is assumed that the population is homogeneously mixing, which is known as the daily contact ratio. The number of people getting infected due to direct contact with an infected individual at a time interval $(t, t+\Delta t)$ is $(\beta SI \Delta t/N)$ (Hethcote 1976):

$$dS/dt = -\beta SI/N \quad dI/dt = \beta SI/N - \mu I \quad dR/dt = \mu I$$

2.2 SIR model with the natural birth rate and the death rate

With the increasing complexity of the epidemic, more parameters are gradually taken into account.

Normally, until now, the SIR model has taken simply the natural birth rate and the natural death rate into account for changing the compartmental class population along with other epidemic parameters. However, now we assign a "disease-related death rate 'd'," which causes the reduction in the infected class of the population:

$$dS/dt = bN - \beta SI/N - wS \quad -dI/dt = \beta SI/N - (\mu + w + d)I$$

$dR/dt = \mu I - wR$ (natural birth rate b and death rate w)

2.3 SIRS model

The SIRS model considers the fact that within the recovered/removed class of the population, a certain part of the population can get cured from the disease and return to the class of the susceptible population, due to the lack of immunity against the disease. Thus, the cycle goes something like this: $S \rightarrow I \rightarrow R \rightarrow S$

Thus, we conceive our first hybrid model, in which we try to combine the characteristics of this SIRS model along with the traits of the model developed so far in this paper.

So, the new model will take into account the following parameters: specific birth rate b and natural death rate w, effective recovery rate f, and disease induced death rate d. (O'Regan 2010)

$$dS/dt = bN - \beta SI/N - wS + fR$$

$$dI/dt = \beta SI/N - (\mu + w + d)I \quad dR/dt = \mu I - wR - fR$$

2.4 SEIR model

In this new model, an incubation period is taken into account, during which the virus (when infects an individual) cultivates within the individual's bloodstream and takes a certain time to do so. After the completion of this process, the individual becomes a part of the infected class. It roughly takes about 4–10 days for incubation, and thus the incubation constant = 1/(incubation period). Since Ebola is a virulent epidemic, we take the lowest case into account and a = 0.25.

With the explanation of this model, we now move on to our next hybridization model in HII, in which we combine the features of HI with the features of the SEIR model, so that the model now takes account of the following parameters:

natural birth rate (b), natural death rate (w), disease-related death rate (d), recovery rate (f), incubation constant (a) along with four different classes S, E, I, and R.

$$dS/dt = bN - \beta SI/N - wS + fR \quad dE/dt = \beta SI/N - wE - aE \quad dI/dt = aE - (\mu + w + d)I \quad dR/dt = \mu I - wR - fR \text{ (Hethcote 2008).}$$

3 RESULTS AND DISCUSSION

3.1 Solution of the differential equations

For the SIR model, $dI/dt = \beta SI/N - \mu I$.

For small t, $S \approx N$. So, $dI/dt \approx I - \mu I$.

Solving this, we obtain the following equation: $I(t) = I(0) \exp(t(\beta - \mu))$.

Here, for the sake of simplicity and as t is very small, $I(0) = 1$(assumed) (Pundir 2010).

Now, it has been already stated that is the average time taken by an individual from being infected to be dead. So, it can be written as

$$I(t) \perp R(t + 1/\mu)$$

$$I(t) = k R(t + 1/\mu)$$

$$\exp \{t(\beta - \mu)\} = k R(t + 1/\mu)$$

where k is a constant and defined as follows:

k = 1/(the fraction of total infected individuals who will die)

Here, calculating from the data, k = 0.64 [6].

Again,

$$t(\beta - \mu) = \ln(1/0.64) + \ln\{R(t + 1/\mu)\}$$

So, the data of the total number of dead individuals at the time t is linearly (taking natural log) fit with the curve.

For t being very small, there is only one data available (after 9 days, total number of dead people = 2) and fitting with this data, we get

$$\beta - \mu = 0.12$$

Now, the average time taken for an infected individual to die varies from 4 to 10 days. So, it can take values from 0.1 to 0.25.

Until now, 8331 people have been infected and 3538 people have died in the Ebola outbreak.

So, the lowest possible value for N(0) will be 8000. In this model, the differential equations have been solved using three different values of N(0): 8000, 10000, and 12000.

3.2 *Simulation of the SIR model*

This is the simplest model used here and many real-life factors have not been considered just to make it simple. So, it is quite possible that in the actual case, the epidemic dynamic may be somewhat different from this model, which has been shown below. Results are illustrated in Figures 1 and 2.

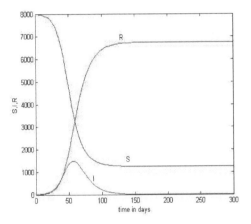

Figure 1. Plotting of S, I, and R. For N(0) = 8000, S(0) = 7990, I(0) = 10, $\beta = 0.22$, $\mu = 0.1$.

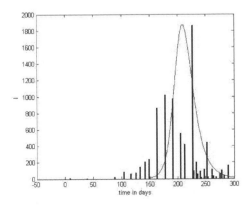

Figure 2. Comparison between the real and modeled data of I. For N(0) = 10000, S(0) = 9990, I(0) = 10, $\beta = 0.22$, $\mu = 0.1$.

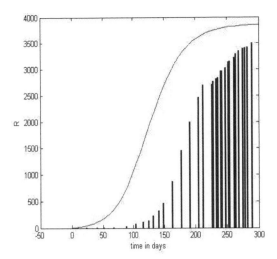

Figure 3. Comparison between the real and modeled data of R when N(0) = 8000, S(0) = 7990, I(0) = 10, $\beta = 0.14$, $\mu = 0.11$.

As shown in Figure 3, in the solution of I(t) from differential equations, the initial time has been taken as t = 150 days. From t = 0 to t = 150 days, the epidemic was still in its early stage, this is a well-guessed alteration to make the values obtained from the model more fit with the actual data.

3.3 *Simulation of the hybrid compartmental SEIRS model*

This is our finalized model, in which we combine the parameters and traits of all the various compartmental models and plot it against the real-time data. Here, the plot is made with three pairs of (β, μ) values to test the variation of the model with reference to these parameters. N(0) = 9000, S(0) = 8200, E(0) = 8, I(0) = 1, R(0) = 0 remains constant for all the three plots.

Interestingly, the basic parameters of the model, i.e., (β, μ) have different values over different values over different periods of time, and this complicates the model even further. After due consideration of the deviations and timelines, the time period of 300 days is roughly divided into three unequal parts and hence three different pairs of value of (β, μ) are used in each of these periods with minute variations in the values of other parameters in the order of $10^{(-3)}$ or $10^{(-4)}$.

Some of the worked out cases are given as follows and the results are shown in the Figures 4–9.

SET-1

300 days is roughly divided into three periods as follows:

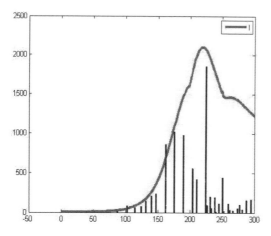

Figure 4. Variation of the class E with time plotted against real-time data.

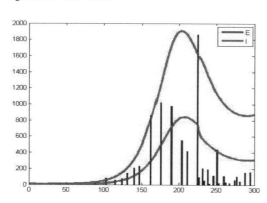

Figure 5. Variation of the classes E and I with time plotted against real-time data.

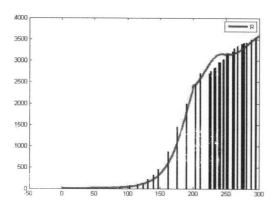

Figure 6. Variation of the class R with time plotted against real-time data.

[0,200] – [200,250] – [250,300]
The values of (β, μ) used in these periods are as follows:

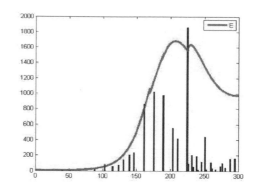

Figure 7. Variation of the class E with time plotted against real-time data.

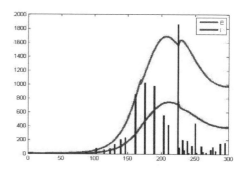

Figure 8. Variation of the classes E and I with time plotted against real-time data.

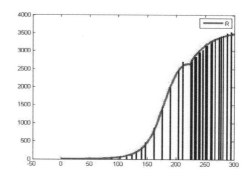

Figure 9. Variation of the class R with time plotted against real-time data.

[0.27,0.15] – [0.33,0.11] – [0.35,0.13]
The results are given below.
SET-2
300 days is roughly divided into three periods:
[0,170] – [170,230] – [230,300]
The values of (β, μ) used in these periods are as follows:

[0.27,0.15] – [0.32,0.11] – [0.39,0.18]
The results are given below.

514

3.4 *Effect of the natural recovery rate on the outbreak of the disease*

Thus, we see that with a variation in the value of the parameters, the curves thus obtained are much closer to the original value after some constant variation.

The value of 'f' or the recovery rate is of critical importance as this is responsible for drastic changes in the trajectory or propagation of the disease as an epidemic. Thus, by adjusting the value of 'f' to a proper value, the course of the epidemic can be adjusted to curb the effect on the victims.

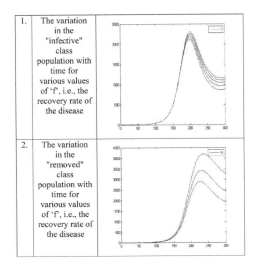

1.	The variation in the "infective" class population with time for various values of 'f', i.e., the recovery rate of the disease
2.	The variation in the "removed" class population with time for various values of 'f', i.e., the recovery rate of the disease

If the calculation is based on the SET-2 of data, we have three different values of f in motion, which are as follows:

f1 = 0.009, f2 = 0.015, f3 = 0.0085

f avg = (f1 + f2 + f3)/3 = 0.01083.

This is the rate of recovery. This amounts to the fact that approximately 11 out of 1000 people recovered only to become susceptible again.

4 CONCLUSION

Thus, throughout this paper, a hybrid model has been developed, and it is evident through this endeavor that the spread of the epidemic Ebola is entirely dependent on the physical parameters that govern the progress of the disease. Therefore, whether the disease will become an epidemic/endemic all depends on the physical parameters such as β, μ, f, and d. If changes occur in these parameters, the entire build-up of the disease is disturbed.

First, changes in the value of 'f' can change the course of the disease. Unlike the other parameters, this parameter is dependent on our choice. Second, the natural recovery rate can be increased by artificial vaccination and proper medication and treatment offered to the victims of the epidemic. This will artificially increase the recovery rate f, which will quickly drive the class population to a lower maximum value and quicker saturation. Third, one of the most noted features of this model is that the basic parameters of the model, i.e., (β, μ) have different values over different values over different periods of time, and this complicates the model even further. These values are assumed based on the real-time data of the previous outbreak of the disease. Finally, with this model, we can be certain that exactly how much the recovery rate should be in order to quickly eradicate Ebola even before it becomes an epidemic. From the above figures, it can be observed that with the increase in the value of the parameter f, the removed class population has a lower maximum value and thus a quicker saturation and lower chances of becoming an epidemic. From the above-mentioned calculations, it amounts to the fact that approximately 11 out of 1000 people recovered only to become susceptible again. Thus, increasing this value will cause a large proportion of people to recover, and with inoculation, these people will gain permanent immunity or at least temporary immunity against the disease, and can thus stop Ebola from becoming a deadly epidemic.

REFERENCES

Astacio Jamie, Bierce Delmer, Guillen Milton, Martinez Josue, Rodriguez Francisco, ValenZuela-Campos Noe (1996), "Mathematical models to study the outbreak of Ebola".

Hethcote, H.W. (1976), "Qualitative analyses of communicable disease models", *Mathematical Biosciences*.

O'Regan Suzanne M., Kelly Thomas C., Korobeinikov Andrei, O'Callaghan Michael J.A., Pokrovskii Alexei V. (2010), "Lyapunov functions for SIR and SIRS epidemic models" *Elsevier*

Hethcote H.W., Herbert W. (2008), "Epidemiology Models with Variable Population Size", Lecture notes series, Institute for Mathematical Sciences National University of Singapore

Pundir Sudhir K., Pundir Rimple (2010), Bio-Mathematics, 2nd ed., Pragati Prakashan, pp. 195–234

http://en.wikipedia.org/wiki/Ebola_virus_epidemic_in_West_Africa

Amin Nikakhtar, Simon M.Hsiang. "Incorporating the dynamics of epidemics in simulation models of healthcare systems",Elsevier.

P.K.Das, S.S.De, "A susceptible-infected removal(SIR) epidemic model", Indian J pure applied Math.

Frontiers in Computer, Communication and Electrical Engineering – Acharyya (Ed.)
© 2016 Taylor & Francis Group, London, ISBN: 978-1-138-02877-7

Lighting design technique of the sports arena

A. Khanra & T. Halder
Supreme Knowledge Foundation Group of Institutions, West Bengal, India

B. Das
Government College of Engineering and Textile Technology, West Bengal, India

ABSTRACT: Sports lighting is a multi—objective tasks and needs to satisfy the user (e.g., players, spectators and referees) requirements, so that their abilities and skills can be justified properly. The lighting ambience can also enhance the spectator's participation. So, the best suitable lighting design is important to enhance the quality of high-resolution pictures. To provide the best lighting design solution, a list of lighting design software is available. For this case study, the lighting design software 'DIALux' is used. Flood lighting is a type of lighting system where the V-H or B-β coordinate system is used. Lighting of sports area is important for providing satisfaction to the players, spectators, and color television viewers. Lighting and lighting arrangement affects the game played, so that the design should be according to the required parameters of particular sports.

1 BASIC DEPENDENT FACTORS

The lighting design of the floodlighting area depends on the factors detailed below.

1.1 *Type of field*

The type of area in which the sports is played is very important. Depending on the shape of the working field, the lighting design technique will be different. The other important consideration regarding a sports area, as far as lighting is concerned, is its dimensions, whether or not it is covered (indoor based or outdoor based), the spectator facilities, and reflectance of boundary surfaces (e.g., ground, walls, and ceiling). The dimension of the sports arena affects the quality requirements of lighting and also the position of luminaires.

1.2 *Type of object*

The type of playing object required for the game is a major influence on the quantity of optical requirement. This is dependent on the physical size of the object (e.g., cricket, football, tennis) and the viewing distance of both players and spectators.

1.3 *Speed of the object*

The quantity of lighting required also depends on the apparent speed of the playing object. This is dependent on the speed and the direction of the movement relative to the direction of view.

1.4 *Class of play*

Class of play depends on which type of game is being played, such as practice, warm up, national, and international game with live coverage.

2 USER OPTICAL REQUIREMENT

Players must be able to clearly see the playing field. Spectators must be able to watch clearly the performance of the players. Players and spectators must be able to see their respective surroundings as well as their immediate neighbors' performance. Lastly, for television or film coverage, lighting should provide necessary conditions to provide good slow motion-based high-resolution pictures, shooting without the production of glare.

3 SPORTS LIGHTING DESIGN PARAMETERS

The lighting criteria for sports area should be such that it can meet the user requirements. For satisfaction, there are some parameters that will have to be taken into consideration. The parameters are as follows:

i. Illuminance (horizontal, vertical, and inclined),
ii. Luminance,
iii. Uniformity on various working planes,
iv. Gradient for horizontal surface,
v. Modeling effect,
  vi. Glare restriction,
   vii. Color properties of the lamp, and
viii. Lamp and luminaire selection.

Simulation results are presented in Figures 1 – 9 and Table 1.

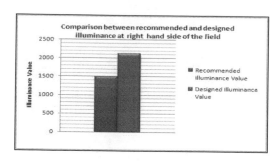

Figure 3.   Recommended and designed illuminance comparison at the right-hand side of the field.

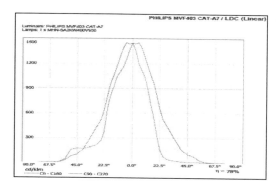

Figure 1.   Wide beam luminaire Cartesian diagram.

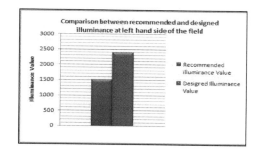

Figure 4.   Recommended and designed illuminance comparison at the left-hand side of the field.

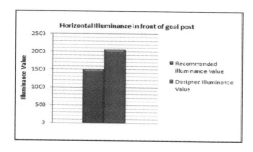

Figure 5.   Recommended and designed illuminance comparison in front of the goal post.

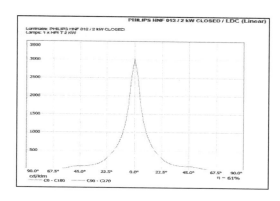

Figure 2.   Narrow beam luminaire Cartesian diagram.

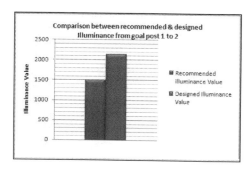

4   COMPARISON BETWEE THE RECOMMENDED AND DESIGNED ILLUMINANCE VALUES OF DIFFERENT CALCULATION SURFACES

Figure 6.   Recommended and designed illuminance comparison from goal post 1 to 2.

PHILIPS HNF 012 / 2 kW CLOSED / Glare Data Sheet

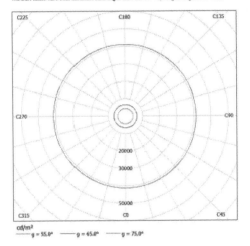

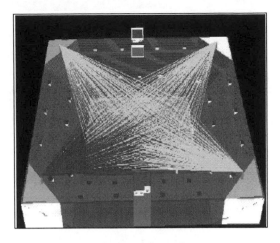

Figure 8.   Ray track view of the stadium.

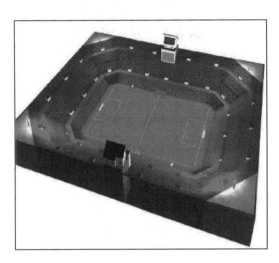

Figure 7.   Glare evaluation of UGR data of the narrow beam luminaire.

Figure 9.   3D simulation view.

Table 1.   Lighting level for football as per FIFA: for televised events.

		Vertical illuminance			Horizontal illuminance		
		E_v Cam Avg.	Uniformity		E_h Avg	Uniformity	
Class	Calculation toward	Lux	U1 (Min/Max)	U2 (Min/Avg)	Lux	U1	U2
International	Slow motion camera	800	0.5	0.7	1500 to 3000	0.6	0.6
	Fixed camera	400	0.5	0.7			
Class V	Mobile camera (at pitch level)	1000	0.3	0.7			

5 CALCULATION OF TOTAL LIGHTING LOAD AND LIGHTING POWER DENSITY

5.1 Lighting load

Power consumed by the lighting system that is used to design the football stadium is called the lighting load, which is expressed in watt or kilo watt.

5.2 Type of luminaire: Philips HNF 012

The number of luminaires of 2 kW(narrow beam) is given by

140*4 = 560

The total wattage of 2 kW (narrow beam) luminaire is given by

140*4*2 kW = 1120 kW

5.3 Type of luminaire: Philips MVF-403 CAT-A7

The number of luminaires of 1 kW (wide beam) is given by

16*2+15*2 = 32+30 = 62

The total wattage of 1 kW (wide beam) luminaire is given by

62*1 kW = 62 kW

The total electrical load of the stadium for lighting is given as follows:

design: = (1120+62)kW = 1182 kW

5.4 Lighting power density

Lighting power density is defined as the lighting power consumed per unit of area.

It technically represents the load of any lighting equipment in any defined area, or the watts per square meter of the lighting equipment.

6 CONCLUSION

For designing the sports lighting area, all the energy-efficient lighting design step parameters are taken into account. The types of lamp and luminaire are selected depending on selection parameters. For sports lighting, color is one important criterion. The lamp is selected depending on the high values of CCT and CRI. For luminaire selection, better optical and distribution characteristics of the luminaire are used depending on the requirement. For finding the layout of the luminaire, DIALux 4.11 software is used. Here, the projection of the luminaire is an important parameter. This is also incorporated in this design. For finding the best possible lighting design solution, several designs are implemented and lighting design parameters are found. Sports lighting is one of the critical working areas where the lighting design is implemented. This design is performed following the National Lighting Code 2010 criteria. The recommended illuminance level is 1500 lux, but the achieved lux level is 2381 lux due to the speed of the game, the presence of electronic media, and the size of the object.

It is worth mentioning that that the average maintained illuminance level is not only the lighting design parameter for the design of indoor and outdoor lighting systems, but also uniformity of illuminance, glare parameters and color are important lighting design parameters. All of these lighting design parameters are considered when designing the sports arena, but are not reported here due to the scarcity of space.

REFERENCES

Ashim Datta, Prof. (Dr) Biswanath Roy, 2009. "*Sports lighting - Fundamentals & Lighting Design concepts*" "*DIALux 4.7 lighting design software manual.*" "*Recommended Practice for Sport Lighting*" IESNA:RP-6-01.

Bhaumik, S. "*2009. Indian Society of Lighting Engineers. Newsletter publication*"; Vol. IX No.I: 21–26.

2005. "*CIE 16x: Technical Report, Practical Design Guidelines for the lighting of sport event for television and filming.*

2005. "*Commission Internationale del' Eclairage, Practice design guidelines for lighting up of sports events for colour television and filming*", CIE169.

2007."*British Standard, Light & Lighting –SportsLighting*", BS EN 12193.

2009. "*The Society of Light and Lighting.Sports Lighting*", SSL LG04.

2012. "*Artificial-sports-lighting-design-guide*".

Frontiers in Computer, Communication and Electrical Engineering – Acharyya (Ed.)
© 2016 Taylor & Francis Group, London, ISBN: 978-1-138-02877-7

Numerical experiment on a modified PLL with the Euler method. Part-1: Locking boundary and stability

S. Roy, S. Ghosh & S. De
Supreme Knowledge Foundation Group of Institutions, Mankundu, Hooghly, West Bengal, India

S. Guha Mallick
ISM, Dhanbad, India

B.N. Biswas
Supreme Knowledge Foundation Group of Institutions, Mankundu, Hooghly, West Bengal, India

ABSTRACT: This paper re-examines the use of the Euler methods (forward and backward) in designing a numerical experiment on a modified phase-locked loop. This paper gives some interesting new results, which have not yet been reported.

1 INTRODUCTION

The origin of the phase-locking technique goes back to the times of Huygens who observed the phenomenon of synchronization of two pendulums hung on a thin wooden plank. The present form of phase locking was observed by H. de Bollesuze in 1932 through an automatic phase control circuit, presently known as a Phase-Locked Loop (PLL).

Unlike the basic PLL structure, incorporating a phase detector, a low-pass filter, and a voltage-controlled oscillator, the present-day PLL incorporates a narrowband bandpass filter for noise. Thus, it introduces a transmission delay τ. The configuration is shown in Figure 1.

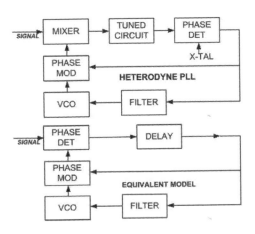

Figure 1.   Block diagram of the PLL with adaptive gain and modulation.

In this paper, we study the locking characteristics of the modified PLL, as shown in Figure 1.

2 MATHEMATICAL APPROACH

Taking the transmission delay into account, the system equation can be written as follows:

$$\frac{d\phi}{dt} = \Omega - K \sin \phi(t - \tau) + \frac{d\psi}{dt}$$
$$- Kp \cos \phi(t - \tau) \frac{d}{dt}[\phi(t - \tau)] \qquad (1)$$

where
Ω = open loop frequency error between the two oscillations;
ϕ = phase difference between the local oscillator and the external signal;
K = open loop gain;
τ = transmission delay;
θ = angle modulation; and
K_p = phase modulation constant

2.1   Forward euler algorithm

$$\phi(k) = \phi(k-1) - \Delta t K \left[\frac{\sin \phi(k-1)}{1 + K_p \cos \phi(k-1)} \right] \qquad (2)$$

Before attempting to arrive at the solution, let us check the numerical stability. Let us assume that the loop parameters are such that they generate an unstable solution. That is, from (2), the successive values of ϕ can be written as $\phi(1), \phi(2), \phi(1), \phi(2)\ldots$. We can write

$$\phi(2) = \phi(1) - K\Delta t \left[\frac{\sin \phi(1)}{1 + K_p \cos \phi(1)} \right] \qquad (3)$$

$$\phi(1) = \phi(2) - K\Delta t \left[\frac{\sin \phi(2)}{1 + K_p \cos \phi(2)} \right] \qquad (4)$$

By adding, we find

$$K\Delta t \left[\frac{\sin \phi(1)}{1 + K_p \cos \phi(1)} + \frac{\sin \phi(2)}{1 + K_p \cos(2)} \right] = 0$$

Therefore, substituting $\phi(1) = -\phi(2)$ in one of the above equations, we obtain

$$2\phi(1) = \frac{K\Delta t \sin \phi(1)}{1 + K_p \cos \phi(1)}$$

$$\frac{2}{K\Delta t} = \frac{\sin \phi(1)}{\phi(1)} \cdot \frac{1}{1 + K_p \cos \phi(1)} \qquad (5)$$

The maximum value of the right-hand side will determine the minimum value of $K\Delta t$ for instability and vice versa for stability. That is,

$$\frac{2}{K\Delta t} > \max \left[\frac{\sin \phi_2}{\phi_2} \cdot \frac{1}{1 + K_p \cos \phi(2)} \right]$$

or

$$K\Delta t < 2.00 + 0.492 Kp + 7.66 Kp^2 - 8.793 Kp^3$$

Variation of the stability zone with variation of the phase modulation constant Kp (forward algorithm) is shown in Figure 2.

2.2 Stability analysis

$$\phi_{R+1} = \phi_R = \phi_s; \; \phi_s = \sin^{-1}(-\Delta\theta)$$

We substitute

$$\phi_{R+1} = \phi_s + x_{R+1} \text{ and } \phi_R = \phi_s + x_R .$$

where x_{R+1} is much smaller than ϕ_{R+1}. Then,

$$x_{R+1} = x_R - K\Delta t \frac{\sin \phi_s + \cos \phi_s \cdot x_R + \Delta\theta}{1 + K_p \cos \phi_s - K_p \sin \phi_s \cdot x_R}$$

$$= x_R - \Delta t \cdot K \frac{-\Delta\theta + \sqrt{1 - (\Delta\theta)^2} \cdot x_R + \Delta\theta}{1 + K_p \sqrt{1 - (\Delta\theta)^2} + K_p \Delta\theta \cdot x_R}$$

$$= x_R - \Delta t K \frac{\sqrt{1 - (\Delta\theta)^2} \cdot x_R}{1 + K_p \sqrt{(1 - \Delta\theta)^2} + K_p \cdot \Delta\theta \cdot x_R}$$

3 BACKWARD EULER ALGORITHM

Numerical results are presented in Figures 3–7.

$$\phi(k+1) \equiv \phi(k)$$

$$- \Delta t \left[\frac{K \sin \left[\phi(k) - \Delta t K \left[\frac{\sin \phi(k) + \Delta\theta}{1 + K_p \cos \phi(k)} \right] \right]}{1 + K_p \cos \left[\phi(k) - \Delta t K \left[\frac{\sin \phi(k) + \Delta\theta}{1 + K_p \cos \phi(k)} \right] \right]} \right]$$

In the case of stability, it can be easily shown that

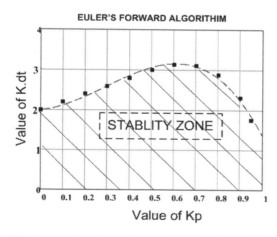

Figure 2. Variation of the stability zone with variation of the phase modulation constant Kp (forward algorithm).

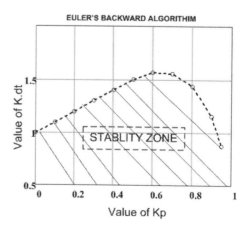

Figure 3. Variation of the stability zone with variation of the phase modulation constant Kp (backward algorithm).

$$\frac{1}{K\Delta t} = \frac{\sin\phi_2}{\phi_2} \cdot \frac{1}{1+K_p \cos\phi(2)}$$

Therefore, for stability,

$$\frac{1}{K\Delta t} > \max\left[\frac{\sin\phi_2}{\phi_2} \cdot \frac{1}{1+K_p \cos\phi(2)}\right] \quad (6)$$

$$K\Delta t < \frac{1}{1.0 - 0.368Kp - 2.069Kp^2 + 2.652Kp^3}$$

3.1 Stability of the backward euler method

$$\Delta x_{k+1} =$$

$$\Delta x_k - \frac{K\Delta t \sin\left[\phi_s + \Delta x_k - \Delta tK \dfrac{\sin\left[\phi_s + \Delta x_k\right]+\Delta\theta}{1+K_p \cos(\phi_s + \Delta x_k)}\right]}{1+K_p \cos\left[\phi_s + \Delta x_k - K\Delta t \dfrac{\sin(\phi_s + \Delta x_k)+\Delta\theta}{1+K_p \cos(\phi_s + \Delta x_k)}\right]}$$

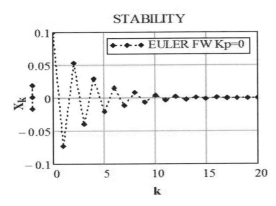

Figure 4. Phase detector output response without the phase modulation constant (forward algorithm).

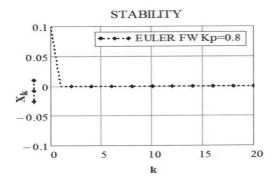

Figure 5. Phase detector output response with the phase modulation constant of 0.8 (forward algorithm).

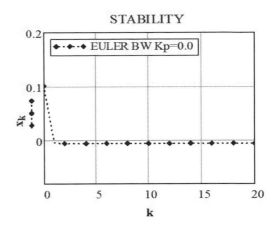

Figure 6. Phase detector output response without the phase modulation constant (backward algorithm).

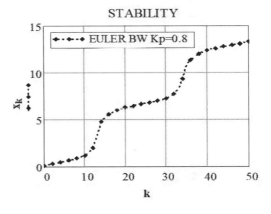

Figure 7. Phase detector output response with the phase modulation constant of 0.8 (backward algorithm).

4 CONCLUSION

The forward Euler algorithm gives a larger locking boundary, but it takes a longer time to settle down from any perturbation. On the other hand, the backward Euler algorithm generates a smaller locking boundary, but it settles down very quickly from any perturbation. The effect of phase modulation produces a strong effect on the forward Euler algorithm, whereas it can run into numerical instability in the backward Euler algorithm. However, the effect of delay has not been considered in this paper, thereby leaving scope for further study.

ACKNOWLEDGMENT

The authors are grateful to Mr B. GuhaMallick, Chairman, SKFGI Management for providing all the assistance for carrying out this work at the Sir J C Bose Creativity Centre of SKFGI. They are also thankful to Mr Somnath Chatterjee, Head of Kanailal Vidyamandir (French section), Chandannagar, Hooghly.

REFERENCES

Biswas, B.N., Chatterjee, S., Mukherjee, S.P.,& Pal, S., (2013) "A Discussion On Euler Method: A Review" *Electronic Journal Of Mathematical Analysis And Applications*, Vol. 1(2) July 2013, pp. 294–317.

Biswas, B.N. (1988), Phase Lock Theories And Application. Oxford & Ibh, New Delhi.

Gardner, Phase Lock Techniques, Third Edition (2005), Wiley.

Numerical experiment on a modified PLL with the Euler method. Part-2: Demodulation criteria

S. De, S. Roy, S. Ghosh & B.N. Biswas
Supreme Knowledge Foundation Group of Institutions, Mankundu, Hooghly, West Bengal, India

ABSTRACT: This paper re-examines the demodulation aspects of a modified Phase-Locked Demodulator (PLD). This paper also takes into account the transmission delay in a commonly used PLD. It also checks the merits and demerits of the modified PLD.

1 INTRODUCTION

Although phase-locked demodulators (PLDs) are commonly used, the problem lies in the signal handling capacity. In order to increase the signal handling capacity, a new modified PLD (MPLD) is proposed, which is depicted in Figure 1. One of the main uses of the PLD is the signal from a moving vehicle.

In this paper, we study the demodulation criteria of a delayed PLL and suggest a method for the elimination of this detection error using an adaptive phase control.

Numerically calculated results are presented in Figures 3–10.

2 ANALYSIS

The system equation governing the delayed PLL system can be expressed as follows (Biswas 1988, Gardner 2005):

$$\frac{d\phi}{dt} = \Omega - K \sin \phi(t - \tau) + \Delta \cdot \cos \omega t$$
$$- Kp \cos \phi(t - \tau)\frac{d}{dt}[\phi(t - \tau)]$$

Ω = open loop frequency error between two oscillations;
ϕ = phase difference between the local oscillator and the external signal;
K = open loop gain;
K_p = phase modulation constant;
τ = transmission delay; and
ω = frequency of the modulating signal.

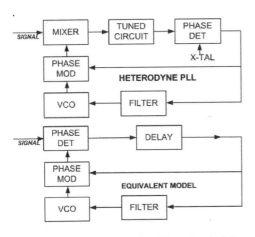

Figure 1. Block diagram of a Phase-Locked Loop (PLL) with a phase modulator and transmission delay.

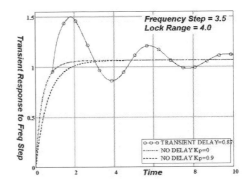

Figure 2. Transient response to the frequency step.

BEAT FREQUENCY

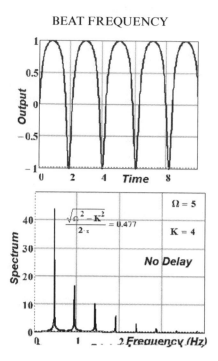

Figure 3. Frequency domain representation of beat variation (with no delay).

BEAT FREQUENCY WITH DELAY

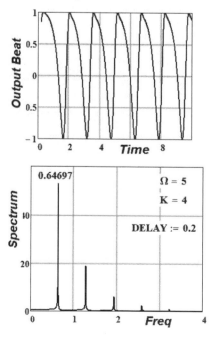

Figure 4. Frequency domain representation of beat variation (with delay Kp = 0).

The study of frequency is necessary because it clearly gives an insight into the locking phenomenon in a PLL.

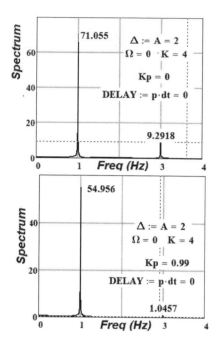

Figure 5. Demodulated signal spectrum of an FM signal with and without the phase control.

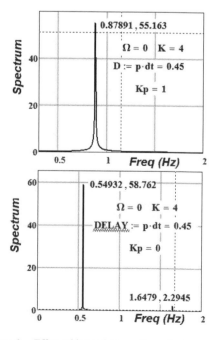

Figure 6. Effect of large delay on the output.

EFFECT OF Kp ON DEMODULATION

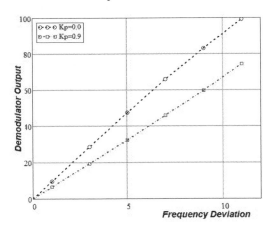

Figure 7.   Modulation characteristics.

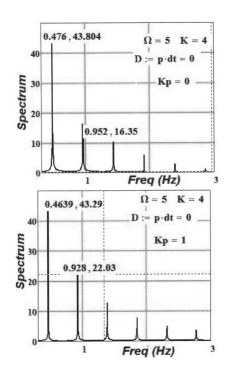

Figure 8.   Beat frequency dependence on phase control with no delay.

The above figure depicts the effect of Kp on the demodulation characteristics of the MPLD. The smaller value of the output with Kp simply indicates better tracking capability.

The above two figures show that the phase control has almost no effect on the beat frequency of the loop when there is no delay.

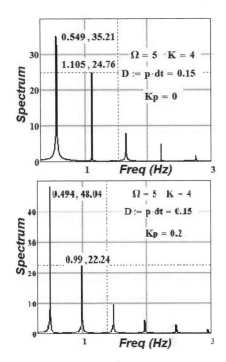

Figure 9.   Effect of delay on beat frequency, showing the effect on the beat frequency of the loop for large delay.

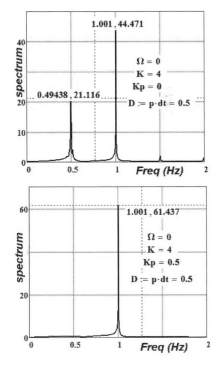

Figure 10.   Effect of large delay on the spectrum when there is no frequency error (instability).

3 CONCLUSION

In the transient response (Figure 2), the delay indicates an under-damped response that is oscillatory, leading to a longer settling time; however, controlling the delay can improve the rise time of the system. The over-damped response is observed in the presence of the phase modulator, making the system slow and sluggish. The signal handling capability of the system increases in the presence of the phase modulator. However, a large value of Kp leads to a new phenomenon called the chaotic phenomenon. This is again observed when the delay is large. This phenomenon has not been reported earlier, and the details of this phenomenon have been left for further study.

ACKNOWLEDGMENT

The authors are grateful to Mr B. GuhaMallick, Chairman, SKFGI Management for providing all the assistance for carrying out this work at the Sir J. C. Bose Creativity Centre of SKFGI. They are also thankful to Dr Somnath Chatterjee, Head of Kanailal Vidyamandir (French section), Chandannagar, Hooghly.

REFERENCES

Biswas, B.N., Chatterjee, S., Mukherjee, S.P., & Pal, S., (2013) "A Discussion On Euler Method: A Review" *Electronic Journal Of Mathematical Analysis And Applications*, Vol. 1(2) July 2013, pp. 294–317.

Biswas, B.N. (1988), Phase Lock Theories And Application. Oxford & Ibh, New Delhi.

Gardner, Phase Lock Techniques, Third Edition (2005), Wiley.

Frontiers in Computer, Communication and Electrical Engineering – Acharyya (Ed.)
© 2016 Taylor & Francis Group, London, ISBN: 978-1-138-02877-7

Design of band pass filter at 13.325 GHz

Souma Guha Mallick & Sushrut Das
Indian School of Mines, Dhanbad, Jharkhand, India

S. Bhanja
SAMEER, Kolkata, India

Tarun Kumar Dey
Supreme Knowledge Foundation Group of Institutions, Mankundu, West Bengal, India

ABSTRACT: A band pass filter having centre frequency 13.32 GHz is to be designed for use in the exciter in a communication system related to an air borne radar. Previously a design was available with PCB of type RO TMM10 provided by the Rogers Corpsubstrate. In this paper RT/DUROID 5880 has been used as the substrate. Further improvements have been obtained using ALUMINA in place of RT/DUROID as the substrate so that it could be integrated with the other components. The design shows results with less error.

1 INTRODUCTION

The advances of telecommunication technology arising hand in hand with the market demands and governmental regulations have given a push to the invention and development of new applications in wireless communication. These new applications offer certain features in telecommunication services that in turn offer three important items to the customers. The first is the coverage, meaning each customer must be supported with a minimal signal level of electromagnetic waves, the second is capacity that means the customer must have sufficient data rate for uploading and downloading of data, and the last is the Quality of Services (QoS) which guarantee the quality of the transmission of data from the transmitter to the receiver with no error. In order to provide additional transmission capacity, a strategy would be to open certain frequency regions for new applications or systems. WiMAX (World Wide Interoperability Microwave Access) which is believed as a key application for solving many actual problems today is an example (Pozar 2012).

In realization of systems like air borne radar a bandpass filter is an important component found in the transmitter orreceiver or in any of its subsystems like the exciter. Bandpass filter is a passive component which is able to select signals inside a specific bandwidth at a certain center frequency and reject signals in another frequency region, espe-

cially in frequency regions, which have the potential to interfere the information signals. In designing the bandpass filter, we face the questions, what is the maximal loss inside the pass region, and the minimal attenuation in the reject/stop regions, and how the filter characteristics must look like in transition regions (Matthaei, Young & Jones 1980).

There are several strategies taken in realization of the filters, for example, the choice of waveguide technology for different applications. The effort to fabricate waveguide filters prevents its application in huge amounts. As an alternative, microstrip filter based on Printed Circuit Board (PCB) offers the advantages easyand cheap in mass production with the disadvantages higher insertion losses and wider transition region (Pozar). Based on (Alaydrus 2010) the authors tried to give a way to conceive, design and fabricate bandpass filter for the exciter of an air borne radar application at Ku-Band with parallel-coupled microstrips.

2 DESIGN OF BAND PASS FILTER

A parallel coupled band pass filter has been designed to work in Ku- Band. The centre frequency selected was 13.32 GHz having a stop band attenuation of 50 dB at 1 GHz offset. The order of the filter calculated was 3 by using the following equation [3].

$$n \geq \frac{\log\left(10^{0.1L_{As}} - 1\right)}{2\log\Omega_s} = 2.16 \qquad (1)$$

where n is the order of the filter, Ω_s is the centre frequency in GHz and L_{As} is the stop band attenuation in dB.

A prototype of low pass filter with Butterworth response of order 3 having element values $g_0 = 1$, $g_1 = 1$, $g_2 = 2$, $g_3 = 1$, $g_4 = 1$, has been initially selected inconsultation with the standard formula and Table (Hong & Lancaster 2001).

Then the characteristic impedances of the J-inverters has been selected from the standard formula (Hong & Lancaster 2001) and hence the electrical parameters for even mode and odd mode have been found out and tabulated (Table 1).

The electrical parameters have been used to obtain the physical parameters through the LIN-CALC tool from the Agilent ADS software. The results have been shown in Table 2 for ROGERS RT/DUROID 5880 and Table 3 for ALUMINA.

Table 1. Electrical parameters of the Band Pass Filter.

i	g_i	$J_{i,i+1}$	$(Z_{0_E})_{i,i+1}$ in ohms	$(Z_{0_O})_{i,i+1}$ in ohms
0	1	0.00686683	73.0612445	38.7270945
1	1			
2	2	0.001667123	54.51522	46.179605
3	1	0.001667123	54.51522	46.179605
4	1	0.00686683	73.0612445	38.7270945

Table 2. Physical parameters of the Band Pass Filter with Rogers Rt/Duroid substrate.

Z_{0_E} in ohms	Z_{0_O} in ohms	Width (mm)	Space (mm)	Length (mm)
73.0612445	38.7270945	0.589124	0.042817	4.194660
54.51522	46.179605	0.764661	0.395741	4.112580
54.51522	46.179605	0.764661	0.395741	4.112580
73.0612445	38.7270945	0.589124	0.042817	4.194660

Table 3. Physical parameters of the Band Pass Filter with Alumina Substrate.

Z_{0_E} in ohms	Z_{0_O} in ohms	Width (mm)	Space (mm)	Length (mm)
73.0612445	38.7270945	0.180634	0.092660	2.275130
54.51522	46.179605	0.239742	0.417861	2.210500
54.51522	46.179605	0.239742	0.417861	2.210500
73.0612445	38.7270945	0.180634	0.092660	2.275130

Automatic Design System (ADS), a CAD based software, from AGILENT TECHNOLOGIES has been used to carry out the simulation study which will be shown next.

3 RESULTS

The results obtained on optimization with the band pass filter, shown in Figure 1, are as under:

A comparison study between simulated value obtained in Figure 1 and desired values are given in Table 4.

This design has some disadvantages stated below:

- Fabrication is difficult and unrealizable
- The passband has ripples even though Butterworth's technique has been incorporated
- The results obtained are not suitable for achieving the goal.

Hence modification had to be done.

So Alumina has been selected as the substrate $(\varepsilon_r = 9.6)$. On using Alumina, performance of the filter has been improved. Once the Ripple has been reduced the fabrication becomes a reality.

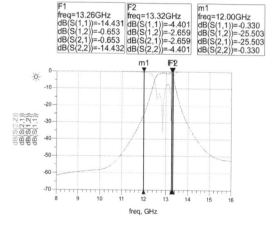

Figure 1. Simulated results for BPF having Rogers Rt/Duroid 5880 as substrate.

Table 4. Comparison study between simulated value obtained in Figure 1 and desired values.

	Desired values	Simulated values
% Bandwidth	7%	6%
Stop band attenuation	>30 db	25.505 Db
Insertion loss	<1 db	2.659 Db
Return loss	>15 db	4.401(S_{11}), 4.401 db(S_{22})

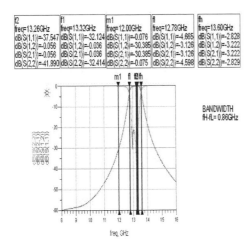

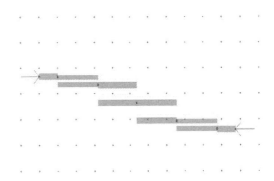

Figure 3. Layout of BPF using alumina as substrate.

Figure 2. Simulation results for BPF having Alumina as substrate.

Table 5. Comparison between the values obtained on simulation as shown in Figure 2 and desired values.

	Desired values	Simulated values
% Bandwidth	7%	6.45%
Stop band attenuation	>30 dB	30.385 dB
Insertion loss	<1 dB	0.036 dB
Return loss	>15 dB	32.124 dB(S_{11}), 32.414 dB(S_{22})

A comparison study between simulated value obtained in Figure 2 and desired values are given in Table 5.

Thus simulated results show that this design with alumina as the substrate may generate the desired results.

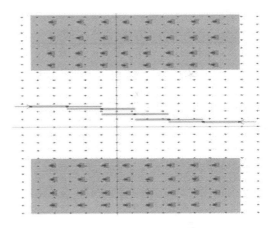

Figure 4. Layout for reduction of unwanted additional coupling in the BPF using alumina as substrate.

4 REMEDY AGAINST COUPLING

Even after these results there was scope for further improvement. Screws can be used for fitting this BPF to the total system but this may create problem due to unwanted coupling. Hence results may show some errors. So some distance has to be maintained in between the structure and the screws. There may be some reception of unwanted electromagnetic waves and pulses which may affect the result. To eliminate this, grounding has been incorporated as shown in figure. The initial and final structures have been shown in Figure 3 and Figure 4 respectively.

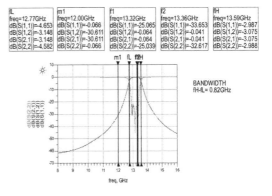

Figure 5. Results on simulation for the design implementing technique to reduce unwanted additional coupling in the BPF.

531

Table 6. Comparison between the values obtained on simulation as shown in Figure 2 and Figure 5.

	Results without taking steps to control unwanted coupling	Results with taking steps to control unwanted coupling
% Bandwidth	6.45%	6.15%
Stop band attenuation	30.385 Db	30.611 Db
Insertion loss	0.036 Db	0.064 Db
Return loss	32.124 db(S_{11}), 32.414 db(S_{22})	25.065 db(S_{11}), 25.039 db(S_{22})

In both the cases, optimization by trial and error method has been adopted. These two Alumina BAND PASS FILTERS are compared in Table 6.

The final result, though deviates from the previous ideal results, suits our requirements. Such design is practically realizable.

5 CONCLUSIONS

i. The designed BPF is to be fitted on the system structure of the transceiver mentioned above.
ii. Such procedure may be applied to different miniaturized structures.
iii. To develop/design a planar subsystem with structures of different substrates such proce-dures may be applied during integration of the structures.
iv. Theoretical results are satisfactory.
v. The actual performance with the practical fittings in the open air test site has not yet been tested. The results may show some deviation from the theoretical data. Hence once again some optimization should be incorporated.

ACKNOWLEDGEMENT

I sincerely thank Sameer—Kolkata Centre for providing the necessary infrastructure to carry out this study. Thanks to Indian School of Mines, Dhanbad for allowing me to undergo this study.

REFERENCES

Alaydrus M. (2010), "Designing Microstrip Bandpass Filter at 3.2 GHz", *International Journal on Electrical Engineering and Informatics - Volume 2, Number 2.*
Hong J.S. & Lancaster M.J. (2001), "Microstrip Filters for RF/Microwave Applications", *Wiley, New York.*
Matthaei G., Young L. & Jones E.M.T. (1980) "Microwave Filters, Impedance-matching Networks, and Coupling Structures", *Artech House, Norwood, MA.*
Pozar D.M. (2012), "Microwave Engineering", *John Wiley & Sons Inc.,* Fourth Edition.

Frontiers in Computer, Communication and Electrical Engineering – Acharyya (Ed.)
© 2016 Taylor & Francis Group, London, ISBN: 978-1-138-02877-7

Bose Einstein Condensation in lithium tantalate ferroelectrics

Arindam Biswas
Department of Electronics and Communication, NSHM Knowledge Campus, Durgapur, West Bengal, India

ABSTRACT: Lithium tantalate is technologically one of the most important ferroelectric materials with a low poling field that has several applications in the field of photonics and memory switching devices. In a Hamiltonian system, such as dipolar system, the polarization behavior of such ferroelectrics can be well-modeled by Klein–Gordon (K-G) equation. For small oscillations, the modal dynamicsis characterized as bound states, revealed via Associated Legendre polynomial. Bose Einstein condensation exclusively takes place around bosonicparticles having different wave functions within the bound states. This paper tried to explore BEC in the realm of bound state for better understanding of switching phenomenon of devices.

1 INTRODUCTION

The most outstanding experimental discovery in recent times is Bose-Einstein Condensation (BEC) in 1995 (Ketterle et al. 2002). Theoretically, BEC is already claimed for Lithium Niobate type ferroelectrics (Biswas et al. 2015). This discovery has triggered both theoretical and experimental works on this fascinating topic of research.The reasons for these vigorous activities are: a) it gives an opportunity to open a new window for a macroscopic view of quantum mechanics, and b) it makes such studies most lucrative in the field of matter-wave relations. On the latter issue, experiments show unusual excitations within the wave that is known as solitons. The unusual properties of such quasi-particles, i.e. both bright and dark solitons (henceforth called bosonic particles or simply particles), in a condensate would allow us to manipulate them in periodic or other potential. There is some evidence for the formation of both dark and bright solitions in the condensate, but the concept of bound state still remains somewhat illusive, despite a lot of activities on nonlinear optical systems that are important for many devices. This gives us motivation to study bound state with a connection to BEC. Further, we find out both lower and upper bounds in relation to frequency and the extent of condensation in the bound state. Although we use data on lithium tantalate ferroelectrics for general theoretical study, it can also be extended to other relevant systems, e.g. in magnon system with two-well Landau potential (Biswas et al. 2015). This effort is made through Klein-Gordon (K-G) equation which on perturbation gives rise to Nonlinear Schrodinger Equation (NLSE) that is a variant of Gross-Pitaevskii Equation (GPE), which is popular and commonly used for studying BEC.

There are excellent reviews on BEC in the vast ocean of literature, notably Ref (Ketterle et al. 2002). An excellent work was done by Kivshar et al on solitons in nonlinear optics (Kivshar et al. 1998);the dynamical generation and control of both bright and dark solitons in matter-wave BEC in optical lattices wereextensively studied by several authors (Kivshar et al. 1998). In this context, a review on dark solitons in atomic BEC by Frantzeskakis and Kevrekedis et al also needs a mention.In atomic optics, the formation of coherent molecular BEC was explained in one-dimension by mean field theories of parametric nonlinearities that convert two solutions to one (and vice-versa) for non-integrable equations (Werner et al. 1996); molecular BEC was invoked for a new type of reaction between molecules (multispecies) to engineer condensates of heavier molecules where macroscopic occupation of single molecular quantum state gives rise to the coherent bosonic stimulation (Werner et al. 1996) BEC was also reported on 10^5 Li_2 molecules in an optical trap with spin mixture of fermionic Li atoms by measuring a collective excitation mode (Jochim et al. 2003). This gives us further motivation to explore BEC in the realm of bound state that might help towards better understanding of switching phenomenon in a vast area of devices including many nano devices. Now, let us look at the potential and the bound states.

2 THEORETICAL DEVELOPMENT

Let us consider an idealized one-dimensional array of N identical rectangular domains along the x direction. Between the neighboring domains, there is domain wall and nearest neighbor coupling (K) is considered. For the mode dynamics of the extended modes and modes that are localized, nonlinear K-G equation relating P against space (x)

and time (t) with a non-dimensional driving field (E_0) is (Kivshar et al. 1998):

$$\frac{\partial^2 P}{\partial t^2} - \bar{K}\frac{\partial^2 P}{\partial x^2} - \bar{\alpha}_1 P + \bar{\alpha}_2 P^3 - E_0 = 0 \qquad (1)$$

K-G equation is a well-known equation of mathematical physics that exhibits a variety of interesting properties with applications in different physical systems (Dauxois et al. 2006). K-G equation is useful for both dark and bright discrete breathers that throw light on quantum localization (Dauxois et al. 2006). e to the localization, the length scale of excitation assumes more significance that obviously drives us to the nano domain, whose importance in the field of solid state physics cannot be denied. Next, let us go for the solutions: In the continuum limit, let P be the solution of Eq. (1) that is replaced by $P = P(x) + f(x,t)$. Here, $P(x)$ and $f(x,t)$ are the functions of x and (x,t) respectively. From physics point of view this combination describes a periodic kink which, a priori, can experience the presence of phonons about its center of mass regardless of its dynamical property. Thus, the resulting eigen value equation will be governed by a linearized problem. Let us write the space dependent equation:

$$-\bar{K}\frac{\partial^2 P(x)}{\partial x^2} - \bar{\alpha}_1 P(x) + \bar{\alpha}_2 P^3(x) = 0 \qquad (2)$$

After some mathematical steps, our Klein-Gordon equation can be expressed in term of wave function ψ [2].

$$(1-z^2)\frac{\partial^2 \psi}{\partial z^2} - 2z\frac{\partial \psi}{\partial z} + \left(n(n+1) - \frac{m^2}{1-z^2}\right)\psi = 0 \qquad (4)$$

where, $m^2 = 2X/\bar{\alpha}_1 = 4 - (2\omega^2/\bar{\alpha}_1)$. The solution of Eq. (3) is: $\psi = p_2^m(z)$. In the bound state, ω is denoted as ω_b and wave function ψ as ψ_b. If $0 \leq \omega_b \leq \sqrt{2\bar{\alpha}_1}$, then '$m$' is real and ALP is only valid if $n = 2$. The existence of different states is considered in the bound state in this limited range of frequency or energy. Here, all our solutions are 'real' and 'stable' in terms of interplay between the mode index and the frequency. The 'lower bound' of the non-degenerate state is at $\omega_b = 0$ for which the wave function (ψ_b) with translation symmetry gives rise to Goldstone Mode (GM) for $m = 2$:

$$\psi_b = P_2^2(z) = 3\sec h^2 qx \qquad (4)$$

To note that bright solitons predominate. This wave function for bosonic particles is not shown here, as it simply shows a typical Gaussian band. The wave functions for other symmetries of GM were not worked out to remain within our main focus on the bound state and BEC formation. As the frequency increases to: $\omega_b > \sqrt{(3\bar{\alpha}_1)/2}$, the system starts showing polarization within a band of $m = \pm 1$, whose wave functions are:

$$\psi_b = P_2^1(z) = 3\tanh qx.\sec hqx \qquad (5)$$

$$\psi_b = P_2^{-1}(z) = -(1/6)\tanh qx.\sec hqx \qquad (6)$$

3 RESULTS AND DISCUSSION

To note that both dark and bright solitons exist, and a pairing or coupling has started in the system. A small number of particles become polarized in opposite directions with the above value of eigen frequency. Wave functions, as per Eq. (5) and (6), are shown in Fig. 1a and Fig. 1b respectively

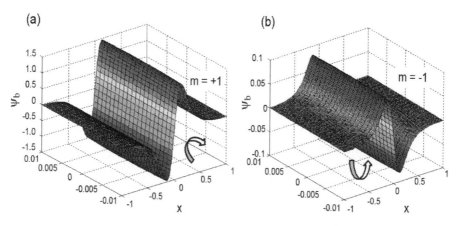

Figure 1. The wave function (ψ_b) for the quasi-particle, (a) As per Eqs. (5), when $m = +1$; (b) As per Eqs. (6), when $m = -1$.

indicating that these behaviors manifest in both + and—directions starting at zero. With the compact operator, the number density of bosons is clearly related to the frequency, which has not been attempted before, and BEC exclusively forms within this bound state. Above critical limit, the phonon is dominant in the system in the unbound state.

4 CONCLUSION

Bose-Einstein condensation is shown in the realm of bound state. This paper tried to explore BEC in the realm of bound state for better understanding of switching phenomenon of devices. This piece of information is considered useful for a future study in this new field of investigation of quantum breathers in ferroelectrics and other applications of BEC in important nonlinear optical materials.

REFERENCES

Biswas, A., Y M Song, J. *Optoelectron. Adv. Mater* 17, (2015).

Dauxois, T., M. Peyrard, Physics of Solitons, *Cambridge Univ. Press, Cambridge, (2006.)*.

Frantzeskakis, D.J., J. Phys. A: *Math. Theor. 43 213001(2010)*.

He, H., M.J. Werner, P.D. Drummond, *Phys. Rev. E 54, 896(1996)*.

Jochim, S., M. Bertenstein, A. Altmeyer, G. Hendl, S. Reidl, C. Chin, J.H. Denschlag, and R. Grimm, *Science.302, 2102(2003)*.

Ketterle, W. Rev. *Mod. Phys. 74*, 1132 (2002).

Kivshar, Y.S., B. Luther-Davis, *Phys. Rep. 298, 81(1998)*.

Frontiers in Computer, Communication and Electrical Engineering – Acharyya (Ed.)
© 2016 Taylor & Francis Group, London, ISBN: 978-1-138-02877-7

Gain and bandwidth enhancement of a microstrip antenna by incorporating air gap

Arindam Biswas
Department of Electronics and Communication, NSHM Knowledge Campus, Durgapur, West Bengal, India

ABSTRACT: This paper presents a design of a rectangular microstrip patch antenna having a large bandwidth using the coaxial feeding method. For this purpose, two substrates were considered. The value of the dielectric constant of the upper substrate was 4.4 and that of the lower substrate was 1 (an air gap was included between the substrate and the ground plane). The height of the upper substrate was considered as 0.158 cm, and the height of the air gap height was also the same. The width and length of the ground plane were also modified. Gain and bandwidth were modified accordingly.

1 INTRODUCTION

Planar antenna designs, such as "microstrip antennas," are often preferred for mobile and satellite communication antennas due to their added advantage of small size, low manufacturing cost, and conformability. In recent years, there has been an increasing demand for square microstrip antennas in the communication scenario due to their polarization diversity, particularly the ability to realize dual or circularly polarized radiation patterns (Bahl *et al.* 2001).

Microstrip patch antennas are widely used because of their many advantages such as low profile, lightweight, low cost, and planar design. However, patch antennas have a disadvantage of a narrow bandwidth, typically 1–5% impedance bandwidth. Researchers have made many efforts to overcome this problem, and many configurations have been presented to increase the bandwidth (Abdelaziz *et al.* 2006). Many configurations have increased the bandwidth, but have also increased the production cost of the antenna due to the misalignment of stacked patches. To overcome the misalignment problem, we use less number of patches. This paper describes the concept of the air gap structure to enhance the antenna's gain and bandwidth (Guha *et al.* 2001). In a normal rectangular patch antenna, we introduce an air gap and find the total field gain ranges from about 3.4 to 4.4 dBi in the bandwidth region. The antenna efficiency ranges from about 67% to 92% within the bandwidth. It can also be seen that the radiation efficiency ranges from about 71% to 97% within the bandwidth.

2 COAXIAL PROBE FEED RECTANGULAR PATCH ANTENNA WITH AN AIR GAP

The drawbacks of the patch antenna are narrow bandwidth and low gain. The bandwidth is affected primarily by substrate height and permittivity. In this section, a rectangular patch antenna is presented, which has a high gain region and a relatively large bandwidth region. The geometry of the antenna is shown in Figure 1.

The antenna is fed with a 50-Ω coax probe. The feed position taken along the x-axis is 0.45 cm and along the y-axis is 0.59 cm from the center of the patch. In this section, the antenna is designed for a larger bandwidth by using the electromagnetic software IE3D. In this simulation, the resonant frequency of the antenna is defined as the frequency at which the maximum resistance occurs. The simulation of the antenna was carried out over a frequency range from 3 GHz to 7 GHz.

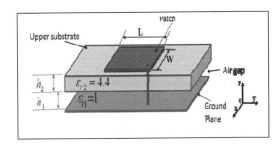

Figure 1. Proposed geometry.

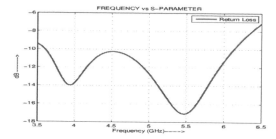

Figure 2. Frequency versus return loss.

Table 1. Return loss and input impedance at different frequencies.

	Output parameters of the antenna		
f_0 GHz	Return loss S[1,1]dB	Real Z[1,1] Ω	Imag Z[1,1] Ω
3.58838	−10.020	52.590	−34.190
3.59	−10.032	52.642	−34.378
3.68	−10.991	53.476	−30.575
3.92	−13.905	62.286	−19.253
4.25	−11.101	79.596	−20.581
4.55	−10.200	78.422	−28.575
5.00	−12.640	67.670	−21.403
5.47872	−17.178	66.110	−2.1020
5.51	−17.125	66.390	−0.7400
5.63	−16.206	67.900	4.700
6.11	−10.214	81.430	27.220
6.12684	−10.049	81.990	27.830

3 SIMULATED RESULTS

In this section, the simulated results of the rectangular patch antenna are presented. The simulated result of the return loss is shown in Figure 2.

It can be seen that it consists of a patch on the top of the substrate and an air gap between the ground plane and the substrate with $\varepsilon r = 4.4$. The height of the substrate with $\varepsilon r = 4.4$ and the air gap is h1 = h2 = 0.158 cm. To improve the bandwidth, the air gap is introduced, and the ground plane width and length are modified as 2.5 cm and 2 cm, respectively. The physical width and length of the patch for the dominant TM01 mode are considered as W = 1.3 cm and L = 1.51 cm, respectively.

The total field gain ranges from about 3.4 to 4.4 dBi in the bandwidth, as given in Table 2. It can also be seen that the directivity obtained within that frequency range is between 4.6 dBi and 5.32 dBi. The total field gain and the directivity versus frequency plots are shown in Figures 4 and 5, respectively.

Table 2. Total field gain, directivity and voltage source gain at different frequencies.

	Output Parameters of the Antenna		
f_0 GHz	Total field gain dBi	Total field directivity dBi	Total field voltage source gain dBi
3.50	3.844580487	4.5040584	1.437796
3.68	4.035360906	4.5767002	1.541965
3.86	4.332925985	4.7265079	1.885397
4.25	4.381467697	5.108824	2.624661
4.55	4.219653487	5.2661915	2.513313
5.00	4.113968424	5.3204146	1.948622
5.63	3.766679843	5.2479775	1.468934
5.93	3.364557553	5.1555993	1.403343
6.11	2.472270124	4.9872449	1.700386

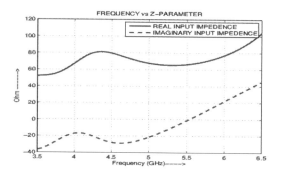

Figure 3. Frequency versus Z-parameter.

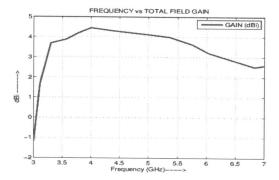

Figure 4. Frequency versus gain.

The antenna resonates at the frequency range of 3.588 GHz to 6.127 GHz with RL ranging between −10.2 dB and 17.2 dB. The bandwidth (S[1,1] ≤ −10 dB) of the antenna is about 2.54 GHz with a maximum gain of 4.38 dBi. Return loss S[1,1] and input impedance Z[1,1] at different frequencies are given in Table 1. The Z-parameter is shown in Figure 3. Real Z[1,1] ranges from 52.6 Ω to 82 Ω within the bandwidth.

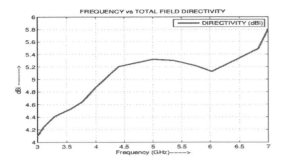

Figure 5. Frequency versus directivity.

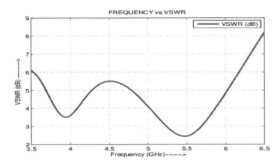

Figure 6. Frequency versus VSWR dB.

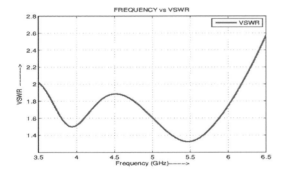

Figure 7. Frequency versus VSWR.

As can be seen from Figure 6 that the VSWR ranges from 1.72 to 1.8, and from Figure 7, it can be seen that the VSWR dB is in the range of 2.7 dB to 4.9 dB within the bandwidth. The antenna efficiency ranges from about 67% to 92% within the bandwidth. It can also be seen that the radiation efficiency ranges from about 71% to 97% within the bandwidth. The 3-dB beam width in the E-plane at $\Phi = 0$ deg is 57°. The beam width in the E-plane of this antenna is narrower than that of the single-layer antenna.

The simulated results are summarized in Tables 2 and 3.

Table 3. Efficiency and VSWR at different frequencies.

| f_0 GHz | Output parameters of the antenna | | | |
	Antenna efficiency %	Radiation efficiency %	VSWR	VSWR dB
3.50	85.9117	96.7158	2.020	6.1092
3.68	88.2808	95.9159	1.798	5.0956
3.86	91.3360	95.5737	1.534	3.7174
4.25	84.5794	91.6948	1.758	4.8981
4.55	78.5862	86.8841	1.883	5.4982
5.00	75.7452	80.1066	1.604	4.1061
5.63	71.1001	72.8449	1.372	2.7472
5.93	66.2058	70.4120	1.653	4.3684
6.11	56.0406	69.4855	2.574	8.2133

4 CONCLUSION

The final result shows that the bandwidth of this antenna is 2.53846 GHz. This antenna works between the frequency range of 3.58838 GHz to 6.12684 GHz. The total field gain obtained is in the range between 3.4 dBi and 4.38 dBi, and the directivity obtained ranges between 4.6 dBi and 5.32 dBi. This concept of bandwidth enhancement will be very useful for antenna applications.

REFERENCES

Abboud F., Damiano J.P., Papiernik A. (1990) A new model for calculating the input impedance of coax-fed circular microstrip antennas with and without air gaps. "*IEEE Transaction on Antenna and Propagation 38: 1882–1885*"

Abdelaziz A.A.;" bandwidth enhancement of microstrip antenna" *Progress In Electromagnetics Research, PIER 63, 311–317, 2006.*

Bahl, I.J.; Bhartia, P., "Microstrip Antennas Design Handbook". *Artech House, USA, 2001.*

Costantine A. Balanis, "Antenna Theory Analysis And Desingn", *Wiley, 2nd edition.*

Guha D. (2001) Resonant frequency of circular microstrip antennas with and without air gaps. *IEEE Transaction on Antenna and Propagation 49: 55–59.*

James, J.R.; Hall, P.S., (Eds.), "Handbook of Microstrip Antennas". *IEE Electromagnetic Waves Series 28, Pete Peregrinus Ltd., United Kingdom*, 198

Nakano, H. and K. Vichien, "Dual frequency square patch antenna with rectangular notch,"*Elec. Letters, Vol. 25, No. 16, 1067–1068, 1989.*

Richards, W.F., S.E. Davidson, and S.A. Long, "Dual band reactively loaded microstrip antenna," *IEEE Trans. Ant. Prop.,Vol. AP-33, No. 5, 556–561, 1985.*

Frontiers in Computer, Communication and Electrical Engineering – Acharyya (Ed.)
© *2016 Taylor & Francis Group, London, ISBN: 978-1-138-02877-7*

A brief study of the ESPRIT Direction-Of-Arrival estimation algorithm in an uncorrelated environment for application in the Smart Antenna System

Dhusar Kumar Mondal
Assistant Director of Operations, Director General of Civil Aviation, Government of India

Biswajit Dian, Chirantan Dutta, Aniket Chatterjee & Abhik Ray
Supreme Knowledge Foundation Group of Institution, Mankundu, Hooghly, West Bengal, India

ABSTRACT: The requirement of today's mobile communication is spectrum efficiency, better coverage, and high quality of service with minimum transmitted power. The smart antenna system promises to solve these technological requirements of wireless communication. The extent up to which the smart antenna system fulfills these requirement depends on the performance of two different algorithms, i.e., the Direction-Of-Arrival (DOA) estimation algorithm and the beamforming algorithm. Among different high-resolution subspaced-based methods of the DOA estimation algorithm, ESPRIT is preferred for analysis in this paper, as it requires less computation and storage space compared with its near competitor MUSIC. This work generates the plots of the error curve for the determination of the signal arrival angle and its dependency on different system parameters from a large number of simulation work performed in MATLAB software. The results enhance the possibility of preferring ESPRIT for practical applications, as it will require less cost compared with MUSIC though its performance is quite acceptable.

1 INTRODUCTION

In the last decade, mobile communication has experienced a rapid growth in demand for the provision of new wireless multimedia services such as internet access from a wireless handset, multimedia services, data transfer, and video call. As the conventional single-input single-output system has limited capacity, it is unable to fulfill this demand. Thus, the use of the Multiple Antenna System (MAS) has been under consideration, as it is able to offer greater capacity than the Single Antenna System. A MAS increases the diversity of communication reliability as well as the data rate by spatial multiplexing (SDMA) [1]. therefore, the Multiple Antenna System offers performance as well as capacity enhancement without additional power. Also, the application of spatial multiplexing increases the frequency reuse factor and thus saves the spectrum bandwidth. The concept of the Smart Antenna System (SAS) arises from the above feature of the Multiple Antenna System.

2 CONCEPT OF THE SAS

The SAS basically utilizes the advance features of the Multiple Antenna System, i.e., an adaptive antenna array where the radiation pattern of the array is controlled by adjusting the amplitude and the relative phase on each element of the array. The radiation pattern of an antenna array can be adjusted by adaptively controlling the amplitude and the phase of the signal feed on different elements of the array. So, the antenna array has the capability to focus the radiation pattern on a desired direction, i.e., steering capability of the radiation pattern. The modern SAS steers the radiation pattern to the desired user and simultaneously forms the null of the radiation pattern in the direction of interference. The complete function of the SAS is performed by two different technologies: (1) determining the direction of the user where the radiation pattern is to be focused, which is performed by the Direction-Of-Arrival (DOA) estimation algorithm; (2) transmitting a focused beam towards the user, which is performed by the beamforming algorithm [1,2]. Both the functions are possible by the combination of software and hardware. The DOA estimation and the steering of the radiation pattern in the same direction are done in a synchronized and adaptive manner, and thus it is also known as the adaptive array smart antenna system. So, the successful design and operation of the adaptive array smart antenna system depends highly on the performance of the DOA estimation algorithm, the beamforming algorithm, and the beam-steering capacity of the antenna array. In this paper, the study of the DOA estimation algorithm is undertaken.

3 DOA ESTIMATION

The performance of the DOA estimation algorithm depends on many parameters such as number of mobile users using the same frequency in the same cell, their spatial separation, number of antenna elements in the array and inter-element distance, number of signal samples taken for calculation, and Signal-to-Noise Ratio (SNR) of the received signal [3]. Many DOA estimation algorithms have been developed in the last few decades. Basically, the non-parametric DOA estimation algorithm can be classified as follows:

1. Conventional DOA estimation method.
2. Subspaced DOA estimation method.

Methods that fall under the class conventional DOA estimation method include Bartlett's method and Capon's method. However, these methods suffer from a lack of angular resolution. For these high-angular resolution subspaced methods, MUSIC and ESPRIT algorithms are mostly used. MUSIC is an acronym that stands for MUltiple SIgnal Classification. The MUSIC algorithm is highly accurate and stable, but requires a large number of computation and storage space, which is an area of concern from the viewpoint of practical implementation costs [4,5].

In this paper, the performance of the ESPRIT algorithm is studied in detail, so that its advantages in terms of its practical application can be judged and simultaneously compared with the MUSIC algorithm. For analyzing its performance, simulations are carried out in MATLAB software.

4 ESPRIT DOA ESTIMATION ALGORITHM

ESPRIT stands for Estimation of Signal Parameters via Rotational Invariance Techniques and was first proposed by Roy and Kailath in 1989. ESPRIT is a computationally efficient and robust method of DOA estimation. It uses two identical arrays in the sense that array elements need to form matched pairs with an identical displacement vector, i.e., the second element of each pair ought to be displaced at the same distance and in the same direction relative to the first element.

The signals induced on each of the arrays from D uncorrelated sources are given by

$$x_1 = A_1 S(k) + n_1(k) \tag{1}$$

$$x_2 = A_2 \phi S(k) + n_2(k) \tag{2}$$

where d is the inter-element distance and k is the wave number and

$$\phi = diag\{e^{jkd \sin \theta_1}, e^{jkd \sin \theta_2}, \ldots \ldots e^{jkd \sin \theta_D}\} \tag{3}$$

$A_i = matrix\ of\ steering\ vector\ for\ subarray\ i = 1, 2$

The complete received signal considering the contributions of both sub arrays is given as follows:

$$x(k) = \begin{bmatrix} x_1(k) \\ x_2(k) \end{bmatrix} = \begin{bmatrix} A_1 \\ A_2 \phi \end{bmatrix} s(k) + \begin{bmatrix} n_1(k) \\ n_2(k) \end{bmatrix} \tag{4}$$

We can now calculate the correlation matrix for either the complete array or for the two sub-arrays. The correlation matrix for the complete array is given by

$$R_{xx} = E\left[xx^H\right] = A\ R_{ss}R^H + \sigma_n^2 I \tag{5}$$

Creating the signal subspace for the entire array results in one signal subspace given by Equation (5). Because of the invariance structure of the array, Equation (5) can be decomposed into the subspaces E1and E2. Next, a 2D × 2D matrix can be formed using the signal subspaces, such that

$$C = \begin{bmatrix} E_1^H \\ E_2^H \end{bmatrix} \begin{bmatrix} E_1 & E_2 \end{bmatrix} = E_C \Lambda E_c^H \tag{6}$$

where E_c is the eigen vector decomposition of C, such that $\{\lambda_1, \lambda_2, \ldots \ldots \lambda_{2D}\}$ and

$$\Lambda = diag\{\lambda_1, \lambda_2, \ldots \ldots \lambda_{2D}\} \tag{7}$$

We then partition E_c into four D × D sub-matrices, such that

$$E_C = \begin{bmatrix} E_{11} & E_{12} \\ E_{21} & E_{22} \end{bmatrix} \tag{8}$$

The rotational operator can be estimated as

$$\psi = -E_{12}E_{22}^{-1} \tag{9}$$

Table 1. The error for each angle of arrival for different values of the SNR in dB.

Signal-to-Noise Ratio	Error for DOA –40 degree	Error for DOA 30 degree	Error for DOA 60 degree
0	0.1667	0.2333	0.3875
5	0.0090	0.0025	0.0473
10	0.0018	0.0015	0.0059
20	0.0022	0.0003	0.0060
30	0.0006	0.0003	0.0006
50	0.0003	0.0001	0.0007
60	0.0001	0.0001	0.0004
70	0.0001	0.0001	0
80	0	0	0
90	0	0	0
100	0	0	0

The angle of arrival can be uniquely determined from the following relation:

$$\theta_i = \sin^{-1}\left(\frac{\arg(\lambda_i)}{k\vec{d}}\right) \qquad i = 1,2,3,.......D \qquad (10)$$

where λ_i is the eigen value of ψ, and $\vec{d}$ is the displacement vector between two sub-arrays [5,6].

5 SIMULATION RESULTS

All the simulations in this work were carried out considering the following parameters:

The number of uncorrelated sources is 3, the type of antenna array is one-dimensional linear array with 12 isotropic elements, and the number of snapshots used is 100.

a. Error in the estimation of DOA using ESPRIT with SNR:

A large number of simulations were carried out in MATLAB software to determine the error in the estimation of DOA by the ESPRIT algorithm. Table 1 presents the error for each angle of arrival for different values of the SNR in dB. Furthermore, the normalized error vs SNR for each angle of arrival as well as the average error for all the angles of arrival vs SNR were plotted individually, as shown in the below figures. Detection error vs. SNR for –40 deg and 30 deg DOA have been swon in Figure 1 and Figure 2 respectively.

b. Error in the estimation of DOA using ESPRIT with number of elements in the array.

Table 2 presents a large number of simulations, from which the effect of the number of elements in the array on the average error for three sources can be observed. Three sources are detectable only

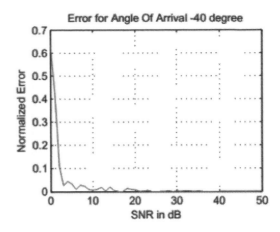

Figure 1. Detection error vs SNR for –40 deg DOA.

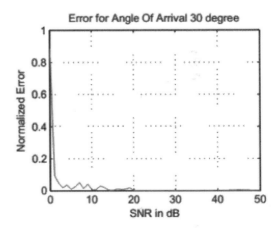

Figure 2. Detection error vs SNR for 30 deg DOA.

Table 2. The effect of the number of elements in the array on the average error for three sources.

Number of elements	Error in the estimation of angle
1	Not detectable
2	Not detectable
3	Not detectable
4	0.0024
5	0.0015
10	0.0004
15	0.0001
20	0.0001
25	0.0001
30	0.0001
35	0.0001
40	0.0002
50	0.0001
100	0

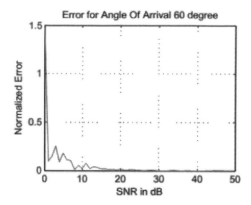

Figure 3a. Detection error vs SNR for 60 deg DOA.

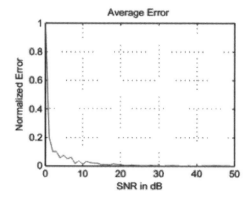

Figure 3b. Average detection error vs SNR.

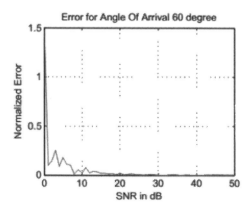

Figure 3c. Average detection error vs number of elements.

when the number of elements is 4 or more and thus the theory is verified. When the number of elements is 5, the detection capability is satisfactory and for beyond 10, it is quite good. However, an increased number in elements increases the array

Table 3. The effect of inter-element spacing on individual DOA estimation.

Inter-element spacing (cm)	Error for DOA -40 degree	Error for DOA 30 degree	Error for DOA 60 degree
2	0.0254	0.5671	0.0712
3	0.0100	0.0523	0.1056
4	0.0149	0.0234	0.0252
5	0.0054	0.0136	0.0321
6	0.0021	0.0041	0.0362
7	0.0020	0.0071	0.0030
8	0.0034	0.0043	0.0022
9	0.0003	0.0048	0.0081
10	0.0015	0.0037	0.0023
11	0.0005	0.0022	0.0009
12	0.0004	0.0002	0.0019
13	0.0004	0.0001	0.0010
14	0.0005	0.0006	0.0006
15	0.0001	0.0009	0.0008

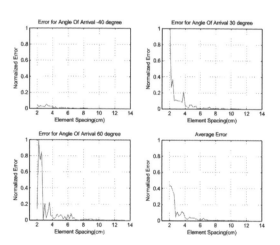

Figure 4. Average error vs inter-element spacing.

size, and thus the miniaturization objective is not achieved. Figure 3 shows variation of normalized error vs number of elements.

c. Error in the estimation of DOA using ESPRIT with inter-element spacing.

Table 3 presents the effect of inter-element spacing on individual DOA estimation. The error decreases with an increase in inter-element spacing. Furthermore, the outcome of the simulation listed in the Table 3 indicates that the DOA detection error depends on the angular separation of the desired source from the undesired one. When two sources are spatially closer to each other, the detection error increases. Average error vs. inter-element spacing is shown in Figure 4.

6 CONCLUSION

In the first author's earlier work [1,2,3], MUSIC has been thoroughly analyzed. Now, the outcome of this paper clearly shows that the performance of ESPRIT is quite acceptable as the performance is close to that of MUSIC, but the implementation cost of MUSIC will be higher as it needs a large number of computation and storage space. Thus, ESPRIT can be implemented for DOA estimation in the SAS.

REFERENCES

[1] Dhusar Kumar Mondal, "Studies of Different Direction of Arrival (DOA) Estimation Algorithm for Smart Antenna in Wireless Communication", International Journal of Electronics & Communication, Vol. 4, Issue-SPL-2, Jan Mar2013.

[2] Dhusar Kumar Mondal, "Study of Spectral Music Direction of Arrival Estimation Algorithm of Smart Antenna Using Half Wavelength Dipole Uniform Linear Array", International Journal of Electronics & Communication, Vol. 5, Issue, SPL-2, Jan-Mar 2014.

[3] Dhusar Kumar Mondal, Aritra Mondal, Pranoy Das, Chayan Seth, B N Biswas, "Analysing Resolution Parameters of Different Direction of Arrival Estimation Algorithm for Un-correlated Environment", International Journal of Electronics & Communication,vol-6–1-spl-1-jan-mar-2015.

[4] Franck B. Gross, "Smart Antenna for Wireless Communication", 2005 by the McGraw-Hill Companies.

[5] Constantine A. Balanis, Panagiotis L, Ioannidcs, "Introduction to Smart Antenna", Synthesis Lecture on Antennas#5. 2007 Morgan & Claypool Publishers.

[6] Lal.C.Godra, "Application of Antenna Arrays to Mobile Communications, Part II: Beam-Forming and Direction of Arrival Considerations", Proceedings of IEEE, Vol. 85, No. 8, August 1997.

Frontiers in Computer, Communication and Electrical Engineering – Acharyya (Ed.)
© 2016 Taylor & Francis Group, London, ISBN: 978-1-138-02877-7

Photonic Integrated Circuit technology for ultra high speed wireless communications

G. Carpintero
Universidad Carlos III de Madrid, Spain

ABSTRACT: We provide an overview of photonic-based schemes to generate signals within the millimeter and terahertz wave range, providing a brief summary of some of the application scenarios where these signals are needed. Photonics techniques present unique advantages, but need to address the cost, size, and reliability. Efforts to address these are directed toward the use of generic integration technology platforms, in which sub-systems can be integrated as a photonic integrated circuit.

1 INTRODUCTION

The future of wireless communications needs to address the increasing need for bandwidth in order to enable transmission of data rates up to 100 Gb/s, aiming to get close to those used in fiber-optic communications [1,2]. An increase of bandwidth can be achieved either through maintaining carrier frequency and using higher order modulation formats or using simple modulation techniques and increasing the carrier frequency into the Millimeter-(MMW) and Terahertz-wave (THz) range [3], from 30 GHz to 10 THz. The latter offers a cost effective solution which has the additional advantage of avoiding coding/decoding latencies.

Particularly, above 275 GHz, there are extremely large bandwidths for "radio" communications available since these frequency bands have not yet been allocated at specific active services. Recent experimental experiments reported data rates up to 48 Gb/s on a 300-GHz carrier frequency [4].

However, one of the bottlenecks for these systems is the generation of the carrier waves at these frequencies with the required signal quality. Photonics-based technologies are at the forefront in the generation of high frequency waves thanks to the availability of telecom-based high-frequency components, lasers, optical modulators, and high-speed photodiodes. Using the photonic approach brings important advantages, especially enabling the use of optical fibers to distribute high-frequency RF signals over long distances. In order to become a competitive technology against electronic alternatives, the photonic approach must address the challenge of realizing small and compact transmitter frontends which are light and have a low cost.

A new road has recently been paved within the framework of different European research projects, developing a generic integration platform for Photonic Integrated Circuit (PIC) technology. The challenge turns into being able to develop the required photonic systems using a small set of standardized Building Blocks (BB) of a given fabrication platform. The building blocks are frequently used basic components, which are brought together in a single integration process that can be optimized for providing high performance. This significantly reduces the design effort, as Process Development Tools can be produced. On top of this, having a unique technological process allows us to establish access to the process through Multi-Project Wafer (MPW) fabrication runs, in which the fabrication costs are shared among several users. In this contribution, we will analyze some results of the integration efforts that have already been conducted.

2 PHOTONIC-BASED GENERATION

2.1 *Photonic-techniques*

Photonic techniques are an alternative approach having some inherent key advantages such as being broadly tunable, having an ultra-wide bandwidth, and being able to be seamlessly connected to wired (fiber optic) networks. As shown in Fig. 1, the two key components for a photonics-based wireless

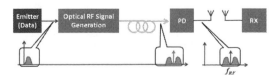

Figure 1. Building blocks of a photonic-based wireless transmitter.

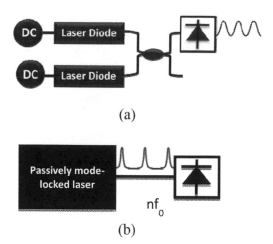

(a)

(b)

Figure 2. Photonic signal generation techniques: (a) optical heterodyne and (b) Mode-locking.

transmitter are the photonic Continuous-Wave (CW) signal generation and a high speed optical-to-electrical (O/E) converter [5]. There are various photonic techniques to generate MMW and THz frequencies, among which optical heterodyning and pulsed techniques are commonly used.

Optical heterodyning is based on mixing two single frequency lasers (λ_1 and λ_2). This technique provides a large tuning range over 10 THz. On the other hand, the performance in terms of phase noise greatly depends on the types of lasers that are used, usually requiring locking the two together. The linewidth of the output millimeter wave is determined by the sum of the linewidth of λ1 and λ2 [6].

Pulsed sources are usually based on mode locked laser structures [7], having recently reported that it can increase the radiated emitted power about 7 dBm above heterodyning schemes [8]. Mode locked sources can either be passive or hybrid, depending on whether the pulse train is locked to an electronic reference. Passive mode locking requires only fixed direct current and reverse voltage to generate the pulsed signal, which lacks stability [9]. The stability can be addressed through hybrid mode locking, in which the reverse voltage is combined with a Continuous Wave (CW) signal at the fundamental frequency. The stability is inherited from the electronic source [10], but is limited by the electrical modulation bandwidth, usually below 30 GHz.

The major drawback of photonic-based approach is that it requires the combination of different photonic components. This requires a costly packaging process, with high impact in the overall cost of the system. Another important factor is that every fiber introduces coupling losses and is a potential source for back-reflections. In order

to achieve higher performance levels and introduce fabrication rationale basic idea behind the work presented here is to investigate photonic integration technology.

2.2 Photonic technology platforms

Currently there are several technology platforms, each based on a different material. The most promising material is Silicon, which allows passive light manipulation [11]. It has many advantages such as having a very small footprint for the components, due to the relative high contrast index of silicon, is a cheap base material, which has mechanical strength so that the wafers can be made larger, and more importantly it is compatible with current CMOS technology, making photonics and electronics compatible. It has one severe drawback which is that there is neither light generation nor amplification. Other silicon-based platforms are available, especially silicon nitride, which allow passive light manipulation, but very importantly have very low transmission loss. This is a key factor for high Q resonators. However, the low refractive index contrast results in large chip sizes.

The most important technology platform is Indium Phosphide, which provides light generation and amplification, and supports modulation and detection, wavelength multiplexing, demultiplexing, optical attenuation, switching, and dispersion compensation. This enables this platform to offer a monolithic integration of all the components into a single chip that performs all the functions.

2.3 Generic integration

Most of the photonic systems can be spelled out using different combinations of a reduced set of components, which are known as basic building blocks [12]. Most of these can be composed of a combination of passive waveguides of different widths and lengths such as straight and curved waveguides, Multi-Mode Interference (MMI) couplers, and so on.

The powerful idea that this brings in is that we do not need to design every component every time, but rather provide these building blocks, investing in developing the technology for a very high performance at the level of the basic building blocks, tested and with a known performance. These basic building blocks are the elements of design software libraries, enabling users to design their own systems using design tools. All this investment is fully justified as the generic integration technology is not application specific and can serve a large market.

The second key ingredient is a low-cost access to the technology, which is achieved through the Multi-Project Wafer (MPW) run approach, combining designs from several users into a single chip mask design. Many small customers are transformed into a bigger one with sufficient volume to get access to the foundry.

3 PHOTONIC INTEGRATED CIRCUIT EXAMPLES

3.1 *Dedicated integration platform*

In order to demonstrate the highest level of integration, we developed a single-chip millimeter-wave wireless transmitter, shown in Fig. 3. The dual wavelength source is based on the monolithic integration of two single mode Distributed Feedback (DFB) lasers combined through a Multimode Interference coupler (MMI). This is done at both ends of the DFB lasers. The left hand side provides an optical output of the dual wavelength for analysis, and the right hand side implements the optical path toward the monolithic high-speed Photodiodes (PD) where the two wavelengths are mixed to generate the millimeter electrical signal passing through Electro-Absorption modulators (EA) for data modulation and Semiconductor Optical Amplifiers (SOA) to boost the optical power within the waveguide. With this approach we have achieved a continuous tuning of wavelength spacing from 5 GHz to 110 GHz. In terms of emitted electrical power, we probed onto the PD with a probe having a DC bias tee and WR08 output, to which we connected a horn antenna. Fig. 4 shows the emitted power detected on an Agilent E4418B EPM series power meter with a W8486 A power sensor (75–110 GHz) when the two DFB lasers were biased at 95 mA (generating a 95.7-GHz beat note frequency). These photodiodes, when diced out, have been proven to have a bandwidth up to 175 GHz and provide up to 0 dBm [13].

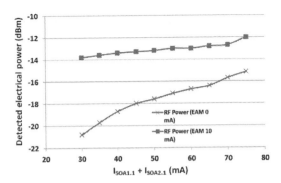

Figure 4. Electrical power emitted by the photodiode at 95,7 GHz, at two different current levels into the EA. Each laser was biased at 95 mA, SOA1.2 at 40 mA, and SOA1.3 at 45 mA.

The main drawbacks of this approach are that it requires a dedicated fabrication process flow and the relatively broad linewidth of the optical modes (around 1 MHz) [14].

3.2 *Photonic Multi-Project Wafer run in a Generic Integration Platform*

An on-chip dual wavelength source suitable for fabrication on a Generic InP-based technology platform to access the cost reduction of a Multi-Project Wafer (MPW) run has been also developed. The design is then restricted to the available building blocks of the platform, which in our case did not include gratings of any kind. The dual wavelength source that was developed, shown in Fig. 5, is based on an Arrayed Waveguide Grating (AWG) laser. The emitted wavelengths originate by forward biasing the SOAs on two channels simultaneously; thus, the spacing is an integer multiple of the AWG channel spacing ($\Delta\lambda$), which usually due to fabrication issues has to be $\Delta\lambda > 100$ GHz. This is the main drawback of this structure, as the wavelength spacing is not tunable because it is fixed by the AWG. The novelty of our design is that it is an on-chip solution through the use of novel Multimode Interference Reflectors (MIR) to create the required Fabry-Perot cavity of the AWG laser without requiring the facets of the chip. The main advantage of this structure is the linewidth of the signal generated from beating the two modes on a high speed photodiode. Fig. 6 shows the linewidth of the optical mode from Channel 2 measured by the self-heterodyne technique. Assuming a Lorentzian shape, optical −3 dB linewidth is 254 KHz, without any stabilization scheme. For this reason, we have selected this structure for a wireless data transmission experiment, described in the next section.

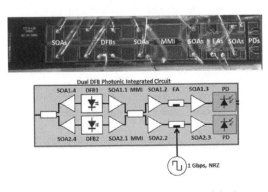

Figure 3. (Top) Picture of the fully integrated dual wavelength source for millimeter wave generation, (bottom) Block diagram of the integrated elements.

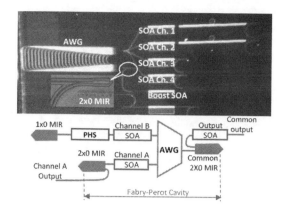

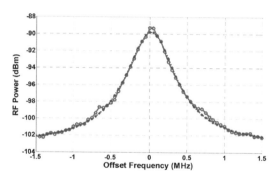

Figure 5. (Top) Picture of the on-chip AWG laser with $\Delta\lambda = 109$ GHz, capable of generating two wavelengths spaced at 109, 218, and 327 GHz (bottom) Block diagram of the integrated elements.

Figure 6. (Red) Measured optical linewidth of the AWG laser modes by self-heterodyning (RBW = 30 kHz), (Blue) lorentzian fit.

REFERENCES

[1] S. Hisatake, G. Carpintero, Y. Yoshimizu, Y. Minamikata, K. Oogimoto, Y. Yasuda, T. Nagatsuma, 2015 "W-band Coherent Wireless Link Using Injection Locked Laser Diodes,". IEEE Photon. Technol. Lett., 27(14), pp. 1565–1568.

[2] A.J. Seeds, H. Shams, M.J. Fice, C.C. Renaud, 2015 "TeraHertz Photonics for Wireless Communications, " J. Lightw. Technol., 33(3), pp. 579–587.

[3] S. Koenig et al., 2013 "Wireless sub-THz communication system with high data rate," Nature Photon. 7, 977–981.

[4] T. Nagatsuma et al., 2013 "Terahertz wireless communications based on photonics technologies," Opt. Exp., 21(20), pp. 23736–23747.

[5] A. Stöhr et al., 2010 "Millimeter-Wave Photonic Components for Broadband Wireless Systems," IEEE Trans. on Microwave Theory and Tech. 58(11), pp. 3071–3082.

[6] G. Carpintero et al., 2012 "95 GHz millimeter wave signal generation using an arrayed waveguide grating dual wavelength semiconductor laser", Optics Letters, 37(17), pp. 3657–3659.

[7] T. Nagatsuma, N. Kukutsu, and Y. Kado, 2007 "Photonic Generation of Millimeter and Terahertz Waves and Its Applications," Applied Electromagnetics and Communications ICECom, Dubrovnik,, pp. 1–4.

[8] L. Moeller, A. Shen, C. Caillaud, and M. Achouche, 2013 "Enhanced THz generation for wireless communications using short optical pulses, " In Infrared, Millimeter, and Terahertz Waves (IRMMW-THz),, pp. 1–3.

[9] K.A. Williams, M.G. Thompson, and I.H. White, 2004 "Long-wavelength mononolithic mode-locked diode lasers," New. J. Phys., vol. 6, p. 179.

[10] G. Carpintero, M.G. Thompson, R.V. Penty, and I.H. White, 2009 "Low noise performance of passively mode-locked 10-GHz quantum-dot laser diode, " IEEE Photon. Technol. Lett., vol. 21, no. 6, pp. 389–391.

[11] C. Doerr 2013 "Integrated Photonic Platforms for Telecomunications: InP and Si" IEICE Trans. Electron. Vol. E96-C, No. 7.

[12] M. Smit et al., 2011, "Generic foundry model for InP-based photonics" IET Optoelectron., Vol. 5, Iss. 5, pp. 187–194.

[13] E. Rouvalis et al., 2012 "High-speed photodiodes for InP-based photonic integrated circuits" Optics Express 20(8), pp. 9172–9177.

[14] G. Carpintero et al., 2014 " Microwave Photonic Integrated Circuits for Millimeter-Wave Wireless Communications," J. Lightwave Technology, 32(20), pp. 511–520.

Author index